GW00402196

Glutamate and GABA Receptors and Transporters

New and Forthcoming Titles in the Pharmaceutical Sciences Series

Forthcoming

Paracetamol (Acetaminophen): A Critical Bibliographic Review
2nd edition
Laurie Prescott
0-415-25845-6 (hbk)

Guide to Cytochromes P450: Structure and Function
David F. V. Lewis
0-7484-0897-5 (pbk)

Guide to Microbiological Control in Pharmaceuticals and Medical Devices
2nd Edition
S. Denyer and *R. Baird (eds)*
0-7484-0615-8 (hbk)

Published

Interindividual Variability in Human Drug Metabolism
Gian Maria Pacifici and *Olavi Pelkonen (eds)*
0-7484-0864-9 (hbk)

Physiological Pharmaceutics: Barriers to Drug Absorption
2nd Edition
Neena Washington, Clive Washington and *Clive Wilson*
0-7484-0562-3 (hbk), 0-7484-0610-7 (pbk)

Drug Therapy for Stroke Prevention
Julien Bogousslavsky (ed.)
0-7484-0934-3 (hbk)

The β3-Adrenoreceptor
A.Donny Strosberg (ed.)
0-7484-0804-5 (hbk)

Handbook of Microbiological Quality Control
R. Baird, N. Hodges and *S. Denyer (eds)*
0-7484-0614-X (hbk)

Pharmaceutical Packaging Technology
Dixie A. Dean, R. Evans and *I. Hall (eds)*
0-7484-0440-6 (hbk)

Particulate Interactions in Dry Powder Formulations for Inhalation
Xian Ming Zeng, Gary Martin and *Christopher Marriot*
0-7484-0960-2 (hbk)

Genetic Polymorphisms and Susceptibility to Disease
M.S. Miller and *M.T. Cronin (eds)*
0-7484-0822-3 (hbk)

Principles and Practice of Bioanalysis
Richard F. Venn (ed.)
0-7484-0842-8 (hbk), 0-7484-0843-6 (pbk)

Handbook of Pre-clinical Continuous Intravenous Infusion
Guy Healing and David Smith (eds)
0-7484-0867-3 (hbk)

Bioactive Compounds from Natural Sources
Corrado Tringali (ed.)
0-7484-0890-8 (hbk)

Pharmaceutical Formulation Development of Peptides and Proteins
Sven Frokjaer and *Lars Hovgaard*
0-7484-0745-6 (hbk)

Pharmaceutical Aspects of Oligonucleotides
Patrick Couvreur and *Claude Malvy (eds)*
0-7484-0841-X (hbk)

Glutamate and GABA Receptors and Transporters

Structure, function and pharmacology

Edited by

Jan Egebjerg, Arne Schousboe and Povl Krogsgaard-Larsen

London and New York

First published 2002 by Taylor & Francis
11 New Fetter Lane, London EC4P 4EE

Simultaneously published in the USA and Canada
by Taylor & Francis Inc,
29 West 35th Street, New York, NY 10001

Taylor & Francis is an imprint of the Taylor & Francis Group

Typeset in 10½/12 Bembo by Wearset, Boldon, Tyne & Wear
Printed and bound in Great Britain by TJ International Ltd, Padstow, Cornwall

British Library Cataloguing in Publication Data
A catalogue record for this book is available from the British Library

Library of Congress Cataloging in Publication Data
Glutamate and gaba receptors and transporters edited by Povl Krogsgaard-Larsen, Jan Egebjerg, Arne Schousboe.
p. cm. — (The Taylor & Francis series in pharmaceutical sciences)
Includes bibliographical references and index.
1. Glutamic acid—Receptors. 2. GABA—Receptors. 3. Carrier proteins. I. Krogsgaard-Larsen, Povl. II. Egebjerg, Jan. III. Schousboe, Arne. IV. Series.

QP364.7 .G567 2001
612.8'042—dc21

2001023961

ISBN 0–7484–0881–9

Contents

Contributors

B. Bettler, Novartis Pharma Inc., Research Department, Therapeutic Area Nervous System, CH 4002, Basel, Switzerland

J. Bockaert, CNRS-UPR 9023, Mecanismes Moleculaires des Communications Cellulaires, CCIPE, 141, rue de la Cardonille, 34094, Montpellier Cedex 5, France

N.G. Bowery, Department of Pharmacology, Medical School, University of Birmingham, Birmingham, B15 2TT

G.L. Collingridge, MRC Centre for Synaptic Plasticity, Department of Anatomy, School of Medical Sciences, University of Bristol, University Walk, Bristol, BS8 1TD

S.G. Cull-Candy, Department of Pharmacology, University College London, Gower Street, London, WC1E 6BT

A.J. Doherty, MRC Centre for Synaptic Plasticity, Department of Anatomy, School of Medical Sciences, University of Bristol, University Walk, Bristol, BS8 1TD

J. Egebjerg, Department of Molecular and Structural Biology, Aarhus University, C.F. Møllers Alle, Building 130, Universitetsparken, DK-8000 Aarhus C, Denmark

B. Ebert, The Royal Danish School of Pharmacy, Department of Medicinal Chemistry, Universitetsparken 2 DK-2100 Copenhagen, Denmark

L. Fagni, CNRS-UPR 9023, Mecanismes Moleculaires des Communications Cellulaires, CCIPE, 141, rue de la Cardonille, 34094, Montpellier Cedex 5, France

N.B. Farber, Department of Psychiatry, Washington University School of Medicine, St. Louis, MO 63110, USA

M. Farrant, Department of Pharmacology, University College London, Gower Street, London, WC1E 6BT

B. Frølund, The Royal Danish School of Pharmacy, Department of Medicinal Chemistry, Universitetsparken 2 SK-2100 Copenhagen, Denmark

J.M. Henley, MRC Center for Synaptic Plasticity, Department of Anatomy, School of Medical Sciences, University of Bristol, University Walk, Bristol, BS8 1TD

D.E. Jane, Department of Pharmacology, School of Medical Sciences, University of Bristol, University Walk, Bristol, BS8 1TD

H.S. Jensen, Department of Molecular and Structural Biology, Aarhus University, C.F. Møllers Allé, Building 130, Universitetsparken, DK-8000 Aarhus C, Denmark

T.N. Johansen, The Royal Danish School of Pharmacy, Department of Medicinal Chemistry, Universitetsparken 2 DK-2100 Copenhagen, Denmark

B. Kanner, Department of Biochemistry, Hadassah Medical School, the Hebrew University, P.O. Box 12272, Jerusalem, Israel, 91120

N. Klix, Novartis Pharma Inc., Research Department, Therapeutic Area Nervous System, CH 4002, Basel, Switzerland

U. Kristiansen, The Royal Danish School of Pharmacy, Department of Medicinal Chemistry, Universitetsparken 2 DK-2100 Copenhagen, Denmark

P. Krogsgaard-Larsen, The Royal Danish School of Pharmacy, Department of Medicinal Chemistry, Universitetsparken 2 DK-2100 Copenhagen, Denmark

L.M. Levy, Department of Physiology, Institute of Basic Medical Sciences, University of Oslo, P.O. Box 1103, Blindern, N-0317, Oslo, Norway

R.L. Macdonald, Departments of Neurology and Physiology, School of Medicine, University of Michigan, Ann Arbor, 48104-1687, MI, USA

U. Madsen, The Royal Danish School of Pharmacy, Department of Medicinal Chemistry, Universitetsparken 2 Dk-2100 Copenhagen, Denmark

G. Meyer, MRC Centre for Synaptic Plasticity, Department of Anatomy, School of Medical Sciences, University of Bristol, University Walk, Bristol, BS8 1TD

H. Möhler, Institute of Pharmacology and Toxicology, ETH and University of Zurich, Winterthurerstr, 1908057, Zurich, Switzerland

J.A. Monn, Lilly Research Laboratories, Eli Lilly and Company, Lilly Corporate Center, drop 0510, Indianapolis, Indiana, 46285, USA

J.W. Newcomer, Department of Psychiatry, Washington University School of Medicine, St. Louis, MO 63110, USA

J.W. Olney, Department of Psychiatry, Washington University School of Medicine, St. Louis, MO 63110, USA

R.W. Olsen, Departments of Molecular and Medical Pharmacology and Anesthesiology, School of Medicine, University of California, Los Angeles, 90095, CA, USA

J.-P. Pin, CNRS-UPR 9023, Mecanismes Moleculaires des Communications Cellulaires, CCIPE, 141, rue de la Cardonille, 34094, Montpellier Cedex 5, France

D.D. Schoepp, Lilly Research Laboratories, Eli Lilly and Company, Lilly Corporate Center, drop 0510, Indianapolis, Indiana, 46285, USA

A. Schousboe, Department of Pharmacology, Royal Danish School of Pharmacy, DK-2100 Copenhagen, Denmark

T.B. Stensbøl, The Royal Danish School of Pharmacy, Department of Medicinal Chemistry, Universitetsparken 2 DK-2100 Copenhagen, Denmark

C. Thomsen, Department for Neurobiology, H. Lundbeck, Ottiliavej 9, DK-2500 Valby, Denmark

W. Wisden, Department of Clinical Neurobiology, University of Heidelberg, Im Neuenheimer Feld 364, D69120 Heidelberg, Germany

Part 1

Ionotropic glutamate receptors

Chapter 1

Ionotropic glutamate receptors

Functional and pharmacological properties in relation to constituent subunits

Stuart G. Cull-Candy

Introduction

Ionotropic glutamate receptors underlie fast excitatory transmission at many central synapses. The fusion of a presynaptic vesicle with the nerve-terminal membrane allows the release of a packet of glutamate into the narrow synaptic cleft. This short-lived chemical signal is converted into an electrical event when neurotransmitter molecules bind to receptors, opening ion channels in the postsynaptic membrane. There is little doubt that many of the principles of transmission, established at the neuromuscular junction (Katz, 1969), also apply at excitatory central synapses. However, important differences do occur. One of the most striking is the variety of functionally distinct postsynaptic receptors that are found in central neurons (reviewed by McBain and Mayer, 1994; Hollmann and Heinemann, 1994; Whiting and Priestley, 1998; Dingledine *et al.*, 1999). The subunits that form glutamate receptors are differentially expressed in the CNS, and their distribution changes during development. Thus individual neuron types express distinct complements of the subunits (Monyer *et al.*, 1994; Akazawa *et al.*, 1995; Geiger *et al.*, 1995; Hollmann and Heinemann, 1994). The resulting heterogeneity in receptor properties allows glutamate to give rise to diverse responses. Information about the mechanisms underlying this diversity in signalling has come from a powerful combination of molecular, functional and pharmacological techniques. This chapter is intended to provide an introductory description of the physiology and pharmacology of glutamate receptors, highlighting some of the key properties of these receptors in relation to their constituent subunits. The following topics will be considered: (1) main classes of receptors, subunits and assemblies; (2) pharmacological properties of AMPA- and kainate-Rs; (3) functional diversity of non-NMDARs; (4) pharmacological and functional characteristics of NMDARs; (5) NMDAR diversity in the CNS.

Classes of ionotropic glutamate receptors, subunits and assemblies

Receptors at the synapse

At many central synapses the transmitter activates a mixed population of non-NMDARs and NMDARs. As illustrated in Figure 1.1, this is reflected in the two component time course of many synaptic currents. Since the NMDAR channel is blocked by Mg^{2+} ions at resting potential, fast transmission is mediated mainly by non-NMDARs. This

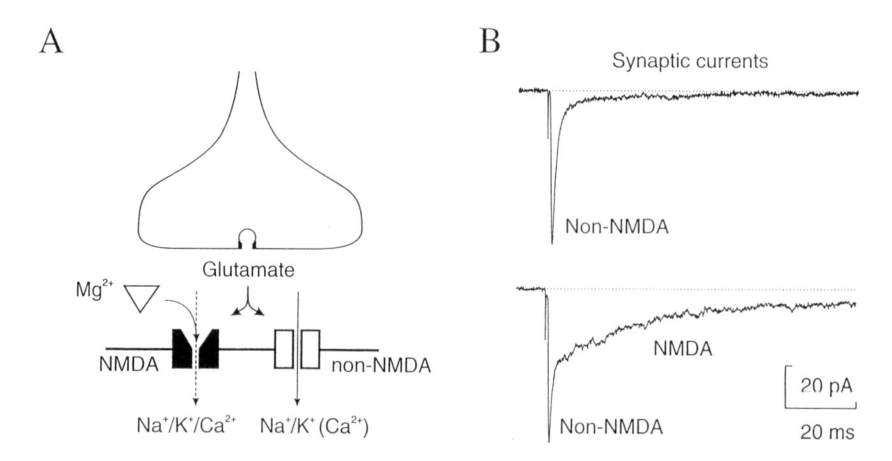

Figure 1.1 Synaptic transmitter activates non-NMDARs and NMDARs, generating a two component EPSC.

(A) Postsynaptic NMDARs are permeable to Na^+, K^+ and Ca^{2+} ions, and are blocked at depolarised potential by Mg^{2+}; non-NMDARs are permeable to Na^+ and K^+ ions. Their permeability to Ca^{2+} depends on the absence of edited subunits (see text). (B) EPSCs recorded under whole-cell clamp (at −60 mV) from the mossy-fibre to granule cell synapse in the cerebellum. *Upper trace:* at the cell's resting potential the EPSC is carried mainly by the non-NMDARs, as NMDAR-channels were blocked by Mg^{2+} ions. *Lower trace:* in the absence of Mg^{2+}-block the EPSC exhibited a two component time course. The fast component was carried by non-NMDARs, the slow component by NMDARs. Modified from Silver *et al.* (1992).

voltage-dependent block is relieved as the postsynaptic cell is depolarized, allowing the NMDAR to function as an 'coincidence detector' of pre- and postsynaptic activity. Thus, the postsynaptic receptor permits an influx of Ca^{2+} and Na^+ ions only at those synapses that receive high frequency presynaptic activity. However, it is worth bearing in mind that the EPSCs at some central synapses are mediated solely by non-NMDARs (e.g. cerebellar Purkinje- and stellate cells: see Momiyama *et al.*, 1996a; Clark and Cull-Candy, 1999).

Identification of AMPA- and kainate receptor subunits

The information revealed by molecular cloning of glutamate receptor subunits (reviewed by Hollmann and Heinemann, 1994; Nakanishi and Masu, 1994; McBain and Mayer, 1994) is central to much of the recent functional work on these receptors. To date, nearly a dozen genes have been cloned encoding non-NMDAR subunits. From their sequence similarities and their pharmacological properties the non-NMDAR subunits fall into two distinct families that constitute the *AMPA-preferring subunits* (activated by α-amino-3-hydroxy-5-methyl-4-isozole propionic acid) and the *kainate-preferring subunits*. As co-assembly does not occur between these two families, individual receptors are composed either of AMPAR- or kainate-R subunits.

From early binding and pharmacological studies it was unclear whether a functional separation existed between kainate- and AMPARs. However, experiments on dorsal root ganglion neurons provided strong evidence that some cell types can possess a 'pure' population of kainate-preferring receptors (see Huettner, 1990, and references therein). Compelling evidence that kainate-preferring subunits formed a separate class of receptors was obtained by Bettler *et al.* (1990) who succeeded in isolating cDNA for GluR5. The low affinity kainate subunits (GluR5, -6 and -7) share only about 20–40 per cent of their 875–890 amino acid sequence with the AMPAR subunits, although their predicted transmembrane structures appear similar.

As illustrated in Table 1.1, the following subunits have been identified: (i) four *AMPA preferring subunits* – GluR1-4 (or GluRA-D), and (ii) five *kainate-preferring sub-*

Table 1.1 Ionotropic glutamate receptor subunits

Non-NMDA

	Splice variants†					*Subfamily*
	i/o	*Q/R*	*R/G*	*M1*	*H*	
GluR1	✓				✓	*AMPAR subunits*
GluR2	✓	✓	✓		✓	
GluR3	✓		✓		✓	
GluR4 (-4c)†	✓		✓		✓	
GluR5 (-1, -2a, b, c)†		✓			✓	*Kainate-R subunits*
GluR6		✓		✓	✓	
GluR7 (-7a, -7b)†					✓	
KA-1						
KA-2						

Key
i/o = possess flip/flop cassette
Q/R = subject to RNA editing at 'Q/R site' in M2
R/G = subject to RNA editing at 'R/G site' preceding flip/flop module
M1 = alternative residues at two (I/V; Y/C) sites in M1
H = subunits that can function as homomeric receptors
† = name of splice variants given in parentheses

NMDA

	Splice variants				
	N-terminal	*C-terminal*			
	N1-cassette (+exon 5)	*C1 (+exon 21)*	*C2 (+exon 22)*	*(C2′)*	
NR1-1a	✗	✓	✓		*NMDAR subunits*
1-1b	✓	✓	✓		
1-2a	✗	✗	✓		
1-2b	✓	✗	✓		
1-3a	✗	✓	✗	✓	
1-3b	✓	✓	✗	✓	
1-4a	✗	✗	✗	✓	
1-4b	✓	✗	✗	✓	
NR2A					
NR2B (splice variants)					
NR2C (splice variants)					
NR2D (splice variants)					
NR3A (splice variants)					
NR3B					

units – GluR5, GluR6 and GluR7 (low affinity kainate subunits), and KA-1 and KA-2 (high affinity kainate subunits). As more kainate subunits have been described, it has become apparent that GluRs 5, 6 and 7 and KA-1, and KA-2 constitute two separate subfamilies with only about 43–46 per cent identity between the groups.

A family of delta subunits (δ1 and δ2) has also been identified (Yamazaki *et al.*, 1992; Lomeli *et al.*, 1993) (not included in Table 1.1). These do not appear to form functional channels either as homomeric assemblies, or in heteromeric combination (when δ subunits are co-expressed with other subunits). However, recent evidence has demonstrated that a single nucleotide change to the δ2 gene gives rise to a channel that is active in the absence of agonist, when expressed in oocytes. This channel appears to be involved in the generation of a large 'constitutively active' current in Purkinje cells of *Lurcher* mice, accounting for neurodegeneration of these cells in the mutant (Zuo *et al.*, 1997).

Isoforms of AMPA- and kainate receptor subunits

As a result of alternative splicing and RNA editing there are additional isoforms of both AMPA- and kainate-R subunits (see Table 1.1 and Figure 1.2). Each AMPAR subunit exists in at least two functionally important splice variants which differ in their 'flip/flop' module – a cassette of 38 amino acids in the extracellular loop between transmembrane domains M3 and M4 (see Hollmann and Heinemann, 1994). Furthermore, certain AMPAR subunits form channels with high Ca^{2+}-permeability. The structural determinant for Ca^{2+}-permeability is a single amino acid residue at the 'Q/R-site' in the pore-lining region (the M2 domain). This site is occupied either by a neutral glutamine (Q) or a positively charged arginine (R). The genes for all AMPA- and kainate-subunits code for a glutamine at this position, the arginine being introduced into GluR2 (but not GluR1, 3, and 4) by site-specific editing of the subunit RNA. GluR2 thus exists *only* in the edited (arginine containing) form *in vivo*, and its presence within a receptor confers low divalent cation permeability (see Jonas and Burnashev, 1995).

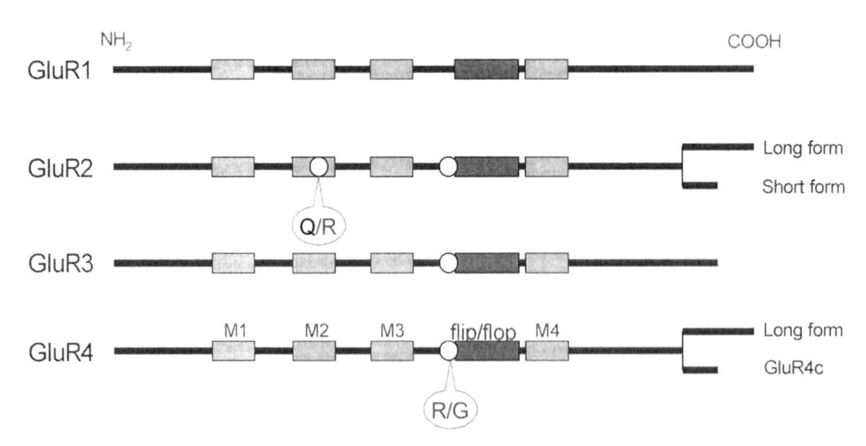

Figure 1.2 Splice and editing variants of the AMPAR subunits.

Sites of RNA editing are indicated. These are the Q/R-site in M2 of GluR2, and the R/G-site immediately preceding the flip/flop cassette of GluR2, 3 and 4. GluR2 and GluR4 exist in two isoforms which differ in their C-terminal lengths. GluR4c is the short form of GluR4, identified in cerebellar granule cells.

In the AMPAR subunits the R/G (arginine/glycine) site, immediately preceding the flip/flop module of GluR2, 3 and 4 (but not GluR1) is also subject to editing (Figure 1.2 and Table 1.1). Editing at this site imparts particularly rapid onset and faster recovery rates from desensitization on those subunits where it occurs. About 80–90 per cent of affected subunits are edited *in vivo* (apart from GluR4 (flip) which is 50 per cent edited) (Sommer *et al.*, 1990; Lomeli *et al.*, 1994). Other splice variants of AMPAR subunits also occur. For example, differential splicing gives rise to a variant of GluR4 (GluR4c) with a short C-terminus. This is expressed at high levels in cerebellar granule cells (Gallo *et al.*, 1992). On the other hand, a small percentage of GluR2 exhibits a long C-terminus (see Hollmann and Heinemann, 1994) The functional relevance of these different C-terminal isoforms is, as yet, unclear.

Variation also occurs within the five kainate subunits, as indicated in Table 1.1 (Sommer *et al.*, 1992; Seeburg, 1996). GluR5 and GluR7 exist in several splice variants (GluR7a, b and GluR5-1, 2a, b, c) while GluR5 and GluR6 (but not GluR7a, b) are subject to Q/R site editing (Bettler *et al.*, 1990; Sommer *et al.*, 1992; Egebjerg and Heinemann, 1993; Seeburg, 1996; Schiffer *et al.*, 1997; reviewed by Hollmann and Heinemann, 1994; Chittajallu *et al.*, 1999, Hollmann, 1999). Q/R site editing of kainate subunits also appears to result in receptors with low divalent cation permeability (Egebjerg and Heinemann, 1993; Swanson *et al.*, 1996). Some initial confusion about this property arose from the fact that homomeric GluR6 edited receptors are not purely cation selective, but also permit the passage of anions (Kohler *et al.*, 1993; see Burnashev, 1996). KA1 and KA2 subunits are not subject to editing (Hollmann and Heinemann, 1994).

In addition to Q/R site editing, GluR6 possesses two further sites in the first transmembrane segment (M1) that are occupied by different amino acids (Köhler *et al.*, 1993). One of these sites (isoleucine/valine: I/V) is encoded by the gene, the other (tyrosine/cysteine: Y/C) by the edited transcript (Seeburg, 1996). However, the functional role of these alternative residues in M1 is not clearly understood. The GluR6 isoform edited in both M1 and M2 constitutes about 65 per cent of the total brain GluR6 mRNA, whereas the fully unedited form amounts to only ~10 per cent. Thus unlike GluR2 which exists only in the edited form *in vivo*, GluR5 and GluR6 can exist as a mixture of edited and unedited subunits. This raises the intriguing possibility that cellular control of editing could play a role in influencing receptor properties at synapses possessing kainate-R subunits. This would contrast with the situation at AMPAR-synapses, where channel properties depend on the *level* of GluR2 subunit expression (Köhler *et al.*, 1993, Geiger *et al.*, 1995). However, even for AMPARs the situation is also likely to be complicated, since subcellular *targeting* of GluR2 also plays an important part in determining Ca^{2+}-permeability of currents (Tóth and McBain, 1998; Liu and Cull-Candy, 2000).

Non-NMDAR subunits form functional homomeric and heteromeric assemblies

Considerable functional diversity arises from the fact that AMPAR subunits can generate functional receptor channels either as homomeric or heteromeric complexes. There is evidence for both of these forms *in vivo*. Similarly, kainate receptors can function as homomeric or heteromeric assemblies. However, while GluR5, 6 and 7 can produce functional homomeric channels (Egebjerg *et al.*, 1991, Sommer *et al.*, 1992; Schiffer *et*

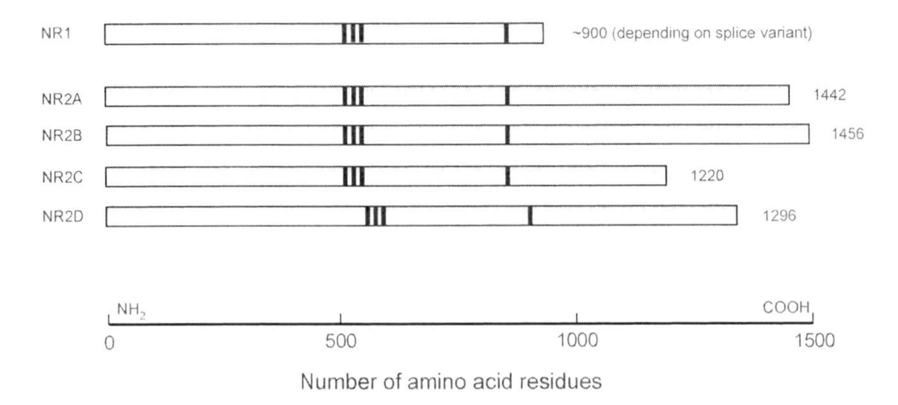

Figure 1.3 Comparison of the lengths of the various NMDAR subunits.

Note the short C-terminus of the NR1 subunit, and the similarity in length of NR2A, NR2B versus that of NR2C, NR2D.

al., 1997), KA-1 and KA-2 form functional receptors *only* when co-assembled with their low affinity (GluR5–7) partners. As described below, the presence of KA-1 and KA-2 in the complex modifies receptor properties in various ways (Herb *et al.*, 1992; Swanson *et al.*, 1996). Recent evidence has demonstrated that GluR5, 6 and 7 can co-assemble with each other to form functional heteromeric receptors (Cui and Mayer, 1999). The fact that KA1 and KA2 cannot form functional receptors without the participation of low affinity subunits makes it very likely that at least some native kainate receptors are composed of two or more subunit types.

Identification of NMDAR subunits

The NMDAR subunits have low homology with the non-NMDAR subunits, with less than 30 per cent identity of amino acid residues. However, the transmembrane topology of the subunits is thought to be similar. At least three families of NMDA receptor subunits have been identified. The relative lengths of the members of two of the main families (NR1 and NR2) are illustrated in Figure 1.3. The C-terminal of the NR1 subunit is shorter than for the NR2 family, and depends on the splice variant. NR2A and NR2B subunits are similar to each other in length, as are NR2C and NR2D.

The NR1 family

This consists of a single family member that exists in at least eight functional isoforms arising from a single gene (Moriyoshi *et al.*, 1991; Sugihara *et al.*, 1992). Combinations of three independent splice variations (see Figure 1.4) give rise to the eight isoforms (Table 1.1). These are generated by the insertion of exon 5 in the N-terminal (referred to as the N1 cassette; Figure 1.4), the deletion of exon 21 in the C-terminal (the C1 cassette), and use of an alternative splice acceptor site in the C-terminal (exon 22). When this is used it results in an alternative C-terminal cassette (C2′, instead of C2).

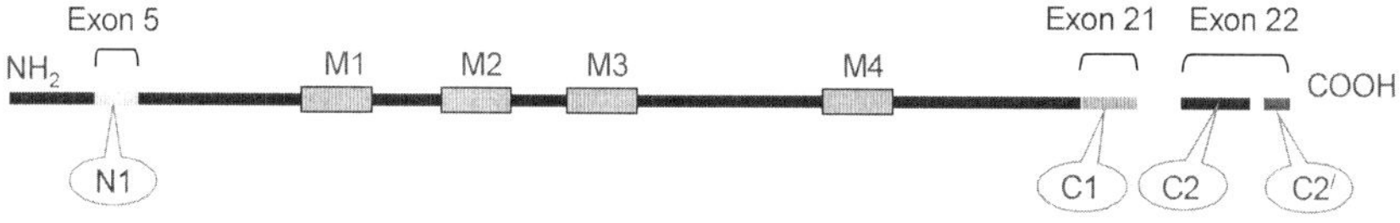

Figure 1.4 Linear depiction of the splice variants of the NR1 NMDAR subunit.

Three independent splice variations produce eight possible isoforms (see also Table 1.1). Insertion of exon 5 in the N-terminal generates the N1 cassette. Deletion of exon 21 in the C-terminal causes the loss of the C1 cassette. Use of an alternative splice acceptor site, exon 22, in the C-terminal generates an alternative C-terminal cassette C2′, instead of C2.

The NR1 splice variants are differential distributed throughout the CNS (Laurie and Seeburg, 1994; reviewed in Dingledine *et al.*, 1999), and vary in several basic properties. The N1 (exon 5) cassette potentiates NMDARs through tonic relief of proton block (Traynelis *et al.*, 1995). Furthermore, splice variants lacking N1 are potentiated by polyamines (Durand *et al.*, 1993; Hollman and Heinemann, 1994; Traynelis *et al.*, 1998) which act by relieving the normal proton block. At physiological pH, ambient protons can reduce the NMDAR-mediated current by as much as ~50 per cent (Traynelis and Cull-Candy, 1990). It has been suggested that exon 5 (the NI cassette) forms a surface loop that is structurally similar to polyamine (Traynelis *et al.*, 1995), acting as a tethered modulator of the pH-sensitive site on the NR1 splice variants. Furthermore, certain C-terminal cassettes interact with intracellular proteins. The C2′ cassette, present in NR1-3 and NR1-4, interacts with the postsynaptic density protein PSD-95, while the C1 cassette, present in NR1-1 and NR1-3 binds to neurofilaments and is involved in NMDAR clustering (see Ehlers *et al.*, 1998; Lin *et al.*, 1998). Domains of the NR1 subunit form the binding site for glycine on the NMDAR-protein (Wafford *et al.*, 1995; Kurytov *et al.*, 1994)

The NR2 family

This consists of four members: NR2A, 2B, 2C and 2D arising from distinct genes (see Moriyoshi *et al.*, 1991; Kutsuwada *et al.*, 1992; Monyer *et al.*, 1994; Ishii *et al.*, 1993; Wisden and Seeburg, 1993; Hollmann and Heinemann, 1994; McBain and Mayer, 1994). There is currently no evidence of alternatively spliced NR2A subunits. However, there is evidence for a variety of splice variants of NR2B, NR2C and NR2D (see Hollmann, 1999). The functional consequences of this are unclear. The *NR2 subunits* form the glutamate binding sites (Laube *et al.*, 1997; Anson *et al.*, 1998)

NR3 subunits (or NMDA-χ)

Two members of this family have been identified, namely NR3A and NR3B (Ciabarra *et al.*, 1995; Sucher *et al.*, 1995; Das *et al.*, 1998). Recent experiments have demonstrated that NR3A subunits can co-assemble with certain NR1/NR2 subunit complexes (Perez-Otana *et al.*, 2001).

NMDAR subunits form heteromeric assemblies

There is overwhelming evidence that NMDARs function as heteromultimeric assemblies, formed from NR1 in combination with at least one type of NR2 subunit. It is known that the NR2 subunits are critical in determining certain key properties of NMDARs. Furthermore, their differential distribution within the CNS gives rise to diversity in the NMDAR subtypes (Monyer *et al.*, 1994; Farrant *et al.*, 1994; Cull-Candy *et al.*, 1995; Momiyama *et al.*, 1996a; Stocca and Vicini, 1998; Wyllie *et al.*, 1998). The type of NR1 isoform involved in the assembly is also of functional importance – although this is less well characterized. The NR3 family, of which only the NR3A subunit has so far received much attention, has been suggested to be a regulatory NMDAR subunit expressed primarily during early development in some regions of the CNS (Das *et al.*, 1998). When co-assembled with NR1 and NR2A it gives rise to NMDARs (NR1/NR2A/NR3A) with reduced Ca^{2+} permeability (Perez-Otana, 2001).

Subunit topology: three membrane spanning domains and a pore loop

Although the identity of many of the subunits is now well established, molecular studies have continued to clarify our picture of the glutamate receptor's transmembrane topology. Recent molecular studies have resulted in a reappraisal of the topology of both non-NMDAR and NMDAR subunits (Wo and Oswald, 1994; Roche *et al.*, 1994, Hollmann *et al.*, 1994; Bennett and Dingledine, 1995; Stern-Bach *et al.*, 1994). The transmembrane architecture, illustrated in Figure 1.5, has been inferred from diverse approaches, including structural studies, examination of ligand binding properties and binding domains of modified subunits, and from the consensus sites for N-linked glycosylation (Wo and Oswald, 1994; Roche *et al.*, 1994, Hollmann *et al.*, 1994; Bennett and Dingledine, 1995). Construction of AMPA-/kainate-R subunit chimeras has revealed that regions in both the amino-terminal (150 amino acids preceding M1) and residues in most of the loop between M3 and M4 are required for agonist selectivity. These extracellular domains, referred to as S1 and S2 respectively, are suggested to show sequence similarities to the bacterial periplasmic amino acid binding proteins (Paas, 1998). The fact that the segments S1 and S2 are important in ligand recognition in AMPA- and kainate receptors is strong support for the idea that the N-terminus and the loop joining M3 and M4 are located extracellularly (Stern-Bach *et al.*, 1994). Based on the bacterial protein structure, models of the glutamate binding site of ionotropic receptors suggest the presence of two lobes separated by a cleft (reviewed by Green *et al.*, 1998). Experiments on a GluR2 S1–S2 soluble fusion protein from *E. coli* have generated constructs that bind kainate and have a bilobed structure (Armstrong *et al.*, 1998). Binding in this construct occurs in the crevice between S1 and S2 from the same subunit, rather that between adjacent subunits.

Considered together, these various approaches provide compelling evidence for a subunit structure that contains an extracellular N-terminus (part of which forms S1), an intracellular C-terminus, and an extracellular loop between M3 and M4 that constitutes S2 and contains the flip-flop cassette (Figure 1.5). Thus only three regions (M1, M3 and M4) completely span the membrane, while the channel lining domain is formed by a re-entrant loop (M2), that dips into the membrane from the cytoplasmic side. The

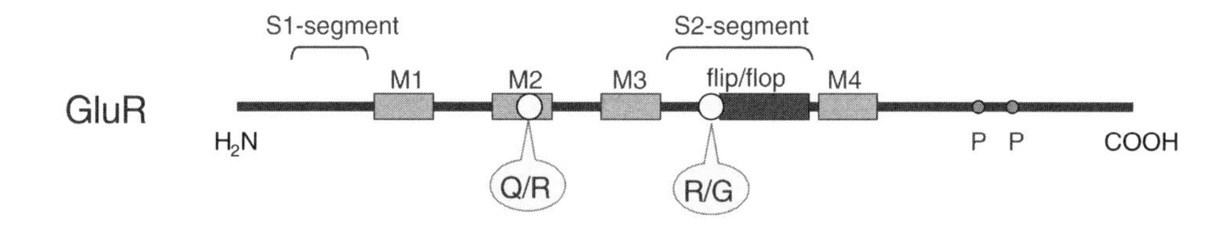

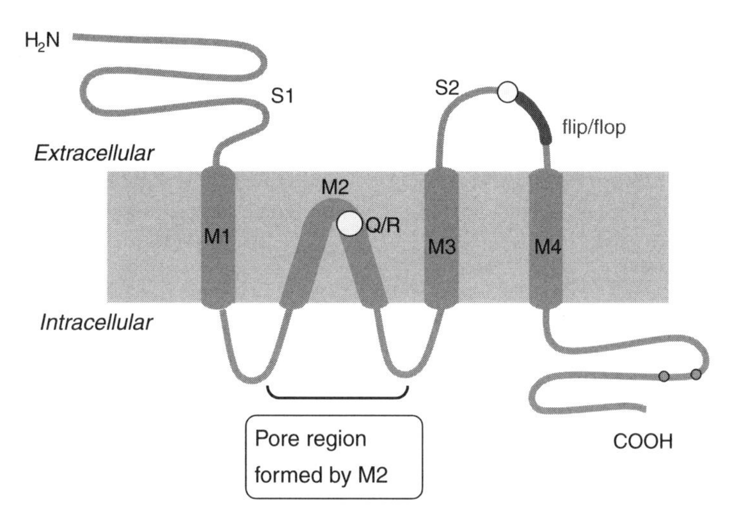

Figure 1.5 Architecture of the AMPAR subunit.

Upper figure shows a linear representation of the GluR subunit; the lower figure indicates transmembrane topology. Only three segments span the membrane: M1, M3 and M4. The pore lining region is generated by M2 which enters the membrane from the intracellular surface forming a 'pore loop' structure. The flip/flop cassette is in the extracellular loop between M3 and M4. The Q/R editing site is in the pore lining region, and the R/G editing site precedes the flip/flop module. Sites of phosphorylation are present on the intracellular C-terminus. The S1 and S2 domains, which are involved in glutamate binding, form part of the N-terminus and the extracellular loop between M3 and M4.

critical 'Q/R site' is located within the re-entrant 'hairpin' loop, while sites on the C-terminus are accessible to a host of intracellular factors, including binding- and signalling-proteins. The consensus phosphorylation sites on the C-terminus may be particularly important in certain receptor functions.

Pharmacological properties of AMPA- and kainate-Rs

We now turn to an overview of the main pharmacological properties of the non-NMDA receptors (summarized in Table 1.2).

Agonists

As expected, AMPA- and kainate-Rs are activated by the transmitter glutamate, but with relatively high EC_{50} values (~500 μM). Homomeric GluR7 receptors are

Table 1.2 Pharmacological and functional properties of non-NMDA receptors

AMPA-receptors	*Kainate-receptors*
Agonists	
• Activated by glutamate ($EC_{50} \sim 500\,\mu M$)[1]	• Activated by glutamate with EC_{50} values of $\sim 500\,\mu M$[2] (GluR7, $EC_{50} \sim 6\,mM$)[3]
• All activated by AMPA[4]	• Some activated by AMPA[5]
• All activated by kainate ($EC_{50} \sim 100\,\mu M$)[6]	• All activated by kainate ($EC_{50} \sim 5–10\,\mu M$)[6]
• Activated by domoate[6]	• Domoate is a high affinity ligand (except for homomeric GluR7)[6,3]
• Activated by SYM 2081 (4-methylglutamate), $EC_{50} = 200–300\,\mu M$[7]	• Activated by SYM 2081(4-methylglutamate) $EC_{50} < 1\,\mu M$.[7] ATPA is an agonist of GluR5[8]
• Rapid desensitization with AMPA or glutamate, *much less* rapid with kainate[9]	• Rapid desensitization with glutamate and kainate, less rapid with domoate[9]
• Desensitization suppressed by diazoxide, cyclothiazide[9] and PEPA[10]	• Desensitization suppressed by concanavalin-A[11] (GluR7 weakly effected)[3]
Antagonists	
• Selectively blocked by GYKI-53655[12] and ATPO[13]	• Insensitive to GYKI compounds[12]
• Unaffected or slightly potentiated by La^{3+}[14]	• Blocked by La^{3+} ($10\,\mu M$)[14]
• Blocked by CNQX, NBQX and related compounds[15,16]	• Blocked by CNQX, NBQX and related compounds[15]
• Not affected by NS-102 or LY294486; weakly antagonized by LY293558[16]	• Recombinant GluR6 blocked by NS-102; recombinant GluR5 blocked by LY294486 and LY293558[16]
Permeability	
• High- or low-Ca^{2+} permeability (depending on editing of subunits)[17]	• High- or low-Ca^{2+} permeability[17]
• Channel conductances ~200fS-30 pS (depending on editing)[18]	• Channel conductances ~200fS-30 pS[19]

Notes
1. Lomeli *et al.* (1994); Partin *et al.* (1996); 2. Heckmann *et al.* (1996); Swanson *et al.* (1996); Traynelis and Wahl, 1997. 3. Schiffer *et al.* (1997); 4. Hollmann *et al.* (1989); Nakanishi *et al.* (1990). 5. Egebjerg *et al.* (1991); Herb *et al.* (1992); Swanson *et al.* (1996). 6. Hollmann *et al.* (1989); Nakanishi *et al.* (1990); Patneau *et al.* (1994). 7. Wilding and Huettner (1997); Zhou *et al.* (1997); Donevan *et al.* (1998). 8. Bleakman and Lodge (1998) 9. Partin *et al.* (1993); Wong and Mayer (1993). 10. Sekaguchi *et al.* (1997). 11. Huettner, (1990); Partin *et al.* (1993). 12. Donevan and Rogawski (1993); Paternain *et al.* (1995); Wilding and Huettner (1995); Bleakman *et al.* (1996). 13. Wahl *et al.* (1998). 14. Reichling and MacDermott (1991); Huettner *et al.* (1998). 15. Honoré *et al.* (1988), Nakanishi *et al.* (1990); Sommer *et al.* 1992. 16. Reviewed by Bleakman and Lodge (1998). 17. Egebjerg and Heinemann (1993); Jonas and Burnashev (1995). 18. Swanson *et al.* (1997a); Wyllie *et al.* (1993). 19. Sahara *et al.* (1997); Howe (1996); Swanson *et al.* (1996);Traynelis and Wahl (1997); Pemberton *et al.* (1998); Huettner (1990).

exceptional in being remarkably insensitive to glutamate, with an EC_{50} in the mM range. This compares with an EC_{50} for glutamate of ~1 μM for some NMDARs (depending on subunit composition). The difference in potency of glutamate on non-NMDARs and NMDARs is likely to relate to their different functional roles.

All AMPARs and a number of kainate-Rs are activated by low concentrations of AMPA; the AMPA-sensitive kainate-R assemblies include heteromeric receptors composed of GluR6/KA-2 and GluR7/KA-2 (or KA-1). Homomeric GluR6 and GluR7 receptors are insensitive to AMPA (see Schiffer *et al.*, 1997), allowing them to be readily distinguished from heteromeric GluR6 and R7 assemblies. However, it is worth noting that the various GluR5 assemblies cannot be distinguished in this way, since both homomeric GluR5 and heteromeric GluR5/KA-2 assemblies respond to AMPA. Remarkably, it has been revealed that a single amino acid residue determines whether the kainate-R subunits GluR5 and GluR6 are AMPA-sensitive (Swanson *et al.*, 1997b).

Kainate is a high affinity ligand for all recombinant kainate receptors, with an EC_{50} of roughly 5–10 μM, and a rapidly desensitizing response. It also activates all AMPARs, although with a lower potency and a relatively large steady-state response (i.e. less marked desensitization). Domoate, a more potent agonist for kainate receptor channels, gives currents that deactivate slowly and desensitize incompletely, displaying a substantial steady-state component (Herb *et al.*, 1992; Sommer *et al.*, 1992). Domoate is an agonist at all kainate receptors, except for homomeric and heteromeric GluR7 assemblies – where it may act as an antagonist, or produce a small current response (Schiffer *et al.*, 1997). Thus, domoate does not *always* identify the presence of kainate receptors; furthermore, at sufficiently high concentrations it can also activate large steady state AMPAR-responses.

Both kainate-receptors and AMPARs rapidly desensitize in response to glutamate. Several amino acid residues appear to be important in this phenomenon. Mutations at the N-terminal end of the S2 region of GluR1 (residue S650; Mano *et al.*, 1996) and the corresponding residue (A689) of GluR6 (Swanson *et al.*, 1997b) have been shown to modify desensitization properties. Furthermore, experiments by Stern-Bach *et al.* (1998) indicate that residue L507 in the S1 region of GluR3 and GluR1 is critical in desensitization. A single mutation (L507Y) at this residue, which is conserved in all AMPAR subunits, abolishes desensitization (Stern-Bach *et al.*, 1998). However, this site does not appear important in controlling desensitization of the kainate-R subunits. The kainate receptor ligand 4-methylglutamate or SYM 2081 (Zhou *et al.*, 1997; Wilding and Huettner, 1997) can be used to activate kainate-Rs at concentrations that produce little detectable response from AMPARs (Pemberton *et al.*, 1998). Furthermore, kainate-R desensitization, which is rapid in response to kainate and glutamate, can be suppressed by pre-treatment with lectins such as concanavalin-A (Con-A). This approach has permitted responses to be detected from various native kainate-Rs, including those in sensory neurons (Huettner, 1990), glial cells of oligodendrocyte lineage (Patneau *et al.*, 1994), and homomeric GluR5 edited recombinant receptors previously thought not to produce functional channels (Swanson *et al.*, 1996). However, the rather *anomalous* kainate subunit GluR7 is potentiated only modestly by Con-A (Schiffer *et al.*, 1997) when compared with other kainate-Rs (see Partin *et al.*, 1993). Recent work, using mutagenesis to investigate the molecular basis for the effect of Con-A, has concluded that lectin binding inhibits the conformational changes required to shift the receptor to its desensitized state (Everts *et al.*, 1999).

Certain drugs are capable of distinguishing between flip and flop isoforms of AMPAR subunits. Cyclothiazide suppresses desensitization only of the *flip* isoform. In contrast the molecule PEPA (4-[2-(phenylsulfonylamino) ethylthio]-2,6-difluorophenoxyacetamide), preferentially potentiates (by up to fifty-fold) *flop* isoforms, compared with only a three-fold potentiation of *flip* isoforms (Sekiguchi *et al.*, 1997).

Antagonists

CNQX, NBQX and DNQX have all proved excellent broad-spectrum blockers of non-NMDA receptors. While these have not usually been effective at distinguishing between AMPA- and kainate-Rs, low concentrations of extracellular La^{3+} blocks kainate-Rs relatively selectively (Reichling and MacDermott, 1991; Huettner *et al.*, 1998). The discovery that the 2,3 benzodiazepines (GYKI compounds: GYKI-52466 and GYKI-53655) act as non-competitive antagonists at non-NMDARs with more selectivity towards AMPARs (Donevan and Rogawski, 1993; Wilding and Huettner, 1995; Paternain *et al.*, 1995), has helped identify kainate-R mediated components at certain synapses (Castillo *et al.*, 1997; Vignes *et al.*, 1997; reviewed by Mody, 1998). Although it has not yet been widely exploited, the AMPAR antagonist ATPO (R,S)-2-amino-3-[5-tert-butyl-3-(phosphonomethoxy)-4-isoxazolyl] propionic acid, is a potent competitive antagonist of AMPARs, with less potency on the GluR4 subunits (Wahl *et al.*, 1998) and with slight antagonist activity on GluR5. A number of newer drugs developed as kainate-R antagonists appear to show subunit selective for GluR5 and AMPARs (LY 293558), or GluR5 alone (LY294486), or GluR6 subunits (NS-102), when examined on recombinant receptors (for further details see Bleakman and Lodge, 1998; Mody, 1998).

Functional diversity of non-NMDARs

RNA editing and Ca^{2+}-permeability

One of the critical developments in the understanding of non-NMDARs has been the discovery of receptors with a high permeability to Ca^{2+} ions (Iino *et al.*, 1996; Ozawa *et al.*, 1998), and the subsequent identification of the underlying molecular determinant (reviewed by Seeburg, 1996). For both AMPA- and kainate-Rs, Ca^{2+}-permeability reflects the *absence* of edited subunits within the assembly (Jonas and Burnashev, 1995). It has become apparent that Q/R site editing influences a variety of functionally important receptor properties (summarized in Figure 1.6).

From early experiments it was clear that Ca^{2+}-permeable non-NMDARs could be distinguished by their characteristic rectifying I–V relationship (Iino *et al.*, 1996; Hollmann and Heinemann, 1994), which contrasted with the ohmic relationship of the Ca^{2+}-impermeable receptors. This rectification resembles that of inwardly rectifying K^+ channels (KIR channels), which possess a 'pore loop' structure. Thus the identification of a 'pore loop' domain (M2 in Figure 1.5) within glutamate receptors prompted several labs independently to investigate whether rectification of these two channel types may have a common origin. As a result, evidence was obtained that rectification of the unedited non-NMDA receptor is not an intrinsic property of the channel, but conferred by an intracellular factor (spermine) (Kamboj *et al.*, 1995; Bowie and Mayer, 1995; Koh *et al.*, 1995a).

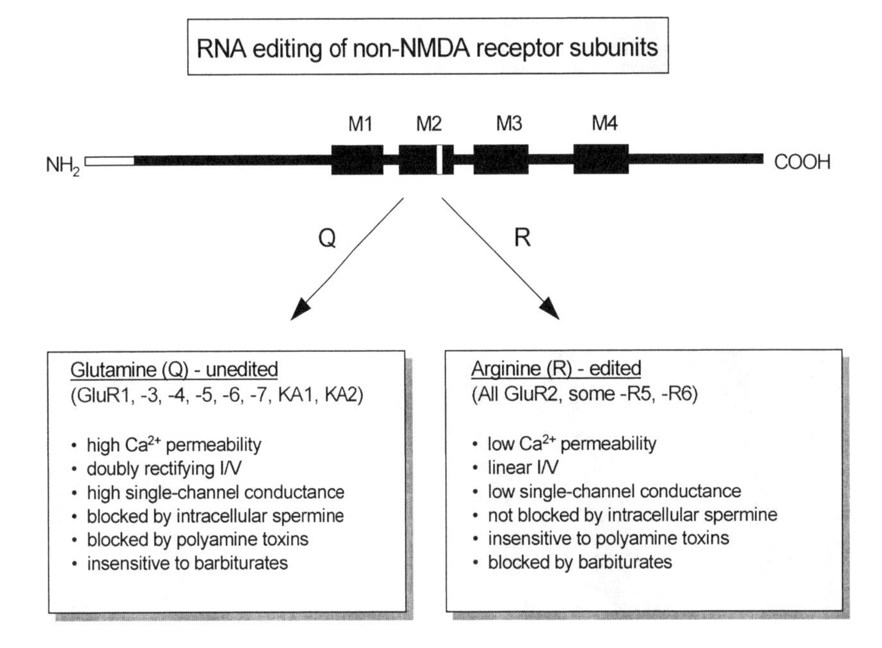

Figure 1.6 Comparison of the main properties displayed by unedited and edited non-NMDA receptors.

The two boxes indicate the subunits which fall into each category, and the main functional and pharmacological and functional properties of the resulting receptors (see text for details). Note that while 'unedited' receptors are composed entirely of subunits with glutamine at the Q/R site, a single edited subunit appears sufficient to confer certain of the characteristic 'edited' properties on a receptor. The characteristic features of edited receptors are likely to reflect the influence of the positively charged arginine, replacing neutral glutamine.

The relative abundance of the native GluR2 subunit has been shown to determine Ca^{2+}-permeability of AMPARs in various neuron types (Wisden and Seeburg, 1993; Jonas *et al.*, 1994; Bochet *et al.*, 1994; Geiger *et al.*, 1995; Otis *et al.*, 1995). However, recent experiments suggest that within individual cells the selective targeting of GluR2 subunits may also play a role in dictating the Ca^{2+}-permeability of synaptic receptors. Thus, many local circuit inhibitory interneurons in the hippocampus express low levels of GluR2 (Geiger *et al.*, 1995; Racca *et al.*, 1996; Leranth *et al.*, 1996). However, while these cells express Ca^{2+}-permeable AMPA receptors at their mossy fibre inputs, the EPSCs at their CA3 inputs are *Ca^{2+}-impermeable* (Tóth and McBain, 1998). Furthermore, in cerebellar stellate cells, high frequency synaptic activity directly influences the targeting of GluR2 subunits, triggering an activity dependent switch in the Ca^{2+} permeability of synaptic receptors (Liu and Cull-Candy, 2000).

Block by polyamines

Spermine produces voltage-dependent block of all the Ca^{2+}-permeable forms of AMPA- and kainate-Rs that have been examined, with little effect on Ca^{2+}-impermeable forms. However, during patch-clamp recording from the whole-cell or isolated patch,

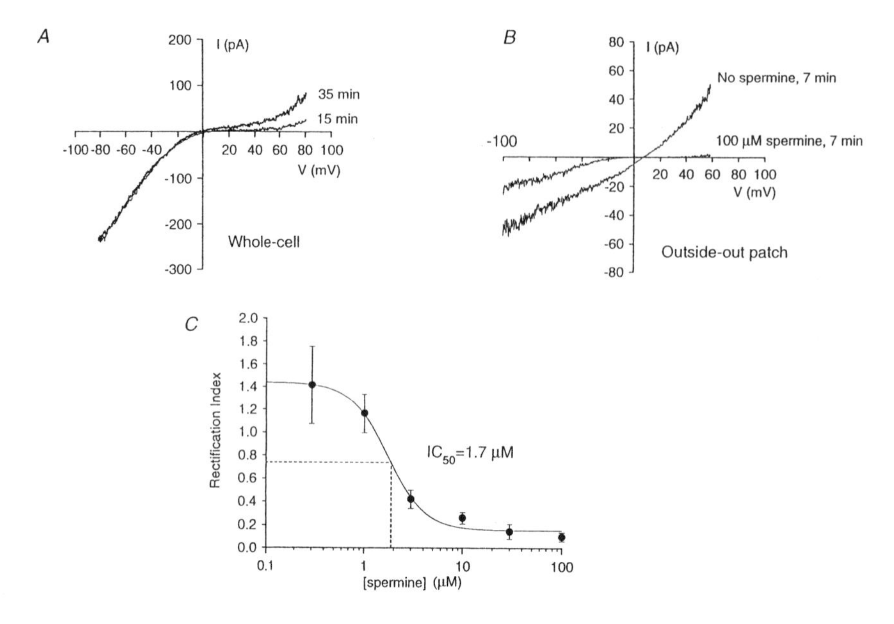

Figure 1.7 Intracellular spermine confers inward rectification on recombinant Ca^{2+}-permeable non-NMDARs receptors.

(A) Plot of agonist evoked current (I) vs membrane potential (V) obtained with whole-cell recording from a single cell transfected with GluR6Q/KA-2 (kainate-Rs). Responses were examined 15 and 35 minutes after the start of whole-cell recording (with no added spermine). The cell shows slight loss of rectification after 35 minutes.
(B) Relationship between agonist evoked current and membrane potential in two outside-out patches. In the absence of added spermine no rectification was seen 7 minutes after patch excision. In a second patch, the inclusion of 100 μM spermine in the pipette solution prevented loss of inward rectification.
(C) Inhibition curve illustrating the relationship between spermine concentration and 'Rectification Index' (a measure of the degree of rectification). The arrow indicates the IC_{50} of 1.7 μM. (Modified from Kamboj *et al.*, 1995)

endogenous intracellular spermine diffuses away and the rectification is gradually lost (Figure 1.7A, B). The speed of dialysis depends on cell size and is less apparent in large cells such as oocytes. As indicated in Figure 1.7B, the rectification of the Ca^{2+}-permeable channels can be readily reinstated by inclusion of spermine (100 μM) in the 'intracellular' pipette solution (Kamboj *et al.*, 1995; Bowie and Mayer, 1995; Koh *et al.*, 1995a). It is thought that polyamines, which are highly charged tetravalent molecules, act as weakly permeable voltage-dependent open channel blockers (see Bähring and Mayer, 1998). Interestingly, the affinity of polyamines differs between GluR subtypes, with kainate receptors having a lower affinity than AMPARs (Bowie and Mayer, 1995).

Surprisingly, it turns out that polyamines also block Ca^{2+}-permeable channels in their closed state (Bowie *et al.*, 1998; Rozov *et al.*, 1999). This block is not voltage-dependent. It has been found that use-dependent relief of intracellular polyamine block occurs mainly when channels are opened at hyperpolarized potentials. As a result, brief repetitive pulses of glutamate applied at the cell's resting potential produce responses that facilitate. This has been ascribed to the fact that the initial response arises from receptors that have become blocked by spermine in their closed state. At negative potentials, a high frequency train of glutamate pulses relieves the polyamine block. As relief occurs more rapidly at negative potentials, a single pulse of glutamate unblocks most channels. Thus, an apparently stronger facilitation is expected to occur at less negative potentials, when a larger pool of blocked channels remains after the initial pulse of glutamate (see Rozov and Burnashev, 1999).

Is spermine the 'intracellular factor' that normally confers rectification on these channels? Spermine is an intracellular constituent that occurs widely in the CNS. Although its concentration in oocytes is ~100–300 μM, buffering by ATP and nucleic acids results in a *free* concentration of only about 8–80 μM (see Kamboj *et al.*, 1995 for refs). As shown in the inhibition curve in Figure 1.7C, experiments on recombinant receptors indicate an IC_{50} of 1.7 μM. Hence, an intracellular concentration of ~2 μM would be sufficient to produce ~50 per cent of the maximum rectification (Kamboj *et al.*, 1995). Consequently, processes that regulate intracellular spermine may influence the degree of rectification. Such processes might include the relative rates of spermine synthesis and degradation, or changes in the general metabolic state that lower ATP levels. This would be expected to result in reduced buffering of free spermine, increasing its free intracellular concentration. Spermine block may therefore be important not only in modifying the normal Ca^{2+}-permeable EPSC (Rozov *et al.*, 1999) but also during ischaemia – when ATP levels are likely to fall dramatically. In these conditions, when cells are depolarized by high levels of extracellular glutamate (Lee *et al.*, 1999), an enhanced block of the Ca^{2+}-permeable non-NMDA channels could serve to limit lethal Ca^{2+} entry.

Identification of Ca^{2+}-permeable channels with polyamine toxins and barbiturates

Experiments with invertebrate toxins have provided further evidence that pores composed entirely of *unedited* subunits have distinctive properties. The Q/R site influences the susceptibility of AMPA- and kainate-Rs to block by extracellular polyamine toxins present in spider and wasp venoms – including Joro spider toxin, argiotoxin and philanthotoxin (Blashke *et al.*, 1993; Herlitze *et al.*, 1993; Brackley *et al.*, 1993; Bahring and Mayer, 1998). The toxin structures consist of a polyamine chain with aromatic amino acid attached at one end. It is likely that the linear polyamine chain penetrates, and blocks, the ion channel while the aromatic group interacts with amino acid residues located above the narrowest region of the channel (see Anis *et al.*, 1990). Perhaps surprisingly, there is some permeation of toxin molecules through the channel (Bahring and Mayer, 1998). The critical Q/R site glutamine residue that allows Ca^{2+} ions to pass also permits block by Joro toxin, argiotoxin and philanthotoxin. This site corresponds to the homologous asparagine (N) residue in M2 of NMDAR subunits, that is involved in block of NMDA channels by Mg^{2+} ions and argiotoxin (Raditsch *et al.*, 1993). This 'Q/R/N site' is thus thought to face the ion permeation pathway, and hence form part of the selectivity filter.

Ca^{2+}-impermeable forms of AMPARs are relatively selectively blocked by pentobarbital, which has less effect on GluR2-lacking AMPARs (Yamakura *et al.*, 1995). Thus, the combined use of these two molecules (polyamine toxins and barbiturates) allows the editing state of native non-NMDARs to be distinguished with reasonable certainty (Liu and Cull-Candy, 2000).

Editing at the Q/R site determines the single channel conductance

Studies of recombinant receptors have demonstrated that assemblies composed entirely of unedited subunits have a high single channel conductance, when compared with their edited counterparts. The low conductance is particularly striking in recombinant homomeric receptors composed of edited subunits e.g. GluR2, or GluR6(R). These give rise to 'femtosiemens' channels that are too small to be directly resolved (Swanson *et al.*, 1996, 1997a; Howe, 1996). In the case of GluR6 subunits, homomeric receptors composed of subunits edited at the Q/R site exhibit a conductance that is ~25–30 times lower than for the equivalent unedited receptor (Swanson *et al.*, 1996; Traynelis and Wahl, 1997).

Figure 1.8 illustrates the marked difference in conductance produced be Q/R site editing. Figure 1.8A,C shows single channels from homomeric kainate GluR6(Q) unedited receptors, giving discrete openings of ~8, 13 and 25 pS. Figure 1.8D illustrates the fact that recombinant receptors composed of GluR6(Q)/KA-2 can be activated by AMPA (unlike homomeric GluR6 receptors), and that this agonist gives rise to discrete opening from the Ca^{2+}-permeable heteromeric channel. The edited form of the GluR6 receptor channels (depicted in Figure 1.8B) gives rise to an inward current

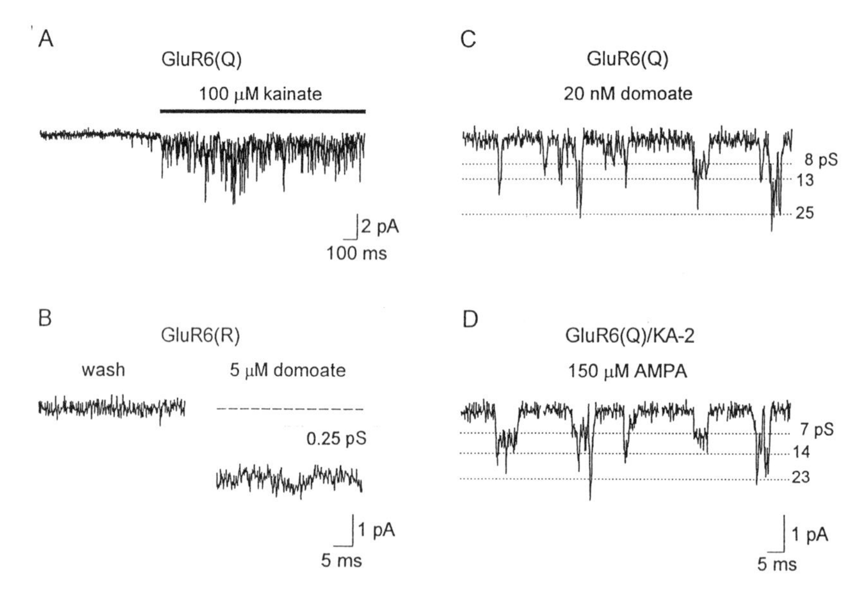

Figure 1.8 Ca^{2+}-permeable non-NMDA channels display high single channel conductances.

(A) Recording (on a slow time-base) from recombinant GluR6(Q) channel openings activated by 100 μM kainate in an outside-out patch (−80 mV).
(B) Response to 5 μM domoate from Ca^{2+}-impermeable GluR6(R) receptors. The current in the absence of agonist is indicated by the dashed line. Note the modest current noise increase associated with the response, and the absence of resolvable single channel openings. The estimated conductance (weighted mean) of channels underlying the noise was 0.25 pS.
(C) Response to 20 nM domoate from Ca^{2+}-permeable GluR6(Q) receptors. Discrete single channel currents are present with conductance levels of 8, 13, and 25 pS.
(D) Response to 150 μM AMPA from Ca^{2+}-permeable GluR6(Q)/KA2 receptors. Discrete single channel currents are present with conductance levels of 7, 14, and 23 pS. (Modified from Swanson *et al.*, 1995)

with no detectable channel events, but with a clear increase in membrane noise. From noise analysis, the conductance of the underlying homomeric edited channel is low – about 250 fS. A reduction in conductance has also been obtained when heteromeric assemblies of kainate-Rs and of AMPARs incorporate edited subunits. Indeed, high single channel conductance appears to be a general hallmark of the Ca^{2+}-permeable recombinant non-NMDA channels, when compared with Ca^{2+}-impermeable forms (see Swanson *et al.*, 1996, 1997a).

Data from a range of cell types appear consistent with the concept that Ca^{2+}-permeable channels exhibit a high channel conductance. Recordings have been obtained from somatic (i.e. extrasynaptic), dendritic and synaptic AMPAR channels. It is notable that Ca^{2+}-permeable AMPA channels (for example in nucleus magnocellularis neurons, and hippocampal basket cells), have single channel conductances of ~25 pS (see Otis *et al.*, 1995; Koh *et al.*, 1995b), compared with ~10 pS for the relatively Ca^{2+}-impermeable channels in hippocampal CA3 cells and some synapses in CA1 pyramidal cells (Spruston *et al.*, 1995) and 4–6 pS for the Ca^{2+}-impermeable synaptic channels at climbing fibre inputs onto cerebellar Purkinje cells (Momiyama *et al.*, 1996b).

Splice variation affects functional and pharmacological properties

Flip (i) isoforms of AMPAR subunits are expressed predominantly in embryonic and postnatal neurons, whereas flop (o) isoforms appear more prevalent in older animals (Monyer *et al.*, 1991). Flip and flop isoforms differ in desensitization time course (Sommer *et al.*, 1990; Mosbacher *et al.*, 1994) and in their sensitivity to certain drugs that suppress desensitization (Partin *et al.*, 1995, 1996; Sekiguchi *et al.*, 1997). Thus, the flip isoforms generate rapidly decaying currents, which fade to a steady-state level in the presence of glutamate. Flop isoforms desensitize more rapidly and profoundly, giving a steady state current that is only ~1 per cent of the peak current (Partin *et al.*, 1994). In particular, GluR3(o) and GluR4(o) undergo very rapid desensitization with time constants of ~1 ms, compared with a value of ~4 ms for GluR3(i) and GluR4(i) forms (Mosbacher *et al.*, 1994).

AMPAR-responses are selectively augmented by cyclothiazide (CTZ), which blocks their desensitization but has little effect on kainate receptors (Partin *et al.*, 1995, 1996). It has been found that flip isoforms are considerably more sensitive to CTZ and exhibit a higher affinity for the drug. Furthermore, in receptor assemblies containing both flip and flop isoforms, the flip form dominates the receptor properties. Thus, the functional channels have desensitization profiles and CTZ sensitivity that resembles homomeric flip-containing receptors (Partin *et al.*, 1994, 1995). In contrast, PEPA (Table 1.2) preferentially modulates flop isoforms of AMPARs (Sekiguchi *et al.*, 1997). Since there is currently little information about which (i/o) splice variants are present at specific synapses, these drugs may prove useful in dissection of the native forms and determining their influence on the transmission process.

Relating observations from recombinant receptors to EPSC properties is clearly a major challenge. While there is little doubt that variation in GluR expression between cell types is functionally important (Jonas and Spruston, 1994), information about the roles played by the various subunits remains scant. A few general rules are starting to emerge, some of which have been outlined above. In addition, differential expression of GluR2(i) and GluR4 subunit isoforms has a well-defined impact on channel gating. Thus, studies on recombinant and native AMPARs suggest slow gating kinetics are conferred by GluR2(i) subunits, while fast gating kinetics are conferred by the presence of GluR4 subunits in heteromeric combinations (Geiger *et al.*, 1995). This raises the obvious issue of what happens when GluR2(i) and GluR4 are co-assembled, as they are at the mossy fibre-granule cell synapse in the young cerebellum. In fact the EPSC time course at this site is fast (fast component time constant ~1 ms; Silver *et al.*, 1996). Furthermore, the Ca^{2+}-permeable variety of non-NMDAR channels generally exhibits a fast deactivation time course (i.e. rapid return to baseline following a brief pulse of glutamate). This is reflected in the fast decay of EPSCs generated by Ca^{2+}-permeable synaptic receptors (Geiger *et al.*, 1995). However, there are also examples of fast time course EPSCs that display low Ca^{2+}-permeability.

Phosphorylation affects non-NMDAR properties

There is extensive evidence that both kainate- and AMPARs are subject to changes in their phosphorylation/dephosphorylation state during long-term potentiation/long-term depression (LTP/LTD) in the hippocampus (Soderling *et al.*, 1994; Kameyama *et al.*, 1998; see Roche *et al.*, 1994). Reversible covalent modification in non-NMDA receptor structure by protein kinase A (PKA), Ca^{2+}/calmodulin-dependent protein kinase II (CAM-kinase II) and calcineurin (phosphatase 2B) is thought to result in changes of basic receptor-channel properties. The co-localization of PKA and calcineurin (phosphatase 2B, activated by CAM-kinase II) by anchoring proteins within dendrites, suggests that these enzymes may have opposite actions on synaptic glutamate receptors. In principle, the phosphorylation/dephosphorylation of the receptors (or associated proteins) could modify channel function by altering various parameters, including single channel conductance, the number of active channels, receptor desensitization and the probability that an agonist-bound receptor will open at the peak of a response. There is evidence that at least two of these parameters can change as a result of phosphorylation/dephosphorylation.

The effects of phosphorylation have been examined in detail on homomeric GluR6 receptors. Mutagenesis studies indicate that PKA phosphorylation can occur at two residues, Ser-684 and Ser-666 (Raymond *et al.*, 1993), causing a substantial enhancement of channel open probability without a measurable change in channel conductance or other receptor properties (Traynelis and Wahl, 1997). On the other hand, dephosphorylation by calcineurin causes a decrease in the probability that agonist-bound GluR6 receptors will open. Paradoxically, existing models of transmembrane topology place these residues on the extracellular surface of the receptor – apparently inaccessible to intracellular PKA.

The effects of phosphorylation on the GluR1 AMPAR subunit are of particular interest because of its probable involvement in long-term changes in transmission (see Soderling *et al.*, 1994). The intracellular C-terminus of GluR1 is a substrate for phosphorylation at two residues: Ser-845 and Ser-831. Recent evidence indicates that LTD is associated with persistent dephosphorylation of Ser-845 (the PKA site on GluR1), while LTP involves an increase in the CaMKII phosphorylation of GluR1 at Ser-831 (see Kameyama *et al.*, 1998). Experiments on recombinant homomeric GluR1 receptors have demonstrated that PKA phosphorylation of GluR1 at Ser-845 increases the probability that agonist-bound receptors will open, while dephosphorylation of this residue, by the phosphatase calcineurin, decreases channel open probability (Banke *et al.*, 1999). Moreover, phosphorylation of Ser-831 by activation of CaM-KII increases the contribution of the higher conductance states of the multiple conductance openings produced by GluR1 channels, without affecting channel open probability or other receptor properties (Derkash *et al.*, 1999). The increased occurrence of higher conductance levels enhances the total current through homomeric GluR1 AMPARs. These observations provide strong support for the finding of Benke *et al.* (1998) that induction of LTP can result in an increase in synaptic single-channel conductance, and fit neatly with the model for bi-directional LTD/LTP proposed by Kameyama *et al.* (1998).

Pharmacological and functional characteristics of NMDARs

We turn next to the main properties of the NMDARs. These receptors play a key role in several aspects of central neurotransmission, including synaptic plasticity and synaptogenesis. They have also been implicated in various disease states such as epilepsy and neurodegeneration following ischaemia (Lee *et al.*, 1999). NMDARs display several features that are highly unusual in ligand-gated receptors. They are permeable to Ca^{2+}, blocked in a voltage-dependent manner by extracellular Mg^{2+}, inhibited by protons (at physiological pH), and require glycine to act as a 'co-agonist'. They are also sensitive to low concentrations of Zn^{2+} and polyamines. These endogenous molecules interact with specific sites on the receptor-channel modulating its behaviour. Furthermore, Mg^{2+} and Zn^{2+} can bind to more than one site, giving rise to diverse effects (see Johnson and Ascher, 1994).

At both a basic and a therapeutic level there is keen interest in the fact that the modulatory binding sites on NMDARs are susceptible to pharmacological manipulation or block. It has been demonstrated that this can be used to alter the NMDAR-response and hence modify pathological changes associated with excessive NMDAR-activation. At a more subtle level, the properties of the modulatory sites have important implications for our understanding of the way that NMDA-channels normally operate – for the amplitude and time course of the synaptic current, and for the membrane potential range over which the channels allow Ca^{2+} to enter the cell. These, in turn, affect one of the main functional roles of NMDARs – the generation of long-term changes in the CNS. Briefly, the main characteristics of the NMDARs are as follows:

Conductance states of the NMDAR channel

Compared with other mammalian glutamate receptors, NMDAR-channels have a relatively large main conductance state (Nowak *et al.*, 1984; Cull-Candy and Ogden, 1985) and exhibit multiple conductance levels (Cull-Candy and Usowicz, 1987; Jahr and Stevens, 1987). Based on their single channel properties, two functionally distinct classes of NMDAR-channels have been identified in neurons (Cull-Candy *et al.*, 1995; Momiyama *et al.*, 1996a). Other conductance levels have also been described, possibly representing further conductance states or perhaps additional NMDAR-channels

(Palecek *et al.*, 1999). The distinct single-channel signatures exhibited by NMDARs yield useful information about the subunit composition of individual receptors.

Mg^{2+}-block, Ca^{2+}-permeability and the Q/R/N site

NMDARs form cation-selective channels with a high permeability to Ca^{2+}, as well as Na^{+} and K^{+} (MacDermott *et al.*, 1986; Ascher and Nowak, 1988; Jahr and Stevens, 1993). Moreover, they are blocked in a voltage-dependent manner by the binding of Mg^{2+} ions to a site within the pore (Nowak *et al.*, 1984; Mayer *et al.*, 1984). The gating of NMDARs is therefore controlled by a combination of ligand(s) and voltage, making them distinct from other types of ligand-gated receptor. Entry of Ca^{2+} through NMDAR-channels is crucial for long-term changes in synaptic transmission. However, this feature appears to be a double-edged sword. The rise in intracellular Ca^{2+}, associated with prolonged NMDAR activation, is a major cause of neuron death following ischaemia. Considerable effort has been made to develop molecules for suppressing excessive NMDAR-activation, with the aim of reducing cell death (Dingledine *et al.*, 1999; Lee *et al.*, 1999). Fundamental to this issue is the need to understand structural features that govern Ca^{2+}-permeability and Mg^{2+}-block of the NMDAR subunits.

The M2 loop of the NMDAR-channel is thought to form a narrow constriction located just over half way across the transmembrane electric field (see Villarroel *et al.*, 1995; Zarei and Dani, 1995). The channel gate appears to be near this constriction, or perhaps slightly cytoplasmic to it (Beck *et al.*, 1999). Since the M2 loop enters the membrane through the intracellular surface, it is clear the lining of the channel vestibule on the *cytoplasmic* side of the membrane is formed from exposed residues of M2 (Beck *et al.*, 1999). However, it is less certain which residues form the lining of the *extracellular* vestibule. Recent data suggests that parts of M3 (at its C-terminal end) and M4 (at its N-terminal end), as well as the region preceding M1, may all contribute.

The cation selectivity of the NMDAR-channel is dependent on a critical residue in the pore-lining region. From alignment of subunit sequences, this critical site is occupied by an asparagine (N) in both NR1 and NR2 (position 598 in NR1, 595 in NR2A, 593 in NR2C) (see Burnashev *et al.*, 1992). The site occupied by this asparagine is directly homologous to the Q/R site in AMPAR subunits. Therefore, the site is collectively referred to as the *Q/R/N site* (or simply the *N-site*, in the case of NMDAR subunits). Although both the NR1 and NR2 subunits contribute to the permeation pathway, they do not contribute equally to the selectivity filter. Thus swapping glutamine for the asparagine (N→Q) in NR1 affects the channel's Ca^{2+}-permeability but not its Mg^{2+}-block, while the same mutation in NR2 affects Mg^{2+}-block, but not Ca^{2+}-permeability (Burnashev *et al.*, 1992).

Experiments of Wollmuth and colleagues (1998a, b) have demonstrated that residues in both NR1 and NR2 contribute to the narrow constriction within the NMDAR-pore. In NR1 the main determinant of the constriction is the N-site asparagine. However, for NR2A the important residue is located adjacent to the N-site, on the C-terminal side (N + 1 site). As it turns out, this is also an asparagine residue. Thus two asparagines – at non-homologous positions in NR1 and NR2 (at the *N-site* and *N+ 1 site*), together constitute the narrow constriction of the NMDAR-channel. Evidence for this idea has also been provided by experiments demonstrating that the site is accessible to methane thiosulphonate from *both* the cytoplasmic and extracellular sides of the membrane. This large molecule cannot itself permeate through the pore (Kuner *et al.*, 1996). It now appears that the two adjacent asparagines (N-site and N + 1 site) in the M2 region of the *NR2-subunit*, form the main blocking site for Mg^{2+} ions arriving from the extracellular side of the membrane. Thus the characteristic voltage-dependent Mg^{2+} block appears to be imparted by binding of Mg^{2+} at this site. However, contrary to expectations, it is the residue at the N + 1 site that makes the strongest contribution to the block. On the other hand, the N-site asparagine of the NR1-subunit appears not to make a major contribution (Wollmuth *et al.*, 1998a, b).

Interestingly, a number of other channel-blocking drugs may also act on, or near to, these residues. Thus, MK801, ketamine and phencyclidine all produce use-dependent block. In the case of MK801, this molecule appears to be a more effective blocker of receptors composed of NR1/NR2A or NR1/NR2B subunits, than those composed of NR1/NR2C or NR1/NR2D (Yamakura *et al.*, 1993; Chazot *et al.*, 1994; Laurie and Seeburg, 1994; Grimwood *et al.*, 1996). From site directed mutagenesis, the N-site residue in NR1 and NR2 also appear to play a role in the binding of these drugs within the ion channel (Burnashev *et al.*, 1992), although it remains to be seen whether both N and N + 1 sites play significant roles here too.

The sensitivity to block by Mg^{2+} varies greatly between NMDAR subtypes. Thus, it has been demonstrated that recombinant NMDARs composed of NR1/NR2A or NR1/NR2B subunits are more sensitive to block by extracellular Mg^{2+} than NMDARs composed of NR1/NR2C or NR1/NR2D (Monyer *et al.*, 1994). As described on p. 000, this variation in Mg^{2+} sensitivity extends to native receptors composed of these various subunits (see Momiyama *et al.*, 1996a), and is therefore of considerable functional interest since it has been suggested that different subtypes of the NMDAR may be active over distinct membrane potential ranges (Momiyama *et al.*, 1996a).

Modulation of NMDARs by glycine: implications for synaptic transmission

Glycine binding appears to be an absolute requirement before NMDARs can be activated (Johnson and Ascher, 1987; Kleckner and Dingledine, 1988). However, the functional significance of this observation has been controversial. Glycine can be detected in the CSF in the μM range, while the EC_{50} for glycine may be as low as ~150 nM (see Johnson and Ascher, 1994), depending on the NMDAR-subtype being examined (see Feldmeyer and Cull-Candy, 1996). Therefore glycine present in the extracellular medium would seem sufficient to activate the binding site maximally at most synapses. For the action of glycine to take on a clear physiological significance, its concentration within the synaptic cleft would need to be kept below a saturating level by some form of control mechanism, such as glycine transport.

Consistent with the idea that the cleft glycine concentration is regulated by an uptake mechanism is the fact that at least one of the glycine transporters has a distribution that closely follows the distribution of NMDARs (Smith *et al.*, 1992). In addition, the NMDAR-mediated component of the synaptic current can be enhanced by externally applied glycine at some synapses (Thomson, 1990; Wilcox *et al.*, 1996; Bergeron *et al.*, 1998; Berger *et al.*, 1998), and by a blocker of the GLYT1 glycine transporter (Bergeron *et al.*, 1998). This seems at odds with the original idea that glycine levels were high at the synapse. Moreover, to produce potentiating effects with externally applied glycine requires concentrations that are greatly in excess of its apparent dissociation constant at NMDARs (by 2–3 orders of magnitude, i.e. >100 μM). Furthermore, D-serine, which also activates the glycine site but is not taken up by transporters, will potentiate the NMDA-component of EPSCs in the concentration range found to be effective on isolated patches (i.e. low μM range).

All of this would appear to be consistent with the idea that glycine transporters act to set the glycine concentration in the cleft below saturation for NMDARs. This then begs the question as to which cells might release glycine (or perhaps D-serine; see Schell *et al.*, 1995) to modulate NMDARs under physiological conditions. The prime suspects are either neighbouring glial cells or glycinergic nerve terminals (see Cull-Candy, 1995). However, despite the accumulating evidence in favour of glycine modulation *in vivo*, its physiological role still remains a puzzle. Nevertheless, the mechanism of glycine action on the NMDAR is clearly of pharmacological and therapeutic interest as simply blocking the glycine site can inactivate NMDARs. Consequently, antagonism at the glycine site has generated much interest because of the possible wider implications (Kemp and Leeson, 1993).

Proton inhibition of NMDARs

Under physiological conditions the NMDARs are subject to tonic inhibition by protons. The IC_{50} coincides with physiological pH (7.3), equivalent to a H^+ concentration of only ~50 nM (see Traynelis and Cull-Candy, 1990). Thus a small shift in pH, above or below the normal level, has a dramatic effect on the size of the NMDAR-current. This inhibition arises mainly through a reduction in the frequency of channel openings, without a decrease in the channel conductance. Moreover, the inhibition is voltage-independent and is not accompanied by a modification in the binding of glycine or NMDA to the receptor (Traynelis and Cull-Candy, 1990, 1991; Traynelis *et al.*, 1995; Vyklicky *et al.*, 1990; see Chesler and Kaila, 1992). This sensitivity to the ambient level of protons is thought to be important during ischaemia and seizures as these conditions are accompanied by acidification of the extracellular medium (Chesler and Kaila, 1992), an effect that could be heightened in the narrow confines of the synaptic cleft (see Traynelis and Cull-Candy, 1991, for discussion). The rise in extracellular $[H^+]$ could therefore suppress activation of NMDARs during the high levels of glutamate that accompany anoxia, serving to reduce ischaemic cell death.

Protons induce their inhibitory effect by interacting with splice variants of the NR1-subunit that lack exon 5 (N1 cassette) in their N-terminus (Figure 1.4; Table 1.1). The tonic proton inhibition can be relieved by application of exogenous polyamines (such as spermine). As described earlier, it has been proposed that this relief from inhibition arises from the fact that exogenous polyamine and the presence of exon 5 can both act to shield the proton sensor of NR1 (Traynelis *et al.*, 1995). Several other molecules also appear capable of potentiating NMDAR-responses by similar, or related, mechanisms. These include Mg^{2+} (acting via the NR2B subunits), aminoglycoside antibiotics, and histamine (Paoletti *et al.*, 1995; Segal and Skolnick, 1998; Vorobjev *et al.*, 1993; Bekkers, 1993). NR1 splice variants that *include* the exon 5 insert are fully active at physiological pH, as these are no longer subject to tonic inhibition. Hence, they are not potentiated by spermine, nor inhibited by Zn^{2+} which appears to give rise to a very similar type of voltage-independent block (Traynelis *et al.*, 1995). Splice variants that lack tonic H^+ inhibition are expected to be more responsive to transmitter which could, in principle, effect their activation during synaptic transmission.

The type of NR2 subunit involved also influences the inhibition of NMDARs by protons and Zn^{2+}. Thus, NMDAR-assemblies containing NR2A or NR2B together with NR1 subunits that lack exon 5, display tonic H^+-inhibition typical of the homomeric NR1 (in zero Mg^{2+}; Paoletti *et al.*, 1995). However, if this variant of NR1 is co-assembled with NR2C or NR2D the resultant receptor is much less sensitive to inhibition by H^+ or Zn^{2+} (Traynelis *et al.*, 1998; Paoletti *et al.*, 1997). This would imply that during ischaemia the overactivation of NR2C- or NR2D-containing NMDARs is not greatly suppressed by the rise in extracellular $[H^+]$. Indeed, it has recently been shown that neuronal death (following vascular occlusion) is reduced in transgenic mice that lack NR2C subunits (Kadotani *et al.*, 1998).

In view of the similarity in the action of polyamines, protons and Zn^{2+} on NR1 subunits lacking the N1-cassette, it is perhaps not surprising that these three molecules share some overlap in the region of the NMDAR to which they bind (Dingledine *et al.*, 1999).

NMDAR-channel kinetics determine the slow decay of NMDAR-EPSCs

The NMDAR-mediated component of the EPSC rises slowly (~10 ms) and decays slowly, compared with most synaptic currents generated by ionotropic receptors. This can be accounted for, in part, by the high affinity of the NMDAR for the transmitter which results in slow unbinding of glutamate. This gives rise to prolonged channel activity. NMDAR channels appear to open, on average, ~10 ms after agonist binding (Dzubay and Jahr, 1996) and continue opening and closing repeatedly until the agonist dissociates from the receptor several hundred milliseconds later (Lester *et al.*, 1990; Clements and Westbrook, 1991; Gibb and Colquhoun, 1992; Lester and Jahr, 1992). It is of considerable interest, from a functional viewpoint, that the offset-decay time itself depends critically on the type of NR2 subunits composing the receptor. Thus, the deactivation time constant is relatively fast for receptors composed of NR1/NR2A subunits (~120 ms), slower for NR1/NR2B- and NR1/NR2C-receptors (400 and 380 ms, respectively), and exceptionally slow for NR1/NR2D containing receptors (4–5 secs) (Monyer *et al.*, 1994; Wyllie *et al.*, 1998; Vicini *et al.*, 1998).

As a point of caution, it is notable that certain properties of the NMDAR-channels are sensitive to recording conditions and the degree of membrane disruption involved in the recording. For example, measurements of single channel conductance (Clark *et al.*, 1997) and probability of channel opening (Rosenmund *et al.*, 1995) are both influenced by the recording method and by the level of extracellular Ca^{2+}. Moreover, quantitative properties of NMDAR-channels can be modified by availability of, or interaction with, intracellular or extracellular Ca^{2+}, ATP and cytoskeletal and other intracellular proteins and enzymes (see Rosenmund and Westbrook, 1993).

NMDAR diversity

There is considerable evidence that a variety of functionally distinct NMDAR subtypes occur in central neurons (Cull-Candy *et al.*, 1987; Howe *et al.*, 1991; Carmignoto and Vicini, 1992; Hestrin, 1992; Kashiwagi *et al.*, 1997; Paoletti *et al.*, 1997). These differ both with respect to their NR1 and their NR2 subunit composition. A number of functionally important biophysical and pharmacological properties of these receptors depend on the type of NR2 subunits present within an assembly (Monyer *et al.*, 1994; Wyllie *et al.*, 1998; Cull-Candy *et al.*, 2001). Thus the various receptor subtypes differ in their response to endogenous ligands and modulators (Feldmeyer and Cull-Candy, 1996).

Distribution of NR2 subunits

Identification of the *native* NMDAR subtypes present within individual cell types has been greatly advanced by studies using *in situ* hybridization (Monyer *et al.*, 1994; Akazawa *et al.*, 1995) and immunocytochemistry (Petralia *et al.*, 1994). These approaches have demonstrated the widespread distribution of NR1 at all stages of development, in keeping with the view that NR1 is indispensable in the formation of functional NMDARs. In contrast, mRNA for NR2A, NR2B, NR2C and NR2D subunits is differentially distributed, and the expression pattern changes during development (Monyer *et al.*, 1994; Akazawa *et al.*, 1994). This clear-cut differential expression provides strong evidence for the presence of functionally diverse types of NMDAR within different cell types.

Recent experiments on the stoichiometry of NMDARs (and non-NMDARs) favour a tetrameric structure (Laub *et al.*, 1998; Rosenmund *et al.*, 1998). NMDARs are suggested to contain two copies of NR1 (Behe *et al.*, 1995; although see Premkumar and Auerbach, 1997) and two copies of NR2 (Premkumar and Auerbach, 1997; Laub *et al.*, 1998). There is compelling evidence to suggest that some native receptors can contain more than one type of NR2 subunit (Sheng *et al.*, 1994; Wafford *et al.*, 1993; Chazot *et al.*, 1994). Therefore cells expressing two types of NR2 subunits might be expected to express three types of receptor assembly, one of which is *heteromeric* in terms of its NR2 subunits. However, the relative contribution of such trimeric (NR1 + two types of NR2) receptors to the native population still remains unclear. Indeed, it has been suggested that the proportion of such native receptors may represent only a minor subset (Chazot and Stephenson, 1997). Recent functional and pharmacological studies have identified native NMDARs that display the properties expected of NR2A-, NR2B-, NR2C- and NR2D-containing receptors. This does not, of course, exclude the presence of other receptor subtypes. In fact, given the large number of NR1 splice variants and the emergence of additional NMDAR subunits (such as the NR3 family), the final picture may well turn out to be complex.

Single channel signatures

Two functionally distinct types of native NMDARs have been distinguished on the basis of their single channel conductances (see Cull-Candy *et al.*, 1995). One of these corresponds to the conventional '50 pS' type of NMDAR that has been widely described. Patch-clamp studies on native (Farrant *et al.*, 1994; Cull-Candy *et al.*, 1995) and recombinant NMDARs (Stern *et al.*, 1992; Brimecombe *et al.*, 1997) have shown these events to be associated with expression of NR2A or NR2B subunits. Experiments on cerebellar slices have demonstrated the presence of a family of native *low conductance* NMDAR channels (Figure 1.9), identified from their characteristic ~35 and 18 pS openings and low sensitivity to Mg^{2+} block (Farrant *et al.*, 1994; Momiyama *et al.*, 1996a).

In cerebellar granule cells these low conductance events arise from NR2C-containing NMDARs (Farrant *et al.*, 1994; Takahashi *et al.*, 1996). Thus, *in situ* hybridization data has revealed that granule cell expression of mRNA for NR2B is replaced by a particularly intense expression of mRNA for NR2C (and NR2A) after postnatal day 10–11 (Monyer *et al.*, 1994; Akazawa *et al.*, 1995). In most other brain regions mRNA for NR2C is relatively sparse. The striking changes in mRNA that occur during granule cell development are accompanied by changes in the single channel currents detected in outside-out patches (see Figure 1.9). Channel openings with a main conductance of ~50 pS are present in granule cells from young animals expressing NR2B subunits. However, in older animals many NMDA channel openings have conductances of ~18 and 38 pS at a stage when cells are expressing mRNA for NR2C (Farrant *et al.*, 1994). Since granule cells from mature mice express *only* the low conductance NMDAR-channels when the NR2A subunit is ablated by gene-knockout (Takahashi *et al.*, 1996), it is apparent that these events arise from NR2C-containing NMDARs. Furthermore, ablation of the NR2C subunit causes loss of the low conductance channels (Ebralidze *et al.*, 1996). It is also notable that recombinant receptors composed of NR1/NR2C subunits give rise to a similar type of low conductance NMDAR channel (Stern *et al.*, 1992).

Single channel recordings from cerebellar Purkinje cells have allowed another distinct type of native low-conductance NMDAR to be identified (Momiyama *et al.*, 1996a; see also Misra *et al.*, 2000b). These cells express mRNA for the NR2D subunit (along with NR1) (see Akazawa *et al.*, 1995), giving rise to a homogeneous population of channels with a low conductance and reduced sensitivity to block by Mg^{2+} (Momiyama *et al.*, 1996a). Cerebellar Golgi and stellate cells also express low conductance NMDAR channels with NR2D characteristics, but in these cells they occur as a mixed population together with high conductance opening (see Misra *et al.*, 2000a). Other cell types that express mRNA for a mixture of NR2 subunits (NR2D, along with other NR2 subunits) also express a 'mixed' population of NMDA channel types. This has been observed in deep cerebellar nuclei neurons and dorsal horn spinal cord neurons (Momiyama *et al.*, 1996a). Confirmation that these *low conductance* NMDAR channels arise from NR2D-containing receptors has been obtained from experiments on recombinant NMDARs, which show striking similarities to the native channels (Wyllie *et al.*, 1998). Interestingly, it has also been noted that, from a structural viewpoint, the NR2 subunits fall into these two groups. NR2A and NR2B subunits possess carboxy tails that are particularly long, and of similar sizes (~630–640 residues) with weak, but clear, identity of structure (Figure 1.4; see Mori and Mishina, 1995).

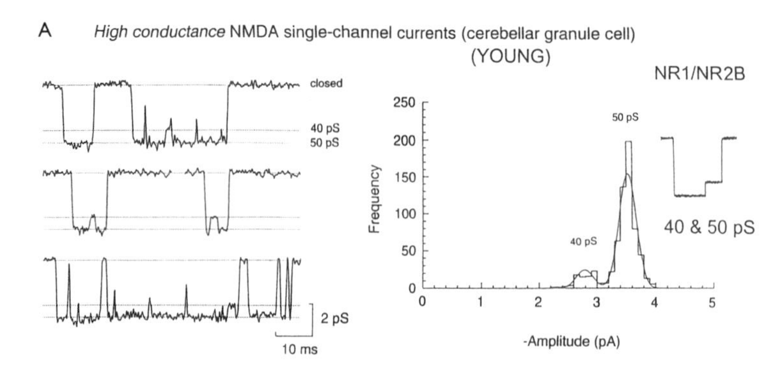

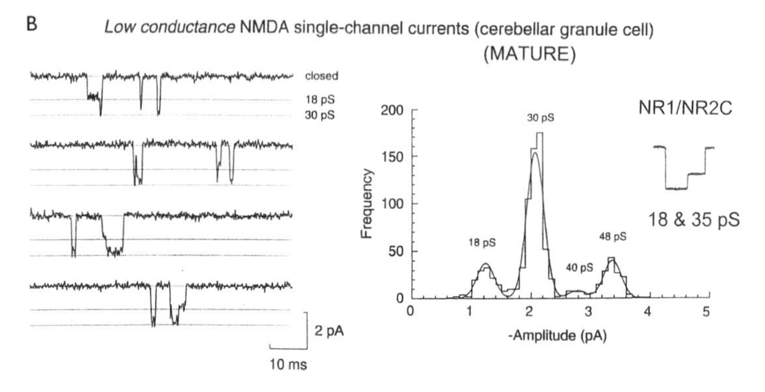

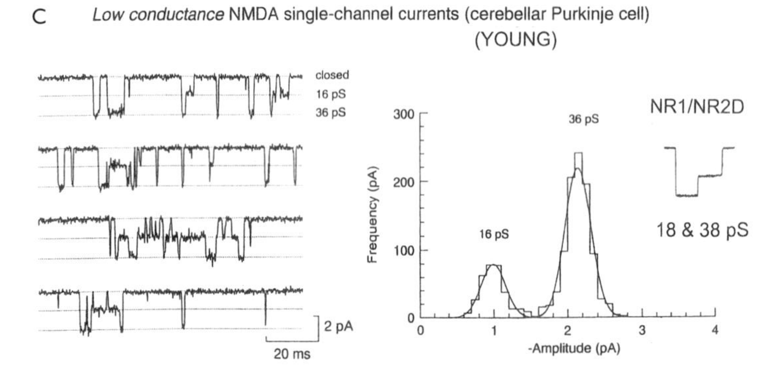

Figure 1.9 Single channel signatures of NMDAR-subtypes.

(A) High conductance '50 pS' NMDAR channel openings recorded from outside-out patches from internal granule cells in a slice from a 12-day-old rat (−70 mV; 1 mM Ca^{2+}). At this stage the cells express predominantly NR1/NR2B subunits. Openings to a subconductance of 40 pS can also be seen. The histogram illustrates the amplitude distribution of single channel currents, with the mean values for the main levels present.

(B) Low conductance NMDAR channel openings to 18 and 30 pS in outside-out patches from internal granule cells in a slice from a 19-day-old rat (−70 mV; 1 mM Ca^{2+}), at a stage when the cells express NR1/NR2C subunits. From the amplitude histogram high conductance openings are also present, as expected since the cells also express NR1/NR2A subunits.

(C) Low conductance NMDAR channel openings to 16 and 36 pS in outside-out patches from a Purkinje cell in a slice from a 5-day-old rat. At this stage the cells express NR1/NR2D subunits. From the amplitude histogram a homogeneous population of low conductance channels is present. Modified from Farrant *et al.* (1994) and Momiyama *et al.* (1996a).

Pharmacological and molecular identification of NMDAR subtypes

A number of drugs, including ifenprodil (Williams, 1993; Kew *et al.*, 1998), its analogue CP101, 606 (Chenard *et al.*, 1995), and haloperidol (Ilyin *et al.*, 1996; Vicini *et al.*, 1998) are all capable of selectively blocking NR2B-containing NMDA receptors. These have been used to distinguish between NR2A- and NR2B-containing NMDA receptors in the synaptic and extrasynaptic membrane (Stocca and Vicini, 1998; Misra *et al.*, 2000a). The best characterized of these compounds, ifenprodil, is an atypical non-competitive antagonist which exhibits approximately four-hundredfold higher apparent affinity for NR2B-, than for NR2A-containing receptors (Williams, 1993; Priestley *et al.*, 1995). Furthermore, the Zn^{2+} chelator TPEN (N,N,N′,N′-tetrakis-(2-pyridylmethyl)-ethylenediamine) enhances responses from NR2A-containing receptors, by sequestering the low levels of contaminating Zn^{2+} which normally suppress responses of NMDARs containing NR2A subunits (Paoletti *et al.*, 1997). Unfortunately, at present little is known about the action of subunit selective drugs on NMDARs receptors that may contain more than one type of NR2 subunit.

Several recent studies have addressed the issue of the NR2 subunits involved in the formation of synaptic receptors by selectively ablating particular NR2 subunits by gene knock out. Experiments using this approach have suggested that differing subtypes of NMDARs are targeted to different synapses in single hippocampal CA3 pyramidal cells (Ito *et al.*, 1997). Examination of the mossy fibre-to-granule cell synapse in the cerebellum of animals lacking NR2C or NR2A has suggested that both of these subunits participate in generating the NMDA-component of the EPSC (Takahashi *et al.*, 1996; Ebralidze *et al.*, 1996). Furthermore, the functional properties of synaptic NMDARs appear to be modified during development, and to reflect the type of NR2 subunit mRNA being expressed (see Flint *et al.*, 1997; Quinlan *et al.*, 1999; Cathala *et al.*, 2000).

Concluding remarks

One of the most striking features of central glutamate synapses is the ability of a single transmitter to generate an impressive array of postsynaptic responses in different nerve cells. It is clear that this can be ascribed, at least in part, to the enormous diversity in glutamate receptor subunits and receptor subtypes expressed in the CNS. However, many basic questions remain unanswered concerning the roles of the different subunits. Furthermore, we remain surprisingly ignorant of the 'molecular identity' of the receptor subtypes present at most central synapses, and of the ground rules which determine why one type of subunit (or subunit combination), rather than another, carries the signal at any given synapse.

Acknowledgements

Work from the author's lab is supported by the Wellcome Trust and the Howard Hughes Medical Institute. I am grateful to my colleagues at UCL for many helpful discussions that have contributed to this chapter.

References

Akazawa, C., Shigemoto, R., Bessho, Y., Nakanishi, S. and Mizuno, N. (1994) 'Differential expression of five NMDA receptor subunit mRNAs in the cerebellum of developing and adult rats', *J. Comp. Neurol.* **347**: 150–160.

Anis, N., Sherby, S., Goodnow, R. Jr, Niwa, M., Konno, K., Kallimopoulos, T., Bukownik, R., Nakanishim K., Usherwood, P. and Eldefrawi, A. (1990) 'Structure–activity relationships of philanthotoxin analogs and polyamines on N-methyl-D-aspartate and nicotinic acetylcholine receptors', *J. Pharmacol. Exp. Ther.* **254**: 764–773.

Anson, L.C., Chen, P.E., Wyllie, D.J.A., Colquhoun, D. and Schoepfer, R. (1998) 'Identification of amino acid residues of the NR2A subunit that control glutamate potency in recombinant NR1/NR2A NMDA receptors', *J. Neurosci.* **18**: 581–589.

Anson, L.C., Schoepfer, R. Colquhoun, D. and Wyllie, D.J.A. (2000) 'Single-channel analysis of an NMDA receptor possessing a mutation in the region of the glutamate binding site', *J Physiol.* **527**: 225–237.

Armstrong, N., Sun, Y., Chen, G.-Q. and Gouaux, E. (1998) 'Structure of a glutamate-receptor ligand-binding core in complex with kainate, *Nature* **395**: 913–917.

Ascher, P. and Nowak, L. (1988) 'The role of divalent cations in the *N*-methyl-D-aspartate responses of mouse central neurones in culture', *J. Physiol.* **399**: 247–266.

Bahring, R. and Mayer, M.L. (1998) 'An analysis of philanthotoxin block for recombinant rat GluR6(Q) glutamate receptor channels', *J. Physiol.* **509**: 635–650.

Banke, T.G., Bowie, D., Lee, H.-K., Huganir, R.L., Schousboe, A. and Traynelis, S.F. (2000) 'Control of GluR1 AMPA receptor function by cAMP-dependent protein kinase', *J. Neurosci.* **20**: 89–102.

Beck, C., Wollmuth, L.P., Seeburg, P.H., Sakmann, B. and Kuner, T. (1999) 'NMDAR channel segments forming the extracellular vestibule inferred from the accessibility of substituted cysteines', *Neuron* **22**: 559–570.

Behe, P., Stern, P., Wyllie, D.J.A., Nassar, M., Schoepfer, R. and Colquhoun, D. (1995) 'Determination of NMDA NR1 subunit copy number in recombinant NMDA receptors', *Proc. R. Soc. Lond. Ser. B.* **262**: 205–213.

Benke, T.A., Lüthi, A., Isaac, J.T.R. and Collingridge, G.L. (1998) 'Modulation of AMPA receptor unitary conductance by synaptic activity', *Nature* **393**: 793–797.

Bekkers, J.M. (1993) 'Enhancement by histamine of NMDA-mediated synaptic transmission in the hippocampus', *Science* **261**: 104–106.

Bennett, J.A. and Dingledine, R. (1995) 'Topology profile for a glutamate receptor: three transmembrane domains and a channel-lining reentrant membrane loop', *Neuron* **14**: 373–384.

Berger, A.J., Dieudonne, S. and Ascher, P. (1998) 'Glycine uptake governs glycine site occupancy at NMDA receptors of excitatory synapses', *J. Neurophysiol.* **80**: 3336–3340.

Bergeron, R., Meyer, T.M., Coyle, J.T. and Greene, R.W. (1998) 'Modulation of N-methyl-D-aspartate receptor function by glycine transport', *Proc. Natl. Acad. Sci. USA* **95**: 15,730–15,734.

Bettler, B., Boulter, J., Hermans-Borgmeyer, I., O'Shea-Greenfield, A., Deneris, E.S., Moll, C., Borgmeyer, U., Hollmann, M. and Heinemann, S. (1990) 'Cloning of a novel glutamate receptor subunit, GluR5: expression in the nervous system during development', *Neuron* **5**: 583–595.

Blaschke, M., Keller, B.U., Rivosecchi, R., Hollmann, M., Heinemann, S. and Konnerth, A. (1993) 'A single amino acid determines the subunit-specific spider toxin block of α-amino-3-hydroxy-5-methylisoxazole-4-propionate/kainate receptor channels', *Proc. Natl. Acad. Sci. USA* **90**: 6528–6532.

Bleakman, D. and Lodge, D. (1998) 'Neuropharmacology of AMPA and kainate receptors', *Neuropharmacology* **37**: 187–204.

Bleakman, D., Ballyk, B.A., Schoepp, D.E., Palmer, A.J., Bath, C.P., Sharpe, E.F., Woolley, M.L., Bufton, H.R., Kamboj, R.K., Tarnawa, I. and Lodge, D. (1996) 'Activity of 2,3-benzodiazepines at native rat and recombinant human glutamate receptors *in vitro*: stereospecificity and selectivity profiles', *Neuropharmacology* **35**: 1689–1702.

Bochet, P., Audinat, E., Lambolez, B., Crépel, F., Rossier, J., Iino, M., Tsuzuki, K. and Ozawa, S. (1994) 'Subunit composition at the single-cell level explains functional properties of a glutamate-gated channel', *Neuron* **12**: 383–388.

Bowie, D. and Mayer, M.L. (1995) 'Inward rectification of both AMPA and kainate subtype glutamate receptors generated by polyamine-mediated ion channel block', *Neuron* **15**: 453–462.

Bowie, D., Lange, G.D. and Mayer, M.L. (1998) 'Activity-dependent modulation of glutamate receptors by polyamines', *J. Neurosci.* **18**: 8175–8185.

Brackley, P.T., Bell, D.R., Choi, S.K., Nakanishi, K. and Usherwood, P.N. (1993) 'Selective antagonism of native and cloned kainate and NMDA receptors by polyamine-containing toxins', *J. Pharmacol. Exp. Ther.* **266**: 1573–1580.

Brimecombe, J.C., Boeckman, F.A. and Aizenman, E. (1997) 'Functional consequences of NR2 subunit composition in single recombinant *N*-methyl-D-aspartate receptors', *Proc. Natl. Acad. Sci. USA* **94**: 11,019–11,024.

Burnashev, N. (1996) 'Calcium permeability of glutamate-gated channels in the central neurons system', *Current Opin. Neurobiol.* **6**: 311–317.

Burnashev, N., Schoepfer, R., Monyer, H., Ruppersberg, J.P., Gunther, W., Seeburgh, P.H. and Sakmann, B. (1992) 'Control by asparagine residues of calcium permeability and magnesium blockade in the NMDA receptor', *Science* **257**: 1415–1419.

Carmignoto, G. and Vicini, S. (1992) 'Activity-dependent decrease in NMDA receptor responses during development of the visual cortex', *Science* **258**: 1007–1011.

Castillo, P.E., Malenka, R.C. and Nicoll, R.A. (1997) 'Kainate receptors mediate a slow post-synaptic current in hippocampal CA3 neurons', *Nature* **388**: 182–186.

Cathala, L., Misra, C. and Cull-Candy, S. (2000) 'Developmental profile of the changing properties of NMDA receptors at cerebellar mossy fiber-granule cell synapses', *Journal of Neuroscience* **20**: 5899–5905.

Chazot, P.L. and Stephenson, F.A. (1997) 'Molecular dissection of native mammalian forebrain NMDA receptors containing the NR1 C2 exon: direct demonstration of NMDA receptors comprising NR1, NR2A, and NR2B subunits within the same complex', *J. Neurochem.* **69**: 2138–2144.

Chazot, P.L., Coleman, S.K., Cik, M. and Stephenson F.A. (1994) 'Molecular characterization of N-methyl-D-aspartate receptors expressed in mammalian cells yields evidence for the coexistence of three subunit types within a discrete receptor molecule', *J. Biol. Chem.* **269**: 24,403–24,409.

Chenard, B.L., Bordner, J., Butler, T.W., Chambers, L.K., Collins, M.A., De Costa, D.L., Ducat, M.F., Dumont, M.L., Fox, C.B., Mena, E.E., Menniti, F.S., Nielsen, J., Pagnozzi, M.J., Richter, K.E.G., Ronau, R.T., Shalaby, I.A., Stemple, J.Z. and White, W.F. (1995) '(1*S*,2*S*)-1-(4-hydroxyphenyl)-2-(4-hydroxy-4-phenylpiperidino)-1-propanol: A potent new neuroprotectant which blocks *N*-methyl-D-aspartate responses', *J. Med. Chem.* **38**: 3138–3145.

Chesler, M. and Kaila, K. (1992) 'Modulation of pH by neuronal activity', *Trends Neurosci.* **15**: 396–402.

Chittajallu, R., Braithwaite, S.P., Clarke, V.R.J. and Henley J.M. (1999) 'Kainate receptors: subunits, synaptic localization and function', *Trends in Pharmacological Sciences* **20**: 26–35.

Ciabarra, A.M., Sullivan, J.M., Gahn, L.G., Pecht, G., Heinemann, S. and Sevarino, K.A. (1995) 'Cloning and characterization of chi-1: A developmentally regulated member of a novel class of the ionotropic glutamate receptor family', *J. Neurosci.* **15**: 6498–6508.

Clark, B.A. and Cull-Candy, S. (1999) 'Frequency-dependent activation of NMDA receptors at an "AMPA receptor only" synapse in the rat cerebellum', *J. Physiol.* **518P**: 156P.

Clark, B.A., Farrany, M. and Cull-Candy, S.G. (1997) 'A direct comparison of the single-channel properties of synaptic and extrasynaptic NMDA receptors', *J. Neurosci.* **17**: 107–116.

Clements, J.D. and Westbrook, G.L. (1991) 'Activation kinetics reveal the number of glutamate and glycine binding sites on the *N*-methyl-D-aspartate receptor', *Neuron* **7**: 605–613.

Cui, C. and Mayer, M.L. (1999) 'Heteromeric kainate receptors formed by the coassembly of GluR5, GluR6, and GluR7', *J. Neurosci.* **19**: 8281–8291.

Cull-Candy, S.G. (1995) 'NMDA receptors: do glia hold the key?', *Current Biology* **5**: 841–843.

Cull-Candy, S.G., Brickley, S.G. and Farrant, M. (2001) 'NMDA receptor subunits: diversity, development and disease', *Current Opinions in Neurobiology* (in press).

Cull-Candy, S.G. and Ogden, D.C. (1985) 'Ion channels activated by L-glutamate and GABA in cultured cerebellar neurons of the rat', *Proc. R. Soc. Lond. B.* **224**: 367–373.

Cull-Candy, S.G. and Usowicz, M.M. (1987) 'Multiple conductance channels activated by excitatory amino acids in cerebellar neurons', *Nature* **325**: 525–528.

Cull-Candy S.G., Farrant M. and Feldmeyer D. (1995) 'NMDA channel conductance: a user's guide', in H. Wheal and A. Thomson (eds) *Excitatory Amino Acids and Synaptic Transmission* (2nd edn), Academic Press. London, pp. 121–132.

Cull-Candy, S.G., Howe, J.R. and Ogden, D.C. (1988) 'Noise and single channels activated by excitatory amino acids in rat cerebellar granule cells', *J. Physiol.* **400**: 189–222.

Cull-Candy, S.G., Howe, J.R. and Usowicz, M.M. (1987) 'Single glutamate-receptor channels in two types of cerebellar neurones', in D. Lodge (ed.) *Excitatory Amino Acids in Health and Disease*, London: Wiley, pp. 165–185.

Cull-Candy, S.G., Momiyama, A., Feldmeyer, D., Misra, C., Brickley, S.G. and Farrant, M. (1998) 'Differential expression of functionally distinct NMDA receptor subtypes in identified cerebellar neurons', *Neuropharmacology* **37**: 1369–1380.

Das, S., Sasaki, Y.F., Rothe, T., Premkumar, L.S., Takasu, M., Crandall, J.E., Dikkes, P., Conner, D.A., Rayudu, P.V., Cheung, W., Vincent Chen, H.-S., Lipton, S.A. and Nakanishi, N. (1998) 'Increased NMDA current and spine density in mice lacking the NMDA receptor subunit NR3A', *Nature* **393**: 377–381.

Derkach, V., Barria, A. and Soderling, T.R. (1999) 'Ca^{2+}/calmodulin-kinase II enhances channel conductance of a-amino-3-hydroxy-5-methyl-4-isoxazolepropionate type glutamate receptors', *Proc. Natl. Acad. Sci.* **96**: 3269–3274.

Dingledine, R., Borges, K., Bowie, D. and Traynelis, S.F. (1999) 'The glutamate receptor ion channels', *Pharmacol. Rev.* **51**: 7–61.

Donevan, S.D. and Rogawski, M.A. (1993) 'GYKI 52466, a 2,3-benzodiazepine, is a highly selective, noncompetitive antagonist of AMPA/kainate receptor responses', *Neuron* **10**: 51–59.

Donevan, S.D., Beg, A., Gunther, J.M. and Twyman, R.E. (1998) 'The methylglutamate, SYM 2081, is a potent and highly selective agonist at kainate receptors', *J. Pharmacol. Exp. Ther.* **285**: 539–545.

Durand, G.M., Bennett, M.V.L. and Zukin, R.S. (1993) 'Splice variants of the *N*-methyl-D-aspartate receptor NR1 identify domains involved in regulation by polyamines and protein kinase C', *Proc. Natl. Acad. Sci. USA* **90**: 6731–6735.

Dzubay, J.A. and Jahr, C.E. (1996) 'Kinetics of NMDA channel opening', *J. Neurosci.* **16**: 4129–4134.

Ebralidze, A.K., Rossi, D.J., Tonegawa, S. and Slater, N.T. (1996) 'Modification of NMDA receptor channels and synaptic transmission by targeted disruption of the NR2C gene', *J. Neurosci.* **16**: 5014–5025.

Egebjerg, J. and Heinemann, S.F. (1993) 'Ca^{2+} permeability of unedited and edited versions of the kainate selective glutamate receptor GluR6', *Proc. Natl. Acad. Sci. USA* **90**: 755–759.

Egebjerg, J., Bettler, B., Hermans-Borgmeyer, I. and Heinemann, S. (1991) 'Cloning of a cDNA for a glutamate receptor subunit activated by kainate but not AMPA', *Nature* **351**: 745–748.

Ehlers, M.D., Fung, E.T., O'Brien, R.J. and Huganir, R.L. (1998) 'Splice variant-specific interaction of the NMDA receptor subunit NR1 with neuronal intermediate filaments', *J. Neurosci.* **18**: 720–730.

Everts, I., Petroski, R., Kizelsztein, P., Teichberg, V.I., Heinemann, S.F. and Hollmann, M.J. (1999) 'Lectin-induced inhibition of desensitization of the kainate receptor GluR6 depends on the activation state and can be mediated by a single native or ectopic N-linked carbohydrate side chain', *J. Neurosci.* **19**: 916–927.

Farrant, M., Feldmeyer, D., Takahashi, T. and Cull-Candy, S.G. (1994) 'NMDA-receptor channel diversity in the developing cerebellum', *Nature* **368**: 335–339.

Feldmeyer, D. and Cull-Candy, S.G. (1994) 'Elusive glutamate receptors', *Current Biology* **4**: 82–84.

Feldmeyer, D. and Cull-Candy, S.G. (1996) 'Functional consequences of changes in NMDA receptor subunit expression during development', *J. Neurocytology* **25**: 857–867.

Flint, A.C., Maisch, U.S., Weishaupt, J.H., Kriegstein, A.R. and Monyer, H. (1997) 'NR2A subunit expression shortens NMDA receptor synaptic currents in developing neocortex', *J. Neurosci.* **17**: 2469–2478.

Gallo, V., Linus, M.U., Hayes, W.P., Vyklicky, Jr, L., Winters, C.A. and Buonanno, A. (1992) 'Molecular cloning and developmental analysis of a new glutamate receptor sub-unit isoform in cerebellum', *J. Neurosci.* **12**: 1010–1023.

Geiger, J.R., Melcher, T., Koh, D.S., Sakmann, B., Seeburg, P.H., Jonas, P. and Monyer, H. (1995) 'Relative abundance of subunit mRNAs determines gating and Ca^{2+} permeability of AMPA receptors in principal neurons and interneurons in rat CNS', *Neuron* **15**: 193–204.

Gibb, A.J. and Colquhoun, D. (1992) 'Activation of *N*-methyl-D-aspartate receptors by L-glutamate in cells dissociated from adult rat hippocampus', *J. Physiol.* **456**: 143–179.

Green, T., Heinemann, S.F. and Gusella, J.F. (1998) 'Molecular neurobiology and genetics: investigation of neural function and dysfunction', *Neuron* **20**: 427–444.

Grimwood, S., Gilbert, E., Ragan, C.I. and Hutson, P.H. (1996) 'Modulation of $_{45}Ca^{2+}$ influx into cells stably expressing recombinant human NMDA receptors by ligands acting at distinct recognition sites', *J. Neurochem.* **66**: 2589–2595.

Heckmann, M., Bufler, J., Franke, C. and Dudel, J. (1996) 'Kinetics of homomeric GluR6 glutamate receptor channels', *Biophys. J.* **71**: 1743–1750.

Herb, A., Burnashev, N., Werner, P., Sakmann, B., Wisden, W. and Seeburg, P.H. (1992) 'The KA-2 subunit of excitatory amino acid receptors shows widespread expression in brain and forms ion channels with distantly related subunits', *Neuron* **8**: 775–785.

Herlitze, S., Raditsch, M., Ruppersberg, J.P., Jahn, W., Monyer, H., Schoepfer, R. and Witzemann, V. (1993) 'Argiotoxin detects molecular differences in AMPA receptor channels', *Neuron* **10**: 1131–1140.

Hestrin, S. (1992) 'Activation and desensitization of glutamate-activated channels mediating fast excitatory synaptic currents in the visual cortex', *Neuron* **9**: 991–999.

Hollmann, M. (1999) 'Structure of ionotropic glutamate receptors', in P. Jonas and H. Monyer (eds) *Ionotropic Glutamate Receptors in the CNS*, Berlin: Springer.

Hollmann, M. and Heinemann, S. (1994) 'Cloned glutamate receptors', *Annu. Rev. Neurosci.* **17**: 31–108.

Hollmann, M., Boulter, J., Maron, C., Beasley, L., Sullivan, J., Pecht, G. and Heinemann, S.F. (1993) 'Zinc potentiates agonist-induced currents at certain splice variants of the NMDA receptor', *Neuron* **10**: 943–954.

Hollmann, M., Maron, C. and Heinemann, S. (1994) '*N*-glycosylation site tagging suggests a three transmembrane domain topology for the glutamate receptor GluR1', *Neuron* **13**: 1331–1343.

Hollmann, M., O'Shea-Greenfield, A., Rogers, S.W. and Heinemann, S. (1989) 'Cloning by functional expression of a member of the glutamate receptor family', *Nature* **342**: 643–648.

Honoré, T., Davies, S.N., Drejer, J., Fletcher, E.J., Jacobsen, P., Lodge, D. and Nielsen, F.E. (1988) 'Quinoxalinediones: potent competitive non-NMDA glutamate receptor antagonists', *Science* **241**: 701–703.

Howe, J.R. (1996) 'Homomeric and heteromeric ion channels formed from the kainate-type subunits GluR6 and KA2 have very small, but different, unitary conductances', *Journal of Neurophysiology* **76**: 510–519.

Howe, J.R., Cull-Candy, S.G. and Colquhoun, D. (1991) 'Currents through single glutamate receptor channels in outside-out patches from rat cerebellar granule cells', *J. Physiol*, **432**: 143–202.

Huettner, J.E. (1990) 'Glutamate receptor channels in rat DRG neurons: activation by kainate and quisqualate and blockade of desensitization by Con A', *Neuron* **5**: 255–266.

Huettner, J.E., Stack, E. and Wilding, T.J. (1998) 'Antagonism of neuronal kainate receptors by lanthanum and gadolinium', *Neuropharmacology* **37**: 1239–1247.

Iino, M., Koike, M., Isa, T. and Ozawa, S. (1996) 'Voltage-dependent blockage of Ca^{2+}-permeable AMPA receptors by joro spider toxin in cultured rat hippocampal neurones', *J. Physiol.* **496**: 431–437.

Ilyin, V.I., Whittemore, E.R., Guastella, J., Weber, E. and Woodward, R.M. (1996) 'Subtype-selective inhibition of *N*-methyl-D-aspartate receptors by haloperidol', *Mol. Pharm.* **50**: 1541–1550.

Ishii, T., Moriyoshi, K., Sugihara, H., Sakurada, K., Kadotani, H., Yokoi, M., Akazawa, C., Shigemoto, R., Mizuno, N., Masu, M. and Nakanishi, S. (1993) 'Molecular characterization of the family of the *N*-methyl-D-aspartate receptor subunits', *J. Biol. Chem.* **268**: 2836–2843.

Ito, I., Futai, K., Katagiri, H., Watanabe, M., Sakimura, K., Mishina, M. and Sugiyana, H. (1997) 'Synapse-selective impairment of NMDA receptor functions in mice lacking NMDA receptor ε1 or ε2 subunit', *J. Physiol.* **500**: 401–408.

Jahr, C.E. and Stevens, C.F. (1987) 'Glutamate activates multiple single channel conductances in hippocampal neurons', *Nature* **325**: 522–525.

Jahr, C.E. and Stevens, C.F. (1993) 'Calcium permeability of the *N*-methyl-D-aspartate receptor channel in hippocampal neurons in culture', *Proc. Natl. Acad. Sci. USA* **90**: 11,573–11,577.

Johnson, J.W. and Ascher, P. (1987) 'Glycine potentiates the NMDA response in cultured mouse brain neurons', *Nature* **325**: 529–531.

Johnson, J.W. and Ascher, P. (1994) 'The NMDA receptor, its channel and its modulation by glycine', in Collingridge and Watkins (eds) *The NMDA Receptor* (2nd edn), Oxford: University Press.

Jonas, P. and Burnashev, N. (1995) 'Molecular mechanisms controlling calcium entry through AMPA-type glutamate receptor channels', *Neuron* **15**: 987–990.

Jonas, P. and Spruston, N. (1994) 'Mechanisms shaping glutamate-mediated excitatory postsynaptic currents in the CNS', *Curr. Opin. Neurobiol.* **4**: 366–372.

Jonas, P., Racca, C., Sakmann, B., Seeburg, P.H. and Monyer, H. (1994) 'Differences in Ca^{2+} permeability of AMPA-type glutamate receptor channels in neocortical neurons caused by differential GluR-B subunit expression', *Neuron* **12**: 1281–1289.

Kadotani, H., Namura, S., Katsuura, G., Terashima, T. and Kikuchi, H. (1998) 'Attenuation of focal cerebral infarct in mice lacking NMDA receptor subunit NR2C', *Neuroreport* **9**: 471–475.

Kamboj, S.K., Swanson, G.T. and Cull-Candy, S.G. (1995) 'Intracellular spermine confers rectification on rat calcium-permeable AMPA and kainate receptors', *J. Physiol.* **486**: 297–303.

Kameyama, K., Lee, H.-K., Bear, M.F. and Huganir, R.L. (1998) 'Involvement of a postsynaptic protein kinase A substrate in the expression of homosynaptic long-term depression', *Neuron* **21**: 1163–1175.

Kashiwagi, K., Pahk, A.J., Masuko, T., Igarashi, K. and Williams, K. (1997) 'Block and modulation of *N*-methyl-D-aspartate receptors by polyamines and protons: Role of amino acid residues in the transmembrane and pore-forming regions of NR1 and NR2 subunits', *Mol. Pharmacol.* **52**: 701–713.

Katz, B. (1969) *The Release of Neural Transmitter Substances*, Liverpool: Liverpool University Press.

Kemp, J.A. and Leeson, P.D. (1993) 'The glycine site of the NMDA receptor – five years on', *Trends Pharmacol. Sci.* **14**: 20–25.

Kew, J.N.C., Richards, J.G., Mutel, V. and Kemp, J.A. (1998) 'Developmental changes in NMDA receptor glycine affinity and ifenprodil sensitivity reveal three distinct populations of NMDA receptors in individual rat cortical neurons', *J. Neurosci.* **18**: 1935–1943.

Kleckner, N.W. and Dingledine, R. (1988) 'Requirement for glycine in activation of NMDA receptors expressed in *Xenopus* oocytes', *Science* **241**: 835–837.

Koh, D.-S., Burnashev, N. and Jonas, P. (1995a) 'Block of native Ca^{2+}-permeable AMPA receptors in rat brain by intracellular polyamines generates double rectification', *J. Physiol.* **486**: 305–312.

Koh, D.-S., Geiger, J.R.P., Jonas, P. and Sakmann, B. (1995b) 'Ca^{2+}-permeable AMPA and NMDA receptor channels in basket cells of rat hippocampal dentate gyrus', *J. Physiol.* **485**: 383–402.

Kohler, M., Burnashev, N., Sakmann, B. and Seeburg, P.H. (1993) 'Determinants of Ca^{2+} permeability in both TM1 an TM2 of high affinity kainate receptor channels: Diversity by RNA editing', *Neuron* **10**: 491–500.

Kuner, T., Wollmuth, L.P., Karlin, A., Seeburg, P.H. and Sakmann, B. (1996) 'Structure of the NMDA receptor channel M2 segment inferred from the accessibility of cysteines', *Neuron* **17**: 343–352.

Kurytov, A., Laube, B., Betz, H. and Kuhse, J. (1994) 'Mutational analysis of the glycine-binding site of the NMDA receptor: Structural similarity with bacterial amino acid-binding proteins', *Neuron* **12**: 1291–1300.

Kutsuwada, T., Kashiwabuchi, N., Mori, H., Sakimura, K., Kushiya, E., Araki, K., Meguro, H., Masaki, H., Kumanishi, T., Arakawa, M. and Mishina, M. (1992) 'Molecular diversity of the NMDA receptor channel', *Nature* **358**: 36–41.

Laube, B., Hirokazu, H., Sturgess, M., Betz, H. and Kuhse, J. (1997) 'Molecular determinants of agonist discrimination by NMDA receptor subunits: analysis of the glutamate binding site on the NR2B subunit', *Neuron* **18**: 493–503.

Laube, B., Kuhse, J. and Betz, H. (1998) 'Evidence for a tetrameric structure of recombinant NMDA receptors', *J. Neurosci.* **18**: 2954–2961.

Laurie, D.J. and Seeburg, P.H. (1994) 'Regional and developmental heterogeneity in splicing of the rat brain MNDAR1 mRNA', *J. Neurosci.* **13**: 3180–3194.

Lee, J.M., Zipfel, G.J. and Choi, D.W. (1999) 'The changing landscape of ischaemic brain injury mechanisms', *Nature* **399**: Suppl. A7–14.

Leranth, C., Szeidemann, Z., Hsu, M. and Buzsáki, G. (1996) 'AMPA receptors in the rat and primate hippocampus: a possible absence of GluR2/3 subunits in most interneurons', *Neuroscience* **70**: 631–652.

Lester, R.A.J. and Jahr, C.E. (1992) 'NMDA channel behavior depends on agonist affinity', *J. Neurosci.* **12**: 635–643.

Lester, R.A.J., Clements, J.D., Westbrook, G.L. and Jahr, C.E. (1990) 'Channel kinetics determine the time course of NMDA receptor-mediated synaptic currents', *Nature* **346**: 565–567.

Lin, J.W., Wyszynski, M., Madhavan, R., Sealock, R., Kim, J.U. and Sheng, M. (1998) 'Yotiao, a novel protein of neuromuscular junction and brain that interacts with specific splice variants of NMDA receptor subunit NR1', *J. Neurosci.* **18**: 2017–2027.

Liu, S.-Q.J. and Cull-Candy, S.G. (2000) 'Synaptic activity at calcium-permeable AMPA receptors induces a switch in receptor subtype', *Nature* **405**: 454–458.

Lomeli, H., Sprengel, R., Laurie, D.J., Kohr, G., Herb, A., Seeburg, P.H. and Wisden, W. (1993) 'The rat delta-1 and delta-2 subunits extend the excitatory amino acid receptor family', *FEBS Lett.* **315**: 318–322.

Lomeli, H., Mosbacher, J., Melcher, T., Höger, T., Geiger, J.R.P., Kuner, T., Monyer, H., Higuchi, M., Bach, A. and Seeburg, P.H. (1994) 'Control of kinetic properties of AMPA receptor channels by nuclear RNA editing', *Science* **266**: 1709–1713.

MacDermott, A.B., Mayer, M.L., Westbrook, G.L., Smith, S.J. and Barker, J.L. (1986) 'NMDA-receptor activation increases cytoplasmic calcium concentration in cultured spinal cord neurones', *Nature* **321**: 519–522.

Mahanty, N.K. and Sah, P. (1998) 'Calcium-permeable AMPA receptors mediate long-term potentiation in interneurons in the amygdala', *Nature* **394**: 683–687.

Mano, I., Lamed, Y. and Teichberg, V.I. (1996) 'A Venus flytrap mechanism for activation and desensitization of a-amino-3-hydroxy-5-methyl-4-isoxazole propionic acid re-ceptors', *J. Biol. Chem.* **271**: 15299–15302.

Mayer, M.L., Westbrook, G.L. and Guthrie, P.B. (1984) 'Voltage-dependent block by Mg^{2+} of NMDA responses in spinal cord neurones', *Nature* **309**: 261–263.

McBain, C.J. and Mayer, M.L. (1994) '*N*-methyl-D-aspartic acid receptor structure and function', *Physiolog. Rev.* **74**: 723–760.

Misra, C., Brickley, S.G., Farrant, M. and Cull-Candy, S.G. (2000a) 'Identification of subunits contributing to synaptic and extrasynaptic NMDA receptors in Golgi cells of the rat cerebellum', *J. Physiol.* **524**: 147–162.

Misra, C., Brickley, S.G., Wyllie, D.J.A. and Cull-Candy, S.G. (2000b) 'Slow deactivation kinetics of NMDA receptors containing NR1 and NR2D subunits in rat cerebellar Purkinje cells', *Journal of Physiology* **525**: 299–305.

Mody, I. (1998) 'Interneurons and the ghost of the sea', *Nat. Neurosci.* **1**: 434–436.

Momiyama, A., Feldmeyer, D. and Cull-Candy, S.G. (1996a) 'Identification of a native low-conductance NMDA channel with reduced sensitivity to Mg^{2+} in rat central neurones. *J. Physiol.* **494**: 479–492.

Momiyama, A., Silver, R.A. and Cull-Candy, S.G. (1996b) 'Conductance of glutamate receptor channels at climbing fibre synapses in rat Purkinje cells in thin slices', *Journal of Physiology* **494**: 86P.

Monyer, H., Burnashev, N., Laurie, D.J., Sakmann, B. and Seeburg, P.H. (1994) 'Developmental and regional expression in the rat brain and functional properties of four NMDA receptors', *Neuron* **12**: 529–540.

Monyer, H., Seeburg, P.H. and Wisden, W. (1991) 'Glutamate-operated channels: Developmentally early and mature forms arise by alternative splicing', *Neuron* **6**: 779–810.

Mori, H. and Mishina, M. (1995) 'Structure and function of the NMDA receptor channel', *Neuropharmacology* **34**: 1219–1237.

Moriyoshi, K., Masu, M., Ishii, T., Shigemoto, R., Mizuno, N. and Nakanishi, S. (1991) 'Molecular cloning and characterization of the rat NMDA receptor', *Nature* **354**: 31–37.

Mosbacher, J., Schoepfer, R., Monyer, H., Burnashev, N., Seeburg, P.H. and Ruppersberg, J.P. (1994) 'A molecular determinant for submillisecond desensitization in glutamate receptors', *Science* **266**: 1059–1061.

Nakanishi, N., Shneider, N.A. and Axel, R. (1990) 'A family of glutamate receptor genes: Evidence for the formation of heteromultimeric receptors with distinct channel properties', *Neuron* **5**: 569–581.

Nakanishi, S. and Masu, M. (1994) 'Molecular diversity and functions of glutamate receptors', *Annu. Rev. Biophys. Biomol. Struct.* **23**: 319–348.

Nowak, L.M., Bregestovski, P., Ascher, P., Herbet, A. and Prochiantz, A. (1984) 'Magnesium gates glutamate-activated channels in mouse central neurones', *Nature* **307**: 462–465.

Otis, T.S., Raman, I.M. and Trussell, L.O. (1995) 'AMPA receptors with high Ca^{2+}-permeability mediate synaptic transmission in the avian auditory pathway', *J. Physiol.* **482**: 309–315.

Ozawa, S., Kamiya, H. and Tsuzuki, K. (1998) 'Glutamate receptors in the mammalian central nervous system', *Prog. Neurobiol.* **54**: 581–618.

Paas, Y. (1998) 'The macro- and microarchitectures of the ligand-binding domain of glutamate receptors', *Trends Neurosci.* **21**: 117–125.

Palecek, J., Abdrachmanova, G., Vlachová, V. and Vyklicky, L. (1999) 'Properties of NMDA receptors in rat spinal cord motoneurons', *Eur. J. Neurosci.* **11**: 827–836.

Paoletti, P., Ascher, P. and Neyton, J. (1997) 'High-affinity zinc inhibition of NMDA NR1-NR2A receptors', *J. Neurosci.* **17**: 5711–5725.

Paoletti, P., Neyton, J. and Ascher, P. (1995) 'Glycine-independent and subunit-specific potentiation of NMDA responses by extracellular Mg^{2+}', *Neuron* **15**: 1109–1120.

Partin, K.M., Bowie, D. and Mayer, M.L. (1995) 'Structural determinants of allosteric regulation in alternatively spliced AMPA receptors', *Neuron* **14**: 833–843.

Partin, K.M., Fleck, M.W. and Mayer, M.L. (1996) 'AMPA receptor flip/flop mutants affecting deactivation, desensitization, and modulation by cyclothiazide, aniracetam, and thiocyanate', *J. Neurosci.* **16**: 6634–6647.

Partin, K.M., Patneau, D.K. and Mayer, M.L. (1994) 'Cyclothiazide differentially modulates desensitization of α-amino-3-hydroxy-5-methyl-4-isoxazolepropionic acid receptor splice variants', *Mol. Pharmacol.* **46**: 129–138.

Partin, K.M., Patneau, D.K., Winters, C.A., Mayer, M.L. and Buonanno, A. (1993) 'Selective modulation of desensitization at AMPA versus kainate receptors by cyclothiazide and concanavalin A', *Neuron* **11**: 1069–1082.

Paternain, A.V., Morales, M. and Lerma, J. (1995) 'Selective antagonism of AMPA receptors unmasks kainate receptor-mediated responses in hippocampal neurons', *Neuron* **14**: 185–189.

Patneau, D.K., Wright, P.W., Winters, C., Mayer, M.L. and Gallo, V. (1994) 'Glial cells of the oligodendrocyte lineage express both kainate- and AMPA-preferring subtypes of glutamate receptor', *Neuron* **12**: 357–371.

Pemberton, K.E., Belcher, S.M., Ripellino, J.A. and Howe, J.R. (1998) 'High-affinity kainate-type ion channels in rat cerebellar granule cells', *J. Physiol.* **510**: 401–420.

Perez-Otano, I., Schulteis, C.T., Contractor, A., Lipton, S.A., Trimmer, J.S., Sucher, N.J. and Heinemann, S.F. (2001) 'Assembly with the NR1 subunit is required for surface expression of NR3A-containing NMDA receptors', *J. Neuroscience* **21**: 1228–1237.

Petralia, R.S., Wang, Y.X. and Wenthold, R.J. (1994) 'The NMDA receptor subunits NR2A and NR2B show histological and ultrastructural localization patterns similar to those of NR1', *J. Neurosci.* **14**: 6102–6120.

Premkumar, L.S. and Auerbach, A. (1997) 'Stoichiometry of recombinant *N*-methyl-D-aspartate receptor channels inferred from single-channel current patterns', *J. Gen. Physiol.* **110**: 485–502.

Priestley, T., Laughton, P., Myers, J., Le Bourdelles, B., Kerby, J. and Whiting, P.J. (1995) 'Pharmacological properties of recombinant human *N*-methyl-D-aspartate receptors comprising NR1a/NR2A and NR1a/NR2B subunit assemblies expressed in permanently transfected mouse fibroblast cells', *Mol. Pharmacol.* **48**: 841–848.

Quinlan, E.M., Philpot, B.D., Huganir, R.L. and Bear, M.F. (1999) 'Rapid, expression-dependent expression of synaptic NMDA receptors in visual cortex in vivo', *Nature Neurosci* **2**: 352–357.

Racca, C., Catania, M.V., Monyer, H. and Sakmann, B. (1996) 'Expression of AMPA-glutamate receptor B subunit in rat hippocampal GABAergic neurons', *Eur. J. Neurosci.* **8**: 1580–1590.

Raditsch, M., Ruppersberg, J.P., Kuner, T., Günther, W., Schoepfer, R., Seeburg, P.H., Jahn, W. and Witzemann, V. (1993) 'Subunit-specific block of cloned NMDA receptors by argiotoxin636', *FEBS Lett.* **324**: 63–66.

Raymond, L.A., Blackstone, C.D. and Huganir, R.L. (1993) 'Phosphorylation and modulation of recombinant GluR6 glutamate receptors by cAMP-dependent protein kinase', *Nature* **361**: 637–641.

Reichling, D.B. and Macdermott, A.B. (1991) 'Lanthanum actions on excitatory amino acid-gated currents and voltage-gated calcium currents in rat dorsal horn neurons', *J. Physiol.* **441**: 199–218.

Roche, K.W., Tingley, W.G. and Huganir, R.L. (1994) 'Glutamate receptor phosphorylation and synaptic plasticity', *Curr. Opin. Neurobiol.* **4**: 383–388.

Rosenmund, C., Stern-Bach, Y. and Stevens, C.F. (1998) 'The tetrameric structure of a glutamate receptor channel', *Science* **280**: 1596–1599.

Rosenmund, C. and Westbrook, G.L. (1993) 'Calcium-induced actin depolymerization reduces NMDA channel activity', *Neuron* **10**: 805–814.

Rosemund, C., Feltz, A., Westbrook, G.L. (1995) 'Synaptic NMDA receptor channels have a low open probability', *J. Neurosci.* **15**: 2788–2795.

Rozov, A. and Burnashev, N. (1999) 'Polyamine-dependent facilitation of postsynaptic AMPA receptors counteracts paired-pulse depression', *Nature* **401**: 594–598.

Sahara, Y., Noro, N., Iida, Y., Soma, K. and Nakamura, Y. (1997) 'Glutamate receptor subunits GluR5 and KA-2 are coexpressed in rat trigeminal ganglion neurons', *J. Neurosci.* **17**: 6611–6620.

Schell, M.J., Molliver, M.E. and Snyder, S. (1995) 'D-Serine, an endogenous synaptic modulator: Localization to astrocytes and glutamate stimulated release', *Proc. Natl. Acad. Sci.* **92**: 3948–3952.

Schiffer, H.H., Swanson, G.T. and Heinemann, S.F. (1997) 'Rat GluR7 and a carboxy-terminal splice variant, GluR7b, are functional kainate receptor subunits with a low sensitivity to glutamate', *Neuron* **19**: 1141–1146.

Seeburg, P.H. (1996) 'The role of RNA editing in controlling glutamate receptor channel properties', *J. Neurochem.* **66**: 1–5.

Segal, J.A. and Skolnick, P. (1998) 'Polyamine-like actions of aminoglycosides and aminoglycoside derivatives at NMDA receptors', *Eur. J. Pharmacol.* **347**: 311–317.

Sekiguchi, M., Fleck, M.W., Mayer, M.L., Takeo, J., Chiba, Y., Yamashita, S. and Wada, K. (1997) 'A novel allosteric potentiator of AMPA receptors: 4-[2(phenylsulfonylami-no) ethylthio]-2,6-difluoro-phenoxyacetamide', *J. Neurosci.* **17**: 5760–5771.

Sheng, M., Cummings, J., Roldan, L.A., Jan, Y.N. and Jan, L.Y. (1994) 'Changing subunit composition of heteromeric NMDA receptors during development of rat cortex', *Nature* **368**: 144–147.

Silver, R.A., Colquhoun, D., Cull-Candy, S.G. and Edmonds, B. (1996) 'Deactivation and desensitization of non-NMDA receptors in patches and the time course of EPSCs in rat cerebellar granule cells', *Journal of Physiology* **493**: 167–173.

Silver, R.A., Traynelis, S.F. and Cull-Candy, S.G. (1992) 'Rapid time course miniature and evoked excitatory currents at cerebellar synapses in situ', *Nature* **355**: 163–166.

Smith, K.E., Boren, L.A., Hartig, P.R., Branchek, T. and Weinshank, R.L. (1992) 'Cloning and expression of a glycine transporter reveal colocalization with NMDA receptors', *Neuron* **8**: 927–935.

Soderling, T.R., Tan, S.E., McGlade-McCulloh, E., Yamamoto, H. and Fukunaga, K. (1994) 'Excitatory interactions between glutamate receptors and protein kinases', *Journal of Neurobiology* **25**: 304–311.

Sommer, B., Burnashev, N., Verdoorn, T.A., Keinanen, K., Sakmann, B. and Seeburg, H. (1992) 'A glutamate receptor channel with high affinity for domoate and kainate', *EMBO J.* **11**: 891–896.

Sommer, B., Keinanen, K., Verdoorn, T.A., Wisden, W., Burnashev, N., Herb, A., Köhler, M., Takagi, T., Sakmann, B. and Seeburg, P.H. (1990) 'Flip and flop: A cell-specific functional switch in glutamate-operated channels of the CNS', *Science* **249**: 1580–1585.

Spruston, N., Jonas, P. and Sakmann, B. (1995) 'Dendritic glutamate receptor channels in rat hippocampal CA3 and CA1 pyramidal neurons', *J. Physiol.* **482**: 325–352.

Stern, P., Behe, P., Schoepfer, R. and Colquhoun, D. (1992) 'Single-channel conductance of NMDA receptors expressed from cloned cDNAs: comparison with native receptors', *Proc. R. Soc. Lond. B.* **250**: 271–277.

Stern-Bach, Y., Bettler, B., Hartley, M., Sheppard, P.O., O'Hara, P.J. and Heinemann, S.F. (1994) 'Agonist selectivity of glutamate receptors is specified by two domains structurally related to bacterial amino acid-binding proteins', *Neuron* **13**: 1345–1357.

Stern-Bach, Y., Russo, S., Neuman, M. and Rosenmund, C. (1998) 'A point mutation in the glutamate binding site blocks desensitization of AMPA receptors', *Neuron* **21**: 907–918.

Stocca, G. and Vicini, S. (1998) 'Increased contribution of NR2A subunit to synaptic NMDA receptors in developing rat cortical neurons', *J. Physiol.* **507**: 13–24.

Sucher, N.J., Akbarian, S., Chi, C.L., Leclerc, C.L., Awobuluyi, M., Deitcher, D.L., Wu, M.K., Yuan, J.P., Jones, E.G. and Lipton, S.A. (1995) 'Developmental and regional expression pattern of a novel receptor-like subunit (NMDAR-L) in the rodent brain', *J. Neurosci.* **15**: 6509–6520.

Sugihara, H., Moriyoshi, K, Ishii, T., Masu, M. and Nakanishi, S. (1992) 'Structures and properties of seven isoforms of the NMDA receptor generated by alternative splicing', *Biochem. Biophys. Res. Commun.* **185**: 826–832.

Swanson, G.T., Feldmeyer, D., Kaneda, M. and Cull-Candy, S.G. (1996) 'Effect of editing and subunit co-assembly on single-channel properties of recombinant kainate receptors', *J. Physiol.* **492**: 129–142.

Swanson, G.T., Kamboj, S.K. and Cull-Candy, S.G. (1997a) 'Single-channel properties of recombinant AMPA receptors depend on RNA editing, splice variation, and subunit composition', *J. Neurosci.* **17**: 58–69.

Swanson, G.T., Gereau, R.W., IV, Green, T. and Heinemann, S.F. (1997b) 'Identification of amino acid residues that control functional behavior in GluR5 and GluR6 kainate receptors', *Neuron* **19**: 913–926.

Takahashi, T., Feldmeyer, D., Suzuki, N., Onodera, K., Cull-Candy, S.G., Sakimura, K. and Mishina, M. (1996) 'Functional correlation of NMDA receptor ϵ subunits expression with the properties of single-channel and synaptic currents in the developing cerebellum', *J. Neurosci.* **16**: 4376–4382.

Thomson, A. (1990) 'Glycine is a coagonist at the NMDA receptor/channel complex', *Progress in Neurobiology* **35**: 53–76.

Thomson, A.M., Walker, V.E. and Flynn, D.M. (1989) 'Glycine enhances NMDA-receptor mediated synaptic potentials in neocortical slices', *Nature* **338**: 422–424.

Tóth, K. and McBain, C. (1998) 'Afferent-specific innervation of two distinct AMPA receptor subtypes on single hippocampal interneurons', *Nat. Neurosci.* **1**: 572–577.

Traynelis, S.F. and Cull-Candy, S.G. (1990) 'Proton inhibition of *N*-methyl-D-aspartate receptors in cerebellar neurons', *Nature* **345**: 347–350.

Traynelis, S.F. and Cull-Candy, S.G. (1991) 'Pharmacological properties and H^+ sensitivity of excitatory amino acid receptor channels in rat cerebellar granule neurones', *J. Physiol.* **433**: 727–763.

Traynelis, S.F. and Wahl, P. (1997) 'Control of rat GluR6 glutamate receptor open probability by protein kinase A and calcineurin', *J. Physiol.* **503**: 513–531.

Traynelis, S.F., Hartley, M. and Heinemann, S.F. (1995) 'Control of proton sensitivity of the NMDA receptor by RNA splicing and polyamines', *Science* **268**: 873–876.

Traynelis, S.F., Silver, R.A. and Cull-Candy, S.G. (1993) 'Estimated conductance of glutamate receptor channels activated during EPSCs at the cerebellar mossy fiber-granule cell synapse', *Neuron* **11**: 279–289.

Traynelis, S.F., Burgess, M.F., Zheng, F., Lyuboslavsky, P. and Powers, J. (1998) 'Control of voltage independent zinc inhibition of NMDA receptors by the NR1 subunit', *J. Neurosci.* **18**: 6163–6175.

Vicini, S., Wang, J.F., Li, J.H., Zhu, W.J., Wang, Y.H., Luo, J.H., Wolfe, B.B. and Grayson, D.R. (1998) 'Functional and pharmacological differences between recombinant *N*-methyl-D-aspartate receptors', *J. Neurophys.* **79**: 555–566.

Vignes, M. and Collingridge, G.L. (1997) 'The synaptic activation of kainate receptors', *Nature* **388**: 179–182.

Villarroel, A., Burnashev, N. and Sakmann, B. (1995) 'Dimensions of the narrow portion of a recombinant NMDA receptor channel', *Biophys. J.* **68**: 866–875.

Vorobjev, V.S., Sharonova, I.N., Walsh, I.B. and Haas, H.L. (1993) 'Histamine potentiates *N*-methyl-D-aspartate responses in acutely isolated hippocampal neurons', *Neuron* **11**: 837–844.

Vyklicky, L., Jr, Vlachová, V. and Krusek, J. (1990) 'The effect of external pH changes on responses to excitatory amino acids in mouse hippocampal neurones', *J. Physiol.* **430**: 497–517.

Vyklicky, L, Patneau, D.K. and Mayer, M.L. (1991) 'Modulation of excitatory synaptic transmission by drugs that reduce desensitization at AMPA/kainate receptors', *Neuron* **7**: 971–984.

Wafford, K.A., Bain, C.J., Le Bourdelles, B., Whiting, P.J. and Kemp, J.A. (1993) 'Preferential coassembly of recombinant NMDA receptors composed of three different subunits', *Neuroreport* **4**: 1347–1349.

Wafford, K.A., Kathoria, M., Bain, C.J., Marshall, G., Le Bourdellès, B., Kemp, J.A. and Whiting, P.J. (1995) 'Identification of amino acids in the *N*-methyl-D-aspartate receptor NR1 subunit that contribute to the glycine binding site', *Mol. Pharmacol.* **47**: 374–380.

Wahl, P., Anker, C., Traynelis, S.F., Egebjerg, J., Rasmussen, J.S., Krogsgaard-Larsen, P. and Madsen, U. (1998) 'Antagonist properties of a phosphono isoxazole amino acid at glutamate R1–4 (*R*,*S*)-2-amino-3-(3-hydroxy-5-methyl-4-isoxazolyl)propionic acid receptor subtypes', *Mol. Pharmacol.* **53**: 590–596.

Wahl, P., Madsen, U., Banke, T., Krogsgaard-Larsen, P. and Schousboe, A. (1996) 'Different characteristics of AMPA receptor agonists acting at AMPA receptors expressed in *Xenopus* oocytes', *Eur. J. Pharmacol.* **308**: 211–218.

Whiting, P.J. and Priestly, T. (1998) 'Molecular biology of *N*-methyl-D-aspartate (NMDA)-type glutamate receptors', in F.A. Stephenson and A.J. Turner (eds) *Amino Acid Neurotransmission*, London: Portland Press.

Wilcox, K.S., Fitzsimond, R.M., Johnson, B. and Dichter, M.A. (1996) 'Glycine regulation of synaptic NMDA receptors in hippocampal neurons', *J. Neurophysiol.* **76**: 3415–3424.

Wilding, T.J. and Huettner, J.E. (1995) 'Differential antagonism of α-amino-3-hydroxy-5-methyl-4-

isoxazolepropionic acid-preferring and kainate-preferring receptors by 2,3-benzodiazepines', *Mol. Pharmacol.* **47**: 582–587.

Wilding, T.J. and Huettner, J.E. (1997) 'Activation and desensitization of hippocampal kainate receptors', *J. Neurosci.* **17**: 2713–2721.

Williams, K. (1993) 'Ifenprodil discriminates subtypes of the *N*-methyl-D-aspartate receptor: Selectivity and mechanisms at recombinant heteromeric receptors', *Mol. Pharmacol.* **44**: 851–859.

Wisden, W. and Seeburg, P.H. (1993) 'A complex mosaic of high-affinity kainate receptors in rat brain', *J. Neurosci.* **13**: 3582–3598.

Wo, Z.G. and Oswald, R.E. (1994) 'Transmembrane topology of two kainate receptor subunits revealed by *N*-glycosylation', *Proc. Natl. Acad. Sci. USA* **91**: 7154–7158.

Wollmuth, L.P., Kuner, T. and Sakmann, B. (1998a) 'Adjacent asparagines in the NR2-subunit of the NMDA receptor channel control the voltage-dependent block by extracellular Mg^{2+}', *J. Physiol.* **506**: 13–32.

Wollmuth, L.P., Kuner, T. and Sakmann, B. (1998b) 'Intracellular Mg^{2+} interacts with structural determinants of the narrow constriction contributed by the NR1-subunit in the NMDA receptor channel', *J. Physiol.* **506**: 33–52.

Wong, L.A. and Mayer, M.L. (1993) 'Differential modulation by cyclothiazide and concanavalin A of desensitization at native alpha-amino-3-hydroxy-5-methyl-4-isoxazolepropionic acid- and kainate-preferring glutamate receptors', *Mol. Pharmacol.* **44**: 504–510.

Wyllie, D.J.A., Behe, P. and Colquhoun, D. (1998) 'Single-channel activations and concentration jumps: Comparison of recombinant NR1a/NR2A and NR1a/NR2D NMDA receptors', *J. Physiol.* **510**: 1–18.

Wyllie, D.J.A., Traynelis, S.F. and Cull-Candy, S.G. (1993) 'Evidence for more than one type of non-NMDA receptor in outside-out patches from the cerebellar granule cells of the rat', *J. Physiol.* **463**: 193–226.

Yamakura, T., Mori, H., Masaki, H., Shimoji, K. and Mishina, M. (1993) 'Different sensitivities of NMDA receptor channel subtypes to non-competitive antagonists', *Neuroreport* **4**: 687–690.

Yamakura, T., Sakimura, K., Mishina, M. and Shimoji, K., (1995) 'The sensitivity of AMPA-selective glutamate receptor channels to pentobarbital is determined by a single amino acid residue of the alpha 2 subunit', *FEBS Lett.* **374**: 412–414.

Yamazaki, M., Araki, K., Shibata, A. and Mishina, M. (1992) 'Molecular cloning of a cDNA encoding a novel member of the mouse glutamate receptor family', *Biochem. Biophys. Res. Com.* **183**: 886–892.

Zarei, M.M. and Dani, J.A. (1995) 'Structural basis for explaining open-channel blockade of the NMDA receptor', *J. Neurosci.* **15**: 1446–1454.

Zhou, L.-M., Gu, Z.-Q., Costa, A.M., Yamada, K.A., Mansson, P.E., Giordano, T., Skolnick, P. and Jones, K.A. (1997) '(2*S*,4*R*)-4-methylglutamic acid (SYM 2081): A selective, high-affinity ligand for kainate receptors', *J. Pharm. Exp. Therap.* **281**: 422–427.

Zuo, J., DeJager, P.L., Takahashi, K.A., Jiang, W., Linden, D.J. and Heintz, N. (1997) 'Neurodegeneration in Lurcher mice caused by mutation in δ2 glutamate receptor gene', *Nature* **388**: 769–773.

Chapter 2

Structure of ionotropic glutamate receptors

Jan Egebjerg and Henrik S. Jensen

Evolution of ionotropic glutamate receptors

The family of glutamate receptor subunits is encoded by at least 17 genes in mammals. The subunits can, based on sequence comparison, be divided into seven groups with amino acid identities higher than 60 per cent within the groups and less than 40 per cent identity between the groups (Figure 2.1). The division correlates with the binding affinities determined for the three key compounds; AMPA, kainate and NMDA, which most commonly have been used to characterize the receptors *in vivo*. The AMPA receptors are formed from the subunits GluR1–GluR4 (also called GluRA–GluRD). The AMPA receptor subunits do not form receptor complexes with subunits from the other groups (Partin *et al.*, 1993; Puchalski *et al.*, 1994). The kainate receptors can be formed of members from two groups: the low affinity kainate receptor subunits, GluR5–GluR7, and the high affinity kainate receptor subunits, KA1 and KA2. The

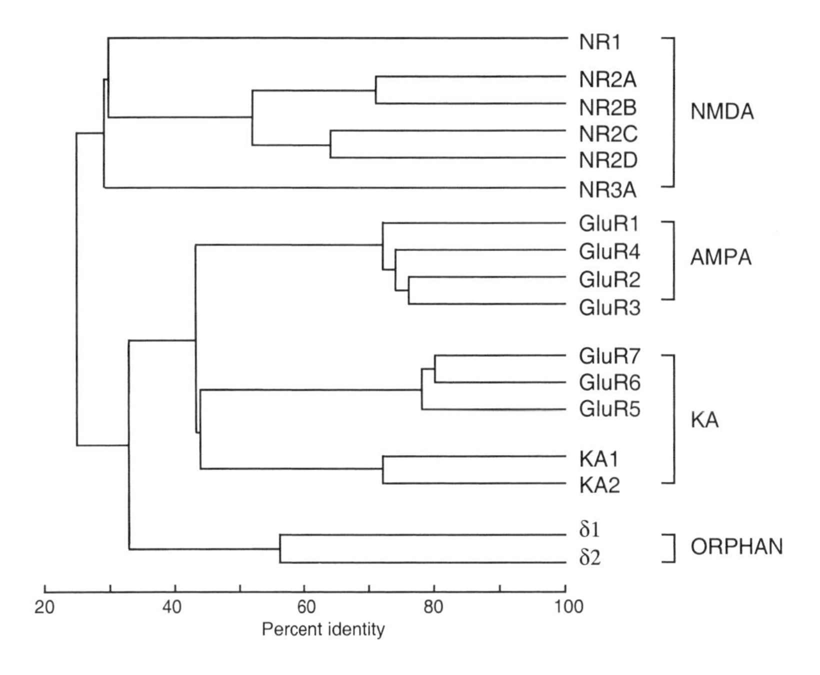

Figure 2.1 Phylogenetic diagram based on amino acid identities.

NMDA receptor subunits fall into three groups: the NR1, the NR2A-NR2D, and the NR3 where the NR1 subunit is required for the formation of a functional channel (see Chapter 1). A seventh related group, consisting of two orphan subunits δ1 and δ2, does not form channels activated by glutamate or assemble with any of the other glutamate receptor subunits (Lomeli *et al.*, 1993). However, several observations suggest that at least δ2 form functional receptors (Kashiwabuchi *et al.*, 1995; Zuo *et al.*, 1997).

The topology of the glutamate receptors has been the topic of many controversies since the first model proposed a subunit topology with four transmembrane segments in analogy with the nicotinic acetylcholine, $GABA_A$ superfamily of receptors. The commonly used nomenclature for the membrane segments (TM1–TM4) is a reminiscence of the early model. However, a large number of different experimental approaches (see Chapter 1; Hollmann *et al.*, 1994; Wo and Oswald, 1994; Stern-Bach *et al.*, 1994; Bennett and Dingledine, 1995; Kuner *et al.*, 1996) and sequence alignments (Nakanishi *et al.*, 1990; O'Hara *et al.*, 1993) with bacterial periplasmatic amino acid binding proteins have provided strong evidence for a three transmembrane topology with a putative pore-forming re-entrant loop between the first and second transmembrane domain. The N-terminal part of the receptor is located extracellularly and is larger (approximately 600 amino acids) than the N-terminal domains of other ligand gated ion channels. Homology searches revealed that the N-terminal part of the domain (X-domain) exhibits similarity to the periplasmatic Leucine-Isoleucine-Valine binding proteins (LIVBP), while the region preceding the first transmembrane and the extracellular domain between the second and third transmembrane region exhibits similarity to the Lysine-Arginine-Ornithine binding proteins (LAOBP, Figure 2.2a; O'Hara *et al.*, 1993; Nakanishi *et al.*, 1990). Both binding proteins have a bi-lobular structure with the amino acid binding site located between the lobes (see Figure 2.3). The pore region, generated by the two first transmembrane regions and the re-entrant loop, resembles the pore structure of the voltage, cyclic-nucleotide gated channels where a number of studies have supported the structural similarity between the re-entrant loop and the P-segment (Figure 2.2). The main difference is that the M2 re-entrant loop enters the membrane from the cytoplasmatic site in the glutamate receptor while the P-element is inserted from the extracellular side (Figure 2.2).

The 'breakdown' of the receptor into modules related to ancestral proteins such as the periplasmatic binding proteins and the P-segment structure have promoted the idea (Wo and Oswald, 1995) that an ancestral form of the glutamate receptors might have evolved as a result of a genetic rearrangement where the gene encoding a K^+-channel-like domain has recombined into a region between the two lobes of the periplasmatic proteins. The third transmembrane region and the intracellular C-terminal might have been associated later in evolution and thereby provided sites for post-translational regulation of the receptor (Chapter 1) and receptor trafficking (Chapter 3).

Evolutionary intermediates lacking the N-terminal LIVBP domain have been identified in frogs and chicken. Recently the evolutionary hypothesis has gained strong support from the cloning of a bacterial precursor, called GluR0, of the eukaryotic GluRs (Chen *et al.*, 1999). GluR0 is a 397 amino acid K^+ selective glutamate gated channel containing only the LAOBP like domain and a K^+ channel segment related to both the eukaryotic GluRs and K^+ channels (Figure 2.2b).

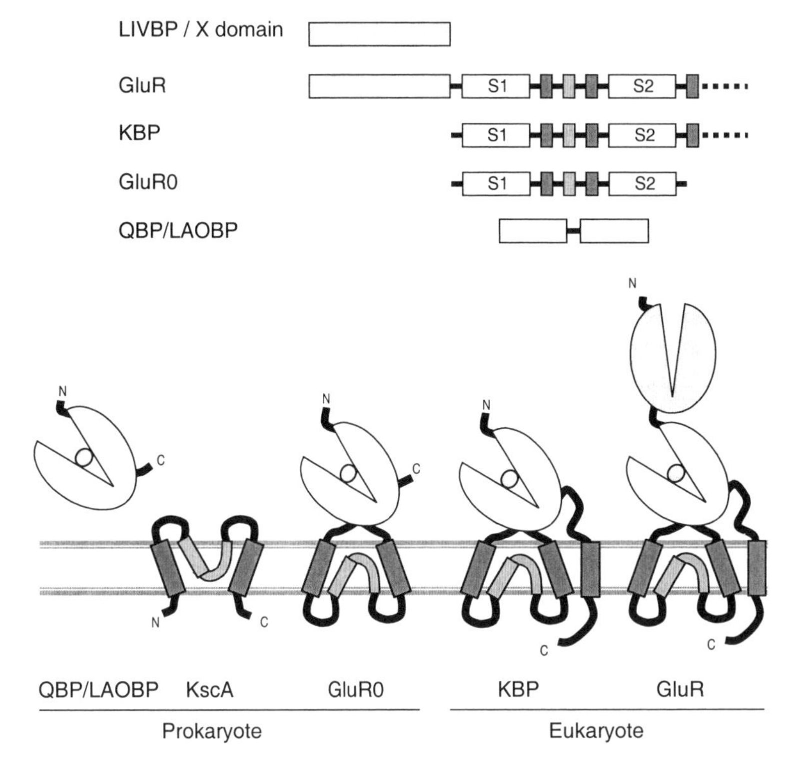

Figure 2.2 (a) Schematic representation of the conserved domains; (b) representation of the structure of the evolutionary related domains between prokaryotic and eukaryotic proteins.

Assembly of ionotropic glutamate receptors

The receptor subunits GluR1-GluR7 can form functional homomeric glutamate receptors, although the potencies vary a thousand fold from a few micromolar to millimolar concentrations for GluR7 (Hollmann and Heinemann, 1994; Schiffer *et al.*, 1997). The efficacy of the channels also varies substantially (e.g. activation of homomeric GluR2 only generates very small currents). The remaining subunits only form functional channels in heteromeric complexes; KA1 and KA2 in complex with GluR5, GluR6 or GluR7 (Chittajallu *et al.*, 1999). Formation of functional NMDA receptors in heterologous expression systems requires co-expression of NR1 and at least one of the NR2 subunits. Reports of activity of homomeric NR1 receptors in *Xenopus* oocytes was most likely due to NR1 assembly with an endogenous oocyte NR2-like subunit (Soloviev and Barnard, 1997). Studies on the assembly of truncated and chimeric subunits indicate that an element in the X-domain and another in the first transmembrane region determine the subunit selective assembly (Leuschner and Hoch, 1999; Kuusinen *et al.*, 1999).

The stoichiometry of the receptor complex is still unclear. Various biochemical approaches, including chemical cross-linking (Brose *et al.*, 1993) and purification of solubilized receptor complex followed by sedimentation analysis (Blackstone *et al.*, 1992;

Wu and Chang, 1994) or gel filtration (Hunter and Wenthold, 1992), have not been able to distinguish between four or five subunits. Neither have careful electrophysiological analysis of either wildtype or mutant subunits reached a final conclusion (for review, see Dingledine *et al.*, 1999). Recent results have been in favor of a tetrameric complex. Firstly, single channel recordings from a non-desensitizing receptor formed by GluR6–GluR3 chimeric subunits suggest that the conductance level of the channel depends on the number of agonists bound to the receptor. The fast transition between these states during receptor activation was overcome by pre-application of the competitive antagonist NBQX before application of the agonist. The rate-limiting release of NBQX permitted resolution of three distinct conductance states. Binding of agonists at two of the subunits was required for activation (with antagonists bound at the remaining subunits), resulting in a low conductance state, while further replacement of the antagonists by agonists resulted in two additional conductance states (Rosenmund *et al.*, 1998) interpreted as consecutive displacement of the antagonist at the remaining two subunits. Secondly, the structural similarities with the tetrameric potassium channels also support the tetrameric composition of the glutamate receptor (Chen *et al.*, 1999).

Agonist binding site

The recognition of the bi-lobe agonist binding site with the structural domains situated on each site of the pore region prompted the construction of a soluble form of the binding domain where the pore region was substituted by a hydrophilic linker and the X-domain and the third transmembrane region removed (Kuusinen *et al.*, 1995). The parts of the protein located at the N-terminal of M1, and the segment from the region between the second and the third transmembrane domains, are referred to as S1 and S2, respectively (Stern-Bach *et al.*, 1994). The soluble protein exhibited a pharmacological profile similar to the receptors, suggesting that the soluble protein folded correctly even in the absence of the transmembrane regions (Kuusinen *et al.*, 1995). Further truncations and optimization of the hydrophobic linker resulted in a soluble form of GluR2 that retained the binding properties, and was suitable for co-crystallization with kainate (Chen and Gouaux, 1997; Armstrong *et al.*, 1998).

The crystal structure showed that most of S1 folded into domain A, except for the C-terminal eight amino acids approaching the pore region folds on domain B. The rest of domain B is formed by the N-terminal part of S2, while the 33 most C-terminal amino acids of S2, including the alternative spliced flip/flop region, are located on the back of the domain A relative to the binding site. The termini connecting the binding domain to the pore domain are located adjacent to each other on the backside (compared to the binding cavity) of S2 (Figure 2.3) (Armstrong *et al.*, 1998).

The co-crystal with kainate showed that the agonist was bound between the two domains stabilizing a closed conformation of the binding domain. The interactions contributing to the stabilization of the closed conformation can be divided into three classes: (1) the amino acids interacting with the glutamate moiety of kainate, (2) a class of receptor-ligand interaction which might exhibit selectivity between ligands, and (3) inter-domain interactions between amino acids brought in close proximity by the ligand induced closure of the binding domain.

The glutamate-like backbone of kainate bridges between the two domains and most likely holds them together, stabilizing a twist between the lobes. Five amino acids (Table 2.1; Figure 2.3b,c, see colour section) were identified in these interactions, which most likely will be involved with all agonist interactions containing a glutamate-like moiety. The α-carboxyl group interacts with lobe A through an ionic interaction with the guanidinium group of Arg485 (GluR2 numbering) and the backbone NH group of Thr480. Thr480 interacts also by the hydroxyl group with the amine on kainate. The amine of kainate is also involved with interactions to lobe B where it interacts with Glu705. Additional interaction with lobe B is directed through the ϵ-carboxyl group which interacts with the backbone amino groups of Ser654 and Thr655, and the hydroxyl group of Thr655 (Figure 2.3d). Mutations in these five residues, at equivalent positions in other glutamate receptors, abolish or greatly reduce agonist affinity or potency (Table 2.2). The amino acids interacting with the glutamate moiety are located deep in the binding cavity, while mutations of a number of proximal amino acid residues located at the lobe interface also influence the ligand interaction, but in a ligand selective manner (Figure 2.3d,e). An example is Tyr450, which forms a wedge between the pyrrolidine ring and the isopropenyl group of kainate and lobe A, thereby preventing full closure of the binding domain. It has been proposed that the gating might relate to the extent of domain closure in a model where an intermediate

Table 2.1 Residues involved in ligand receptor interactions

R2 number	*R1–R4*	*R5–R7*	*KA1/KA2*	*NR1*	*NR2A–2D*	*NR3A*	*δ1/δ2*
Residues interacting with the glutamate backbone of kainate							
480	Thr	Thr/Ala/Tyr	Tyr	Thr	Thr	Ser	Thr
485	Arg	Arg	Arg	Arg	Arg	Arg	Arg
654	Ser	Ser/Ala/Ala	Ser	Ser	Ser	Ser	Ala
655	Thr	Thr	Ser/Thr	Val	Thr	Ala	Val
705	Glu	Glu	Glu	Asp	Asp	Asp	Asp
Residues proposed to interact selectively with drugs							
405	Tyr	Tyr/Tyr/Phe	Tyr	Phe	Phe	Phe	Phe
407	Met	Met/Leu/Met	Met	Tyr	Ile	Phe	Met
447	Asp	Asp	Asp	Asp	Asn	Asp	Asp
450	Tyr	Tyr	Tyr	Phe	His	Tyr	Tyr
478	Pro	Pro	Gly/Ala	Pro	Ser	Ser	Ala
483	Leu	Tyr/Tyr/His	Ala	Asn	Glu	Thr	Pro
649	Thr	Ala	Thr	Thr	Thr	Thr	Thr
650	Leu	Val	Ile	Val	Val	Val	Val
653	Gly	Gly	Gly	Ser	Gly	Ser	Ser
708	Met	Ser/Thr/Thr	Met	Val	Val	Leu	Val
727	Leu	Ile	Leu	Phe	Phe	Phe	Ile/Val
728	Asp	Asp	Asp	Phe	Ala	Ala	Ser/Ala
Residues involved in inter domain interactions							
402	Glu	Glu	Glu	Gln	Glu	Glu	Glu
449	Lys	Lys	Val	Lys	Lys	Lys	Arg/Lys
651	Glu/Asp*	Arg/Glu/Lys	His	Lys	Pro	Arg	Arg/Leu
652	Ala/Ser†	Asp	Glu/Ala	Gln	Asn	Glu	Asp
686	Thr	Ser/Asn/Asn	Thr	Ala	Val	Thr	Pro/Ser

*Asp for GluR2–4
†Ser for GluR2–4

Table 2.2 Effect of mutations in residues critical for receptor–ligand interaction

Residue	Effect
Residues interacting with the glutamate backbone of kainate	
480T	GluR3 (T→A) EC_{50}: Glu inc. 134X
	cKBP (T→A) Affinity: Glu dec. 100X, Kai dec. 58X
485R	NR2B (R→K) no response
	GluR1(R→E, Q or K) no response
	cKBP (R→S) no Kai binding
654S	NR2B (S→G) EC_{50}: Glu inc. 100X
	cKBP (S→A) Affinity: Glu WT, Kai dec. 6X
655T	NR2A (T→A) EC_{50}: Glu inc. 1000X (Gly WT)
	cKBP (T→A) no Kai binding
705E	NR1 (D→E, N, A or G) EC_{50}: Gly inc. >4000X; Glu, DCQX and 5,7.DCK WT
	cKBP (E→Q) no Kai binding
Residues proposed to interact selectively with drugs	
405Y	NR1 (F→S) EC_{50}: Gly inc. 63X; IC_{50} 7-Chlorokyrinic acid inc. 48X; Glu WT
	NR2B (F→S) EC_{50}: Glu inc. 50X
	cKBP (Y→I) Affinity: Kai dec. 10X; Glu dec. 30X; CNQX no binding
407M	NR1 (Y→A) EC_{50}: Gly inc. 12X; Glu WT
447D	NR1 (D→N) EC_{50}: Gly inc. 7X; Glu WT
	NR2A (N→A) EC_{50}: Glu inc. 6X; Gly WT
	GluR1 (D→K) EC_{50}: Glu inc. 5X
450Y	NR1 (F→A) EC_{50}: Gly inc. 6300X; Glu inc. 10X; IC_{50} 7-Chlorokyrinic acid inc. >300X
	NR2A (H→F) EC_{50}: Glu inc. 9X
	NR2B (H→A) EC_{50}: Glu inc. 220X; Gly WT
	cKBP (Y→I) Affinity: Glu dec. 90X; Kai dec. 10X
478P	NR2B (S→A) EC_{50}: Glu inc. 40X
	cKBP (P→A) Affinity: Kai WT; CNQX dec. 10X
483L	GluR3 (L→Y) no desensitization
649T	NR2A (T→A) EC_{50}: Glu inc. 7X
650L	NR1 (V→A) EC_{50}: Gly inc. 13X; Glu WT
	NR2B (V→A) EC_{50}: Glu inc. 20X
	NR2A (V→A) EC_{50}: Glu inc. 11X
	GluR1 (L→T) EC_{50}: Glu inc. 20X
653G	NR1 (S→G) EC_{50}: Gly inc. 25X; Glu WT
	NR2A (G→A) EC_{50}: Glu inc. 320X
	cKBP (S→A) Affinity: Kai 5X; Glu WT
708M	NR2B (V→A) EC_{50}: Glu inc. 30X
727L	NR1 (F→A) EC_{50}: Gly inc. 15X
728D	NR1 (S→G) EC_{50}: Gly inc. 28X
Residues involved in inter-domain interactions	
402E	NR1 (Q→K) EC_{50}: Gly inc. 14000X; Glu inc. 13X
	NR2B (E→A) EC_{50}: Glu inc. 240X; Gly WT
	GluR1 (E→L) EC_{50}: Kai inc. 3–4X
	GluR1 (E→K) EC_{50}: Glu inc. 10^5X; Kai inc. 27X; AMPA inc. 21X
	cKBP (E→V) Affinity: Kai 110X; Glu WT
449K	NR1 (K→Q) EC_{50}: Gly inc. 130X
	NR2A (K→E) EC_{50}: Glu inc. 10X
	NR2B (K→E) EC_{50}: Glu inc. 180X; Gly WT
	GluR1 (K→Q) EC_{50}: Glu inc. 3X; AMPA inc. 51X
651	No mutant
652	No mutant
686	GluR6 (N→S) can be activated by AMPA and iodo-wilardine

References
Anson *et al.* (1998), Kuryatov *et al.* (1994), Hirai *et al.* (1996), Wafford *et al.* (1995), Williams *et al.* (1996), Laube *et al.* (1997), Mano *et al.* (1996), Uchino *et al.* (1992), Stern-Bach *et al.* (1998), Swanson *et al.* (1997a, 1998), Paas *et al.* (1996).

closure will correspond to the gating state while further closure will induce the desensitized state. Thus, the steric clashes between kainate and Tyr450 would result in a partially closed form with a low level (or no) of desensitization and a low binding affinity, while AMPA and glutamate easily can accommodate Tyr450 and allow further closure which results in desensitization and high binding affinity (Armstrong *et al.*, 1998). Alternative models have proposed that gating is induced by changes within lobe B (Mano *et al.*, 1996).

The proposed agonist mediated closure of the lobes is further stabilized by inter-lobe interaction which does not participate directly in the kainate interaction, such as Glu402-Thr686 and Lys449 interaction with Asp651 and Ser652, respectively (Armstrong *et al.*, 1998). However the residues might interact with other ligands since the position equivalent to Thr686 in GluR6 (Asn721) prevents AMPA binding to GluR6 (Swanson *et al.*, 1997a).

The current model for the composition of the NMDA receptor predicts that the glycine binding site is formed by the NR1 subunit(s) and the glutamate site by the NR2 subunit(s). The model is based on the binding properties of the soluble agonist binding domain of NR1 (Ivanovic *et al.*, 1998) and detailed mutagenesis studies which show that mutations in the NR1 subunit greatly affect the glycine interaction (Kuryatov *et al.*, 1994). In particular the mutants Gln387 and Phe466 (equivalent to 402 and 450 in GluR2) in S1 reduce the potency for glycine more that two-thousand fold, but also mutants Ser669 (653 in GluR2) in S2 affect the glycine interaction (Kuryatov *et al.*, 1994) (Table 2.2). Mutations in NR2 greatly alter the glutamate potency but only impose minor changes in the glycine potency (Laube *et al.*, 1997; Anson *et al.*, 1998). The model suggests that subtype-specific competitive glycine antagonists might not be expected while the molecular diversity of the glutamate site remains a potential target for competitive subtype specific drugs. However, examining the key residues in the binding pocket (Table 2.1) determined from comparisons with the GluR2 crystal structure reveals a highly conserved binding pocket between the NR2 subtypes. The differences in potency for NMDA and glutamate between the subtypes is also less than fourfold, with order of potency NR1/NR2D > NR1/NR2B = NR1/NR2C > NR1/NR2A. Other agonists such as homoquinolinate exhibit the same relative subunit selectivity but the order was NR1/NR2A > NR1/NR2B > NR1/NR2C > NR1/NR2D (Buller *et al.*, 1994). Interestingly, antagonists exhibit some subunit selectivity, as AP-5 and CCP which inhibit NR1/NR2A with more than tenfold higher potency than NR1/NR2D receptors. Mutagenesis data show that AP-5 inhibition is affected by mutations which do not affect glutamate interaction, suggesting that the AP-5 binding site exceeds the glutamate binding pocket (Table 2.2; Laube *et al.*, 1997).

Interestingly, only the mutations in S1 of NR1 affect 7-chlorokynurenic acid antagonism of glycine activity, suggesting that 7-chlorokynurenic acid exerts its effect by selective binding to one of the binding domains (Kuryatov *et al.*, 1994). Similarly (*R*)-CPP inhibition is only affected by mutants in lobe A of NR2, while AP-5 is sensitive to mutations in both binding domains (Laube *et al.*, 1997), suggesting that competitive antagonists might inhibit agonist interaction either by selective shielding of the agonist binding site on one lobe (as CCP or 7-chlorokynurenic) or by stabilizing the binding domain in a partially closed state unable to activate gating (as AP-5). The AMPA receptor antagonist NBQX might also stabilize a partial closed form since the activity of a constitutive active GluR1 mutant is greatly increased by NBQX (Taverna *et al.*, 2000).

Most of the compound acting on the AMPA receptors exhibit a low degree of inter-subunit selectivity, which is reflected in the high conservation of the key residues (Table 2.1). A few compounds, such as Br-HIBO (Coquelle *et al.*, 2000) and the currently most potent AMPA receptor agonist 1-methyl-tetrazole-AMPA (Chapter 5; Vogensen *et al.*, 2000), exhibit 10-fold differences in potency between the subunits. However, the maximal responses elicited by these agonists are much smaller on the receptors exhibiting the highest potency, suggesting a larger degree of desensitization. This raises two concerns: first, using functional studies to estimate binding affinities can be misleading due to different contributions from different functional conformations (Colquhoun, 1998); second, highly potent agonists which induce a large degree of desensitization might functionally act as inhibitors of the receptor response.

The kainate receptor family exhibits the most distinct inter-subunit pharmacology, which is also reflected in the variations of the key residues (Table 2.1). Homomeric GluR5 is activated by AMPA, while GluR6 receptors are inert to AMPA activation. The subunit selectivity resides in the amino acid at position 721 in GluR6 (686 in GluR2) (Swanson *et al.*, 1997a). Substitutions at the 5 position of the isoxazole ring in AMPA with bulky hydrophobic substituents, as *tert*-butyl in ATPA, increased the potency at GluR5 significantly, still without activation of GluR6 (Bleakman *et al.*, 1996; Stensbol *et al.*, 1999). The increase in potency for ATPA on GluR5 receptors is difficult to explain structurally since there is no obvious hydrophobic pocket that can accommodate the *tert*-butyl group and facilitate the binding.

Modulatory sites

All the glutamate receptor subtypes are targets for a number of non-competitive drugs and metal ion interactions that modulate the functionality of the receptor with a high degree of subtype specificity (Chapter 1). One of the best-described "natural" modulators is the pH influence on the NMDA receptor activity (Chapter 1). The molecular determinants for the pH sensor are not well defined, but the action of a number of other modulators seems to converge to the pH sensor as a common structural element. For example, the presence of exon 5 (e.g. NR1b), or spermine, abolishes the pH inhibition, while the drug ifenprodil (at NR1a/NR2B receptors), or nano-molar Zn^{2+} (in NR1a/NR2A receptors), enhances the pH inhibition (Choi and Lipton, 1999; Traynelis *et al.*, 1998; Mott *et al.*, 1998). Zn^{2+} acts both in a voltage dependent manner, most likely as a channel blocker, and in a voltage independent manner. The voltage independent block shows a very high selectivity for NR2A containing receptors. Interestingly Zn^{2+} seems to exert its action through allosteric modulations, since a number of residues contributing to the high affinity Zn^{2+} binding sites are located in the X-domain (Choi and Lipton, 1999; Low *et al.*, 2000; Fayyazuddin *et al.*, 2000).

The most selective compounds acting on the AMPA receptors are the 2,3-benzodiazepines GYKI 52466 and GYKI 53655, which act as inhibitors, and the potentiating compounds such as cyclothiazide, aniracetam and PEPA referred to as ampakines (Chapter 5). The benzodiazepines and cyclothiazides bind at different sites distinct from the agonist binding site. Cyclothiazides preferentially reduce desensitization in the flip splice forms. The molecular mechanism underlying cyclothiazide potentiation is not clearly understood. However, the structural determinant for difference between the flip and flop splice variants depends on the amino acid at position 750 in GluR1 (754 in GluR2) (Partin *et al.*, 1995). Examination of the GluR2 crystal structure shows that residue 754 (in GluR2) is located adjacent to a remarkably hydrophobic solvent-exposed surface, suggesting that the region might be involved in subunit–subunit interaction in the assembled receptor (Armstrong *et al.*, 1998). Other mutagenesis studies have shown that aromatic substitutions of Leu497 in the S1 domain in GluR1 completely relive desensitization (Stern-Bach *et al.*, 1998), suggesting either that these regions might interact or that desensitization might result from allosteric transitions involving different parts of the receptor subunits.

The pore region

The initial electrophysiological characterization of the recombinant AMPA receptor subunits clearly demonstrated that the presence of the GluR2 subunit within the receptor complex changed the channel properties; for example, homomeric GluR1 exhibited a rectifying current–voltage (I/V) relationship while the I/V relation of GluR1GluR2 heteromeric complexes is linear (Boulter *et al.*, 1990). Later it was also shown that GluR2-containing receptors were impermeable to Ca^{2+} and, that the differences depend on one amino acid difference in the putative transmembrane region 2, where GluR2 contain an arginine (R) while the other AMPA receptor subunits contains a glutamine (Q). The R codon (CGG) is not encoded by the GluR2 gene but occurs as a result of an RNA editing process (Sommer *et al.*, 1991). Similar RNA editing is observed in GluR5 and GluR6 pre-mRNA, resulting in Q to R changes (Q/R site)(see Chapter 1).

A structural model for the pore region has to account for the basic channel properties, which are to form a pore through the hydrophobic membrane that permits permeation of ions at a reasonable rate and generates an environment which selects between different ions. Furthermore, there should also be a gating mechanism, which opens the channel in the presence of a ligand but also ensures a closed or ion-impermeable condition in the absence of a ligand. The channel formed by the four transmembrane receptor family (nACh, GABA, etc) is traditionally viewed as an hourglass, where the narrow constriction in the middle forms the gate. The hourglass-like shape is formed by bend α-helices contributed from each of the five subunits. The main functions of the transmembrane part of the channel are the gating mechanism and the formation of an ion-permeable hydrophilic path, while the ion selective filter is not well defined but most likely located outside the transmembrane region. The channel conductance of the nACh receptor is determined by the charge of the residues located at the ends of the α-helix (Green *et al.*, 1998). Many of the early permeation studies were interpreted under the assumption that the glutamate receptor topology resembled the nACh receptor. However, as already discussed, a structural similarity to the voltage-gated, potassium-gated channels is more likely despite a low sequence identity. Combination of the recently published three-dimensional structure of the KcsA channel (Doyle *et al.*, 1998) with the extensive mutagenesis studies, including the substituted cysteine accessibility method (SCAM), has provided some insight on the architecture of the glutamate receptor pore.

The assumed similarity to the KcsA channel suggests that the transmembrane helices form a cone-like structure where the TM2 region is inserted from the cytoplasmatic side into the base of the cone (Figure 2.3a). SCAM analysis showed that the C-terminal part of TM3 (11 residues) and the N-terminal part of TM4 were reactive to extracellular applied reagents, while TM1 was inaccessible (Beck *et al.*, 1999). In addition, the sequence preceding TM1 (preTM1) was reactive, suggesting that these three reactive regions form the extracellular vestibule or the tip of the cone. Structural changes might occur in the preTM1 region during receptor activation since the amino acid located just before TM1 is only accessible in the presence of a ligand, and two residues within the preTM1 region (A555 and S556) of NR2A are critical for the slow component of the desensitization (Beck *et al.*, 1999; Krupp *et al.*, 1998). The fast component depends on residues in the binding domain.

Changes in the highly conserved (YTANLAAF) motif in the C-terminal part of TM3 affect gating more directly. The mutant A634C in NR1 (the third last) exhibits, in the absence of an agonist, a strong current after exposure to the reagent MTS (Beck *et al.*, 1999). In addition, recent studies on the Lurcher mutant revealed that a constitutive active mutant of δ2 was caused by an alanine to threonine change (at the second-last position) (Kashiwabuchi *et al.*, 1995). Introduction of the same mutation in the NMDA receptors and the GluR1 receptor also resulted in a constitutive activated phenotype (Taverna *et al.*, 2000; Kohda *et al.*, 2000). More studies are needed to determine if the C-terminal part of TM3 indeed participates in the formation of the gate (in analogy to KscA) or whether the mutation merely changes the connection between the ligand binding domain and the pore segment.

Mutagenesis studies support a similar structure of the TM2 region and the P-element of KcnA. TM2 might form a re-entrant loop where the N-terminal part forms a α-helical structure located parallel with the walls of the cone formed by the transmembrane elements. The α-helical structure is followed by a random coiled structure pointing toward the center of the pore (Doyle *et al.*, 1998). The narrow constriction of the channel has been determined by the differential reactivity in SCAM analysis when the cysteine reacting reagents were applied from either the extracellular or the intracellular side. The Q/R/N site (referred to as position 0) in NR1 or NR2C is only accessible from the extracellular surface, while the residues located at the positions +2 to +4 in the putative random coiled region also are accessible – but only from the cytoplasmatic site (Kuner *et al.*, 1996). This suggests that the Q/R/N site is located at the tip of the TM2 re-entrant loop. The size of the pore has been investigated using different size organic cations as charge carriers (Wollmuth *et al.*, 1996; Villarroel *et al.*, 1995). This showed that the pore size for the organic ions in the NMDA receptor was determined by the residue at the 0 position in NR1 and the +1 and +2 positions in NR2. However, the pore size of the kainate receptors was independent of the Q/R site (Burnashev *et al.*, 1996).

The selective filter in the potassium channel is formed in the random coil region, where the geometry of the pore selectively can accommodate two potassium ions. A similar mechanism is difficult to imagine for the rather unselective glutamate receptor channel. Many observations suggest that the selectivity between monovalent and divalent ions, at least in the NMDA receptor, might result from different modes of permeation where divalent and monovalent ions bind different binding sites within the channel (Antonov *et al.*, 1998). The difference in permeability between Mg^{2+} and Ca^{2+} in the NMDA channel has been attributed to the larger size of hydrated Mg^{2+} compared to the dehydrated Ca^{2+} ion. The different size, and also the difference in voltage block, suggests different binding sites for the two divalent ions. However, most of the residues which contribute to the voltage dependent Mg^{2+} block of the NMDA receptor also affect the Ca^{2+} permeation (Burnashev *et al.*, 1992; Kawajiri and Dingledine, 1993; Sakurada *et al.*, 1993; Wollmuth *et al.*, 1996; Schneggenburger, 1998; Behe *et al.*, 1995; Premkumar *et al.*, 1997; Ferrer-Montiel *et al.*, 1996). The Mg^{2+} binding site is difficult to determine with the standard methods, but mutations in the polar residues at the 0 and +2 positions in NR1 and the +1 position in NR2, as well as trypthophans at position −5 and −8 in NR2, affects the IC_{50}. The former sites overlap with the residues involved in the formation of the narrowest part of the pore, supporting the notion that the hydrated Mg^{2+} ion is excluded from permeation due to size.

The Ca permeability of the AMPA/kainate receptors depends mainly on the residue at the Q/R site, while mutations of other sites have minor effects (Hume *et al.*, 1991; Dingledine *et al.*, 1992). The pore size cannot account for the selectivity since it is independent of the residue at the Q/R site. It is likely that the ion selectivity at the AMPA/kainate receptors is due simply to a larger electrostatic repulsion of Ca^{2+} by the arginines. In agreement with the observed chloride permeability of the homomeric GluR6R and GluR2 receptors (Burnashev *et al.*, 1996).

The putative Ca^{2+} binding sites in the pore of the NMDA receptor might partially account for the very high fractional Ca permeability of the NMDA receptor compared to the homomeric "Q-form" of the AMPA/kainate receptors. However, the major contribution might originate from a high local Ca concentration in the vestibule due to additional Ca binding sites located in the C-terminal to the TM3 region (Wollmuth *et al.*, 1998).

The Q/R/N site is a pivotal site for understanding many of the channel properties of normal signal transduction, but the site is also involved in interaction with a number of drugs, such as the philantotoxins on the 'Q-form' AMPA/kainate receptors (Blaschke *et al.*, 1993), and in the binding of the uncompetitive blockers as MK801 and PCP on the NMDA receptor (Ferrer-Montiel *et al.*, 1998).

References

Anson, L.C., Chen, P.E., Wyllie, D.J.A., Colquhoun, D. and Schoepfer, R. (1998) 'Identification of amino acid residues of the NR2A subunit that control glutamate potency in recombinant NR1/NR2A NMDA receptors', *J. Neurosci.* **18**: 581–589.

Antonov, S.M., Gmiro, V.E. and Johnson, J.W. (1998) 'Binding sites for permeant ions in the channel of NMDA receptors and their effects on channel block', *Nat. Neurosci.* **1**: 451–461.

Armstrong, N., Sun, Y., Chen, G.Q. and Gouaux, E. (1998) 'Structure of a glutamate-receptor ligand-binding core in complex with kainate', *Nature* **395**: 913–917.

Beck, C., Wollmuth, L.P., Seeburg, P.H., Sakmann, B. and Kuner, T. (1999) 'NMDAR channel segments forming the extracellular vestibule inferred from the accessibility of substituted cysteines', *Neuron* **22**: 559–570.

Behe, P., Stern, P., Wyllie, D.J., Nassar, M., Schoepfer, R. and Colquhoun, D. (1995) 'Determination of NMDA NR1 subunit copy number in recombinant NMDA receptors', *Proc. R. Soc. Lond. B. Biol. Sci.* **262**: 205–213.

Bennett, J.A. and Dingledine, R. (1995) 'Topology profile for a glutamate receptor: three transmembrane domains and a channel-lining reentrant membrane loop', *Neuron* **14**: 373–384.

Blackstone, C.D., Moss, S.J., Martin, L.J., Levey, A.I., Price, D.L. and Huganir, R.L. (1992)

'Biochemical characterization and localization of a non-N-methyl-D-aspartate glutamate receptor in rat brain', *J. Neurochem.* **58**: 1118–1126.

Blaschke, M., Keller, B.U., Rivosecchi, R., Hollmann, M., Heinemann, S. and Konnerth, A. (1993) 'A single amino acid determines the subunit-specific spider toxin block of alpha-amino-3-hydroxy-5-methylisoxazole-4-propionate/kainate receptor channels', *Proc. Natl. Acad. Sci. USA* **90**: 6528–6532.

Bleakman, R., Schoepp, D.D., Ballyk, B., Bufton, H., Sharpe, E.F., Thomas, K., Ornstein, P.L. and Kamboj, R.K. (1996) 'Pharmacological discrimination of GluR5 and GluR6 kainate receptor subtypes by (3S,4aR,6R,8aR)-6-[2-(1(2)H-tetrazole-5-yl)ethyl]decahyd roisdoquinoline-3 carboxylic-acid', *Mol. Pharmacol.* **49**: 581–585.

Boulter, J., Hollmann, M., O'Shea-Greenfield, A., Hartley, M., Deneris, E., Maron, C. and Heinemann, S. (1990) 'Molecular cloning and functional expression of glutamate receptor subunit genes', *Science* **249**: 1033–1037.

Brose, N., Gasic, G.P., Vetter, D.E., Sullivan, J.M. and Heinemann, S.F. (1993) 'Protein chemical characterization and immunocytochemical localization of the NMDA receptor subunit NMDA R1', *J. Biol. Chem.* **268**: 22,663–22,671.

Buller, A.L., Larson, H.C., Schneider, B.E., Beaton, J.A., Morrisett, R.A. and Monaghan, D.T. (1994) 'The molecular basis of NMDA receptor subtypes: native receptor diversity is predicted by subunit composition', *J. Neurosci.* **14**: 5471–5484.

Burnashev, N., Schoepfer, R., Monyer, H., Ruppersberg, J.P., Gunther, W., Seeburg, P.H. and Sakmann, B. (1992) 'Control by asparagine residues of calcium permeability and magnesium blockade in the NMDA receptor', *Science* **257**: 1415–1419.

Burnashev, N., Villarroel, A. and Sakmann, B. (1996) 'Dimensions and ion selectivity of recombinant AMPA and kainate receptor channels and their dependence on Q/R site residues', *J. Physiol. (Lond.)* **496**: 165–173.

Chen, G.Q., Cui, C., Mayer, M.L. and Gouaux, E. (1999) 'Functional characterization of a potassium-selective prokaryotic glutamate receptor', *Nature* **402**: 817–821.

Chen, G.Q. and Gouaux, E. (1997) 'Overexpression of a glutamate receptor (GluR2) ligand binding domain in Escherichia coli: application of a novel protein folding screen', *Proc. Natl. Acad. Sci. USA* **94**: 13,431–13,436.

Chittajallu, R., Braithwaite, S.P., Clarke, V.R. and Henley, J.M. (1999) 'Kainate receptors: subunits, synaptic localization and function', *Trends Pharmacol. Sci.* **20**: 26–35.

Choi, Y.B. and Lipton, S.A. (1999) 'Identification and mechanism of action of two histidine residues underlying high-affinity Zn2+ inhibition of the NMDA receptor', *Neuron* **23**: 171–180.

Colquhoun, D. (1998) 'Binding, gating, affinity and efficacy: the interpretation of structure–activity relationships for agonists and of the effects of mutating receptors', *Br. J. Pharmacol.* **125**: 924–947.

Coquelle, T., Christensen, J.K., Banke, T.G., Madsen, U., Schousboe, A. and Pickering, D.S. (2000) 'Agonist discrimination between AMPA receptor subtypes', *Neuroreport* **11**: 2643–2648.

Dingledine, R., Borges, K., Bowie, D. and Traynelis, S.F. (1999) 'The glutamate receptor ion channels', *Pharmacol. Rev.* **51**: 7–61.

Dingledine, R., Hume, R.I. and Heinemann, S.F. (1992) 'Structural determinants of barium permeation and rectification in non-NMDA glutamate receptor channels', *J. Neurosci.* **12**: 4080–4087.

Doyle, D.A., Morais Cabral, J., Pfuetzner, R.A., Kuo, A., Gulbis, J.M., Cohen, S.L., Chait, B.T. and MacKinnon, R. (1998) 'The structure of the potassium channel: molecular basis of K+ conduction and selectivity', *Science* **280**: 69–77.

Fayyazuddin, A., Villarroel, A., Le Goff, A., Lerma, J. and Neyton, J. (2000) 'Four residues of the extracellular N-terminal domain of the NR2A subunit control high-affinity Zn2+ binding to NMDA receptors', *Neuron* **25**: 683–694.

Ferrer-Montiel, A.V., Merino, J.M., Planells-Cases, R., Sun, W. and Montal, M. (1998) 'Structural determinants of the blocker binding site in glutamate and NMDA receptor channels', *Neuropharmacology* **37**: 139–147.

Ferrer-Montiel, A.V., Sun, W. and Montal, M. (1996) 'A single tryptophan on M2 of glutamate receptor channels confers high permeability to divalent cations', *Biophys. J.* **71**: 749–758.

Green, T., Heinemann, S.F. and Gusella, J.F. (1998) 'Molecular neurobiology and genetics: investigation of neural function and dysfunction', *Neuron* **20**: 427–444.

Hirai, H., Kirsch, J., Laube, B., Betz, H. and Kuhse, J. (1996) 'The glycine binding site of the N-methyl-D-aspartate receptor subunit NR1: identification of novel determinants of co-agonist potentiation in the extracellular M3–M4 loop region', *Proc. Natl. Acad. Sci. USA* **93**: 6031–6036.

Hollmann, M. and Heinemann, S. (1994) 'Cloned glutamate receptors', *Annu. Rev. Neurosci.* **17**: 31–108.

Hollmann, M., Maron, C. and Heinemann, S. (1994) 'N-glycosylation site tagging suggests a three transmembrane domain topology for the glutamate receptor GluR1', *Neuron* **13**: 1331–1343.

Hume, R.I., Dingledine, R. and Heinemann, S.F. (1991) 'Identification of a site in glutamate receptor subunits that controls calcium permeability', *Science* **253**: 1028–1031.

Hunter, C. and Wenthold, R.J. (1992) 'Solubilization and purification of an alpha-amino-3-hydroxy-5-methylisoxazole-4-propionic acid binding protein from bovine brain', *J. Neurochem.* **58**: 1379–1385.

Ivanovic, A., Reilander, H., Laube, B. and Kuhse, J. (1998) 'Expression and initial characterization of a soluble glycine binding domain of the N-methyl-D-aspartate receptor NR1 subunit', *J. Biol. Chem.* **273**: 19,933–19,937.

Kashiwabuchi, N., Ikeda, K., Araki, K., Hirano, T., Shibuki, K., Takayama, C., Inoue, Y., Kutsuwada, T., Yagi, T., Kang, Y., *et al.* (1995) 'Impairment of motor coordination, Purkinje cell synapse formation, and cerebellar long-term depression in GluR delta 2 mutant mice', *Cell* **81**: 245–252.

Kawajiri, S. and Dingledine, R. (1993) 'Multiple structural determinants of voltage-dependent magnesium block in recombinant NMDA receptors', *Neuropharmacology* **32**: 1203–1211.

Kohda, K., Wang, Y. and Yuzaki, M. (2000) 'Mutation of a glutamate receptor motif reveals its role in gating and delta2 receptor channel properties', *Nat. Neurosci.* **3**: 315–322.

Krupp, J.J., Vissel, B., Heinemann, S.F. and Westbrook, G.L. (1998) 'N-terminal domains in the NR2 subunit control desensitization of NMDA receptors', *Neuron* **20**: 317–327.

Kuner, T., Wollmuth, L.P., Karlin, A., Seeburg, P.H. and Sakmann, B. (1996) 'Structure of the NMDA receptor channel M2 segment inferred from the accessibility of substituted cysteines', *Neuron* **17**: 343–352.

Kuryatov, A., Laube, B., Betz, H. and Kuhse, J. (1994) 'Mutational analysis of the glycine-binding site of the NMDA receptor: structural similarity with bacterial amino acid-binding proteins', *Neuron* **12**: 1291–1300.

Kuusinen, A., Abele, R., Madden, D.R. and Keinanen, K. (1999) 'Oligomerization and ligand-binding properties of the ectodomain of the alpha-amino-3-hydroxy-5-methyl-4-isoxazole propionic acid receptor subunit GluRD', *J. Biol. Chem.* **274**: 28,937–28,943.

Kuusinen, A., Arvola, M. and Keinanen, K. (1995) 'Molecular dissection of the agonist binding site of an AMPA receptor', *Embo. J.* **14**: 6327–6332.

Laube, B., Hirai, H., Sturgess, M., Betz, H. and Kuhse, J. (1997) 'Molecular determinants of agonist discrimination by NMDA receptor subunits: analysis of the glutamate binding site on the NR2B subunit', *Neuron* **18**: 493–503.

Leuschner, W.D. and Hoch, W. (1999) 'Subtype-specific assembly of alpha-amino-3-hydroxy-5-methyl-4-isoxazole propionic acid receptor subunits is mediated by their n-terminal domains', *J. Biol. Chem.* **274**: 16,907–16,916.

Lomeli, H., Sprengel, R., Laurie, D.J., Kohr, G., Herb, A., Seeburg, P.H. and Wisden, W. (1993) 'The rat delta-1 and delta-2 subunits extend the excitatory amino acid receptor family', *FEBS Lett.* **315**: 318–322.

Low, C.M., Zheng, F., Lyuboslavsky, P. and Traynelis, S.F. (2000) 'Molecular determinants of coordinated proton and zinc inhibition of N-methyl-D-aspartate NR1/NR2A receptors', *Proc. Natl. Acad. Sci. USA* **97**: 11,062–11,067.

Mano, I., Lamed, Y. and Teichberg, V.I. (1996) 'A venus flytrap mechanism for activation and desensitization of alpha-amino-3-hydroxy-5-methyl-4-isoxazole propionic acid receptors', *J. Biol. Chem.* **271**: 15,299–15,302.

Mott, D.D., Doherty, J.J., Zhang, S., Washburn, M.S., Fendley, M.J., Lyuboslavsky, P., Traynelis, S.F. and Dingledine, R. (1998) 'Phenylethanolamines inhibit NMDA receptors by enhancing proton inhibition', *Nat. Neurosci.* **1**: 659–667.

Nakanishi, N., Shneider, N.A. and Axel, R. (1990) 'A family of glutamate receptor genes: evidence for the formation of heteromultimeric receptors with distinct channel properties', *Neuron* **5**: 569–581.

O'Hara, P.J., Sheppard, P.O., Thogersen, H., Venezia, D., Haldeman, B.A., McGrane, V., Houamed, K.M., Thomsen, C., Gilbert, T.L. and Mulvihill, E.R. (1993) 'The ligand-binding domain in metabotropic glutamate receptors is related to bacterial periplasmic binding proteins', *Neuron* **11**: 41–52.

Paas, Y., Eisenstein, M., Medevielle, F., Teichberg, V.I. and Devillers-Thiery, A. (1996) 'Identification of the amino acid subsets accounting for the ligand binding specificity of a glutamate', *Neuron* **17**: 979–990.

Partin, K.M., Bowie, D. and Mayer, M.L. (1995) 'Structural determinants of allosteric regulation in alternatively spliced AMPA receptors', *Neuron* **14**: 833–843.

Partin, K.M., Patneau, D.K., Winters, C.A., Mayer, M.L. and Buonanno, A. (1993) 'Selective modulation of desensitization at AMPA versus kainate receptors by cyclothiazide and concanavalin A', *Neuron* **11**: 1069–1082.

Premkumar, L.S., Qin, F. and Auerbach, A. (1997) 'Subconductance states of a mutant NMDA receptor channel kinetics, calcium, and voltage dependence', *J. Gen. Physiol.* **109**: 181–189.

Puchalski, R.B., Louis, J.C., Brose, N., Traynelis, S.F., Egebjerg, J., Kukekov, V., Wenthold, R.J., Rogers, S.W., Lin, F., Moran, T., *et al.* (1994) 'Selective RNA editing and subunit assembly of native glutamate receptors', *Neuron* **13**: 131–147.

Rosenmund, C., Stern-Bach, Y. and Stevens, C.F. (1998) 'The tetrameric structure of a glutamate receptor channel', *Science* **280**: 1596–1599.

Sakurada, K., Masu, M. and Nakanishi, S. (1993) 'Alteration of Ca2+ permeability and sensitivity to Mg2+ and channel blockers by a single amino acid substitution in the N-methyl-D-aspartate receptor', *J. Biol. Chem.* **268**: 410–415.

Schiffer, H.H., Swanson, G.T. and Heinemann, S.F. (1997) 'Rat GluR7 and a carboxy-terminal splice variant, GluR7b, are functional kainate receptor subunits with a low sensitivity to glutamate', *Neuron* **19**: 1141–1146.

Schneggenburger, R. (1998) 'Altered voltage dependence of fractional Ca2+ current in N-methyl-D-aspartate channel pore mutants with a decreased Ca2+ permeability', *Biophys. J.* **74**: 1790–1794.

Soloviev, M.M. and Barnard, E.A. (1997) 'Xenopus oocytes express a unitary glutamate receptor endogenously', *J. Mol. Biol.* **273**: 14–18.

Sommer, B., Kohler, M., Sprengel, R. and Seeburg, P.H. (1991) 'RNA editing in brain controls a determinant of ion flow in glutamate-gated channels', *Cell* **67**: 11–19.

Stensbol, T.B., Borre, L., Johansen, T.N., Egebjerg, J., Madsen, U., Ebert, B. and Krogsgaard-Larsen, P. (1999) 'Resolution, absolute stereochemistry and molecular pharmacology of the enantiomers of ATPA', *Eur. J. Pharmacol.* **380**: 153–162.

Stern-Bach, Y., Bettler, B., Hartley, M., Sheppard, P.O., O'Hara, P.J. and Heinemann, S.F. (1994) 'Agonist selectivity of glutamate receptors is specified by two domains structurally related to bacterial amino acid-binding proteins', *Neuron* **13**: 1345–1357.

Stern-Bach, Y., Russo, S., Neuman, M. and Rosenmund, C. (1998) 'A point mutation in the glutamate binding site blocks desensitization of AMPA receptors', *Neuron* **21**: 907–918.

Swanson, G.T., Feldmeyer, D., Kaneda, M. and Cull-Candy, S.G. (1996) 'Effect of RNA editing and subunit co-assembly single-channel properties of recombinant kainate receptors', *J. Physiol. (Lond.)* **492**: 129–142.

Swanson, G.T., Gereau, R.W.T., Green, T. and Heinemann, S.F. (1997a) 'Identification of amino acid residues that control functional behavior in GluR5 and GluR6 kainate receptors', *Neuron* **19**: 913–926.

Swanson, G.T., Kamboj, S.K. and Cull-Candy, S.G. (1997b) 'Single-channel properties of recombinant AMPA receptors depend on RNA editing, splice variation, and subunit composition', *J. Neurosci.* **17**: 58–69.

Swanson, G.T., Green, T. and Heinemann, S.F. (1998) 'Kainate receptors exhibit differential sensitivities to (S)-5-iodowillardiine', *Mol. Pharmacol.* **53**: 942–949.

Taverna, F., Xiong, Z.G., Brandes, L., Roder, J.C., Salter, M.W. and MacDonald, J.F. (2000) 'The Lurcher mutation of an alpha-amino-3-hydroxy-5-methyl-4-isoxazolepropionic acid receptor subunit enhances potency of glutamate and converts an antagonist to an agonist', *J. Biol. Chem.* **275**: 8475–8479.

Traynelis, S.F., Burgess, M.F., Zheng, F., Lyuboslavsky, P. and Powers, J.L. (1998) 'Control of voltage-independent zinc inhibition of NMDA receptors by the NR1 subunit', *J. Neurosci.* **18**: 6163–6175.

Uchino, S., Sakimura, K., Nagahari, K. and Mishina, M. (1992) 'Mutations in a putative agonist binding region of the AMPA-selective glutamate receptor channel', *FEBS Lett.* **308**: 253–257.

Villarroel, A., Burnashev, N. and Sakmann, B. (1995) 'Dimensions of the narrow portion of a recombinant NMDA receptor channel', *Biophys. J.* **68**: 866–875.

Vogensen, S.B., Jensen, H.S., Stensbol, T.B., Frydenvang, K., Bang-Andersen, B., Johansen, T.N., Egebjerg, J. and Krogsgaard-Larsen, P. (2000) 'Resolution, configurational assignment, and enantiopharmacology of 2-amino-3-[3-hydroxy-5-(2-methyl-2H-tetrazol-5-yl)isoxazol-4-yl]propionic acid, a potent GluR3- and GluR4-preferring AMPA receptor agonist', *Chirality* **12**: 705–713.

Wafford, K.A., Kathoria, M., Bain, C.J., Marshall, G., Le Bourdelles, B., Kemp, J.A. and Whiting, P.J. (1995) 'Identification of amino acids in the N-methyl-D-aspartate receptor NR1 subunit that contribute to the glycine binding site', *Mol. Pharmacol.* **47**: 374–380.

Williams, K., Chao, J., Kashiwagi, K., Masuko, T. and Igarashi, K. (1996) 'Activation of N-methyl-D-aspartate receptors by glycine: role of an aspartate residue in the M3–M4 loop of the NR1 subunit', *Mol. Pharmacol.* **50**: 701–708.

Wo, Z.G. and Oswald, R.E. (1994) 'Transmembrane topology of two kainate receptor subunits revealed by N-glycosylation', *Proc. Natl. Acad. Sci. USA* **91**: 7154–7158.

Wo, Z. and Oswald, R. (1995) 'Unraveling the modular design of glutamate-gated ion channels', *Trends Neurosci.* **18**: 161–168.

Wollmuth, L.P., Kuner, T. and Sakmann, B. (1998) 'Adjacent asparagines in the NR2-subunit of the NMDA receptor channel control the voltage-dependent block by extracellular Mg2+', *J. Physiol. (Lond.)* **506**: 13–32.

Wollmuth, L.P., Kuner, T., Seeburg, P.H. and Sakmann, B. (1996) 'Differential contribution of the NR1- and NR2A-subunits to the selectivity filter of recombinant NMDA receptor channels', *J. Physiol. (Lond.)* **491**: 779–797.

Wu, T.Y. and Chang, Y.C. (1994) 'Hydrodynamic and pharmacological characterization of putative alpha-amino-3-hydroxy-5-methyl-4-isoxazolepropionic acid/kainate-sensitive L-glutamate receptors solubilized from pig brain', *Biochem. J.* **300**: 365–371.

Zuo, J., De Jager, P.L., Takahashi, K.A., Jiang, W., Linden, D.J. and Heintz, N. (1997) 'Neurodegeneration in Lurcher mice caused by mutation in delta2 glutamate receptor gene', *Nature* **388**: 769–773.

Chapter 3

Glutamate receptor trafficking

Guido Meyer and Jeremy M. Henley

Introduction

The precise trafficking of ionotropic receptors to the postsynaptic membrane is a prerequisite of synaptic transmission. Therefore processes must exist for targeting the appropriate receptors from the cell body to specific synapses. While it is clear that protein–protein interactions provide the underlying mechanisms for the targeting, transport, clustering and anchoring of receptors the details of these processes are only now beginning to be elucidated. Within the ionotropic glutamate receptor family *N*-methyl-D-aspartate receptors (NMDA receptors) and α-amino-3-hydroxy-5-methylisoxazolepropionate receptors (AMPA receptors) are mainly co-localised in postsynaptic membranes. However, there are significant differences in the synaptic expression mechanisms for NMDA and AMPA receptors. For example, functional NMDA receptors are present and functional at synapses at earlier developmental time points than AMPA receptors (Rao *et al.*, 1998). Recent advances in the cell biology of receptors have now revealed that separate sets of proteins interact with these two types of glutamate receptors and that these interactions orchestrate their trafficking and functional surface expression. A good deal of what has been learned has been the result of the application of the yeast two-hybrid assay (Y2H: Nishimune *et al.*, 1996) using the intracellular C-terminal domains of individual NMDA receptor or AMPA receptor subunits.

Proteins interacting with NMDA receptors

The majority of glutamatergic synapses are located on dendritic spines and possess a submembraneous protein network known as postsynaptic density (PSD). A prominent feature of the core proteins of the PSD is their insolubility in detergents like Triton X-100 (Kennedy, 1997). Functional NMDA receptors are integral membrane proteins and, as such, are contained in the postsynaptic membrane. However, NMDA receptors are also insoluble in Triton X-100 and NR2 subunits appear as core components in preparations of the PSD (Moon *et al.*, 1994; Allison *et al.*, 1998). Thus, NMDA receptors are tightly anchored in the PSD. NMDA receptor subunits exhibit very long cytoplasmic C-terminal tails compared to other glutamate receptor subtypes, and the anchoring of NMDA receptors in the PSD is likely to be the result of an interaction between the C-terminal tails of the NMDA receptor subunits and PSD proteins. Indeed, a surprisingly large variety of intracellular proteins have been identified that interact with the cytoplasmic portion of NMDA receptors (Table 3.1).

Table 3.1 Proteins that interact with NMDA receptors

Primary interactor	*Secondary interactor*	*Tertiary interactor*
PSD-95 (Kornau *et al.*, 1995)	GKAP (Kim *et al.*, 1997; Takeuchi *et al.*, 1997)	Shank (Naisbitt *et al.*, 1999)
	Neuroligin (Irie *et al.*, 1997)	Neurexin (Ichtchenko *et al.*, 1995)
	Neuronal Nitric Oxide Synthase (Brenman *et al.*, 1996)	
	SynGAP (Kim *et al.*, 1998)	
	CRIPT (Niethammer *et al.*, 1998)	Tubulin (Niethammer *et al.*, 1998)
	Citron (Zhang *et al.*, 1999)	
	BEGAIN (Deguchi *et al.*, 1998)	
	Fyn, Yes, Sre (Tezuka *et al.*, 1999)	
Chapsyn-110 (Kim *et al.*, 1996)		
SAP102 (Muller *et al.*, 1996)		
Mammalian LIN-7 (Jo *et al.*, 1999)	CASK (Kaech *et al.*, 1998; Butz *et al.*, 1998)	Mint (Kaech *et al.*, 1998; Butz *et al.*, 1998)
		Protein 4.1 (Cohen *et al.*, 1998)
		Syndecan (Cohen *et al.*, 1998)
		Neurexin (Hata *et al.* 1996)
S-SCAM (Hirao *et al.*, 1998)	GKAP (Hirao *et al.*, 1998)	
	NPRAP/δ-catenin (Ide *et al.*, 1999)	
CIPP (Kurschner *et al.*, 1998)		
α-actinin (Wyszinski *et al.*, 1997)	Actin (Beggs *et al.*, 1992)	
Tubulin (van Rossum *et al.*, 1999)		
Spectrin (Wechsler and Teichberg, 1998)		
Neurofilament-L (Ehlers *et al.*, 1998)		
Calmodulin (Ehlers *et al.*, 1996)		
Calcium/Calmodulin-dependent protein kinase II (Strack and Colbran, 1998; Gardoni *et al.*, 1998)		

Interestingly, it appears that some of the interactions occur only in certain types of neurons (Zhang *et al.*, 1999) and it is likely that specific interactions are restricted to subcellular compartments such that NMDA receptors follow a consecutive sequence of such protein interactions during trafficking. Nonetheless, a large number of protein interactions with NMDA receptors have been reported to occur at synapses and an important area for future research is to determine which of these NMDA receptor–protein interactions at the synapse are competitive and which can occur simultaneously. The complexity of the NMDA receptor-associated protein network at synapses is further increased by proteins that interact indirectly with NMDA receptors via primary NMDA receptor binding partners (Table 3.1). The identification of NMDA receptor-associated proteins is complemented by systematic biochemical approaches to elucidate the components of the PSD (reviewed in (Langnaese *et al.*, 1996; Kennedy, 1998) and the identification of proteins that become induced by synaptic activity (Lanahan and Worley, 1998). Some of the potentially most important currently known proteins that interact directly with NMDA receptors are described below.

NMDA receptor interaction with PSD-95

NMDA receptors interact with members of a family of membrane-associated guanylate kinases (MAGUKs). The synaptically localised members of this family comprise PSD-95 (also named SAP90), Chapsyn 110 (also named PSD-93), SAP102 and SAP97 (also named hdlg) (for review see Kennedy, 1998; Kim and Huganir, 1999). The guanylate kinase domain of MAGUKs is apparently non-functional and seems to act as a protein interaction cassette. MAGUKs are scaffolding proteins that contain four additional protein interaction modules, three of which are PDZ domains and one is a src-homology 3 domain. The PDZ domain is a protein–protein interaction module of approximately ninety amino acids, containing a conserved motif Gly-Leu-Gly-Phe. Its name is derived from three proteins initially shown to possess this module, namely PSD-95 (Cho *et al.*, 1992), its *Drosophila* homologue discs-large tumour suppressor gene product (DLG protein; Woods and Bryant, 1991), and ZO-1, a tight junction protein (Itoh *et al.*, 1993). NMDA receptor subunits 2A and 2B as well as NR1 splice variants 1d and 1e interact with the second PDZ domain of PSD-95 (Kornau *et al.*, 1995). The extreme C-terminus of these subunits conforms to the consensus motif for interaction with PDZ domains. A classification of type I PDZ domains which interact with proteins having a serine or threonine at the −2 position, and an isoleucine, valine or other hydrophobic residue at the 0 position (Songyang *et al.*, 1997), accounts for these forms of interaction. Type 2 PDZ domains interact with proteins containing a hydrophobic amino acid in both the −2 and 0 positions. Proteins may also bind to PDZ domains by other mechanisms which do not involve the extreme C-terminus but an internal 'finger' structure (Hillier *et al.*, 1999).

PSD-95 co-localises with NMDA receptors at synapses in cultured hippocampal neurons, but is absent from GABAergic synapses (Kornau *et al.*, 1995). During synaptogenesis in neuronal cultures postsynaptic aggregation of PSD-95 precedes the accumulation of glutamate receptors (Rao *et al.*, 1998), and PSD-95 aggregates are stable over a time course of hours (Okabe *et al.*, 1999). Thus, synaptic PSD-95 can serve as a marker for glutamatergic synapses even in the absence of glutamate receptors. The recruitment of PSD-95 to synapses depends on its N-terminal palmitoylation, as well as on yet unknown protein interactions of the first two PDZ domains and a C-terminal fragment (Craven *et al.*, 1999). Despite the specific localisation of PSD-95 at glutamatergic synapses evidence for a role of PSD-95 in synaptic anchoring of NMDA receptors remains indirect. A Drosophila member of the MAGUK family, DLG protein, interacts with Shaker-type potassium channels (Tejedor *et al.*, 1997). Genetic mutations of the DLG protein or the binding motif on Shaker channels abolish the clustering and synaptic localisation of the Shaker channels (Budnik, 1996). However, similar genetic mutations of PSD-95 (Migaud *et al.*, 1998) or NMDA receptors (Sprengel *et al.*, 1998) in mice have no effect on basic NMDA receptor-mediated synaptic transmission, suggesting that clustering and synaptic localisation of NMDA receptors is unimpaired. Instead, the mutations targeting the NMDA receptor/PSD-95 interaction disturb NMDA receptor-dependent plastic changes in synaptic strength (Migaud *et al.*, 1998; Sprengel *et al.*, 1998). Although the analysis of these transgenic mice does not exclude an involvement of PSD-95 in the synaptic anchoring of NMDA receptors, they are more in line with a role of PSD-95 in the recruitment of signalling molecules to NMDA receptors.

NMDA receptor interaction with α-actinin

A more direct link of NMDA receptors to the submembranous cytoskeleton might be mediated by the actin-binding protein α-actinin. NR1 subunits bind directly to α-actinin-2 (Wyszynski *et al.*, 1997). This interaction is competitive with the binding of calmodulin to the NRI subunit (Ehlers *et al.*, 1996; Wyszynski *et al.*, 1997). Displacement of α-actinin by calmodulin results in a functional inactivation of NMDA receptors (Ehlers *et al.*, 1996; Zhang *et al.*, 1998; Krupp *et al.*, 1999). α-actinin is well placed to act as a postsynaptic anchor for NMDA receptors at dendritic spines (Wyszynski *et al.*, 1998). However, although actin depolymerisation using latrunculin A disperses α-actinin aggregates, NMDA receptors remain clustered (Allison *et al.*, 1998). Thus, it appears α-actinin is not a necessary requirement for NMDA receptor aggregation. Nonetheless, it is possible that it may play a role in synaptic localisation of NMDA receptor clusters, since some of the receptor aggregates move away from synapses after displacement of α-actinin (Allison *et al.*, 1998).

NMDA receptor interaction with mammalian LIN-7

In *C. elegans* a protein containing a single PDZ domain named LIN-7 is necessary for the synaptic localisation of an AMPA-type glutamate receptor GLP-1 (Rongo *et al.*, 1998). Although the extreme C-terminus of GLR-1 conforms to a consensus sequence for binding of type 1 PDZ domains, no direct interaction partner of the type 1 PDZ domain of LIN-7 has yet been identified in *C. elegans*. Surprisingly, mammalian homologues of LIN-7 (MALS or Velis) interact via their PDZ domain with NR2B subunits of NMDA receptors (Jo *et al.*, 1999).

While the significance of this interaction for the synaptic expression of NMDA receptors remains to be evaluated, it is known that mammalian LIN-7 binds independently of its PDZ domain to a member of the MAGUK family, CASK, which is a mammalian homologue of LIN-10 in *C. elegans* (Butz *et al.*, 1998; Kaech *et al.*, 1998). CASK in turn recruits yet another PDZ domain containing protein, the mammalian homologue of LIN-2 named Mint or X11 (Butz *et al.*, 1998; Kaech *et al.*, 1998; Borg *et al.*, 1999). In *C. elegans* all three components of this complex of LIN-2/LIN-7/LIN-10 are necessary for the correct localisation of the receptor tyrosine kinase LET-23 (Kim, 1997). The mammalian correlate of this complex is mainly localised in perinuclear regions and the Golgi apparatus of neurons, but is also present at synapses (Borg *et al.*, 1999). Thus, it could play an important role in the sorting of NMDA receptors for transport to synapses (Bredt, 1998).

Proteins interacting with AMPA receptors

AMPA receptors are coexpressed with NMDA receptors at the majority of glutamatergic synapses in the adult brain. In contrast to NMDA receptors, AMPA receptors are readily extracted from the postsynaptic membrane by detergents like Triton X-100 (Allison *et al.*, 1998). This suggests that AMPA receptors are either anchored to the PSD by protein interaction(s) that are sensitive to mild detergents, or that AMPA receptors become concentrated in the postsynaptic membrane by means other than anchoring to the PSD. In addition, it has been shown recently that AMPA receptors undergo a rapid recycling at the postsynaptic membrane with a functional half-life of a proportion of AMPA receptors being in the order of 10 min (Lüthi *et al.*, 1999; Noel *et al.*, 1999). No such rapid recycling has been reported for NMDA receptors. These differences in synaptic expression of AMPA receptors to that of NMDA receptors is reflected in the contingent of potential binding partners. None of the proteins reported to interact with NMDA receptors (Table 3.1) bind to AMPA receptors (Table 3.2) and vice versa. The specific postsynaptic surface localisation of AMPA receptors at glutamatergic synapses, and their functional expression on the postsynaptic membrane, is tightly controlled in a developmental and activity dependent profile. In general terms, the GluR2 subunit of AMPA receptors appears to be the prime target for protein interactions, and proteins that interact fall into two groups: PDZ-containing and non-PDZ-containing interactors (Braithwaite *et al.*, 1999). The main proteins currently known to interact with AMPA receptors are discussed below.

Glutamate receptor interacting protein (GRIP) and AMPA receptor binding proteins (ABPs)

GRIP is a 130 kDa protein that contains seven PDZ domains, of which domains 4 and 5 mediate binding to the extreme C-terminal ESVKI motif of GlurR2 and GluR3 (Dong *et al.*, 1997). GRIP does not bind to GluR1 or GluR4. The function of GRIP

Table 3.2 Proteins that interact with AMPA receptors

Primary	*Secondary*
GRIP (Dong *et al.*, 1997)	Ephrins and Eph receptors (Torres *et al.*, 1998, Bruckner *et al.*, 1999)
ABP/GRIP (Srivastava *et al.*, 1998)	Ephrins and Eph receptors (Bruckner *et al.*, 1999)
Pick1 (Xia *et al.*, 1999; Dev *et al.*, 1999)	Protein kinase Cα (Staundinger *et al.*, 1995)
	Ephrins and Eph receptors (Torres *et al.*, 1998; Bruckner *et al.*, 1999)
NSF (Nishimune *et al.*, 1998; Osten *et al.*, 1998; Song *et al.*, 1998)	
α and β-SNAP (Osten *et al.*, 1998)	
SAP97 (Leonard *et al.*, 1998)	
Lyn (Hayashi *et al.*, 1999)	
Guanine-nucleotide-binding protein αi 1 (Wang *et al.*, 1997)	
Calnexin (Rubio and Wenthold, 1999)	
Immunoglobin binding protein (Rubio and Wenthold, 1999)	

is unclear since it does not cluster AMPA receptors, although multiple PDZ domains would suggest it may function similarly to PSD95 (Dong *et al.*, 1997). Thus the PDZ domains other than 4 and 5 could provide a mechanism to link AMPA receptors to other binding partners of GRIP. The subcellular distribution of GRIP was initially reported to be synapse specific; however, it has now been shown to be in the cytosol (Wyszynski *et al.*, 1998) and present in post-Golgi vesicles (Xia *et al.*, 1999), as well as at inhibitory GABAergic synapses on interneurones (Xia *et al.*, 1999).

GRIP has recently been shown to interact with the family of Eph-receptors and their membrane-bound ligands the ephrins (Hsueh and Sheng, 1998; Torres *et al.*, 1998). Eph-receptors are receptor tyrosine kinases that bind the PDZ domains 6 and 7 of GRIP (as opposed to 4 and 5 for GluR2/3). It is believed that interactions between Eph-receptors and their ligands located on adjacent cells are important for processes involved in neurite extension and axonal guidance (Gale and Yancopoulos, 1997). No direct linkage of AMPA receptors to Eph-receptors via GRIP has yet been reported.

AMPA receptor binding proteins (ABPs) are related to GRIP in structure and share 64–93 per cent homology in their PDZ domains. ABP exists in two isoforms: one of 130 kDa which also exhibits seven PDZ domains, and a shorter 98 kDa isoform which contains only six PDZ domains, of which domains 3, 5 and 6 are capable of binding to the VKI region of the C-terminal of the GluR2/3 (Srivastava *et al.*, 1998). The PDZ domain 2 of the 98 kDa isoform of ABP mediates homodimerization as well as heterodimerization with GRIP, thus ABP can form multimeric complexes with itself as well as heteromeric complexes with GRIP (Srivastava *et al.*, 1998). Thus, analogous to PSD-95, GRIP and ABPs have the potential to form large synaptic scaffolds capable of binding AMPA receptors to other proteins.

Protein interacting with C-kinase (PICK1)

Protein interacting with C-kinase (PICK1) is a single PDZ-containing protein that was originally identified through 2-hybrid screening with the catalytic subunit of PKCα (Staudinger *et al.*, 1995). Two groups have shown subsequently that PICK1 also binds to AMPA receptor subunits, including GluR2 (Dev *et al.*, 1999; Xia *et al.*, 1999). Since both PKCα and GluR2 can bind to PICK1 via its PDZ domain it is possible that the interactions are competitive, although this has not yet been demonstrated.

It has also been reported that PICK1 can dimerise at a site different from the PKCα/GluR2 binding region (Staudinger *et al.*, 1997). This suggests several possible functions for PICK1. For example, it could act to cluster AMPA receptors and provide a mechanism for concentrating AMPA receptors in puncta at synapses similar to the functions suggested for other, albeit multi-PDZ-containing proteins (O'Brien *et al.*, 1998). Alternatively, PICK1 could be involved in the targeting of PKCα to AMPA receptors and thereby provide a mechanism for the selective phosphorylation of AMPA receptors. AMPA receptors are heterooligomeric assemblies, either tetrameric or pentameric, the majority of which contain GluR1 which has been shown to be directly phosphorylated by PKCα (Roche *et al.*, 1994). Indeed, there is evidence that the catalytic domain of PKCα can regulate the function of hippocampal AMPA receptors (Wang *et al.*, 1994; Benke *et al.*, 1996). A related but different possibility is that PICK1 could target AMPA receptors to PKCα bound to intracellular membranes. In this scenario PICK1 would be involved in AMPA receptor post-translational modification prior to membrane insertion and functional expression. This last hypothesis has some support from the observed distribution of flag-PICK1 in transiently transfected COS cells which display a predominantly perinuclear localisation, with staining also occurring in the rough endoplasmic reticulum and Golgi apparatus. Some isoforms of PKC have also been shown to translocate to the perinuclear region but whether PICK1 is responsible in localizing and interacting with specific perinuclear targeted PKCα or GluRs is unclear.

SAP97

Using conventional biochemical techniques, GluR1 has been shown to bind to the PDZ-containing synapse associated protein SAP97 (Leonard *et al.*, 1998). As mentioned above, other members of the SAP family have been reported to interact with NMDA receptor subunits. The interaction between GluR1 and SAP97 occurs via the extreme C-terminus of the GluR1 and the PDZ domain of SAP97.

N-*ethylmaleimide sensitive fusion protein (NSF)*

NSF is an ATPase which plays an essential role in transporting proteins through the Golgi apparatus; it is also an essential for vesicular release of neurotransmitters at the presynaptic membrane (Rothman, 1994). NSF is also abundant in the postsynaptic density (Walsh and Kuruc, 1992), and it has been reported recently that transient cerebral ischaemia leads to an accumulation of postsynaptic NSF (Hu *et al.*, 1998). Furthermore, electrophysiological experiments have shown that the induction of long-term potentiation (LTP) is blocked by the potent inhibitor of NSF *N*-ethylmaleimide (NEM) (Lledo *et al.*, 1998).

Recent studies have shown that NSF, in addition to its other roles within the cell, binds to a specific non-PDZ-type recognition site on the intracellular C-terminal domain of the GluR2 subunit of AMPA receptors (Henley *et al.*, 1997; Nishimune *et al.*, 1998; Osten *et al.*, 1998; Song *et al.*, 1998; Noel *et al.*, 1999). Specific blockade of the interaction between NSF and GluR2 with a peptide corresponding to the binding domain of GluR2 (pep2m) or an anti-NSF antibody caused a rapid and substantial decrease in evoked AMPA receptor-mediated synaptic transmission (Nishimune *et al.*, 1998). These data suggest that NSF may regulate the membrane insertion/stabilisation, and thus the functional expression of GluR2-containing AMPA receptors. In addition, adenoviral expression of pep2m resulted in a dramatic loss in the number of AMPA receptor puncta on the cell surface (Noel *et al.*, 1999), suggesting that blocking the interaction between NSF and GluR2 does not prevent receptor synthesis or passage through the ER/Golgi systems but rather that it has an effect at the synapse.

Lyn

Lyn is a non-receptor protein tyrosine kinase of the Src-family that is particularly abundant in brain and which has been reported to interact directly with AMPA receptors (Hayashi *et al.*, 1999). Lyn binds to AMPA receptors via its Src homology region but the precise site(s) of interaction on the GluR subunit(s) have not been established

(Hayashi *et al.*, 1999). Activation of Lyn by AMPA receptor stimulation results in initiation of the mitogen-activated protein kinase (MAPK) signalling cascade which leads to increased transcription of RNA encoding brain-derived neurotrophic factor (BDNF). Since BDNF plays a role in synaptic plasticity in the neonatal and adult CNS (Lo, 1995; Thoenen, 1995) it has been proposed that AMPA receptors could serve as a postsynaptic signal transducer capable of converting short-term plasticity to long-term plasticity by stimulating BDNF production.

LIN-10

In *C. elegans* the GLR-1 AMPA-type glutamate receptors are expressed postsynaptically at target cells of the sensory neuron ASH and are required for ASH-mediated touch-sensitivity (Hart *et al.*, 1995; Maricq *et al.*, 1995). The postsynaptic targeting of these receptors is dependent on the interaction with the PDZ-containing protein LIN-10 (Rongo *et al.*, 1998). LIN-10 is the *C. elegans* orthologue of the mammalian protein X-11 (Borg *et al.*, 1996), also known as Mint (Okamoto and Südhof, 1997), a protein of unknown function but which is abundant in brain (Okamoto and Südhof, 1997). Based on the work from *C. elegans* where it is known that X-11 can restore the correct localisation of GLR-1 in mutant *C. elegans* lacking LIN-10 (Rongo *et al.*, 1998), it has been suggested that X-11 could play a role in the postsynaptic localisation of GluRs at mammalian synapses. As yet, however, the direct involvement of X-11 in the synaptic localisation of mammalian GluRs has not been demonstrated.

Concluding remarks

The increasing number of proteins that interact with NMDA and AMPA receptors provides some insight into the complexity of the mechanisms which regulate the trafficking of these receptors. Investigation of the cellular organisation and orchestration of these interactions should keep researchers occupied for many years to come.

References

Allison, D.W., Gelfand, V.I., Spector, I. and Craig, A.M. (1998) 'Role of actin in anchoring postsynaptic receptors in cultured hippocampal neurons: differential attachment of NMDA versus AMPA receptors', *J. Neurosci.* **18**: 2423–2436.

Beggs, A.H., Byers, T.J., Knoll, J.H., Boyce, F.M., Bruns, G.A. and Kunkel, L.M. (1992) 'Cloning and characterization of two human skeletal muscle alpha-actinin genes located on chromosomes 1 and 11', *J. Biol. Chem.* **267**: 9281–9288.

Benke, T.A., Bresink, I., Collett, V.J., Doherty, A.J., Henley, J.M. and Collingridge, G.L. (1996) 'Post-translational mechanisms which could underlie the postsynaptic expression of LTP and LTD', in M.S. Fazeli and G.L. Collingridge (eds) *Cortical Plasticity* Oxford: BIOS Scientific, pp. 83–104.

Borg, J., Ooi, J., Levy, E. and Margolis, B. (1996) 'The phosphotyrosine interaction domains of X11 and FE65 bind distinct sites on the YENPTY motif of amyloid precursor protein', *Mol. Cell. Biol.* **16**: 6229–6241.

Borg, J.-P., Lopez-Figueroa, M.O., de Taddeo-Borg, M., Kroon, D.E., Turner, R.S., Watson, S.J. and Margolis, B. (1999) 'Molecular analysis of the X11-mLin-2/CASK complex in brain', *J. Neurosci.* **19**: 1307–1316.

Braithwaite, S.P., Meyer, G. and Henley, J.M. (1999) 'Synaptic targeting and anchoring surface expressed of AMPA receptors', *Neuropharmacol.* **39**: 919–930.

Bredt, D.S. (1998) 'Sorting out genes that regulate epithelial and neuronal polarity', *Cell* **94**: 691–694.

Brenman, J.E., Chao, D.S., Gee, S.H., McGee, A.W., Craven, S.E., Santillano, D.R., Wu, Z., Huang, F., Xia, H., Peters, M.F., Froehner, S.C. and Bredt, D.S. (1996) 'Interaction of nitric oxide synthase with the postsynaptic density protein PSD-95 and α1-syntrophin mediated by PDZ domains', *Cell* **84**: 757–767.

Bruckner, K., Pablo Labrador, J., Scheiffele, P., Herb, A., Seeburg, P.H. and Klein, R. (1999) 'Ephrinβ ligands recruit GRIP family PDZ adaptor proteins into raft membrane nucrodomams', *Neuron* **22**: 511–524.

Budnik, V. (1996) 'Synapse maturation and structural plasticity at Drosophila neuromuscular junctions', *Curr. Op. Neurobiol.* **6**: 858–867.

Butz, S., Okamoto, M. and Südhof, T.C. (1998) 'A tripartite protein complex with the potential to couple synaptic vesicle exocytosis to cell adhesion in brain', *Cell* **94**: 773–782.

Cho, K.O., Hunt, C.A. and Kennedy, U.B. (1992) 'The rat brain postsynaptic density fraction contains a homolog of the Drosophila discs-large tumor suppressor protein', *Neuron* **9**: 929–942.

Cohen, A.R., Woods, D.F., Marfatia, S.M., Walther, Z., Chishti, A.H., Anderson, J.M. and Woods, D.F. (1998) 'Human CASK/LIN-2 binds syndecan-2 and protein 4.1 and localizes to the basolateral membrane of epithelial cells', *J. Cell Biol.* **142**: 129–138.

Craven, S.E., El-Husseini, A. and Bredt, D.S. (1999) 'Synaptic targeting of the postsynaptic density protein PSD-95 mediated by lipid and protein motifs', *Neuron* **22**: 497–509.

Degushi, M., Hata, Y., Takeuchi, M., Ide, N., Hirao, K., Yao, I., Irie, M., Toyoda, A. and Takai, Y. (1998) 'BEGAIN (Brain-enriched guanylate kinase-associated protein), a novel neuronal PSD-95/SAP90-binding protein', *J. Biol. Chem.* **273**: 26269–26272.

Dev, K.K., Nishimune, A., Henley, J.M. and Nakanishi, S. (1999) 'The protein kinase Ca binding protein PICK1 interacts with short but not long form alternative splice variants of AWA receptor subunits', *Neuropharmacology* **38**: 635–644.

Dong, H., O'Brien, R.J., Fung, E.T., Lanahan, A.A., Worley, P.F. and Huganir, R.L. (1997) 'GRIP: a synaptic PDZ domain-containing protein that interacts with AMPA receptors', *Nature* **386**: 279–284.

Ehlers, M.D., Zhang, S., Bernhardt, J.P. and Huganir, R.L. (1996) 'Inactivation of NMDA receptors by direct interaction of calmodulin with the NRI subunit', *Cell* **84**: 745–755.

Ehlers, M.D., Fung, E.T., O'Brien, R.J. and Huganir, R.L. (1998) 'Splice variant-specific interaction of the NMDA receptor subunit NR1 with neuronal intermediate filaments', *J. Neurosci.* **18**: 720–730.

Gale, N.W. and Yancopoulos, Q.D. (1997) 'Ephrins and their receptors: a repulsive topic?', *Cell Tissue Res.* **290**: 227–241.

Gardoni, F., Caputi, A., Cimino, M., Pastrorino, L., Cattabeni, F. and Di Luca, M. (1998) 'Calcium/calmodulin-dependent protein kinase 11 is associated with NR2A/B subunits of NMDA receptor in postsynaptic densities', *J. Neurochem.* **71**: 1733–1741.

Hart, A., Sims, S. and Kaplan, J. (1995) 'A synaptic code for sensory modalities revealed by analysis of the *C. elegans* GLR-1 glutamate receptor', *Nature* **378**: 82–85.

Hata, Y., Butz, S. and Südhof, T.C. (1996) 'CASK: a novel dig/PSD-95 homolog with a N-terminal calmodulin-dependent protein kinase domain identified by interaction with neurexins', *J. Neurosci.* **16**: 2488–2494.

Hayashi, T., Umemori, K., Mishina, M. and Yamamoto, T. (1999) 'The AWA receptor interacts with and signals through the protein tyrosine kinase Lyn', *Nature* **397**: 72–76.

Henley, J.K., Nishimune, A., Nash, S.R. and Nakanishi, S. (1997) 'Use of the two-hybrid system to find novel proteins that interact with AMPA receptor subunits', *Biochem. Soc. Trans.* **25**: 838–841.

Hillier, B.J., Christopherson, K.S., Prehoda, K.E., Bredt, D.S. and Lim, W.A. (1999) 'Unexpected modes of PDZ domain scaffolding revealed by structure of nNOS-syntrophin complex', *Science* **284**: 812–815.

Hirao, K., Hata, Y., Ide, N., Takeuchi, M., Irie, M., Yao, I., Deguchi, M., Toyoda, A., Sudhof, T.C.

and Takai, Y. (1998) 'A novel multiple PDZ domain-containing molecule interacting with N-methyl-D-aspartate receptors and neuronal cell adhesion proteins', *J. Biol. Chem.* **273**: 21,105–21,110.

Hsueh, Y.P. and Sheng, M. (1998) 'Eph receptors, ephrins, and PDZs gather in neuronal synapses', *Neuron* **21**: 1227–1229.

Hu, B.-R., Park, M., Martone, M.E., Fischer, W.H., Ellisman, M.H. and Zivin, J.A. (1998) 'Assembly of proteins to postsynaptic densities after transient cerebral ischemia', *J. Neurosci.* **18**: 625–633.

Ichtchenko, K., Hata, Y., Nguyen, T., Ullrich, B., Missler, M., Moomaw, T. and Südhof, T.C. (1995) 'Neuroligin1: a splice site-specific ligand for β-neurexins', *Cell* **81**: 435–443.

Ide, N., Hata, Y., Deguchi, M., Hirao, K., Yao, I. and Takai, Y. (1999) 'Interaction of S-SCAM with neural plakophilin-related Armadillo-repeat protein/delta-catenin', *Biochem. Biophys. Res. Commun.* **256**: 456–461.

Irie, M., Hata, Y., Takeuchi, M., Ichtchenko, K., Toyoda, A., Hirao, K., Takai, Y., Rosahl, T.W. and Sudhof, T.C. (1997) 'Binding of neuroligins to PSD-95', *Science* **277**: 1511–1515.

Itoh, M., Nagafuchi, A., Yonemura, S., Kitani, Y.T., Tsukita, S. and Tsukita, S., (1993) 'The 220-kD protein colocalising with cadherins in non-epithelial cells is identical to ZO-1, a tight junction-associated protein in epithelial cells: cDNA cloning and immunoelectron microscopy', *J. Cell. Biol.* **121**: 491–502.

Jo, K., Derin, R., Li, M. and Bredt, D.S. (1999) 'Characterization of MALS/Velis-1, -2, and -3: a family of mammalian LIN-7 homologs enriched at brain synapses in association with the postsynaptic density-95/NMDA receptor postsynaptic complex', *J. Neurosci.* **19**: 4189–4199.

Kaech, S.M., Whitfield, C.W. and Kim, S.K. (1998) 'The LIN-2ALIN-7ALINIO complex mediates basolateral membrane localization of the C. elegans EGF receptor LET-23 in vulval epithelial cells', *Cell* **94**: 761–771.

Kennedy, M.B. (1997) 'The postsynaptic density at glutamatergic synapses', *Trends Neurosci.* **20**: 264–268.

Kennedy, M.B. (1998) 'Signal transduction molecules at the glutamatergic postsynaptic membrane', *Brain Res. Rev.* **26**: 243–257.

Kim, E., Cho, K.-O., Rothschild, A. and Sheng, M. (1996) 'Heteromultimerization and receptor-clustering activity of −110, a member of the PSD-95 of proteins', *Neuron* **17**: 103–113.

Kim, E., Naisbitt, S., Hsueh, Y.-P., Rao, A., Rothschild, A., Craig, A.M. and Sheng, M. (1997) 'GKAP, a novel synaptic protein that interacts with the guanylate kinase-like domain of the PSD-95/SAP90 family of channel clustering molecules', *J. Cell Biol.* **136**: 669–678.

Kim, J.H. and Huganir, R.L. (1999) 'Organisation and regulation of proteins at synapses', *Curr. Opin. Cell Biol.* **11**: 248–254.

Kim, J.H., Liao, D., Lau, L.-F. and Huganir, R.L. (1998) 'SynGAP: a synaptic RasGAP that associates with the PSD-95/SAP90 protein family', *Neuron* **20**: 683–691.

Kim, S.K. (1997) 'Polarized signaling: basolateral receptor localization in epithelial cells by PDZ-containing proteins', *Curr. Opin. Cell Biol.* **9**: 853–859.

Kornau, H.-C., Schenker, L.T., Kennedy, U.B. and Seeburg, P.H. (1995) 'Domain interaction between NMDA receptor subunits and the postsynaptic density protein PSD-95', *Science* **269**: 1737–1740.

Krupp, J.J., Vissel, B., Thomas, C.G., Heinemann, S.F. and Westbrook, G.L. (1999) 'Interactions of calmodulin and α-actinin with the NRI subunit modulate Ca^{2+}-dependent inactivation of NMDA receptors', *J. Neurosci.* **19**: 1165–1178.

Kurschner, C., Mermelstein, P.G., Holden, W.T. and Surmeier, D.J. (1998) 'CIPP, a novel multivalent PDZ domain protein, selectively interacts with Kir4.0 family members, NMDA receptor subunits, neurexins, and neuroligins', *Mol. Cell. Neurosci.* **11**: 161–172.

Lanahan, A. and Worley, P. (1998) 'Immediate-early genes and synaptic function', *Neurobiology of Learning and Memory* **70**: 37–43.

Langnaese, K., Seidenbecher, C., Wex, K., Seidel, B., Hartung, K., Appeltauer, U., Garner, A., Voss,

B., Mueller, B., Garner, C.C. and Gundelfinger, E.D. (1996) 'Protein components of a rat brain synaptic junctional protein preparation', *Mol. Brain Res.* **42**: 118–122.

Leonard, A.S., Davare, M.A., Horne, M.C., Garner, C.C. and Hell, J.W. (1998) 'SAP97 is associated with the AMPA receptor GluR1 subunit', *J. Biol. Chem.* **273**: 19518–19524.

Lledo, P.-M., Zhang, X., Südhof, T.C., Malenka, R.C. and Nicoll, R.A. (1998) 'Postsynaptic membrane fusion and long-term potentiation', *Science* **279**: 399–403.

Lo, D.C. (1995) 'Neurotrophic factors and synaptic plasticity', *Neuron* **15**: 979–981.

Lüthi, A., Chittajallu, R., Duprat, F., Palmer, M.J., Benke, T.A., Kidd, F.L., Henley, J.M., Isaac, J.T.R. and Collingridge, G.L. (1999) 'Hippocampal LTD expression involves a pool of AMPARs regulated by the NSF-GluR2 interaction', *Neuron* **24**: 389–399.

Maricq, A.V., Peckol, E., Driscoll, M. and Bargmann, C. (1995) 'glr-1, a C. elegans glutamate receptor that mediates mechanosensory signaling', *Nature* **378**: 78–81.

Migaud, M., Charlesworth, P., Dempster, M., Webster, L.C., Watabe, A.M., Makhinson, M., He, Y., Ramsay, M.F., Morris, R.G., Morrison, J.H., O'Dell, T.J. and Grant, S.G. (1998) 'Enhanced long-term potentiation and impaired learning in mice with mutant postsynaptic density-95 protein', *Nature* **296**: 433–439.

Moon, I.S., Apperson, M.L. and Kennedy, M.B. (1994) 'The major tyrosine-phosphorylated protein in the postsynaptic density fraction is n-methyl-D-aspartate receptor subunit 2B', *Proc. Natl. Acad. Sci. USA* **91**: 3954–3958.

Muller, B.M., Kistner, U., Kindler, S., Chung, W.J., Kuhlenddahl, S., Fenster, S.D., Lau, L.-F., Veh, R.W., Huganir, R.L., Gundelfinger, E.D. and Garner, C.C. (1996) 'SAP102, a novel postsynaptic protein that interacts with NMDA receptor conplexes in vivo', *Neuron* **17**: 255–265.

Naisbitt, S., Kim, E., Tu, J.C., Xiao, B., Sala, C., Valtschanoff, J., Weinberg, R.J., Worley, P.F. and Sheng, M. (1999) 'Shank, a novel family of postsynaptic density proteins that binds to the NMDA receptor/PSD-95/GKAP complex and cortactin', *Neuron* **23**: 569–582.

Niethammer, M., Valtschanoff, J.G., Kapoor, T.M., Allison, D.W., Weinberg, R.J., Craig, A.M. and Sheng, M. (1998) 'CRIPT, a novel postsynaptic protein that binds to the third PDZ domain of PSD95/SAP90', *Neuron* **20**: 693–707.

Nishimune, A., Isaac, J.T.R., Molnar, E., Noel, J., Nash, S.R., Tagaya, M., Collingridge, G.L., Nakanishi, S. and Henley, J.M. (1998) 'NSF binding to GluR2 regulates synaptic transmission', *Neuron* **21**: 87–97.

Nishimune, A., Nash, S.R., Nakanishi, S. and Henley, J.M. (1996) 'Detection of protein–protein interactions in the nervous system using the two-hybrid system', *Trends Neurosci.* **19**: 261–266.

Noel, J., Ralph, G.S., Pickard, L., Williams, J., Molnar, E., Uney, J.B., Collingridge, G.L. and Henley, J.M. (1999) 'Surface expression of AMPA receptors in hippocampal neurons is regulated by an NSF-dependent mechanism', *Neuron* **23**: 365–376.

O'Brien, R.J., Kamboj, S., Ehlers, M.D., Rosen, K.R., Fischbach, G.D. and Huganir, R.L. (1998) 'Activity-dependent modulation of synaptic AMPA receptor accumulation', *Neuron* **21**: 1067–1078.

Okabe, S., Kim, H.-D., Miwa, A., Kuriu, T. and Okado, H. (1999) 'Continual remodeling of postsynaptic density and its regulation by synaptic activity', *Nature Neurosci.* **2**: 804–811.

Okamoto, M. and Südhof, T. (1997) 'Mints, Munc18-interacting proteins in synaptic vesicle exocytosis', *J. Biol. Chem.* **272**: 31,459–31,464.

Osten, P., Srivastava, S., Inman, G.J., Vilim, F.S., Khatri, L., Lee, L.M., States, B.A., Einheber, S., Milner, T.A., Hanson, P.I. and Ziff, E.B. (1998) 'The AMPA receptor GluR2 C terminus can mediate a reversible, ATP-dependent interaction with NSF and α-and β-SNAPs', *Neuron* **21**: 99–110.

Rao, A., Kim, E., Sheng, M. and Craig, A.M. (1998) 'Heterogeneity in the molecular composition of excitatory sites during development of hippocampal neurons in culture', *J. Neurosci.* **18**: 1217–1229.

Roche, K.W., Tingley, W.G. and Huganir, R.L. (1994) 'Glutamate receptor phosphorylation and synaptic plasticity', *Curr. Opin. Neurobiol.* **4**: 383–388.

Rongo, C., Whitfield, C.W., Rodal, A., Kim, S.K. and Kaplan, J.M. (1998) 'LIN-10 is a shared component of the polarized protein localisation pathways in neurons and epithelia', *Cell* **94**: 751–759.

Rothman, J.E. (1994) 'Intracellular membrane fusion', *Adv. Second Messenger Phosphoprotein Res.* **29**: 81–96.

Rubio, M.E. and Wenthold, R.J. (1999) 'Calnexin and immunoglobulin binding protein (BiP) coimmunoprecipitate with AMPA receptors', *J. Neurochem.* **73**: 942–948.

Song, I., Kamboj, S., Xia, J., Dong, H., Liao, D. and Huganir, R.L. (1998) 'Interaction of the N-ethylmaleimide-sensitive factor with AMPA receptors', *Neuron* **21**: 393–400.

Songyang, Z., Fanning, A.S., Fu, C., Xu, J., Marfatia, S.M., Chishti, A.H., Crompton, A., Chan, A.C., Anderson, J.M. and Cantley, L.C. (1997) 'Recognition of unique carboxl-terminal motifs by distinct PDZ domains', *Science* **275**: 73–77.

Sprengel, R., Suchanek, B., Amico, C., Brusa, R., Burnashev, N., Rozov, A., Hvalby, O., Jensen, V., Paulsen, O., Andersen, P., Kim, J.J., Thompson, R.F., *et al.* (1998) 'Importance of the intracellular domain of NR2 subunits for NMDA receptor function in vivo', *Cell* **92**: 279–289.

Srivastava, S., Osten, P., Vilim, F.S., Khatri, L, Inman, G.J., States, B.A., Daly, C., DeSouza, S., Abagyan, R., Valtschanoff, J.G., Weinberg, R.J. and Ziff, E.B. (1998) 'Novel anchorage of GluR2/3 to postsynaptic density by an AMPA receptor binding protein ABP', *Neuron* **21**: 581–591.

Staudinger, J., Lu, J. and Olson, E.N. (1997) 'Specific interaction of the PDZ domain protein PICK1 with the COOH terminus of protein kinase C', *J. Biol. Chem.* **272**: 32,019–32,024.

Staudinger, J., Zhou, J., Burgess, R., Elledge, S.J. and Olson, E.N. (1995) 'PICK1: A perinuclear binding protein and substrate for protein kinase C isolated by the yeast two-hybrid system', *J. Cell Biol.* **128**: 263–271.

Strack, S. and Colbran, R.J. (1998) 'Autophosphorylation-dependent targeting by the NR2B subunit of the N-methyl-D-aspartate receptor', *J. Biol. Chem.* **273**: 20,689–20,692.

Takeuchi, M., Hata, Y., Hirao, K., Toyoda, A., Irie, M. and Takai, Y. (1997) 'SAPAPS. A family of PSD-95/SAP90-associated proteins localized at postsynaptic density', *J. Biol. Chem.* **272**: 11,943–11,951.

Tejedor, F.J., Bokhar, A., Rogero, O., Gorczyca, N.L., Zhang, J., Kim, E., Sheng, M. and Budnik, V. (1997) 'Essential role for dlg in synaptic clustering of Shaker K+ channels in vivo', *J. Neurosci.* **17**: 152–159.

Tezuka, T., Umemori, H., Akiyama, T., Nakanishi, S. and Yamamoto, T. (1999) 'PSD-95 promotes Fyn-mediated tyrosine phosphorylation of the N-methyl-D-aspartate receptor subunit NR2A', *Proc. Natl. Acad. Sci. USA* **96**: 435–440.

Thoenen, H. (1995) 'Neurotrophins and neuronal plasticity', *Science* **270**: 593–598.

Torres, R., Firestein, B.L., Dong, H., Staudinger, J., Olson, E.N., Huganir, R.L., Bredt, D.S., Gale, N.W. and Yancopoulos, G.D. (1998) 'PDZ proteins bind, cluster, and synaptically colocalize with Eph receptors and their ephrin ligands', *Neuron* **21**: 1453–1463.

Van Rossum, D., Kuhse, J. and Betz, M. (1999) 'Dynamic interaction between soluble tubulin and C-terminal domains of N-methyl-D-aspartate receptor subunits', *J. Neurochem.* **72**: 962–973.

Walsh, M.J. and Kuruc, N. (1992) 'The Postsynaptic Density – constituent and associated proteins characterized by electrophoresis, immunoblotting, and peptide sequencing', *J. Neurochem.* **59**: 667–678.

Wang, L.Y., Dudek, E.M., Browning, M.D. and Macdonald, J.F. (1994) 'Modulation of AMPA/kainate receptors in cultured murine hippocampal neurones by protein kinase-C', *J. Physiol. (London)* **475**: 431–437.

Wang, Y.Z., Small, D.L., Stanimirovic, D.B., Morley, P. and Durkin, J.P. (1997) 'AMPA receptor-mediated regulation of a Gi-protein in cortical neurons', *Nature* **389**: 502–504.

Wechsler, A. and Teichberg, V.I. (1998) 'Brain spectrin binding to the NMDA receptor is regulated by phosphorylation, calcium and calmodulin', *EMBO J.* **17**: 3931–3939.

Woods, D.F. and Bryant, P.J. (1991) 'The discs-large tumor suppressor gene Drosophila encodes a guanylate kinase homolog localised at septate junctions', *Cell* **66**: 451–464.

Wyszynski, M., Kim, E., Yang, F.-C. and Sheng, M. (1998) 'Biochemical and immunocytochemical characterisation of GRIP, a putative ANPA receptor anchoring protein in rat brain', *Neuropharmacol.* **37**: 1335–1344.

Wyszynski, M., Lin, J., Rao, A., Nigh, E., Beggs, A.H., Craig, A.M. and Sheng, M. (1997) 'Competitive binding of α-actin and clamodulin to the NMDA receptor', *Nature* **385**: 439–442.

Xia, J., Zhang, N., Staudinger, J. and Huganir, R.L. (1999) 'Clustering of AMPA receptors by the synaptic PDZ domain-containing protein PICK1', *Neuron* **22**: 179–187.

Zhang, S., Ehlers, M.D., Bernhardt, J.P., Su, C.-T. and Huganir, R.L. (1998) 'Calmodulin mediates calcium-dependent inactivation of N-methyl-D-aspartate receptors', *Neuron* **21**: 443–453.

Zhang, W., Vazquez, L., Apperson, M. and Kennedy, M.B. (1999) 'Citron binds to PSD-95 at glutamatergic synapses on inhibitory neurons in the hippocampus', *J. Neurosci.* **19**: 96–108.

Chapter 4

Pharmacology of NMDA receptors

David Jane

Introduction

N-methyl-D-aspartic acid (NMDA, Figure 4.1) was initially synthesised as part of a structure–activity study aimed at obtaining aspartic (asp) and glutamic acid (glu) (Figure 4.1) analogues, which showed a degree of stereoselectivity for activation of glutamate receptors (Curtis and Watkins, 1960; Watkins, 1962). Indeed, NMDA was the most potent excitant tested at that time and, unlike asp, showed stereoselectivity, the D-form being the active isomer, while the L-form was much less potent (Curtis and Watkins, 1963). The increase in potency observed with NMDA compared to asp and glu was later ascribed to the low affinity of NMDA for the transporter responsible for the uptake of glu and asp (Balcar and Johnston, 1972). Since this time structure–activity studies have revealed agonists with greater potency than NMDA (see pp. 71–73).

The search for selective antagonists led to the discovery of three likely candidates: magnesium ions (Evans *et al.*, 1977), α,ϵ-diaminopimelate (DAP, Figure 4.1) (Biscoe

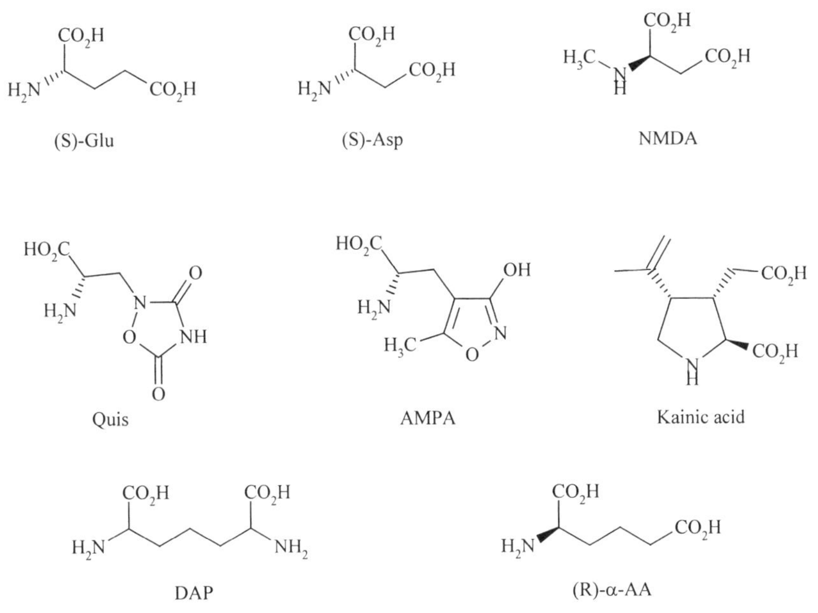

(S)-Glu (S)-Asp NMDA

Quis AMPA Kainic acid

DAP (R)-α-AA

Figure 4.1 Structures of ligands used to classify excitatory amino acid receptors.

et al., 1977a) and 3-amino-1-hydroxypyrrolidin-2-one (HA-966, see p. 80 and Figure 4.9) (Biscoe *et al.*, 1977a; Evans *et al.*, 1978) each of which antagonised NMDA-induced but were a lot loss effective against quisqualate- or kainate-induced depolarisations of spinal motoneurons. At the same time it was reported that (RS)-α-aminoadipate (α-AA) antagonised (S)-glu-induced responses whereas (S)-α-aminoadipate had excitatory effects, implying that the R isomer possessed the antagonist activity (Hall *et al.*, 1977). It was later reported that (R)-α-AA (Figure 4.1) antagonised NMDA-induced but not quisqualate- or kainate-induced depolarisations of spinal motoneurons (Biscoe *et al.*, 1977b). These observations led to the initial classification of ionotropic glutamate receptors (iGluRs) into NMDA and non-NMDA receptors, which was later refined to the present system of NMDA, AMPA and kainate receptors (for a review see Monaghan *et al.*, 1989).

Further refinement of the structure of (R)-α-AA by replacement of the ω-carboxyl group with a phosphonate moiety led to the potent and selective antagonist (R)-2-amino-5-phosphonopentanoate ((R)-AP5, Figure 4.4), and chain extension led to (R)-2-amino-7-phosphonoheptanoate ((R)-AP7, Figure 4.4).

The observation that NMDA and other EAA receptor agonists such as kainic acid could cause neuronal degeneration led to the realisation that antagonists may be useful neuroprotectants for therapeutic intervention into a number of neurodegenerative disorders. This led to an explosion of interest in NMDA receptor antagonists, many based on the (R)-AP5 and (R)-AP7 structure (for reviews see Jane *et al.*, 1994; Johnson and Ornstein, 1996). A review of the structure–activity studies carried out on the way to obtaining potent and selective NMDA receptor antagonists acting at the glutamate recognition site is given on pp. 73–79.

More recently, it has been demonstrated that glycine is a necessary co-agonist for NMDA receptor channel opening (Kleckner and Dingledine, 1988), and that glycine binds to a site distinct from the glutamate recognition site (Henderson *et al.*, 1990). Thus, the NMDA receptor is distinct from other iGluRs in requiring the binding of two agonists for optimal functioning. This gave medicinal chemists another target for producing antagonists that interfere with the normal functioning of the NMDA receptor (for a review see Leeson and Iversen, 1994), and studies aimed at producing such antagonists are reviewed on pp. 80–84.

The NMDA receptor complex has a number of other sites that have become targets for drug design, such as the PCP site within the channel pore (Lodge and Johnson, 1990) and the polyamine binding sites (Ransom and Stec, 1988) (see pp. 84–89).

Agonists interacting with the glutamate binding site

Structure–activity studies have demonstrated that an α-amino, α-carboxyl and a terminal acidic group are essential for NMDA receptor agonist activity (Olverman *et al.*, 1988; Jane *et al.*, 1994), consistent with the three-point attachment pharmacophore model previously proposed by Curtis and Watkins (1960). A number of structural modifications of asp and glu have been made in order to optimise NMDA receptor agonist activity including:

Replacement of the terminal carboxyl group

The terminal carboxyl group of either asp or glu has been replaced by a number of different acidic groups leading to a marked variation in agonist potency at the NMDA receptor (Jane *et al.*, 1994). Irregardless of stereochemistry at the α-amino acid centre the rank order of potency for asp analogues with different ω-acidic groups is $CO_2H \geq SO_2H > SO_3H > PO_3H_2$, whereas the rank order of potency for (S)-glu analogues is $CO_2H > SO_2H > SO_3H \gg PO_3H_2$ and that for (R)-glu analogues is $SO_2H \geq SO_3H > CO_2H \gg PO_3H_2$. In the case of both asp and (S)-glu analogues (see Figure 4.2 for structures) a terminal carboxyl group is the preferred substituent, while an ω-phosphonate group is detrimental to agonist activity. Racemic asp and glu analogues bearing a terminal phosphinic acid group, APIA and APIBA respectively (for structures, see Figure 4.2), were found to be only tenfold less potent than the parent compounds in a [^{3}H]-glu binding assay (Fagg and Baud, 1988). A functional assay confirmed that APIA and APIBA (Figure 4.2) were agonists, which contrasts with the weak antagonist properties of the corresponding phosphonates, (RS)-AP3 and (R)-AP4 (Fagg and Baud, 1988) (Figure 4.2). The effect of ω-tetrazole substitution has been examined; an asp analogue, (RS)-tetrazolylglycine (α-Tetgly) (Figure 4.2), with a planar tetrazole moiety as a terminal acidic group proved to be more potent than NMDA as an NMDA receptor agonist and has potent convulsant and excitotoxic activity (Schoepp *et al.*, 1991; Lunn *et al.*, 1992). The structure–activity data would appear to suggest that a monobasic terminal acidic group is favoured for optimal agonist activity at NMDA receptors and that such a group should be planar rather than tetrahedral.

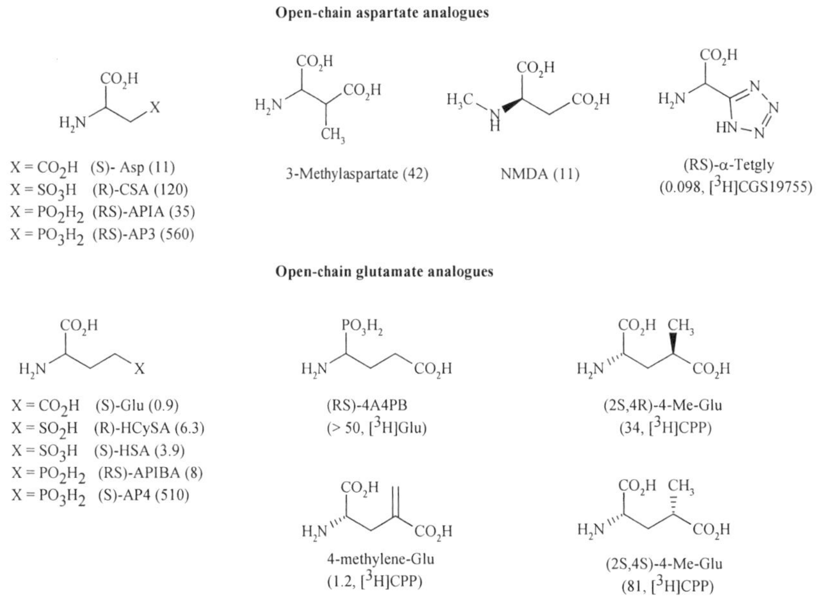

Open-chain aspartate analogues

$X = CO_2H$ (S)-Asp (11)
$X = SO_3H$ (R)-CSA (120)
$X = PO_2H_2$ (RS)-APIA (35)
$X = PO_3H_2$ (RS)-AP3 (560)

3-Methylaspartate (42)

NMDA (11)

(RS)-α-Tetgly
(0.098, [^{3}H]CGS19755)

Open-chain glutamate analogues

$X = CO_2H$ (S)-Glu (0.9)
$X = SO_2H$ (R)-HCySA (6.3)
$X = SO_3H$ (S)-HSA (3.9)
$X = PO_2H_2$ (RS)-APIBA (8)
$X = PO_3H_2$ (S)-AP4 (510)

(RS)-4A4PB
(> 50, [^{3}H]Glu)

(2S,4R)-4-Me-Glu
(34, [^{3}H]CPP)

4-methylene-Glu
(1.2, [^{3}H]CPP)

(2S,4S)-4-Me-Glu
(81, [^{3}H]CPP)

Figure 4.2 Structures of open chain aspartate and glutamate analogues. IC_{50} or K_i (μM) values for displacement of [^{3}H](R)-AP5 binding to rat brain membranes given in parenthesis (unless otherwise stated). Values taken from Jane *et al.* (1994) and Bräuner-Osborne *et al.* (1997).

Replacement of the α-carboxyl group:

Substitution of the α-carboxyl group of glutamate by an α-phosphono group to give (RS)-4A4PB (Figure 4.2) leads to a marked reduction in affinity for NMDA receptors (Jane *et al.*, 1994). This is perhaps due to the presence of the additional hydroxyl group or due to steric considerations imposed by the tetrahedral geometry of the phosphono group. To date, the effect of replacement of the α-carboxyl by acidic groups other than phosphonate on NMDA receptor agonist activity has not been investigated.

N-alkylation

Monoalkylation of the α-amino group of either asp or glu is generally detrimental to NMDA receptor agonist activity (Jane *et al.*, 1994). An exception to this rule is the *N*-methyl analogue of (R)-asp (NMDA) which displays almost the same affinity as asp for the NMDA receptor.

Substitution on the inter-acidic group chain

For both asp and glu, α-methylation leads to loss of affinity for the NMDA receptor (Olverman *et al.*, 1988; Jane *et al.*, 1994, 1996). Although the α-ethyl analogue of (S)-glu has no NMDA receptor agonist activity, it is a moderately potent group II metabotropic glutamate (mGlu) receptor antagonist (Jane *et al.*, 1996). Substitution of a methyl group at the 3-position is highly detrimental to NMDA receptor activity for glu but not asp. Indeed, racemic 3-methylaspartate (Figure 4.2) retains high affinity for the NMDA receptor (Olverman *et al.*, 1988; Jane *et al.*, 1994). Affinity for the NMDA receptor is somewhat reduced on substitution of glu at the 4-position with an amino, fluoro, hydroxyl, methyl or methylene group, and markedly reduced upon substitution with a carboxyl group (Olverman *et al.*, 1988). The (2S,4R)-isomer of 4-methylglu (SYM 2081; see Figure 4.2) displaced [^{3}H]CPP binding with a lower potency than (S)-4-methylene-glu (Figure 4.2) but potently displaced [^{3}H]kainate binding and displayed weak agonist activity on recombinant mGlu2 receptors (Bräuner-Osborne *et al.*, 1997). The (2S,4S)-isomer of 4-methylglu (Figure 4.2) was a moderately potent agonist at mGlu1α and mGlu2 and a weak agonist at mGlu4a; it also displaced [^{3}H]CPP and [^{3}H]kainate binding, but with lower affinity than the (2S,4R)-isomer (Bräuner-Osborne *et al.*, 1997). Thus, the isomers of 4-methylglutamate show a complex pharmacology activating NMDA, kainate (Donevan *et al.*, 1998) and metabotropic glutamate receptors.

Conformational restriction

Conformational restriction of asp and glu has been achieved by incorporation of some or all of the inter-acidic group chain into a carbocyclic or heterocyclic ring or by incorporation of the α-amino group and the inter-acidic group chain into a heterocyclic ring. Specific examples of conformationally restricted analogues with an affinity for the NMDA receptor that is comparable to or greater than asp or glu include (RS)-α-Tetgly (Lunn *et al.*, 1992), *trans*-ACBD (Allan *et al.*, 1990; Lanthorn *et al.*, 1990; Kyle *et al.*, 1992), (2S,1′R,2′S)-CCG, (2R,1′S,2′R)-CCG, (2R,1′S,2′S)-CCG, (Monahan *et al.*,

1990; Kawai *et al.*, 1992), homoquinolinic acid (HQA) (Jane *et al.*, 1994), (1R,3R)-ACPD (Sunter *et al.*, 1991; Jane *et al.*, 1994), (RS)-ibotenic acid (Kyle *et al.*, 1992) and (RS)-AMAA (Kyle *et al.*, 1992) (for structures and binding data see Figure 4.3). Of these it would appear that (2S,1′R,2′S)-CCG has the highest affinity for the NMDA receptor, being approximately seventeenfold more potent than (S)-glu (Kawai *et al.*, 1992). The conformationally restricted glu analogue homoquinolinic acid (HQA) is unusual in that the α-amino group and part of the inter-acidic group chain is incorporated into a planar pyridine ring. This may account for the observed selectivity of HQA for NMDA receptors containing the NR2B subunit as determined on recombinant and native NMDA receptor subtypes (Monaghan and Beaton, 1992; Buller *et al.*, 1994; Buller and Monaghan, 1997). A tritiated form of HQA has been shown to label two sites in the brain, a sub-population of NMDA receptors containing the NR2B subunit and a novel binding site (Brown *et al.*, 1998). The two sites can be distinguished by the use of an additional ligand such as 2-carboxy-3-carboxymethylquinoline (CCMQ), which binds only to the novel binding site. The enhanced potency of certain of the conformationally restricted glu analogues such as TetGly, *trans*-ACBD and (2S,1′R,2′S)-CCG suggests that glu binds in a folded conformation at the agonist recognition site of the NMDA receptor. A number of computer-aided molecular modelling studies have also suggested that this is the case (for a review, see Jane *et al.*, 1994).

Competitive NMDA receptor antagonists

Major advances in the design of NMDA receptor antagonists stemmed from the discovery that chain extended analogues of glu bearing ω-phosphono groups such as (R)-AP5 and (R)-AP7 (Evans *et al.*, 1982) (Figure 4.4) were moderately potent and highly selective NMDA receptor antagonists (for reviews, see Jane *et al.*, 1994; Johnson and Ornstein, 1996). Two general requirements for optimal NMDA receptor antagonist activity arose from early structure–activity studies: firstly, R absolute stereochemistry at the α-amino acid stereogenic centre; secondly, a chain length of either 3 or 5 methylenes

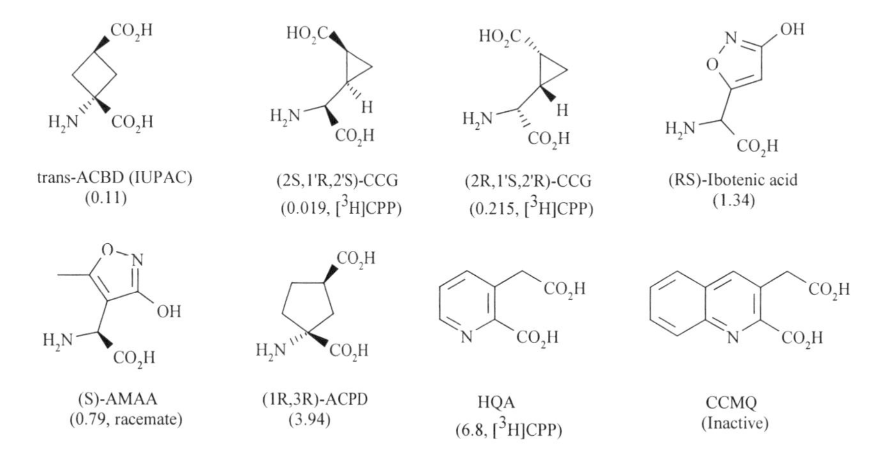

Figure 4.3 Structures of conformationally restricted glutamate analogues. IC_{50} or K_i (μM) values for displacement of [^{3}H](S)-glutamate binding to rat brain membranes given in parenthesis (unless otherwise stated). Values taken from Jane *et al.* (1994).

between the glycine moiety and the phosphonate group (Evans *et al.*, 1982; Jane *et al.*, 1994). However, examples do exist of NMDA receptor antagonists with the S stereochemistry at the α-amino acid centre (particularly so for AP7 analogues) or with inter-acidic group chain lengths that are at variance with the general rule.

Although antagonists such as (R)-AP5 and (R)-AP7 were useful for establishing the physiological roles of the NMDA receptor (Collingridge and Lester, 1989), and its involvement in epilepsy (Croucher *et al.*, 1982) and neurodegeneration (Meldrum, 1985), poor blood–brain barrier penetration was a major limitation to their therapeutic application. Further structure–activity studies were therefore undertaken in order to identify structural features necessary for potent antagonist activity and good blood–brain barrier penetration. Structural modifications of the basic AP5 and AP7 structure that have been made, alone or in combination, include bioisosteric replacement of the terminal phosphonate group, conformational restriction and introduction of lipophilic substituents onto, or heteroatoms into, the inter-acidic group chain. Replacement of the ω-phosphono group of either AP5 or AP7 with various acid groups led to the following rank order of antagonist potency: $PO_3H_2 >$ tetrazole $> CO_2H \gg SO_3H$ (Jane *et al.*, 1994). Thus, for optimal antagonist potency a phosphonate moiety is preferred, though a tetrazole group is well tolerated, in contrast to the situation for NMDA receptor agonists where for potent activity either a carboxyl or a sulphono group is favoured. A methylphosphinate group, although less well tolerated than a phosphonate group, has been shown to be an effective terminal acidic group (Hays *et al.*, 1990). However, an AP5 analogue with a ω-phosphinate group displayed only weak NMDA receptor antagonist activity (Fagg and Baud, 1988), and phosphonate monoesters are weaker antagonists than the corresponding doubly charged phosphonate analogues (Jane *et al.*, 1994).

This structure–activity data has been interpreted as suggesting that the two hydroxyl groups of the ω-phosphono moiety bind to two interaction points within the glu recognition site of the receptor, thereby enhancing binding (Ortwine *et al.*, 1992). As one of the hydroxyl groups of the phosphono moiety (pKa < 1 and 7.8) is fully ionised, and the other is somewhat less than half ionised at physiological pH, it has been suggested that it could be acting as both a charge donor and acceptor within the binding site. However, the 3.7-fold increase in antagonist potency of AP7 observed on increasing the pH from 7.3 to 8.2 has been suggested as evidence that for optimal NMDA receptor antagonism the doubly ionised form of the phosphonate group is favoured (Mayer *et al.*, 1994).

Replacement of the α-carboxyl of AP5 with an α-phosphono group has been shown to be detrimental to antagonist activity (Monahan and Michel, 1987), as is esterification of the α-carboxyl group (Jane *et al.*, 1994).

By far the best method of obtaining potent NMDA receptor antagonists has been to synthesise conformationally restricted analogues of AP5 and AP7 (Jane *et al.*, 1994; Johnson and Ornstein, 1996). Potent antagonists have been obtained by conformational restriction of AP5 by inclusion of a double bond (e.g. CGP 37849, the active 2R-isomer CGP 40116 and CGP 39653 (Figure 4.4); see Fagg *et al.*, 1990) or a keto group (e.g. MDL 100,453 (Figure 4.4); see Whitten *et al.*, 1990) into the inter-acidic group chain. Unlike the parent compound, the ethyl ester of CGP 37849, CGP 39551 (Figure 4.4) displayed oral bioavailability as an anticonvulsant presumably acting as a prodrug form of CGP 37849, thus allowing effective penetration of the blood–brain barrier (Fagg *et al.*, 1990).

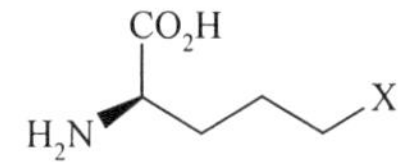

X = CO_2H, (R)-α-AA (13)
X = PO_3H_2, (R)-AP5 (0.62)

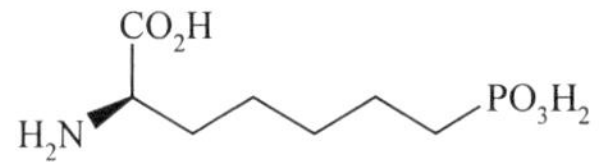

(R)-AP7 (1.7)

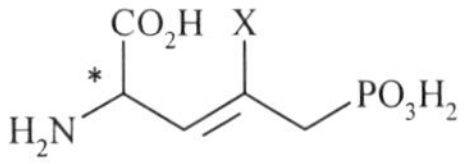

X = CH_3 CGP 37849 (0.035)
* = R, CGP 40116 (0.019)
X = n-C_3H_7 CGP 39653
(0.007, [^{3}H]CGP 39653)

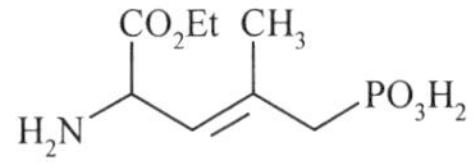

CGP 39551 (0.31)

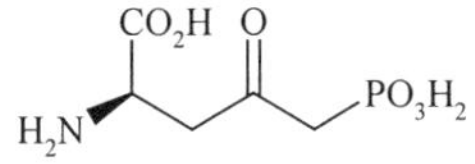

MDL 100,453 (0.109)

Figure 4.4 Structures of open chain AP5 and AP7 analogues. IC_{50} or K_i (μM) values for displacement of [^{3}H](R)-AP5 binding to rat brain membranes given in parenthesis (unless otherwise stated). Values taken from Jane *et al.* (1994).

Another common strategy for conformationally restricting either AP5 or AP7 is to incorporate the α-amino group and part of the inter-acidic group chain into a heterocycle such as a piperazine (e.g. CPP; see Davies *et al.*, 1986; Aebischer *et al.*, 1989) or piperidine (e.g. CGS19755 and CPPP; see Lehmann *et al.*, 1988; Hutchison *et al.*, 1989) ring (for structures see Figure 4.5). In order to improve potency within the series

CGS19755 (0.183)

CPPP (0.179)

I (0.075, [^{3}H]CGS19755)

R = $(CH_2)_3PO_3H_2$ (R)-CPP (0.14)
R = CH_2-CH=CH_2-PO_3H_2
(R)-CPP-ene (0.044)

MDL 100,925
(0.04)

LY 235959
(0.025, [^{3}H]CGS19755)

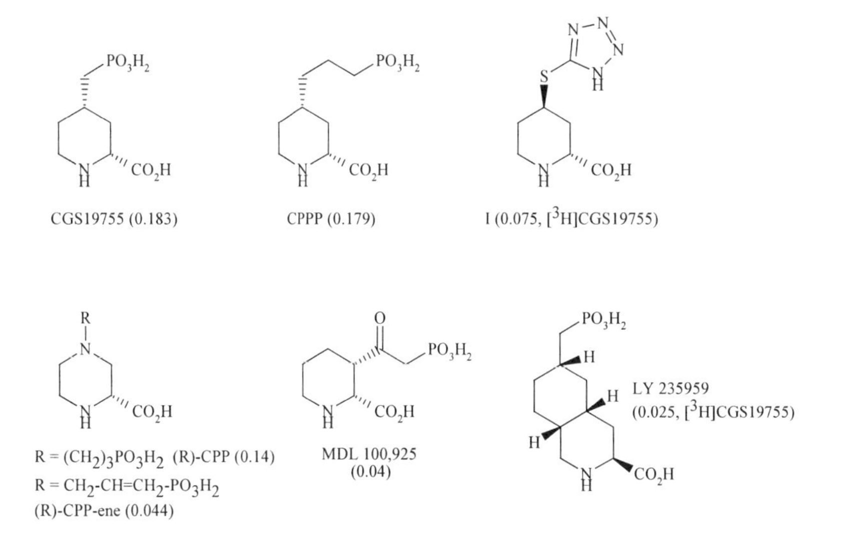

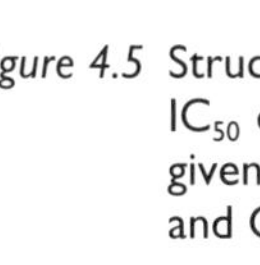

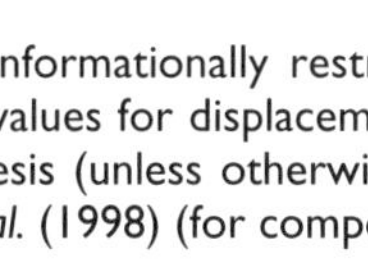

Figure 4.5 Structures of conformationally restricted heterocyclic analogues of AP5 and AP7. IC_{50} or K_i (μM) values for displacement of [^{3}H]CPP binding to rat brain membranes given in parenthesis (unless otherwise stated). Values taken from Jane *et al.* (1994) and Ornstein *et al.* (1998) (for compound I).

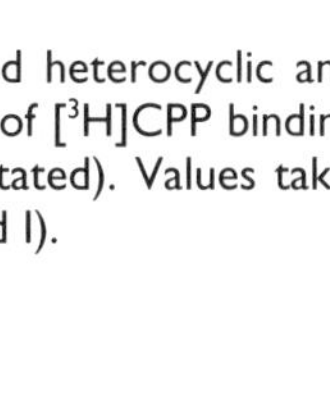

of heterocyclic AP7 analogues either a double bond (e.g. CPP-ene; see Aebischer *et al.*, 1989) or a keto group (e.g. MDL 100,925; see Whitten *et al.*, 1991) has been introduced into the phosphonoalkyl side-chain (Figure 4.5). Further conformational restriction of piperidine analogues by incorporating part of the phosphonoalkyl side-chain into a second ring leading to the formation of a bicyclic decahydroisoquinoline ring system (e.g. LY 235959, Figure 4.5) improved antagonist potency still further (Ornstein and Klimkowski, 1992; Ornstein *et al.*, 1992). In contrast to the vast majority of AP5 or AP7 analogues, in the series of decahydroisoquinolines the isomers with the S configuration at the α-amino acid stereogenic centre have the potent NMDA receptor antagonist activity. Like their open chain counterparts, for decahydroisoquinolines and other heterocyclic analogues of AP5 or AP7, a ω-phosphono or ω-tetrazole group is the preferred distal acid moiety for optimal NMDA receptor antagonist activity (Ornstein and Klimkowski, 1992; Ornstein *et al.*, 1992). Recently it has been reported that appropriate substitution of a heteroatom for a methylene group in the tetrazolylmethyl side-chain of some heterocyclic AP5-AP7 analogues (e.g. compound I, Figure 4.5) led to an increase in affinity for the NMDA receptor (Ornstein *et al.*, 1998).

Conformationally restricted carbocyclic analogues of AP5-AP7 such as NPC 17742 (Figure 4.7) (Ferkany *et al.*, 1993a, b) have also been synthesised and shown to be potent, systemically active, NMDA receptor antagonists. In some cases it is the conformationally restricted carbocyclic analogues of AP6 which posses the potent NMDA receptor antagonist activity (e.g. the cyclobutane, ACPEB; see Figure 4.6 (Gaoni *et al.*, 1994) and the quinoxalin-2-yl-alanine (compound II; see Figure 4.6 (Baudy *et al.*, 1993)). The latter is one of the most potent competitive NMDA receptor antagonists yet reported; however, it also possesses high affinity for the glycine binding-site on the NMDA receptor. Although these AP6 analogues are exceptions to the rule that potent activity resides in compounds with AP5 or AP7 chain length, molecular modelling studies have confirmed that both the cyclobutane (Gaoni *et al.*, 1994) and quinoxaline (Baudy *et al.*, 1993) analogues of AP6 gave the best fit to a template consisting of the conformationally restricted AP5 analogue CGS19755.

Incorporation of an aryl ring into the inter-acidic group chain of AP7 led to a series of phenylalanine analogues (e.g. SDZ-EAB-515, SDZ-215–439, SDZ-220–040, SDZ-220–581) (Müller *et al.*, 1992; Urwyler *et al.*, 1996a; see Figure 4.7). Like the decahydroisoquinoline series of AP7 analogues it is the S enantiomers of the phenylalanines which possess the potent NMDA receptor antagonist activity. It was noted in structure–activity studies that *m*-phenyl substitution of 3-phosphonophenylalanine to give

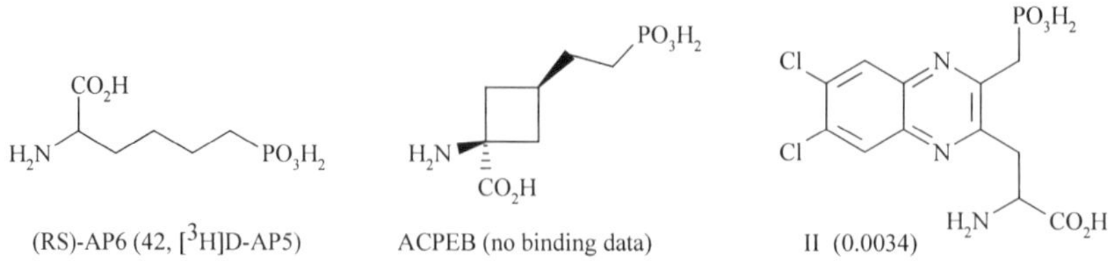

(RS)-AP6 (42, [^{3}H]D-AP5) — ACPEB (no binding data) — II (0.0034)

Figure 4.6 Structures of conformationally restricted analogues of AP6. IC_{50} or K_i (μM) values for displacement of [^{3}H]CPP binding to rat brain membranes given in parenthesis (unless otherwise stated). Values taken from Jane *et al.* (1994). No binding data is available for ACPEB, but in an electrophysiological assay it is approximately equipotent with (RS)-CPP (Gaoni *et al.*, 1994).

SDZ-EAB-515 led to a marked increase in NMDA receptor antagonist potency (Müller *et al.*, 1992). A further improvement in the potency of SDZ-EAB-515 was obtained by hydroxy-substitution at R^1, and substitution of the second phenyl ring in the para-position (see SDZ-215–439, Figure 4.7), and especially the ortho-position (optimally with a chloro group; see SDZ-220–040, Figure 4.7), led to increased NMDA receptor antagonist potency (Urwyler *et al.*, 1996a). Ortho-substitution of the second phenyl ring is thought to increase antagonist potency by increasing the torsion angle between the two phenyl rings to the optimal arrangement for interaction with the NMDA receptor binding site (Urwyler *et al.*, 1996a). The enhanced potency of SDZ-220–040 may also be explained by the stabilisation of the optimal folded conformation for interaction with the NMDA receptor binding site (Ortwine *et al.*, 1992) through intramolecular hydrogen bonding of the phenolic hydroxyl group with the phosphonomethyl and alanine side-chains. The biphenyl-AP7 analogue, SDZ-220–581 has been shown to be neuroprotective with a potency comparable to that of (R)-CPP-ene and was more effective than the latter upon oral administration (Urwyler *et al.*, 1996b). The oral bioavailability of SDZ-220–581 is likely explained by the increased lipophilicity of this class of compound and it has also been suggested that it may be actively transported from the intestine and across the blood brain barrier (Li *et al.*, 1995; Urwyler *et al.*, 1996b).

Bioisosteric replacement of the glycine unit by a squaric acid amide moiety is possible, as demonstrated by the almost equal affinity of the *N*-phosphonoethyl analogue (compound III, Figure 4.8) and (RS)-AP7 for the NMDA receptor (Kinney *et al.*, 1992). A greater than tenfold increase in affinity for the NMDA receptor was obtained upon introducing a degree of conformational restriction by formation of the bicyclic analogue (IV, Figure 4.8) of compound II (Figure 4.8) (Kinney *et al.*, 1992; Kinney and Garrison, 1992).

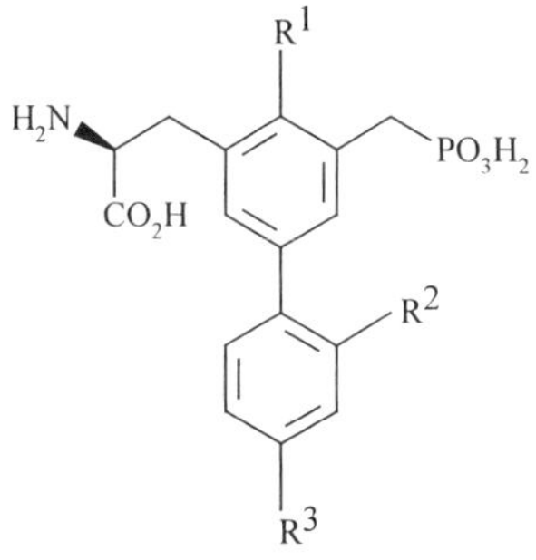

R^1 H_2N PO_3H_2 CO_2H R^2 R^3

pK_i values for displacement of $[^3H]$CGP39653 given in parenthesis:

$R^1 = R^2 = R^3 =$ H SDZ-EAB-515 (6.6)
R^1 = H, R^2 = H, R^3 = Ph SDZ-215-439 (7.8)
R^1 = OH, $R^2 = R^3$ = Cl SDZ-220-040 (8.5)
R^1 = H, R^2 = Cl, R^3 = H SDZ-220-581 (7.7)

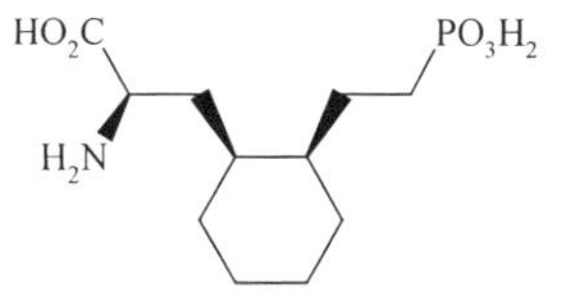

HO_2C PO_3H_2 H_2N

NPC 17742 (0.148, $[^3H]$CGS19755)

Figure 4.7 Structures of phenylalanine analogues and a phenylalanine with a saturated phenyl ring, conformationally restricted AP7 analogues. IC_{50} or K_i (μM) values for displacement of radioligand binding to rat brain membranes given in parenthesis. Values taken from Jane *et al.* (1994) and Urwyler *et al.* (1996a).

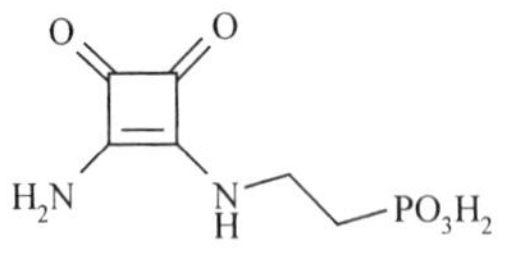

III (0.47)

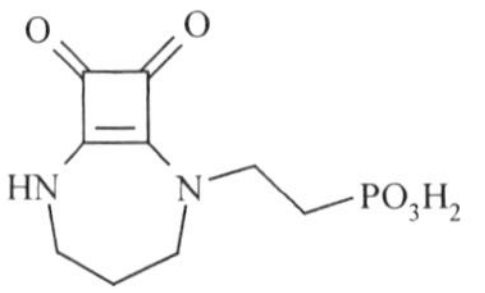

IV (0.030)

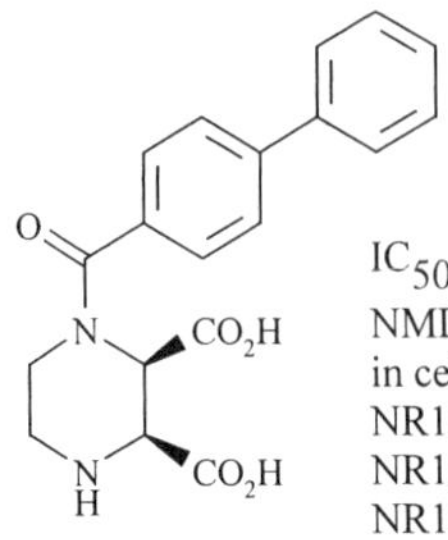

IC_{50} (μM) for antag. of
NMDA-induced depols
in cells expressing:
NR1/NR2A 15.79
NR1/NR2B 5.01
NR1/NR2C 8.98
NR1/NR2D 4.29

PBPD

Figure 4.8 Structures of two squaric acid amide analogues and the conformationally restricted aspartate analogue PBPD. IC_{50} or K_i (μM) values for displacement of [^{3}H]CPP binding to rat brain membranes given in parenthesis (unless otherwise stated). Values taken from Jane *et al.* (1994), Johnson and Ornstein (1996) (for IV) and Buller and Monaghan (1997) (for PBPD).

The development of the potent NMDA receptor antagonists described above was aided by the introduction of radioligands such as [^{3}H]AP5 (Olverman *et al.*, 1984, 1988), and the higher affinity ligands, [^{3}H]CGS19755 (Murphy *et al.*, 1988), [^{3}H]CPP (Olverman *et al.*, 1986) and [^{3}H]CGP 39653 (Sills *et al.*, 1991). The high affinity of [^{3}H]CGP 39653 (K_D value 7 nM) allowed the use of this ligand in a filtration binding assay facilitating compound throughput. A photoaffinity probe has been developed based on the structure of CGP 39653 (Heckendorn *et al.*, 1993).

Pharmacophore modelling of competitive NMDA receptor antagonists

A number of computer-aided molecular modelling studies have been carried out in order to define pharmacophore models for NMDA receptor antagonists (for a review of the types of approaches used in these studies see Jane *et al.*, 1994). In the most comprehensive modelling study (Ortwine *et al.*, 1992) it was concluded that antagonists bind in a folded conformation to a primary interaction point on the receptor, proximal to the distal NMDA receptor agonist binding site, thus allowing both the ω-phosphonate and α-carboxyl groups to interact simultaneously with this site. It was also suggested that the ω-phosphonate, but not an ω-carboxyl group, interacts with a distal secondary site giving an explanation for the increased affinity observed with phosphonates over the corresponding monobasic carboxy-substituted analogues. In order to estimate the exclusion volume for the NMDA receptor a large number of structurally dissimilar compounds were included in this study. Excluded volume was noted around the α-amino group and the methylene group adjacent to the ω-phosphonate, while a degree of bulk is tolerated around the 4-6-positions of the piperidine ring of CGS19755 (Ortwine *et al.*, 1992). When SDZ-EAB-515 was added to the model it was noted that the biphenyl moiety of (S)-isomer did not conflict with known excluded volume, whereas this was not the case for the (R)-isomer (Bigge, 1993; Li *et al.*, 1995). This observation provides an explanation for the stereoselectivity of the NMDA receptor antagonist activity of SDZ-EAB-515. Problems exist with all the modelling studies carried out to date, the most serious of which include problems associated with modelling charged molecules, the unknown environment of the receptor binding site (with respect to the presence of water molecules and metal ions such as Ca^{2+}) and the likely existence of NMDA receptor subtypes (see Jane *et al.*, 1994). The binding data referred to in these modelling studies comes from forebrain assays, an area rich in NMDA receptors containing mainly NR2A and/or NR2B subunits (Buller *et al.*, 1994). Thus, it is likely that the antagonist pharmacophore models to date are describing the binding site found in NMDA receptors containing either NR2A or NR2B but not NR2C or NR2D subunits.

Subtype selectivity of competitive NMDA receptor antagonists

Molecular biologists have isolated genes encoding for six subunits termed $NR1_{a\text{-}h}$ (eight splice variants are known), NR2A-D and NR3A (Monyer *et al.*, 1992; Dingledine *et al.*, 1999). NMDA receptors are thought to be either tetrameric or pentameric complexes comprised of at least two NR1 and two NR2A-D subunits (Laube *et al.*, 1998; Dingledine *et al.*, 1999). The glutamate recognition site is located on the NR2 subunit (Laube *et al.*, 1997), whilst the glycine co-agonist binding site is located on the NR1 subunit (Hirai *et al.*, 1996). The NR3A subunit is thought to play a regulatory role in NMDA receptor function, likely controlling Ca^{2+} ion influx through the channel (Das *et al.*, 1998). As NR2 subunits can assemble in a variety of combinations, it is likely that NMDA receptor subtypes exist in native tissue. The heterogeneous anatomical distribution of mRNA encoding for NR2A-D subunits in the CNS provides evidence for the presence of subtypes in native tissue (Buller *et al.*, 1994). In addition, some NMDA receptor agonists and antagonists show a differential binding to ventral lateral thalamus (NR2A and NR2B containing), medial striatum (NR2B enriched), cerebellum (NR2C enriched) or midline thalamus (NR2D enriched), thus indicating a degree of subtype selectivity for such compounds (Buller and Monaghan, 1997; Brown *et al.*, 1998).

This can be illustrated with (R)-CPP-ene, which shows higher affinity for forebrain NMDA receptors than for receptors found in the cerebellum and midline thalamus, whereas SDZ-EAB-515 is unable to discriminate between receptors found in the forebrain and midline thalamus (Buller and Monaghan, 1997). This selectivity was also observed on recombinant NMDA receptors individually expressed in *Xenopus* oocytes, where (R)-CPP-ene had higher affinity for NR1/NR2A or NR1/NR2B containing receptors than for those containing NR1/NR2C or NR1/NR2D, while SDZ-EAB-515 had high affinity for receptors comprised of either NR1/NR2D or NR1/NR2B subunits (Buller *et al.*, 1994; Buller and Monaghan, 1997). The pattern of selectivity displayed by SDZ-EAB-515 on recombinant and native NMDA receptors was also shown by the biphenyl analogue PBPD (Figure 4.8) (Buller and Monaghan, 1997). Whether this is due to the presence of a binding site in the NR2C or NR2D subunit with which the biphenyl moiety of PBPD or SDZ-EAB-515 can interact remains to be determined.

Ligands binding to the glycine co-agonist site of the NMDA receptor

A number of side-effects such as motor impairment (ataxia), impairment of spatial learning, psychotomimetic effects and the production of vacuoles in rat cortical neurons have been noted for competitive NMDA receptor antagonists tested in animal models (Herrling, 1997). The observation that glycine is a necessary co-agonist for activation of the NMDA receptor (Johnson and Ascher, 1987) has led to an interest in antagonists and partial agonists which interact with this site as an alternative therapeutic strategy (for a review see Leeson and Iversen, 1994). The neutral amino acids (R)-alanine and (R)-serine (Figure 4.9) display high affinity for the glycine site and behave as full agonists, the effect is stereoselective as the S enantiomers have much lower affinity (Leeson and Iversen, 1994). A number of NMDA receptor glycine site partial agonists have been reported, including HA-966 (Figure 4.9) which was one of the first compounds identified as an NMDA receptor antagonist (Evans *et al.*, 1978). A number of analogues of HA-966 have since been synthesised and of these L-687,414 (Figure 4.9) was more potent as an NMDA receptor antagonist than the parent compound. Recently, it has been demonstrated that L-687,414 (Leeson *et al.*, 1993) does not prevent hippocampal LTP *in vivo* at plasma levels known to have neuroprotective effects, suggesting that NMDA receptor glycine site partial agonists may have therapeutic advantages over full antagonists (Priestly *et al.*, 1998). An analogue of HA-966, compound V (Figure 4.9), has recently been synthesised (Cordi *et al.*, 1999), which has lower affinity for the glycine site than L-687,414 but higher affinity than the previously reported partial agonist (R)-cycloserine (Figure 4.9) (Hood *et al.*, 1989a). A series of cycloalkyl glycine analogues bind to the glycine site, with the cyclopropyl analogue ACC (Figure 4.9) being almost a full agonist (Marvizon *et al.*, 1989; Watson and Lanthorn, 1990), while the cyclobutane analogue ACBC (Figure 4.9) is a low efficacy partial agonist (Hood *et al.*, 1989b).

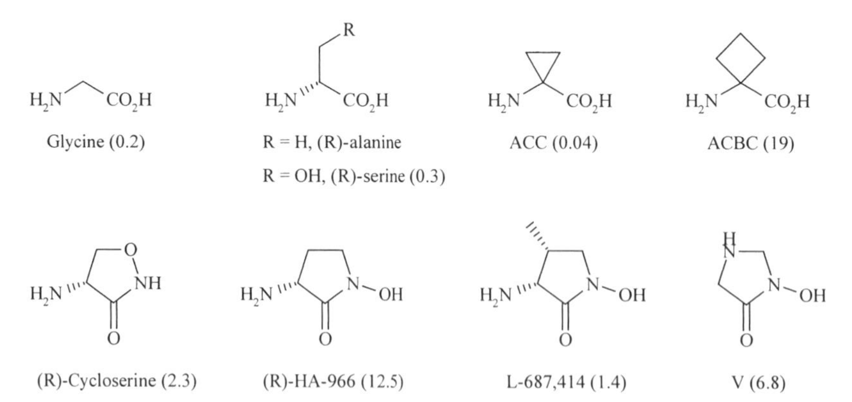

Figure 4.9 Structures of NMDA receptor glycine binding site agonists and partial agonists. IC_{50} (μM) values for displacement of [^{3}H]glycine binding to rat brain membranes given in parenthesis. Values taken from Leeson and Iversen (1994) and Cordi *et al.* (1999) (for compound V; in the same assay (R)-cycloserine had an IC_{50} value of 7.37 μM).

The development of antagonists acting at the glycine binding site associated with the NMDA receptor and the therapeutic potential of such compounds has been reviewed (Leeson and Iversen, 1994). A number of heterocyclic templates have been used for the development of potent antagonists, including kynurenic acid (e.g. L-683,344, Leeson *et al.*, 1991; L-689,560, Leeson *et al.*, 1992), quinoline-2-ones (e.g. L-701,324, Kulagowski *et al.*, 1994), quinoxaline-2,3-diones (e.g. compound VII, Figure 4.10, Nagata *et al.*, 1994; ACEA-1021, Cai *et al.*, 1997), and 2-carboxyindoles (GV150526A, Di Fabio, 1997) (for structures see Figures 4.10–4.13).

Further elaboration of compounds based on the quinoxaline-2,3-dione structure led to a range of phosphonoalkylamino-substituted analogues with selectivity for either the glycine binding site of the NMDA receptor or the AMPA receptor (Auberson *et al.*, 1999). The S isomer, compound VI (Figure 4.10), displayed high affinity and 500-fold selectivity for the glycine binding site relative to AMPA receptors. A structure–activity study on this series of compounds revealed that a combination of a 7-bromo- and (S)-α-phosphoalanine substituent led to optimal potency and selectivity for the glycine binding site. The phosphono group conferred good water solubility on these compounds thus solving one of the problems associated with the earlier generation of antagonists based on the quinoxalinedione structure. It was proposed that effective blood–brain barrier penetration, aided by an active transport system, explained the excellent *in vivo* anticonvulsant activity of these compounds. Recently, a number of non-planar analogues of the previously reported AMPA and NMDA receptor glycine site antagonist, PNQX (Figure 4.10), have been reported (Nikam *et al.*, 1999). The sarcosine analogue (VIII, Figure 4.10) retained high affinity for the AMPA receptor and the NMDA receptor glycine site and possessed improved water solubility when compared to PNQX. The rationale for the increased water solubility was that the

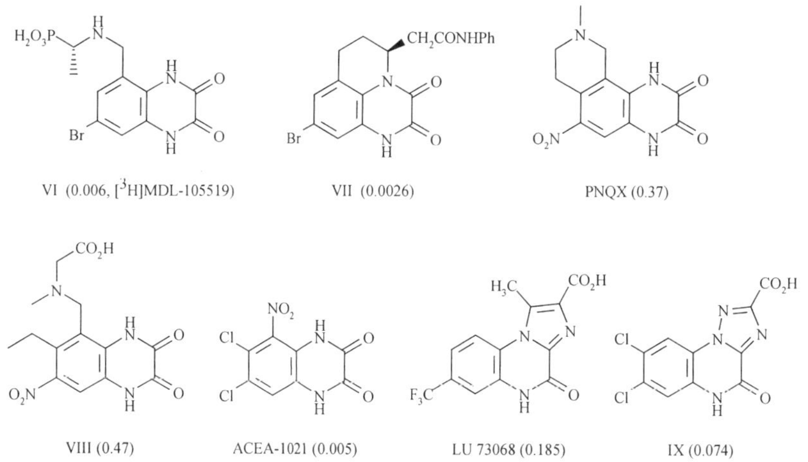

VI (0.006, [^{3}H]MDL-105519) VII (0.0026) PNQX (0.37)

VIII (0.47) ACEA-1021 (0.005) LU 73068 (0.185) IX (0.074)

Figure 4.10 Structures of NMDA receptor glycine binding site antagonists based on the quinoxaline-2,3-dione structure. IC_{50} (μM) values for displacement of [^{3}H]glycine binding to rat brain membranes given in parenthesis (unless otherwise stated). Values taken from Auberson *et al.* (1999) (for VI), Nagata *et al.* (1994) (for VII), Nikam *et al.* (1999) (for PNQX and VIII), Potschka *et al.* (1998) (for LU 73068) and Catarzi *et al.* (1999) (for IX). K_B value (μM) for ACEA-1021 is from a functional assay on cortical neurons (Leeson and Iversen, 1994).

non-planarity of compound VIII (Figure 4.10) results in a reduction in packing efficiency in the solid state. A detailed comparison of pharmacophore models for the NMDA receptor glycine site and the AMPA receptor has been reported (Bigge *et al.*, 1995; Nikam *et al.*, 1999). The model for glycine site antagonists identified an H-bond acceptor site for the NH group in the quinoxaline ring, an acceptor site for Coulombic interaction with the dione functionality on the quinoxaline ring, a size limited electropositive H-bonding region with which substituents at the 4 and 5 positions of the quinoxaline ring can interact and a site with which electron-withdrawing groups and, in particular, halogens can interact. It has been proposed that in order to achieve significant *in vivo* potency in animal models of stroke and anticonvulsant activity a more balanced affinity for AMPA, kainate and NMDA receptor glycine binding-sites may be necessary (Bigge *et al.*, 1995; Potschka *et al.*, 1998). A novel quinoxalinedione analogue, LU 73068 (Figure 4.10), has equally high affinity for the glycine site of the NMDA receptor and the AMPA receptor, as well as moderate affinity for some kainate receptor subunits, and has been shown to be a potent and effective anticonvulsant agent (Potschka *et al.*, 1998). In order for this combination therapy approach to be successful, the therapeutic benefits will need to outweigh the adverse effects on motor and respiratory function.

A recent report has disclosed a range of tricyclic quinoxalinedione analogues with mixed antagonist activity at AMPA receptors and the glycine site of the NMDA receptor (Catarzi *et al.*, 1999). In particular, the dichloro-substituted analogue (compound IX, Figure 4.10) displayed high affinity for the NMDA receptor glycine site and ~18-fold selectivity for this site compared to the AMPA receptor. Structural features necessary for activity at the NMDA receptor glycine site included a free carboxylic acid substituent at the 2-position and at least one electron-withdrawing substituent no larger than a chlorine atom on the benzo-fused ring.

Elaboration of the structure of kynurenic acid, the first glycine site antagonist to be reported (Birch *et al.*, 1988; Watson *et al.*, 1988), has led to a number of potent antagonists such as L-683,344 and L-689,560 (Leeson *et al.*, 1991, 1992) (for structures see Figure 4.11). Based on a detailed structure–activity study on a range of kynurenic acid analogues, a glycine site antagonist pharmacophore model has been proposed in which regions of bulk tolerance and intolerance and size-limited hydrophobic regions for the chloro substituents on the aromatic ring have been identified (Leeson and Iversen, 1994). Interaction sites for charged residues were also identified, including an H-bond acceptor site for the amine group, a size-limited polar region with which the 2-carboxylate group interacts, and an H-bond donor site with which the carbonyl group of the 4-position substituent interacts. Recently, a photoaffinity label, [^{3}H]CGP 61594 (Figure 4.11), based on the structure of L-689,560, has been developed for the NMDA receptor glycine site (Benke *et al.*, 1999). The protein photolabelled by [^{3}H]CGP 61594 in brain membrane preparations was identified as the NR1 subunit of the NMDA receptor, and future studies are likely to reveal the nature of the amino acid residues lining the glycine binding site. It was previously reported that CGP 61594 displays higher affinity for the NR1/NR2B receptor subtype over those containing NR2A, NR2C or NR2D subunits (Honer *et al.*, 1998) and thus represents a novel lead for the development of NR2B-selective antagonists. The dependency of the affinity of glycine receptor agonists for the glycine site of the NMDA receptor on the NR2 subtype present, despite the glycine binding site being located on the NR1 subunit, has previously been noted (Kutsuwada *et al.*, 1992; Buller *et al.*, 1994).

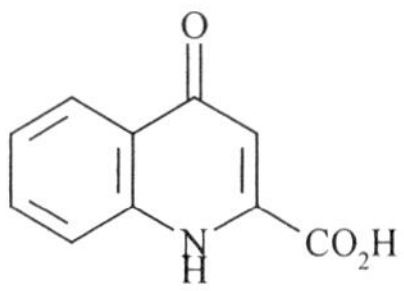

Kynurenic acid (41)

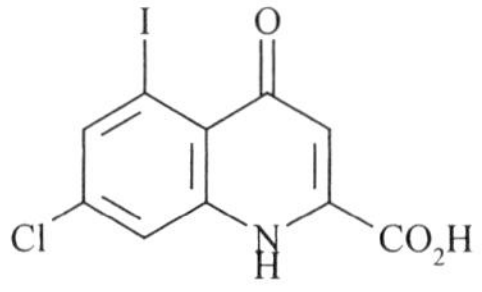

L-683,344 (0.032)

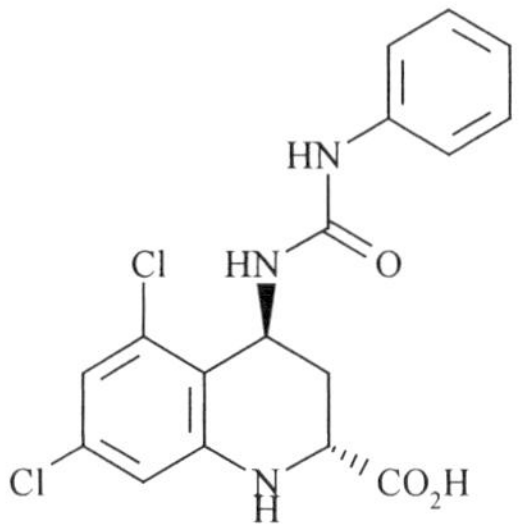

L-689,560 (0.0078)

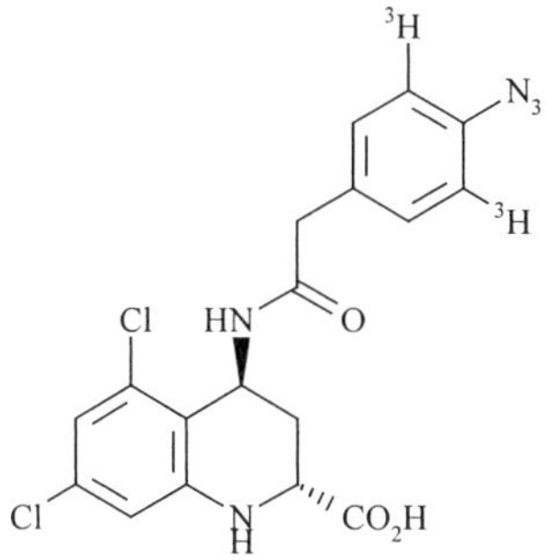

[^{3}H]CGP 61594 (K_D 23 nM)

Figure 4.11 Structures of NMDA receptor glycine binding site antagonists based on the kynurenic acid structure. IC_{50} (μM) values for displacement of [^{3}H]glycine binding to rat brain membranes given in parenthesis (unless otherwise stated). Values taken from Leeson and Iversen (1994) and Benke *et al.* (1999) (for [^{3}H]CGP 61594).

The 4,6-dichloroindole-2-carboxylate nucleus has been used successfully in the development of a number of potent NMDA receptor glycine site antagonists (Leeson and Iversen, 1994). Substitution of a 2-carboxyethyl group at C-3 of the indole ring (see compound X, Figure 4.12) improved potency (Salituro *et al.*, 1992) and this was further increased by substitution at C-3 of the indole with a β-unsaturated amide leading to GV150526A (Figure 4.12) (Di Fabio *et al.*, 1997). Further elaboration of GV150526A led to GV196771A (Figure 4.12) which potently antagonised recombinant

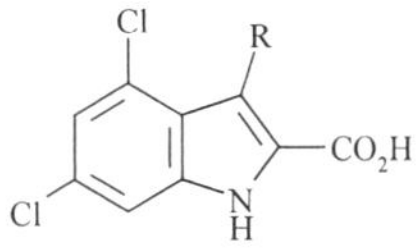

X, R = -$(CH_2)_2CO_2H$ (0.14)
GV150526A, R = -CH=CHC(O)NHPh (pKi value 8.5)

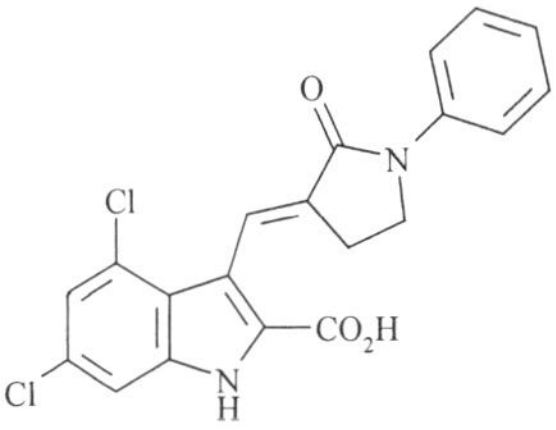

GV196771A (pK_B 8.04)

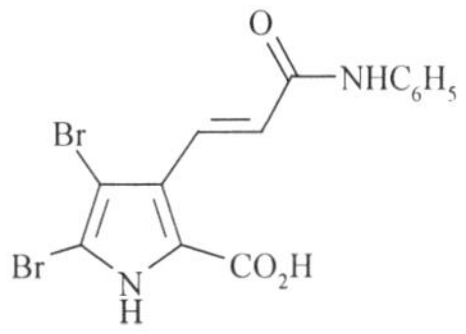

XI (pK_i value 7.95)

Figure 4.12 Structures of NMDA receptor glycine binding site antagonists based on the indole-2-carboxylic acid template. IC_{50} (μM) or pK_i values for displacement of [^{3}H]glycine binding to rat brain membranes given in parenthesis. Values taken from Leeson and Iversen (1994) (for X), Di Fabio *et al.* (1997) (for GV150526A) and Balsamini *et al.* (1998) (for XI). pK_B value for GV196771A is from an electrophysiological assay (Carignani *et al.*, 1998).

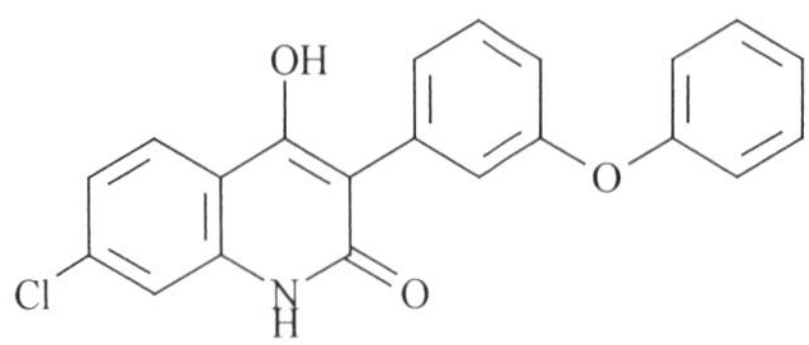

L-701,324 (0.002)

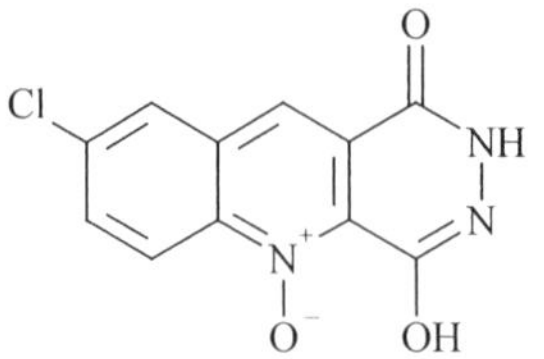

MRZ 2/576 (0.090, [^{3}H]5,7-DCKA)

Figure 4.13 Structures of L-701,324 and MRZ 2/576, two potent NMDA receptor glycine binding site antagonists. IC_{50} (μM) values for displacement of [^{3}H]glycine binding to rat brain membranes given in parenthesis (unless otherwise stated). Values taken from Leeson and Iversen (1994) and Parsons *et al.* (1996) (MRZ 2/576).

NMDA receptors, showing some selectivity for those containing either NR2A or NR2B subunits, and antagonised NMDA receptors in embryonic spinal neurones with a similar potency (Carignani *et al.*, 1998). GV196771A blocked spinal cord wind-up (the augmented response in C-fibres following an initial burst of repetitive stimulation), and was also shown to be efficacious in two animal models of neuropathic pain (Bordi *et al.*, 1998; Carignani *et al.*, 1998). Pyrrole analogues of GV150526A retained potent glycine site antagonist activity as long as substituents were present on the 4- and 5-position of the pyrrole ring, with 4,5-dibromo substitution (e.g. compound XI, Figure 4.12) being optimal (Balsamini *et al.*, 1998). A quantitative structure–activity study suggested a correlation between affinity for the glycine site and electron-withdrawing ability, bulk and lipophilicity of the 4,5-substituents.

A series of pyrido-phtalazindiones (e.g. MRZ 2/576, Figure 4.13) have been reported to have high affinity for the glycine binding site (Parsons *et al.*, 1997) and displayed neuroprotective (Wenk *et al.*, 1998) and anti-nociceptive (Williams *et al.*, 1999) properties in animal models.

Channel blocking uncompetitive NMDA receptor antagonists

The dissociative anaesthetics phencyclidine (PCP) and ketamine (Figure 4.14) block NMDA receptor mediated responses (Anis *et al.*, 1983) by a use (i.e. initial activation of the receptor by an agonist is required to open the channel) and voltage-dependent mechanism (Huettner and Bean, 1988; Lodge and Johnson, 1990). Due to the slow kinetics of dissociation from the binding site, the antagonist can become trapped upon channel closure and subsequent recovery from this trapped closed state is slow. This mechanism of action led to the suggestion that these antagonists are blocking the NMDA receptor channel by binding to a site deep within it. Indeed, site-specific mutagenesis has revealed that an asparagine residue (N598) in the pore-lining M2 segment of the NMDA receptor is important for antagonist binding (Sakurada *et al.*, 1993).

The use-dependent mechanism of action of open channel blockers led to the suggestion that such compounds may have therapeutic utility as neuroprotective agents for the treatment of ischaemia resulting from over-activation of NMDA receptors and excessive Ca^{2+} entry (Lipton, 1993). A number of potent NMDA receptor open channel

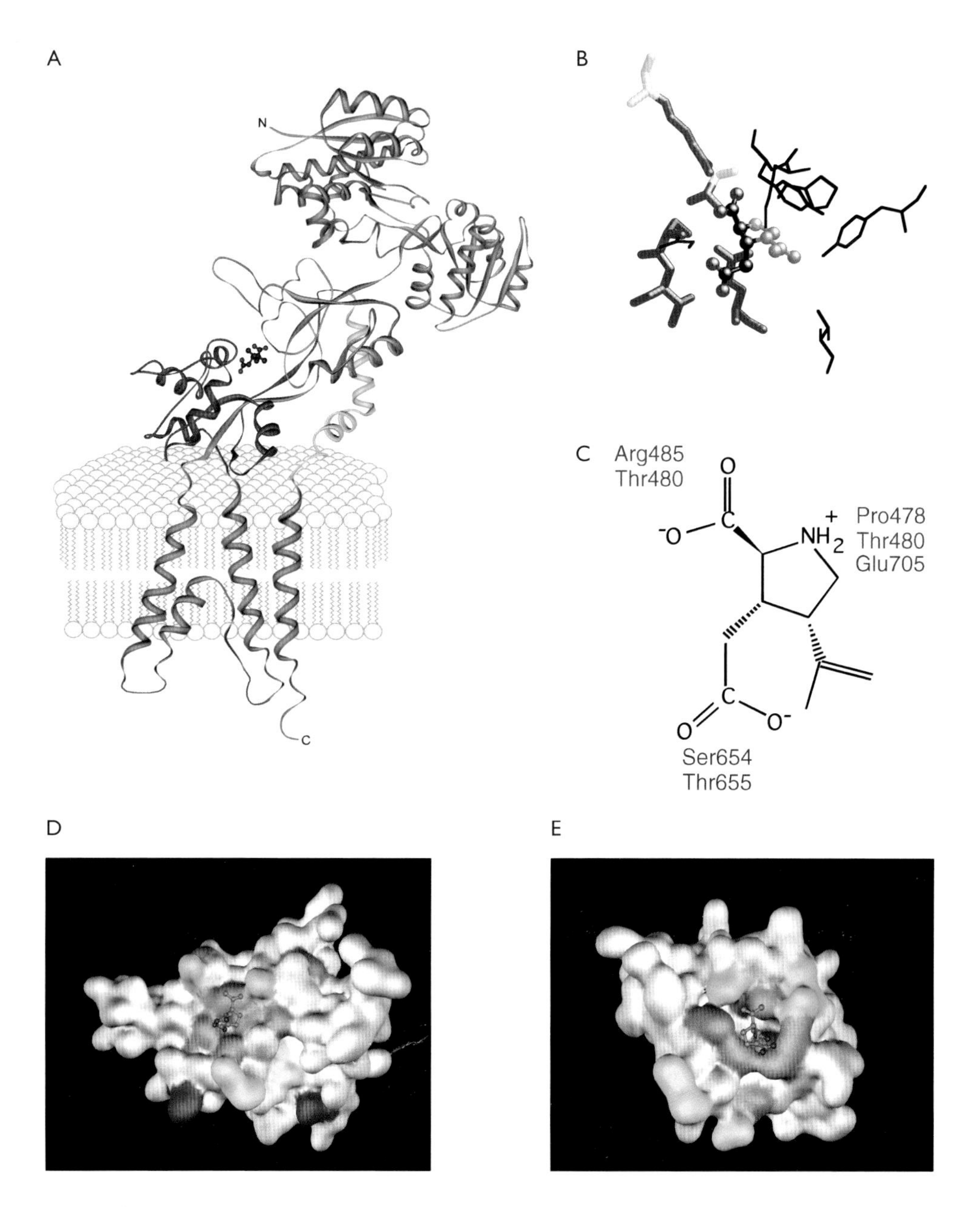

Figure 2.3 (a) Model of the glutamate receptor. The X-domain is folded as a LIVBP protein (gray), the binding domain is folded as the GluR2 structure determined from the soluble binding domain (Armstrong, 1998 #366), lobe A is blue with the flip/flop region in green, and lobe B is depicted in red. (b) Kainate bridging between the lobe A (blue) and lobe B (red). (c) Residues interacting with the glutamate like moiety of kainate in the GluR2 subunit. (d, e) Surface plots of a "cut-open" binding domain viewing the ligand binding surface of lobe A (d) and lobe B (e). Residues interacting with the glutamate-moiety of kainate are shown in pink, residues which affect interactions in a ligand-specific manner (Table 2.2) (orange), inter-lobe interacting residues which stabilize the closed conformation (green) and residues where the lobes are covalent attached (cyan).

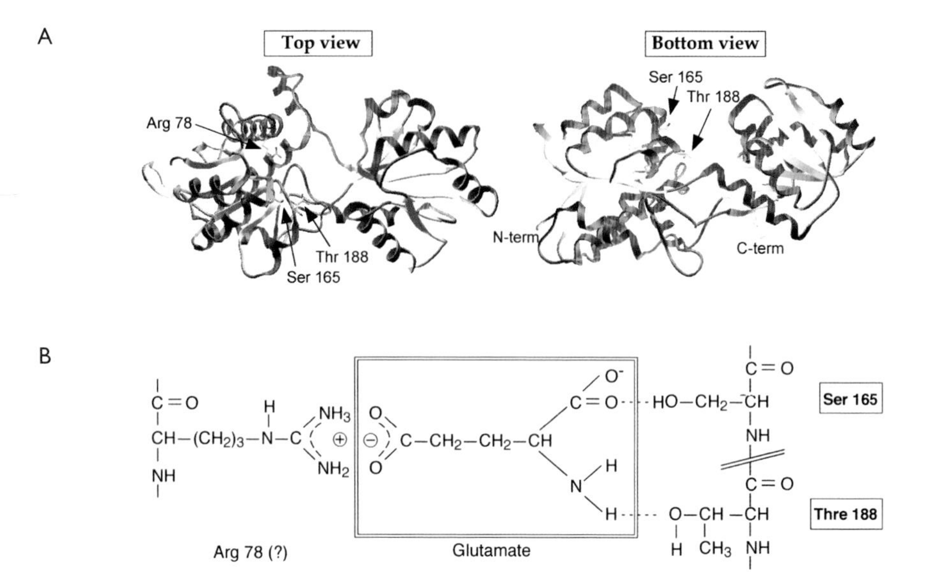

Figure 6.3 Glutamate binding site of mGluR1.

(a) the three-dimensional model of the open state of mGluR1 binding domain. This model has been generated according to the sequence alignment of the extracellular domain of mGluRs with periplasmic binding proteins as proposed by O'Hara *et al.* (1993) and using the program modeller (Molecular stimulation Inc. San Diego USA). The three large insertions not found in the bacterial periplasmic proteins have not been included in the model. Helices are indicated in red and sheets in yellow. Residues likely to be involved in binding glutamate are indicated Arg^{78} (see text), Ser^{165} and Thr^{188}.

(b) Putative chemical interactions between glutamate and the binding site:

- a salt bridge is formed between the positively charged guanidium group of Arg^{78} and the γ carboxyl group of glutamate[78];
- the C=O of the α-carboxyl group of glutamate is hydrogen bound to the OH of Ser^{165};
- the N—H of the α-amino-group of glutamate is hydrogen bound to the —OH of Thr^{188}.

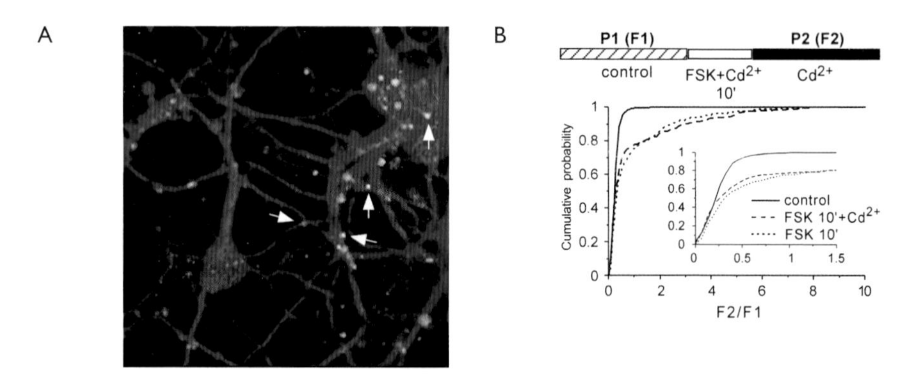

Figure 6.7 Increase in cAMP stimulated the vesicular release of glutamate by a direct activation of synaptic release machinery, an effect blocked by low μM concentrations of L-AP4. The ratio F2/F1 represented the ratio between the uptake of two antibodies against the same epitope of synaptotagmin I during a control (P1) and a test (P2) period at each vesicular release site of cerebellar granular cells. The uptake of the two antibodies, one from goat (during P1) and one from rabbit (during P2), was quantified using secondary antibodies labelled with Texas red (anti-goat antibody) and fluorescein (anti-rabbit antibody).

C

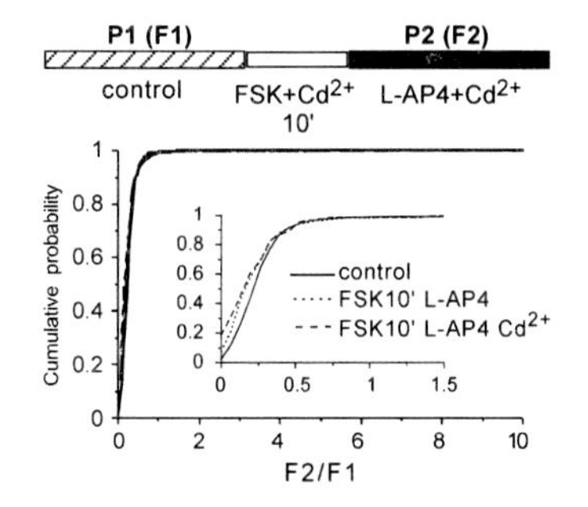

(a) Yellow dots indicate that the release sites were active during the control (P1) and the test periods, whereas the green dots indicate 'silent' release sites which become active after the treatment. (b) Treatment with forskoline (FSK, 10 μM) in the presence of Cd^{+} to block the VSCC increased vesicular release (right shift of the curve) and revealed activity in previously 'silent' sites (blue sites in (a). Inset is an extension of the first part of the curves. (c) Treatment with L-AP4, at a concentration which stimulated mGluR4 receptors, blocked the effect of FSK in the absence or the presence of Cd^{2+}. Modified from Chavis *et al.* (1998b).

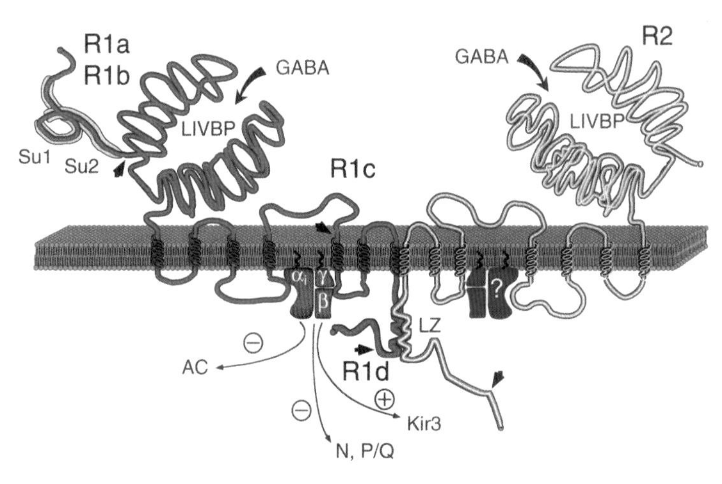

Figure 11.1 Structural model and major effector systems of the functional heteromeric $GABA_B$ receptor. $GABA_B$R1 and $GABA_B$R2 assemble as subunits to produce a single pharmacologically and functionally defined receptor *in vivo*. The amino-terminal extracellular domain is roughly limited to a region with homology to bacterial periplasmatic proteins like LIVBP. Based on its crystal structure, one predicts that this domain is constituted of two lobes which close upon ligand binding (Sack *et al.*, 1989; Galvez *et al.*, 1999). The mature R1b protein (815 amino acids) differs from R1a (944 amino acids) in that the amino-terminal 147 residues are replaced by 18 different residues. Additional splice variants, designated $GABA_B$R1c and $GABA_B$R1d, generate isoforms with sequence differences in presumed extracellular and intracellular domains (alternative splice sides are marked by an arrow head). Two C-terminal splice variants of the human $GABA_B$R2 have been reported (Ng *et al.*, 1999) The R1a-specific region contains two copies of short consensus repeats (SCRs) about 60 amino acids each, also known as sushi repeats (Su1, Su2) or complement control protein (CCP). $GABA_B$R1 and R2 may heterodimerize through an interaction between their intracellular carboxyl terminal tails using a coiled–coil structure, a dimerization signal that is also used by leucine zipper (LZ) transcription factors. Specificity of G-protein coupling of $GABA_B$R1 is likely to be provided by the second intracellular loop. The activated α-subunit of $G_{i/o}$ type proteins inhibits adenylate cyclase (AC). Presynaptically, $\beta\gamma$-subunits negatively modulate voltage-dependent Calcium channels of the N, P/Q type, yielding a reduced neurotransmitter and neuropeptide release. At postsynaptic sites, $GABA_B$ receptors activate Kir3.1+3.2 type K^{+} channels resulting in a hyperpolarization of the neuron. The G-protein coupling status of $GABA_B$R2 is unclear.

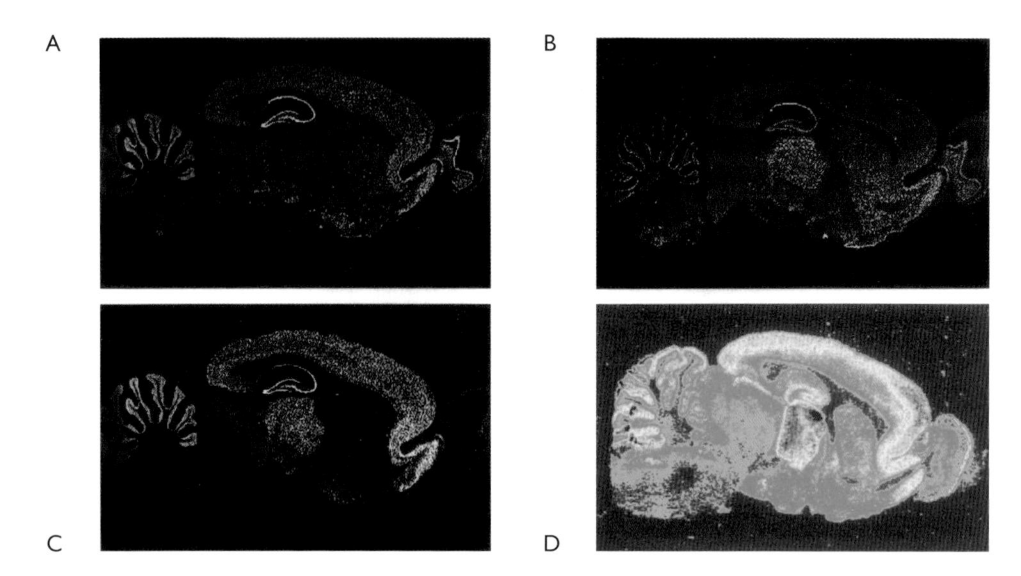

Figure 11.2 Distribution of $GABA_B$ receptor mRNA and binding sites on adjacent sagittal rat brain sections. (a), (b) and (c) are darkfield photomicrographs representing the spatial distribution of $GABA_BR1a$ (a), $GABA_BR1b$ (b) and $GABA_BR2$ (c) transcripts obtained by *in situ* hybridization with specific ^{35}S-labelled riboprobes. (d) shows the distribution of $GABA_BR1$ binding sites labelled with the specific and high affinity $GABA_BR1$ receptor antagonist [^3H]CGP 54626A. Binding was assessed by quantitative autoradiography on cryosection and shown as specific binding (methods described in detail in Bischoff *et al.*, 1999).

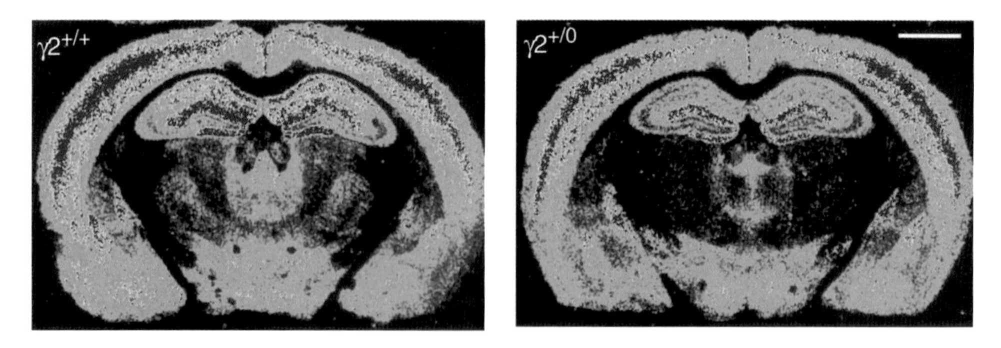

Figure 16.1 The reduction of the gene dosage of the γ2 subunit results in a region-specific decrease of [^3H]-flumazenil binding and a concomitant reduction of $GABA_A$-receptor clusters in $\gamma2^{+/0}$ mice (Crestani *et al.*, 1999).

Color-coded autoradiograms of [^3H]-flumazenil binding in transverse sections of wt and $\gamma2^{+/0}$ brain. Red indicates maximal density, followed by yellow, green and blue. Note the profound decrease in the hippocampus CA1 and CA3 regions compared to the moderate reduction in the dentate gyrus (DG). The average decrease in the entire forebrain was −20% of wt, as assessed by densitometry; regions with a greater than average decrease include cingulate (−25%), frontal (Fr) (−23%), and piriform cortex (Pi) (−25%), the CA1 (−35%) and CA3 (−28%) regions of the hippocampus, as well as the lateral septum (−30%) and several thalamic nuclei. Conversely, regions with a lesser than average decrease include the striatum (Cpu) (−6%), globus pallidus (GP) (−13%), dentate gyrus (DG) (−15%) and most of the amygdala. Abbreviations: BL, basolateral nucleus of the amygdala; Ce, central nucleus of the amygdala; S1, primary somatosensory cortex. Scale bar, 1 mm.

blockers have since been developed, such as MK-801 (Figure 4.14) which has been widely used to study the therapeutic utility of this type of antagonist (Wong *et al.*, 1986; Gill *et al.*, 1987). A pharmacophore model has been proposed for the MK-801 recognition site (Leeson *et al.*, 1990). High affinity channel blockers such as PCP and MK-801 display a number of adverse side effects such as ataxia, impairment of memory and learning, neuronal vacuolisation and psychotomimetic effects (Iversen and Kemp, 1994; Parsons *et al.*, 1999) and thus they have not been developed for use in the clinic. However, low affinity blockers such as dextromethorphan, memantine, amantadine and remacemide appear to display a better therapeutic ratio (Parsons *et al.*, 1995, 1999; Williams *et al.*, 1999) (for structures see Figure 4.14). The relationship between the promising therapeutic profile and the mechanism of action of memantine has been thoroughly reviewed (Parsons *et al.*, 1999). In common with the aforementioned low affinity channel blockers memantine shows faster open channel blocking and unblocking kinetics than MK-801. It has been suggested that the combination of fast kinetics and strong voltage dependency allows memantine to leave the channel rapidly upon transient activation by glutamate but blocks sustained activation by glutamate as occurs during chronic excitotoxic insults (Parsons *et al.*, 1999). Recently it has been proposed

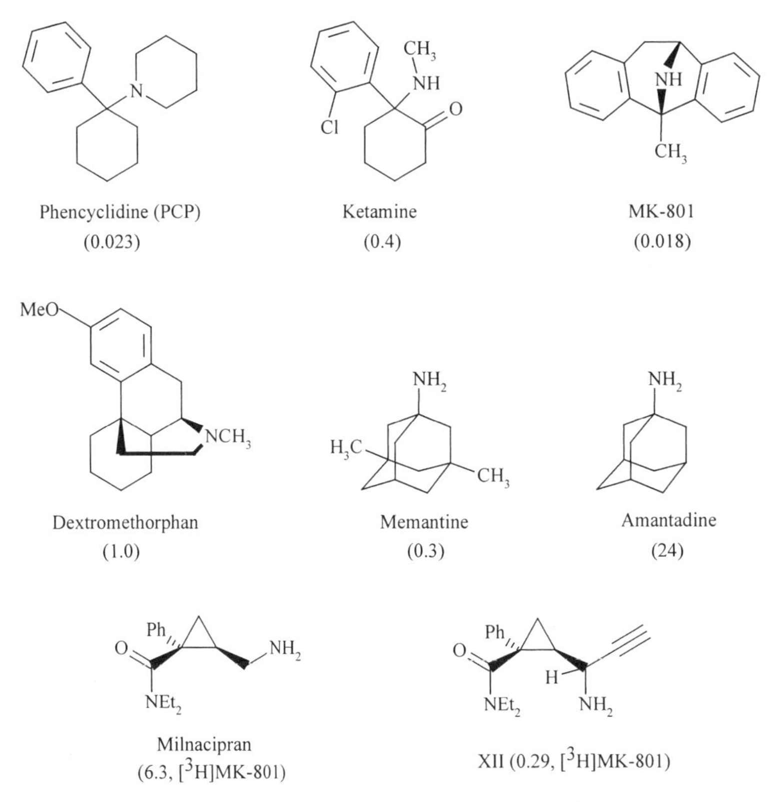

Figure 4.14 Structures of open channel blockers of the NMDA receptor. Values in parenthesis represent the concentration at which the compound inhibits half of the maximal agonist activated current (Ferrer-Montiel *et al.*, 1998). Values in parenthesis for (±)-milnacipran and analogue (XII) represent IC_{50} (μM) values for displacement of [^{3}H]MK-801 binding to rat cerebral cortical synaptic membranes (Shuto *et al.*, 1998).

that memantine and amantadine are only partially trapped in closed NMDA receptor channels whereas high affinity blockers such as MK-801 (Blanpied *et al.*, 1997) stay trapped for longer due to slow unblocking kinetics. This mechanism may explain the improved therapeutic profile of memantine as in the absence of agonist a significant proportion of channels would unblock and therefore be available for subsequent activation under normal physiological conditions. However, under pathological conditions prolonged receptor activation by over-release of glutamate would result in strong block of the NMDA receptor by memantine (Parsons *et al.*, 1999).

In some studies, memantine shows moderate subunit selectivity with two- to threefold greater potency for NR2B over NR2A-containing receptors. However, this is not a consistent finding as other workers have found almost equal potency for receptors containing either NR2A or NR2B subunits (see Parsons *et al.*, 1999). The three- to fivefold greater potency for NR2C over NR2A receptors exhibited by memantine is a more consistent finding, but the relevance of this selectivity to the improved therapeutic profile of memantine has yet to be established. Other channel blockers such as dextromethorphan also display higher affinity for NR2C-containing receptors over NR2A or NR2B-containing receptors (Monaghan and Larsen, 1997). The high affinity channel blocker MK-801 shows slower kinetics for channel block/unblock at NR2C-containing receptors to those containing NR2A or NR2B (Monaghan and Larsen, 1997).

Thus a number of factors may contribute to the improved therapeutic ratio observed with low affinity channel blockers, but whatever the mechanism such compounds may have the desired properties to be therapeutically useful drugs.

Recently, a series of analogues of (±)-milnacipran (Figure 4.14), a clinically effective antidepressant, have been shown to be open channel blockers of NMDA receptors (Shuto *et al.*, 1998). One of these, with an ethynyl substituent α to the primary amine (compound XII, Figure 4.14), displayed high affinity for the NMDA receptor but was also a potent 5-HT uptake inhibitor (K_i value 0.19 μM). This series of compounds may represent a new target for drug development since the parent compound, milnacipran, shows no serious side effects and is able to cross the blood–brain barrier.

Non-competitive NMDA receptor antagonists

The endogenous polyamines spermine and spermidine modulate both native and recombinant NMDA receptor function by at least three mechanisms: glycine-dependent potentiation, glycine- and voltage-independent potentiation and voltage-dependent inhibition (for reviews see Rock and Macdonald, 1995; Dingledine *et al.*, 1999). Potentiation of NMDA receptor function by polyamines is dependent on glycine concentration being greater at low concentration than at saturating concentrations (Benveniste and Mayer, 1993). Glycine-independent potentiation of NMDA receptor function is dependent on the NR1 subunit (Durand *et al.*, 1993) as well as on the NR2B, but not the NR2A, NR2C or NR2D subunits (Williams *et al.*, 1994; Williams, 1995). However, glycine-dependent potentiation of the NMDA receptor does not involve the NR1 subunit but does occur at both NR2A- and NR2B-containing receptors. The open channel block of the NMDA receptor by polyamines is weakly voltage dependent (Rock and Macdonald, 1992; Benveniste and Mayer, 1993) and appears to display the same selectivity as Mg^{2+} for NMDA receptors containing the

NR2C subunit (Williams *et al.*, 1994; Williams, 1995). Thus, it would appear that polyamines interact with at least three sites on the NMDA receptor complex.

The phenylethanolamine, ifenprodil (Figure 4.15), which was originally developed as an α_1 adrenoreceptor antagonist, has since been shown to be a potent non-competitive NMDA receptor antagonist (Carter *et al.*, 1989). The glycine-dependent high affinity block of the NMDA receptor by ifenprodil has been reported to require *N*-terminal residues on the NR2B subunit (Williams, 1993; Gallagher *et al.*, 1996). Indeed, ifenprodil displays remarkable selectivity for the NR2B subunit showing a greater than a hundredfold higher affinity for NR2B over NR2A (Williams, 1993) and very low affinity for NR2C or NR2D (Williams, 1995). A range of 1,4-disubstituted piperidine analogues of ifenprodil such as eliprodil (Avenet *et al.*, 1997), Ro 25–6981 (Fischer *et al.*, 1997; Mutel *et al.*, 1998), CP-101,606 (Chenard *et al.*, 1995), Co 101676 (Tamiz *et al.*, 1998a) and Co 101526 (Tamiz *et al.*, 1998b) (for structures see Figure 4.15) have also been reported to be NR2B selective NMDA receptor antagonists. A structure–activity study on a series of ifenprodil analogues suggested that the piperidine ring and alkyl chain substituents are not necessary for potent antagonist activity but that a phenolic hydroxyl group, a basic nitrogen atom and the distance between the two phenyl rings were the main factors that determined potency (Tamiz *et al.*, 1998a). A number of studies have concentrated on further conformational restriction of ifenprodil, and this approach has led to potent NR2B-selective antagonists such as CP-283,097 (Butler *et al.*, 1998), hydroxypropyne (compound XIII, Figure 4.15) and hydroxybutyne (compound XIV, Figure 4.15) analogues (Wright *et al.*, 1999) and a tetrahydropyridoindole (compound XV, Figure 4.15) (Tamiz *et al.*, 1999). Ifenprodil and analogues have been shown to possess anticonvulsant (Kotlinska and Liljequist, 1996), neuroprotective (Chenard *et al.*, 1995; Fischer *et al.*, 1997) and antinociceptive (Bernardi *et al.*, 1996) properties without showing the psychotomimetic effects associated with competitive antagonists and some channel blockers. One explanation that has been suggested for the improved therapeutic ratio is the NR2B subunit selectivity displayed by these compounds. Recently, it has been suggested that ifenprodil binding leads to a shift of the pKa for proton block of the NMDA receptor to more alkaline values, and thus at physiological pH a larger proportion of the receptors are inhibited (Mott *et al.*, 1998; Dingledine *et al.*, 1999). This has important consequences for the use of ifenprodil analogues as neuroprotective agents as in ischaemic tissue the pH is lower than 7 and low pH increases the potency of some ifenprodil analogues when tested on recombinant receptors or in NMDA-induced toxicity assays (see Dingledine *et al.*, 1999). It is conceivable that ifenprodil analogues could be designed that are inactive at physiological pH but block NMDA receptor function when the pH lowers after an ischaemic event.

Although ifenprodil is selective for NR2B-containing NMDA receptors it is also a potent σ ligand (Schoemaker *et al.*, 1990) and a moderately potent blocker of L-, N- and P-type neuronal calcium channels (Church *et al.*, 1994). Newer analogues of ifenprodil are more selective, Ro 25–6981 does not bind to σ receptors, α_1-adrenergic, or serotonergic binding sites but binds to NR2B-containing NMDA receptors with high affinity (Mutel *et al.*, 1998). Radiolabelled Ro 25–6981 has been used to probe the distribution of NR2B-containing NMDA receptors in the CNS and to characterise the binding of a range of ifenprodil analogues to rat brain membranes (Mutel *et al.*, 1998).

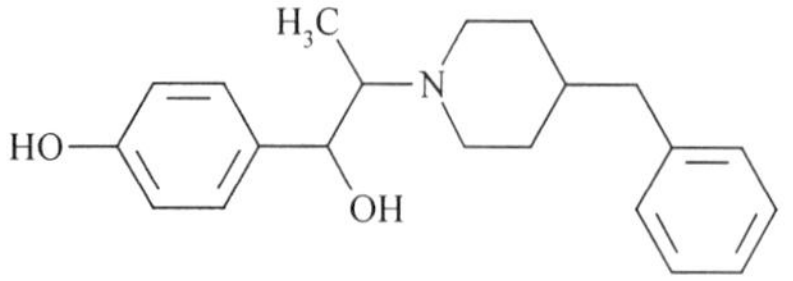

Ifenprodil (0.02)

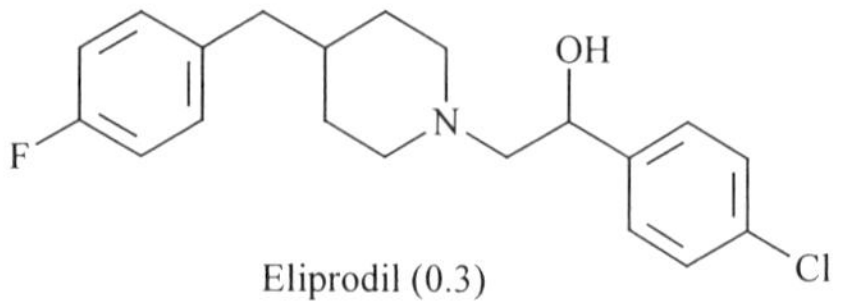

Eliprodil (0.3)

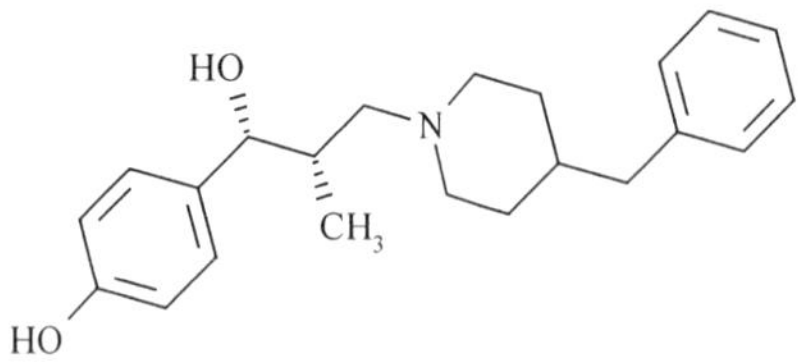

Ro 25-6981 (0.006)

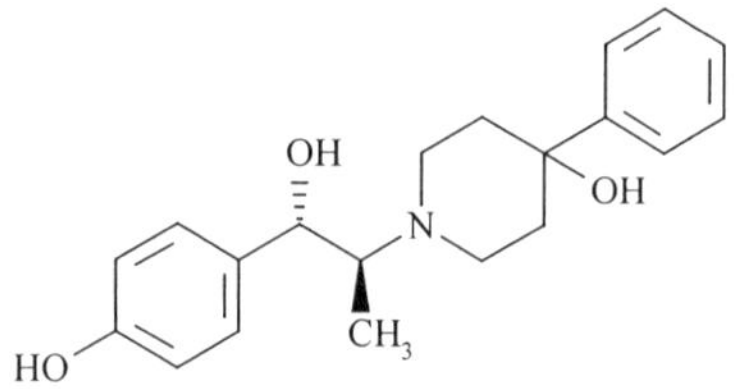

CP-101,606 (0.008)

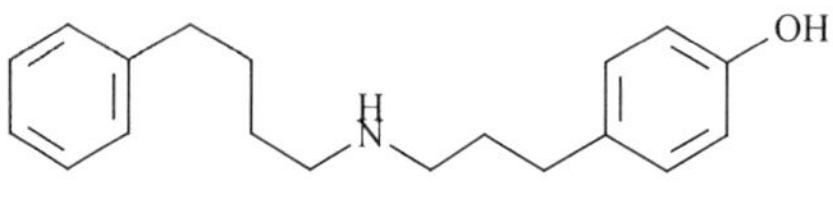

Co 101676 (NR2B IC_{50} 0.020)

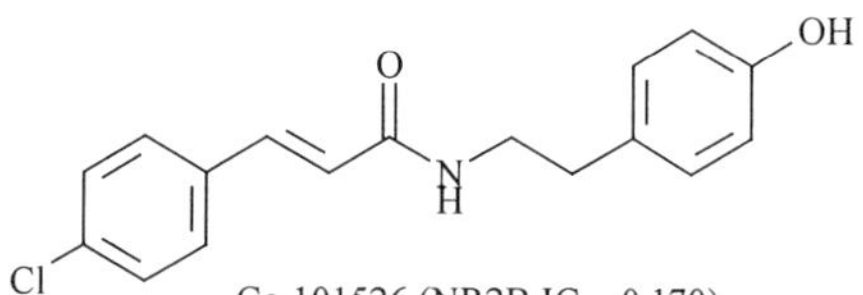

Co 101526 (NR2B IC_{50} 0.170)

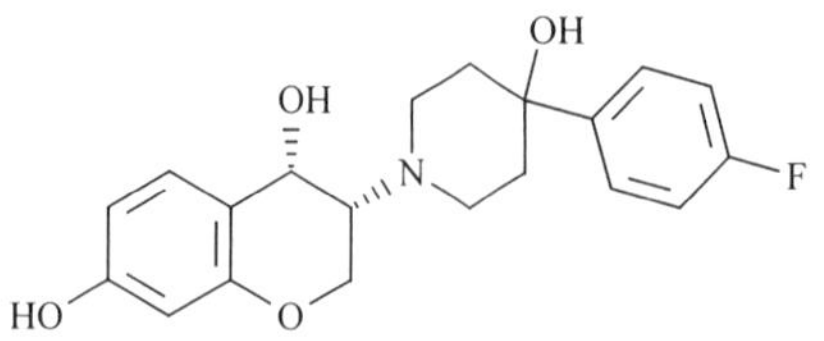

CP-283,097 (0.018, [^{3}H] CP-101,606)

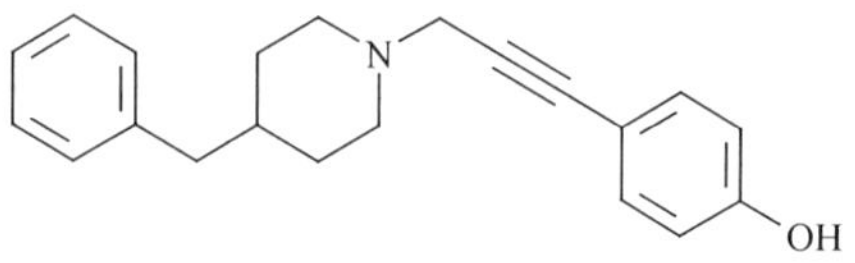

XIII (NR1a/NR2B IC_{50} 0.10)

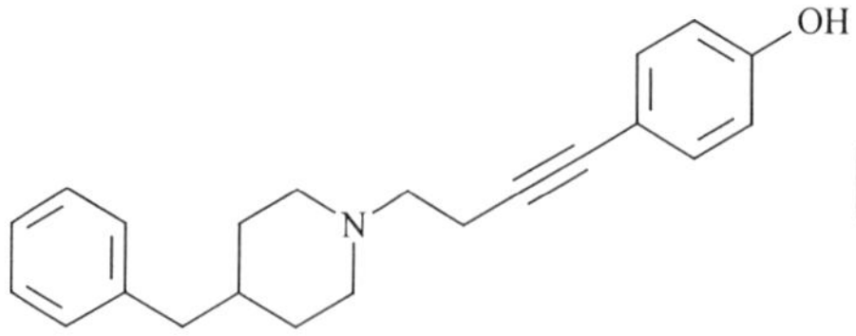

XIV (NR1a/NR2B IC_{50} 0.17)

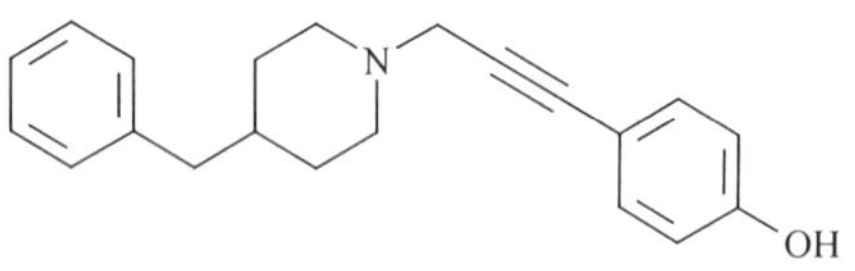

XIII (NR1a/NR2B IC_{50} 0.10)

Figure 4.15 Structures of non-competitive NMDA receptor antagonists based on ifenprodil. K_i values (μM) for displacement of [^{3}H]Ro 25-6981 from rat brain membranes are given in parenthesis (unless otherwise stated). Values taken from Mutel *et al.* (1998). IC_{50} (μM) value for Co 101676 and CP-283,097 taken from Tamiz *et al.* (1999) and Butler *et al.* (1998), respectively. IC_{50} (μM) values for Co 101526 (Tamiz *et al.*, 1998b), XIII, XIV (Wright *et al.*, 1999) and XV (Tamiz *et al.*, 1999) are from electrophysiological assays on *Xenopus* oocytes expressing NR1a/NR2B receptors.

Two peptides, Conantokins-T and G (Con-T and Con-G respectively) (Haack *et al.*, 1990; Mena *et al.*, 1990), isolated from the venom of marine snails of the genus *Conus*, have been reported to be non-competitive inhibitors of the positive modulatory effects of polyamines at NMDA receptors (Skolnick *et al.*, 1992; Chandler *et al.*, 1993). Structure–activity studies in which individual amino acid residues in the 17-amino acid polypeptide Con-G were replaced suggested that γ-carboxyglutamate (Gla) residues at positions 3 and 4 are required for antagonist activity (Zhou *et al.*, 1996). In addition, a synthetic analogue of Con-G in which the amino acid at position 7 of the polypeptide chain was replaced by an alanine residue (Con-G[A7] was fourfold more potent than the parent peptide as an inhibitor of spermine-stimulated [^{3}H]MK-801 binding (IC_{50} value ~45 nM). A recent study has confirmed that the Gla-Ala-7 substitution increases potency and has also shown that Leu5 is important for antagonist activity (Nielsen *et al.*, 1999). Binding studies on human NMDA receptors confirmed the potent inhibition of polyamine enhancement of [^{3}H]MK-801 seen in rat brain. A novel con-G analogue, Con-G[K7], displayed differential binding in different brain areas, showing greater affinity for NMDA receptors in the superior temporal gyrus than for the mid-frontal gyrus. It was therefore concluded that the Con-G binding site may be a target for the pharmacological discrimination of NMDA receptor subtypes (Nielsen *et al.*, 1999).

Concluding remarks

Since the discovery of the first selective NMDA receptor agonists and competitive antagonists by Watkins and co-workers much progress has been made in the design of potent and selective systemically active competitive antagonists. The lack of success of such antagonists in clinical trials prompted researchers to look for alternative strategies of inhibiting NMDA receptor function. This has led to an explosion of interest in the design of antagonists binding to other sites on the NMDA receptor complex such as glycine site antagonists, low affinity open channel blockers and ifenprodil analogues. The outcome of clinical trials involving such compounds is eagerly awaited. Another strategy for limiting side effects has been to identify subtype selective NMDA receptor antagonists, such as ifenprodil analogues, which are selective for NR2B-containing NMDA receptors. Subtype selective competitive NMDA receptor antagonists are starting to emerge and it will be interesting to see if these compounds display an improved side-effect profile.

References

Aebischer, B., Frey, P., Haerter, H.P., Herrling, P.L., Mueller, W.A., Olverman, H.J. and Watkins, J.C. (1989) 'Synthesis and NMDA antagonist properties of the enantiomers of 4-(3-phosphonopropyl)piperazine-2-carboxylic acid (CPP) and of the unsaturated analogue (E)-4-(3-phosphonoprop-2-enyl)piperazine-2-carboxylic acid (CPP-ene)', *Helv. Chim. Acta* **72**: 1043–1051.

Allan, R.D., Hanrahan, J.R., Hambley, T.W., Johnston, G.A.R., Mewett, K.N. and Mitrovic, A.D. (1990) 'Synthesis and activity of a potent *N*-methyl-D-aspartic acid agonist, *trans*-1-aminocyclobutane-1,3-dicarboxylic acid, and related phosphonic and carboxylic acids', *J. Med. Chem.* **33**: 2905–2915.

Anis, N.A., Berry, S.C., Burton, N.R. and Lodge, D. (1983) 'The dissociative anaesthetics, ketamine and phencyclidine, selectively reduce excitation of central mammalian neurons by N-methyl-D-aspartate', *Br. J. Pharmacol.* **79**: 565–575.

Auberson, Y.P., Acklin, P., Bischoff, S., Moretti, R., Ofner, S., Schmutz, M. and Veenstra, S.J. (1999) 'N-phosphonoalkyl-5-aminomethylquinoxaline-2,3-diones: *in vivo* active AMPA and NMDA (glycine) antagonists', *Bioorg. Med. Chem. Lett.* **9**: 249–254.

Avenet, P., Léonardon, J., Besnard, F., Graham, D., Depoortere, H. and Scatton, B. (1997) 'Antagonist properties of eliprodil and other NMDA receptor antagonists at rat NR1A/NR2A and NR1A/NR2B receptors expressed in *Xenopus* oocytes', *Neurosci. Lett.* **223**: 133–136.

Balcar, V.J. and Johnston, G.A.R. (1972) 'The structural specificity of the high affinity uptake of L-glutamate and L-aspartate by rat brain slices' *J. Neurochem.* **19**: 2657–2666.

Balsamini, C., Bedini, A., Diamantini, G., Spadoni, G., Tontini, A., Tarzia, G., Di Fabio, R., Feriani, A., Reggiani, A., Tedesco, G. and Valigi, R. (1998) '(*E*)-3-(2-(*N*-Phenylcarbamoyl)vinyl)pyrrole-2-carboxylic acid derivatives. A novel class of glycine site antagonists', *J. Med. Chem.* **41**: 808–820.

Baudy, R.B., Greenblatt, L.P., Jirkovsky, I.L., Conklin, M., Russo, R.J., Bramlett, D.R., Emrey, T.A., Simmonds, J.T., Kowal, D.M., Stein, R.P. and Tasse, R.P. (1993) 'Potent quinoxaline-spaced phosphono α-amino acids of the AP6 type as competitive NMDA antagonists: synthesis and biological evaluation', *J. Med. Chem.* **36**: 331–342.

Benke, D., Honer, M., Heckendorn, R., Pozza, M.F., Allgeier, H., Angst, C. and Mohler, H. (1999) '[^{3}H]CGP 61594, the first photoaffinity ligand for the glycine site of NMDA receptors', *Neuropharmacology* **38**: 233–242.

Benveniste, M. and Mayer, M.L. (1993) 'Multiple effects of spermine on *N*-methyl-D-aspartic acid receptor responses of rat cultured hippocampal neurons', *J. Physiol.* **464**: 131–163.

Bernardi, M., Bertolini, A., Szczawinska, K. and Genedani, S. (1996) 'Blockade of the polyamine site of NMDA receptors produces antinociception and enhances the effect of morphine, in mice', *Eur. J. Pharmacol.* **298**: 51–55.

Bigge, C.F. (1993) 'Structural requirements for the development of potent *N*-methyl-D-aspartic acid (NMDA) receptor antagonists', *Biochemical Pharmacology* **45**: 1547–1561.

Bigge, C.F., Malone, T.C., Boxer, P.A., Nelson, C.B., Ortwine, D.F., Schelkun, R.M., Retz, D.M., Lescosky, L.J., Borosky, S.A., Vartanian, M.G., Schwarz, R.D., Campbell, G.W., Robichaud, L.J. and Wätjen, F. (1995) 'Synthesis of 1,4,7,8,9,10-hexahydro-9-methyl-6-nitropyrido[3,4-*f*]-quinoxaline-2,3-dione and related quinoxalinediones: characterisation of α-amino-3-hydroxy-5-methyl-4-isoxazolepropionic acid (and N-methyl-D-aspartate) receptor and anticonvulsant activity', *J. Med. Chem.* **38**: 3720–3740.

Birch, P.J., Grossman, C.J. and Hayes, A.G. (1988) 'Kynurenic acid antagonises responses to NMDA via an action at the strychnine-insensitive glycine receptor', *Eur. J. Pharmacol.* **154**: 85–87.

Biscoe, T.J., Davies, J, Dray, A., Evans, R.H., Francis, A.A., Martin, M.R. and Watkins, J.C. (1977a) 'Depression of synaptic excitation and of amino acid-induced excitatory responses of spinal neurones by D-α-aminoadipate, α,ε-diaminopimelic acid and HA-966', *Eur. J. Pharmacol.* **45**: 315–316.

Biscoe, T.J., Evans, R.H., Francis, A.A., Martin, M.R., Watkins, J.C., Davies, J. and Dray, A. (1977b) 'D-α-aminoadipate as a selective antagonist of amino acid-induced and synaptic excitation of mammalian spinal neurones', *Nature*, **270**: 743–745.

Blanpied, T.A., Boeckman, F.A., Aizenman, E. and Johnson, J.W. (1997) 'Trapping channel block of NMDA-activated responses by amantadine and memantine', *J. Neurophysiol.* **77**: 309–323.

Bordi, F., Quartaroli, M., Reggiani, A. and Trist, D.G. (1998) 'Suppression of noxious stimulus-evoked activity by the glycine site antagonist GV196771A in the posterolateral nucleus of the thalamus of rats with an experimental peripheral mononeuropathy', *Naunyn-Schmeidbergs Archives of Pharmacology* **358**: P1120.

Bräuner-Osborne, H., Nielsen, B., Stensbøl, T.B., Johansen, T.N., Skjærbæk, N. and Krogsgaard-Larsen, P. (1997) 'Molecular pharmacology of 4-substituted glutamic acid analogues at ionotropic and metabotropic excitatory amino acid receptors', *Eur. J. Pharmacol.* **335**: R1–R3.

Brown III, J.C., Tse, H.W., Skifter, D.A., Christie, J.M., Andaloro, V.J., Kemp, M.C., Watkins, J.C., Jane, D.E. and Monaghan, D.T. (1998) '[^{3}H]Homoquinolinate binds to a subpopulation of NMDA receptors and to a novel binding site', *J. Neurochem.* **71**: 1464–1470.

Buller, A.L., Larson, H.C., Schneider, B.E., Beaton, J.A., Morrisett, R.A. and Monaghan, D.T. (1994) 'The molecular basis of NMDA receptor subtypes: native receptor diversity is predicted by subunit composition', *J. Neurosci.* **14**: 5471–5484.

Buller, A.L. and Monaghan, D.T. (1997) 'Pharmacological heterogeneity of NMDA receptors: characterization of NR1a/NR2D heteromers expressed in Xenopus oocytes', *Eur. J. Pharmacol.* **320**: 87–94.

Butler, T.W., Blake, J.F., Bordner, J., Butler, P., Chenard, B.L., Collins, M.A., DeCosta, D., Ducat, M.J., Eisenhard, M.E., Menniti, F.S., Pagnozzi, M.J., Sands, S.B., Segelstein, B.E., Volberg, W., White, W.F. and Zhao, D. (1998) '(3R,4S)-3-[4-(4-Fluorophenyl)-4-hydroxypiperidibn-1-yl]chroman-4,7-diol: a conformationally restricted analogue of the NR2B subtype-selective NMDA antagonist (1S,2S)-1-(4-hydroxyphenyl)-2-(4-hydroxy-4-phenylpiperidino)-1-propanol', *J. Med. Chem.* **41**: 1172–1184.

Cai, S.X., Kehr, S.M., Zhou, Z.-L., Ilyin, V., Espitia, S.A., Tran, M., Hawkinson, J.E., Woodward, R.M., Weber, E. and Keana, J.F.W. (1997) 'Structure–activity relationships of alkyl- and alkoxy-substituted 1,4-dihydroquinoxaline-2,3-diones: potent and systemically active antagonists for the glycine site of the NMDA receptor', *J. Med. Chem.* **40**: 730–738.

Carignani, C., Ugolini, A., Pinnola, V., Belardetti, F., Trist, D.G. and Corsi, M. (1998) 'NMDA receptor subunit characterization of the glycine site antagonist GV196771A and its action on the spinal cord wind-up', *Naunyn-Schmeidbergs Archives of Pharmacology* **358**: P1119.

Carter, C., Rivy, J.P. and Scatton, B. (1989) 'Ifenprodil and SL-82.0715 are antagonists at the polyamine site of the N-methyl-D-aspartate (NMDA) receptor', *Eur. J. Pharmacol.* **164**: 611–612.

Catarzi, D., Colotta, V., Varano, F., Cecchi, L., Filacchioni, G., Galli, A. and Costagli, C. (1999) '4,5-Dihydro-1,2,4-triazolo[1,5-a]quinoxalin-4-ones: excitatory amino acid antagonists with combined glycine/NMDA and AMPA receptor affinity', *J. Med. Chem.* **42**: 2478–2484.

Chandler, P., Pennington, M., Maccecchini, M.-L., Nashed, N.T. and Skolnick, P. (1993) 'Polyamine-like action of peptides derived from conantokin-G, an *N*-methyl-D-aspartate (NMDA) antagonist', *J. Biol. Chem.* **268**: 17,173–17,178.

Chenard, B.L., Bordner, J., Butler, T.W., Chambers, L.K., Collins, M.A., Decosta, D.L., Ducat, M.F., Dumont, M.L., Fox, C.B., Mena, E.E., Menniti, F.S., Nielsen, J., Pagnozzi, M.J., Richter, K.E.G., Ronau, R.T., Shalaby, I.A., Stemple, J.Z. and White, W.F. (1995) '(1S,2S)-1-(4-Hydroxyphenyl)-2-(4-hydroxy-4-phenylpiperidino)-1-propanol – a potent new neuroprotectant which blocks N-methyl-D-aspartate responses', *J. Med. Chem.* **38**: 3138–3145.

Church, J., Fletcher, E.J., Baxter, K. and Macdonald, J.F. (1994) 'Blockade by ifenprodil of high voltage-activated Ca^{2+} channels in rat and mouse cultured hippocampal pyramidal neurons – comparison with N-methyl-D-aspartate receptor antagonist actions', *Br. J. Pharmacol.* **113**: 499–507.

Collingridge, G.L. and Lester, R.A.J. (1989) 'Excitatory amino acid receptors in the vertebrate central nervous system', *Pharmacological Reviews* **40**: 143–210.

Cordi, A., Lacoste, J.-M., Audinot, V. and Millan, M. (1999) 'Design, synthesis and structure–activity relationships of novel strychnine-insensitive glycine receptor ligands', *Bioorg. Med. Chem. Lett.* **9**: 1409–1414.

Croucher, M.J., Collins, J.F. and Meldrum, B.S. (1982) 'Anticonvulsant action of excitatory amino acid antagonists', *Science* **216**: 899–901.

Curtis, D.R. and Watkins, J.C. (1960) 'The excitation and depression of spinal neurones by structurally related amino acids', *J. Neurochem.* **6**: 117–141.

Curtis, D.R. and Watkins, J.C. (1963) 'Acidic amino acids with strong excitatory actions on mammalian neurones', *J. Physiol.* **166**: 1–14.

Das, S., Sasaki, Y.F., Rothe, T., Premkumar, L.S., Takasu, M., Crandall, J.E., Dikkes, P., Conner, D.A., Rayudu, P.V., Cheung, W., Vincent Chen, H.-S., Lipton, S.A. and Nakanishi, N. (1998) 'Increased NMDA current and spine density in mice lacking the NMDA receptor subunit NR3A', *Nature (Lond.)* **393**: 377–381.

Davies, J., Evans, R.H., Herrling, P.L., Jones, A.W., Olverman, H.J., Pook, P. and Watkins, J.C.

(1986) 'CPP a new and selective NMDA antagonist. Depression of central neuron responses, affinity for [^{3}H]D-AP5 binding sites on brain membranes and anticonvulsant activity', *Brain Research* **382**: 169–173.

Di Fabio, R., Capelli, A.M., Conti, N., Cugola, A., Donati, D., Feriani, A., Gastaldi, P., Gaviraghi, G., Hewkin, C.T., Micheli, F., Missio, A., Mugnaini, M., Pecunioso, A., Quaglia, A.M., Ratti, E., Rossi, L., Tedesco, G., Trist, D.G. and Reggiani, A. (1997) 'Substituted indole-2-carboxylates as *in vivo* potent antagonists acting at the strychnine-insensitive glycine binding site', *J. Med. Chem.* **40**: 841–850.

Dingledine, R., Borges, K., Bowie, D. and Traynelis, S.F. (1999) 'The glutamate receptor ion channels', *Pharmacological Reviews* **51**: 7–61.

Donevan, S.D., Beg, A., Gunther, J.M. and Twyman, R.E. (1998) 'The methylglutamate, SYM 2081, is a potent and highly selective agonist at kainate receptors', *J. Pharm. Exp. Ther.* **285**: 539–545.

Durand, G.M., Benett, M.V. and Zukin, R.S. (1993) 'Splice variants of the N-methyl-D-aspartate receptor NR1 identify domains involved in regulation by polyamines and protein kinase C', *Proc. Natl. Acad. Sci. USA* **90**: 6731–6735.

Evans, R.H., Francis, A.A., Jones, A.W., Smith, D.A.S. and Watkins, J.C. (1982) 'The effects of a series of ω-phosphonic α-carboxylic amino acids on electrically evoked and amino acid induced responses in isolated spinal cord preparations', *Br. J. Pharmacol.* **75**: 65–75.

Evans, R.H., Francis, A.A. and Watkins, J.C. (1977) 'Selective antagonism by Mg^{2+} of amino acid-induced depolarization of spinal neurones', *Experientia* **33**: 489–491.

Evans, R.H., Francis, A.A. and Watkins, J.C. (1978) 'Mg^{2+}-like selective antagonism of excitatory amino acid-induced responses by α,ε-diaminopimelic acid, D-α-aminoadipate and HA-966 in isolated spinal cord of frog and immature rat', *Brain Research* **148**: 536–542.

Fagg, G.E. and Baud, J. (1988) 'Characterisation of NMDA receptor-ionophore complexes in the brain', in D. Lodge (ed.) *Excitatory Amino Acids in Health and Disease*, Chichester: John Wiley, pp. 63–90.

Fagg, G.E., Olpe, H.-R., Pozza, M.F., Baud, J., Steinmann, M., Schmutz, M., Portet, C., Baumann, P., Thedinga, K., Bittiger, H., Allgeier, H., Heckendorn, R., Angst, C., Brundish, D. and Dingwall, J.G. (1990) 'CGP 37849 and CGP 39551: novel and competitive *N*-methyl-D-aspartate receptor antagonists with oral activity', *Br. J. Pharmacol.* **99**: 791–797.

Ferkany, J.W., Hamilton, G.S., Patch, R.J., Huang, Z., Borosky, S.A., Bednar, D.L., Jones, B.E., Zubrowski, R., Willetts, J. and Karbon, E.W. (1993a) 'Pharmacological profile of NPC 17742 [2R,4R,5S-(2-amino-4,5-(1,2-cyclohexyl)-7-phosphonoheptanoic acid)], a potent, selective and competitive *N*-methyl-D-aspartate receptor antagonist', *J. Pharm. Exp. Ther.* **264**: 256–264.

Ferkany, J.W., Willetts, J., Borosky, S.A., Clissold, D.B., Karbon, E.W. and Hamilton, G.S. (1993b) 'Pharmacology of (2R,4R,5S)-2-amino-4,5-(1,2-cyclohexyl)-7-phosphonoheptanoic acid (NPC 17742); a selective, systemically active, competitive NMDA antagonist', *Bioorg. Med. Chem. Lett.* **3**: 33–38.

Ferrer-Montiel, A.V., Merino, J.M., Planells-Cases, R., Sun, W. and Montal, M. (1998) 'Structural determinants of the blocker binding site in glutamate and NMDA receptor channels', *Neuropharmacology* **37**: 139–147.

Fischer, G., Mutel, V., Trube, G., Malherbe, P., Kew, J.N.C., Mohaesi, E., Heitz, M.P. and Kemp, J.A. (1997) 'Ro 25-6981, a highly potent and selective blocker of NMDA receptors containing the NR2B subunit. Characterization *in vitro*', *J. Pharmacol. Exp. Ther.* **283**: 1285–1292.

Gallagher, M.J., Huang, H., Pritchett, D.R. and Lynch, D.R. (1996) 'Interactions between ifenprodil and the NR2B subunit of the *N*-methyl-D-aspartate receptor', *J. Biol. Chem.* **271**: 9603–9611.

Gaoni, Y., Chapman, A.G., Parvez, N., Pook, P.C.-K., Jane D.E. and Watkins, J.C. (1994) 'Synthesis, NMDA receptor antagonist activity, and anticonvulsant action of 1-aminocyclobutanecarboxylic acid derivatives', *J. Med. Chem.* **37**: 4288–4296.

Gill, R., Foster, A.C. and Woodruff, G.N. (1987) 'Systemic administration of MK-801 protects against ischaemia-induced hippocampal neurodegeneration in the gerbil', *J. Neurosci.* **7**: 3343–3349.

Haack, J.A., Rivier, J., Parks, T.N., Mena, E.E., Cruz, L.J. and Olivera, B.M. (1990) 'Conantokin-T: a gamma-carboxyglutamate containing peptide with *N*-methyl-D-aspartate antagonist activity', *J. Biol. Chem.* **265**: 6025–6029.

Hall, J.G., Mclennan, H. and Wheal, H.V. (1977) 'The actions of certain amino acids on neuronal excitants', *J. Physiol.* **272**: 52–53P.

Hays, S.J., Bigge, C.F., Novak, P.M., Drummond, J.T., Bobovski, T.P., Rice, M.J., Johnson, G., Brahce, L.J. and Coughenour, L.L. (1990) 'New and versatile approaches to the synthesis of CPP-related competitive NMDA antagonists. Preliminary structure–activity relationships and pharmacological evaluation', *J. Med. Chem.* **33**: 2916–2924.

Heckendorn, R., Allgeier, H., Baud, J., Gunzenhauser, W. and Angst, C. (1993) 'Synthesis and binding properties of 2-amino-5-phosphono-3-pentenoic acid photoaffinity ligands as probes for the glutamate recognition site of the NMDA receptor', *J. Med. Chem.* **36**: 3721–3726.

Henderson, G., Johnson, J.W. and Ascher, P. (1990) 'Competitive antagonists and partial agonists at the glycine modulatory site of the mouse *N*-methyl-D-aspartate receptor', *J. Physiol. (Lond.)* **430**: 189–212.

Herrling, P.L. (1997) *Excitatory Amino Acids – Clinical Results with Antagonists*, London: Academic Press.

Hirai, H., Kirsch, J., Laube, B., Betz, H. and Kuhse, J. (1996) 'The glycine binding site of the N-methyl-D-aspartate receptor subunit NR1: identification of novel determinants of co-agonist potentiation in the extracellular M3–M4 loop region', *Proc. Natl. Acad. Sci. USA* **93**: 6031–6036.

Honer, M., Benke, D., Laube, B., Kuhse, J., Heckendorn, R., Allgeier, H., Angst, C., Monyer, H., Seeburg, P.H., Betz, H. and Mohler, H. (1998) 'Differentiation of glycine antagonist sites of *N*-methyl-D-aspartate receptor subtypes: Preferential interaction of CGP 61594 with NR1/2B receptors', *J. Biol. Chem.* **273**: 11,158–11,163.

Hood, W.F., Compton, R.P. and Monahan, J.B. (1989a) 'D-cycloserine: a ligand for the N-methyl-D-aspartate coupled glycine receptor has partial agonist characteristics', *Neurosci. Lett.* **98**: 91–95.

Hood, W.F., Sun, E.T., Compton, R.P. and Monahan, J.B. (1989b) '1-Aminocyclobutane-1-carboxylate (ACBC) a specific antagonist of the N-methyl-D-aspartate receptor coupled glycine receptor', *Eur. J. Pharmacol.* **161**: 281–282.

Huettner, J.E. and Bean, B.P. (1988) 'Block of *N*-methyl-D-aspartate-activated current by the anticonvulsant MK-801: selective binding to open channels', *Proc. Natl. Acad. Sci. USA* **85**: 1307–1311.

Hutchison, A.J., Williams, M., Angst, C., de Jesus, R., Blanchard, L., Jackson, R.H., Wilusz, E.J., Murphy, D.E., Bernard, P.S., Schneider, J., Campbell, T., Guida, W. and Sills, M.A. (1989) '4-(Phosphonoalkyl)- and 4-(Phosphonoalkenyl)-2-piperidinecarboxylic acids: synthesis, activity at *N*-methyl-D-aspartic acid receptors, and anticonvulsant activity', *J. Med. Chem.* **32**: 2171–2178.

Iversen, L.L. and Kemp, J.A. (1994) 'Non-competitive NMDA antagonists as drugs', in G.L. Collingridge. and J.C. Watkins (eds) *The NMDA Receptor*, Oxford: Oxford University Press, pp. 469–486.

Jane, D.E., Olverman, H.J. and Watkins, J.C. (1994) 'Agonists and competitive antagonists: structure–activity and molecular modelling studies', in G.L. Collingridge and J.C. Watkins (eds) *The NMDA Receptor*, Oxford: Oxford University Press, pp. 31–104.

Jane, D.E., Thomas, N.K., Tse, H.-W. and Watkins, J.C. (1996) 'Potent antagonists at the L-AP4- and (1S,3S)-ACPD-sensitive presynaptic metabotropic glutamate receptors in the neonatal rat spinal cord', *Neuropharmacology* **35**: 1029–1035.

Johnson, G. and Ornstein P.L. (1996) 'Competitive NMDA antagonists – a comprehensive analysis of molecular biological, structure activity and molecular modelling relationships', *Current Pharmaceutical Design* **2**: 331–356.

Johnson, J.W. and Ascher, P. (1987) 'Glycine potentiates the NMDA response in cultured mouse brain neurons', *Nature (Lond.)* **325**: 529–531.

Kawai, M., Horikawa, Y., Ishihara, T., Shimamoto, K. and Ohfune, Y. (1992) '2-(Carboxycyclopropyl)glycines: binding, neurotoxicity and induction of intracellular free Ca^{2+} increase', *Eur. J. Pharmacol.* **211**: 195–202.

Kinney, W.A. and Garrison, D.C., European Patent Application, 1992, 496,561 A2.

Kinney, W.A., Lee, N.E., Garrison, D.T., Podlesney Jr., E.J., Simmonds, J.T., Bramlett, D., Notvest, R.R., Kowal, D.M. and Tasse, R.P. (1992) 'Bioisosteric replacement of the alpha-amino carboxylic-acid functionality in 2-amino-5-phosphonopentanoic acid yields unique 3,4-diamino-3-cyclobutene-1,2-dione containing NMDA antagonists', *J. Med. Chem.* **35**: 4720–4726.

Kleckner, N.W. and Dingledine, R. (1988) 'Requirement for glycine in activation of NMDA receptors expressed in Xenopus oocytes', *Science* **241**: 835–837.

Kotlinska, J. and Liljequist, S. (1996) 'Oral administration of glycine and polyamine receptor antagonists blocks ethanol withdrawal seizures', *Psychopharmacology* **127**: 238–244.

Kulagowski, J.J., Baker, R., Curtis, N.R., Leeson, P.D., Mawer, I.M., Moseley, A.M., Ridgill, M.P., Rowley, M., Stansfield, I., Foster, A.C., Grimwood, S., Hill, R.G., Kemp, J.A., Marshall, G.R., Saywell, K.L. and Tricklebank, M.D. (1994) '3′-(Arylmethyl)- and 3′-(Aryloxy)-3-phenyl-4-hydroxyquinolin-2(1H)-ones: orally active antagonists of the glycine site of the NMDA receptor', *J. Med. Chem.* **37**: 1402–1405.

Kutsuwada, T., Kashiwabuchi, N., Mori, H., Sakimura, K., Kushiya, E., Araki, K., Meguro, H., Masaki, H., Kumanishi, T., Arakawa, M. and Mishina, M. (1992) 'Molecular diversity of the NMDA receptor channel', *Nature (Lond.)* **358**: 36–41.

Kyle, D.J., Patch, R.J., Karbon, E.W. and Ferkany, J.W. (1992) 'NMDA receptors: heterogeneity and agonism', in P. Krogsgaard-Larsen and J.J. Hansen (eds) *Excitatory Amino Acid Receptors. Design of Agonists and Antagonists*, Chichester: Ellis Horwood Ltd, pp. 121–162.

Lanthorn, T.H., Hood, W.F., Watson, G.B., Compton, R.P., Rader, R.K., Gaoni, Y. and Monahan, J.B. (1990) '*cis*-2,4-methanoglutamate is a potent and selective *N*-methyl-D-aspartate receptor agonist', *Eur. J. Pharmacol.* **182**: 397–404.

Laube, B., Hirai, H., Sturgess, M., Betz, H. and Kuhse, J. (1997) 'Molecular determinants of agonist discrimination by NMDA receptor subunits: analysis of the glutamate binding site on the NR2B subunit', *Neuron* **18**: 493–503.

Laube, B., Kuhse, J. and Betz, H. (1998) 'Evidence for a tetrameric structure of recombinant NMDA receptors', *J. Neurosci.* **18**: 2954–2961.

Leeson, P.D., Baker, R., Carling, R.W., Curtis, N.R., Moore, K.W., Williams, B.J., Foster, A.C., Donald, A.E., Kemp, J.A. and Marshall, G.R. (1991) 'Kynurenic acid derivatives. Structure–activity relationships for excitatory amino acid antagonism and identification of potent and selective antagonists at the glycine site on the *N*-methyl-D-aspartate receptor', *J. Med. Chem.* **34**: 1243–1252.

Leeson, P.D., Carling, R.W., James, K., Smith, J.D., Moore, K.W., Wong, E.H.F. and Baker, R. (1990) 'Role of hydrogen bonding in ligand interaction with the *N*-methyl-D-aspartate receptor ion channel', *J. Med. Chem.* **33**: 1296–1305.

Leeson, P.D., Carling, R.W., Moore, K.W., Moseley, A.M., Smith, J.D., Stevenson, G., Chan, T., Baker, R., Foster, A.C., Grimwood, S., Kemp, J.A., Marshall, G.R. and Hoogsteen, K. (1992) '4-Amido-2-carboxytetrahydroisoquinolines. Structure–activity relationships for antagonism of the glycine site of the NMDA receptor', *J. Med. Chem.* **35**: 1954–1968.

Leeson, P.D. and Iversen, L.L. (1994) 'The glycine site on the NMDA receptor: structure–activity relationships and therapeutic potential', *J. Med. Chem.* **37**: 4053–4067.

Leeson, P.D., Williams, B.J., Rowley, M., Moore, K.W., Baker, R., Kemp, J.A., Priestly, T., Foster, A.C. and Donald, E.A. (1993) 'Derivatives of 1-hydroxy-3-aminopyrrolidin-2-one (HA-966). Partial agonists at the glycine site of the NMDA receptor', *Bioorg. Med. Chem. Lett.* **3**: 71–76.

Lehmann, J., Hutchison, A.J., McPherson, S.E., Mondadori, C., Schmutz, M., Sinton, C.M., Tsai, C., Murphy, D.E., Steel, D.J., Williams, M., Cheney, D.L. and Wood, P.L. (1988) 'CGS 19755, a selective and competitive *N*-methyl-D-aspartate-type excitatory amino acid receptor antagonist', *J. Pharmacol. Exp. Ther.* **246**: 65–75.

Li, J.-H., Bigge, C.F., Williamson, R.M., Borosky, S.A., Vartanian, M.G. and Ortwine, D.F. (1995) 'Potent, orally-active, competitive *N*-methyl-D-aspartate (NMDA) receptor antagonists are substrates for a neutral amino-acid-uptake system in Chinese-hamster ovary cells', *J. Med. Chem.* **38**: 1955–1965.

Lipton, S.A. (1993) 'Prospects for clinically tolerated NMDA antagonists: open channel blockers and alternative redox states of nitric oxide', *Trends Neurosci.* **16**: 527–532.

Lodge, D. and Johnson, K.M. (1990) 'Noncompetitive excitatory amino acid antagonists', *Trends Pharm. Sci.* **11**: 81–86.

Lunn, W.H.W., Schoepp, D.D., Calligaro, D.O., Vasileff, R.T., Heinz, L.J., Salhoff, C.R. and O'Malley, P.J. (1992) 'D,L-Tetrazol-5-ylglycine, a highly potent NMDA agonist: its synthesis and NMDA receptor efficacy', *J. Med. Chem.* **35**: 4608–4612.

Marvizon, J.C.G., Lewin, A.H. and Skolnick, P. (1989) '1-Aminocyclopropane carboxylic acid: a potent and selective ligand for the glycine modulatory site of the N-methyl-D-aspartate receptor complex', *J. Neurochem.* **52**: 992–994.

Mayer, M.L., Benveniste, M. and Patneau, D.K. (1994) 'NMDA receptor agonists and competitive antagonists', in G.L. Collingridge and J.C. Watkins (eds) *The NMDA Receptor*, Oxford: Oxford University Press, pp. 132–146.

Meldrum, B.S. (1985) 'Possible therapeutic applications of antagonists of excitatory amino acid neurotransmitters', *Clinical Science* **68**: 113–122.

Mena, E.E., Gullak, M.F., Pagnozzi, M.J., Richter, K.E., Rivier, J., Cruz, L.J. and Olivera, B.M. (1990) 'Conantokin-G: a novel peptide antagonist to the *N*-methyl-D-aspartic acid (NMDA) receptor', *Neurosci. Lett.* **118**: 241–244.

Monaghan, D.T. and Beaton, J.A. (1992) 'Pharmacologically-distinct NMDA receptor populations of the cerebellum, medial thalamic nuclei, and forebrain', *Mol. Neuropharm.* **2**: 71–75.

Monaghan, D.T., Bridges, R.J. and Cotman, C.W. (1989) 'The excitatory amino acid receptors: their classes, pharmacology and distinct properties in the function of the central nervous system', *Ann. Rev. Pharmacol. Toxicol.* **29**: 365–402.

Monaghan, D.T. and Larsen, H. (1997) 'NR1 and NR2 subunit contributions to *N*-methyl-D-aspartate receptor channel blocker pharmacology', *J. Pharmacol. Exp. Ther.* **280**: 614–620.

Monahan, J.B., Hood, W.F., Compton, R.P., Cordi, A.A., Snyder, J.P., Pellicciari, R. and Natalina, B. (1990) 'Characterisation of D-3,4-cyclopropylglutamates as *N*-methyl-D-aspartate receptor agonists', *Neuroscience Letters* **112**: 328–332.

Monahan, J.B. and Michel, J. (1987) 'Identification and characterisation of an N-methyl-D-aspartate specific L-[^{3}H]glutamate recognition site in synaptic plasma membranes', *J. Neurochem.* **48**: 1699–1708.

Monyer, H., Sprengel, R., Schoepfer, R., Herb, A., Higuchi, M., Lomeli, H., Burnashev, N., Sakmann, B. and Seeburg, P.H. (1992) 'Heteromeric NMDA receptors: molecular and functional distinction of subtypes', *Science* **256**: 1217–1221.

Mott, D.D., Doherty, J.J., Zhang, S., Washburn, M.S., Fendley, M.J., Lyuboslavsky, P., Traynelis, S.F. and Dingledine, R. (1998) 'Enhancement of proton inhibition: A novel mechanism of inhibition of NMDA receptors by phenylethanolamines', *Nat. Neurosci.* **1**: 659–667.

Müller, W., Lowe, D.A., Neijt, H., Urwyler, S., Herrling, P.L., Blaser, D. and Seebach, D. (1992) 'Synthesis and N-methyl-D-aspartate (NMDA) antagonist properties of the enantiomers of α-amino-5-(phosphonomethyl)[1,1′-biphenyl]-3-propanoic acid. Use of a new chiral glycine derivative', *Helv. Chim. Acta* **75**: 855–864.

Murphy, D.E., Hutchison, A.J., Hurt, S.D., Williams, M. and Sills, M.A. (1988) 'Characterization of the binding of [^{3}H]-CGS 19755: a novel *N*-methyl-D-aspartate antagonist with nanomolar affinity in rat brain', *Br. J. Pharmacol.* **95**: 932–938.

Mutel, V., Buchy, D., Klingelschmidt, A., Messer, J., Bleurel, Z., Kemp, J.A. and Richards, J.G. (1998) 'In vitro binding properties in rat brain of [^{3}H]Ro 25–6981, a potent and selective antagonist of NMDA receptors containing NR2B subunits', *J. Neurochem.* **70**: 2147–2155.

Nagata, R., Tanno, N., Kodo, T., Ae, N., Yamaguchi, H., Nishimura, T., Antoku, F., Tatsuno, T., Kato, T., Tanaka, Y. and Nakamura, M. (1994) 'Tricyclic quinoxalinediones: 5,6-dihydro-1*H*-

pyrrolo[1,2,3-*de*]quinoxaline-2,3-diones and 6,7-dihydro-1*H*,5*H*-pyrido[1,2,3-*de*]quinoxaline-2,3-diones as potent antagonists for the glycine binding site of the NMDA receptor', *J. Med. Chem.* **37**: 3956–3968.

Nielsen, K.J., Skjærbæk, N., Dooley, M., Adams, D.A., Mortensen, M., Dodd, P.R., Craik, D.J., Alewood, P.F. and Lewis, R.J. (1999) 'Structure–activity studies of conantokins as human N-methyl-D-aspartate receptor modulators', *J. Med. Chem.* **42**: 415–426.

Nikam, S.S., Cordon, J.J., Ortwine, D.F., Heimbach, T.H., Blackburn, A.C., Vartanian, M.G., Nelson, C.B., Schwarz, R.D., Boxer, P.A. and Rafferty, M.F. (1999) 'Design and synthesis of novel quinoxaline-2,3-dione AMPA/Gly_N receptor antagonists: amino acid derivatives', *J. Med. Chem.* **42**: 2266–2271.

Olverman, H.J., Jones, A.W., Mewett, K.N. and Watkins, J.C. (1988) 'Structure–activity relations of NMDA receptor ligands as studied by their inhibition of [^{3}H]-D-AP5 binding in rat brain membranes', *Neuroscience* **26**: 17–31.

Olverman, H.J., Jones, A.W. and Watkins, J.C. (1984) 'L-glutamate has higher affinity than other amino acids for [^{3}H]-D-AP5 binding sites in rat brain membranes', *Nature (Lond.)* **307**: 460–462.

Olverman, H.J., Monaghan, D.T., Cotman, C.W. and Watkins, J.C. (1986) '[^{3}H]-CPP, a new competitive ligand for NMDA receptors', *Eur. J. Pharmacol.* **131**: 161–162.

Ornstein, P.L., Arnold, M.B., Lunn, W.H.W., Heinz, L.J., Leander, J.D., Lodge, D. and Schoepp, D.D. (1998) 'Heteroatom-substitution as a strategy for increasing the potency of competitive NMDA antagonists', *Bioorg. Med. Chem. Lett.* **8**: 389–394.

Ornstein, P.L. and Klimkowski, V.J. (1992). 'Competitive NMDA receptor antagonists', in P. Krogsgaard-Larsen and J.J. Hansen (eds) *Excitatory Amino Acid Receptors. Design of Agonists and Antagonists*, Chichester: Ellis Horwood Ltd, pp. 183–201.

Ornstein, P.L., Schoepp, D.D., Arnold, M.B., Augenstein, N.K., Lodge, D., Millar, J.D., Chambers, J., Campbell, J., Paschal, J.W., Zimmerman D.M. and Leander, J.D. (1992) '6-substituted decahydroisoquinoline-3-carboxylic acids as potent and selective conformationally constrained NMDA receptor antagonists', *J. Med. Chem.* **35**: 3547–3560.

Ortwine, D.F., Malone, T.C., Bigge, C.F., Drummond, J.T., Humblet, C., Johnson, G. and Pinter, G.W. (1992) 'Generation of N-methyl-D-aspartate agonist and competitive antagonist pharmacophore models. Design and synthesis of phosphonoalkyl-substituted tetrahydroisoquinolines as novel antagonists', *J. Med. Chem.* **35**: 1345–1370.

Parsons, C.G., Danysz, W. and Quack, G. (1999) 'Memantine is a clinically well tolerated *N*-methyl-D-aspartate (NMDA) receptor antagonist – a review of preclinical data', *Neuropharmacology* **38**: 735–767.

Parsons, C.G., Danysz, W., Hartmann, S., Bartmann, A., Gold, M., Kalvinch, I., Piskunova, I. and Rozhkov, E. (1996) 'Novel antagonists of the glycine site of the NMDA receptor: electrophysiological and biochemical characterisation', Poster presented at the 26th Annual Meeting of the Society for Neuroscience, Washington, USA, 16–21 November.

Parsons, C.G., Danysz, W., Quack, G., Hartmann, S., Lorenz, B., Baran, L., Przegalinski, E., Kostowski, W., Krzascik, P., Headley, P.M., Chizh, B. (1997) 'Novel antagonists of the glycine site of the NMDA receptor. Electrophysiological, biochemical and behavioural characterisation', *J. Pharmacol. Exp. Ther.* **283**: 1264–1275.

Parsons, C.G., Quack, G., Bresink, I., Baran, L., Przegalinski, E., Kostowski, W., Krzascik, P., Hartmann, S. and Danysz, W. (1995) 'Comparison of the potency, kinetics and voltage-dependency of a series of uncompetitive NMDA receptor antagonists *in vitro* with anticonvulsive and motor impairment activity *in vivo*', *Neuropharmacology* **34**: 1239–1258.

Potschka, H., Löscher, W., Wlaz, P., Behl, B., Hofmann, H.P., Treiber, H.-J. and Szabo, L. (1998) 'LU 73068, a new non-NMDA and glycine/NMDA receptor antagonist: pharmacological characterization and comparison with NBQX and L-701,324 in the kindling model of epilepsy', *Br. J. Pharmacol.* **125**: 1258–1266.

Priestly, T., Marshall, G.R., Hill, R.G. and Kemp, J.A. (1998) 'L-687,414, a low efficacy NMDA receptor glycine site partial agonist in vitro, does not prevent hippocampal LTP in vivo at plasma levels known to be neuroprotective', *Br. J. Pharmacol.* **124**: 1767–1773.

Ransom, R.W. and Stec, N.L. (1988) 'Cooperative modulation of [^{3}H]MK-801 binding to the *N*-methyl-D-aspartate receptor-ion channel complex by L-glutamate, glycine, and polyamines', *J. Neurochem.* **51**: 830–836.

Rock, D.M. and Macdonald, R.L. (1992) 'Spermine and related polyamines produce a voltage-dependent reduction of *N*-methyl-D-aspartate receptor single channel conductance', *Mol. Pharmacol.* **42**: 157–164.

Rock, D.M. and Macdonald, R.L. (1995) 'Polyamine regulation of *N*-methyl-D-aspartate receptor channels', *Annu. Rev. Pharmacol. Toxicol.* **35**: 463–482.

Sakurada, K., Masu, M. and Nakanishi, S. (1993) 'Alteration of Ca^{2+} permeability and sensitivity to Mg^{2+} and channel blockers by a single amino acid substitution in the N-methyl-D-aspartate receptor', *J. Biol. Chem.* **268**: 410–415.

Salituro, F.G., Harrison, B.L., Baron, B.M., Nyce, P.L., Stewart, K.T., Kehne, J.H., White, H.S. and McDonald, I.A. (1992) '3-(2-Carboxyindol-3-yl)propionic acid-based antagonists of the N-methyl-D-aspartic acid receptor associated glycine binding site', *J. Med. Chem.* **35**: 1791–1799.

Schoemaker, H., Allen, J. and Langer, S.Z. (1990) 'Binding of [^{3}H]ifenprodil, a novel NMDA antagonist, to a polyamine-sensitive site in the rat cerebral cortex', *Eur. J. Pharmacol.* **176**: 249–250.

Schoepp, D.D., Smith, C.L., Lodge, D., Millar, J.D., Leander, J.D., Sacaan, A.I. and Lunn, W.H.W. (1991) 'D,L-Tetrazol-5-ylglycine: a novel and highly potent NMDA receptor agonist', *Eur. J. Pharmacol.* **203**: 237–243.

Shuto, S., Ono, S., Imoto, H., Yoshi, K. and Matsuda, A. (1998) 'Synthesis and biological activity of conformationally restricted analogues of milnacipran: (1S,2R)-1-phenyl-2-[(R)-1-amino-2-propynyl]-N,N-diethylcyclopropanecarboxamide is a novel class of NMDA receptor channel blocker', *J. Med. Chem.* **41**: 3507–3514.

Sills, M.A., Fagg., G., Pozza, M., Angst, C., Brundish, D.E., Hurt, S.D., Wilusz, E.J. and Williams, M. (1991) '[^{3}H]CGP 39653: a new *N*-methyl-D-aspartate antagonist radioligand with low nanomolar affinity in rat brain', *Eur. J. Pharmacol.* **192**: 19–24.

Skolnick, P., Boje, K., Miller, R., Pennington, M. and Maccecchini, M.-L. (1992) 'Noncompetitive inhibition of *N*-methyl-D-aspartate by conantokin-G: evidence for an allosteric interaction at polyamine sites', *J. Neurochem.* **59**: 1516–1521.

Sunter, D.C., Edgar, G.E., Pook, P.C.-K., Howard, J.A.K., Udvarhelyi, P.M. and Watkins, J.C. (1991) 'Actions of the four isomers of 1-aminocyclopentane-1,3-dicarboxylate (ACPD) in the hemisected spinal cord of the neonatal rat', *Br. J. Pharmacol. (Proc. Suppl.)* **104**: 377P.

Tamiz, A.P., Whittemore, E.R., Zhou, Z.-L., Huang, J.-C., Drewe, J.A., Chen, J.-C., Cai, S.-X., Weber, E., Woodward, R.M. and Keana, J.F.W. (1998a) 'Structure–activity relationships for a series of bis-(phenylalkyl)amines: potent subtype-selective inhibitors of *N*-methyl-D-aspartate receptors', *J. Med. Chem.*, **41**: 3499–3506.

Tamiz, A.P., Whittemore, E.R., Schelkun, R.M., Yuen, P.-W., Woodward, R.M., Cai, S.-X., Weber, E. and Keana, J.F.W. (1998b) '*N*-(2-(4-hydroxyphenyl)ethyl)-4-chlorocinnamide: a novel antagonist at the 1A/1B NMDA receptor subtype', *Bioorg. Med. Chem. Lett.* **8**: 199–200.

Tamiz, A.P., Whittemore, E.R., Woodward, R.M., Upasani, R.B. and Keana, J.F.W. (1999) 'Structure–activity relationship for a series of 2-substituted 1,2,3,4-tetrahydro-9H-pyrido[3,4-*b*]indoles: potent subtype-selective inhibitors of *N*-methyl-D-aspartate (NMDA) receptors', *Bioorg. Med. Chem. Lett.* **9**: 1619–1624.

Urwyler, S., Laurie, D., Lowe, D.A., Meier, C.L. and Müller, W. (1996a) 'Biphenyl-derivatives of 2-amino-7-phosphonoheptanoic acid, a novel class of potent competitive N-methyl-D-aspartate receptor antagonists – I. Pharmacological characterization *in vitro*', *Neuropharmacology* **35**: 643–654.

Urwyler, S., Campbell, E., Fricker, G., Jenner, P., Lemaire, M., McAllister, K.H., Neijt, H.C., Park, C.K., Perkins, M., Rudin, M., Sauter, A., Smith, L., Wiederhold, K.H. and Müller, W. (1996b) 'Biphenyl-derivatives of 2-amino-7-phosphonoheptanoic acid, a novel class of potent competitive

N-methyl-D-aspartate receptor antagonists – II. Pharmacological characterization *in vivo*', *Neuropharmacology* **35**: 655–669.

Watkins, J.C. (1962) 'The synthesis of some acidic amino acids possessing neuropharmacological activity', *J. Med. Pharm. Chem.* **5**: 1187–1199.

Watson, G.B., Hood, W.F., Monahan, J.B. and Lanthorn, T.H. (1988) 'Kynurenate antagonizes actions of N-methyl-D-aspartate through a glycine sensitive receptor', *Neurosci. Res. Commun.* **2**: 169–174.

Watson, G.B. and Lanthorn, T.H. (1990) 'Pharmacological characteristics of cyclic homologues of glycine at the N-methyl-D-aspartate receptor-associated glycine site', *Neuropharmacology* **29**: 727–730.

Wenk, G.L., Baker, L.M., Stoehr, J.D., Hauss-Wegrzyniak, B. and Danysz, W. (1998) 'Neuroprotection by novel antagonists at the NMDA receptor channel and $glycine_B$ sites', *Eur. J. Pharmacol.* **347**: 183–187.

Whitten, J.P., Baron, B.M., Miller, D.M.F., White, H.S. and McDonald, I.A. (1990) '(R)-4-Oxo-5-phosphononorvaline: A new competitive glutamate antagonist at the NMDA receptor complex', *J. Med. Chem.* **33**: 2961–2963.

Whitten, J.P., Muench, D., Cube, R.V., Nyce, P.L., Baron, B.M. and McDonald, I.A. (1991) 'Synthesis of 3(S)-phosphonoacetyl-2(R)-piperidine-carboxylic acid, a conformationally-restricted glutamate antagonist', *Bioorg. Med. Chem. Lett.* **1**: 441–444.

Williams, K. (1993) 'Ifenprodil discriminates subtypes of the *N*-methyl-D-aspartate receptor: polyamine-like and high-affinity antagonist actions', *J. Pharmacol. Exp. Ther.* **266**: 231–236.

Williams, K. (1995) 'Pharmacological properties of recombinant *N*-methyl-D-aspartate (NMDA) receptors containing the epsilon 4 (NR2D) subunit', *Neurosci. Lett.* **184**: 181–184.

Williams, K., Zappia, A.M., Pritchett, D.B., Shen, Y.M. and Molinoff, P.B. (1994) 'Sensitivity of the *N*-methyl-D-aspartate receptor to polyamines is controlled by NR2 subunits', *Mol. Pharmacol.* **45**: 803–809.

Williams, M., Kowaluk, E.A. and Arneric, S.P. (1999) 'Emerging molecular approaches to pain therapy', *J. Med. Chem.* **42**: 1481–1500.

Wong, E.H.F., Kemp, J.A., Priestley, T., Knight, A.R., Woodruff, G.N. and Iversen, L.L. (1986) 'The anticonvulsant MK-801 is a potent N-methyl-D-aspartate antagonist', *Proc. Nat. Acad. Sci. USA* **83**: 7104–7108.

Wright, J.L., Gregory, T.F., Bigge, C.F., Boxer, P.A., Serpa, K., Meltzer, L.T., Wise, L.D., Cai, S.X., Hawkinson, J.E., Konkoy, C.S., Whittemore, E.R., Woodward, R.M. and Zhou, Z.-L. (1999) 'Subtype-selective *N*-methyl-D-aspartate receptor antagonists: synthesis and biological evaluation of 1-(arylalkynyl)-4-benzylpiperidines', *J. Med. Chem.* **42**: 2469–2477.

Zhou, L.-M., Szendrei, G.I., Fossom, L.H., Maccecchini, M.-L., Skolnick, P. and Otvos, Jr., L. (1996) 'Synthetic analogues of conantokin-G: NMDA antagonists acting through a novel polyamine-coupled site', *J. Neurochem.* **66**: 620–628.

Chapter 5

Pharmacology of AMPA/kainate receptors

Ulf Madsen, Tommy N. Johansen, Tine B. Stensbøl and Povl Krogsgaard-Larsen

Introduction

For approximately twenty years ionotropic glutamate (iGlu) receptors have been divided into the three classes NMDA, AMPA (originally named quisqualate receptors) and kainate receptors, named after three agonists (McLennan and Lodge, 1979; Watkins and Evans, 1981; Watkins *et al.*, 1990). This receptor classification was primarily based on the selective pharmacology observed for the three agonists and the action of a few antagonists. The pharmacological characterization of the NMDA receptors was developed fairly quickly, due to the relatively high degree of selectivity of NMDA itself and the early availability of a large number of selective and potent competitive NMDA receptor antagonists (see Chapter 4). For AMPA and kainate receptors there has been and, to some extent, still is a shortage of especially selective antagonists, thus these two receptor classes are often collectively named non-NMDA receptors. However, a number of compounds with selective action at AMPA receptors have been described, whereas much fewer compounds acting at kainate receptors have been published.

The cloning of the different subunits of iGlu receptors was a major breakthrough concerning the pharmacology of AMPA and kainate receptors. Nine different subunits of non-NMDA receptors have been cloned (see Chapter 1); four AMPA-preferring subunits, GluR1-4, and five kainate-preferring subunits, GluR5-7 and KA1-2. The classification based on sequence similarity has supported the original pharmacological classification, though the AMPA and kainate-preferring subunits are structurally related. This structural similarity manifests itself in the difficulty in obtaining selective compounds suitable for pharmacological differentiation of the two receptor classes.

The development of selective ligands is indispensable in order to obtain tools for pharmacological characterization and as potential therapeutic agents, and the need for such ligands is far from being fulfilled. However, therapeutic administration of Glu agonists as well as Glu antagonists presents potential problems to be overcome concerning potential or observed adverse effects (see Chapter 17). The use of partial agonists, with an appropriately balanced agonist/antagonist profile, may be an alternative for future therapeutic applications.

Naturally occurring AMPA/kainate receptor agonists

A number of naturally occurring compounds have been identified, which show potent, but nonselective, agonist activity at AMPA/kainate receptors (Figure 5.1). Many of these

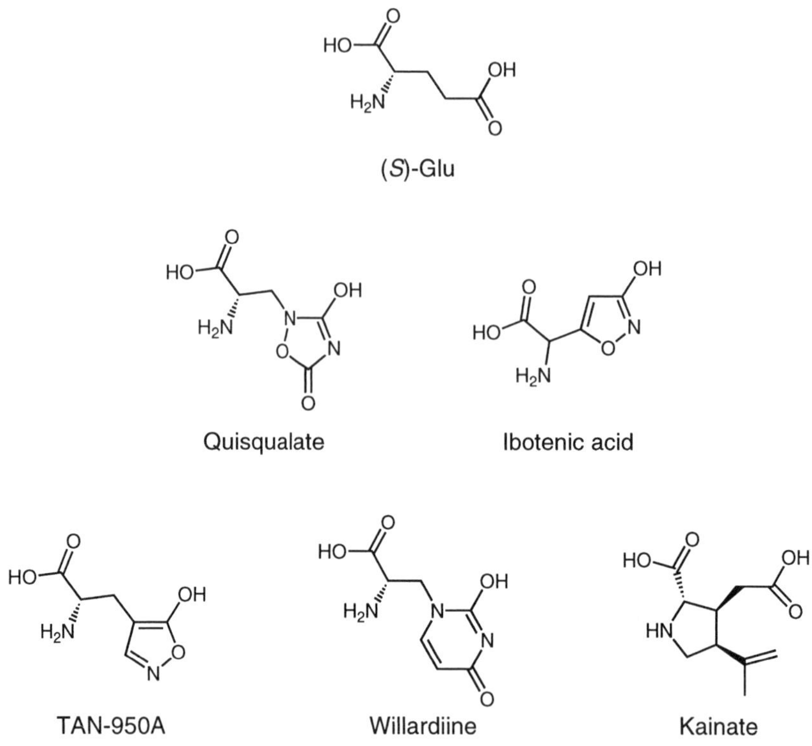

Figure 5.1 Structure of (*S*)-Glu and some naturally occurring excitatory amino acid receptor agonists.

compounds are bioisosterically related to Glu and they are generally potent neurotoxins. Quisqualate has been isolated from *Quisqualis* species and is a potent AMPA receptor agonist. Later multiple actions of quisqualate have been discovered, notably activity at metabotropic Glu (mGlu) receptors and interaction with Glu uptake mechanisms (Sladeczek *et al.*, 1985; Watkins *et al.*, 1990). Ibotenic acid, isolated from the fly agaric mushroom *Amanita muscaria*, was originally shown primarily to have activity at NMDA receptors, and subsequent potent interaction with mGlu receptors was disclosed, whereas ibotenic acid only shows weak interaction with AMPA/kainate receptors (Krogsgaard-Larsen *et al.*, 1980; Bräuner-Osborne *et al.*, 1998). The use of ibotenic acid as a lead structure has, however, given rise to the development of a large number of compounds with selective action at different receptors, notably AMPA receptor ligands, including AMPA itself, and more recently also kainate receptor ligands.

TAN-950A, isolated from *Streptomycetes* species, is structurally related to ibotenic acid and AMPA, and TAN-950A also shows AMPA agonist activity (Iwama *et al.*, 1991). Another naturally occurring Glu bioisostere is willardiine, found in *Acacia* and *Mimosa* species (Evans *et al.*, 1980). Willardiine has been used as a lead for the design of selective AMPA or kainate agonists (see pp. 101 and 111). Kainate, isolated from the red algae *Digenea simplex*, gave its name to the kainate receptors (Shinozaki, 1992). Kainate is not only a potent kainate receptor agonist, but also shows fairly potent and nondesensitizing activity at AMPA receptors, and is widely used as an experimental neurotoxin. These and other naturally occurring amino acids have been, and still are being, used as lead structures in the design of selective agonists or antagonists at different subtypes of Glu receptors.

Ibotenic acid as a lead structure

The 3-isoxazolol moiety of ibotenic acid has proved very versatile as a bioisostere to the distal carboxyl group of Glu. Thus, ibotenic acid has been used extensively as a lead structure in the search for potent and selective compounds acting at iGlu or mGlu receptors. These structural manipulations of ibotenic acid have provided compounds with potent NMDA agonist activity (e.g. AMAA; Madsen *et al.*, 1990), AMPA agonists with different pharmacological profiles (e.g. AMPA, ACPA and HIBO; Krogsgaard-Larsen *et al.*, 1980; Madsen and Wong, 1992), functional partial agonists (e.g. (*R*)- and (*S*)-APPA; Ebert *et al.*, 1994a), AMPA receptor antagonists (e.g. AMOA; Krogsgaard-Larsen *et al.*, 1991), the GluR5 receptor agonist ATPA (Lauridsen *et al.*, 1985; Clarke *et al.*, 1997) and the selective mGluR6 agonist Homo-AMPA (Bräuner-Osborne *et al.*, 1996) (Figure 5.2). The pharmacology of some of these and related compounds will be described in the following sections.

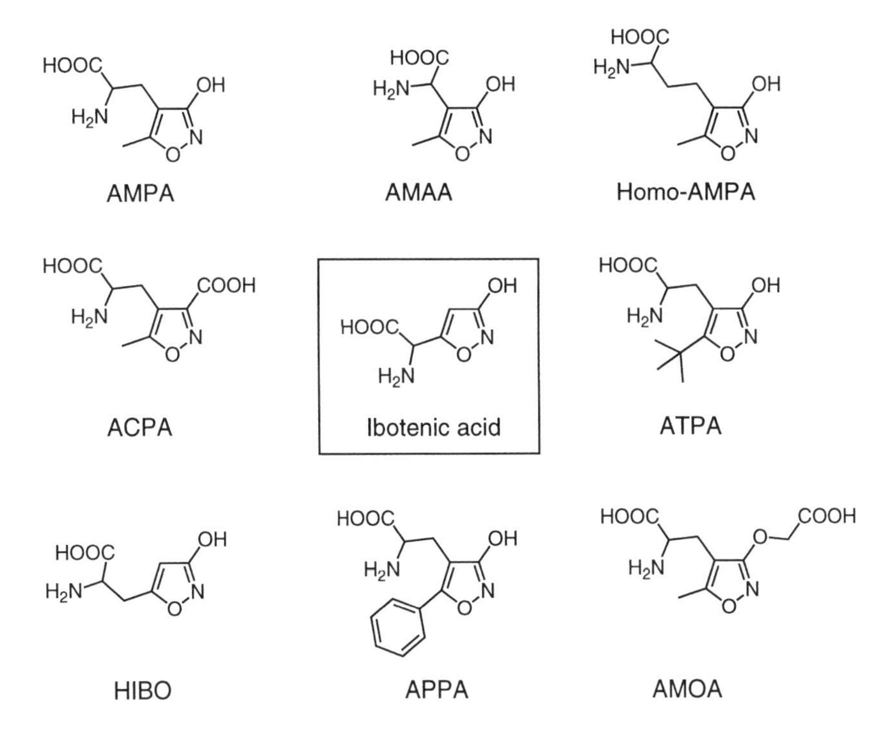

Figure 5.2 Structure of ibotenic acid and a number of excitatory amino acid receptor ligands developed using ibotenic acid as a lead structure.

AMPA receptor agonists

The synthesis and initial pharmacological characterization of AMPA were published in 1980, and later [^{3}H]AMPA was introduced for binding studies of AMPA receptors (at that time named quisqualate receptors) (Hansen and Krogsgaard-Larsen, 1980; Krogsgaard-Larsen *et al.*, 1980; Honoré and Nielsen, 1985). Enzymatic resolution of AMPA showed (*S*)-AMPA to be the more potent enantiomer, having almost a four thousand times higher affinity for AMPA receptors compared to the (*R*)-form (Hansen *et al.*, 1983; Nielsen *et al.*, 1993). The selectivity of AMPA was markedly higher than that of quisqualate, and this afforded grounds for the change of name from quisqualate receptors to AMPA receptors (Collingridge and Lester, 1989; Monaghan *et al.*, 1989; Watkins *et al.*, 1990). AMPA is the agonist of choice, used not only for binding studies but also as a standard agonist in electrophysiological studies. Kainate has much lower affinity for AMPA receptors compared to AMPA, but in contrast to AMPA kainate does not desensitize the AMPA receptors; thus a much larger steady-state current is obtained with kainate. Therefore kainate is often used as an agonist for studies of AMPA receptors (Fletcher and Lodge, 1996).

In Figure 5.3 a number of other AMPA receptor agonists of importance are depicted. ACPA has been shown to possess very potent AMPA receptor activity – more potent than AMPA itself (Madsen and Wong, 1992). Studies on homomeric GluR1 receptors expressed in oocytes have shown ACPA to produce much larger steady state currents than AMPA, similar to the findings for kainate (see Figure 5.4) (Wahl *et al.*, 1996; Banke and Lambert, 1999). In analogy with (*S*)-AMPA, (*S*)-ACPA has been shown to be the more potent enantiomer (Johansen *et al.*, 2001).

A series of willardiine analogues containing the pyrimidine-2,4-dione moiety functioning as a carboxylic acid bioisostere, have shown potent AMPA agonist activity. Different 5-substituted willardiines were shown to produce highly varying degrees of desensitization, depending on the substituent, (*S*)-fluorowillardiine being the most potent AMPA agonist (Patneau *et al.*, 1992). Like AMPA, (*S*)-fluorowillardiine produced a high degree of receptor desensitization in hippocampal neurons, whereas the (*S*)-iodo analogue showed much less desensitization, thus having kainate-like activity (see p. 111). In a later study performed on cells expressing human homomeric AMPA-preferring receptors (hGluR1, 2 or 4), similar degrees of desensitization were shown for the two analogues (Jane *et al.*, 1997). In this study (*S*)-Cl-azawillardiine displayed very

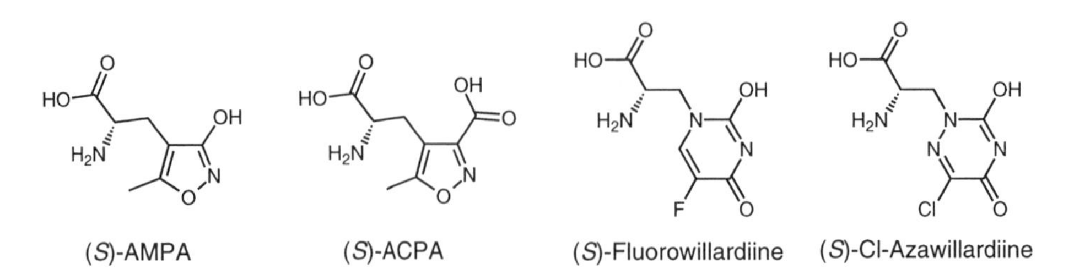

Figure 5.3 Structure of (*S*)-AMPA and other potent and selective AMPA receptor agonists.

high potency at AMPA-preferring receptors and, in contrast to AMPA and (*S*)-fluorowillardiine, a particularly high affinity for hGluR4. Thus, high activity at AMPA-preferring receptor subtypes was obtained with willardiine analogues containing small electron-withdrawing substituents in the 5-position, whereas larger lipophilic substituents afforded high affinity for the kainate-preferring GluR5 receptor. ACPA and (*S*)-fluorowillardiine (Hawkins *et al.*, 1995) have been synthesized in tritiated forms for binding studies, and comparison with [^{3}H]AMPA does show distinct binding characteristics for the three agonists (Stensbøl *et al.*, 1999b).

Future ligand design and structure–activity studies on AMPA receptors may be based on structural information about the receptor proteins. Recently, the crystal structure of the GluR2 ligand binding domain, containing kainate in the agonist binding site, has been published (Armstrong *et al.*, 1998). From this structure, interaction points between ligand and receptor can be deduced, and the steadily increasing knowledge about the structure of ligand binding sites should facilitate design of new compounds.

AMPA analogues

Many AMPA analogues with different substituents in the 5-position of the isoxazole ring have been synthesized and characterized as AMPA receptor ligands. These compounds have been prepared as part of structure–activity studies at AMPA receptors in the search for agonists, partial agonists and antagonists. From Table 5.1 is it evident that small substituents such as methyl, ethyl, propyl, cyclopropyl and isopropyl are preferred for potent AMPA agonist activity (Madsen *et al.*, 1993b; Skjærbæk *et al.*, 1995; Sløk *et al.*, 1997). De-Methyl-AMPA does show fairly high affinity in the [^{3}H]AMPA binding assay, but poor activity in electrophysiological recordings. This may reflect that De-Methyl-AMPA, in contrast to AMPA, is a substrate for Glu uptake systems. The more lipophilic *tert*-butyl analogue ATPA has limited potency at AMPA receptors, but, in contrast to AMPA, ATPA is pharmacologically active after systemic administration to animals (Turski *et al.*, 1992; Arnt *et al.*, 1995). Introduction of larger substituents leads to very weak or inactive compounds (Sløk *et al.*, 1997).

The acidity of the 3-isoxazolol moiety of Trifluoro-AMPA is significantly higher than for AMPA itself, whereas the agonist potencies of these two compounds are similar. Thus, enhanced acidity of the distal acidic group does not *per se* lead to enhanced activity (Madsen *et al.*, 1992). The compound ABPA was designed as a chemically reactive analogue, but the pharmacology of ABPA showed potent AMPA receptor agonist activity with no sign of irreversible receptor interaction (Krogsgaard-Larsen *et al.*, 1985; Ebert *et al.*, 1992). Compound **1**, containing a 2-methyl-5-tetrazolyl substituent, shows remarkably high AMPA receptor agonist activity, being more potent than AMPA and with a potency similar to that of ACPA (Bang-Andersen *et al.*, 1997). The phenyl analogue APPA was found to be a rather weak agonist, but displaying partial agonist activity (Christensen *et al.*, 1989). Resolution of the two enantiomers has been performed by diastereomeric salt formation using the enantiomers of α-phenylethylamine as chiral auxiliaries, and the absolute configuration of (*S*)- and (*R*)-APPA was established by an X-ray analysis. Pharmacological examination of the two enantiomers revealed (*S*)-APPA to be a full AMPA agonist, whereas (*R*)-APPA turned out to be a competitive AMPA antagonist, thus explaining the apparent partial agonist activity of racemic APPA (Ebert *et al.*, 1994a). This finding gave rise to the

Table 5.1 Structure, receptor binding affinity and *in vitro* electrophysiological activity of a number of AMPA analogues.

(structure: HOOC, H_2N, OH, N, O, R)	—R	$[^3H]AMPA$ IC_{50} (μM)	Electrophys. EC_{50} (μM)
De-Methyl-AMPA	—H	0.27	900
AMPA	$-CH_3$	0.04	3.5
Ethyl-AMPA	$-CH_2CH_3$	0.03	2.3
Propyl-AMPA	$-CH_2CH_2CH_3$	0.09	5.0
Cyclopropyl-AMPA	$-CH(CH_2)_2$	0.04	5.5
Isopropyl-AMPA	$-CH(CH_3)_2$	0.19	9.0
Butyl-AMPA	$-CH_2CH_2CH_2CH_3$	1.0	32
Isobutyl-AMPA	$-CH_2CH(CH_3)_2$	0.61	23
Isopentyl-AMPA	$-CH_2CH_2CH(CH_3)_2$	>100	>1000
ATPA	$-C(CH_3)_3$	11	48
Neopentyl-AMPA	$-CH_2C(CH_3)_3$	55	420
4-Heptyl-AMPA	$-CH(CH_2CH_2CH_3)_2$	99	>1000
Trifluoro-AMPA	$-CF_3$	0.08	2.3
ABPA	$-CH_2Br$	0.03	13
I	(2-methyltetrazol-5-yl: N, N, N, N, CH_3)	0.03	0.9
APPA	(phenyl)	35	390
2-Py-AMPA	(2-pyridyl: N)	0.57	7.4
5-HPCA	(bicyclic structure: HOOC, HN, OH, N, O)	1.8	70

development of the concept of functional partial agonism, which implies that partial agonism can be established at any desired level of efficacy by co-administration of agonist and competitive antagonist at appropriate ratios. This can be established at AMPA receptors and at any other ionotropic or metabotropic receptors by the use of appropriate mixtures of an agonist and a competitive antagonist (Ebert *et al.*, 1994b).

The pyridyl analogue, 2-Py-AMPA, shows a remarkably increased agonist activity compared to APPA, being approximately fifty times more potent (Falch *et al.*, 1998). The bicyclic AMPA analogue, 5-HPCA, has very limited structural flexibility, but 5-HPCA still is a fairly potent AMPA agonist (Krogsgaard-Larsen *et al.*, 1985; Ebert *et al.*, 1992). Interestingly 5-HPCA desensitizes recombinant GluR1 receptors to an even higher degree than AMPA (Figure 5.4) (Wahl *et al.*, 1996).

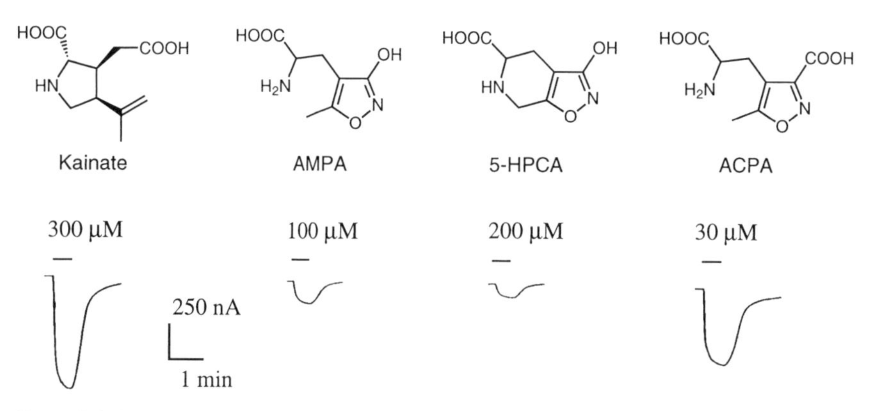

Figure 5.4 Inward currents activated by near saturating concentrations of kainate, AMPA, 5-HPCA and ACPA recorded from a single oocyte injected with GluR1-flop receptor cRNA.

Homoibotenic acid analogues

Another series of ibotenic acid derivatives with AMPA receptor agonist activity is the homoibotenic acid (HIBO) analogues (Table 5.2). In parallel to the AMPA series, the analogues with small substituents in the 4-position of the isoxazole ring are the more potent ones, Methyl-HIBO and Br-HIBO being the most potent AMPA agonists in this series. Butyl-HIBO is somewhat weaker and Octyl-HIBO is inactive (Krogsgaard-Larsen *et al.*, 1980; Hansen *et al.*, 1992; Christensen *et al.*, 1992). Analogous to De-Methyl-AMPA, the unsubstituted analogue HIBO is fairly weak in electrophysiological experiments, reflecting that HIBO is probably a substrate for Glu uptake (Bischoff *et al.*, 1995). These aspects are under investigation using recombinant Glu transporters. Resolution of HIBO analogues has been carried out using different methods: asymmetric synthesis (HIBO; see Bischoff *et al.*, 1985), enzymatic resolution (Methyl-HIBO and Br-HIBO; see Hansen *et al.*, 1989, 1992) and diastereomeric salt formation (Butyl-HIBO; see Johansen *et al.*, 1998). For all four HIBO analogues the (*S*)-form has been found to be the more potent AMPA receptor agonist.

Within the series of HIBO analogues effects at mGlu receptors are also observed, notably as mGluR1 and 5 (Group I) antagonists (Bräuner-Osborne *et al.*, 1998). The metabotropic activity of the HIBO analogues also reside in the (*S*)-enantiomers. In contrast to this, both the (*S*)- and the (*R*)-forms of the HIBO analogues generally show affinity for $CaCl_2$-dependent [^{3}H]Glu binding (Hansen *et al.*, 1989, 1992) and potent activity after quisqualate priming in electrophysiological experiments (Madsen *et al.*, 1993a, and unpublished results), suggesting interaction with transport mechanisms, which may be associated with neurotransmitter vesicles.

Electrophysiological experiments have shown (*S*)-Butyl-HIBO to be strongly potentiated by simultaneous application with an otherwise inactive concentration of (*R*)-Butyl-HIBO (Figure 5.5A) (Johansen *et al.*, 1998). This potentiation is virtually eliminated by lowering the temperature to 2–4°C (Figure 5.5B) in agreement with the general temperature dependence observed for uptake/transport mechanisms. The bicyclic ibotenic acid analogue 6-HPCA (Table 5.2) is inactive both as an AMPA

Table 5.2 Structure, receptor binding affinity and *in vitro* electrophysiological activity of a number of HIBO analogues

	—*R*	*[³H] AMPA* IC_{50} *(μM)*	*Electrophys.* EC_{50} *(μM)*
HIBO	—H	1.5	370
Methyl-HIBO	$—CH_3$	0.61	20
Br-HIBO	—Br	0.60	23
Butyl-HIBO	$—CH_2CH_2CH_2CH_3$	1.8	37
Octyl-HIBO	$—(CH_2)_7CH_3$	>100	>1000
6-HPCA		>100	>1000

agonist, as a mGlu receptor ligand, and also after quisqualate priming (Madsen *et al.*, 1986 and unpublished results).

AMPA receptor antagonists

Different series of competitive AMPA receptor antagonists have been developed. Early pharmacological studies on AMPA and kainate receptors were hampered by the lack of selective and potent antagonists. CNQX and related compounds (Figure 5.6) offered a breakthrough in this respect, being quite potent antagonists, though nonselective (Fletcher *et al.*, 1988; Honoré *et al.*, 1988). Subsequently, NBQX was published, which

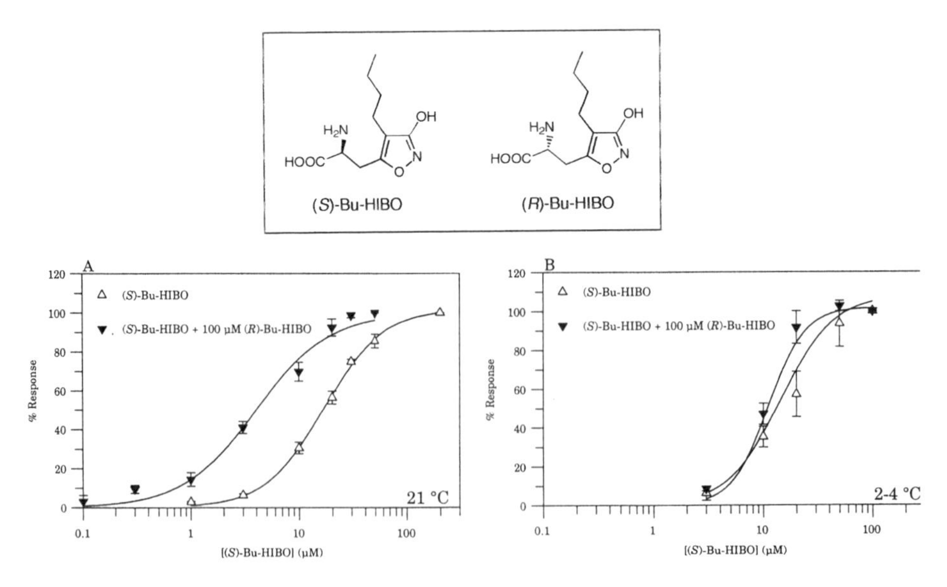

Figure 5.5 Structure of (*S*)- and (*R*)-Bu-HIBO and below dose–response curves obtained from the rat cortical slice preparation for (*S*)-Bu-HIBO alone and in combination with 100 μM (*R*)-Bu-HIBO at 21 °C (a) or at 2–4 °C (b).

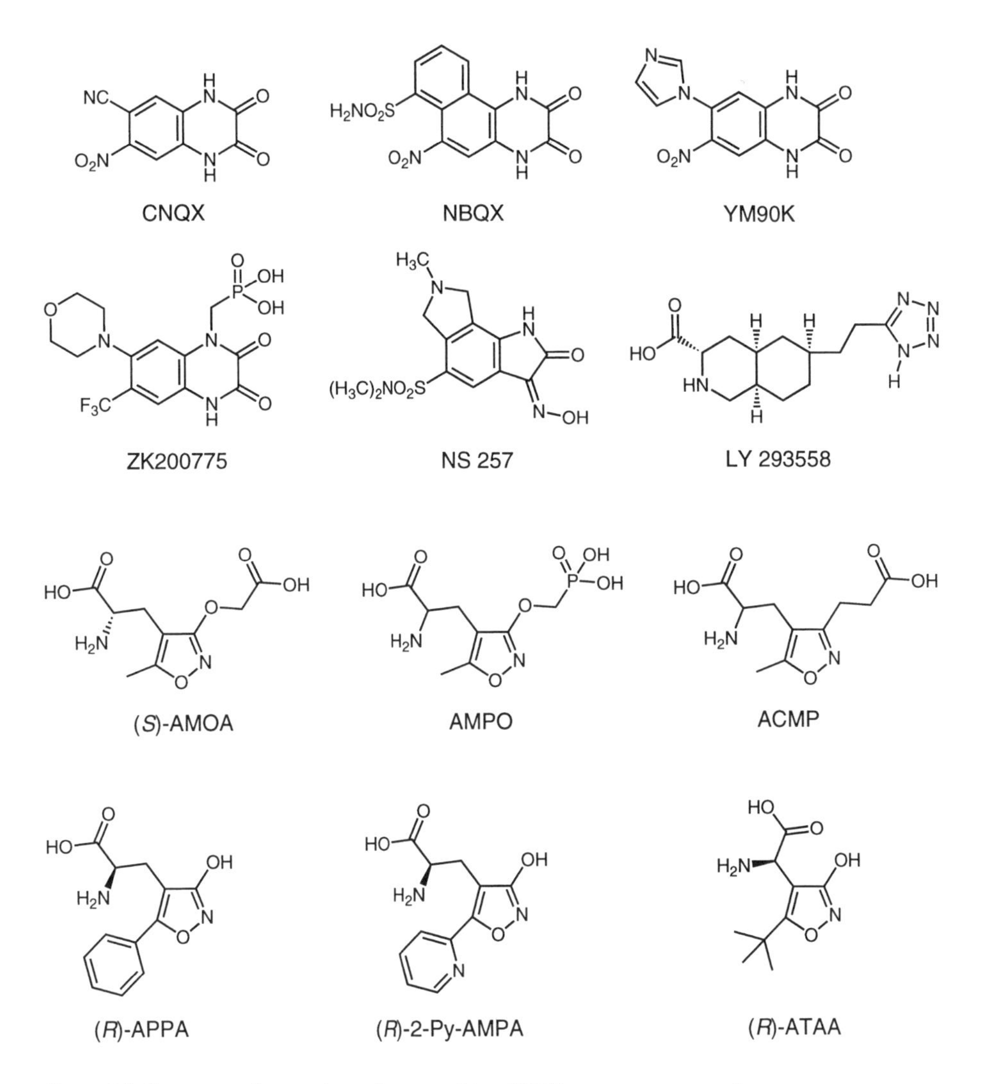

Figure 5.6 Structure of a number of competitive AMPA receptor antagonists.

had improved AMPA receptor selectivity compared to CNQX (Sheardown *et al.*, 1990). A large number of compounds have been developed with the quinoxalinedione structure, not only as AMPA receptor antagonists but also as compounds showing effects at other Glu receptor sites, notably glycine antagonists. YM90K shows high potency compared to NBQX as an AMPA receptor antagonist and is systemically active (Ohmori *et al.*, 1994). Introduction of a phosphonate moiety afforded ZK200775, which is a potent and selective AMPA receptor antagonist with improved water solubility and a longer duration of action (Turski *et al.*, 1998). A number of isatin oximes, such as NS 257, have shown antagonist activity and these compounds show higher water solubility than NBQX, and they have been shown to be systemically active in anticonvulsive tests (Wätjen *et al.*, 1993, 1994).

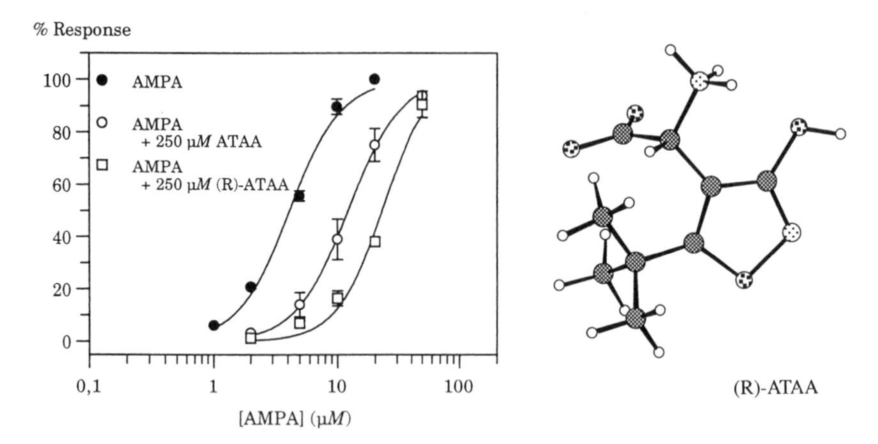

Figure 5.7 Dose–response curves obtained from the rat cortical slice preparation for AMPA alone and in combination with 250 μM ATAA or 250 μM (*R*)-ATAA. A ball-and-stick model of (*R*)-ATAA is shown at the right. The enantiomers of ATAA were obtained by diastereomeric salt formation and the absolute stereochemistry established by X-ray crystallography.

One group of AMPA receptor antagonists contains the acidic amino acid structure generally found in Glu agonists, whereas another group of antagonists is composed of acidic amino acids containing a longer backbone. One member of the latter group is the tricyclic compound LY 293558 (Ornstein *et al.*, 1993). Similarly, the AMPA receptor antagonists AMOA and AMPO, which have been developed using AMPA as a lead structure (Krogsgaard-Larsen *et al.*, 1991; Madsen *et al.*, 1996), have carbon backbones longer than those normally found in AMPA agonist molecules. For LY 293558 and AMOA, the (*S*)-form has been shown to be the more potent enantiomer (Wahl *et al.*, 1992; Ornstein *et al.*, 1993). Recently, the carbon analogue of AMOA, ACMP, has shown improved AMPA antagonist potency compared to AMOA (Madsen *et al.*, 1999).

Within the group of AMPA antagonists having agonist chain lengths, the (*R*)-form is generally the active enantiomer, as exemplified by (*R*)-APPA (Ebert *et al.*, 1994a), (*R*)-Z-Py-AMPA (Johansen *et al.*, 1997a; Falch *et al.*, 1998) and (*R*)-ATAA (Figures 5.6 and 5.7) (Johansen *et al.*, 1997b). These antagonists are remarkable because, in spite of their "agonist-like" structure and AMPA antagonist activity, they show very low receptor affinity in the [^{3}H]AMPA binding assay.

AMOA analogues

A number of AMOA analogues have been synthesized in the search for more potent compounds as part of ongoing structure–activity studies on AMPA receptor antagonists (Madsen *et al.*, 1996). Two series were prepared, carboxylic acid analogues and analogues containing a phosphonate group. Compounds of the latter group generally show higher potency than those belonging to the former group. For both classes, the *tert*-butyl analogue shows the highest potency, ATPO being the most potent compound (Table 5.3). This finding is surprising in light of the structure–activity relationships

Table 5.3 Structure, receptor binding affinity and *in vitro* electrophysiological activity of a number of AMOA analogues

	—R	—X	[^{3}H] AMPA IC_{50} (μM)	Electrophys. EC_{50} (μM)
AMOA	$-CH_3$	—COOH	90	320
ATOA	$-C(CH_3)_3$	—COOH	33	150
AMPO	$-CH_3$	$-PO_3H_2$	31	60
ATPO	$-C(CH_3)_3$	$-PO_3H_2$	35	28

observed in the agonist series, where AMPA is significantly more potent than the agonist with a *tert*-butyl group, ATPA (Table 5.1). Compounds containing larger substituents did, however, show lower potency in analogy with the findings for the agonist series (Madsen *et al.*, 1996).

ATPO has been shown to have selective antagonist effect at recombinant homomeric and heteromeric AMPA receptors expressed in HEK293 cells relative to homomeric and heteromeric NMDA or kainate receptors (Wahl *et al.*, 1998). Recently, separation of (*R*)- and (*S*)-ATPO by chiral HPLC has shown the (*S*)-form to be the active enantiomer (Møller *et al.*, 1999). In this study the selective antagonist effect at AMPA-preferring receptors, GluR1–4 was confirmed, though (*S*)-ATPO did show some antagonist effect at the kainate-preferring receptor GluR5, though markedly weaker (Figure 5.8).

AMPA receptor modulators

A number of compounds with noncompetitive antagonist effects at AMPA receptors have been identified (Figure 5.9). GYKI 52466 (Tarnawa *et al.*, 1989) and other 2,3-benzodiazepines such as GYKI 53655 show potent and selective antagonist effect at

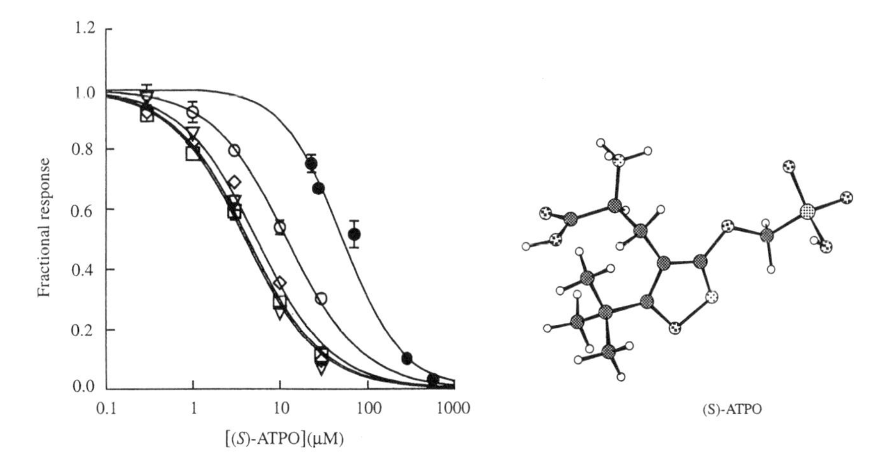

Figure 5.8 Inhibition by (*S*)-ATPO of currents evoked by 25 μM kainate in oocytes expressing GluR1 (□), GluR1-2 (▽), GluR3 (◇), GluR4 (○) or GluR5 (●). A ball-and-stick model of (*S*)-ATPO is shown at the right. The absolute configuration of (*R*)- and (*S*)-ATPO was established by X-ray crystallography.

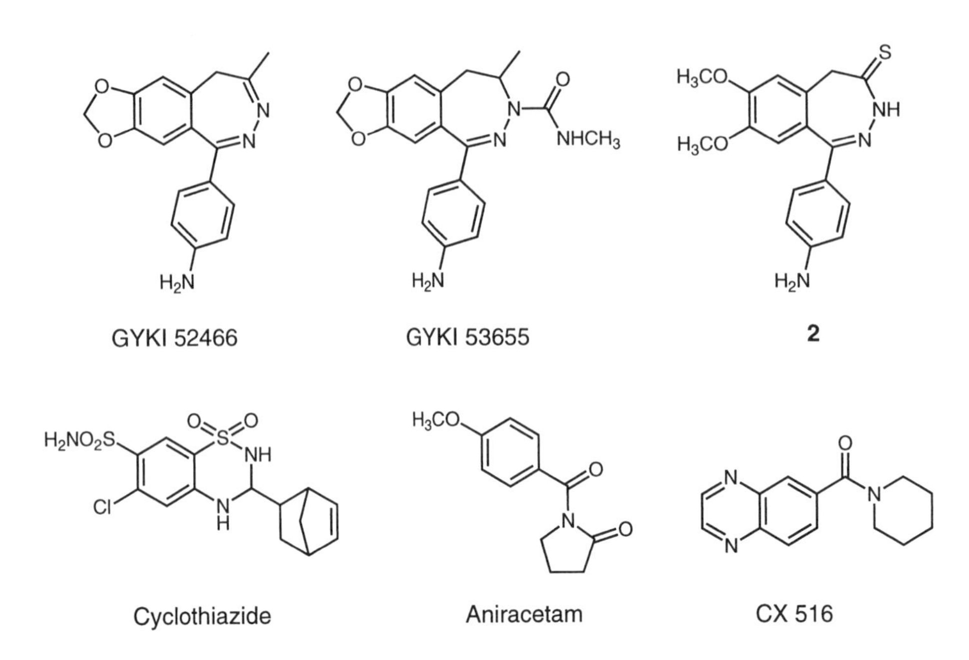

Figure 5.9 Structure of modulators of AMPA receptor activity with noncompetitive antagonist activity or inhibitory activity towards AMPA receptor desensitization.

AMPA receptors by interaction with an allosteric site (Wilding and Huettner, 1995; Donevan and Rogawski, 1993). The benzodiazepine-4-thione **2** has been shown to produce higher potency, lower toxicity, and longer-lasting anticonvulsant activity than GYKI 52466 (Chimirri *et al.*, 1998).

The rapid desensitization observed after application of agonists at AMPA receptors can be modulated by compounds such as the diuretic cyclothiazide and the nootropic agent aniracetam (Isaacson and Nicoll, 1991; Patneau *et al.*, 1993, Zorumski *et al.*, 1993). These compounds essentially block desensitization to agonists, thereby enhancing the excitatory activity, in some cases severalfold, depending on the initial level of desensitization observed for the individual agonist. The recognition sites for cyclothiazide and aniracetam appear to be different, and are also distinct from both the AMPA agonist binding site and the site for noncompetitive antagonists such as GYKI 53655 (Desai *et al.*, 1995; Fallerino *et al.*, 1995). Aniracetam is of rather low potency, but more potent analogues have been developed, such as CX 516 (Arai *et al.*, 1994).

Potassium thiocyanate is used to enhance agonist affinity in the [^{3}H]AMPA binding assay (Honoré and Nielsen, 1985). Thiocyanate has been described as a chaotropic agent, in the absence of knowledge about the actual mechanism. Later studies have shown that thiocyanate ions block AMPA receptors by converting the receptors into the desensitized state (Bowie and Smart, 1993). Obviously, the desensitized state has a higher affinity for AMPA than the sensitized state, and it has been suggested that the sites of interaction for cyclothiazide and thiocyanate are identical, though with opposite effects on desensitization (Donevan and Rogawski, 1998). Furthermore, the interaction seems to take place only on AMPA receptors comprising the GluR2 subunit (Eugene *et al.*, 1996). Interestingly, binding of [^{3}H]ACPA, an agonist showing weak desensitization of AMPA receptors (see Figure 5.4), is not affected by thiocyanate ions (Stensbøl *et al.*, 1999b).

Kainate receptors

Kainate receptor agonists

Studies of kainate receptors have for many years been hampered by the lack of selective ligands, both agonists and in particular antagonists. Kainate has been the standard agonist in spite of its non-selective action, and [^{3}H]kainate is the ligand of choice for studies of high and low affinity kainate binding sites (Foster and Fagg, 1984; Young and Fagg, 1990). Kainate shows relatively potent interaction with AMPA receptors as well, and is often used as the agonist for studies on AMPA receptors, because kainate, in contrast to AMPA itself, does not desensitize AMPA receptors (see p. 102).

Other naturally occurring kainoids (Figure 5.10) also interact potently with kainate receptors. Domoic acid and acromelic acid A, found in a seaweed and in a poisonous mushroom, respectively, have been identified. They are both highly potent kainate receptor agonists, more potent than kainate and also very effective neurotoxins (Ishida and Shinozaki, 1988; Shinozaki, 1988; Patneau and Mayer, 1990). (2*S*,4*R*)-4-methyl-Glu has been reported to have selective affinity for [^{3}H]kainate binding sites (Gu *et al.*, 1995), and potent agonist activity at recombinant homomeric GluR5 and GluR6 receptors (Jones *et al.*, 1997; Donevan *et al.*, 1998). Significant potency of (2*S*,4*R*)-4-methyl-Glu has, however, also been reported at mGlu receptors as well (Bräuner-Osborne *et al.*, 1997). Potent and selective GluR5 agonist activity has also been shown for (*S*)-iodowillardiine, whereas very weak or no activity were observed at GluR6 or GluR7 (Jane *et al.*, 1997; Swanson *et al.*, 1998). Potent GluR5 agonist activity has been observed for the AMPA analogue ATPA, which show high selectivity towards GluR5

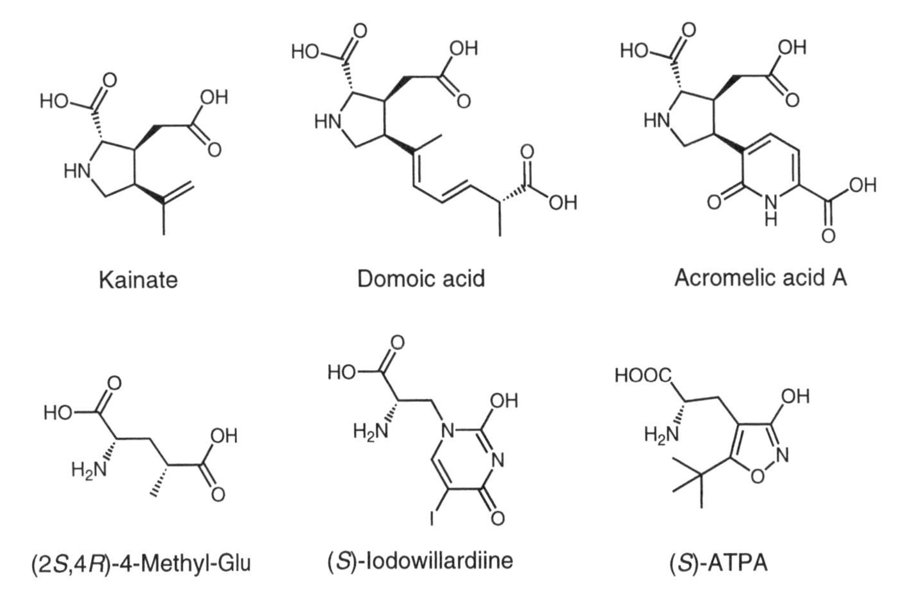

Figure 5.10 Structure of kainate and other kainate receptor agonists.

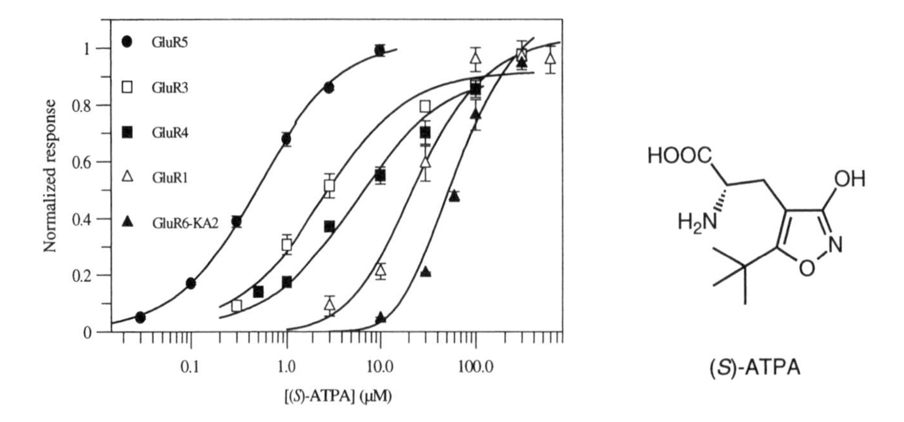

Figure 5.11 Normalized dose–response curves for (*S*)-ATPA on GluR1, GluR3, GluR4, GluR5 or GluR6-KA2 receptors expressed in oocytes. The structure of (*S*)-ATPA is shown at the right.

relative to mGluR6 and AMPA-preferring receptors, both in binding experiments and in electrophysiological studies (Clarke *et al.*, 1997). The activity at GluR5 as well as the weak AMPA receptor activity have been shown to reside in (*S*)-ATPA (Figure 5.11), whereas the (*R*)-form turned out to be a weak but selective antagonist only at AMPA-preferring receptors (Stensbøl *et al.*, 1999a).

Activity at kainate receptors can be modulated by lectins, such as Concanavalin A, which attenuates the desensitization of kainate receptors and thus enhances the excitatory activity (Huettner, 1990; Egebjerg *et al.*, 1991).

Kainate receptor antagonists

The number of selective kainate receptor antagonists is still very small and the characterization of these is fairly limited. NS 102 (Figure 5.12) does show selective affinity for the low affinity [^{3}H]kainate binding site (Johansen *et al.*, 1993), and has been described to have selective antagonist effect at homomeric GluR6 (Verdoorn *et al.*, 1994), although low water solubility may limit the pharmacological utility of NS 102. LY 294486 has been shown to have competitive antagonist activity with some selectivity towards GluR5, relative to AMPA-preferring subunits, and without activity at other kainate-pre-

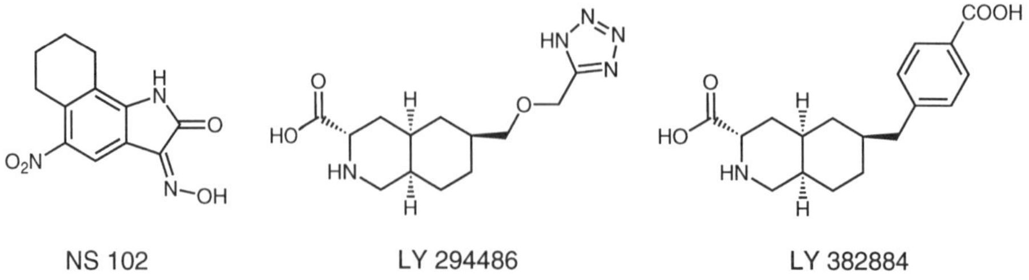

NS 102 LY 294486 LY 382884

Figure 5.12 Structure of the kainate receptor antagonists NS 102, LY 294486 and LY 382884.

ferring subunits (Clarke *et al.*, 1997). Similarly, LY 382884 has been described to have promising GluR5 selective antagonist activity relative to GluR2 (O'Neill *et al.*, 1998).

Future studies on receptors of known subunit combination will be of importance for the development of kainate receptor ligands. The AMPA receptor antagonist LY 293558 (Figure 5.6) also shows some antagonist activity at GluR5 receptors, but no antagonism of GluR6 receptors (Bleakman *et al.*, 1996).

Conclusion

A steadily growing number of potent and selective agonists and antagonists have been described for AMPA receptors, whereas the selection of kainate-preferring compounds is more limited. In particular, more selective kainate receptor antagonists are needed. Today recombinant receptors, homomeric as well as heteromeric, play a key role in the pharmacological characterization of Glu receptor ligands. Many questions concerning subunit composition of native receptors are, however, unanswered. This means that some of the already known compounds may become useful subtype selective ligands for *in vivo* studies. Certainly, the use of molecular biology techniques has disclosed a large number of receptor subtypes and has emphasized the demand for new types of subtype-selective ligands.

References

Arai, A., Kessler, M., Xiao, P., Ambros-Ingerson, J., Rogers, G. and Lynch, G.A. (1994) 'A centrally active drug that modulates AMPA receptor gated currents', *Brain Res.* **638**: 343–346.

Armstrong, N., Sun, Y., Chen, G.-Q. and Gouaux, E. (1998) 'Structure of a glutamate-receptor ligand-binding core in complex with kainate', *Nature* **395**: 913–917.

Arnt, J., Sánchez, C., Lenz, S.M., Madsen, U. and Krogsgaard-Larsen, P. (1995) 'Differentiation of in vivo effects of AMPA and NMDA receptor ligands using drug discrimination methods and convulsant/anticonvulsant activity', *Eur. J. Pharmacol.* **285**: 289–297.

Bang-Andersen, B., Lenz, S.M., Skjærbæk, N., Søby, K.K., Hansen, H.O., Ebert, B., Bøgesø, K.P. and Krogsgaard-Larsen, P. (1997) 'Heteroaryl analogues of AMPA. Synthesis and quantitative structure–activity relationships', *J. Med. Chem.* **40**: 2831–2842.

Banke, T.G. and Lambert, J.D.C. (1999) 'Novel potent AMPA analogues differentially affect desensitisation of AMPA receptors in cultured hippocampal neurons', *Eur. J. Pharmacol.* **367**: 405–412.

Bischoff, F., Johansen, T.N., Ebert, B., Krogsgaard-Larsen, P. and Madsen, U. (1995) 'Excitatory amino acid receptor ligands: asymmetric synthesis, absolute stereochemistry and pharmacology of (*R*)- and (*S*)-homoibotenic acid', *Bioorg. Med. Chem.* **3**: 553–558.

Bleakman, D., Schoepp, D.D., Ballyk, B., Bufton, H., Sharpe, E.F., Thomas, K., Ornstein, P.L. and Kamboj, R.K. (1996) 'Pharmacological discrimination of GluR5 and GluR6 kainate receptor subtypes by (3*S*,4a*R*,6*R*,8a*R*)-6-[2-(1(2)*H*-tetrazole-5-yl)ethyl]decahydroisoquinoline-3-carboxylic acid', *Mol. Pharmacol.* **49**: 581–585.

Bowie, D. and Smart, T.G. (1993) 'Thiocyanate ions selectively antagonize AMPA-evoked responses in *Xenopus laevis* oocytes microinjected with rat brain mRNA', *Br. J. Pharmacol.* **109**: 779–787.

Bräuner-Osborne, H., Sløk, F.A., Skjærbæk, N., Ebert, B., Sekiyama, N., Nakanishi, S. and Krogsgaard-Larsen, P. (1996) 'A new highly selective metabotropic excitatory amino acid agonist: 2-amino-4-(3-hydroxy-5-methyl-4-yl)butyric acid', *J. Med. Chem.* **39**: 3188–3194.

Bräuner-Osborne, H., Nielsen, B., Stensbøl, T.B., Johansen, T.N., Skjærbæk, N. and Krogsgaard-Larsen, P. (1997) 'Molecular pharmacology of 4-substituted glutamic acid analogues at ionotropic and metabotropic excitatory amino acid receptors', *Eur. J. Pharmacol.* **335**: R1–3.

Bräuner-Osborne, H., Nielsen, B. and Krogsgaard-Larsen, P. (1998) 'Molecular pharmacology of homologues of ibotenic acid at cloned metabotropic glutamic acid receptors', *Eur. J. Pharmacol.* **350**: 311–316.

Chimirri, A., DeSarro, G., DeSarro, A., Gitto, R., Quartarone, S., Zappala, M., Constanti, A. and Libri, V. (1998) '3,5-Dihydro-4H-2,3-benzodiazepine-4-thiones: a new class of AMPA receptor antagonists', *J. Med. Chem.* **41**: 3409–3416.

Christensen, I.T., Reinhardt, A., Nielsen, B., Ebert, B., Madsen, U., Nielsen, E.Ø., Brehm, L. and Krogsgaard-Larsen, P. (1989) 'Excitatory amino acid agonists and partial agonists', *Drug Des. Del.* **5**: 57–71.

Christensen, I.T., Ebert, B., Madsen, U., Nielsen, B., Brehm, L. and Krogsgaard-Larsen, P. (1992) 'Excitatory amino acid receptor ligands. Synthesis and biological activity of 3-isoxazolol amino acids structurally related to homoibotenic acid', *J. Med. Chem.* **35**: 3512–3519.

Clarke, V.R.J., Ballyk, B.A., Hoo, K.H., Mandelzys, A., Pellizzari, A., Bath, C.P., Thomas, J., Sharpe, E.F., Davies, C.H., Ornstein, P.L., Schoepp, D.D., Kamboj, R.K., Collingridge, G.L., Lodge, D. and Bleakman, D. (1997) 'A hippocampal GluR5 kainate receptor regulating inhibitory synaptic transmission', *Nature* **389**: 599–603.

Collingridge, G.L. and Lester, R.A.S. (1989) 'Excitatory amino acid receptors in the vertebrate central nervous system', *Pharmacol. Rev.* **40**: 145–195.

Desai, M.A., Burnett, J.P., Ornstein, P.L. and Schoepp, D.D. (1995) 'Cyclothiazide acts at a site on the α-amino-3-hydroxy-5-methyl-4-isoxazole propionic acid receptor complex that does not recognize competitive or noncompetitive AMPA receptor antagonists', *J. Pharmacol. Exp. Ther.* **272**: 38–43.

Donevan, S.D. and Rogawski, M.A. (1993) 'GYKI 52466, a 2,3-benzodiazepine, is a highly selective, noncompetitive antagonist of AMPA/kainate receptor responses', *Neuron* **10**: 51–59.

Donevan, S.D. and Rogawski, M.A. (1998) 'Allosteric regulation of α-amino-3-hydroxy-5-methyl-4-isoxazole-propionate receptors by thiocyanate and cyclothiazide at a common modulatory site distinct from that of 2,3-benzodiazepines', *Neuroscience* **87**: 615–629.

Donevan, S.D., Beg, A., Gunther, J.M. and Twyman, R.E. (1998) 'The methylglutamate, SYM 2081, is a potent and highly selective agonist at kainate receptors', *J. Pharmacol. Exp. Ther.* **285**: 539–545.

Ebert, B., Madsen, U., Lund, T.M., Holm, T. and Krogsgaard-Larsen, P. (1992) 'Molecular pharmacology of cortical and spinal AMPA receptors', *Mol. Neuropharmacol.* **2**: 47–49.

Ebert, B., Lenz, S., Brehm, L., Bregnedal, P., Hansen, J.J., Frederiksen, K., Bøgesø, K.P. and Krogsgaard-Larsen, P. (1994a) 'Resolution, absolute stereochemistry, and pharmacology of the (*S*)-(+)- and (*R*)-(−)-isomers of the apparent partial AMPA receptor agonist (*R,S*)-2-amino-3-(3-hydroxy-5-phenylisoxazol-4-yl)propionic acid [(*R,S*)-APPA]', *J. Med. Chem.* **37**: 878–884.

Ebert, B., Madsen, U., Lund, T.M., Lenz, S.M. and Krogsgaard-Larsen, P. (1994b) 'Molecular pharmacology of the AMPA agonist (*S*)-2-amino-3-(3-hydroxy-5-phenyl-4-isoxazolyl)propionic acid [(*S*)-APPA] and the AMPA antagonist, (*R*)-APPA', *Neurochem. Int.* **24**: 507–515.

Egebjerg, J., Bettler, B., Hermans, B.I. and Heinemann, S. (1991) 'Cloning of a cDNA for a glutamate receptor subunit activated by kainate but not AMPA', *Nature* **351**: 745–748.

Eugene, D., Moss, S.J. and Smart, T.G. (1996) 'Thiocyanate ions inhibit AMPA-activated currents in recombinant non-NMDA receptors expressed in Xenopus laevis oocytes: the role of the GluR2 subunit', *Eur. J. Neurosci.* **8**: 1983–1993.

Evans, R.H., Jones, A.W. and Watkins, J.C. (1980) 'Willardiine: a potent quisqualate-like excitant', *J. Physiol. (London)* **308**: 71P-72P.

Falch, E., Brehm, L., Mikkelsen, I., Johansen, T.N., Skjærbæk, N., Nielsen, B., Stensbøl, T.B., Ebert, B. and Krogsgaard-Larsen, P. (1998) 'Heteroaryl analogues of AMPA. 2. Synthesis, absolute stereochemistry, photochemistry, and structure–activity relationships', *J. Med. Chem.* **41**: 2513–2523.

Fallarino, F., Genazzani, A.A., Silla, S., Lepiscopo, M.R., Camici, O., Corazzi, L., Nicoletti, F. and Fioretti, M.C. (1995) '[^{3}H]Aniracetam binds to specific recognition sites in brain membranes', *J. Neurochem.* **65**: 912–918.

Fletcher, E.J. and Lodge, D. (1996) 'New developments in the molecular pharmacology of α-amino-3-hydroxy-5-methyl-4-isoxazole propionate and kainate receptors', *Pharmacol. Ther.* **70**: 65–89.

Fletcher, E.J., Martin, D., Aram, J.A., Lodge, D. and Honoré, T. (1988) 'Quinoxalinediones selectively block quisqualate and kainate receptors and synaptic events in rat neocortex and hippocampus and frog spinal cord in vitro', *Br. J. Pharmacol.* **95**: 585–597.

Foster, A.C. and Fagg, G.E. (1984) 'Acidic amino acid binding sites in mammalian neuronal membranes: their characteristics and relationship to synaptic receptors', *Brain Res. Rev.* **7**: 103–164.

Gu, Z.-Q., Hesson, D.P., Pelletier, J.C., Maccecchini, M.-L., Zhou, L.M. and Skolnick, P. (1995) 'Synthesis, resolution, and biological evaluation of the four stereoisomers of 4-methylglutamic acid: selective probes of kainate receptors', *J. Med. Chem.* **38**: 2518–2520.

Hansen, J.J. and Krogsgaard-Larsen, P. (1980) 'Isoxazole amino acids as glutamic acid agonists. Synthesis of some analogues and homologues of ibotenic acid', *J. Chem. Soc. Perkin Trans. I:* 1826–1833.

Hansen, J.J., Lauridsen, J., Nielsen, E. and Krogsgaard-Larsen, P. (1983) 'Enzymic resolution and binding to rat brain membranes of the glutamic acid agonist α-amino-3-hydroxy-5-methyl-4-isoxazolepropionic acid', *J. Med. Chem.* **26**: 901–903.

Hansen, J.J., Nielsen, B., Krogsgaard-Larsen, P., Brehm, L., Nielsen, E.Ø. and Curtis, D.R. (1989) 'Excitatory amino acid agonists. Enzymic resolution, X-ray structure and enantioselective activities of (R)- and (S)-bromohomoibotenic acid', *J. Med. Chem.* **32**: 2254–2260.

Hansen, J.J., Jørgensen, F.S., Lund, T.M., Nielsen, B., Reinhardt, A., Breum, I., Brehm, L. and Krogsgaard-Larsen, P. (1992) 'AMPA receptor agonists: structural, conformational and stereochemical aspects', in P. Krogsgaard-Larsen and J.J. Hansen (eds) *Excitatory Amino Acid Receptors: Design of Agonists and Antagonists*, Chichester: Ellis Horwood, pp. 216–245.

Hawkins, L.M., Beaver, K.M., Jane, D.E., Taylor, P.M., Sunter, D.C. and Roberts, P.J. (1995) 'Characterization of the pharmacology and regional distribution of (*S*)-[^{3}H]-5-fluorowillardiine in rat brain', *Br. J. Pharmacol.* **116**: 2033–2039.

Honoré, T. and Nielsen, M. (1985) 'Complex structure of quisqualate-sensitive glutamate receptors in rat cortex', *Neurosci. Lett.* **54**: 27–32.

Honoré, T., Davies, S.N., Drejer, J., Fletcher, E.J., Jacobsen, P., Lodge, D. and Nielsen, F.E. (1988) 'Quinoxalinediones: potent competitive non-NMDA glutamate receptor antagonists', *Science* **241**: 701–703.

Huettner, J.E. (1990) 'Glutamate receptor channels in rat DRG neurons: activation by kainate and quisqualate and blockade of desensitization by Con A', *Neuron* **5**: 255–266.

Isaacson, J.S. and Nicoll, R.A. (1991) 'Aniracetam reduces glutamate receptor desensitization and slows the decay of fast excitatory synaptic currents in the hippocampus', *Proc. Natl. Acad. Sci. USA* **88**: 10,936–10,940.

Ishida, M. and Shinozaki, H. (1988) 'Acromelic acid is a much more potent excitant than kainic acid or domoic acid in the isolated rat spinal cord', *Brain Res.* **474**: 386–389.

Iwama, T., Nagai, Y., Tamura, N., Harada, S. and Nagaoka, A. (1991) 'A novel glutamate agonist, TAN-950 A, isolated from streptomycetes', *Eur. J. Pharmacol.* **197**: 187–192.

Jane, D.E., Hoo, K., Kamboj, R., Deverill, M., Bleakman, D. and Mandelzys, A. (1997) 'Synthesis of willardiine and 6-azawillardiine analogs: pharmacological characterization on cloned homomeric human AMPA and kainate receptor subtypes', *J. Med. Chem.* **40**: 3645–3650.

Johansen, T.H., Drejer, J., Wätjen, F. and Nielsen, E.Ø. (1993) 'A novel non-NMDA antagonist shows selective displacement of low-affinity [^{3}H]kainate binding', *Eur. J. Pharmacol. Mol. Pharmacol. Sect.* **246**: 195–204.

Johansen, T.N, Ebert, B., Falch, E. and Krogsgaard-Larsen, P. (1997a) 'AMPA receptor agonists: resolution, configurational assignment, and pharmacology or (+)-(S)- and (−)-(R)-2-amino-3-[3-hydroxy-5-(2-pyridyl)-isoxazol-4]-propionic acid (2-Py-AMPA)', *Chirality* **9**: 274–280.

Johansen, T.N., Frydenvang, K., Ebert, B., Madsen, U. and Krogsgaard-Larsen, P. (1997b) 'Excitatory amino acid receptor antagonists: resolution, absolute stereochemistry, and pharmacology of

(S)- and (R)-2-amino-2-(5-*tert*-butyl-3-hydroxyisoxazol-4-yl)acetic acid (ATAA)', *Chirality* **9**: 529–536.

Johansen, T.N., Ebert, B., Bräuner-Osborne, H., Didriksen, M., Christensen, I.T., Søby, K.K., Madsen, U., Krogsgaard-Larsen, P. and Brehm, L. (1998) 'Excitatory amino acid receptor ligands: resolution, absolute stereochemistry, and enantiopharmacology of 2-amino-3-(4-butyl-3-hydroxy-isoxazolyl-5-yl)propionic acid', *J. Med. Chem.* **41**: 930–939.

Johansen, T.N., Stensbøl, T., Neilsen, B., Vogensen, S.B., Frydenvang, K., Sløk, F., Bräuner-Osborne, H., Madsen, U. and Krogsgaard-Larsen (2001) 'Resolution, configurational assignment and enantiopharmacology at glutamate receptors of 2-amino-3-(3-carboxy-5-methyl-4-isoxazolyl)propionic acid (ACPA) and Demethyl-ACPA', *Chirality* (in press).

Jones, K.A., Wilding, T.J., Huettner, J.E. and Costa, A.-M. (1997) 'Desensitization of kainate receptors by kainate, glutamate and diastereomers of 4-methylglutamate', *Neuropharmacology* **36**: 853–863.

Krogsgaard-Larsen, P., Honoré, T., Hansen, J.J., Curtis, D.R. and Lodge, D. (1980) 'New class of glutamate agonist structurally related to ibotenic acid', *Nature* **284**: 64–66.

Krogsgaard-Larsen, P., Brehm, L., Johansen, J.S., Vinzents, P., Lauridsen, J. and Curtis, D.R. (1985) 'Synthesis and structure–activity studies on excitatory amino acids structurally related to ibotenic acid', *J. Med. Chem.* **28**: 673–679.

Krogsgaard-Larsen, P., Ferkany, J.W., Nielsen, E.Ø., Madsen, U., Ebert, B., Johansen, J.S., Diemer, N.H., Bruhn, T., Beattie, D.T. and Curtis, D.R. (1991) 'Novel class of amino acid antagonists at non-N-methyl-D-aspartic acid excitatory amino acid receptors. Synthesis, in vitro and in vivo pharmacology, and neuroprotection', *J. Med. Chem.* **34**: 123–130.

Lauridsen, J., Honoré, T. and Krogsgaard-Larsen, P. (1985) 'Ibotenic acid analogues. Synthesis, molecular flexibility, and in vitro activity of agonists and antagonists at central glutamic acid receptors', *J. Med. Chem.* **28**: 668–672.

Madsen, U., Schaumburg, K., Brehm, L., Curtis, D.R. and Krogsgaard-Larsen, P. (1986) 'Ibotenic acid analogues. Synthesis and biological testing of two bicyclic 3-isoxazolol amino acids', *Acta Chem. Scand.* **B40**: 92–97.

Madsen, U., Brehm, L., Schaumburg, K., Jørgensen, F.S. and Krogsgaard-Larsen, P. (1990) 'Relationship between structure, conformational flexibility, and biological activity of agonists and antagonists at the N-methyl-D-aspartic acid subtype of excitatory amino acid receptors', *J. Med. Chem.* **33**: 374–380.

Madsen, U., Ebert, B., Krogsgaard-Larsen, P. and Wong, E.H.F. (1992) 'Synthesis and pharmacology of (RS)-2-amino-3-(3-hydroxy-5-flouromethyl-4-isoxazolyl)propionic acid, a potent AMPA receptor agonist', *Eur. J. Med. Chem.* **27**: 479–484.

Madsen, U. and Wong, E.H.F. (1992) 'Heterocyclic excitatory amino acids. Synthesis and biological activity of novel analogues of AMPA', *J. Med. Chem.* **35**: 107–111.

Madsen, U., Ebert, B., Hansen, J.J. and Krogsgaard-Larsen, P. (1993a) 'The non-depolarizing D-form of bromohomoibotenic acid enhances depolarizations evoked by the L-form or quisqualate', *Eur. J. Pharmacol.* **230**: 383–386.

Madsen, U., Frølund, B., Lund, T.M., Ebert, B. and Krogsgaard-Larsen, P. (1993b) 'Design, synthesis and pharmacology of model compounds for indirect elucidation of the topography of AMPA receptor sites', *Eur. J. Med. Chem.* **28**: 791–800.

Madsen, U., Bang-Andersen, B., Brehm, L., Christensen, I.T., Ebert, B., Kristoffersen, I.T.S., Lang, Y. and Krogsgaard-Larsen, P. (1996) 'Synthesis and pharmacology of highly selective carboxy and phosphono isoxazole amino acid AMPA receptor antagonists', *J. Med. Chem.* **39**: 1682–1691.

Madsen, U., Sløk, F.A., Stensbøl, T.B., Bräuner-Osborne, H., Lützhøft, H.-C.H., Poulsen, M., Eriksen, L. and Krogsgaard-Larsen, P. (2000) 'Ionotropic excitatory amino acid receptor ligands. Synthesis and pharmacology of a new amino acid AMPA antagonist', *Eur. J. Med. Chem.* **35**: 69–76.

McLennan, H. and Lodge, D. (1979) 'The antagonism of amino acid-induced excitation of spinal neurons in the cat', *Brain Res.* **169**: 83–90.

Monaghan, D.T., Bridges, R.J. and Cotman, C.W. (1989) 'The excitatory amino acid receptors: their classes, pharmacology, and distinct properties in the function of the central nervous system', *Ann. Rev. Pharmacol. Toxicol.* **29**: 365–402.

Møller, E.H., Egebjerg, J., Brehm, L., Stensbøl, T.B., Johansen, T.N., Madsen, U. and Krogsgaard-Larsen, P. (1999) 'Resolution, absolute stereochemistry, and enantiopharmacology of the GluR1–4 and GluR5 antagonist 2-amino-3-[5-*tert*-butyl-3-(phosphonomethoxy)-4-isoxazolyl]propionic acid', *Chirality* **11**: 752–759.

Nielsen, B., Fisker, H., Ebert, B., Madsen, U., Curtis, D.R., Krogsgaard-Larsen, P. and Hansen, J.J. (1993) 'Enzymatic resolution of AMPA by use of α-chymotrypsin', *Bioorg. Med. Chem. Lett.* **3**: 107–114.

Ohmori, J., Sakamoto, S., Kubota, H., Shimizu-Sasamata, M., Okada, M., Kawasaki, S., Hidaka, K., Togami, J., Furuya, T. and Murase, K. (1994) '6-(1*H*-Imidazol-1-yl)-7-nitro-2,3-(1*H*,4*H*)-quinoxalinedione hydrochloride (YM90K) and related compounds: structure–activity relationships for the AMPA-type non-NMDA receptor', *J. Med. Chem.* **37**: 467–475.

O'Neill, M.J., Bond, A., Ornstein, P.L., Ward, M.A., Hicks, C.A., Hoo, K., Bleakman, D. and Lodge, D. (1998) 'Decahydroisoquinolines: novel competitive AMPA/kainate antagonists with neuroprotective effects in global cerebral ischaemia', *Neuropharmacology* **37**: 1211–1222.

Ornstein, P.L., Arnold, M.B., Augenstein, N.K., Lodge, D., Leander, D.J. and Schoepp, D.D. (1993) '(3*SR*,4a*RS*,6*RS*,8a*RS*)-6-[2-(1*H*-Tetrazol-5-yl)ethyl)decahydroisoquinoline-3-carboxylic acid: a structurally novel, systematically active, competitive AMPA receptor antagonist', *J. Med. Chem.* **36**: 2046–2048.

Paternain, A.V., Morales, M. and Lerma, J. (1995) 'Selective antagonism of AMPA receptors unmasks kainate receptor-mediated responses in hippocampal neurons', *Neuron* **14**: 185–189.

Patneau, D.K. and Mayer, M.L. (1990) 'Structure–activity relationships for amino acid transmitter candidates acting at N-methyl-D-aspartate and quisqualate receptors', *J. Neurosci.* **10**: 2385–2399.

Patneau, D.K., Mayer, M.L., Jane, D.E. and Watkins, J.C. (1992) 'Activation and desensitization of AMPA/kainate receptors by novel derivatives of willardiine', *J. Neurosci.* **12**: 595–606.

Patneau, D.K., Vyklicky, L.J. and Mayer, M.L. (1993) 'Hippocampal neurons exhibit cyclothiazide-sensitive rapidly desensitizing responses to kainate', *J. Neurosci.* **13**: 595–606.

Sheardown, M.J., Nielsen, E.Ø., Hansen, A.J., Jacobsen, P. and Honoré, T. (1990) '2,3-Dihydroxy-6-nitro-7-sulfamoyl-benzo(F)quinoxaline. A neuroprotectant for cerebral ischemia', *Science* **247**: 571–574.

Shinozaki, H. (1988) 'Pharmacology of the glutamate receptor', *Prog. Neurobiol.* **30**: 399–435.

Shinozaki, H. (1992) 'Kainic acid receptor agonists', in P. Krogsgaard-Larsen and J.J. Hansen (eds) *Excitatory Amino Acid Receptors: Design of Agonists and Antagonists,* Chichester: Ellis Horwood, pp. 261–291.

Skjærbæk, N., Ebert, B., Falch, E., Brehm, L. and Krogsgaard-Larsen, P. (1995) 'Excitatory amino acids. Synthesis of (*RS*)-2-amino-3-(5-cyclopropyl-3-hydroxyisoxazol-4-yl)propionic acid, a new potent and specific AMPA receptor agonist', *J. Chem. Soc. Perkin Trans.* **I**: 221–225.

Sladeczek, F., Pin, J.P., Récasens, M., Bockaert, J. and Wiess, S. (1985) 'Glutamate stimulates inositol phosphate formation in striatal neurons', *Nature* **317**: 717–719.

Sløk, F.A., Ebert, B., Lang, Y., Krogsgaard-Larsen, P., Lenz, S.M. and Madsen, U. (1997) 'Excitatory amino acid receptor agonists. Synthesis and pharmacology of analogues of 2-amino-3-(3-hydroxy-5-methylisoxazol-4-yl)propionic acid (AMPA)', *Eur. J. Med. Chem.* **32**: 329–338.

Stensbøl, T.B., Borre, L., Johansen, T.N., Egebjerg, J., Madsen, U., Ebert, B. and Krogsgaard-Larsen, P. (1999a) 'Resolution, absolute stereochemistry and molecular pharmacology of the enantiomers of ATPA', *Eur. J. Pharmacol.* **380**: 153–162.

Stensbøl, T.B., Sløk, F.A., Trometer, J., Hurt, J., Ebert, B., Kjøller, C., Egebjerg, J., Madsen, U., Diemer, N.H. and Krogsgaard-Larsen, P. (1999b) 'Characterization of a new AMPA receptor radioligand, [^{3}H]2-amino-3-(3-carboxy-5-methyl-4-isoxazolyl)propionic acid', *Eur. J. Pharmacol.* **373**: 251–262.

Swanson, G.T., Green, T. and Heinemann, S.F. (1998) 'Kainate receptors exhibit differential sensitivities to (*S*)-5-iodowillardiine', *Mol. Pharmacol.* **53**: 942–949.

Tarnawa, I., Farkas, S., Berzsenyi, P., Pataki, A. and András i, F. (1989) 'Electrophysiological studies with a 2,3-benzodiazepine muscle relaxant: GYKI52466', *Eur. J. Pharmacol.* **167**: 193–199.

Turski, L., Jacobsen, P., Honoré, T. and Stephens, D.N. (1992) 'Relief of experimental spasticity and anxiolytic/anticonvulsant actions of the α-amino-3-hydroxy-5-methyl-4-isoxazolepropionate antagonist 2,3-dihydroxy-6-nitro-7-sulfamoylbenzo(F)quinoxaline', *J. Pharmacol. Exp. Ther.* **260**: 742–747.

Turski, L., Huth, A., Sheardown, M., McDonald, F., Neuhaus, R., Schneider, H.H., Dirnagl, U., Wiegand, F., Jacobsen, P. and Ottow, E. (1998) 'ZK200775: A phosphonate quinoxazolinedione AMPA antagonist for neuroprotection in stroke and trauma', *Proc. Natl. Acad. Sci. USA* **95**: 10,960–10,965.

Verdoorn, T.A., Johansen, T.H., Drejer, J. and Nielsen, E.Ø. (1994) 'Selective block of recombinant GluR6 receptors by NS-102, a novel non-NMDA receptor antagonist', *Eur. J. Pharmacol.* **269**: 43–49.

Wahl, P., Nielsen, B., Krogsgaard-Larsen, P., Hansen, J.J., Schousboe, A. and Miledi, R. (1992) 'Stereoselective effects of AMOA on non-NMDA receptors expressed in Xenopus oocytes', *J. Neurosci. Res.* **33**: 392–397.

Wahl, P., Madsen, U., Banke, T., Krogsgaard-Larsen, P. and Schousboe, A. (1996) 'Different characteristics of AMPA receptor agonists acting at AMPA receptors expressed in *Xenopus* oocytes', *Eur. J. Pharmacol.* **308**: 211–218.

Wahl, P., Anker, C., Traynelis, S.F., Egebjerg, J., Rasmussen, J.S., Krogsgaard-Larsen, P. and Madsen, U. (1998) 'Antagonist properties of a phosphono isoxazole amino acid at glutamate R1–4 (R,S)-2-amino-3-(3-hydroxy-5-methyl-4-isoxazolyl)propionic acid receptor subtypes', *Mol. Pharmacol.* **53**: 590–596.

Wätjen, F., Nielsen, E.Ø., Drejer, J. and Jensen, L.H. (1993) 'Isatin oximes – a novel series of bioavailable non-NMDA antagonists', *Bioorg. Med. Chem. Lett.* **3**: 105–106.

Wätjen, F., Bigge, C.F., Jensen, L.H., Boxer, P.A., Lescosky, L.J., Nielsen, E.Ø., Malone, T.C., Campbell, G.W., Coughenour, L.L., Rock, D.M., Drejer, J. and Marcoux, F.W. (1994) 'NS 257 (1,2,3,6,7,8-hexahydro-3-(hydroxyimino)-N,N,7-trimethyl-2-oxobenzo[2,1-b:3,4-c′]dipyrrole-5-sulfonamide) is a potent, systemically active AMPA receptor antagonist', *Bioorg. Med. Chem. Lett.* **4**: 371–376.

Watkins, J.C. and Evans, R.H. (1981) 'Excitatory amino acid transmitters', *Annu. Rev. Pharmacol. Toxicol.* **21**: 165–204.

Watkins, J.C., Krogsgaard-Larsen, P. and Honoré, T. (1990) 'Structure–activity relationships in the development of excitatory amino acid receptor agonists and competitive antagonists', *Trends Pharmacol. Sci.* **11**: 25–33.

Wilding, T.J. and Huettner, J.E. (1995) 'Differential antagonism of α-amino-3-hydroxy-5-methyl-4-isoxazolepropionic acid-preferring and kainate-preferring receptors by 2,3-benzodiazepines', *Mol. Pharmacol.* **47**: 582–587.

Young, A.B. and Fagg, G.E. (1990) 'Excitatory amino acid receptors in the brain: membrane binding and receptor autoradiographic approaches', *Trends Pharmacol. Sci.* **11**: 126–133.

Zorumski, C.F., Yamada, K.A., Price, M.T. and Olney, J.W. (1993) 'A benzodiazepine recognition site associated with the non-NMDA glutamate receptor', *Neuron* **10**: 61–67.

Part 2

Metabotropic glutamate receptors

Chapter 6

Metabotropic glutamate receptors (mGluRs)

Structure and function

J. Bockaert, L. Fagni and J.-P. Pin

Introduction and general presentation

Glutamate mediates transmission at most central excitatory synapses. The modulation of synaptic activity of those synapses is responsible for several forms of plasticity involved in learning and memory (Nakanishi, 1992). Excessive release of glutamate has been implicated in acute, as well as in slow, neuronal death associated with pathologies such as ischaemia, epilepsy, Parkinson's disease, amyotrophic lateral sclerosis, Huntington's chorea and Alzheimer's disease (Choi, 1988; Olney, 1991). Fifteen years ago, our conception of synaptic activation of glutamate receptors was simply explained by fast postsynaptic currents generated by AMPA receptor channels and a slower response generated by calcium-permeable NMDA receptor channels. Today, the glutamatergic synaptic transmission appears more complex.

First, the synaptic ionotropic glutamate receptors have recently been joined by kainate receptors (Lerma, 1997; Mulle *et al.*, 1998). All these ionotropic receptors are composed of a large number of subunits and spliced variants. Their subsynaptic localization and function can be modified following their interaction with associated proteins (such as SAP 95) and phosphorylation, respectively (for a review see Dingledine *et al.*, 1999).

Second, the discovery of glutamate metabotropic receptors (mGluRs) localized at pre- and postsynaptic sites of both glutamatergic and GABAergic neurons further add to the complexity of these synapses (Pin and Bockaert, 1995; Pin and Duvoisin, 1995; Conn and Pin, 1997). mGluRs were first characterized as phospholipase C(PLC)-coupled receptors (Sladeczek *et al.*, 1985; Nicoletti *et al.*, 1986), which are now classified as group I mGluRs composed of two members, mGluR1 and mGluR5. They are generally localized postsynaptically (Baude *et al.*, 1993; Shigemoto *et al.*, 1997), with some exceptions (Gereau and Conn, 1995a; Manzoni and Bockaert, 1995; Lu *et al.*, 1997). Additional mGluR subtypes, negatively coupled to adenylyl cyclase (AC), have been identified both pharmacologically and via the cloning of their cDNA. They constitute groups II (mGluR2, mGluR3 and DmGluRA, a Drosophila cloned mGluR) and III (mGlu R4, 6, 7 and 8) (Conn and Pin, 1997). They are mostly localized presynaptically (Shigemoto *et al.*, 1996, 1997; Masugi *et al.*, 1999).

The transduction mechanisms of mGluRs are much more complex than activation of PLC or inhibition of AC, as we will discuss in this chapter. The diversity of mGluRs is increased by the generation of spliced variants, all differing in their C-terminal domain (Figure 6.1). The interest in mGluRs has also been boosted by their structure,

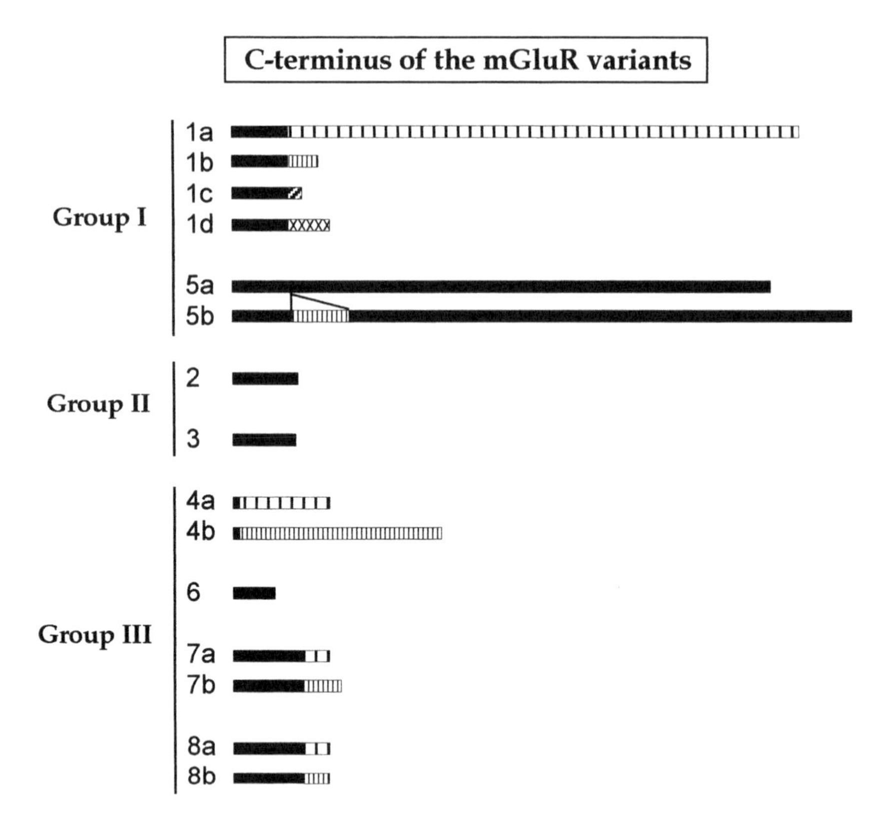

Figure 6.1 mGluR spliced variants. Schema representing the different C-terminals of mGluR spliced variants. In black are the identical sequences within one gene. Each of the other sequences is unique for a given variant.

which is completely different from that of all other G protein-coupled receptors (GPCRs), as it will be largely analysed here.

Due to their ability to control second messenger production, kinases, ionic channels, as well as ionotropic glutamate receptor activities, and to their pre- and postsynaptic localizations, mGluRs are ideal candidates to finely tune synaptic glutamatergic (but also GABAergic) synaptic transmissions.

In addition, their diversity and their functional properties make mGluRs target candidates for drugs designed to treat a variety of psychiatric and neurological disorders (Pin *et al.*, 1999).

mGluRs are members of family 3 GPCRs

GPCRs have a common basic domain structure with seven transmembrane helices (TM I–VII, called TMD here) connected by three intracellular (i1, i2, i3) and three extracellular (e1, e2, e3) loops. Although no high resolution structure of GPCRs remains to be determined, a low resolution (9Å) electron diffraction structure of rhodopsin has revealed the orientation of the transmembrane α-helices (Unger *et al.*,

1997). Three main families (1, 2, 3) of GPCRs are recognized. The total absence of sequence similarities between receptors from these families suggests that we are in the presence of a remarkable example of a molecular convergence (Bockaert and Pin, 1999).

Family 1 contains most GPCRs, including rhodopsin, receptors for odorants, catecholamines, peptides, but also for some high molecular weight glycoproteins such as LH and TSH. Family 2 GPCRs include high molecular weight hormone receptors such as glucagon, secretine, VIP-PACAP, PTH and also receptors for latrotoxin, the Black widow spider toxin. No endogenous ligand is known yet for latrotoxin receptors. Family 3 (Figure 6.2) has been the most intriguing one for a long time because it only contains metabotropic glutamate receptors (mGluRs) and the Ca^{2+}-sensing receptors in charge of sensing Ca^{2+} in parathyroid glands, kidney and brain (Brown *et al.*, 1993; Ruat *et al.*, 1996). Over the past two years, novel members, such as the GABA-B receptors and a group of putative pheromone receptors coupled to the G protein Go (termed VRs and Go-VN), have joined the family (reviewed in Bockaert and Pin, 1999). This year, distantly related putative taste receptors having 30 per cent identity with Ca^{2+}-sensing receptors and 20–30 per cent identity with mGluRs and pheromone receptors have also been described (Honn *et al.*, 1999). The intracellular loops of family 3 GPCRs are short compared to those of other GPCRs. They all possess a very large N-terminal domain (450–600 residues) which includes the binding site having some sequence similarity with periplasmic bacterial amino-acid binding proteins (PBP, such as leucine binding protein (LBP) and leucine, isoleucine, valine, binding protein (LIVBP) (O'Hara *et al.*, 1993). Downstream of the binding site, mGluRs and the Ca^{2+}-sensing receptor possess a domain rich in cysteines which is absent in the GABA-B receptors. This domain is necessary for the binding of glutamate (Figure 6.2).

In vertebrates, two other modestly represented GPCRs families do exist: family 4 comprising pheromones receptors (VNs) associated with the Gi protein (Dulac and Axel, 1995), and family 5 comprising the 'frizzled' and the 'smoothened' (Smo) receptors involved in embryonic development and in particular in cell polarity and segmentation (Bockaert and Pin, 1999).

The binding domain of mGluRs

As already stated, the binding site of mGluRs is localized at the N-terminal domain. This was demonstrated by studying the pharmacological and functional characteristics of chimeras. These include mGluR2/mGluR1 (Takahashi *et al.*, 1993), mGluR3/mGluR1 (Wroblewska *et al.*, 1997), mGluR4/mGluR1 (Tones *et al.*, 1995), DmGluRA/mGluR1 (Parmentier *et al.*, 1998) and even Ca^{2+}-sensing receptor/mGluR1 chimeras (Hammerland *et al.*, 1999). The association of the *N*-terminal domain of mGluR1 with the TMD of DmGluRA gave a functional receptor having the pharmacology of the mGluR1 (Parmentier *et al.*, 1998). Similarly, the association of the N-terminal domain of the Ca^{2+}-sensing receptor with the TMD of mGluR1 gave a functional receptor responding to Ca^{2+} (Hammerland *et al.*, 1999). Finally, production of the extracellular domain of mGluR1 and mGluR4 as soluble proteins has been found to be sufficient for binding mGluR1 and mGluR4 ligands respectively (Okamoto *et al.*, 1998; Han and Hampson, 1999). Similarly, the GABA-BR1 binding site is localized in the extracellular domain and some specific residues

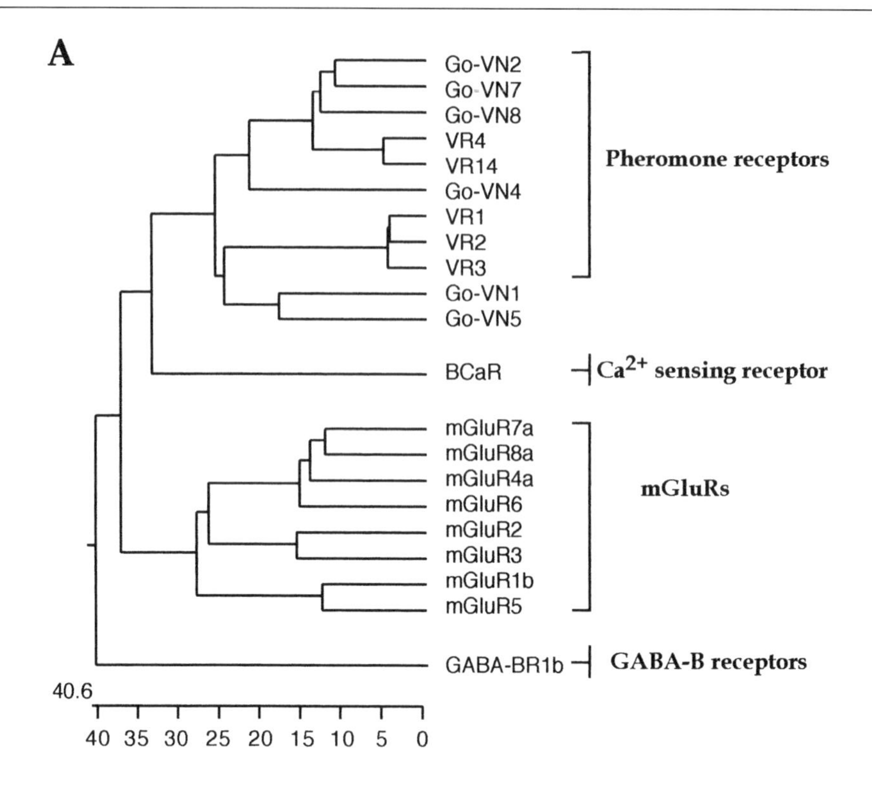
A
Go-VN2
Go-VN7
Go-VN8
VR4
VR14
Go-VN4
VR1
VR2
VR3
Go-VN1
Go-VN5
Pheromone receptors
BCaR
Ca2+ sensing receptor
mGluR7a
mGluR8a
mGluR4a
mGluR6
mGluR2
mGluR3
mGluR1b
mGluR5
mGluRs
GABA-BR1b
GABA-B receptors
40.6
40 35 30 25 20 15 10 5 0

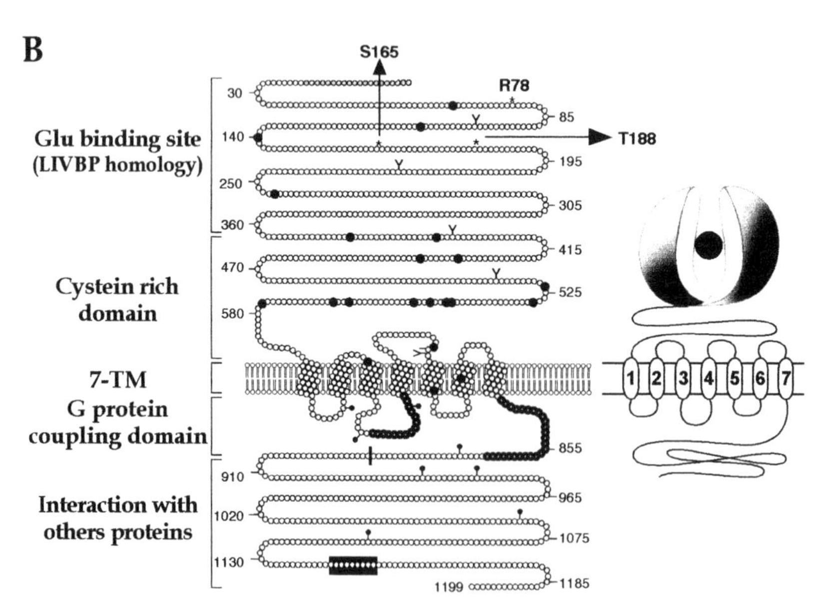
B
S165
R78
T188
Glu binding site
(LIVBP homology)
Cystein rich
domain
7-TM
G protein
coupling domain
Interaction with
others proteins
30
85
140
195
250
305
360
415
470
525
580
855
910
965
1020
1075
1130
1185
1199
1
2
3
4
5
6
7

Figure 6.2 (a) Dendrogram of family 3 GPCRs. The dendrogram was established with Clustal W. Four subfamilies can be recognized: the pheromone receptors coupled to Go, the Ca^{2+}-sensing receptor, the mGluRs and the GABA-B receptors. (b) Schematic representation of the mGluR1a.

The cysteine residues conserved among all mGluRs are indicated by black circles. In the N-terminal domain:

- the first N-terminal stretch of residues (in grey) corresponds to the signal peptide;
- the putative LIVBP domain contains the glutamate binding site;
- the three putative main residues interacting with glutamate (R78 (see text), S165 and R188) are indicated by a star;
- the putative glycosylation sites are indicated by Y.

In the intracellular loops and the N-terminal part of the C-terminal domain:

- the domains which determines the G protein coupling specificity are in bold. The putative intracellular phosphorylation sites are indicated by small black dots attached to the amino-acid.

In the C-terminal domain:

- the position of the common splicing site of mGluR1 is indicated by the vertical bar;
- the black rectangle indicates the localization of the Homer domain (PPXXFR) within the domain supposed to interact with other proteins.

On the right part, the 'fly trap' binding domain is represented in the closed state containing glutamate (ball).

involved in the binding of some antagonists and agonists have been identified (Galvez *et al.*, 1999). It was Patrick O'Hara who first proposed that the extracellular domain of mGluR1 shares some structural similarities with some bacterial periplasmic binding proteins (PBP) such as LBP (leucine binding protein) and LIVBP (leucine, isoleucine binding protein) (O'Hara *et al.*, 1993). A similar structure has also been proposed for the binding site of ionotropic glutamate receptors and verified by resolving the crystal structure of this domain in GluR2, an AMPA receptor subunit (Armstrong *et al.*, 1998). According to this model, the binding site of family 3 GPCRs is constituted of two lobes interconnected by 3 linkers (Figure 6.3, see colour section). The agonist is supposed to bind to lobe 1 within the cleft between the two lobes and according to the Venus fly trap model of PBP, this induces the closure of the two lobes. Homology modelling and mutagenesis suggest that Ser^{165} and Thr^{188} of mGluR1, two residues conserved in the entire mGluR family (including Drosophila and *C. elegans*), play a crucial role in glutamate binding. They may be forming a hydrogen bond with the α-carboxyl and α-amino-group of glutamate, respectively (Figure 6.3). The γ-carboxyl may bind to an Arg residue (Figure 6.3) which has not been localized. The most likely one is a conserved Arg – likely Arg^{61} in mGluR2, Arg^{78} in mGluR1. This residue occupies the expected position in mGluR1, mGluR2 and mGluR8 models that we (unpublished data) and others (Constantino and Pelliciari, 1996) constructed. However, this hypothesis remains to be validated by site directed mutagenesis. Arg^{57} and Ser^{167} have been shown to be important factors for the binding of a group II tritiated antagonist ([^{3}H]-LY341495), but they are not conserved in mGluRs of other groups and therefore are not likely to be implicated in glutamate binding (Yang *et al.*, 1998). We have also constructed a 3D model of the LBP-like domain of the GABA-BR1 receptor based on known structures of LBP. This model predicts that four cysteines, two on each lobe, are important for the folding of the protein, likely through a formation of disulfide bridges. This receptor mutation of Ser^{269} was found to affect the affinity of various GABA analogs differently, indicating that this residue is involved in the selectivity of recognition of GABA-B ligands. The mutation of two other residues, Ser^{247} and Gln^{312}, was found to increase the affinity of agonists and decrease the affinity of antagonists, an effect which can be explained by the Venus flytrap model (Galvez *et al.*, 1999). In Ca^{2+}-sensing receptors nothing is known about the binding site, but numerous mutations within the PBP domains activate mutations, suggesting that structural changes of that region lead to receptor activation (Brown and Herbert, 1997).

The TMD of mGluRs

We know almost nothing about the topology of TM helices of family 3 GPCRs. Three-dimensional models of family 1 GPCRs have been constructed based on the projection map of bacteriorhodopsin and rhodopsin, and they generally fit with the data obtained from site-directed mutagenesis studies. The absence of sequence similarities between family 1 and family 3 GPCRs, as well as the absence of mutagenesis studies on the TMD region of family 3 GPCRs, prevents the construction of similar models for mGluRs.

Interestingly, the TMD which is likely to be involved in the transduction of the structural changes, triggered within the N-terminal domain by the ligand to the intracellular domains, has been found to be the site of interaction of CPCCOEt, a specific mGluR1 antagonist (Litschig *et al.*, 1999). This compound has no amino-acid-like structure and is a non-competitive antagonist. Recently, two residues localized on top of TMVII of mGluR1, Thr^{815} and Ala^{818}, have been shown to be responsible for the binding of CPCCOEt (Litschig *et al.*, 1999). It is likely that MPEP, a non-competitive and specific mGluR5 antagonist, also interacts with the TMD (Gasparini *et al.*, 1999). It is possible, but not demonstrated, that NPS-R-467 and NPS-R-568, two allosteric positive modulators of the Ca^{2+}-sensing receptors, also interact with the TMD (Nemeth *et al.*, 1998). Moreover, Gd^{3+} has been shown to activate a truncated Ca^{2+}-sensing receptor lacking the N-terminal domain (Gasparini *et al.*, 1999). These data indicate that the TMD of family 3 GPCRs, like the TMD of family 1, can adopt an active state in the absence of the N-terminal domain.

Activation and coupling to G proteins

For family 1 GPCRs, and particularly rhodopsin, the structural modifications of the TMD following receptor activation are correctly documented. 'Site directed spin labelling' experiments done with rhodopsin in which the spin marker has been introduced at the junction between TMIII and i2 (Cys^{139}) on the one hand and the C-terminal part of i3 on the other (Cys^{248}), clearly indicated that activation results in a 30 degree clockwise rotation of TMVI (viewed from the cytoplasmic face) which moves apart from TMIII. The TMII, III, VI and VII move apart, whereas TM I, IV and V are more stable (for a review see Bockaert and Pin, 1999). In family 1 GPCR, one residue (Asp) in TMII, and a tripeptide (DRY or ERW) at the interface of TMIII and i2 are important for receptor activation. Since these residues are not conserved in the other GPCR families, one may conclude that either the change in conformation of the core domain or the molecular events leading to these changes are not conserved between members of these different families.

These changes in conformation of the 'core domain' affect the conformation of the i2 and i3 intracellular loops (which are directly linked to TMIII and TMVI, respectively). These loops constitute one of the key sites for G-protein recognition and activation. They certainly provide a way of fishing for specific domains of the G proteins (in particular their C-terminal domain) within the 'core domain'.

In family 3 GPCRs the first problem is to understand how the Venus flytrap, once the glutamate is inside, can induce the structural changes of the TMD. Several hypotheses can be proposed:

1 The flytrap interacts with the TMD and delivers glutamate to the TMD. Glutamate activates the TMD as small ligands of family 1. This is unlikely because chimera having the N-terminal domain of mGluR1 and the TMD of the Ca^{2+}-sensing receptor respond to glutamate, whereas the reciprocal chimera responds to Ca^{2+} (Hammerland *et al.*, 1999).
2 The flytrap, with its ligand, interacts with the TMD and behaves as an agonist.
3 The ligand modifies the structural relationship between two partners of a dimer, a situation which can be analogous to that of tyrosine kinase receptors of the receptor dimer. Indeed, as for other GPCRs, family 3 GPCRs form dimers. In mGluR and Ca^{2+}-sensing receptors the homodimers are formed (in the absence of ligand) via a disulfide bridge between their N-terminal domains (Romano *et al.*, 1996; Fan *et al.*, 1998; Ward *et al.*, 1998; Pace *et al.*, 1999) (Figure 6.4). In GABA-B, GABA-BR_1, GABA-BR_2 heterodimers are formed via coiled-coil interaction of their C-terminal domain (White *et al.*, 1998; Kuner *et al.*, 1999) (Figure 6.4). In the case of GABA-B receptors, the heterodimer is an absolute requirement for an efficient coupling to G proteins.

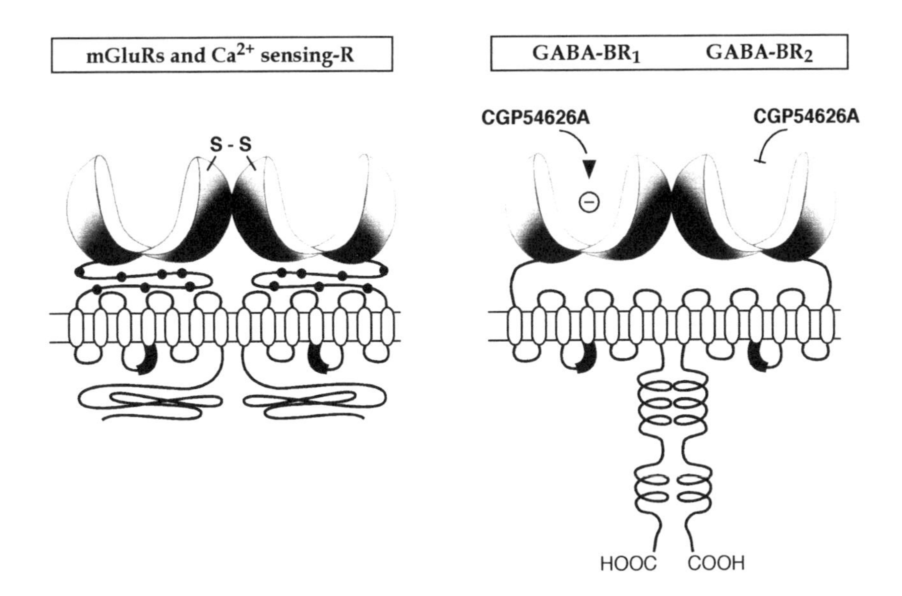

Figure 6.4 Homodimers and heterodimers of family 3 GPCRs. mGluRs and Ca^{2+}-sensing receptors form homodimers. The two subunits are likely to be linked by a S-S bridge connecting their two 'flytrap' domains. GABA-BR1 and GABA-BR2 form heterodimers. The two subunits are likely to be linked by coiled–coil interaction of their C-terminal domains. The pharmacology of GABA-BR1 (which binds the antagonist CGP-5466A) differs from that of GABA-BR2 (which does not bind CGP-5466A).

In the case of Ca^{2+}-sensing receptors, the covalent dimerization via the S-S bridge facilitates the receptor activity which remains present (Pace *et al.*, 1999). Dimerization via the TMD also occurs in Ca^{2+}-sensing receptors (Bai *et al.*, 1998). The next question is to know which intracellular domains are involved in G-protein coupling. The chimera generated using receptors with different G-protein selectivity revealed that the i2 loop of family 3 GPCRs plays a critical role in the specificity of G-protein recognition, whereas the other loops (in particular i3) and the C-terminal domain are important for the efficacy of coupling. Using a series of Gαq chimera in which the last five residues have been replaced by those of Gαi, Gαo or Gαz, as well as Gα15 and Gα16 (which have identical C-terminal domains), we found that groups II and III mGluRs have different profiles of specificity when considering their coupling to G proteins. We also found that the difference in specificity between mGluR2 and mGluR4 for their coupling to Gαqo and Gαqz is due to residue −4 (C in Gαqo and I in Gαqz) (Figure 6.5) (Blahos *et al.*, 1998; Parmentier *et al.*, 1998). In contrast, the difference in the coupling of groups II and III to Gα15 (which was activated by both groups) and Gα16 (which was activated only by group II) cannot be due to a difference in the C-terminal of these Gα subunits because they are identical. Recent data from our laboratory indicate that the L9 loop of these Gα proteins is one of the most important domains determining the differential specificity of coupling to these mGluRs.

G protein coupling specificity

A

		Gαq	Gαqi	Gαqo	Gαqz	Gαqo(CI)	Gαqz(IC)	Gα15	Gα16
Group II	mGluR2	0	+	+	0	0	+	+	+
	DmGluRA	0	+	+	0	-	-	+	+
Group III	mGluR4a	0	+	+	+	+	+	+	0
	mGluR7a	0	+	+	+	-	+	+	0
	mGluR8a	0	+	+	+	-	+	+	0
Chimera	mGluR2/4i2				+			+	0

B

	-5	-4	-3	-2	-1
αqo	G	C	G	L	Y
αqz	G	I	G	L	C

	-5	-4	-3	-2	-1
α15	E	I	N	L	L
α16	E	I	N	L	C

Figure 6.5 G-protein coupling specificity.

(a) The coupling of group II and III mGluRs (including $DmGluR_A$, a Drosophila mGluR with high primary sequence homologies and close pharmacology to group II mGluRs (Parmentier *et al.*, 1996)) and of a chimera in which the i2 loop of mGluR2 has been replaced by that of mGluR4 (named mGluR2/4i2) to:

1 natural (Gαq, Gα15 and Gα16);
2 chimera of Gαq in which the last five residues have been replaced by those of Gαo (named Gαqo) or those of Gαz (named Gαqz);
3 Gαqo in which Cys residue at position −4 has been replaced by Ile (named Gαqo(CI);
4 Gαqz in which the Ile residue at position −4 has been replaced by Cys.

(b) Amino-acid sequence of the C-terminal domain of Gαo, Gαz, Gα15, Gα16 used in (a).

As in other GPCRs families, the C-terminal domain of mGluRs is important for their fine-tuning to G proteins. The long C-terminal variants (more than 350 residues) of mGluRs such as mGluR1a and mGluR5 have a spontaneous coupling to G proteins, i.e. agonist-independent when expressed in heterologous systems (Prézeau *et al.*, 1996). In contrast, no such coupling was observed with the short C-terminal tail splice variants like mGluR1b,c,d. Mutagenesis indicated that the absence of agonist-independent activity of these short receptors resulted from the inhibitory action of a cluster of four basic residues (KKRR) located close to their extreme carboxy terminus (Mary *et al.*, 1998). Since this cluster is also found in mGluR1a, we proposed that the long carboxy-terminal sequence of mGluR1a prevents the action of this short inhibitory sequence. The PKC phosphorylation of a single threonine (T840) residue in the C-terminal domain of mGluR5 (not present in mGluR1) plays a crucial role in generating Ca^{2+} oscillations associated with mGluR5, but not with mGluR1 activation (Kawabata *et al.*, 1996). Ca^{2+} oscillations have been observed in transfected cells as well as in astrocytes (Nakahara *et al.*, 1997) and express mGluR5 rather than mGluR1. PKC phosphorylation of mGluRs is also involved in their desensitization (Alaluf *et al.*, 1995b; Gereau and Heinemann, 1998). Interestingly, mGluR5 desensitization can be reversed by low (not high) NMDA receptor activation, likely via the activation of Ca^{2+}-dependent serine/threonine phosphatases (Alagarsamy *et al.*, 1999). Since group I mGluRs potentiate NMDA receptor activation (Ben-Ari *et al.*, 1992), this reciprocal potentiation may be highly important in amplification and induction of NMDA-dependent processes.

Transduction mechanisms

Several extensive reviews have discussed in detail the transduction mechanisms associated to mGluR activation (Pin and Bockaert, 1995; Pin and Duvoisin, 1995; Conn and Pin, 1997). We will only deal here with the recent aspects of this question and try to show the highly pleiotropic aspects of these transductions.

Transduction mechanisms of group I mGluRs: more than a simple activation of PLC

The classical coupling of group I mGluRs to PLC is responsible for the production of InsP3 and for the activation of PKC. Both are responsible for a wide variety of cellular events. The most classical one is an InsP3-induced release of Ca^{2+} from intracellular Ca^{2+} stores localized in different cellular compartments, including postsynaptic spines (Finch and Augustine, 1998; Takechi *et al.*, 1998). This release of Ca^{2+} triggers a variety of biochemical events and modulation of Ca^{2+}-dependent channels (Fagni *et al.*, 1991; Fiorillo and Williams, 1998). Among PKC-mediated events we can quote their own phosphorylation (Alaluf *et al.*, 1995a; Gereau and Heinemann, 1998) and the potentiation of NMDA and AMPA ionotropic glutamate receptors (Ben-Ari *et al.*, 1992). In addition, a series of group I mGluR-mediated cellular events do not seem to result from a simple activation of PLC.

Mobilization of InsP3 and ryanodine intracellular Ca^{2+} stores

In Purkinje cell spines, the intracellular Ca^{2+} stores mobilized by mGluR1 are clearly InsP3-sensitive stores. This pathway is directly implicated in inducing long-term depression (LTD) (Finch and Augustine, 1998; Daniel *et al.*, 1999) recorded at synapses between parallel fibres and Purkinje cells. In cerebellar granular cells (Chavis *et al.*, 1996), in midbrain dopaminergic neurons (Fiorillo and Williams, 1998), or in presynaptic reticulospinal axons of lamprey (Cochilla and Alford, 1998), activation of mGluR1 mobilized essentially the intracellular ryanodine/caffeine sensitive pools. It is not clear whether or not activation of mGluR1 first causes InsP3-induced Ca^{2+} release and then a Ca^{2+}-induced Ca^{2+} release through ryanodine receptors (Irving *et al.*, 1992). Some recent data from our laboratory suggest a more complex relationship between mGluR1 and ryanodine receptors, as now discussed.

In cerebellar granular cells, mGluR1 activates L-type Ca^{2+} channels by triggering their tight association with ryanodine receptors

We have found that mGluR1 can trigger Ca^{2+} entry through L-type Ca^{2+} channels in a ryanodine-dependent manner even after blockade of InsP3 receptors with heparin (Chavis *et al.*, 1996). This effect was mimicked by caffeine, but not by InsP3, and involved a Pertussis toxin-resistant G protein. Stimulation of mGluR1 enhanced L-type Ca^{2+} channels recorded in cell-attached. This stimulation resists after excision of the patch into the inside-out configuration (Figure 6.6). Then, it can be blocked by ryanodine applied to the internal surface of the patch, suggesting a close functional interaction between the ryanodine receptor and plasma membrane Ca^{2+} channels. This is reminiscent of the cross-talk between ryanodine receptor type 1 and L-type Ca^{2+} channels in skeletal muscle cells.

mGluR1, but not muscarinic receptor-mediated activation of Ca^{2+}-sensitive K^+ channels, is mediated via the ryanodine-L-type Ca^{2+} channel interaction

Interestingly, in cerebellar granular cells, the activation of Ca^{2+}-dependent K^+ channels by mGluR1, but not by muscarinic receptors, was blocked by ryanodine and nitrendipine, an L-type Ca^{2+} channel blocker (Chavis *et al.*, 1998a). This indicated that when mGluR1 were activated, the Ca^{2+} entry via L-type Ca^{2+} channels was responsible for the activation of Ca^{2+}-dependent K^+ channels. Similarly, in midbrain dopaminergic neurons and hippocampal CA3 pyramidal neurons, the activation of Ca^{2+}-sensitive K^+ channels by mGluR1 was ryanodine-sensitive (Fiorillo and Williams, 1998).

In contrast, when muscarinic receptors were activated in cerebellar granular cells, Ca^{2+} was released from InsP3 receptors responsible for the activation of Ca^{2+}-dependent K^+ channels (Figure 6.6).

Activation of a Na^+/Ca^{2+} exchanger

Several data indicate that group I mGluRs can directly activate, or in coincidence with AMPA receptor activation, a Na^+/Ca^{2+} exchanger generating non-selective cation currents (Staub *et al.*, 1992) or an increase in phospholipase A2 (Dumuis *et al.*, 1990, 1993). It is not known whether the mechanism involves activation of PLC.

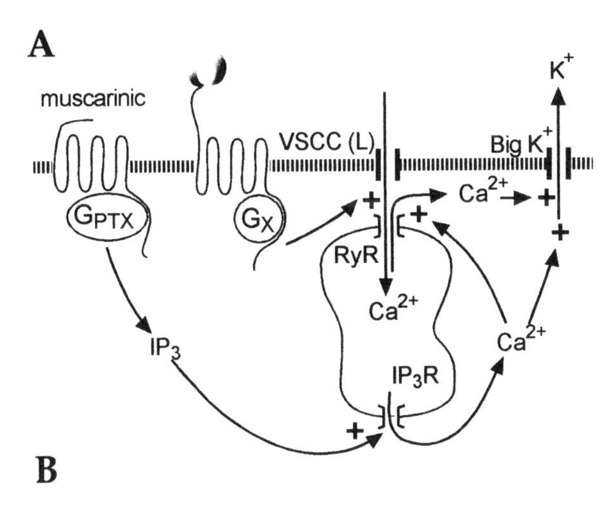

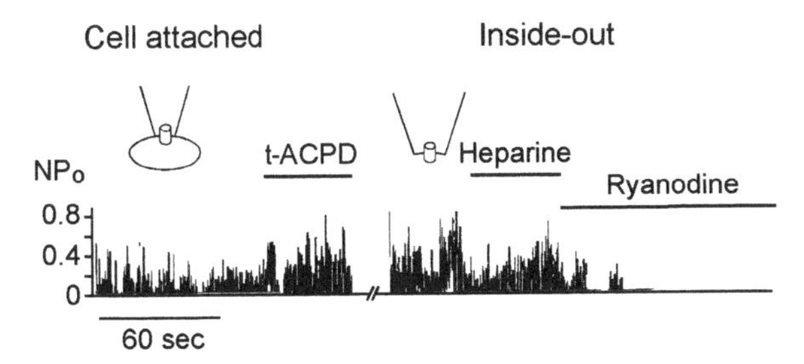

Figure 6.6 Relationships between mGluRs, intracellular Ca^{2+} stores and plasma membrane ionic channels.

(a) Schematic representation of putative interactions between group I mGluRs, intracellular Ca^{2+} stores and membrane voltage-sensitive Ca^{2+} channels (VSCC) of type LGX and GPTX are G protein-insensitive and G-protein-sensitive to Pertussis toxin, respectively. Big K^+ is the big K^+ channel. RyR and IP3R are the ryanodine and InsP3 sensitive receptors on the intracellular Ca^{2+} stores which are represented here by a unique reservoir. (b) L-type VSCC activity was recorded on cerebellar granular cells using the cell-attached mode of patch clamp technique. The channel open probability (NPo) is monitored. Addition of trans-ACPD (400 μM), a mGluR agonist, increased the open probability. This effect was maintained after excision of the patch into the inside-out configuration. Heparin did not attenuate the activation. In contrast, ryanodine completely inhibited the Ca^{2+} channel activity. These experiments (see Chavis *et al.*, 1996, 1998a) suggest that mGluR1 induced a tight coupling between RyR and L-type VSCC leading to channel activation and an increase in intracellular Ca^{2+}. The latter activated the big K^+ channel. Muscarinic receptors activated the same channel but via the IP3 sensitive Ca^{2+} store.

Inhibition of the IAHP current, activation of cation conductance in hippocampal neurons by group I mGluRs

One of the classical effects of mGluRs is the inhibition of the IAHP (inward after-hyperpolarization) current following action potentials (Charpak *et al.*, 1990). Generally, this current is partly mediated by inhibition of apamine-sensitive Ca^{2+}-dependent K^+ channels (SK channels) but also by apamine-insensitive channels. In CA3 neurons, group I-mediated inhibition (Gereau and Conn, 1995b) was Ca^{2+}-independent and not mediated by PKC and PKA (Gerber *et al.*, 1992). In CA3 neurons, glutamate receptors also activate cationic channels in a G-protein-independent manner (Guérineau *et al.*, 1995). However, the mGluRs nature of the receptors involved in the latter effect is not clear. In CA1 neurons, group I mGluRs activated a non-selective cationic channel (Congar *et al.*, 1997).

mGluR1 heterologously expressed in sympathetic neurons inhibited Ca^{2+} currents and M-type K^+ (IM) currents in a complex manner

The nature of coupling of mGluR1a obviously depends on the nature of the neurons in which they are expressed. It also certainly depends on their subcellular localizations. For example, it has recently been shown that mGluR1a (1) inhibits, in a voltage-dependent manner, N-type Ca^{2+} channels via $G\beta\gamma$ released from Gi/Go and Gq/G_{11}, and (2) inhibits, in a voltage-independent manner, likely via Gq and G_{11}, N-type Ca^{2+} channels and M-type K^+ channels. This pathway probably involves PLC activation (Kammermeier and Ikeda, 1999).

Time-dependent modulation of group I mGluRs coupling

Fiorillo and Williams observed that glutamatergic synaptic transmission resulted in postsynaptic midbrain dopaminergic neurons, in the generation of a rapid and transient IPSPs followed by a slow (excitatory postsynaptic potential) (EPSP Fiorillo and Williams, 1998). Both postsynaptic potentials were mediated by mGluRs. The IPSP was due, as already described, to the activation of Ca^{2+}-dependent K^+ channels by Ca^{2+} released from the ryanodine-sensitive stores, whereas the EPSP was due to blockade of voltage-dependent K^+ channels. Thus, stimulation of group I mGluR in the same postsynaptic neuron resulted in two temporally distinct events. It has been proposed that mGluR1 switched from a coupling leading to activation of Ca^{2+}-dependent K^+ channels to a coupling leading to inhibition of other K^+ channels (Pin, 1998). The former coupling would be transient because of a rapid desensitization of this pathway. Indeed, only a fast application of mGluR1 agonists can generate IPSP. A possible molecular basis for such a switch can be proposed based on mechanisms described for β2-adrenergic receptors. Indeed, this receptor switches from a coupling to Gs (stimulation of adenylyl cyclase) to a coupling to Gi (inhibition of adenylyl cyclase) after it has been phosphorylated by PKA (Daaka *et al.*, 1997). Another switch between positive and negative effects mediated by group I mGluRs has been reported at the presynaptic level. Sanchez-Prieto's laboratory found that group I mGluRs potentiated glutamate release but that the same receptors inhibit glutamate release when their coupling to PLC (and particularly to diacylglycerol production) has been desensitized (Herrero *et al.*, 1998).

Homer proteins modulate transduction mechanisms of group I mGluRs

The discovery that mGluR1a and mGluR5 interact via a specific Homer binding sequence (PPXXFR) in their long C-terminal domain with a domain of a series of Homer proteins, homologous to the EVH1 domain of the Ena/VASP protein family, adds to the complexity of the transduction machinery (Brakeman *et al.*, 1997). Homer proteins (Homer 1a/b/c, Homer 2 and Homer 3) have been cloned from mouse brain,

Drosophila and human. With the exception of Homer 1a, they can all form dimers or multimers via coiled–coil interactions of their C-terminal domains. They also interact via their EVH1 domain with a Homer sequence of InsP3 and ryanodine receptors (Tu *et al.*, 1998; Xiao *et al.*, 1998). Therefore, one can imagine that mGluR1a and mGluR5 can form clusters via Homer proteins with intracellular Ca^{2+} stores, a feature which may explain some observations described above. Homer 1a is a C-terminal truncated form of Homer which does not form coiled–coil interaction with other Homer proteins and yet interacts with mGluR1a. Thus, Homer 1a can disrupt the bridge between mGluR1a and InsP3 receptors assembled by Homer dimers (Tu *et al.*, 1998; Xiao *et al.*, 1998). This would reduce the speed of mGluR-induced Ca^{2+} release from intracellular stores. Since Homer 1a is an immediate early gene expressed under epileptic activity or during LTP induction (Brakeman *et al.*, 1997; Kato *et al.*, 1998), its expression slows down mGluR1-induced intracellular Ca^{2+} release.

Transduction mechanisms of groups II and III mGluRs: much more than inhibition of adenylyl cyclase (AC)

The best characterized transduction mechanism associated with groups II and III mGluRs is the inhibition of AC. Easily demonstrated in transfected cells but also in neurons, the physiological importance of this inhibition is not always evident (Pin and Duvoisin, 1995; Conn and Pin, 1997). Since groups II and III mGluRs are mainly implicated in the inhibition of neurotransmitter release, the inhibition of cAMP production may be part of the many additional mechanisms involved. For example, at interneuron–Purkinje cell synapses, it has been demonstrated that glutamate released from Purkinje dendrites acts as a retrograde messenger on presynaptic terminals of interneurons. There, glutamate inhibits GABA released via a group II mGluR. Since this effect was reduced by forskolin, an AC activator (Glitsch *et al.*, 1996), mGluR inhibition of cAMP production is likely to be involved. However, most of groups II and III mGluR functions do not require inhibition of AC.

Inhibition of Ca^{2+} channels

Numerous studies have shown that groups II and III mGluRs inhibit the L, N and P/Q type Ca^{2+} channels, depending on the neuron considered (for review, see Conn and Pin, 1997). When tested, these inhibitions were blocked by Pertussis toxin indicating that a Gi or a Go protein is involved. The inhibition of the N and P/Q type Ca^{2+} channels by mGluRs is certainly due to a direct interaction between $G\beta\gamma$ released from Gi/Go with the channels as demonstrated in other systems. Intracellular messengers are therefore not required (Bourinet *et al.*, 1996; Herlitze *et al.*, 1997). However, we have found in cerebellar granular cells that the inhibition of L-type Ca^{2+} channels by group II mGluRs can be observed in the cell-attached configuration of the patch clamp technique, indicating that a diffusible messenger should exist between the receptors localized outside the patch and the channels, or that $\beta\gamma$ (or $G\alpha o$ because the L-type Ca^{2+} channels have no consensus binding site for $\beta\gamma$) can diffuse under the patch. Since the inhibition was observed in the presence of IBMX and saturating concentrations of cAMP, an inhibition of cAMP production could not explain the results (Chavis *et al.*, 1994).

Activation of K^+ channels

The mechanisms by which groups II and III mGluRs inhibit glutamate release are diverse. Among them, activation of sensitive K^+ channels have been demonstrated in the reticulospinal cord of the lamprey (Cochilla and Alford, 1998).

Direct inhibition of the neurotransmitter release machinery

Another way of inhibiting glutamate release is to control the mechanism of vesicular release at a step downstream of Ca^{2+} entry in the presynaptic terminals. It has been shown in CA1 neurons (Gereau and Conn, 1995a), CA3 neurons (Scanziani *et al.*, 1995), cortico-striatal co-cultures (Tyler and Lovinger, 1995) and neurons from accumbens (Manzoni *et al.*, 1997), that the mEPSCs, frequency, but not their amplitude, is reduced by group II and III agonists even in the presence of voltage-dependent Ca^{2+} channels and cAMP. Using direct visualization of synaptic activity, we have shown that in cerebellar granular cells L-AP4, a group III specific agonist, reduced the cAMP/PKA-mediated enhancement of synaptic glutamate release by acting downstream of Ca^{2+} channels (Chavis *et al.*, 1998b) (Figure 6.7; see colour section). All these effects were Gi/Go mediated events since they were blocked by Pertussis toxin.

Activation of cAMP production

It is known that AC of group II is activated by coincident interaction with $G\alpha s$ of Gs and $\beta\gamma$ generally released from Gi/Go (Taussig and Gilman, 1995). It is certainly the reason why activation of groups II and III mGluRs potentiate cAMP production mediated by Gs-activating receptors, such as β-adrenergic in brain slices. Evidence suggests that such an interaction can occur in glial cells, and this generates local cAMP-dependent production of adenosine. Adenosine may then inhibit glutamate release via presynaptic adenosine A1 receptors (for a review, see Conn and Pin, 1997).

Activation of cGMP phosphodiesterase

mGluR6, the L-AP4-sensitive mGluR of ON-bipolar cells of retina, has been proposed to activate a cGMP phosphodiesterase (Nawy and Jahr, 1990). However, a direct demonstration of this effect is still lacking. The mechanism is proposed to be the same as the one occurring in rod cells following light activation. As in rods, the G protein involved is both PTX-sensitive and cholera toxin-sensitive, but has not been fully characterized (Shiells and Falk, 1990).

Roles of mGluRs in synaptic transmission

Most of the transduction mechanisms reviewed above lead either to the modulation of rapid synaptic ionotropic (glutamatergic or GABAergic) transmission or to the generation of slow postsynaptic currents.

Presynaptic events: modulation of neurotransmitter release

Activation of transmitter release by group I mGluRs

Pharmacological evidence: In cortical synaptosomes, activation of group I mGluRs increased 4-AP-induced glutamate release, an effect mediated by PKC (Herrero *et al.*, 1992; Moroni *et al.*, 1998; Reid *et al.*, 1999). This effect is transient and followed by an inhibition of glutamate release (Herrero *et al.*, 1994, 1998). In hippocampal interneurons, group I mGluR agonists stimulate GABA release which generates IPSCs in postsynaptic CA3 neurons (Poncer *et al.*, 1995).

Physiological evidence: At the reticulospinal axon synapses of the lamprey, repetitive firing (five action potentials at 50 Hz) generated EPSCs of increasing intensity. Application of CPCCOEt, a mGluR1-specific antagonist, did not modify the presynaptic action potential but massively reduced the gradual increase in postsynaptic EPSCs. These data, plus the demonstration that group I mGluRs are localized presynaptically leading to an increase in presynaptic Ca^{2+} (released from ryanodine-sensitive pools), suggest that presynaptic mGluRs potentiated glutamate release (Cochilla and Alford, 1998).

Inhibition of transmitter release by group I mGluRs

Pharmacological evidence suggests that group I mGluRs inhibit glutamate release at the Schaeffer collateral-CA1 (Gereau and Conn, 1995a; Manzoni and Bockaert, 1995) and GABA release at interneurons-CA1 synapses in adult rats (Gereau and Conn, 1995a; Poncer *et al.*, 1995). In Schaeffer collateral-CA1 synapses, the receptors involved are mGluR5 since the effect was absent in mGluR5 knock-out mice (Lu *et al.*, 1997).

Inhibition of transmitter release by group II and III mGluRs

Pharmacological evidence: There are numerous reports which show that application of group II and III agonists inhibit glutamatergic transmission in many synapses, and this has been reviewed (Conn and Pin, 1997). This is consistent with the localization of group II mGluRs at the periphery of the presynaptic terminals and of group III (mGluR7) within the synaptic grid (Shigemoto *et al.*, 1996, 1997). Some reports indicate that mGluR7 can also be present within the synaptic grid of GABAergic synapses (Lujan *et al.*, 1998), and that group II and III agonists can inhibit GABAergic transmission (for a review, see Conn and Pin, 1997). The problem concerning mGluR7 is that its affinity is so low (mM) that one wonders where glutamate comes from acting within the GABAergic synapses.

Physiological evidence: Scanziani *et al.*, have shown at the hippocampal mossy fibre synapses, that glutamate that accumulates within the synapse following high frequency stimulation activates presynaptic mGluRs (likely group II), which inhibits glutamatergic transmission. This was not observed at low frequency stimulation (Scanziani *et al.*, 1997). This use-dependent activation of presynaptic mGluRs represents a negative feedback mechanism to control the strength of synaptic transmission. As mentioned above, the GABAergic-interneuron-Purkinje cell synapses, glutamate, released from the post-synaptic Purkinje cell (which is GABAergic but contains mM concentrations of glutamate) acts as a retrograde messenger to activate a presynaptic group II mGluR which inhibits GABA release (Glitsch *et al.*, 1996).

Postsynaptic excitatory events

Pharmacological evidence: Most of the experiments showing that group I mGluRs generated postsynaptic excitations use application of agonists. Generally, excitations are the result of an inhibition of K^+ channels and activation of non-selective cation currents (for a review, see Pin and Duvoisin, 1995; Conn and Pin, 1997).

Physiological evidence: High frequency stimulation of neurons making synapses on midbrain dopaminergic neurons elicited EPSPs (via inhibition of K^+ channels), which were blocked by MCPG and mimicked by application of group I specific agonists (Fiorillo and Williams, 1998). Similarly, in the cerebellum, stimulation of presynaptic parallel fibres resulted, at the postsynaptic Purkinje cell level, in the generation of slow EPSPs, blocked by MCPG (Batchelor and Garthwaite, 1997). Interestingly, these EPSPs were potentiated for a long period (2 minutes) by a coincident stimulation of climbing fibres making synapses on the same Purkinje cell. A slow synaptic inward current evoked by high frequency stimulation, in the presence of K^+ channel blockers, ionotropic glutamate and GABA-A receptors antagonists, has been isolated in CA1 neurons and resulted from activation of non-selective cationic synaptic current by group I mGluRs (Congar *et al.*, 1997).

Postsynaptic inhibitory events

L-AP4-sensitive mGluR6 is restrictively expressed on ON-bipolar cells of the retina. In the dark, glutamate released from photoreceptor cells activates mGluR6; mGluR6 activates the hydrolysis of cGMP, leading to the closure of cGMP cationic channels and cellular hyperpolarization. In the presence of light, glutamate is no longer released from photoreceptors, mGluR6 is not activated, cationic channels are opened leading to cellular depolarization. In mGluR6-deficient mice, the synaptic transmission in the ON pathway was almost completely abolished, establishing clear evidence for a synaptic role for these receptors (Masu *et al.*, 1995). The animals have reduced sensitivity of pupilary responses to light stimulus and impaired ability to drive optokinetic nystagmus (Iwakabe *et al.*, 1997).

For a short period of time, high frequency stimulation of neurons making synapses on midbrain dopaminergic neurons, elicited IPSPs (via activation of Ca^{2+}-dependent K^+ channels). This effect was blocked by MCPG (Fiorillo and Williams, 1998).

Long-term potentiation (LTP) and long-term depression (LTD)

The involvement of group I mGluRs in LTP induction in CA1 hippocampal neurons has been a matter of controversial debate. In this respect, group I knock-out mice have not been much help (Aiba *et al.*, 1994a; Conquet *et al.*, 1994; Lu *et al.*, 1997). It is beyond the scope of this chapter to review this issue, which will be discussed in other chapters. It is enough to say that many groups have indeed shown that activation of mGluRs are able to facilitate NMDA receptor mediated LTP (Aniksztejn *et al.*, 1992; O'Connor *et al.*, 1994), but that the poor pharmacological antagonists available so far lead to a controversy on the physiological importance of mGluRs in LTP induction as well as the nature of group I mGluRs which could be involved (Bortolotto and Collingridge, 1992, 1993; Chinestra *et al.*, 1993; Bortolotto *et al.*, 1994; Manzoni *et al.*,

1994). Recently, Fitzjohn *et al.*, proposed that a not yet cloned receptor could be implicated (Fitzjohn *et al.*, 1998), whereas Wilsch *et al.* (Fitzjohn *et al.*, 1998), using the non-selective antagonist MCPG, showed that the physiological role of group I mGluRs may be confined to certain types of LTP which are induced by weak tetanization. During strong tetanization protocols the mGluR intracellular Ca^{2+} release will be bypassed by Ca^{2+} going through VDCCs.

The situation is much simpler for the induction of LTD at parallel fibres/Purkinje cell synapses. Here, clear physiological and genetic (knock-out mGluR1 mice) evidence suggests a role for mGluR1 in LTD induction (Ito, 1989; Aiba *et al.*, 1994b; Conquet *et al.*, 1994; Finch and Augustine, 1998; Takechi *et al.*, 1998). The clear Ca^{2+} in Purkinje neuron spines, triggered following mGluR1 activation, is likely to be the critical cue that determines the input specificity of LTD (Finch and Augustine, 1998; Takechi *et al.*, 1998). LTD induced by low frequency stimulation at the mossy fibre CA3 synapses was almost abolished in mice lacking mGluR2 (Yokoi *et al.*, 1996).

Pathologies which may benefit from specific drugs acting on mGluRs

As already stated, only a few drugs acting on ionotropic glutamate receptors are used in therapy or are under development, generally for pathologies relating to excitotoxic neuronal death such as Parkinson's disease, epilepsy or dementia (Danysz *et al.*, 1995). MK-801, the most potent NMDA receptor antagonist, produced side-effects, including neuronal death (Olney *et al.*, 1991) at doses lower than those which have clinical efficacies (Dingledine *et al.*, 1999). The activation or inhibition of mGluRs which have mainly regulatory roles at glutamatergic neurotransmission could be of more therapeutic value.

Group I mGluRs: potential therapeutic applications and new available drugs

Although group I mGluRs were the first to be described, it is only recently that specific ligands have been synthesized. The most selective agonist is 3′,5′-DHPG, although it acts on NMDA at high doses. CHPG has been proposed as a mGluR5 specific agonist, but its low affinity excludes its use without caution. LY393675 is a non-specific group I antagonist (for a review, see Pin *et al.*, 1999).

Specific mGluR1 antagonists such as (S)-CBPG and LY367385 (S-2-methyl-4-carboxy-phenylglycine) are available and are competitive antagonists. CPCCOEt is non-competitive. A highly potent specific and non-competitive antagonist of mGluR5 (MPEP) has recently been described (Gasparini *et al.*, 1999).

Role in neurodegeneration

Group I mGluRs as targets for therapy: Group I mGluRs can potentiate NMDA and AMPA/kainate-induced neuronal death by two mechanisms, the first one is a potentiation of ionotropic receptor activities (Ben-Ari *et al.*, 1992; Bruno *et al.*, 1995), the second one is an increase in intracellular Ca^{2+} release and PKC activation. Although no very specific agonists and antagonists have been used, experiments on cortical neurons, on hippocampal slices and *in vivo* suggest that NMDA, kainate and hypoxia/

hypoglycaemia-induced neuronal death may have a group I mGluR component (Nicoletti *et al.*, 1996; Strasser *et al.*, 1998). Therefore, potent group I antagonists may be useful in these acutely triggered neuronal death. It is interesting to recall that Homer 1a is synthesized during epileptic neuronal activity and may be involved in a negative feedback loop leading to a reduction in group I transduction (Xiao *et al.*, 1998). A benefit action of group I mGluR antagonists in pentylenetetrazole (PTZ)-induced seizures has also been reported (Thomsen and Dalby, 1998).

It is also possible that activation of group I mGluRs can be useful in some situations to inhibit apoptosis, certainly by triggering a gentle elevation in Ca^{2+} (Copani *et al.*, 1995; Anneser *et al.*, 1998).

Role in nociception

Nociceptive fibres (A and C) terminate primarily in the superficial dorsal horn, which comprises the lamina I and II (also called *substantia gelatinosa*). mGluR1 are concentrated in lamina II and are poorly expressed in the motor neuron area (Corsi *et al.*, 1998). mGluR5 are expressed only in lamina II (see discussion in Corsi *et al.*, 1998).

Using a protocol of stimulation providing a way of recruiting dorsal root C-fibres, the late polysynaptic phase of the ventral root potential (VRP) was blocked by MCPG (Figure 6.8) and the mGluR1-specific antagonist 4-CPG (Corsi *et al.*, 1998). MCPG also blocks the cumulative depolarization 'wind-up' of the VRP evoked by repetitive stimulation of the dorsal root (Boxal *et al.*, 1996) (Figure 6.8). These observations, plus those to follow, clearly demonstrate an indication of group I mGluRs in treating pain: (1) group I mGluRs potentiated the glutamate ionotropic responses in spinal cord (Ugolini *et al.*, 1997); (2) intrathecal injections of group I agonists (such as DHPG), but neither group II nor III agonists, induced spontaneous nociceptive behaviour (Fischer and Coderre, 1996a); (3) group I antagonists reduced formalin-induced pain (Fischer and Coderre, 1996b), and AP3, a poor mGluR antagonist, reduced neuronal spinal activity resulting from inflammatory knee (Neugebauer *et al.*, 1994); (4) antisense ablation of mGluR1 in the spinal cord developed marked analgesia as measured by an increase in the latency to tail-flick (55°C) over a period of 4–7 days (Young *et al.*, 1998); (5) mGluR1 −/− mice have a higher paw-linking latency time from a hot plate than wild type (Figure 6.8), and 4-CPG also increasing this latency (Corsi *et al.*, 1998). The most convincing evidence of a role for mGluR5 in inflammatory pain has been reported using MPEP. Whereas group I mGluR agonists induce the development of mechanical hyperalgia in inflamed rat hind paw, MPEP dose-dependently inhibits this effect. MPEP was active when given intraperitoneally but not intrathecally, suggesting a periperheral action (Bowes *et al.*, 1999).

At the thalamic level, the thalamic relay cells (RC) receive information from sensory afferents, send information to the layer 6 of the somato-sensory area, and receive information back from the same cortical area. These neurons synaptically activate RC neurons via NMDA but also via mGluR1 receptors. Therefore, cortical inactivation reduces the nociceptive responses recorded on RC cells. Noxious stimuli, like heat, generated action potentials which can be recorded on RC cells. These are inhibited by mGluR1 antagonists, and in particular LY367385, which also block the action potentials generated by iontophoretic application of ACPD, but not NMDA nor AMPA (Salt and Turner, 1998b; see also Salt and Turner, 1998a; Figure 6.9).

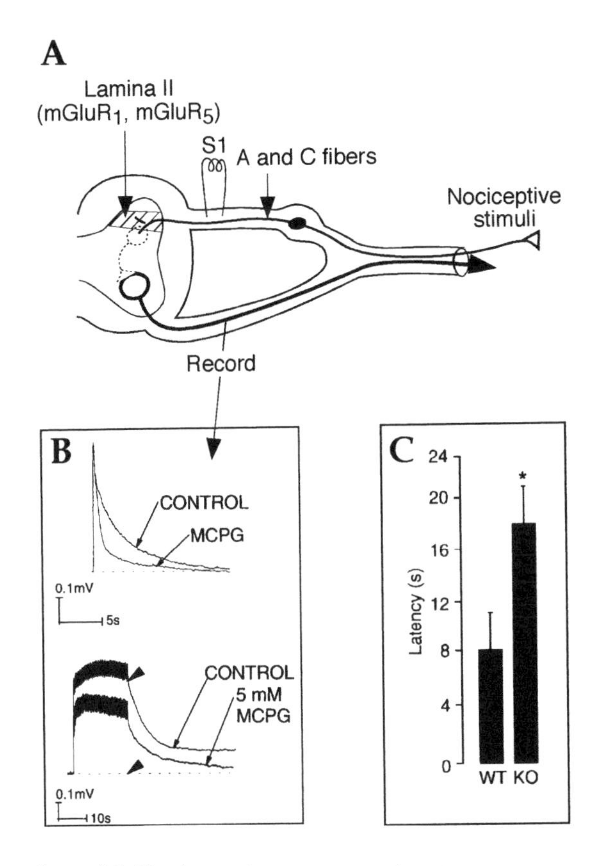

Figure 6.8 Metabotropic receptors and nociception.

(a) Schematic representation of spinal nociceptive reflex. mGluR1 and mGluR5 are localized in lamina II. (b) MCPG reduced the prolonged phase of the single shock C-fibre-evoked ventral root depolarization (upper record). Low-frequency, high-intensity stimulation of the dorsal root evoked a wind-up response in the ventral horn, the intensity of which was attenuated by MCPG. Modified from (Boxal *et al.*, 1996). (c) The latency time before the animal licks its hind paw was measured. Control and mGluR1 knock-out (KO) mice were confined to a hot (55°C) plate. Modified from (Corsi *et al.*, 1998).

Altogether, these results suggest that potent group I antagonists may be useful in pain therapy.

Group II mGluRs: potential therapeutic applications and new available drugs

Following the synthesis of agonists with mGluR potencies (such as (1*S*,3*S*)-ACPD and (2*R*,4*R*)-ACPD), two very potent specific group II agonists (nM or less) have been described (LY354740, Schoepp *et al.*, 1997) LY379268, Moon *et al.*, 1998). Thanks to these agonists, and to very potent competitive antagonists such as LY307452 (Wermuth *et al.*, 1996), the possible use of group II mGluRs as targets for therapeutic drugs has started to be analysed.

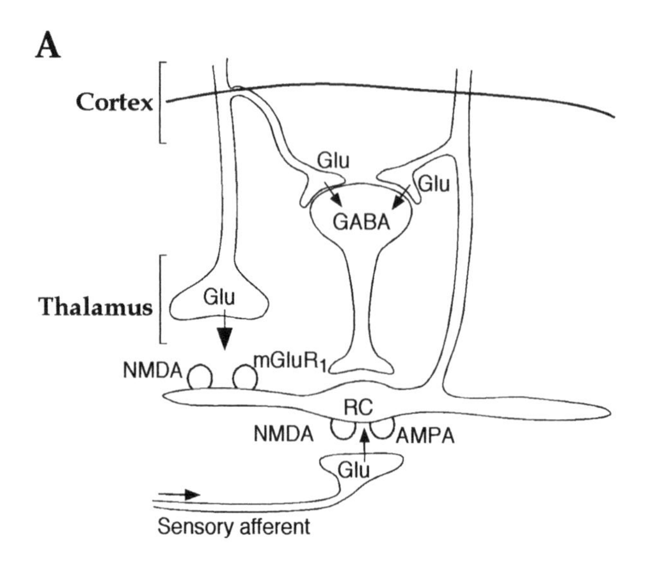

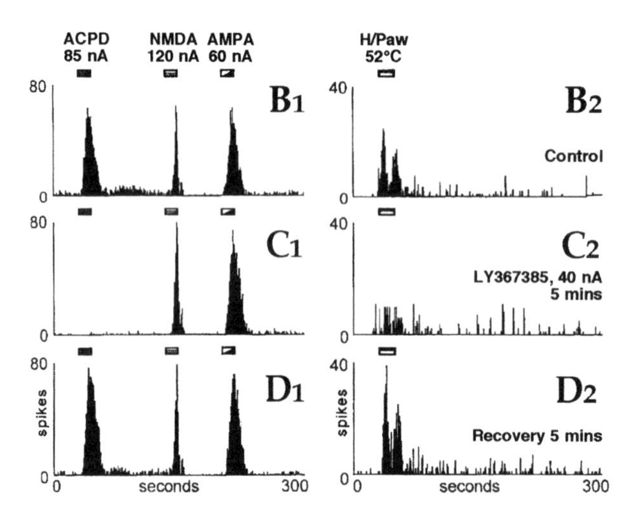

Figure 6.9 Thalamic mGluR1 and nociceptive responses.

(a) Simplified schema of sensory transmission in the cortico-thalamic circuit. The glutamate sensory afferents make synapses on thalamic relay cells (RC). These cells send information to the cortex and inhibitory GABA neurons. The nociceptive response coming from the cortex goes back to RC cells and generates action potentials mediated via mGluR1 receptors. Modified from Salt and Turner (1998a). (b) A single nociceptive thalamic neuron was recorded extracellularly after iontophoretic application of ACPD, NMDA, AMPA (B1), or after a noxious stimulation of the contralateral hindpaw (52°C) (B2). Histograms show action potential spikes counted into 100 ms epochs. C1 and C2, the same experiment as in B1 and B2, respectively, but in the presence of LY367385 (S-2-methyl-4-carboxy-phenylglycine), a mGluR1 antagonist. D1 and D2 is a similar experiment to B1 and B2, respectively, done 5 mins after C1 and C2 experiments (recovery). Modified from Salt and Turner (1998b).

Neuroprotection

DCG-IV or 4C3HPG are two group II mGluR agonists having some selectivity problems. The first one is the most potent group III antagonist described so far (Brabet *et al.*, 1998), the second one is a group I antagonist. *In vitro*, these compounds have been shown to reduce neuronal death induced by NMDA, kainate and hypoxia combined with hypoglycaemia (for a review, see Nicoletti *et al.*, 1996). The mechanism involved in neuroprotection remains to be identified, but it could involve a reduction of insult and post-insult release of glutamate. Indeed, group II mGluRs are mainly localized presynaptically (Shigemoto *et al.*, 1997). A neuroprotective effect mediated via Gi/Goα or G$\beta\gamma$ triggered events, such as the activation of MAP kinase, cannot be excluded. Another mechanism has recently been proposed by Bruno *et al.* (1998). These authors have shown that activation of group II (presumably mGluR3) of astrocytes leads to an increased release of TGFβ, known to be neuroprotective against NMDA insult *in vitro*. Finally, it has been shown that LY354740 is neuroprotective against a short (3 minute) period of global ischaemia in the gerbil model (Bond *et al.*, 1998).

Anti-parkinsonian properties

In Parkinson's disease, muscle rigidity is certainly due to an over-stimulation of the striatopallidal neurons following a decrease in their inhibition by dopamine D2 receptors and the prevalence of the cortico-striatal excitation. This is certainly why NMDA antagonists have been shown to exhibit anti-parkinsonism properties in different animal models, including haloperidol-induced muscle rigidity (Ossowska, 1994). Similarly, LY354740, presumably via a reduction in cortico-striatal glutamate release, has been shown to reduce haloperidol-induced muscle rigidity (Konieczny *et al.*, 1998).

Anxiolytic properties

In two tests, the fear potentiated startle and elevated plus maze tests, LY354740 (0.3 to 10 mg/kg per os.) was as potent as diazepam. In addition, the anxiolytic activity of LY354740 was not associated with unwanted secondary pharmacology of diazepam, such as sedation and deficits in neuromuscular coordination memory impairment (Helton *et al.*, 1998a).

Anti-addictive properties

In humans, nicotine withdrawal results in a number of symptoms which include anxiety and irritability. In rodents, the symptoms are more difficult to analyse. A model which can be used is the increase in startle response during 5 days following withdrawal. Readministration of nicotine, but also of LY354740, greatly reduces this withdrawal-induced behaviour (Helton *et al.*, 1998b).

Anti-schizophrenic properties

Phencyclidine (PCP) treatment induces a series of symptoms which are considered to be similar to those observed in schizophrenic patients. Some of them may be due to an increase in glutamatergic transmission and, indeed, PCP increases glutamate release in pre-frontal cortex and nucleus accumbens. LY354740, at doses which have no effect on dopamine neurotransmission, was able to attenuate the disruptive effects of PCP on working memory, stereotypy, locomotion and glutamate efflux (Moghaddam and Adams, 1998). Thus, targeting group II mGluRs may be of use as a non-dopaminergic therapeutic treatment of schizophrenia or related disorders.

Group III mGluRs: potential therapeutic applications

Neuroprotection

Group III mGluRs are mainly presynaptically localized (Shigemoto *et al.*, 1997; Masugi *et al.*, 1999), and mGluR7 has been localized in the presynaptic grid of glutamatergic synapses (Shigemoto *et al.*, 1994). Therefore, group III agonists are expected to have multiple therapeutic applications. However, there are no high affinity specific agonists or antagonists available to test their possible pertinence as therapeutic targets *in vivo* (for reviews, see Conn and Pin, 1997; Pin *et al.*, 1999). *In vitro*, it has been found that μM doses of L-AP4, which likely activate mGluR4 or 8, were neuroprotective on cortical neurons (Bruno *et al.*, 1996). High concentrations of L-AP4 (mM), likely to act on mGluR7, have been found to be neuroprotective against NMDA-induced toxicity on glutamatergic cerebellar granular cells *in vitro* (Lafon-Cazal *et al.*, 1999). This was certainly due to a reduction in glutamate release during and after the insult. In contrast, mM concentrations of L-AP4 (activating mGluR7-like receptors) potentiated NMDA-induced cell death in striatal neurons which are mainly GABAergic neurons (unpublished observations). Again this was certainly due to a reduction in GABA release, which has a neuroprotective effect. It would be surprising to find mGluR7-like receptors on terminals of GABAergic neurons, knowing their low affinity for glutamate (mM). However, such a localization has been described using electron microscopic analysis (Lujan *et al.*, 1998).

Anti-epileptic properties

A predominant role for mGluR7 in controlling excessive glutamate release is also suggested by the observation that 12-week-old mGluR7 −/− knockout mice developed epilepsy (Masugi *et al.*, 1999). Potent and selective mGluR7 ligands are urgently needed to be tested in epilepsy.

Conclusion

The mGluR saga, started 14 years ago, has provided an amazing amount of research and discoveries. First member of a unique family of GPCRs, our knowledge on their structure and coupling to G proteins has still to be developed in parallel with similar efforts on other members of the family, such as GABA-B receptors. Hopefully, one extracellular binding site of these receptors will soon be crystallized. In addition, the physiology of mGluRs reveals their essential role in synaptic transmission. Finally, we all, especially pharmaceutical companies, hope for therapeutic drugs to act on mGluRs. We can be optimistic about this matter; chronic pain, anxiety and schizophrenia may be the first main pathologies treated with mGluRs-related drugs. Of course, optimism and hard work is certainly not enough – luck also has something to do with it.

References

Aiba, A., Chen, C., Herrup, K., Rosenmund, C., Stevens, C.F. and Tonegawa, S. (1994a) 'Reduced hippocampal long-term potentiation and context-specific deficit in associative learning in mGluR1 mutant mice', *Cell* **79**: 365–375.

Aiba, A., Kano, M., Chen, C., Stanton, M.E., Fox, G.D., Herrup, K., Zwingman, T.A. and Tonegawa, S. (1994b) 'Deficient cerebellar long-term depression and impaired motor learning in mGluR1 mutant mice', *Cell* **79**: 377–388.

Alagarsamy, S., Marino, M.J., Rouse, S.T., Gereau, R.W., Heinemann, S.F. and Conn, P.J. (1999) 'Activation of NMDA receptors reverses desensitization of mGluR5 in native and recombinant systems', *Nature Neurosci.* **2**: 234–240.

Alaluf, S., Mulvihill, E.R. and McIlhinney, R.A.J. (1995a) 'The metabotropic glutamate receptor mGluR4: but not mGluR1a, is palmitoylated when expressed in BHK cells', *J. Neurochem.* **64**: 1548–1555.

Alaluf, S., Mulvihill, E.R. and McIlhinney, R.A.J. (1995b) 'Rapid agonist-mediated phosphorylation of the metabotropic glutamate receptor1a by protein kinase C in permanently transfected BHK cells', *FEBS Lett.* **367**: 301–305.

Aniksztejn, L., Otani, S. and Ben-Ari, Y. (1992) 'Quisqualate metabotropic receptors modulate NMDA currents and facilitate induction of long-term potentiation through protein kinase C', *Eur. J. Neurosci.* **4**: 500–505.

Anneser, J.M.H., Horstmann, S., Weydt, P. and Borasio, G.D. (1998) 'Activation of metabotropic glutamate receptors delays apoptosis of chick embryonic motor neurons in vitro', *NeuroReport* **9**: 2039–2043.

Armstrong, N., Sun, Y., Chen, G. and Gouaux, E. (1998) 'Structure of the glutamate receptor ligand-binding core in complex with kainate', *Nature* **395**: 913–917.

Bai, M., Trivedi, S. and Brown, E. (1998) 'Dimerization of the extracellular calcium-sensing receptor (CaR) on the cell surface of CaR-transfected HEK293 cells', *J. Biol. Chem.* **273**: 23,605–23,610.

Batchelor, A.M. and Garthwaite, J. (1997) 'Frequency detection and temporally dispersed synaptic signal association through a metabotropic glutamate receptor pathway', *Nature* **385**: 74–77.

Baude, A., Nusser, Z., Roberts, J.D.B., Mulvihill, E., McIlhinney, R.A.J. and Somogyi, P. (1993) 'The metabotropic glutamate receptor (mGluR1a) is concentrated at perisynaptic membrane of neuronal subpopulations as detected by immunogold reaction', *Neuron* **11**: 771–787.

Ben-Ari, Y., Aniksztejn, L. and Bregestovski, P. (1992) 'Protein kinase C modulation of NMDA currents: an important link for LTP induction', *Trends Neurosci.* **9**: 333–339.

Blahos II, J., Mary, S., Perroy, J., de Colle, C., Brabet, I., Bockaert, J. and Pin, J.-P. (1998) 'Extreme C terminus of G protein a-subunits contains a site that discriminates between G_i-coupled metabotropic glutamate receptors', *J. Biol. Chem.* **273**: 25,765–25,769.

Bockaert, J. and Pin, J.-P. (1999) 'Molecular tinkering of G protein-coupled receptors: an evolutionary success', *EMBO J.* **18**: 1723–1729.

Bond, A., O'Neil, M.J., Hicks, C.A., Moon, J.A. and Lodge, D. (1998) 'Neuroprotective effects of a systemically active group II metabotropic glutamate receptor agonist LY354740 in a gerbil model of global ischemia', *NeuroReport* **9**: 1191–1193.

Bortolotto, Z.A., Bashir, Z.I., Davies, C.H. and Collingridge, G.L. (1994) 'A molecular switch activated by metabotropic glutamate receptors regulates induction of long-term potentiation', *Nature* **368**: 740–743.

Bortolotto, Z.A. and Collingridge, G.L. (1992) 'Activation of glutamate metabotropic receptors induces long-term potentiation', *Eur. J. Pharmacol.* **214**: 297–298.

Bortolotto, Z.A. and Collingridge, G.L. (1993) 'Characterization of LTP induced by the activation of glutamate metabotropic receptors in area CA1 of the hippocampus', *Neuropharmacology* **32**: 1–9.

Bourinet, E., Soong, T., Stea, A. and Snutch, T. (1996) 'Determinants of the G protein-dependent opioid modulation of neuronal calcium channels', *Proc. Natl. Acad. Sci. USA* **93**: 1486–1491.

Bowes, M., Panesar, M., Gentry, C., Urban, L., Gasparini, F., Kuhn, R. and Walker, K. (1999) 'Anti-hyperalgesic effects of the novel metabotropic glutamate receptor 5 antagonist, methylphenylethynylpyridine, in rat models of inflammatory pain', *Proceeding Supplement to the Br. J. Pharmacol.* **126**: 250.

Boxal, S.J., Thompson, S.W.N., Dray, A., Dickenson, A.H. and Urban, L. (1996) 'Metabotropic glutamate receptor activation contributes to nociceptive reflex activity in the rat spinal cord in vitro', *Neuroscience* **74**: 13–20.

Brabet, I., Parmentier, M.-L., De Colle, C., Bockaert, J., Acher, F. and Pin, J.-P. (1998) 'Comparative effect of L-CCG-I, DCG-IV and g-carboxy-L-glutamate on all cloned metabotropic glutamate receptor subtypes', *Neuropharmacology* **37**: 1043–1051.

Brakeman, P.R., Lanahan, A.A., O'Brien, R., Roche, K., Barnes, C.A., Huganir, R.L. and Worley, P.F. (1997) 'Homer: a protein that selectively binds metabotropic glutamate receptors', *Nature* **386**: 284–288.

Brown, E.M., Gamba, G., Riccardi, D., Lombardi, M., Butters, R., Kifor, O., Sun, A., Hediger, M.A., Lytton, J. and Hebert, S.C. (1993) 'Cloning and characterization of an extracellular Ca^{2+}-sensing receptor from bovine parathyroid', *Nature* **366**: 575–580.

Brown, E.M. and Herbert, S.C. (1997) 'Calcium-receptor-regulated parathyroid and renal function', *Bone* **20**: 303–309.

Bruno, V., Battaglia, G., Casabona, G., Copani, A., Caciagli, F. and Nicoletti, F. (1998) 'Neuroprotection by glial metabotropic glutamate receptors is mediated by transforming growth factor-b', *J. Neurosci.* **18**: 9594–9600.

Bruno, V., Copani, A., Bonanno, L., Knoepfel, T., Kuhn, R., Roberts, P.J. and Nicoletti, F. (1996) 'Activation of group III metabotropic receptors is neuroprotective in cortical cultures', *Eur. J. Pharmacol.*, **310**: 61–66.

Bruno, V., Copani, A., Knopfel, T., Kuhn, R., Casabona, G., Dell' Albani, P., Condorelli, D. and Nicoletti, F. (1995) 'Activation of metabotropic glutamate receptors coupled to inositol phospholipid hydrolysis amplifies NMDA-induced neuronal degeneration in cultured cortical cells', *Neuropharmacology* **34**: 1089–1098.

Charpak, S., Gähwiler, B.H., Do, K.Q. and Knöpfel, T. (1990) 'Potassium conductances in hippocampal neurons blocked by excitatory amino-acid transmitters', *Nature* **347**: 765–767.

Chavis, P., Fagni, L., Lansman, J.B. and Bockaert, J. (1996) 'Functional coupling between ryanodine receptors and L-type calcium channels in neurons', *Nature* **382**: 719–722.

Chavis, P., Ango, F., Michel, J.-M., Bockaert, J. and Fagni, L. (1998a) 'Modulation of big K^+ channel activity by ryanodine receptors and L-type Ca^{++} channels in neurons', *Eur. J. Neurosci.* **10**: 2322–2327.

Chavis, P., Mollard, P., Bockaert, J. and Manzoni, O. (1998b) 'Visualization of cyclic-AMP-regulated presynaptic activity at cerebellar granule cells', *Neuron* **4**: 773–781.

Chavis, P., Shinozaki, H., Bockaert, J. and Fagni, L. (1994) 'The metabotropic glutamate receptor types 2/3 inhibit L-type calcium channels via a Pertussis toxin-sensitive G-protein in cultured cerebellar granule cells', *J. Neurosci.* **14**: 7067–7076.

Chinestra, P., Aniksztejn, L., Diabira, D. and Ben-Ari, Y. (1993) '(RS)-a-Methyl-4-carboxyphenylglycine neither prevents induction of LTP nor antagonizes metabotropic glutamate receptors in CA1 hippocampal neurons', *J. Neurophysiol.* **70**: 2684–2689.

Choi, D.W. (1988) 'Glutamate neurotoxicity and diseases of the nervous system', *Neuron* **1**: 623–634.

Cochilla, A. and Alford, S. (1998) 'Metaboropic glutamate receptor-mediated control of neurotransmitter release', *Neuron* **20**: 1007–1016.

Congar, P., Leinekugel, X., Ben-Ari, Y. and Crepel, V. (1997) 'A long lasting calcium-activated nonselective cationic current is generated by synaptic stimulation or exogenous activation of group I metabotropic glutamate receptors in CA1 pyramidal neurons', *J. Neurosci.* **17**: 5366–5379.

Conn, P.J. and Pin, J.P. (1997) 'Pharmacology and functions of metabotropic glutamate receptors', *Annu. Rev. Pharmacol. Toxicol.* **37**: 205–237.

Conquet, F., Bashir, Z.I., Davies, C.H., Daniel, H., Ferraguti, F., Bordi, F., Franz-Bacon, K., Reggiani, A., Matarese, V., Condé, F., Colingridge, G.L. and Crépel, F. (1994) 'Motor deficit and impairment of synaptic plasticity in mice lacking mGluR1', *Nature* **372**: 237–243.

Constantino, G. and Pelliciari, G. (1996) 'Homology modeling of metabotropic glutamate receptors (mGluRs) structural motifs affecting binding modes and pharmacological profile of mGluR1 agonists and competitive antagonists', *J. Med. Chem.* **39**: 3998–4006.

Copani, A., Bruno, V.F., Barresi, V., Battaglia, G., Conderelli, D.F. and Nicoletti, F. (1995) 'Activation of metabotropic glutamate receptors prevents neuronal apoptosis in culture', *J. Neurochem.* **64**: 101–108.

Corsi, M., Ugolini, A., Quartaroli, M., Chiamulera, C., Corti, C., Maraia, G., Conquet, F. and Ferraguti, F. (1998) Phospholipase C-coupled metabotropic glutamate receptors modulate nociceptive transmission, in F. Nicoletti and D. Pellegrini-Giampietro (eds) *Metabotropic Glutamate Receptors and Brain Function*, London: Portland Press, pp. 37–47.

Daaka, Y., Luttrell, L.M. and Lefkowitz, R.J. (1997) 'Switching of the coupling of the b_2-adrenergic receptor to different G proteins by protein kinase A', *Nature* **390**: 88–91.

Daniel, H., Levenes, C., Fagni, L., Conquet, F., Bockaert, J. and Crepel, F. (1999) 'InsP3-mediated rescue of cerebellar long-term depression in mGluR1 mutant mouse', *Neuroscience* **92**: 1–6.

Danysz, W., Parsons, C.G., Bresink, I. and Quack, G. (1995) 'Glutamate in CNS disorders', *Drug News and Perspectives* **8**: 261–276.

Dingledine, R., Borges, K., Bowie, D. and Traynelis, S.F. (1999) 'The glutamate receptor ion channels', *Pharmacol. Rev.* **51**: 7–61.

Dulac, C. and Axel, R. (1995) 'A novel family of genes encoding putative pheromone receptors in mammals', *Cell* **83**: 195–206.

Dumuis, A., Pin, J.P., Oomagari, K., Sebben, M. and Bockaert, J. (1990) 'Arachidonic acid released from striatal neurons by joint stimulation of ionotropic and metabotropic quisqualate receptors', *Nature*, **347**: 182–184.

Dumuis, A., Sebben, M., Fagni, L., Prézeau, L., Manzoni, O., Cragoe Jr., E.J. and Bockaert, J. (1993) 'Stimulation of arachidonic acid release by glutamate receptors depends on Na^+/Ca^{2+} exchanger in neuronal cells', *Mol. Pharmacol.* **43**: 976–981.

Fagni, L., Bossu, J.-L. and Bockaert, J. (1991) 'Activation of a large-conductance Ca^{2+}-dependent K^+-channel by stimulation of glutamate phosphoinositide-coupled receptors in cultured cerebellar granule cells', *Eur. J. Neurosci.* **3**: 778–789.

Fan, G.F., Ray, K., Zhao, X.M., Goldsmith, P.K. and Spiegel, A.M. (1998) 'Mutational analysis of the cysteines in the extracellular domain of the human Ca^{2+} receptor: effects on cell surface expression, dimerization and signal transduction', *FEBS Lett.* **346**: 353–356.

Finch, E.A. and Augustine, G.J. (1998) 'Local calcium signaling by inositol-1,4,5,-trisphosphate in Purkinje cell dendrites', *Nature* **396**: 753–756.

Fiorillo, C.D. and Williams, J.T. (1998) 'Glutamate mediates an inhibitory postsynaptic potential in dopamine neurons', *Nature* **394**: 78–82.

Fischer, K. and Coderre, T.J. (1996a) 'Comparison of the nociceptive effects produced by intrathecal administration of mGluR agonists', *NeuroReport* **7**: 2743–2747.

Fischer, K. and Coderre, T.J. (1996b) 'The contribution of metabotropic glutamate receptors (mGluRs) to formalin-induced nociception', *Pain* **68**: 255–263.

Fitzjohn, S.M., Bortolotto, Z.A., Palmer, M.J., Doherty, A.J., Ornstein, P.L., Schoepp, D.D., Kingston, A.E., Lodge, D. and Collingridge, G.L. (1998) 'The potent mGlu receptor antagonist LY341495 identifies roles for both cloned and novel mGlu receptors in hippocampal synaptic plasticity', *Neuropharmacology* **37**: 1445–1458.

Galvez, T., Joly, C., Parmentier, M.-L., Malitschek, B., Kaupmann, K., Kuhn, R., Bittiger, H., Froestl, W., Bettler, B. and Pin, J.-P. (1999) 'Mutagenesis and modeling of the $GABA_B$ receptor extracellular domain support a Venus Flytrap mechanism for ligand binding', *J. Biol. Chem.* **247**: 13,362–13,369.

Gasparini, F., Linggenhoehl, K., Flor, P.J., Munier, N., Heirich, M., Vranesic, F., Biollaz, M., Heckendron, R., Allgeir, H., Varney, M.A., Johnson, E., Hess, S.D., Velicelebi, G. and Kuhn, R. (1999) '2-methyl-6-(phenylethyl)-pyridine (MPEP): a novel potent, subtype-selective and systemically active antagonist at metabotropic glutamate receptor subtype 5', *Br. J. Pharmacol.* BPS meeting abstract **126**: 249.

Gerber, U., Sim, J.A. and Gähwiler, B.H. (1992) 'Reduction of potassium conductances mediated by metabotropic glutamate receptors in rat CA3 pyramidal cells does not require protein kinase C or protein kinase A', *Eur. J. Neurosci.* **4**: 792–797.

Gereau IV, R. and Heinemann, S.F. (1998) 'Role of protein kinase C phosphorylation in rapid desensitization of metabotropic glutamate receptor 5', *Neuron* **20**: 143–151.

Gereau, R. and Conn, P. (1995a) 'Multiple presynaptic metabotropic glutamate receptors modulate excitatory and inhibitory synaptic transmission in hippocampal area CA1', *J. Neurosci.* **15**: 6879–6889.

Gereau, R.W. and Conn, P.J. (1995b) 'Roles of specific metabotropic glutamate receptor subtypes in regulation of hippocampal CA1 pyramidal excitability', *J. Neurophysiol.* **74**: 122–129.

Glitsch, M., Llano, I. and Marty, A. (1996) 'Glutamate as a candidate retrograde messenger at interneurone-Purkinje cell synapses of rat cerebellum', *J. Physiol.* **497**: 531–537.

Guérineau, N.C., Bossu, J.-L., Gähwiler, B.H. and Gerber, U. (1995) 'Activation of a non selective cationic conductance by metabotropic glutamatergic and muscarinic agonists in CA3 pyramidal neurons of rat hippocampus', *J. Neurosci.* **15**: 4395–4407.

Hammerland, L., Garrett, J., Hung, B., Levinthal, C. and Nemeth, E. (1999) 'Domains determining ligand specificity for Ca^{2+} receptors', *Mol. Pharmacol.* **53**: 1083–1088.

Han, G. and Hampson, D.R. (1999) 'Ligand binding to the amino-terminal domain of the mGluR4 subtype of metabotropic.

glutamate receptor', *J. Biol. Chem.* **274**: 10,008–10,013.

Helton, D.R., Tizzano, J.P., Moon, J.A., Scoepp, D.D. and Kallman, J. (1998a) 'Anxiolytic and side-effect profile of LY354740: a potent, highly selective, orally active agonist for group II metabotropic glutamate receptors', *J. Pharmacol. Exp. Ther.* **284**: 651–660.

Helton, D.R., Tizzano, J.P., Noon, J.A., Schoep, D.D. and Kallman, M.J. (1998b) 'LY354740: a metabotropic glutamate receptor agonist which ameliorates symptoms of nicotine withdrawal in rats', *Neuropharmacology* **36**: 1511–1516.

Herlitze, S., Hockerman, G., Scheuer, T. and Catterall, W. (1997) 'Molecular determinants of inactivation and G protein modulation in the intracellular loop connecting domains I and II of the calcium channel a_{1A} subunit', *Proc. Natl. Acad. Sci. USA* **94**: 1512–1516.

Herrero, I., Miras-Portugal, T. and Sanchez-Prieto, J. (1992) 'Positive feedback of glutamate exocytosis by metabotropic presynaptic receptor stimulation', *Nature* **360**: 163–166.

Herrero, I., Miras-Portugal, M.T. and Sanchez-Prieto, J. (1994) 'Rapid desensitization of the metabotropic glutamate receptor that facilitates glutamate release in rat cerebrocortical nerve terminals', *Eur. J. Neurosci.* **6**: 115–120.

Herrero, I., Miras-Portugal, M.T. and Sanchez-Prieto, J. (1998) 'Functional switch from facilitation to inhibition in the control of glutamate release by metabotropic glutamate receptors', *J. Biol. Chem.* **273**: 1951–1958.

Honn, M.A., Adler, E., Lindemeir, J., Battey, J.F., Ryba, J.P. and Zucker, C.S. (1999) 'Putative mammalian taste receptors: a class of taste-specific GPCRs with distinct topographic selectivity', *Cell* **96**: 541–551.

Irving, A.J., Collingridge, G.L. and Schofield, J.G. (1992) 'Interactions between Ca^{2+} mobilizing mechanisms in cultured rat cerebellar granule cells', *J. Physiol. (Lond.)* **456**: 667–680.

Ito, M. (1989) 'Long-term depression', *Ann. Rev. Neurosci.* **12**: 85–102.

Iwakabe, H., Katsuura, G., Ishibashi, C. and Nakanishi, S. (1997) 'Impairment of pupillary responses and optokinetic nystagmus in the mGluR6-deficient mouse', *Neuropharmacology* **36**: 135–143.

Kammermeier, P. and Ikeda, S. (1999) 'Expression of RGS2 alters the coupling of metabotropic glutamate receptor 1a to M-type K^+ and N-type Ca^{2+} channels', *Neuron* **22**: 819–829.

Kato, A.F., Ozawa, F., Saitoh, Y., Fukazawa, H., Sugiyama, H. and Inokuchi, K. (1998) 'Novel members of the Vesl/Homer family of PDZ proteins that bind metabotropic glutamate receptors', *J. Biol. Chem.* **273**: 23,969–23,975.

Kawabata, S., Tsutsumi, R., Kohara, A., Yamaguchi, T., Nakanishi, S. and Okada, M. (1996) 'Control of calcium oscillation of metabotropic glutamate receptors', *Nature* **383**: 89–92.

Konieczny, J., Ossowska, K., Wolfarth, S. and Pilc, A. (1998) 'LY354740: a group II metabotropic glutamate receptor agonist with potential anti-parkinsonian properties in rats', *Naunyn-Schmiedeberg's Arch. Pharmacol.* **358**: 500–502.

Kuner, R., Grunewald, K.H., Eisenhardt, G., Bach, A. and Kornau, H.C. (1999) 'Role of heteromer formation in GABA B receptor function', *Science* **283**: 74–77.

Lafon-Cazal, M., Fagni, L., Guiraud, M.-J., Mary, S., Lerner-Natoli, M., Pin, J.-P., Shigemoto, R. and Bockaert, J. (1999) 'mGluR7-like metabotropic glutamate receptors inhibit NMDA-mediated excitotoxicity in cultured cerebellar granule neurons', *Eur. J. Neurosci.* **11**: 663–672.

Lerma, J. (1997) 'Kainate reveals its target', *Neuron* **19**: 1155–1158.

Litschig, S., Gasparini, F., Rueegg, D., Munier, N., Flor, P.J., Vranesic, I., Prézeau, L., Pin, J.P., Thomsen, C. and Kuhn, R. (1999) 'CPCCOEt, a non-competitive mGluR1 antagonist, inhibits receptor signaling without affecting glutamate binding', *Mol. Pharmacol.* **55**: 453–461.

Lu, Y.M., Jia, Z., Janus, C., Henderson, J.T., Gerlai, R., Wotjtowicz, J.M. and Roder, J.C. (1997) 'Mice lacking metabotropic glutamate receptor 5 show impairment learning and reduced CA1 long-term potentiation (LTP) but normal CA3 LTP', *J. Neurosci.* **17**: 5196–5205.

Lujan, R., Shigemoto, R. and Somogyi, P. (1998) 'A presynaptic metabotropic glutamate receptor (mGluR7) is located at the vesicle release site on GABAergic terminals in the rat hippocampus', *Eur. J. Neurosci.* **10**: 128 (abstract Berlin FENS, 1998).

Manzoni, O. and Bockaert, J. (1995) 'Metabotropic glutamate receptors inhibiting excitatory synapses in CA1 area of rat hippocampus', *Eur. J. Neurosci.*, **7**: 2518–2523.

Manzoni, O., Michel, J.M. and Bockaert, J. (1997) 'Metabotropic glutamate receptors modulate excitatory transmission in the rat nucleus accumbens', *Eur. J. Neurosci.* **9**: 1514–1523.

Manzoni, O.J., Weisskop, M.G. and Nicoll, R.A. (1994) 'MCPG antagonizes metabotropic glutamate receptors but not long-term potentiation in the hippocampus', *Eur. J. Neurosci.* **6**: 1050–1054.

Mary, S., Gomeza, J., Prézeau, L., Bockaert, J. and Pin, J.-P. (1998) 'A cluster of basic residues in the carboxyl-terminal tail of the short metabotropic glutamate receptor 1 variants impairs their coupling to phospholipase C', *J. Biol. Chem.* **273**: 425–432.

Masu, M., Iwakabe, H., Tagawa, Y., Miyoshi, T., Yamashita, M., Fukuda, Y., Sasaki, H., Hiro, K., Nakamura, Y., Shigemoto, R., Takada, M., Nakamura, K., Nakao, K., Katsuki, M. and Nakanishi, S. (1995) 'Specific deficit of ON response in visual transmission by targeted disruption of the mGluR6 gene', *Cell* **80**: 757–765.

Masugi, M., Yokoi, M., Shigemoto, R., Muguruma, K., Watanabe, Y., Sansig, G., van der Putten, H. and Nakanishi, S. (1999) 'Metabotropic glutamate receptor subtype 7 ablation causes deficit in fear response and conditioned taste aversion', *J. Neurosci.* **19**: 955–963.

Moghaddam, B. and Adams, B.W. (1998) 'Reversal of phencyclidine effects by group II metabotropic glutamate receptor agonist in rats', *Science* **281**: 1349–1352.

Moon, J.A., Valli, M.J., Andis, S.L., Wright, R.A., Johnson, B.G., Tomlinson, R., Kingston, A.E. and Schoepp, D.D. (1998) '*In vitro* characterization of LY379268: a highly potent and selective agonist for group-II metabotropic glutamate receptors', *Soc. Neurosci. Abs.* **24**: 583.

Moroni, F., Cozzi, A., Lombardi, G., Sourtcheva, S., Leonardi, P., Carfi, M. and Pellicciari, R. (1998) 'Presynaptic mGlu1 type receptors potentiate transmitter output in the rat cortex', *Eur. J. Pharmacol.* **347**: 189–195.

Mulle, C., Sailer, A., Perez-Otano, I., Dickinson-Anson, H., Castillo, P.E., Bureau, I., Maron, C., Gage, F.H., Mann, J.R., Bettler, B. and Heineman, S.F. (1998) 'Altered synaptic physiology and reduced susceptibility to kainate-induced seizures in GluR6-deficient mice', *Nature* **392**: 601–605.

Nakahara, K., Okada, M. and Nakanishi, S. (1997) 'The metabotropic glutamate receptor mGluR5 induces calcium oscillations in cultured astrocytes via protein kinase C phosphorylation', *J. Neurochem.* **69**: 1467–1475.

Nakanishi, S. (1992) 'Molecular diversity of glutamate receptors and implications for brain function', *Science* **258**: 597–603.

Nawy, S. and Jahr, C.E. (1990) 'Suppression by glutamate of cGMP-activated conductance in retina bipolar cells', *Nature* **346**: 269–271.

Nemeth, E., Steffey, M., Hammerland, L., Hung, B., Van Wagenen, B., DelMar, E. and Balandrin, M. (1998) 'Calcimimetics with potent and selective activity on parathyroid calcium receptor', *Proc. Natl. Acad. Sci. USA* **95**: 4040–4045.

Neugebauer, V., Lucke, T. and Schaible, H.G. (1994) 'Requirement of metabotropic glutamate receptors for the generation of inflammation-evoked hyperexcitability in rat spinal cord neurons', *Eur. J. Neurosci.* **6**: 1179–1186.

Nicoletti, F., Bruno, V., Copani, A., Casabona, G. and Knöpfel, T. (1996) 'Metabotropic glutamate receptors: a new target for the therapy of neurodegenerative disorders', *Trends Neurosci.* **19**: 267–271.

Nicoletti, F., Wroblewski, J.T., Novelli, A., Alho, H., Guidotti, A. and Costa, E. (1986) 'The activation of inositol phospholipid metabolism as a signal-transduction system for excitatory amino acids in primary cultures of cerebellar granule cells', *J. Neurosci.* **6**: 1905–1911.

O'Connor, J.J., Rowan, M.J. and Anwyl, R. (1994) 'Long-lasting enhancement of NMDA receptor-mediated synaptic transmission by metabotropic glutamate receptor activation', *Nature* **367**: 557–559.

O'Hara, P.J., Sheppard, P.O., Thogersen, H., Venezia, D., Haldeman, B.A., McGrane, V., Houamed, K.M., Thomsen, C., Gilbert, T.L. and Mulvihill, E.R. (1993) 'The ligand-binding domain in metabotropic glutamate receptors is related to bacterial periplasmic binding proteins', *Neuron* **11**: 41–52.

Okamoto, T., Sekiyama, N., Otsu, M., Shimada, Y., Sato, A., Nakanishi, S. and Jingami, H. (1998) 'Expression and purification of extracellular ligand binding region of metabotropic glutamate receptor subtype', *J. Biol. Chem.* **273**: 13,089–13,096.

Olney, J.M., Labruyere, J., Wang, G., Wozniak, D.F., Price, M.T. and Sesma, M.A. (1991) 'NMDA antagonist neurotoxicity-mechanism and prevention', *Science* **254**: 1515–1518.

Olney, J.W. (1991) Excitotoxicity and neuropsychiatric disorders, in P. Ascher, D.W. Choi and Y. Christen (eds) *Glutamate, Cell Death and Memory*, New York: Springer-Verlag, pp.77–101.

Ossowska, K. (1994) 'The role of excitatory amino-acids in experimental models of Parkinson's disease', *J. Neural. Transm.* **8**: 39–71.

Pace, A., Gama, L. and Breitwieser, G.E. (1999) 'Dimerization of the calcium-sensing receptor occurs within the extracellular domain and is eliminated by Cys-Ser mutations at Cys101 and Cys236', *J. Biol. Chem.* **274**: 11,629–11,634.

Parmentier, M.-L., Joly, C., Restituito, S., Bockaert, J., Grau, Y. and Pin, J.-P. (1998) 'The G-protein coupling profile of the metabotropic glutamate receptors, as determined with exogenous proteins, is independent of their ligand recognition domain', *Mol. Pharmacol.* **53**: 778–786.

Parmentier, M.-L., Pin, J.-P., Bockaert, J. and Grau, Y. (1996) 'Cloning and functional expression of a Drosophila metabotropic glutamate receptor expressed in the embryonic CNS', *J. Neurosci.* **16**: 6687–6694.

Pin, J.-P. (1998) 'The two faces of glutamate (News & Views)', *Nature* **394**: 19–21.

Pin, J.-P. and Bockaert, J. (1995) 'Get receptive to metabotropic glutamate receptors', *Curr. Opin. Neurobiol.* **5**: 342–349.

Pin, J.-P. and Duvoisin, R. (1995) 'The metabotropic glutamate receptors: structure and functions', *Neuropharmacology* **34**: 1–26.

Pin, J.-P., De Colle, C., Bessis, A. and Acher, F. (1999) 'New perspectives for development of selective metabotropic glutamate receptor ligands', *Eur. J. Pharmacol.* **375**, 1–3: 277–294.

Poncer, J.C., Shinozaki, H. and Miles, R. (1995) 'Dual modulation of synaptic inhibition by distinct metabotropic glutamatereceptors in the rat hippocampus', *J. Physiol.* **485**: 121–134.

Prézeau, L., Gomeza, J., Ahern, S., Mary, S., Galvez, T., Bockaert, J. and Pin, J.-P. (1996) 'Changes in the carboxyl-terminal domain of metabotropic glutamate receptor 1 by alternative splicing generate receptors with differing agonist-independent activity', *Mol. Pharmacol.* **49**: 422–429.

Reid, M., Toms, N., Bedingfield, J. and Roberts, P. (1999) 'Group I mGlu receptors potentiate synaptosomal [^{3}H]-glutamate release independently of exogenously applied arachidonic acid', *Neuropharmacology* **38**: 477–485.

Romano, C., Yang, W.-L. and O'Malley, K.L. (1996) 'Metabotropic glutamate receptor 5 is a disulfide-linked dimer', *J. Biol. Chem.* **271**: 28,612–28,616.

Ruat, M., Snowman, A.M., Hester, L.D. and Snyder, S.H. (1996) 'Cloned and expressed rat Ca^{2+}-sensing receptor. Differential cooperative responses to calcium and magnesium', *J. Biol. Chem.* **271**: 5972–5975.

Salt, T.E. and Turner, J.P. (1998a) Functions of metabotropic glutamate receptors in sensory processing in the thalamus, in F. Moroni, F. Nicoletti and D.E. Pellegrini-Giampietro (eds) *Metabotropic Glutamate Receptors and Brain Function*, London: Portland Press, pp. 49–58.

Salt, T.E. and Turner, J.P. (1998b) 'Reduction of sensory and metabotropic glutamate receptor responses in the thalamus by novel metabotropic glutamate receptor-1-selective antagonist S-2-methyl-4-carboxy-phenylglycine', *Neuroscience* **85**: 655–658.

Scanziani, M., Gahwiller, B.H. and Thompson, S.M. (1995) 'Presynaptic inhibition of the excitatory synaptic transmission by muscarinic and metabotropic glutamate receptor activation in hippocampus: are Ca^{2+} channels involved?', *Neuropharmacology* **34**: 1549–1557.

Scanziani, M., Salin, P.A., Vogt, K.E., Malenka, R.C. and Nicoll, R.A. (1997) 'Use-dependent increases in glutamate concentration activate presynaptic metabotropic glutamate receptors', *Nature* **385**: 630–634.

Schoepp, D.D., Johnson, B.G., Wright, R.A., Salhoff, C.R., Mayne, N.G., Wu, S., Cockerham, S.L., Burnett, J.P., Belegaje, R., Bleakman, D. and Monn, J.A. (1997) 'LY354740 is a potent and highly selective group II metabotropic glutamate receptor agonist in cells expressing human glutamate receptors', *Neuropharmacology* **36**: 1–11.

Shiells, R.A. and Falk, G. (1990) 'Glutamate receptors of rod bipolar cells are linked to a cyclic GMP cascade via a G-protein', *Proc. R. Soc. Lond. (Biol.)* **242**: 91–94.

Shigemoto, R., Abe, T., Nomura, S., Nakanishi, S. and Hirano, T. (1994) 'Antibodies inactivating mGluR1 metabotropic glutamate receptor block long-term depression in cultured Purkinje cells', *Neuron* **12**: 1245–1255.

Shigemoto, R., Kinoshita, A., Wada, E., Nomura, S., Ohishi, H., Takada, M., Flor, P.J., Neki, A., Abe, T., Nakanishi, S. and Mizuno, N. (1997) 'Differential presynaptic localization of metabotropic glutamate receptor subtypes in the rat hippocampus', *J. Neurosci.* **17**: 7503–7522.

Shigemoto, R., Wada, E., Oshini, H., Takada, M., Mizumo, M., Roberts, J. and Somogyi, P. (1996) 'Target cell specific concentration of a metabotropic glutamate receptor in the presynaptic active zone', *Nature* **381**: 523–525.

Sladeczek, F., Pin, J.-P., Récasens, M., Bockaert, J. and Weiss, S. (1985) 'Glutamate stimulates inositol phosphate formation in striatal neurones', *Nature* **317**: 717–719.

Staub, C., Vranesic, I. and Knöpfel, T. (1992) 'Responses to metabotropic glutamate receptor activation in cerebellar Purkinje cells: Induction of an inward current', *Eur. J. Neurosci.* **4**: 832–839.

Strasser, U., Lobner, D., Margaritta Behrens, M., Canzoniero, L.M.T. and Choi, D.W. (1998) 'Antagonists for group I mGluRs attenuate excitotoxic neuronal death in cortical cultures', *Eur. J. Neurosci.* **10**: 2848–2855.

Takahashi, K., Tsuchida, K., Tanabe, Y., Masu, M. and Nakanishi, S. (1993) 'Role of the large extracellular domain of the metabotropic glutamate receptors in agonist selectivity determination', *J. Biol. Chem.* **268**: 19,341–19,345.

Takechi, H., Eilers, J. and Konnerth, A. (1998) 'A new class of synaptic response involving calcium release in dentritic spines', *Nature* **396**: 757–760.

Taussig, R. and Gilman, A.G. (1995) 'Mammalian membrane-bound adenylyl cyclases', *J. Biol. Chem.* **270**: 1–4.

Thomsen, C. and Dalby, N.O. (1998) 'Roles of metabotropic glutamate receptor subtypes in modulation of pentylenetetrazole-induced seizure activity', *Neuropharmacology* **37**: 1465–1473.

Tones, M.A., Bendali, H., Flor, P.J., Knopfel, T. and Kuhn, R. (1995) 'The agonist selectivity of a class III metabotropic glutamate receptor, human mGluR4a, is determined by the N-terminal extracellular domain', *NeuroReport* **7**: 117–120.

Tu, J.C., Xiao, B., Yuan, J.P., Lanahan, A.A., Leoffert, K., Li, M., Linden, D.J. and Worley, P.F. (1998) 'Homer binds a novel proline-rich motif and links group 1 metabotropic glutamate receptors with IP3 receptors', *Neuron* **21**: 717–726.

Tyler, E.C. and Lovinger, D.M. (1995) 'Metabotropic glutamate receptor modulation of synaptic transmission in cortico-striatal co-cultures: role of calcium influx', *Neuropharmacology* **34**: 939–952.

Ugolini, A., Corsi, M. and Bordi, F. (1997) 'Potentiation of NMDA and AMPA responses by group I mGluR in spinal cord motoneurons', *Neuropharmacology* **36**: 1047–1055.

Unger, V.M., Hargrave, P.M., Baldwin, J.M. and Schertler, G.F. (1997) 'Arrangement of rhodopsin transmembrane alpha-helices', *Nature* **389**: 203–206.

Ward, D.T., Brown, E.M. and Harris, H.W. (1998) 'Disulfide bonds in the extracellular calcium-plyvalent cation-sensing receptor correlate with dimer formation and its response to divalent cations in vitro', *J. Biol. Chem.* **273**: 14,476–14,483.

Wermuth, C.G., Mann, A., Schenfelder, A., Wright, R.A., Johnson, B.G., Burnett, J.P., Mayne, N.G. and Schoepp, D.D. (1996) '(2S,4S)-2-amino-4-(4,4-diphenylbut-1-yl)-pentane-1.5-dioic acid: a potent and selective antagonist for metabotropic glutamate receptors negatively linked to adenylate cyclase', *J. Med. Chem.* **39**: 814–816.

White, J.H., Wise, A., Main, M.J., Green, A., Fraser, N.J., Disney, G.H., Barnes, A.A., Emson, P., Foord, S.M. and Marshall, F.H. (1998) 'Heterodimerization is required for the formation of a functional $GABA_B$ receptor', *Nature* **396**: 679–682.

Wroblewska, B., Wroblewski, J.T., Pshenichkin, S., Surin, A., Sullivan, S.E. and Neale, J.H. (1997) 'N-acetylaspartylglutamate selectively activates mGluR3 receptors in transfected cells', *J. Neurochem.*, **69**: 174–181.

Xiao, B., Tu, J.C., Petralia, R.S., Yuan, J.P., Doan, A., Breder, C.D., Ruggiero, A., Lanahan, A.A., Wenthold, R.J. and Worley, P.F. (1998) 'Homer regulates the association of group 1 metabotropic receptors with multivalent complexes of Homer-related, synaptic proteins', *Neuron* **21**: 701–716.

Yang, P., Chaney, M. and Baez, M. (1998) 'Mutagenesis analysis of a human mGluR2 binding site model using a novel tritiated antagonist, LY341495: as a probe', *Soc. Neurosci. Abs.* **24**: 1343.

Yokoi, M., Kobayashi, K., Manabe, T., Takahashi, T., Sakaguchi, I., Katsuura, G., Shigemoto, R., Ohishi, H., Nomura, S., Nakamura, K., Nakao, K., Katsuki, M. and Nakanishi, S. (1996) 'Impairment of hippocampal mossy fiber LTD in mice lacking mGluR2', *Science* **273**: 645–650.

Young, M.R., Blackburn-Munro, G., Dickinson, T., Johnson, M.J., Anderson, H., Nakalembe, I. and Fleetwood-Walker, S.M. (1998) 'Antisense ablation of type I metabotropic glutamate receptor mGluR1 inhibits spinal nociceptive transmission', *J. Neurosci.* **18**: 10,180–10,188.

Chapter 7

Pharmacology of metabotropic glutamate receptors

Darryle D. Schoepp and James A. Monn

Background

Ion channel linked (ionotropic) glutamate receptors were recognized decades ago to play a direct role in fast excitatory synaptic transmission within the majority of synapses of the central nervous system (see Watkins and Evans, 1981; Watkins, 1986). More recently, "metabotropic" glutamate (mGlu) receptors are recognized as a novel heterogeneous family of G-protein coupled receptor proteins which function to modulate glutamate transmission via pre-synaptic, post-synaptic, and glial mechanisms (see Pin and Duvoisin, 1995; Conn and Pin, 1997; Anwyl, 1999). Currently, eight subtypes of mGlu receptors (mGlu1–8) have been cloned; their structure and function have been described in the preceding chapter (Pin and Bockaert, 1999). Briefly, group I mGlu receptors are coupled to activation of phospholipase C and include mGlu1 and mGlu5 (and their slice variants). Both group II mGlu receptors (mGlu2 and mGlu3) and group III (mGlu4, mGlu6, mGlu7, and mGlu8) are coupled via Gi to inhibition of adenylate cyclase. The basis for describing mGlu receptors into these three groups was originally described by Nakanishi (1992) and included (1) structural homology of about 70 percent within groups and about 40 percent between groups, (2) similar transduction mechanisms or G-protein coupling, and (3) shared pharmacological properties within a group. This classification of mGlu receptors into groups has been very useful for describing the pharmacological actions of many novel agents. Many current agents are selective for mGlu receptor groups but they do not distinguish between subtypes within a group. Thus, this classification has been very useful for understanding the relative role of each mGlu receptor group in modulation of excitatory synaptic transmission. However, better pharmacological tools are needed to distinguish clearly the relative role of a single mGlu receptor subtype. The recent development of agents with subtype selectivity within an mGlu receptor group will be very useful. Also, a clearer understanding of the actions of known agents across all eight known receptors is needed. With this in mind, we have recently reviewed the literature in detail on the effects of known pharmacological agents across all eight mGlu receptor subtypes (Schoepp *et al.*, 1999a).

In this chapter we review the actions of current agonists and antagonists, with some attempt to understand the relationship between the actions of agents in cloned versus native systems and their usefulness as pharmacological tools. Only compounds with high potency and/or useful (or potentially useful) selectivity for mGlu receptors as a class are discussed. In many cases, the actions of compounds across all eight mGlu subtypes are not known. Thus, the effects of these agents in native tissues and their usefulness as pharmacological tools to study functions of mGlu receptor subtypes should be interpreted with caution. One presumably mGlu receptor mediated native tissue response which has been difficult to link to a given subtype of mGlu receptor is the activation of phospholipase D. For the sake of brevity, the effect of mGlu active agents on phospholipase D activity are not detailed in this chapter. For more detailed information on this aspect of mGlu receptor pharmacology, see Pin and Duvoisin (1995), Conn and Pin (1997) and Schoepp *et al.* (1999a).

For the purposes of this chapter, we have grouped pharmacological agents based on shared structural features and relative selectivity (as known to date) for mGlu receptor groups. There have been a number of different cell lines used for expression, various assay techniques (e.g. binding versus functional assays), species versions (e.g. rat versus human), and splice variants of mGlu receptor subtypes used by different investigators to characterize mGlu receptor activities of pharmacological agents. Therefore, the selectivity of certain agents at a given concentration should not be assumed to be the same in different tissue systems, as factors such as receptor reserve and the presence of endogenous glutamate in the media have been shown to influence agonist/antagonist potencies (see Schoepp *et al.*, 1999a). Whenever relevant, these factors are brought out in our discussions below.

Agonist pharmacology

Group I selective agonists

Quisqualate and related compounds

QUISQUALATE

Prior to the cloning of mGlu receptor subtypes, the study mGlu receptors generally involved the activation of the phosphoinositide hydrolysis second messenger system or subsequent intracellular calcium mobilization in a number of preparations, including cultured neurons, brain slices, cultured glia, and *Xenopus* oocytes injected with rat brain mRNA. The pharmacology of this response was clearly distinguished from known ionotropic glutamate receptor subtypes at that time (reviewed by Schoepp *et al.*, 1990; Schoepp and Conn, 1993). Quisqualate (Figure 7.1) was the most potent mGlu receptor agonist in eliciting these metabotropic responses, having submicromolar potency in many tissues. In cloned mGlu receptor subtypes, quisqualate remains the most potent and, at least across the mGlu subtypes, selective group I (mGlu1 and mGlu5) receptor agonist (see Schoepp *et al.*, 1999a). However, its high potency as an AMPA receptor agonist generally has precluded, or at least complicates, the use of quisqualate to study mGlu receptors in native tissues. Nevertheless, quisqualate has been useful as a group I selective agonist in recombinant systems and rat brain tissues. For example, ^{3}H-quisqualate was used as a high affinity ligand to study structure/function of cloned mGlu receptors in a recombinant system where its AMPA receptor activity would not interfere. In that study, Okamoto *et al.* (1998) showed that a soluble protein encoding the N-terminal extracellular domain of the rat mGlu1 receptor retained, when compared to the full length receptor, high affinity ^{3}H-quisqualate binding with a similar rank order of agonist inhibitions. Thus, ^{3}H-quisqualate appears to be a useful ligand for

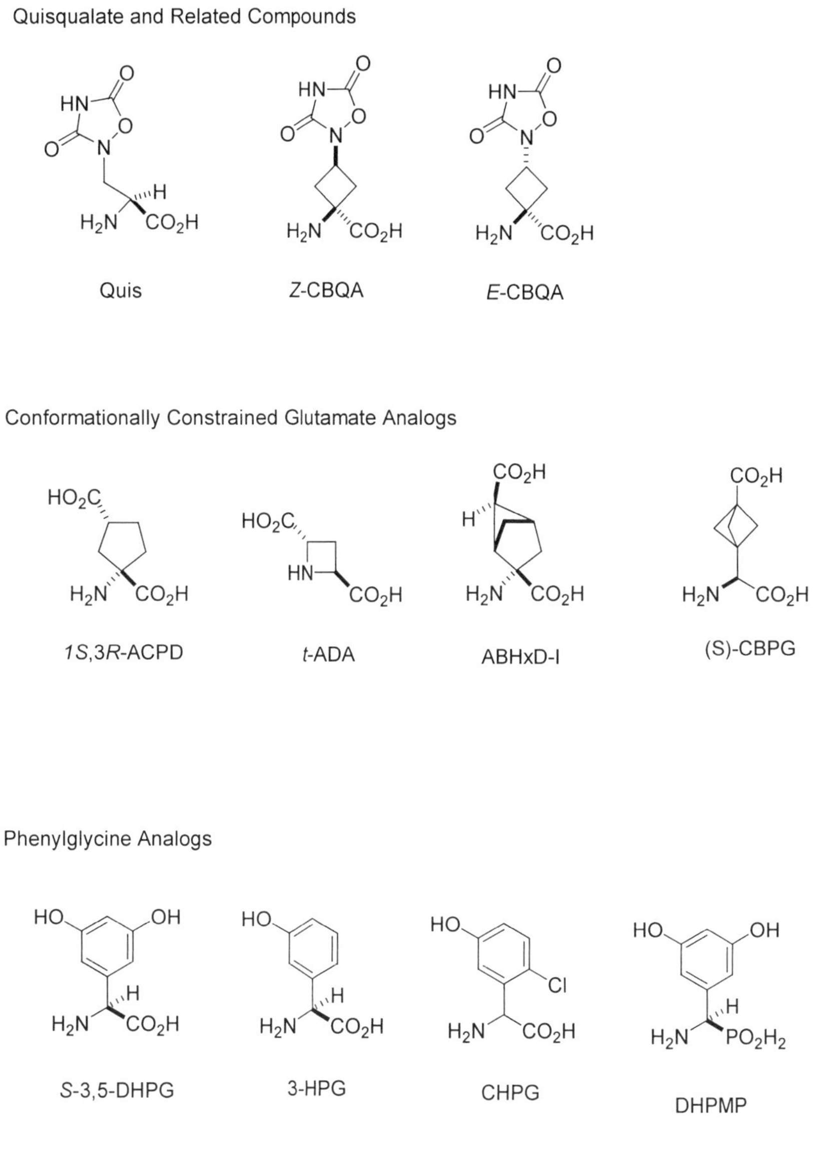

Quisqualate and Related Compounds

Quis   *Z*-CBQA   *E*-CBQA

Conformationally Constrained Glutamate Analogs

1S,3R-ACPD   *t*-ADA   ABHxD-I   (S)-CBPG

Phenylglycine Analogs

S-3,5-DHPG   3-HPG   CHPG   DHPMP

Figure 7.1 Group I selective agonists.

studies of mGlu1 (and possibly mGlu5) in recombinant systems. The usefulness of ^{3}H-quisqualate as a ligand for native tissue group I receptors needs to be investigated. Theoretically, ^{3}H-quisqualate binding to AMPA receptors in tissues could be prevented by including non-labeled AMPA or other such compounds in the incubation media, and then study other quisqualate-sensitive sites, which may include group I mGlu receptors. This approach has been successfully used for ^{3}H-glutamate binding to mGlu receptors. In this case, non-labeled AMPA, NMDA, and kainate are included in the incubation media and specific mGlu receptor binding was defined as binding sensitive to displacement by the group I/II mGlu receptor agonist compound 1*S*,3*R*-ACPD (Cha *et al.*, 1990; Schoepp and True, 1992). Interestingly, in these studies displacement of ACPD-sensitive ^{3}H-glutamate binding to rat brain exhibited both quisqualate-sensitive and quisqualate-insensitive binding sites that likely correspond to displacement of group I and group II mGlu receptors, respectively (Catania *et al.*, 1993; Wright *et al.*, 1994). More recently, the chemical structure of quisqualate, particularly the presence of the distal isoxazole acidic bioistere, was found to serve as a starting platform for other novel mGlu selective molecules (see CBQA isomers below).

CBQA ISOMERS

(*Z*)- and (*E*)-1-amino-3-[2′-(3′,5′-dioxo-1′,2′,4′-oxadiazolidinyl)] cyclobutane-1-carboxylic acid (CBQA) are conformationally constrained cyclobutane analogs of quisqualic acid (Littman *et al.*, 1999) (see Figure 7.1). Both (*Z*)- and (*E*)-CBQA stimulated phosphoinositide hydrolysis in cells expressing rat mGlu5a with low micromolar potencies (EC_{50}s of 18 and 53 μM, respectively), while having no effect at mGlu1a, mGlu2 or mGlu4a receptors. Consistent with being a potent mGlu5a selective agonist, (*Z*)- and (*E*)-CBQA also potently stimulated phosphoinositide hydrolysis in the rat hippocampus, a tissue which expresses high levels of mGlu5 receptors relative to mGlu1 receptors (Abe *et al.*, 1992; Lujan *et al.*, 1996). In contrast, neither compound stimulated phosphoinositide hydrolysis in the rat cerebellum, a tissue which expresses high levels of mGlu1a receptors (Masu *et al.*, 1991; Fotuhi *et al.*, 1993). Also, these compounds do not block uptake of ^{3}H-glutamate into rat brain synaptosomes, indicating that they are not inhibitors or substrates for glutamate uptake sites in the rat brain. Overall, (*Z*)- and (*E*)-CBQA appear to be valuable new compounds to explore the selective activation of mGlu5 receptors, as they appear reasonably potent and unlike most other group I agonists (e.g. see 3,5-DHPG) they would appear useful to discriminate between the group I receptor subtypes in native tissues.

Conformationally constrained glutamate analogs

1*S*,3*R*-ACPD

The conformationally constrained cyclopentane derivative of glutamate, 1-amino-cyclopentane-1,3-dicarboxylic acid (ACPD), gives rise to four different isomers: 1*S*,3*R*, 1*R*,3*S*-, 1*S*,3*S*- and 1*R*,3*R*- (see Schoepp *et al.*, 1990). Racemic (±)*trans*-ACPD (or *cis*-ACPD according to IUPAC nomenclature) was the first mGlu receptor selective agonist, as it activated phosphoinositide hydrolysis in rat hippocampus at concentrations having no effects on ionotropic glutamate receptor ligand binding (Palmer *et al.*, 1989; Desai and Conn, 1990). Separation of (+)*trans*-ACPD into its 1*S*,3*R*- and 1*R*,3*S*-isomers, demonstrated that 1*S*,3*R*-ACPD (Figure 7.1) was the active component responsible for stimulation of phosphoinositide hydrolysis in rat brain tissue (Schoepp *et al.*, 1991). Also, consistent with its metabotropic glutamate receptor activity in rat brain slices, 1*S*,3*R*-ACPD (not 1*R*,3*S*-ACPD) was found to stimulate intracellular calcium mobilization in neurons (Irving *et al.*, 1990). However, it is now recognized that 1*S*,3*R*-ACPD (at μM concentrations) has been reported to act as an agonist at multiple mGlu receptor subtypes, including mGlu1, mGlu2, mGlu3, mGlu4, mGlu5, mGlu6 and mGlu8 receptor subtypes (see Schoepp *et al.*, 1999a). This non-subtype selective mGlu receptor agonist profile likely explains why a range of pharmacological actions have been reported for 1*S*,3*R*-ACPD. These include increases in phosphoinositide hydrolysis and calcium mobilization in neurons and glia, increases in phospholipase D activity in rat brain, suppression of cAMP formation in multiple brain tissues, enhancement of cAMP formation in brain slices, and modulation of multiple ion currents or conductances in neurons (I_{AHP}, I_M, calcium, potassium, ionotropic NMDA and AMPA responses) (see Schoepp and Conn, 1993; Pin and Duvoisin, 1995; Anwyl, 1999). Thus, although it is considered by many investigators to be "diagnostic" for the involvement of mGlu receptors in native systems, data with 1*S*,3*R*-ACPD do not indicate the specific subtypes of mGlu receptors involved, and thus the usefulness of this tool for studying mGlu subtypes is limited.

t-ADA

Trans-azetidine-2,4-dicarboxylic acid (*t*-ADA, Figure 7.1) is a conformationally constrained glutamate analog which was initially reported to stimulate phosphoinositide hydrolysis in cultured cerebellar neurons at 1 mM concentration, but was devoid of mGlu1a receptor activity (Favaron *et al.*, 1993). Therefore, it was hypothesized in their paper that activation of the other known phosphoinositide coupled mGlu receptor subtype, namely the mGlu5 receptor, may have been responsible for this result. Subsequently, Kozikowski *et al.* (1993) resolved the (−) and (+) isomers of *t*-ADA and showed that both isomers activated phosphoinositide hydrolysis in cerebellar granule cells in culture. However, later work by Knöpfel *et al.* (1995a) reported that neither isomer of *t*-ADA (2*S*,4*S*- or 2*R*,4*R*-tested at 500 μM) activated recombinant rat mGlu5a receptors. In that study, *t*-ADA also had no agonist (or antagonist) activities at rat mGlu1b or mGlu4a receptors, but it only weakly suppressed forskolin-stimulated cAMP formation in rat mGlu2 expressing cells. More recently, Klein *et al.* (1997) demonstrated that *t*-ADA potently ($EC_{50} \leq 10\,\mu M$) increased phosphoinositide hydrolysis in slices of the neonatal (8-day-old) rat hippocampus, and it weakly (at 1 mM) increased phospholipase D activity in the same tissue. Therefore, at present it is difficult to reconcile the mGlu agonist-like activity of this compound in the cerebellar neurons and rat hippocampal tissues with a corresponding response in cloned mGlu receptors. Until this aspect is better understood, *t*-ADA does not appear to be a highly useful agent to study the known mGlu receptor subtypes in native mGlu receptor systems.

ABHxD-I

2-aminobicyclo[2.1.1]hexane-2,5-dicarboxylic acid-I (ABHxD-I) is a conformationally constrained glutamate analog in which the cyclopentane moiety of 1*S*,3*R*-ACPD is further rigidified in an extended conformation by a CH_2 bridge (Figure 7.1) (Kozikowski *et al.*, 1998). ABHxD-I exhibited varying mGlu receptor activities across group I (mGlu1a and mGlu5a), group II (mGlu2 and mGlu3), and group III (mGlu4a and mGlu6), with a relative potency order of mGlu2 > mGlu5 > mGlu1a ≥ mGlu3 > mGlu6 > mGlu4a. At mGlu2, ABHxD-I exhibited an EC_{50} value of 0.3 μM, but activation of mGlu4a receptors required about 1–2 orders of magnitude higher concentrations (EC_{50} 23 μM). Concentrations up to 1 mM did not exhibit ionotropic glutamate receptor activities. Thus, work with ABHxD-I indicates that broad mGlu receptor agonist selectivity can be produced by a fully extended form of glutamate. However, the usefulness of ABHxD-I as a tool to explore the functions of specific mGlu receptor subtypes is limited.

(S)-CBPG

(*S*)-(+)-2-(3′-carboxybicyclo[1.1.1]pentyl)-glycine (CBPG) is a conformationally constrained glutamate analog (Figure 7.1) which was initially reported as a mGlu1 selective antagonist ($IC_{50} = 25$ μM), with partial agonist activity (54 percent maximal effect versus glutamate and $EC_{50} = 103$ μM) at mGlu5 receptors, and no appreciable activities at mGlu2 or mGlu4 receptors (Pellicciari *et al.*, 1996a). A more recent study (Mannaioni *et al.*, 1999) has shown that CBPG activates phosphoinositide hydrolysis in rat hippocampus with a comparable potency to its effects on mGlu5, while having no effects on cAMP formation in that tissue. Like other selective and non-selective mGlu5 agonists such as CHPG and DHPG, CBPG depolarized CA1 pyramidal neurons and blocked spike frequency adaption and after hypolarizations in these cells. These studies indicate that mGlu5 receptors mediate the direct excitatory actions of group I agonists in CA1 hippocampal cells and indicate that the CBPG is a useful new tool for dissecting the relative role of mGlu1 versus mGlu5 in modulation of excitatory synaptic transmission.

Phenylglycine analogs

3,5-DHPG

(*RS*)3,5-dihydroxyphenylglycine (3,5-DHPG) was initially described by Ito *et al.* (1992) as a more potent activator of rat mGlu1a receptors expressed in *Xenopus* oocytes than (±)*trans*-ACPD. At that time, (±)*trans*-ACPD was one of the few known selective mGlu receptor agonists. Also, in the original paper on this compound, the putative mGlu receptor effects of 3,5-DHPG were studied electrophysiologically on native neurons (rat hippocampal CA3 pyramidal cells), where 3,5-DHPG, like ACPD, produced a slow excitation resulting from depression of inhibitory currents (I_{AHP} and I_M) (Ito *et al.*, 1992). This suggested that structurally novel phenylglycines, such as 3,5-DHPG, could mimic the actions of more ‘glutamate-like’ diacidic amino acids at mGlu receptors.

Subsequent second messenger studies in the rat hippocampus showed that 3,5-DHPG was a μM potent group I mGlu receptor selective agonist, as it activated phosphoinositide hydrolysis but, unlike 1*S*,3*R*-ACPD, had no effect on forskolin-stimulated cAMP formation (Schoepp *et al.*, 1994). This group I selective activity of 3,5-DHPG on the phosphoinositide hydrolysis pathway in the rat hippocampus resided in the *S*-isomer (Baker *et al.*, 1995). At cloned mGlu receptor subtypes, 3,5-DHPG exhibits about equal agonist potencies at mGlu1 and mGlu5 receptor subtypes but has no agonist or antagonist activities at recombinant rat and/or human group II (mGlu2 and mGlu3) or group III (mGlu4, 6, 7, or 8) mGlu receptors (see Desai *et al.*, 1995; Gereau and Conn, 1995; Brabet *et al.*, 1995; Wu *et al.*, 1998).

The mGlu1/5 selectivity of 3.5-DHPG has made this compound the most widely used compound to date to explore the relative role of group I mGlu receptors in glutamatergic excitation in native tissues. For example, it has been shown in hippocampal and striatal neurons that 3,5-DHPG, although not directly exciting neurons, enhance neuronal excitations by the potentiation of the ionotropic glutamate receptor agent NMDA (Fitzjohn *et al.*, 1996; Pisani *et al.*, 1997). Consistent with these pro-excitatory actions, central administration of 3,5-DHPG has been shown to lead the induction of seizures and/or neurotoxicity (Tizzano *et al.*, 1995a; Ong and Balcar, 1997; Camon *et al.*, 1998) and pain-like states (Fisher and Coderre, 1996, 1998) in animals. Studies such as these are consistent with the potential anti-convulsant and analgesic actions of selective group I receptor antagonists in animals (see Bordi and Ugolini, 1999).

3-HYDROXYPHENYL GLYCINE

Like 3,5-DHPG, the monohydroxy-substituted phenyl glycine analog 3-hydroxyphenyl glycine (3-HPG) (Figure 7.1) also selectively activates group I mGlu receptors (see Schoepp *et al.*, 1999a; Birse *et al.*, 1993; Hayashi *et al.*, 1994; Brabet *et al.*, 1995). The work of Brabet *et al.* (1995) compared the activities of these two compounds in the same test system, and found that the removal of the hydroxy group from the 5-position of 3,5-DHPG reduced the group I mGlu receptor activity of 3-HPG about one-order of magnitude (3-HPG EC_{50} values for mGlu1 and mGlu5 were 97 μM and 14 μM, respectively). The relative potency order of 3,5-DHPG > 3-HPG is a useful approach to understanding group I mGlu receptor involvement in cellular and behavioral processes. For example, in the guinea pig hippocampus, both 3,5-DHPG and 3-HPG were shown to convert picrotoxin-induced intra-ictal bursts into prolonged discharges (Merlin and Wong, 1997). The concentrations required for these effects were 50–100 μM of 3,5-DHPG, but 250–500 μM of 3-HPG. Thus, the effects of these compounds and their relative potency order in this system strongly indicate a potential role for group I mGlu receptors in inducing epileptiform activity *in vitro*, and are consistent with *in vivo* studies showing that 3,5-DHPG is more potent than 3-HPG in inducing limbic seizures in mice (Tizzano *et al.*, 1995a, 1995b).

CHPG

(*RS*)-2-chloro-5-hydroxyphenylglycine (CHPG) is a phenylglycine analog where a choro-substituent was added on the phenyl ring para to the hydroxy group of 3-HPG (see Figure 7.1). This resulted in a compound with measurable mGlu5 receptor activity (EC_{50} 750 μM), but no measurable mGlu1 receptor activity at up to 10,000 μM (Doherty *et al.*, 1997). Thus, CHPG has been suggested to be useful for examining the relative role of mGlu5 versus mGlu1 receptors. Consistent with the actions of the more potent mGlu5 (and mGlu1) receptor agonists such as 3,5-DHPG and 1*S*,3*R*-ACPD, CHPG (1 mM) was shown to potentiate NMDA induced depolarizations in the rat hippocampus (Doherty *et al.*, 1997). These data indicate that mGlu5, not mGlu1 is responsible for this excitatory action of group I agonists. However, the low potency of CHPG and its unknown activities across all metabotropic and ionotropic glutamate receptors at the mM concentrations needed to activate mGlu5 receptors somewhat limits the usefulness of this compound to study mGlu5 receptor activation definitively.

DHPMP

(*RS*)-amino (3,5-dihydroxyphenyl)methylphosphinic acid (DHPMP) (see Figure 7.1) has been reported as a potent and selective group I mGlu receptor agonist (Boyd *et al.*, 1996). In slices of the guinea pig cerebral cortex, DHPMP was about equi-potent (EC_{50} 28 μM) with 3,5-DHPG in inducing phosphoinositide hydrolysis, but had no effect at up to 1000 μM on forskolin-stimulated cAMP formation. However, the group I mGlu receptor selectivity of DHPMP needs to be confirmed across recombinant mGlu receptor subtypes.

Group II selective agonists

Conformationally constrained glutamate analogs

CYCLOPROPYLGLYCINE ANALOGS

2-(carboxycyclopropyl)glycines (CCGs) are conformationally restricted glutamate analogs in which the proximal and distal carboxy groups can exist in various conformations where they are relatively extended or folded. Studies with the eight different isomers of CCG indicate that the extended conformation is preferred for mGlu receptor agonist potency and selectivity, as L-CCG-I (Figure 7.2) effectively activates multiple mGlu receptor subtypes, while having no effects on ionotropic glutamate receptors (Nakagawa *et al.*, 1990; Hayashi *et al.*, 1992; Ohfune *et al.*, 1993). L-CCG-I was one of the most potent (nM activity) compounds for group II (mGlu2 or mGlu3) receptors. However, it is now recognized that L-CCG-I can also activate mGlu1, mGlu5, mGlu, mGlu4, mGlu6, and mGlu8 receptors at about 10–30 times higher (μM) concentrations (see Brabet *et al.*, 1998; Schoepp *et al.*, 1999a). Consistent with this, in slices of the rat hippocampus, L-CCG-I was about ten times more potent in suppressing forskolin-stimulated cAMP formation than it was in activating phosphoinositide hydrolysis (Schoepp *et al.*, 1995a). As there are now compounds with much greater mGlu2/3 selectivity than L-CCG-I for group II mGlu receptors (see below), the usefulness of L-CCG-I as a pharmacological tool is somewhat limited. In this regard, combining carboxycyclopropylglycine pharmacophoric elements that led to the mGlu selective agonist L-CCG-I (extended glutamate conformation) and the potent NMDA agonist L-CCG-IV (folded glutamate conformation) resulted in the 2′,3′-

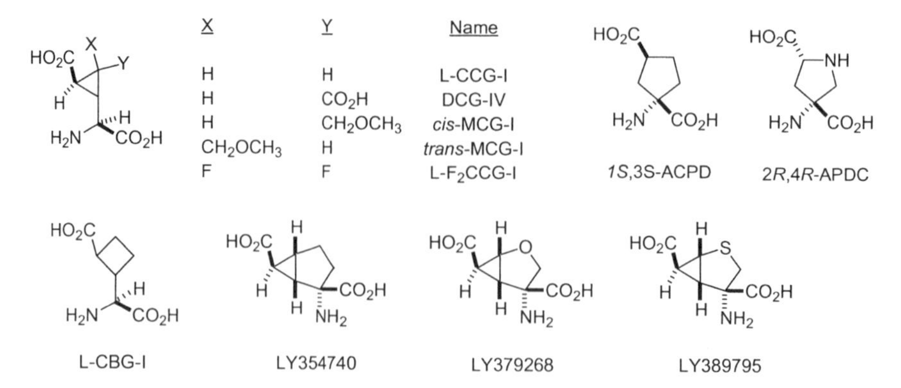

Figure 7.2 Group II selective agonists.

dicarboxycyclopropylglycine analog DCG-IV (Figure 7.2) (Ohfune *et al.*, 1993). Interestingly, DCG-IV is highly potent and relatively selective for group II mGlu receptors. Its potency for mGlu2 and mGlu3 is low nM (Hayashi *et al.*, 1993; Brabet *et al.*, 1998); however, at higher μM concentrations it is an antagonist at group I and group III mGlu receptor subtypes (Brabet *et al.*, 1998). Also, these higher concentrations have been reported to activate NMDA receptors in native tissues (Wilsch *et al.*, 1994; Breakwell *et al.*, 1997; Uyama *et al.*, 1997). In any case, when used at appropriate sub-micromolar concentrations, DCG-IV appears to be a highly useful agent for the selective activation of group II mGlu receptors. Recently, ^{3}H-DCG-IV has also been shown to bind to CHO cell membranes expressing recombinant rat mGlu2 receptors with relatively high affinity (Kd = 160 nM) (Cartmell *et al.*, 1998). The pharmacology of this binding (e.g. potent displacement by other group II agonists such as LY354740) is consistent with the selective labeling of mGlu2 receptors. In rat brain cortex homogenates, ^{3}H-DCG-IV is bound to a single site with a Kd = 180 nM and B_{max} of 780 fmol/mg protein, and this binding was potently displaced by the group II antagonist LY341495 and group II agonist LY354740 (Mutel *et al.*, 1998). This indicates that ^{3}H-DCG-IV is a useful ligand for studies of mGlu2 (and likely mGlu3) in native tissues. However, the selectivity of ^{3}H-DCG-IV binding across the mGlu receptor subtypes needs to be further explored.

Further substitutions of L-CCG-I by placement of a methoxymethyl substituent on the cyclopropane ring (see Figure 7.2) have led to additional compounds with reported group II mGlu receptor selectivity. These include *cis*- and *trans*-MCG-I which, like other group II mGlu agonists, potently suppress monosynaptic excitations in the newborn rat spinal cord and/or forskolin-stimulated cAMP formation in cultured rat cortical neurons, with the *trans*-MCG-I isomer being more potent (Shimamoto and Ohfune, 1993; Ishida *et al.*, 1994, 1995). Likewise, the difluoro analog of L-CCG-I, L-F_2CCG-I (Figure 7.2) was shown to be about three times more active that L-CCG-I in suppressing reflexes in the rat spinal cord (Shibuya *et al.*, 1998). However, little is known about the actions of these agents across cloned mGlu receptor subtypes (see Schoepp *et al.*, 1999a).

1*S*,3*S*-ACPD

Among the isomers of ACPD, 1*S*,3*S*-ACPD (Figure 7.2) is considered to be a relatively selective group II mGlu receptor agonist (see Schoepp *et al.*, 1999a). At cloned mGlu receptor subtypes, 1*S*,3*S*-ACPD activates mGlu2 and mGlu3 receptors with low μM potency (Kozikowski *et al.*, 1998); however, high μM or mM concentrations are required to activate cloned group I mGlu receptors (Thomsen *et al.*, 1994a). In rat hippocampal slices ≥100 μM 1*S*,3*S*-ACPD is needed to activate phosphoinositide hydrolysis (Schoepp *et al.*, 1995a). However, in the neonatal rat spinal cord 10 μM of 1*S*,3*S*-ACPD has been shown to greatly depress the monosynaptic component of the dorsal root-evoked ventral root potential (Pook *et al.*, 1992), an action consistent with its more potent group II mGlu agonist activity in cloned receptors. Although apparently a relatively selective group II agonist in that preparation, its full profile across all mGlu receptors (particularly group III receptors) has not been reported. This, along with its group I mGlu receptor activity at higher concentrations, suggests that data with 1*S*,3*S*-ACPD in less defined preparations needs to interpreted with caution.

2*R*,4*R*-APDC

2*R*,4*R*-4-aminopyrrolidine-2,4-dicarboxylate (2*R*,4*R*-APDC; LY314593; see Figure 7.2) is a conformationally constrained glutamate analog in which a pyrrolidine ring structure is present rather than the cyclopentane group in ACPD (Monn *et al.*, 1996). The presence of this nitrogen in the ring produced a group II selective compound with about the same potency as 1*S*,3*S*-ACPD (see Schoepp *et al.*, 1999a). In human mGlu2 or mGlu3 expressing cells 2*R*,4*R*-APDC suppresses forskolin-stimulated cAMP formation with EC_{50} values of ~400 nM, but there were no appreciable effects on group I (mGlu1 or mGlu5) or group III (mGlu4, mGlu7, mGlu6, or mGlu8) receptors at up to 100 μM (Schoepp *et al.*, 1996; Wu *et al.*, 1998; Tückmantel *et al.*, 1997). In rat brain slices, 2*R*,4*R*-APDC selectively suppresses forskolin-stimulated cAMP formation at concentrations (1–100 μM) having no effect on phosphoinositide hydrolysis *per se* (Schoepp *et al.*, 1995b). These studies suggest that 2*R*,4*R*-APDC is a highly useful agent for examining the role of group II mGlu receptors in native tissues. In this regard, 2*R*,4*R*-APDC, like the other group II selective agonist 1*S*,3*S*-ACPD, was shown to inhibit the development of kindled seizures in rats (Attwell *et al.*, 1995, 1998), suggesting a role for group II mGlu receptors in epilepsy. In contrast, Fisher and Coderre (1996) reported that, unlike the group I agonist 3,5-DHPG, 2*R*,4*R*-APDC does not induce spontaneous nociceptive behaviors in rats. These *in vivo* studies are consistent with the highly selectivity of 2*R*,4*R*-APDC for group II mGlu receptors *in vitro*.

LY354740

As described above, a number of selective agonists for group II mGlu receptors have now been described. However, their usefulness to probe the CNS for possible therapeutic uses is limited by either low potency and/or lack of activity following systemic administration. The conformationally restricted bicycloamino acid compound, LY354740 ((1*S*,2*S*,5*R*,6*S*)-2-aminobicyclo[3.1.0.]hexane-2,6-dicarboxylate monohydrate) (Figure 7.2) (Monn *et al.*, 1997), represents the first group II selective agonist which has been shown to have potent pharmacological effects in animals, following systemic administration (for a detailed review on this compound see Schoepp *et al.*, 1999b). *In vitro* studies have shown that LY354740 potently activates mGlu2 (EC_{50} = 5 nM) and mGlu3 (EC_{50} = 24 nM), but had no appreciable effects on other mGlu subtypes or ionotropic glutamate receptors at concentrations 2–4 orders of magnitude higher (Schoepp *et al.*, 1997; Wu *et al.*, 1998). In rat hippocampus, LY354740 suppresses forskolin-stimulated cAMP formation with about the same potency as was observed in mGlu3 expressing cells (EC_{50} = 22 nM), but LY354740 has no effect on phosphoinositide hydrolysis *per se* (Schoepp *et al.*, 1998). These studies indicate that LY354740 is a highly potent and selective mGlu2/3 agonist at both recombinant human and native rat receptors. In rat brain tissue, ^{3}H-LY354740 has been shown to bind with high affinity (Kd = 8 nM). Specific ^{3}H-LY354740 in the rat brain exhibited a pharmacology and distribution consistent with the selective labeling of group II mGlu receptors (Schauffhauser *et al.*, 1998). A number of studies have now shown that LY354740 is active *in vivo*. Consistent with the role of group II mGlu receptors in suppression of glutamate transmission *in vitro*, systemic LY354740 has been shown to prevent veratridine-evoked glutamate and aspartate release in free moving rats undergoing microdialysis of the striatum (Battaglia *et al.*, 1997). Orally administered LY354740 potently suppressed fear-potentiated startle responding in rats, indicating it has anxiolytic activity, but unlike other anxiolytics (e.g. diazepam), LY354740 did not produce CNS depression at any dose (Helton *et al.*, 1998a). Acute LY354740 administration will block withdrawal reactions to multiple agents, including morphine (Vandergriff and Rasmussen, 1999), nicotine (Helton *et al.*, 1997), and diazepam (Helton *et al.*, 1998b). However, subchronic administration of LY354740 itself does not produce withdrawal. More recently, LY354740 has been reported to suppress the *in vivo* actions of phencyclidine in rats (Moghaddam and Adams, 1998; Cartmell *et al.*, 1999). These data suggest that LY354740 may be useful for the treatment of various psychiatric disorders. Studies in models of neurological disorders have shown the LY354740 in neuroprotective in the gerbil global ischemia model (Bond *et al.*, 1998) and traumatic neuronal injury *in vitro* and *in vivo* (Allen *et al.*, 1999). However, LY354740 was not active following permanent focal ischemia in the rat (Lam *et al.*, 1998). Thus, the role of group II mGlu receptor agonists as novel neuroprotectants requires additional investigation.

More recently, the conformationally constrained heterocyclic compounds with either oxygen (LY379268) or sulfur (LY389795) in the 2-position of the bicyclohexane ring have been described as highly potent and systemically active mGlu2/3 agonists (see Figure 7.2) (Monn *et al.*, 1999). In functional assays of human mGlu2 or mGlu3 expressing cells, LY379268 and LY389795 were 3–8 times more potent than LY354740. In a model of limbic seizures in mice induced by 1*S*,3*R*-ACPD (Tizzano *et al.*, 1995a), as shown previously for LY354740 (Monn *et al.*, 1997), both LY379268 and LY389795 were anticonvulsant in this test when given parenterally (i.p.) (Monn *et al.*, 1999). In another study (Cartmell *et al.*, 1999), LY379268 was compared to LY354740 in the phencyclidine (PCP) model of psychosis in rats. When administered by the subcutaneous route, both compounds blocked motor activations induced by 5 mg/kg PCP; however, LY379268 was about three times more potent than LY354740. These studies demonstrate that these heterobicyclic compounds, like LY354740, are useful agents to investigate the therapeutic prospects of mGlu2/3 agonists following systemic administration.

Group III selective agonists

Acidic bioisosteres of glutamate

Simple bioisosteric replacement of the distal carboxylic acid moiety on glutamic acid has lead to novel and interesting group III mGlu selective agonists. The most well-studied and notable of these compounds are *S*-4-phosphono-2-aminobutyric acid (*S*-AP4) and *S*-serine-O-phosphate (*S*-SOP) (see Figure 7.3). Prior to the recognition that mGlu receptors existed, (*S*)-AP4 was reported to suppress excitatory transmission in multiple excitatory synapses, including the lateral perforant path (Koerner and Cotman, 1981), mossy fiber pathway (Yamamoto *et al.*, 1983; Lanthorn *et al.*, 1984), lateral olfactory tract (Collins, 1982), and retina (Shiells *et al.*, 1981). Ganong and Cotman (1982) also showed that this presynaptic inhibitory action of *S*-AP4 in the lateral perforant path was also produced by *S*-SOP. This so called "L-AP4" receptor remained something of an enigma until the cloning of mGlu receptor subtypes, when it was recognized that these compounds act as relatively potent and somewhat selective agonists for mGlu4, mGlu6, mGlu7, and mGlu8 receptors (see Thomsen, 1997). As

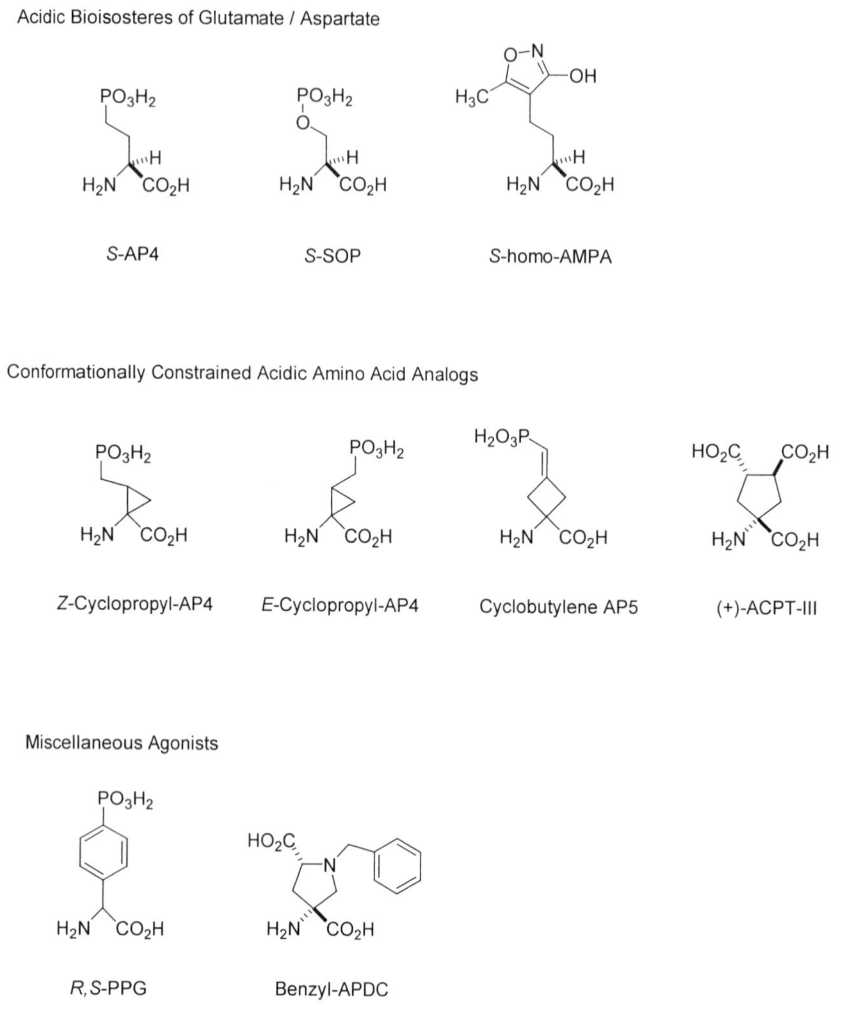

Acidic Bioisosteres of Glutamate / Aspartate

S-AP4 *S*-SOP *S*-homo-AMPA

Conformationally Constrained Acidic Amino Acid Analogs

Z-Cyclopropyl-AP4 *E*-Cyclopropyl-AP4 Cyclobutylene AP5 (+)-ACPT-III

Miscellaneous Agonists

R,S-PPG Benzyl-APDC

Figure 7.3 Group III selective agonists.

each of the *S*-AP4-sensitive members of the cloned group III mGlu receptors (e.g. mGlu4, mGlu7, and mGlu8) are found to exist presynaptically in *S*-AP4 sensitive excitatory pathways, it is possible that in a given synapse one or more of the group III receptors function as negative feedback receptors that regulate glutamate release. Studies have attempted to understand which *S*-AP4-sensitive subtype(s) are responsible for suppression of excitatory transmission in specific preparations (see Thomsen, 1997; Johansen *et al.*, 1995). For example, the L-AP4 receptor in the retina has been suggested to be mGlu6, as this receptor shows selective expression in this tissue (Nakajima *et al.*, 1993). *S*-AP4 and *S*-SOP are each nM to low μM potent agonists for multiple group III mGlu receptor subtypes, including mGlu4, mGlu6 and mGlu8 (see Thomsen, 1997; Schoepp *et al.*, 1999a). As *S*-AP4 is relatively inactive at cloned group I or group II mGlu receptors, this has made this compound "diagnostic" for the involvement of group III receptors in cellular functions. However, mM concentrations of *S*-AP4 are required to activate mGlu7 receptors, and these concentrations have also been shown to activate cloned mGlu2 and mGlu3 receptors (see Schoepp *et al.*, 1999a). Furthermore, the selectivity of *S*-SOP across all cloned mGlu receptors needs to be better established. Clearly, better and more subtype selective agonists for group III mGlu receptors than *S*-AP4 or *S*-SOP are desirable to study the role of specific group III mGlu receptors in regulation of excitatory synaptic transmission. In this regard, additional agents targeted at advancing group III mGlu receptor pharmacology have recently been described and are discussed below.

S-homo-AMPA

The compound 2-amino-4-(3-hydroxy-5-methylisoxazol-4-yl)butyric acid (homo-AMPA), the 3-isoxazole bioisostere of 2-aminoadipic acid (see Figure 7.3), was reported to activate rat mGlu6 receptors with about four times less potency than glutamate ($EC_{50} = 82\,\mu M$) (Bräuner-Osborne *et al.*, 1996). Importantly homo-AMPA had no ionotropic glutamate receptor affinity, or mGlu1, mGlu2, or mGlu4 receptor activities, at concentrations which activated mGlu6. A later study (Ahmadian *et al.*, 1997) showed that the mGlu6 agonist activity of homo-AMPA resides in the *S*-isomer ($EC_{50} = 52\,\mu M$), and concentrations of up to $1000\,\mu M$ had no effect on other mGlu receptor subtypes including mGlu1, mGlu2, mGlu3, mGlu4, mGlu5, or mGlu7. Although its activity at mGlu8 receptors needs to be determined, based on data in cloned receptors, *S*-homo-AMPA appears to be a highly useful compound for studies of mGlu6 receptors in native tissues.

Conformationally constrained acidic amino acid analogs

Several conformationally constrained acidic amino acid analogs have been described which potentially enhance, or at least retain, the group III potency of other known group III agonists such as *S*-AP4. The cyclopropyl analog of AP4, (*Z*)-2-amino-2,3-methano-4-phosphonobutanoic acid (*Z*-cyclopropyl-AP4) (Figure 7.3) has been reported to retain the potency of *S*-AP4 at the mGlu4a receptor ($EC_{50} = 580\,nM$); however, *E*-cyclopropyl-AP4 (Figure 7.3) was about ten times less potent ($EC_{50} = 8\,\mu M$) (see Figure 7.3 for structures) (Johansen *et al.*, 1995). Likewise, the constrained phospho-amino acid compound cyclobutylene AP5 (Figure 7.3) was reported as a relatively potent ($EC_{50} = 4.4\,\mu M$) mGlu4a receptor agonist (Johansen *et al.*, 1995; Peterson *et al.*, 1992). However, the actions of these compounds across other mGlu subtypes and ionotropic receptors have not been reported. Nevertheless, this pharmacological information has been useful to attempt to identify mGlu receptors involved in *S*-AP4 sensitive pathways in the rat brain. For example, although *Z*- and *E*-cyclopropyl-AP4 have different potencies at cloned mGlu4a receptors, they were equipotent in suppressing excitatory transmission in the lateral perforant pathway (Kroona *et al.*, 1991; Johansen *et al.*, 1995). Furthermore, although cyclobutylene AP5 was two times more potent than *E*-cyclopropyl-AP4 as an mGlu4a agonist, it was 2–3 times less potent in suppressing excitatory transmission in the lateral perforant pathway (Johansen *et al.*, 1995). Based on these data, mGlu4 would not appear to represent the L-AP4 receptor in the lateral perforant pathway.

ACPTS

Acher *et al.* (1997) have reported on the four stereoisomers of 1-aminocyclopentane-1,3,4-tricarboxylic acid (ACPT). Two of these isomers, ACPT-I and ACPT-III (Figure 7.3) were μM potent agonists for mGlu4a, with weak antagonist activities at mGlu1 and mGlu2. These compounds may be useful to explore group III mGlu receptors, but their activities across other glutamate receptors needs to be established.

Miscellaneous agonists

(*R,S*)-PPG

The phosphono-substituted phenylglycine compound (*R,S*)4-phosphonophenylglycine ((*R,S*)-PPG, see Figure 7.3) has been shown to be a potent and selective mGlu receptor agonist across each cloned group III receptor subtype, with the following order of decreasing agonist potencies: mGlu8 ($EC_{50} = 0.2\,\mu M$) > mGlu6 = mGlu4 (EC_{50}s = 5 μM) > mGlu7 ($EC_{50} = 185\,\mu M$). Importantly, (*R,S*)-PPG is relatively inactive at cloned group I and group II mGlu receptors or inotropic glutamate receptors (Gasparini *et al.*, 1999a). However, similar to other phosphono-amino acids, (*R,S*)-PPG showed micromolar potency at Ca^{2+}/Cl^{-} dependent binding sites in the rat brain. Consistent with other group III agonists such as (*S*)-AP4, (*R,S*)-PPG was neuroprotective agonist NMDA mediate excitoxicity in cultured neurons *in vitro* and when co-injected with NMDA into the rat striatum *in vivo*. Nevertheless, the usefulness of (*R,S*)-PPG as a new tool is limited by the fact that it was not pharmacologically active when given systemically, as it was fully protective against maximal electroshock seizures in mice when administered intracerebral ventricular (173 nmol) but did not protect when given parenterally (100 mg/kg intraperitoneal, or 10 mg/kg intravenous) in the same test. In any case, these studies with (*RS*)-PPG further support a role for group III agonists as novel agents to treat neurodegenerative disorders (for a review on this subject see Nicoletti *et al.*, 1996).

BENZYL-APDC

The 1-benzyl derivative of the conformationally contrained glutamate analog APDC, benzyl-APDC (see Figure 7.3), has been reported as a selective mGlu6 receptor agonist ($EC_{50} = 20\,\mu M$), with no activity at mGlu1 receptors and relatively weak antagonist activity at mGlu5 and mGlu2 receptors (Tückmantel *et al.*, 1997). The actions of this compound on other mGlu receptors subtypes and native mGlu receptor responses have not been reported.

Antagonist pharmacology

Group I selective antagonists

Phenyglycine analogs

Before the 1990s the only selective antagonist for mGlu receptor responses, in this case phosphoinositide hydrolysis in rat brain tissues, was L-2-amino-3-phosphonopropanoate (L-AP3). In general, L-AP3 is only weakly active in many preparations, requiring high μM–mM concentrations, and in some systems (e.g. brain slices) its inhibitory actions are non-competitive in nature (see Schoepp *et al.*, 1990, 1999a). As discussed above, the phenylglycine analog 3,5-DHPG was shown selectively to activate group I mGlu receptors with comparable potency to *trans*-ACPD and glutamate. The search for more potent competitive mGlu receptor antagonists led to the recognition that various other phenylglycine derivatives possess antagonist, agonist, and even mixed agonist/antagonist activities across multiple subtypes of mGlu receptors. The various phenylglycine derivatives have been reviewed in detail elsewhere (see Roberts, 1995; Watkins and Collingridge, 1994; Schoepp *et al.*, 1999a). The focus here will be on select compounds of this chemical class with unique and useful properties for investigating mGlu receptor pharmacology and function. The system where many of these compounds have been tested is the neonatal spinal cord preparation (Pook *et al.*, 1992; Cao *et al.*, 1995). In this preparation, group I agonists (e.g. 1*S*,3*R*-ACPD, 3,5-DHPG) depolarize motoneurons, while group II selective agonists (e.g. 1*S*,3*S*-ACPD) and group III selective agonists (e.g. L-AP4) depress the monosynaptic component of the dorsal root evoked-ventral root potential (mDR-VRP), presumably via a presynaptic mechanism. The specific mGlu receptor subtypes mediating these agonist responses have not been completely established, so such data in the absence of clone data needs to interpreted with caution. Many phenylglycines have been examined at representative recombinant receptors, confirming to a certain extent their mGlu subtype selectivities.

(*S*)-4CPG

Substitution of the para-hydroxy group in 3,5-dihydroxyphenylglycine with a carboxy group, as in *S*-4-carboxy phenylglycine (*S*-4CPG, Figure 7.4), produced a mGlu1 selective competitive group I antagonist (IC_{50}s 10–100 μM) with weak agonist activity at mGlu2 ($EC_{50} \sim 500\,\mu$M), but no activity at mGlu4 receptors (Thomsen *et al.*, 1994b; Hayashi *et al.*, 1994; see Schoepp *et al.*, 1999a).

(*S*)-MCPG

The incorporation of an α-methyl group into *S*-4CPG, as in *S*-α-methyl-4-carboxyphenylglycine ((*S*)-MCPG) maintains the group I antagonist potency, and leads to about equivalent antagonist potency at mGlu2 receptors (Jane *et al.*, 1993; Hayashi *et al.*, 1994; Thomsen *et al.*, 1994b; see Schoepp *et al.*, 1999a). Although inactive at mGlu4 receptors, (*S*)-MCPG had significant antagonist activity at mGlu8 receptors (Saugstad *et al.*, 1997; Wu *et al.*, 1998). As (*S*)-MCPG has been reported to block the actions of L-AP4 in the neonatal spinal cord (Kemp *et al.*, 1994), it suggests that mGlu8 receptors mediate the depressant effects of L-AP4 in that preparation. The competitive nature of (*S*) MCPG inhibitions, along with the commercial availability of this compound, has made it a widely used pharmacological tool for probing the functions of mGlu receptors in cellular processes (see Watkins and Collingridge, 1994). However, the recent discovery of more potent and selective phenylglycine and non-phenylglycine antagonist compounds currently limits the usefulness of (*S*)-MCPG.

(*RS*)-PeCPG

As (*S*)-MCPG is a competitive mGlu antagonist, additional substitutions of the α-carbon of (*S*)4CPG have been explored. These include the *n*-pentyl derivative

Phenylglycine Analogs

S-4-CPG

S-MCPG

R,S-PeCPG

AIDA

S-4C3HPG

LY367385
(+)-4C2MPG

Miscellaneous Competitive Antagonists

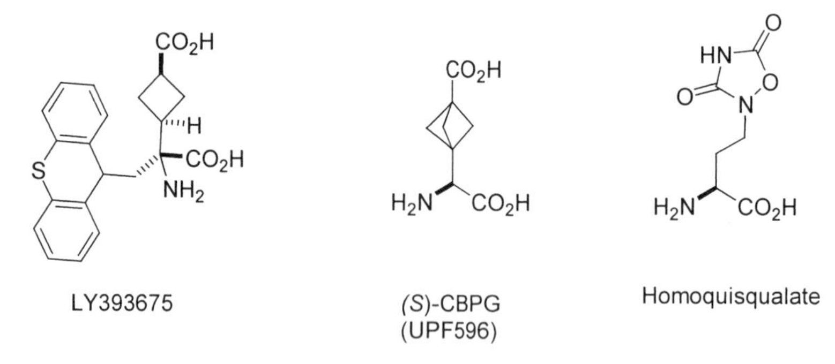

Figure 7.4 Group I selective competitive antagonists.

(*RS*)-α-pentyl-4-carboxyphenylglycine ((*RS*)-PeCPG, see Figure 7.4), which appears to be equipotent at mGlu1 when compared to (*S*)4CPG but somewhat more potent at mGlu5 (Doherty *et al.*, 1999). Likewise, the thioxanthyl-analog (Figure 7.4), is a low micromolar potent group I antagonist with about equal activities at mGlu1 and mGlu5 (Clark *et al.*, 1998).

LY367385

Modifications at the ortho position of the phenyl ring in *S*-4CPG include (+)-2-methyl-4-carboxylphenylglycine (LY367385, Figure 7.4) (Clark *et al.*, 1997). LY367385 was about four times more potent than (*S*)-4CPG as an antagonist at human mGlu1a receptors ($IC_{50} = 9\,\mu M$), with no appreciable mGlu5 receptor activity at up to $100\,\mu M$.

AIDA

The compound (*RS*)-1-aminoindan-1,5-dicarboxylic acid (AIDA) can be considered a conformationally constrained *S*-4CPG analog, with a cyclopentane bridge between the ortho position of the phenyl and the α-carbon (Figure 7.4) (Pellicciari *et al.*, 1995). AIDA is a reported selective antagonist of mGlu1 (pA2 = 4.21), with weaker antagonist action at mGlu5, and was inactive at mGlu2 receptors (Moroni *et al.*, 1997). Consistent with this group I antagonist profile, AIDA reversed 1*S*,3*R*-ACPD stimulated phosphoinositide hydrolysis with somewhat greater potency than (*S*)-4CPG or (*S*)-MCPG, but had no effect on cAMP responses in the same tissue (Moroni *et al.*, 1997). Recently, intracerebral injection of either LY367385 or AIDA were shown to be anticonvulsant in three different rodent models, including DBA/2 mice, lethargic mice and genetically epilepsy prone rats (Chapman *et al.*, 1999). This suggests a role for mGlu1 antagonists in the treatment of epilepsy. Likewise, LY367385 selectively reduced noxious excitatory responses in the rat thalamus (Salt and Turner, 1998), also suggesting a role for mGlu1 receptors in nociceptive processes of the brain.

(S)-4C3HPG

The 3-hydroxy substituted analog of the mGlu1 selective antagonist (*S*)4CPG, namely (*S*)-4-carboxy-3-hydroxyphenylglycine ((*S*)-4C3HPG) has provided unique insights into mGlu pharmacology. (*S*)-4C3HPG is an antagonist at mGlu1 receptors, but is about equipotent as an agonist at mGlu2, with no appreciable activities at mGlu4 (Thomsen *et al.*, 1994b; Hayashi *et al.*, 1994). As both group I antagonists and group II agonists have been shown to be anticonvulsant in certain animal models of epilepsy, it is interesting that (*S*)-4C3HPG combines both of these activities in one molecule. Thomsen *et al.* (1994c) found that (*S*)-4C3HPG was relatively potent following intracerebral injection as an anticonvulsant against audiogenic seizures in DBA/2 mice. Thus, the possibility exists that compounds with a group II agonist/group I antagonist profile might be more effective than either mGlu activity alone. However, this hypothesis requires additional investigations with other more potent "mixed" compounds, and/or possibly administrations of combinations of more potent and systemically active subtype selective group II agonists or group I antagonists.

Miscellaneous competitive antagonists

LY393675

A series of α-substituted-3-carboxycyclobutylglycines were studies for activity at human mGlu1a or mGlu5a receptors (Baker *et al.*, 1998). In this study, the cis-α-thioxanth-9-yl-methyl analog, LY393675 (Figure 7.4), was the most potent, showing submicromolar potency at mGlu1a ($IC_{50} = 0.35\,\mu M$) and mGlu5a ($IC_{50} = 0.35\,\mu M$). However, the activity of this compound across mGlu subtypes needs to be established. In any case, its submicromolar potency may make this compound a more useful tool for *in vivo* studies.

(*S*)-CBPG (UPF596)

This is a micromolar potent mGlu1 antagonist, but an agonist/partial agonist at mGlu5 (Figure 7.4; Pellicciari *et al.*, 1996a; Mannaioni *et al.*, 1999) (see group I agonists described above). As such this compound may be useful for delineating the relative role for mGlu1 versus mGlu5 in tissues.

(*S*)-HOMOQUISQUALATE

Likewise, (*S*)-2-amino-4-(3-hydroxy-5-methylisoxazol-4-yl)butyric acid ((*S*)-homoquisqualate) (Figure 7.4) has both mGlu agonist and antagonist properties. (*S*)-homoquisqualate was an antagonist at mGlu1 (K_B 184 μM), but had full agonist activities at mGlu5 ($EC_{50} = 36\,\mu M$) and mGlu2 ($EC_{50} = 23\,\mu M$) receptors, with no activity at mGlu4 receptors (Bräuner-Osborne and Krogsgaard-Larsen, 1998). The unique receptor profile of this compound needs to be further explored across all clones and then possibly investigated in native tissue systems.

Non-competitive antagonists

Recently, a number of novel group I mGlu receptor inhibitors with non-competitive mechanisms that appear to bind at domains other the glutamate binding regions have been described and are discussed below. Interestingly, these compounds have high selectivities for single mGlu receptor subtypes within the group I class, making them potentially useful agents for exploring mGlu1 versus mGlu5 receptors.

CPCCOEt

7-hydroxyiminocyclopropan[b]chromene-1α-carboxylic acid (Figure 7.5) exhibits low μM potency as a non-competitive inhibitor of functional responses at human mGlu1 receptors, but did not exhibit any appreciable activities at mGlu2, mGlu4a, mGlu5a, mGlu7b, or mGlu8a receptors (Annoura *et al.*, 1996; Hermans *et al.*, 1998; Litschig *et al.*, 1999). CPCCOEt does not directly interact at the glutamate binding domain, as it did not displace ^{3}H-glutamate binding to the receptor (Litschig *et al.*, 1999). Interestingly, it appears to interact at a region downstream from the N-terminal extracellular domain, specifically involving residues Thr^{815} and Ala^{818} of mGlu1, as introduction of these residues into the mGlu5a receptor introduced sensitivity for CPCCOEt inhibition to this receptor. Likewise, substitution of the corresponding mGlu5a residues for the critical residues on mGlu1 resulted in loss of CPCCOEt inhibition (Litschig *et al.*, 1999). These studies with CPCCOEt illustrate a highly specific molecular interaction of this agent at a site distal to the glutamate binding region.

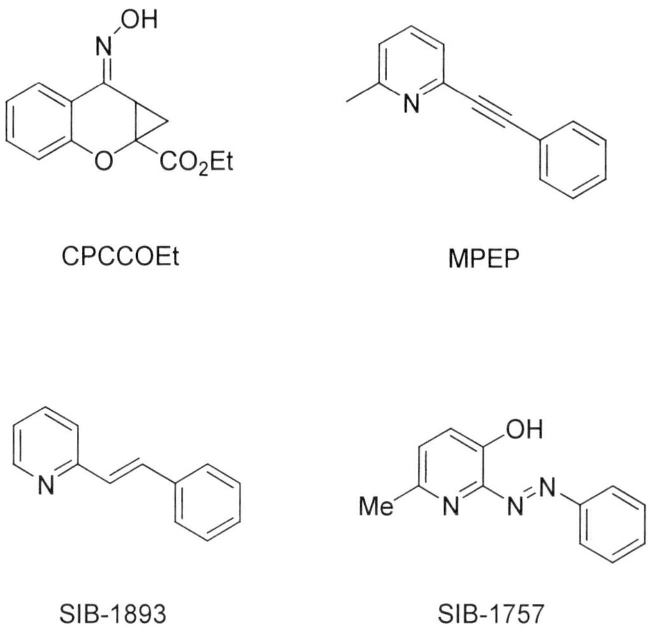

CPCCOEt MPEP

SIB-1893 SIB-1757

Figure 7.5 Group I non-competitive antagonists.

SIB-1757

6-methyl-2-(phenylazo)-3-pyrindol (Figure 7.5) is a structurally novel, potent and highly selective mGlu5 receptor antagonist. This compound was discovered by high throughput screening of compounds for their ability to inhibit glutamate induced intracellular calcium mobilization in a cell line expressing human mGlu5a receptors (Varney *et al.*, 1999). SIB-1757 potently inhibited glutamate responses at human mGlu5a receptors ($IC_{50} = 0.37\,\mu M$) in a noncompetitive manner. No activities of this compound were noted when tested up to $100\,\mu M$ on human versions of mGlu1b, mGlu2, mGlu3, mGlu4a, mGlu6, mGlu7b, mGlu8a. At recombinant AMPA, kainate or NMDA receptors, SIB-1757 was inactive when tested up to $30\,\mu M$. Consistent with this profile at human recombinant receptors, SIB-1757 potently antagonized 3,5-DHPG-induced phosphoinostide hydrolysis in rat brain regions expressing high levels of mGlu5 (hippocampus and striatum), but had no effect in the rat cerebellum, a region primarily expressing mGlu1 receptors.

SIB-1893

(*E*)-2-methyl-6-(2-phenylethenyl)pyridine (Figure 7.5), a second compound in this series (Varney *et al.*, 1999), had similar potency at human mGlu5a receptors ($IC_{50} = 0.29\,\mu M$), and a similar profile (high selectivity for mGlu5) across all other glutamate receptor subtypes (metabotropic and ionotropic). Based on these data, SIB-1757 and SIB-1893 represent highly useful compounds for probing the functions of mGlu5 receptors in animals. However, as these compounds were derived from random screening efforts of chemical libraries their usefulness may depend on the yet to be demonstrated lack of any non-specific biological actions in complex native tissues. Additional data on these compounds and on their effects in native tissues, such as measuring parameters of mGlu5 receptor mediated excitability (e.g. CA1 pyramidal excitation by DHPG), would help to address this issue.

MPEP

2-methyl-6-(phenylethynyl)-pyridine was derived from chemical derivatization around SIB-1757 and SIB-1893, in which the ethylene group was replaced by an acetylene linkage (Figure 7.5). MPEP is about ten times more potent than either SIB-1757 or SIB-1893 at either human recombinant or native rat mGlu5 receptors, with an IC_{50} values of 32 nM in human mGlu5a expressing cells and 15 nM against 3,5-DHPG stimulated phosphoinositide hydrolysis in the neonatal rat hippocampus. It was highly selective for mGlu5 receptors, with high μM concentrations having no effects across other cloned mGlu receptor and ionotropic glutamate receptor subtypes (Gasparini *et al.*, 1999b). In rat CA1 pyramidal neurons, i.v. administration of MPEP selectively blocked depolarization induced by 3,5-DHPG, but had no effects on AMPA depolarizations. MPEP, administered by either oral (10, 30, 100 mg/kg) or intraplantar (10–100 nmol) routes, was shown to prevent mechanical hyperalgesis induced by Freund's complete adjuvant (Bowes *et al.*, 1999). However, intrathecal MPEP (10–100 nmol) was inactive in this test. These data indicate a role for peripheral mGlu5 receptors in inflammatory pains states, and further suggest that MPEP is a useful pharmacological agent to perform *in vivo* studies on the role of mGlu5 receptors in physiological and pathological conditions.

Competitive antagonists for group II mGlu receptors

Substituted glutamate analogs

ADBD

Subtype selective competitive mGlu receptor antagonist activity has been shown to exist in a series of 4-substituted glutamate analogs. One of the most selective, potent, and well characterized group II antagonist compounds from the series includes the diphenylbutyl-substituted glutamate analog, (2*S*,4*S*)-2-amino-4-(4,4-diphenylbut-1-yl)pentane-1,5-dioic acid (ADBD or LY307452) (Figure 7.6) (Wermuth *et al.*, 1996; Escribano *et al.*, 1998). LY307452 was a relatively potent antagonist against 1*S*,3*R*-ACPD induced inhibitions of forskolin-stimulated cAMP formation at human mGlu2 (IC_{50} = 50 μM) and mGlu3 (IC_{50} = 30 μM) expressing non-neuronal cells, but had no activities at human mGlu1a, mGlu5a, mGlu4a, mGlu7a, or mGlu8a at up to 300 μM. In rat brain homogenates, LY307452 does not appreciably displace ^{3}H-AMPA, ^{3}H-kainate, or ^{3}H-CGS-19755 binding (IC_{50}s > 100 μM) (Wermuth *et al.*, 1996). However, consistent with its group II antagonist activities in cell lines, LY307452 potently displaced 1*S*,3*R*-ACPD sensitive ^{3}H-glutamate binding to rat brain (IC_{50} = 4.2 μM). In slices of the rat hippocampus, LY307452 (100 μM) did not block 3,5-DHPG stimulation of phosphoinositide hydrolysis *per se*, but it prevents potentia-

Substituted Glutamate Analogs

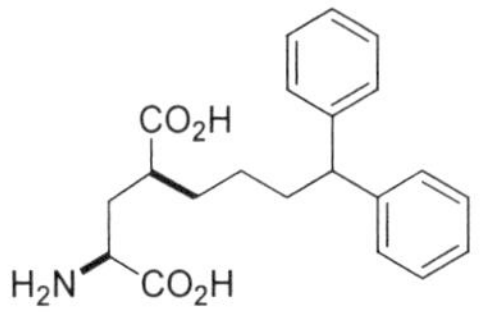

LY307452
(ADBD)

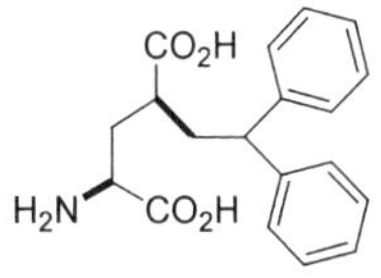

LY310225
(ADED)

Substituted Carboxycyclopropylglycine Analogs

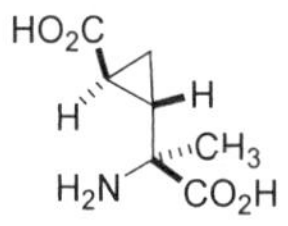

MCCG

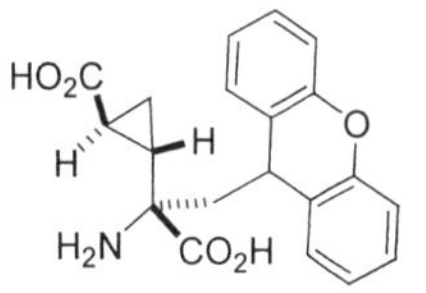

LY341495

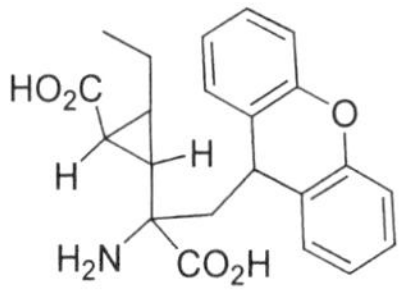

CECXG

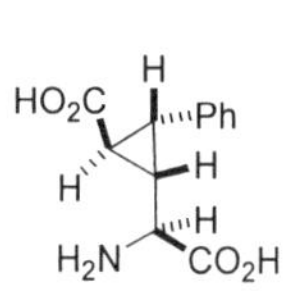

PCCG-4

Figure 7.6 Antagonists for group II mGlu receptors.

tion of this response by the group II selective agonist 2*R*,4*R*-APDC (Schoepp *et al.*, 1996). In the song-control nucleus HVc of the zebra finch, group II mGlu receptor agonists (ACPD, L-CCG-I, and 2*R*,4*R*-APDC) hyperpolarize these neurons via the coupling to GIRK (G-protein coupled inward-rectifying K^+) channels. The effect of these group II agonists on these neurons was selectively blocked by LY307452, but the hyperpolarizing effect of the group III agonist L-AP4 was not (Dutar *et al.*, 1999). These data indicate the LY307452 is a useful pharmacological tool for investigation of the role of group II mGlu receptors in native rat tissues.

ADED

(2*S*,4*S*)-2-amino-4-(2,2-diphenylethyl)pentane-1,5-dioic acid (ADED or LY310225) (Figure 7.6), a shorter chain version of LY307452, was shown to be more potent than LY307452 as a functional antagonist for mGlu2 or mGlu3 receptors, and in its displacement of 1*S*,3*R*-ACPD sensitive glutamate binding to rat brain ($IC_{50} = 0.38\,\mu M$). Like LY307452, LY310225 did not activate or inhibit other mGlu subtypes or displace ionotropic glutamate receptor binding to rat brain (Escribano *et al.*, 1998).

Substituted cyclopropyl glutamate analogs

MCCG

(2*S*,1′*S*,2′*S*)-2-methyl-2-(2′-carboxycyclopropyl)glycine (Figure 7.6) is the α-methyl derivative of the potent group II mGlu receptor agonist L-CCG-I. Pharmacological studies have demonstrated that this simple substitution produces an antagonist for group II mGlu receptors with high μM affinity. In the neonatal spinal cord preparation, MCCG blocked the presynaptic depressant effects of 1*S*,3*S*-ACPD and L-CCG-I, while having minimal effects on the depressant effects of L-AP4. MCCG had no effect on the depolarizing response to 1*S*,3*R*-ACPD, indicating no appreciable group I mGlu receptor activity (Jane *et al.*, 1994). This group II selective antagonist profile has also been observed in recombinant systems expressing mGlu1, mGlu2, and mGlu4 cloned receptors (Knöpfel *et al.*, 1995b; see Schoepp *et al.*, 1999a). However, in adult rat cerebral cortex slices, MCCG acted as agonist/partial agonist, as indicated by suppression of forskolin-stimulated cAMP formation to about 80 percent of that observed with maximally effective concentration of L-CCG-I (Kemp *et al.*, 1996). The reasons for this are not clear, but the actions of MCCG on mGlu3 receptors have not been reported, and mGlu3 would likely also contribute along with mGlu2 receptors in cAMP response in the rat brain.

LY341495

This, the α-xanth-9yl-methyl analog of L-CCG-I (Figure 7.6), represented the most potent/optimal substituent from a series of phenylethyl or diphenylethyl substituted compounds (Ornstein *et al.*, 1998a, 1998b). Full characterization of LY341495 across human mGlu receptor subtypes indicated that LY341495 is a competitive antagonist with low nM potency at mGlu2 and mGlu3 receptors, but at higher concentrations (high nM–μM) LY341495 will antagonize mGlu receptor responses in cloned group I and group III receptors (Kingston *et al.*, 1998; see Schoepp *et al.*, 1999a). LY341495 had no measurable affinity for ionotropic glutamate receptors. The high potency and selectivity for group II receptors when used at submicromolar concentrations, but measurable antagonist actions at other receptors at higher (μM) concentrations, have made LY341495 a highly useful concentration-dependent agent for investigating the potential role of multiple mGlu receptor subtypes. For example, ^{3}H-LY341495 selectively binds to recombinant human and native rat mGlu2 and mGlu3 receptors (K_D values of 1.7 and 0.75 nM, respectively), with a pharmacology (at 1 nM) indicating specific labeling of these receptors in either cells expressing cloned mGlu2/3 subtypes (Johnson *et al.*, 1999) or in the rat brain (Ornstein *et al.*, 1998c). This makes ^{3}H-LY341495 a highly useful ligand (under these conditions) for quantifying mGlu2/3 receptor expressing, investigation of affinity of other compounds for mGlu2 or mGlu3 binding sites, and to investigate the physiological/pathological regulation of these binding sites.

However, higher concentrations of LY341495 which block all mGlu receptor subtypes have been used to rule out mGlu receptor involvement in physiological processes. Fitzjohn *et al.* (1998) showed that higher concentrations of LY341495 which block all recombinant mGlu receptor subtypes and multiple mGlu second messenger response in the rat hippocampus also block NMDA receptor-independent depotentiation and setting of the molecular switch involved in long-term potentiation (LTP). This indicates, as shown previously with MCPG, that this effect is mediated by mGlu receptors. In contrast, LY341495 under these conditions did not affect NMDA receptor dependent homosynaptic long-term depression (LTD) or LTP, suggesting that these processes do not involve any mGlu receptor dependent processes.

LY341495 has also been used *in vivo*, where relatively low parenteral doses (0.3–1 mg/kg) have been shown to block the pharmacological actions of selective group II mGlu receptor agonists. In the elevated plus maze test for anxiety in mice, 0.3 mg/kg i.p. LY341495 reversed the anxiolytic actions of LY354740 while having no effect on its own (Ornstein *et al.*, 1998b). In the phencyclidine (PCP) model of psychosis in rats, LY341495 (1 mg/kg s.c.) alone had no effect on PCP-evoked motor activations; however, the inhibitory actions of LY379268 against PCP were completely reversed (Cartmell *et al.*, 1999). These potent and selective actions of LY341495 make it a useful pharmacological agent to study group II mGlu receptor functions.

CECXG

Interestingly, the 3′-ethyl analog of LY341495 (tested as a diastereomeric mixture), (2*SR*)- and (2*RS*)-2-(1′*SR*,2′*SR*,3′*SR*)-2′-carboxy-3′-ethylcyclopropyl)-2-(9-xanthylmethyl)glycine (CECXG) (Figure 7.6), was about five times less potent than LY341495 on mGlu3 receptors ($IC_{50} = 0.076\,\mu M$), but it was thirty-eight times more potent at mGlu3 than mGlu2 ($IC_{50} = 2.9\,\mu M$). CECXG was also considerably less potent at other mGlu receptor subtypes, including mGlu1a, mGlu5a, mGlu4a, mGlu7a, and mGlu8a (Collado *et al.*, 1998). Thus, CECXG might be useful to investigate the role of mGlu3 versus mGlu2 in native tissues.

PCG-IV

Of the sixteen different isomers of 2-(2′-carboxy-3′-phenylcyclopropyl)glycine (PCCGs), PCCG-4 (Figure 7.6) was found to be a relatively potent antagonist ($IC_{50} = 8\,\mu M$) at rat mGlu2 receptors, with no activity at a representative group I (mGlu1), and weak agonist activity at a representative group III (in this case mGlu4a) receptor ($EC_{50} = 156\,\mu M$) (Pellicciari *et al.*, 1996b). Consistent with this mixed pharmacological profile, PCCG-4 was a relatively potent compound in reversing the *in vitro* neuroprotective actions of the group II agonist DCG-IV in mouse cortical neurons. However, higher concentrations of PCCG-4 *per se* were neuroprotectant, like L-AP4, suggesting that the native mGlu4 receptors (or another group III receptor) are activated by these higher concentrations (Thomsen *et al.*, 1996).

Competitive antagonists for group III mGlu receptors

In general, group III mGlu receptor antagonists is the least developed area of mGlu receptor pharmacology. Many agents with group III antagonist activity also possess group II mGlu receptor antagonist activity, and/or their profile across all four group III receptors has not been described. Nevertheless, the actions of these agents which block group III mGlu subtypes can be compared to the effects of more selective group II agonists, to get an indication of the possible involvement of group III mGlu receptors in native tissues. Furthermore, the ability of a compound to reverse the actions of selective group III agonists such as L-AP4 in a native tissue is useful information, suggesting group III involvement. With these caveats, certain compounds targeted at these receptors have emerged as useful agents and are discussed below. For a more detailed discussion of the structure–activity relationship of other compounds related to the ones described below, see Schoepp *et al.* (1999a).

Substituted phosphono/isoxazole amino acids

MAP4, MSOP

(*S*)-α-methyl-2-amino-4-phosphonobutanoic acid (MAP4) and (*RS*)-α-methylserine-O-phosphate (MSOP) are the α-methyl analogs of the group III selective agonists L-AP4 and L-SOP, respectively (see Figure 7.7). In the rat neonatal spinal cord preparation where these compounds were initially described, both were moderately potent group III selective competitive antagonists in that they selectively reversed the depressant effects of L-AP4 (Jane *et al.*, 1994, 1996; Thomas *et al.*, 1996; Cao *et al.*, 1997). Likewise, in the rat thalamus *in vivo* (Salt and Eaton, 1995), CA1 region of the

Substituted Phosphono-Amino Acids

PO_3H_2 CH_3 H_2N CO_2H

MAP4

PO_3H_2 O CH_3 H_2N CO_2H

MSOP

Phenylglycine Analogs

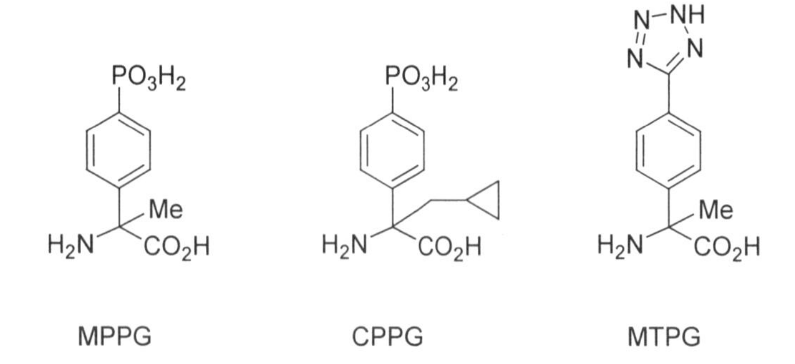

Figure 7.7 Antagonists for group III mGlu receptors.

hippocampus (Vignes *et al.*, 1995), and the hippocampal lateral perforant path (Bushell *et al.*, 1995), MAP4 selectively blocked the actions of L-AP4. Thus, at least in these preparations, there appears to be a group III L-AP4 activated mGlu receptor subtype that is sensitive to MAP4 and/or MSOP. Studies indicate that MAP4 and MSOP are antagonists at mGlu4 receptors. However, the profile of MAP4 and MSOP across all eight receptors is not known, and there are some discrepancies in their activities across different laboratories (see Schoepp *et al.*, 1999a). For example, data in certain preparations suggest that MAP4 may have group II and/or group II mGlu receptor agonist/partial agonist activities. In adult rat cortical slices MAP4 potently suppressed forskolin-stimulated cAMP formation ($EC_{50} = 0.11\,\mu M$) (Kemp *et al.*, 1996), and in contrast to others (Johansen and Robinson, 1995), Knöpfel *et al.* (1995b) found MAP4 to be a partial agonist at rat mGlu4a receptors. Thus, until these discrepancies are resolved the antagonist actions of these agents should not be considered "diagnostic" for the involvement of a group III mGlu receptor in physiological or pathological processes involving glutamate.

Phenylglycine analogs

MPPG

(*RS*)-α-methyl-4-phosphonophenylglycine (Figure 7.7) is the phospho-substituted version of MCPG. As the substitution of the distal carboxy group in glutamate with a phosphono acidic bioisostere (producing L-AP4) leads to group III activity, one could suggest that a similar change to MCPG might produce group III antagonist activity. Indeed, MPPG has been shown to be a group III mGlu receptor antagonist in a number of test systems, as indicated by selective reversal (generally at 50–200 μM concentrations) of the actions of L-AP4 in rat spinal cord (Jane *et al.*, 1995), in the ventrobasal thalamus (Salt and Turner, 1996), in the lateral perforant path of the hippocampus (Bushell *et al.*, 1996), and in the L-AP4 suppression of forskolin-stimulated cAMP formation in adult rat cortex (Bedingfield *et al.*, 1996). However, MPPG appears to also act on group II receptors, as at similar or higher concentrations it has been shown to also reverse the actions 1*S*,3*S*-ACPD in the spinal cord (Jane *et al.*, 1996) and L-CCG-I suppression of forskolin-stimulated cAMP formation in the rat cortex (Bedingfield *et al.*, 1996). Studies in recombinant mGlu receptors have shown that MPPG is both a group II and group III mGlu receptor antagonist. It is a moderately potent antagonist at rat mGlu2 and mGlu4 receptors (Gomeza *et al.*, 1996; Thomsen *et al.*, 1996; Ma *et al.*, 1997; Kowal *et al.*, 1998). Wu *et al.* (1998) compared its actions across human group III receptors in the same cell line and found that MPPG most potently blocked mGlu8 receptors ($IC_{50} = 47\,\mu M$) followed by mGlu7 ($IC_{50} = 528\,\mu M$) and mGlu4 ($IC_{50} = 540\,\mu M$). Thus, in general the usefulness of MPPG is limited by lack of good selectivity and potency for the group III mGlu receptors.

CPPG

(*RS*)-α-cyclopropyl-4-phosphonophenylglycine differs from MPPG by the incorporation of a cyclopropyl group at the α-carbon rather than a methyl (Figure 7.7). With this change, the potency of CPPG as an antagonist for both group II and group III mGlu receptors was greatly increased in the rat cerebral cortex. CPPG antagonized

L-AP4 and L-CCG-I inhibitions of forskolin-stimulated cAMP formation with IC_{50} values of 2.2 nM and 46 nM, respectively. However, it only weakly ($K_B = 650\,\mu M$) blocked group I agonist stimulations of phosphoinositide hydrolysis (Toms *et al.*, 1996). In membranes prepared from rat mGlu4a expressing cells, CPPG displaced ^{3}H-AP4 binding ($Ki = 23\,\mu M$) with about threefold greater potency than MPPG or MAP4 (Han and Hampson, 1999). However, the micromolar potency at mGlu4a receptors is at odds with nM antagonist potency observed against L-AP4 second messengers in rat brain. Thus, one of the other group III receptors would appear to be responsible for this effect of L-AP4 in the rat brain. Additional data on recombinant receptors is needed to fully understand the actions and usefulness of this compound.

Past/future perspectives

In the past decade it became recognized that multiple subtypes for mGlu receptors exist (eight currently). Compared to the early 1990s when the first cloned receptor was elucidated, much progress has been made on the pharmacology of mGlu receptors. Advances in the discovery of new agents now include many nM potent compounds with high selectivity for certain mGlu receptor groups such as group I (mGlu1/5) and group II (mGlu2/3) receptors. Some of these compounds (i.e. DCG-IV, LY354740, LY341495) have been used as high affinity radioligands, useful to perform receptor modeling, investigations of mGlu receptor expression levels and receptor regulation, and to determine the relative affinities of compounds for mGlu receptor binding sites in cells expressing cloned receptors and in the rat brain. Non-competitive antagonists for each group I receptor subtype (CPCCOEt for mGlu1 and MPEP for mGlu5) have been described. In the case of CPCCOEt, very specific and well defined molecular interactions have been elucidated which increase understanding of mGlu receptor structure/function. Systemically active pharmacological tools (e.g. LY354740, LY379268, MPEP) have been described that have enabled the exploration in animals of the potential therapeutic applications of mGlu receptor ligands. In the future, the field will likely be propelled by availability of cloned mGlu receptors for structure activity studies, as well as additional high-throughput screening efforts. In this manner, many more potent, selective, and drug-like molecules – in addition to those described here – are likely to emerge.

References

Abe, T., Sugihara, H., Nawa, H. Shigemoto, R., Mizuno, N. and Nakanishi, S. (1992) 'Molecular characterization of a novel metabotropic glutamate receptor mGluR5 coupled to inositol phosphate/Ca^{2+} signal transduction', *J. Biol. Chem.* **267**: 13,361–13,368.

Acher, F.C., Tellier, F.J., Azerad, R., Brabet, I.N., Fagni, L. and Pin, J.P. (1997) 'Synthesis and pharmacological characterisation of aminocyclopentanetricarboxylic acids: New tools to discriminate between metabotropic glutamate receptor subtypes', *J. Med. Chem.* **40**: 3119–3129.

Ahmadian, H., Nielsen, B., Bräuner-Osborne, H., Johansen, T.N., Stensbøl, T.B., Sløk, F.A., Sekiyama, N., Nakanishi, S., Krogsgaard-Larsen, P. and Madsen, U. (1997) '(S)-Homo-AMPA, a specific agonist at the mGlu6 subtype of metabotropic glutamic acid receptors', *J. Med. Chem.* **40**: 3700–3705.

Allen, J.W., Ivanova, S.A., Fan, L., Espey, M.G., Basile, A.S. and Faden, A.I. (1999) 'Group II metabotropic glutamate receptor activation attenuates traumatic neuronal injury and improves neurological recovery after traumatic brain injury', *J. Pharmacol. Exp. Ther.* **290**: 112–128.

Annoura, H., Fukunaga, A., Uesugi, M., Tatsuoka, T. and Horikawa, Y. (1996) 'A novel class of antagonists for metabotropic glutamate receptors, 7-(hydroxyimino)cyclopropa[b]chromen-1a-carboxylates', *Bioorg. Med. Chem. Lett.* **6**: 763–766.

Anwyl, R. (1999) 'Metabotropic glutamate receptors: electrophysiological properties and role in plasticity', *Brain Res. Reviews* **29**: 83–120.

Attwell, P.J.E., Kaura, S., Sigala, G., Bradford, H.F., Croucher, M.J., Jane, D.E. and Watkins, J.C. (1995) 'Blockade of both epileptogenesis and glutamate release by (1S,3S)-ACPD, a presynaptic glutamate receptor agonist', *Brain Res.* **698**: 155–162.

Attwell, P.J.E., Koumentaki, A., Croucher, M.J. and Bradford, H.F. (1998) 'Specific group II metabotropic glutamate receptor activation inhibits the development of kindled epilepsy in rats', *Brain Res.* **787**: 286–291.

Baker, S.R., Goldsworthy, J., Harden, R.C., Salhoff, C.R. and Schoepp, D.D. (1995) 'Enzymatic resolution and pharmacological activity of the enantiomers of 3,5-dihydroxyphenylglycine, a metabotropic glutamate receptor agonist', *Bioorg. Med. Chem. Lett.* **5**: 223–228.

Baker, S.R., Clark, B.P., Harris, J.R., Griffy, K.I., Kingston, A.E. and Tizzano, J.P. (1998) 'LY393675: an α-substituted-cyclobutylglycine, is a potent group I metabotropic glutamate receptor antagonist', *Soc. Neurosci. Abstr.* **24**: 576.

Battaglia, G., Monn, J.A. and Schoepp, D.D. (1997) 'In vivo inhibition of veratridine-evoked release of striatal excitatory amino acids by the group II metabotropic glutamate receptor agonist LY354740', *Neurosci. Lett.* **229**: 161–164.

Birse, E.F., Eaton, S.A., Jane, D.E., Jones, P.L.St.J., Porter, R.H.P., Pook, P.C.-K., Sunter, D.C., Udvarhelyi, P.M., Wharton, B., Roberts, P.J., Salt, T.E. and Watkins, J.C. (1993) 'Phenylglycine derivatives as new pharmacological tools for investigating the role of metabotropic glutamate receptors in the central nervous system', *Neuroscience* **52**: 481–488.

Bond, A, O'Neill, M.J., Hicks, C.A., Monn, J.A. and Lodge, D. (1998) 'Neuroprotective effects of a systemically active group II metabotropic glutamate receptor agonist LY354740 in a gerbil model of global ischaemia', *NeuroReport* **9**: 1191–1193.

Bordi, F. and Ugolini, A. (1999) 'Group I metabotropic glutamate receptors: implications for brain diseases', *Prog. Neurobiol.* **59**: 55–79.

Bowes, M., Panesar, M., Gentry, C., Urban, L., Gasparini, F., Kuhn, R. and Walker, K. (1999) 'Anti-hyperalgesic effects of the novel metabotropic glutamate receptor 5 antagonist, methylphenylethynylpyridine, in rat models of inflammatory pain', *Br. J. Pharmacol.* **126** (suppl.): P154.

Boyd, E.A., Alexander, S.P.H., Kendall, D.A. and Loh, V.M. Jr. (1996) 'DHPMP: a novel group I specific metabotropic glutamate receptor agonist', *Bioorg. Med. Chem. Lett.* **6**: 2137–2140.

Brabet, I, Mary, S., Bockaert, J. and Pin, J.-P. (1995) 'Phenylglycine derivatives discriminate between mGluR1- and mGluR5-mediated responses', *Neuropharmacology* **34**: 895–903.

Brabet, I., Parmentier, M.-L., De Colle, C., Bockaert, J., Acher, F. and Pin, J.-P. (1998) 'Comparative effect of L-CCG-I, DCG-IV and γ-carboxy-L-glutamate on all cloned metabotropic glutamate receptor subtypes', *Neuropharmacology* **37**: 1043–1051.

Bräuner-Osborne, H, Sløk, F.A., Skjærbæk, N., Ebert, B., Sekiyama, N., Nakanishi, S. and Krogsgaard-Larsen, P. (1996) 'A new highly selective metabotropic excitatory amino acid agonist: 2-amino-4-(3-hydroxy-5-methylisoxazol-4-yl)butyric acid', *J. Med. Chem.* **39**: 3188–3194.

Bräuner-Osborne, H. and Krogsgaard-Larsen, P. (1998) 'Pharmacology of (S)-homoquisqualic acid and (S)-2-amino-5-phosphonopentanoic acid [(S)-AP5] at cloned metabotropic glutamate receptors', *Br. J. Pharmacol.* **123**: 269–274.

Bedingfield, J.S., Jane, D.E., Kemp, M.C., Toms, N.J. and Roberts, P.J. (1996) 'Novel potent selective phenylglycine antagonists of metabotropic glutamate receptors', *Eur. J. Pharmacol.* **309**: 71–78.

Breakwell, N.A., Huang, L.-Q., Rowan, M.J. and Anwyl, R. (1997) 'DCG-IV inhibits synaptic transmission by activation of NMDA receptors in area CA1 of rat hippocampus', *Eur. J. Pharmacol.* **322**: 173–178.

Bushell, T.J., Jane, D.E., Tse, H.-W., Watkins, J.C., Davies, C.H., Garthwaite, J. and Collingridge, G.L. (1995) 'Antagonism of the synaptic depressant actions of L-AP4 in the lateral perforant path by MAP4', *Neuropharmacology* **34**: 239–241.

Bushell, T.J., Jane, D.E., Tse, H.-W., Watkins, J.C., Garthwaite, J. and Collingridge, G.L. (1996) 'Pharmacological antagonism of the actions of group II and III mGluR agonists in the lateral perforant path of rat hippocampal slices', *Br. J. Pharmacol.* **117**: 1457–1462.

Camon, L., Vives, P., de Vera, N. and Martinez, E. (1998) 'Seizures and neuronal damage induced in the rat by activation of group I metabotropic glutamate receptors with their selective agonist 3,5-dihydroxyphenylglycine', *J. Neurosci.* **51**: 339–348.

Cao, C.Q., Evans, R.H., Headley, P.M. and Udvarhelyi, P.M. (1995) 'A comparison of the effects of selective metabotropic glutamate receptor agonists on synaptically evoked whole cell currents of rat spinal ventral horn neurones *in vitro*' *British J. Pharmacol.* **115**: 1469–1474.

Cao, C.Q., Tse, H.-W., Jane, D.E., Evans, R.H. and Headley, P.M. (1997) 'Metabotropic glutamate receptor antagonists, like $GABA_B$ antagonists potentiate dorsal root-evoked excitatory synaptic transmission at neonatal rat spinal motoneurones in vitro', *Neuroscience* **78**: 243–250.

Cartmell, J., Adam, G., Chaboz, S., Henningsen, R., Kemp, J.A., Klingelschmidt, A., Metzler, V., Monsma, F., Schaffhauser, H., Wichmann, J. and Mutel, V. (1998) 'Characterization of [^{3}H]-(2S,2′R,3′R)-(2′,3′-dicarboxycyclopropyl)glycine ([^{3}H]-DCG IV) binding to metabotropic mGlu2 receptor-transfected cell membranes', *Br. J. Pharmacol.* **123**: 497–504.

Cartmell, J., Monn, J.A. and Schoepp, D.D. (1999) 'The mGlu2/3 receptor agonists, LY354740 and LY379268: selectively attenuate phencyclidine versus d-amphetamine motor behaviors in rats', *J. Pharmacol. Exp. Ther.* **291**: 161–170.

Catania, M.V., Hollingsworth, Z., Penny, J.B. and Young, A.B. (1993) 'Quisqualate resolves two distinct metabotropic [^{3}H]glutamate binding sites', *NeuroReport* **4**: 311–313.

Cha, J.-H., Makowiec, R.L., Penny, J.B. and Young, A.B. (1990) 'L-[^{3}H]Glutamate labels the metabotropic excitatory amino acid receptor in rodent brain', *Neurosci. Lett.* **113**: 78–83.

Chapman, A.G., Yip, P.K., Yap, J.S., Quinn, L.P., Tang, E., Harris, J.R. and Meldrum, B.S. (1999) 'Anticonvulsant actions of LY367385 ((+)-2-methyl-4-carboxyphenylglycine) and AIDA ((RS)-1-aminoindan-1,5-dicarboxylic acid)', *Eur. J. Pharmacol.* **368**: 17–24.

Clark, B.P., Baker, S.R., Goldsworthy, J., Harris, J.R. and Kingston, A.E. (1997) '(+)-2-Methyl-4-carboxyphenylglycine (LY367385) selectively antagonises metabotropic glutamate mGluR1 receptors', *Bioorg. Med. Chem. Lett.* **7**: 2777–2780.

Clark, B.P., Harris, J.R., Kingston, A.E. and McManus, D. (1998) 'α-substituted phenylglycines as group I metabotropic glutamate receptor antagonists', XVth European International Symposium on Medicinal Chemistry, Edinburgh, 6–10 September 1998.

Collado, I, Ezquerra, J., Mazon, A., Pedregal, C., Yruretagoyena, B., Kingston, A.E., Tomlinson, R., Wright, R.A., Johnson, B.G. and Schoepp, D.D. (1998) '2,3′-disubstituted-2-(2′-carboxycyclopropyl)glycines as potent and selective antagonists of metabotropic glutamate receptors', *Bioorg. Med. Chem. Lett.* **8**: 2849–2854.

Collins, G.G.S. (1982) 'Some effects of excitatory amino acid receptor antagonists on synaptic transmission in the rat olfactory cortex slice', *Brain Res.* **244**: 311–318.

Conn, P.J. and Pin, J.-P. (1997) 'Pharmacology and functions of metabotropic glutamate receptors', *Annu. Rev. Pharmacol. Toxicol.* **37**: 205–237.

Desai, M.A. and Conn P.J. (1990) 'Selective activation of phosphoinositide hydrolysis by a rigid analogue of glutamate', *Neurosci. Lett.* **109**: 157–162.

Desai, M.A., Burnett, J.P., Mayne, N.G. and Schoepp, D.D. (1995) 'Cloning and expression of a human metabotropic glutamate receptor 1α: enhanced coupling on co-transfection with a glutamate transporter', *Mol. Pharmacol.* **48**: 648–657.

Doherty, A.J., Palmer, M.J., Henley, J.M., Collingridge, G.L. and Jane, D.E. (1997) '(RS)-2-Chloro-5-hydroxyphenylglycine (CHPG) activates mGlu5: but not mGlu1: receptors expressed in CHO cells and potentiates NMDA responses in the hippocampus', *Neuropharmacol.* **36**: 265–267.

Doherty, A.J., Collingridge, G.L. and Jane, D.E. (1999) 'Antagonist activity of α-substituted 4-carboxyphenylglycine analogues at group I metabotropic glutamate receptors expressed in CHO cells', *Br. J. Pharmacol.* **126**: 205–210.

Dutar, P., Vu, H.M. and Perkel, D.J. (1999) 'Pharmacological characterization of an unusual mGluR-evoked neuronal hyperpolarization mediated by activation of GIRK channels', *Neuropharmacology* **38**: 467–475.

Escribano, A., Ezquerra, J., Pedregal, C., Rubio, A., Yruretagoyena, B., Baker, S.R., Wright, R.A., Johnson, B.G. and Schoepp, D.D. (1998) '(2*S*,4*S*)-2-amino-4-(2,2-diphenylethyl)pentanedioic acid selective group 2 metabotropic glutamate receptor antagonist', *Bioorg. Med. Chem. Lett.* **8**: 765–770.

Favaron, M., Manev, R.M., Candeo, P., Arban, R., Gabellini, N., Kozikowski, A.P. and Manev, H. (1993) 'Trans-azetidine-2,4-dicarboxylic acid activates neuronal metabotropic receptors', *NeuroReport* **4**: 967–970.

Fisher, K. and Coderre, T.J. (1996) 'Comparison of nociceptive effects produced by intrathecal administration of mGluR agonists', *NeuroReport* **7**: 2743–2747.

Fisher, K. and Coderre, T.J. (1998) 'Hyperalgesia and allodynia induced by the intrathecal (RS)-dihydroxyphenylglycine in rats', *NeuroReport* **9**: 1169–1172.

Fitzjohn, S.M., Irving, A.J., Palmer, M.J., Harvey, J., Lodge, D. and Collingridge, G.L. (1996) 'Activation of group I mGluRs potentiates NMDA responses in rat hippocampal slices', *Neurosci. Lett.* **203**: 211–213.

Fitzjohn, S.M., Bortolotto, Z.A., Palmer, M.J., Doherty, A.J., Ornstein, P.L., Schoepp, D.D., Kingston, A.E., Lodge, D. and Collingridge, G.L. (1998) 'The potent mGlu receptor antagonist LY341495 identifies roles for both cloned and novel mGlu receptors in hippocampal synaptic plasticity', *Neuropharmacology* **37**: 1445–1458.

Fotuhi, M., Sharp, A.H., Glatt, C.E., Hwang, P.M., von Krosigk, M., Synder, S.H. and Dawson, T.M. (1993) 'Differential localization of phosphoinositide-linked metabotropic glutamate receptor (mGluR1) and inositol 1,4,5-trisphosphate receptor in rat brain', *J. Neurosci.* **13**: 2001–2012.

Ganong, A.H. and Cotman, C.W. (1982) 'Acidic amino acid antagonists of lateral perforant path synaptic transmission: agonist–antagonist interactions in the dentate gyrus', *Neurosci. Lett.* **34**: 195–200.

Gasparini, F., Bruno, V., Battaglia, G., Lukic, S., Leonhardt, T., Inderbitzin, W., Laurie, D., Sommer, B., Varney, M.A., Hess, S.D., Johnson, E.C., Kuhn, R., Urwyler, S., Sauer, D., Portet, C., Schmutz, M., Nicoletti, F. and Flor, P.J. (1999a) '(*R*,*S*)-4-phosphonophenyglycine, a potent and selective group III metabotropic glutamate receptor agonist, is anticonvulsant and neuroprotective in vivo', *J. Pharmacol. Exp. Ther.* **290**: 1678–1687.

Gasparini, F., Lingenhoehl, K., Flor, P.J., Munier, N., Heinrich, M., Vranesic, I., Biollaz, M., Heckendorn, R., Allgeier, H., Varney, M., Johnson, E., Hess, S.D., Veliçelebi, G. and Kuhn, R. (1999b) 'Methylphenylethynylpyridine (MPEP): a novel potent, subtype-selective and systemically active antagonist at metabotropic glutamate receptor subtype 5', *Br. J. Pharmacol.* **126** (suppl): P154.

Gereau, R.W. and Conn, J.P. (1995) 'Roles of specific metabotropic glutamate receptor subtypes in regulation of hippocampal CA1 pyramidal cell excitability', *J. Neurophysiology* **74**: 122–129.

Gomeza, J., Mary, S., Brabet, I., Parmentier, M.-L., Restituito, S., Bockaert, J. and Pin, J.-P. (1996) 'Coupling of metabotropic glutamate receptors 2 and 4 to Gα15: Gα16: and chimeric Gαq/i proteins: characterization of new antagonists', *Mol. Pharmacol.* **50**: 923–930.

Han, G. and Hampson, D.R. (1999) 'Ligand binding of the amino-terminal domain of the mGluR4 subtype of metabotropic glutamate receptor', *J. Biol. Chem.* **274**: 10,008–10,013.

Hayashi, Y., Tanabe, Y., Aramori, I., Masu, M., Shimamoto, K., Ohfune, Y. and Nakanishi, S. (1992) 'Agonist analysis of 2-(carboxycyclopropyl)glycine isomers for cloned metabotropic glutamate receptor subtypes expressed in Chinese hamster ovary cells', *Br. J. Pharmacol.* **107**: 539–543.

Hayashi, Y., Momiyama, A., Takahashi, T., Ohishi, H., Ogawa-Meguro, R., Shigemoto, R., Mizuno, N. and Nakanishi, S. (1993) 'Role of a metabotropic glutamate receptor in synaptic modulation in the accessory olfactory bulb', *Nature* **366**: 687–690.

Hayashi, Y., Sekiyama, N., Nakanishi, S., Jane, D.E., Sunter, D.C., Birse, E.F., Udvarhelyi, P.M. and Watkins, J.C. (1994) 'Analysis of agonist and antagonist activities of phenylglycine derivatives for different cloned metabotropic glutamate receptor sub-types', *J. Neurosci.* **14**: 3370–3377.

Helton, D.R., Tizzano, J.P., Monn, J.A., Schoepp, D.D. and Kallman, M.J. (1997) 'LY354740: A potent, orally active, highly selective metabotropic glutamate receptor agonist which ameliorates symptoms of nicotine withdrawal', *Neuropharmacology* **36**: 1511–1516.

Helton, D.R., Tizzano, J.P., Monn, J.A., Schoepp, D.D. and Kallman, M.J. (1998a) 'Anxiolytic and side-effect profile of LY354740: A potent, highly selective, orally active agonist for group II metabotropic glutamate receptors', *J. Pharmacol. Exp. Ther.* **284**: 651–660.

Helton, D.R., Schoepp, D.D., Monn, J.A., Tizzano, J.P. and Kallman, M.J., (1998b) 'A role for metabotropic glutamate receptors in drug withdrawal states', in F. Moroni, F. Nicoletti and D.E. Pellegrini-Giampietro, (eds) *Metabotropic Glutamate Receptors and Brain Function*, London: Portland Press Ltd, pp. 305–314.

Hermans, E., Nahorski, S.R. and Chaliss, R.A.J. (1998) 'Reversible and non-competitive antagonist profile of CPCCOEt at the human type 1a metabotropic glutamate receptor', *Neuropharmacology* **37**: 1645–1647.

Irving, A.J., Schofield, J.G., Watkins, J.C., Sunter, D.C. and Collingridge, G.L. (1990) '1S,3R-ACPD stimulates and L-AP3 blocks Ca^{2+} mobilization in rat cerebellar neurons', *Eur. J. Pharmacol.* **186**: 363–365.

Ishida, M., Saitoh, T., Nakamura, Y., Kataoka, K. and Shinozaki, H. (1994) 'A novel metabotropic glutamate receptor agonist: (2*S*,1′*S*,2′*R*,3′*R*)-2-(2-carboxy-3-methoxymethylcyclopropyl)glycine (cis-MCG-I)', *Eur. J. Pharmacol. Mol. Pharmacol.* **268**: 267–270.

Ishida, M., Saitoh, T., Tsuji, K., Nakamura, Y., Kataoka, K. and Shinozaki, H. (1995) Novel agonists for metabotropic glutamate receptors: trans- and cis-2-(2-carboxy-3-methoxymethylcyclopropyl)glycine (trans- and cis-MCG-I)', *Neuropharmacol.* **34**: 821–827.

Ito, I., Kohda, A., Tanabe, S., Hirose, E., Hayashi, M., Mitsunaga, S. and Sugiyama, H. (1992) '3,5-dihydroxyphenylglycine: a potent agonist of metabotropic glutamate receptors', *NeuroReport* **3**: 1013–1016.

Jane, D.E., Jones, P.L.St.J., Pook, P.C.-K., Salt, T.E., Sunter, D.C. and Watkins, J.C. (1993) 'Stereospecific antagonism by (+)-α-methyl-4-carboxyphenylglycine (MCPG) of (1S,3R)-ACPD-induced effects in neonatal rat motoneurones and rat thalamic neurones', *Neuropharmacology* **32**: 725–727.

Jane, D.E., Jones, P.L.St.J., Pook, P.C.-K., Tse, H.-W. and Watkins, J.C. (1994) 'Actions of two new antagonists showing selectivity for different sub-types of metabotropic glutamate receptor in the neonatal rat spinal cord', *Br. J. Pharmacol.* **112**: 809–816.

Jane, D.E., Pittaway, K., Sunter, D.C., Thomas, N.K. and Watkins, J.C. (1995) 'New phenylglycine derivatives with potent and selective antagonist activity at presynaptic glutamate receptors in neonatal rat spinal cord', *Neuropharmacology* **34**: 851–856.

Jane, D.E., Pittaway, K., Sunter, D.C., Thomas, N.K. and Tse, H.-W. (1996) 'Phosphono substituted amino acids as selective metabotropic glutamate receptor antagonists', *Phosphorus, Sulphur, Silicon and Related Elements* **109–110**: 313–316.

Johansen, P.A., Chase, L.A., Sinor, A.D., Koerner, J.F., Johnson, R.L. and Robinson, M.B. (1995) 'Type 4a metabotropic glutamate receptor: identification of new potent agonists and differentiation from the L-(+)-2-amino-4-phosphonobutanoic acid-sensitive receptor in the lateral perforant pathway in rats', *Mol. Pharmacol.* **48**: 140–149.

Johansen, P.A. and Robinson, M.B. (1995) 'Identification of 2-amino-2-methyl-4-phosphonobutanoic acid as an antagonist at the mGlu(4a) receptor', *Eur. J. Pharmacol. Mol. Pharmacol.* **290**: R1–R3.

Johnson, B.G., Wright, R.A., Arnold, M.B., Wheeler, W.J., Ornstein, P.L. and Schoepp, D.D. (1999) '[^{3}H]-LY341495 as a novel rapid filtration antagonist radioligand for group II metabotropic glutamate receptors: Characterization of binding to membranes of mGlu receptor subtype expressing cells', *Neuropharmacology* **38**: 1519–1529.

Kemp, M.C., Roberts, P.J., Pook, P.C.-K., Jane, D.E., Jones, A.W., Jones, P.L.St.J., Sunter, D.C., Udvarhelyi, P.M. and Watkins, J.C. (1994) 'Antagonism of presynaptically mediated depressant responses and cyclic AMP-coupled metabotropic glutamate receptors', *Eur. J. Pharmacol. Mol. Pharmacol.* **266**: 187–192.

Kemp, M.C, Jane, D.E., Tse, H.-W. and Roberts, P.J. (1996) 'Agonists of cyclic AMP-coupled metabotropic glutamate receptors in adult-rat cortical slices', *Eur. J. Pharmacol.* **309**: 79–85.

Kingston, A.E., Ornstein, P.L., Wright, R.A., Johnson, B.G., Mayne, N.G., Burnett, J.P., Belagaje, R., Wu, S. and Schoepp, D.D. (1998) 'LY341495 is a nanomolar potent and selective antagonist of group II metabotropic glutamate receptors', *Neuropharmacology* **37**: 1–12.

Klein, J., Reymann, K.G. and Riedel, G. (1997) 'Activation of phospholipases C and D by the novel metabotropic glutamate receptor agonist t-ADA', *Neuropharmacology* **36**: 261–263.

Knöpfel, T., Sakaki, J., Flor, P.J., Baumann, P., Sacaan, A.I., Veliçelebi, G., Kuhn, R. and Allgeier, H. (1995a) 'Profiling of trans-azetidine-2,4-dicarboxylic acid at the human metabotropic glutamate receptors mGlu1b, -2, -4a and -5a', *Eur. J. Pharmacol.* **288**: 389–392.

Knöpfel, T., Lukic, S., Leonardt, T., Flor, P.J., Kuhn, R. and Gasparini, F. (1995b) 'Pharmacological characterization of MCCG and MAP4 at the mGluR1b, mGluR2 and mGluR4a human metabotropic glutamate receptor subtypes', *Neuropharmacology* **34**: 1099–1102.

Koerner, J.F. and Cotman, C.W. (1981) 'Micromolar L-2-amino-4-phosphonobutyric acid selectively inhibits perforant path synapses from lateral entorhinal cortex', *Brain Res.* **216**: 192–198.

Kowal, D.M., Hsiao, C., Wardwell-Swanson, J. and Tasse, J.R. (1998) 'Correlation between mGluR ligand activities observed in CHO cells transfected with hmGluR4a using [^{3}H]L-AP4 and [^{35}S]GTPγS binding assays', *Soc. Neurosci. Abstr.* **24**: 576.

Kozikowski, A.P., Tückmantel, W., Liao, Y., Manev, H., Ikonomovic, S. and Wroblewski, J.T. (1993) 'Synthesis and metabotropic receptor activity of the novel rigidified glutamate analogues (+)- and (−)-*trans*-azetidine-2,4-dicarboxylic acid and their N-methyl derivatives', *J. Med. Chem.* **36**: 2706–2708.

Kozikowski, A.P., Steensma, D., Luca-Araldi, G., Tückmantel, W., Wang, S., Pshenichkin, S., Surina, E. and Wroblewski, J.T. (1998) 'Synthesis and biology of the conformationally restricted ACPD analog, 2-aminobicyclo[2.1.1]hexane-2,5-dicarboxylic acid-I, a potent mGluR agonist', *J. Med. Chem.* **41**: 1641–1650.

Kroona, H.B., Peterson, N.L., Koerner, J.F. and Johnson, R.L. (1991) 'Synthesis of the 2-amino-4-phosphonobutanoic acid analogues (E)- and (Z)-2-amino-2,3-methano-4-phosphonobutanoic acid and their evaluation as inhibitors of hippocampal excitatory neurotransmission', *J. Med. Chem.* **34**: 1692–1699.

Lam, A.G.M., Soriano, M.A., Monn, J.A., Schoepp, D.D., Lodge, D. and McCulloch, J. (1998) 'Effects of the selective metabotropic glutamate agonist LY354740 in a rat model of permanent ischaemia', *Neurosci. Lett.* **254**: 121–123.

Lanthorn, T.H., Ganong, A.H. and Cotman, C.W. (1984) '2-amino-4-phosphonobutyrate selectively blocks mossy fiber-CA3 responses in guinea pig but not rat hippocampus', *Brain Res.* **290**: 174–178.

Litschig, S., Gasparini, F., Rueegg, D., Stoehr, N., Flor, P.J., Vranesic, I., Prézeau, L., Pin, J.-P., Thomsen, C. and Kuhn, R. (1999) 'CPCCOEt, a noncompetitive metabotropic glutamate receptor 1 antagonist, inhibits receptor signaling without affecting glutamate binding', *Mol. Pharmacol.* **55**: 453–461.

Littman, L., Tokar, C., Venkatraman, S., Roon, R.J., Koerner, J.F., Robinson, M.B. and Johnson, R.L. (1999) 'Cyclobutane quisqualic acid analogues as selective mGluR5a metabotropic glutamic acid receptor ligands', *J. Med. Chem.* **42**: 1639–1647.

Lujan, R., Nusser, Z., Roberts, J.D.B., Shigemoto, R. and Somogi, P. (1996) 'Perisynaptic location of metabotropic glutamate receptors mGluR1 and mGluR5 on dendrites and dendritic spines in the rat hippocampus', *Eur. J. Neurosci.* **8**: 1488–1500.

Ma, D., Tian, H., Sun, H., Kozikowski, A.P., Pschenichkin, S. and Wroblewski, J.T. (1997)

'Synthesis and biological activity of cyclic analogues of MPPG and MCPG as metabotropic glutamate receptor antagonists', *Bioorg. Med. Chem. Lett.* **9**: 1195–1198.

Mannaioni, G., Attucci, S., Missanelli, A., Pellicciari, R., Corradetti, R. and Moroni, F. (1999) 'Biochemical and electrophysiological studies on (*S*)-(+)-2-(3′-carboxybicyclo[1.1.1]pentyl)-glycine (CBPG), a novel $mGlu_5$ receptor agonist endowed with $mGlu_1$ receptor antagonist activity', *Neuropharmacology* **38**: 917–926.

Masu, M., Tanabe, Y., Tsuchida, K., Shigemoto, R. and Nakanishi, S. (1991) 'Sequence and expression of a metabotropic glutamate receptor', *Nature* **349**: 760–765.

Merlin, L.R. and Wong, R.K.S. (1997) 'Role of group I metabotropic glutamate receptors in the patterning of epileptiform activities in vitro', *J. Neurophysiol.* **78**: 539–544.

Moghaddam, B. and Adams, B.W. (1998) 'Reversal of phencyclidine effects by group II metabotropic glutamate receptor agonist in rats', *Science* **281**: 1349–1352.

Monn, J.A., Valli, M.J., Johnson, B.G., Salhoff, C.R., Wright, R.A., Howe, T., Bond, A., Lodge, D., Spangle, L.A., Paschal, J.W., Campbell, J.B., Griffey, K., Tizzano, J.P. and Schoepp, D.D. (1996) 'Synthesis of the four isomers of 4-aminopyrrolidine-2,4-dicarboxylate: identification of a potent, highly selective, and systemically active agonist for metabotropic glutamate receptors negatively coupled to adenylate cyclase', *J. Med. Chem.* **39**: 2990–3000.

Monn, J.A., Valli, M.J., Massey, S.M., Wright, R.A., Salhoff, C.R., Johnson, B.G., Howe, T., Alt, C.A., Rhodes, G.A., Robey, R.L., Griffey, K.R., Tizzano, J.P., Kallman, M.J., Helton, D.R. and Schoepp, D.D. (1997) 'Design, synthesis and pharmacological characterization of (+)-2-aminobicyclo[3.1.0]hexane-2,6-dicarboxylic acid (LY354740): A potent, selective and orally active group 2 metabotropic glutamate receptor agonist possessing anticonvulsant and anxiolytic properties', *J. Med. Chem.* **40**: 528–537.

Monn, J.A., Valli, M.J., Massey, S.M., Hansen, M.M., Kress, T.J., Wepsiec, J.P., Harkness, A.R., Grutsch, J.L. Jr., Wright, R.A., Johnson, B.G., Andis, S.L., Kingston, A.E., Tomlinson, R., Lewis, R., Griffey, K.R., Tizzano, J.P. and Schoepp, D.D. (1999) 'Synthesis, pharmacological characterization and molecular modeling of heterobicyclic amino acids related to LY354740: Identification of LY379268 and LY389795: Two new potent, selective and systemically active agonists for group II metabotropic glutamate receptors', *J. Med. Chem.* **42**: 1027–1040.

Moroni, F., Lombardi, G., Thomsen, C., Leonardi, P., Attucci, S., Peruginelli, F., Torregrossa, S.A., Pellegrini-Giampietro, D.E., Luneia, R. and Pellicciari, R. (1997) 'Pharmacological characterization of 1-aminoindan-1,5-dicarboxylic acid, a potent mGluR1 antagonist', *J. Pharmacol. Exp. Ther.* **281**: 721–729.

Mutel, V., Adam, G., Chaboz, S., Kemp, J.A., Klingelschmidt, A., Messer, J., Wichmann, J., Woltering, T. and Richards, J.G. (1998) 'Characterization of (2*S*,2′*R*,3′*R*)-2–2′,3′-[^{3}H]-dicarboxycyclopropyl)glycine binding in rat brain', *J. Neurochem.* **71**: 2558–2564.

Nakagawa, Y., Saitoh, K., Ishihara, T., Ishida, M. and Shinozaki, H. (1990) '(2S,3S,4S)-α-(carboxycyclopropyl)glycine is a novel agonist of metabotropic glutamate receptors', *Eur. J. Pharmacol.* **184**: 205–206.

Nakajima, Y., Iwakabe, H., Akazawa, C., Nawa, H., Shigemoto, R., Mizuno, R. and Nakanishi, S. (1993) 'Molecular characterization of a novel retinal metabotropic glutamate receptor mGluR6 with a high agonist selectivity for L-2-amino-4-phosphonobutyrate', *J. Biol. Chem.* **268**: 11,868–11,873.

Nakanishi, S. (1992) 'Molecular diversity of glutamate receptors and implications for brain function', *Science* **258**: 597–603.

Nicoletti, F., Bruno, V., Copani, A., Casabona, G., Knopfel, T. (1996) 'Metabotropic glutamate receptors: a new target for the therapy of neurodegenerative disorders?', *Trends in Neurosci.* **19**: 267–271.

Ohfune, Y., Shimamoto, K., Ishida, M. and Shinozaki, H. (1993) 'Synthesis of L-2-(2,3-dicarboxycyclopropyl)glycines. Novel conformationally restricted glutamate analogues', *Bioorg. Med. Chem. Lett.* **3**: 15–18.

Okamoto, T., Sekiyama, N., Otsu, M., Shimada, Y., Sato, A., Nakanishi, S. and Jingami, H. (1998) 'Expression and purification of the extracellular ligand binding region of metabotropic glutamate receptor subtype 1', *J. Biol. Chem.* **273**: 13,089–13,096.

Ong, W.Y. and Balcar, V.J. (1997) 'Group I metabotropic glutamate receptor agonist causes neurodegeneration in rat hippocampus', *J. Brain Res.* **3**: 317–322.

Ornstein, P.L., Bleisch, T.J., Arnold, M.B., Wright, R.A., Johnson, B.G. and Schoepp, D.D. (1998a) '2-substituted (2*SR*)-2-amino-2-((1*SR*,2*SR*)-2-carboxycycloprop-1-yl)glycines as potent and selective antagonists of group II metabotropic glutamate receptors. 1. Effects of alkyl, arylalkyl, and diarylalkyl substitution', *J. Med. Chem.* **41**: 346–357.

Ornstein, P.L., Bleisch, T.J., Arnold, M.B., Kennedy, J.H., Wright, R.A., Johnson, B.G., Tizzano, J.P., Helton, D.R., Kallman, M.J., Schoepp, D.D. and Hérin, M. (1998b) '2-substituted (2*SR*)-2-amino-2-((1*SR*,2*SR*)-2-carboxycycloprop-1-yl)glycines as potent and selective antagonists of group II metabotropic glutamate receptors. 2. Effects of aromatic substitution, pharmacological characterization, and bioavailability', *J. Med. Chem.* **41**: 358–378.

Ornstein, P.L., Arnold, M.B., Bleisch, T.J., Wright, R.A., Wheeler, W.J. and Schoepp, D.D. (1998c) '[^{3}H]LY341495: a highly potent, selective and novel radioligand for labeling group II metabotropic glutamate receptors', *Bioorg. Med. Chem. Lett.* **8**: 1919–1922.

Palmer, E., Monaghan, D.T. and Cotman, C.W. (1989) 'Trans-ACPD, a selective agonist of the phosphoinositide-coupled excitatory amino acid receptor', *Eur. J. Pharmacol.* **166**: 585–587.

Pellicciari, R., Luneia, R., Constantino, G., Marinozzi, M., Natalini, B., Jakobsen, P., Kanstrup, A., Lombardi, G., Moroni, F. and Thomsen, C. (1995) '1-aminoindan-1,5-dicarboxylic acid: a novel antagonist of phospholipase C-linked metabotropic glutamate receptors', *J. Med. Chem.* **38**: 3717–3719.

Pellicciari, R., Raimondo, M., Marinozzi, M., Natalini, B., Costantino, G. and Thomsen, C. (1996a) '(*S*)-(+)-2-(3′-carboxybicyclo[1,1,1]pentyl)glycine, a structurally new group I metabotropic glutamate receptor antagonist', *J. Med. Chem.* **39**: 2874–2876.

Pellicciari, R., Marinozzi, M., Natalini, B., Costantino, G., Luneia, R., Giorgi, G., Moroni, F. and Thomsen, C. (1996b) 'Synthesis and pharmacological characterisation of all sixteen stereoisomers of 2-(2′-carboxy-3′-phenylcyclopropyl)glycine. Focus on (2*S*,1′*S*,2′*S*,3′*R*)-2-(2′-carboxy-3′-phenylcyclopropyl)glycine, a novel and selective group II metabotropic glutamate receptor antagonist', *J. Med. Chem.* **39**: 2259–2269.

Peterson, N.L., Kroona, H.B., Johnson, R.L. and Koerner, J.F. (1992) 'Activity of the conformationally rigid 2-amino-4-phosphonobutanoic acid (AP4) analogue (*RS*)-1-amino-3-(phosphonomethylene)cyclobutane-1-carboxylic acid (cyclobutylene AP5) on evoked responses in the perforant pathway of rat hippocampus', *Brain Res.* **571**: 162–164.

Pin, J.-P. and Bockaert J. (1999) 'Structure and function of metabotropic glutamate receptors', in J. Egebjerg, A. Schousboe and P. Krogsgarrd-Larsen, P. (eds) *Glutamate and GABA Receptors and Transporters: Structure, Function and Pharmacology*, London: Taylor & Francis, pp. 134–165.

Pin, J-P. and Duvoisin, R. (1995) 'The metabotropic glutamate receptors: structure and functions', *Neuropharmacology* **34**: 1–26.

Pisani, A., Calabresi, P., Centonze, D. and Bernardi, G. (1997) 'Enhancement of NMDA responses by group I metabotropic glutamate receptor activation in striatal neurones', *Brit. J. Pharmacol.* **120**: 1007–1014.

Pook, P.C.-K., Sunter, D.C., Udvarhelyi, P.M. and Watkins, J.C. (1992) 'Evidence for pre-synaptic depression of monosynaptic excitation in neonatal rat motoneurons by (1*S*,3*S*)-ACPD and (1*S*,3*R*)-ACPD', *Experimental Physiology* **77**: 529–532.

Roberts, P.J. (1995) 'Pharmacological tools for the investigation of metabotropic glutamate receptors (mGluRs): phenylglycine derivatives and other selective antagonists – an update', *Neuropharmacology* **34**: 813–819.

Salt, T.E. and Eaton, S.A. (1995) 'Distinct presynaptic metabotropic receptors for L-AP4 and CCG1 on GABAergic terminals: pharmacological evidence using novel α-methyl derivative mGluR antagonists, MAP4 and MCCG, in the rat thalamus in vivo', *Neuroscience* **65**: 5–13.

Salt, T.E. and Turner, J.P. (1996) 'Antagonism of the presumed presynaptic action of L-AP4 on GABAergic transmission in the ventrobasal thalamus by the novel mGluR antagonist MPPG', *Neuropharmacology* **35**: 239–241.

Salt, T.E. and Turner, J.P. (1998) 'Reduction of sensory and metabotropic glutamate receptor responses in the thalamus by the novel metabotropic glutamate receptor-1-selective antagonist S-2-methyl-4-carboxy-phenylglycine', *Neuroscience* **85**: 655–658.

Saugstad, J.A., Kinzie, J.M., Shinohara, M.M., Segerson, T.P. and Westbrook, G.L. (1997) 'Cloning and expression of rat metabotropic glutamate receptor 8 reveals a distinct pharmacological profile', *Mol. Pharmacol.* **51**: 119–125.

Schauffhauser, H., Richards, J.G., Cartmell, J., Chaboz, S., Kemp, J.A., Klingelschmidt, A., Messer, J., Stadler, H., Woltering, T. and Mutel, V. (1998) 'In vitro binding characteristics of a new selective group II metabotropic glutamate receptor radioligand, [^{3}H]LY354740: in rat brain', *Mol. Pharmacol.* **53**: 228–233.

Schoepp, D.D. and Conn, P.J. (1993) 'Metabotropic glutamate receptors in brain function and pathology', *Trends in Pharmacol. Sci.* **14**: 13–20.

Schoepp, D.D. and True, R.A. (1992) '1S,3R-ACPD-sensitive (metabotropic) [^{3}H]glutamate receptor binding in membranes', *Neurosci. Lett.*, **145**: 100–104.

Schoepp, D.D., Bockaert, J. and Sladeczek, F. (1990) 'Pharmacological and functional characteristics of metabotropic excitatory amino acid receptors', *Trends Pharmacol. Sci.* **11**: 508–515.

Schoepp, D.D., Johnson, B.G., True, R.A. and Monn, J.A. (1991) 'Comparison of (1S,3R)-1-aminocyclopentane-1,3-dicarboxylic acid (1S,3R-ACPD)- and 1R,3S-ACPD-stimulated brain phosphoinositide hydrolysis', *Eur. J. Pharmacol. Mol. Pharmacol.* **207**: 351–353.

Schoepp, D.D., Goldsworthy, J., Johnson, B.G., Salhoff, C.R. and Baker, S.R. (1994) '3,5-dihydroxyphenylglycine is a highly selective agonist for phosphoinositide-linked metabotropic glutamate receptors in the rat hippocampus', *J. Neurochem.* **63**: 769–772.

Schoepp, D.D., Johnson, B.G., Salhoff, C.R., Wright, R.A., Goldsworthy, J.S. and Baker, S.R. (1995a) 'Second-messenger responses in brain slices to elucidate novel glutamate receptors', *J. Neurosci. Methods* **59**: 105–110.

Schoepp, D.D., Johnson, B.G., Salhoff, C.R., Valli, M.J., Desai, M.A., Burnett, J.P., Mayne, N.G. and Monn, J.A. (1995b) 'Selective inhibition of forskolin-stimulated cyclic AMP formation in rat hippocampus by a novel mGluR agonist, 2*R*,4*R*-4-aminopyrrolidine-2,4-dicarboxylate', *Neuropharmacology* **34**: 843–850.

Schoepp, D.D., Salhoff, C.R., Wright, R.A., Johnson, B.G., Burnett, J.P, Mayne, N.G., Belagaje, R., Wu, S. and Monn, J.A. (1996) 'The novel metabotropic glutamate receptor agonist 2R,4R-APDC potentiates stimulation of phosphoinositide hydrolysis in the rat hippocampus by 3,5-dihydroxyphenylglycine: evidence for a synergistic interaction between group 1 and group 2 receptors', *Neuropharmacology* **35**: 1661–1672.

Schoepp, D.D., Johnson, B.G., Wright, R.A., Salhoff, C.R., Mayne, N.G., Wu, S., Cockerham, S.L., Burnett, J.P., Belagaje, R., Bleakman, D. and Monn, J.A. (1997) 'LY354740 is a potent and highly selective group II metabotropic glutamate receptor agonist in cells expressing human glutamate receptors', *Neuropharmacol.* **36**: 1–11.

Schoepp, D.D., Johnson, B.G., Wright, R.A., Salhoff, C.R. and Monn, J.A. (1998) 'Potent, stereoselective, and brain region selective modulation of second messengers in the rat brain by (+)LY354740: a novel group II metabotropic glutamate receptor agonist', *Naunyn-Schmiedeberg's Arch. Pharm.* **358**: 175–180.

Schoepp, D.D., Jane, D.E. and Monn, J.A. (1999a) 'Pharmacological agents acting at subtypes of metabotropic glutamate receptors', *Neuropharmacology* **38**: 1431–1476.

Schoepp, D.D., Jane, D.E. and Monn, J.A. (1999b) 'LY354740: A systemically active mGlu2/mGlu3 receptor agonist', *CNS Drug Reviews* **5**: 1–13.

Shibuya, A., Sato, A. and Taguchi, T. (1998) 'Preparation of difluoro analogs of CCGs and their pharmacological evaluations', *Bioorg. Med. Chem. Lett.* **8**: 1979–1984.

Shiells, R.A., Falk, G. and Naghshinch, S. (1981) 'Action of glutamate and aspartate analogs on rod horizontal and bipolar cells', *Nature* **294**: 592–594.

Shimamoto, K. and Ohfune, Y. (1993) 'Inversion of cis-substituted α-cyclopropyl acyl anion. Stereoselective entry to the synthesis of a potent metabotropic glutamate agonist, (2S,1′S,2′S)-2-(carboxycyclopropyl)glycine (L-CCG-I), and its 3′-substituted analogues', *SynLett.* **N12**: 919–920.

Thomas, N.K., Jane, D.E., Tse, H.-W. and Watkins, J.C. (1996) 'α-methyl derivatives of serine-O-phosphate as novel selective competitive metabotropic glutamate receptor antagonists', *Neuropharmacology* **35**: 637–642.

Thomsen, C. (1997) 'The L-AP4 receptor', *Gen. Pharmac.* **29**: 151–158.

Thomsen, C., Hansen, L. and Suzdak, P.D. (1994a) 'L-glutamate uptake inhibitors may stimulate phosphoinositide hydrolysis in baby hamster kidney cells expressing mGluR1a via heteroexchange with L-glutamate without direct activation of mGluR1a', *J. Neurochem.* **63**: 2038–2047.

Thomsen, C., Boel, E. and Suzdak, P.D. (1994b) 'Actions of phenylglycine analogs at subtypes of the metabotropic glutamate receptor family', *Eur. J. Pharmacol. Mol. Pharmacol.* **267**: 77–84.

Thomsen, C., Klitgaard, H., Sheardown, M., Jackson, H.C., Eskesen, K., Jacobsen, P., Treppendahl, S. and Suzdak, P.D. (1994c) '(*S*)-4-carboxy-3-hydroxyphenylglycine, an antagonist of metabotropic glutamate receptor (mGluR)1a and an agonist of mGlu2: protects against audiogenic seizures in DBA/2 mice', *J. Neurochem.* **62**: 2492–2495.

Thomsen, C., Bruno, V., Nicoletti, F., Marinozzi, M. and Pellicciari, R. (1996) '(2*S*,1′*S*,2′*S*,3′*R*)-2-(2′-carboxy-3′-phenylcyclopropyl)glycine, a potent and selective antagonist of type 2 metabotropic glutamate receptors', *Mol. Pharmacol.* **50**: 6–9.

Tizzano, J.P., Griffey, K.I. and Schoepp, D.D. (1995a) 'Induction of protection of limbic seizures in mice by mGluR subtype selective agonists', *Neuropharmacol.* **8**: 1063–1067.

Tizzano, J.P., Griffey, K.I., Valli, M.J., Helton, D.R., Kallman, M.J., Schoepp, D.D. and Monn, J.A. (1995b) 'Intracerebral injections of metabotropic glutamate receptor agonists can induce or protect against limbic seizures in mice', *Behavioral Pharmacology* **6**: 633.

Toms, N.J., Jane, D.E., Kemp, M.C., Bedingfield, J.S. and Roberts, P.J. (1996) 'The effects of (*RS*)-α-cyclopropyl-4-phosphonophenylglycine ((*RS*)-CPPG), a potent and selective metabotropic glutamate receptor antagonist', *Br. J. Pharmacol.* **119**: 851–854.

Tückmantel, W., Kozikowski, A.P., Wang, S., Pshenichkin, S. and Wroblewski, J.T. (1997) 'Synthesis, molecular modeling, and biology of the 1-benzyl derivative of APDC – an apparent mGluR6 selective ligand', *Biomed. Chem. Lett.* **7**: 601–606.

Uyama, Y., Ishida, M. and Shinozaki, H. (1997) 'DCG-IV, a potent metabotropic glutamate receptor agonist, as an NMDA receptor agonist in the rat cortical slice', *Brain Res.* **752**: 327–330.

Vandergriff, J. and Rasmussen, K. (1999) 'The selective mGlu2/3 receptor agonist LY354740 attenuates morphine-withdrawal-induced activation of locus coerulus neurons and behavioral signs of morphine withdrawal', *Neuropharmacology* **38**: 217–222.

Varney, M.A., Cosford, N., Jachec, C., Rao, S.P., Sacaan, A., Lin, F.-F., Bleicher, L., Santori, E.M., Flor, P.J., Allgeier, H., Gasparini, F., Kuhn, R., Hess, S.D., Velicelebi, G. and Johnson, E.C. (1999) 'SIB-1757 and SIB-1893: Selective, non-competitive antagonists of metabotropic glutamate receptor type 5', *J. Pharmacol. Exp. Ther.* **290**: 170–181.

Vignes, M., Clarke, V.R.J., Davies, C.H., Chambers, A., Jane, D.E., Watkins, J.C. and Collingridge, G.L. (1995) 'Pharmacological evidence for an involvement of group II and group III mGluRs in the presynaptic regulation of excitatory synaptic responses in the CA1 region of rat hippocampal slices', *Neuropharmacology* **34**: 973–982.

Watkins, J.C. (1986) 'Twenty-five years of excitatory amino acid research', in P.J. Roberts, J. Storm-Mathisen and H.F. Bradford (eds) *Excitatory Amino Acids*, London: Macmillan, pp. 1–39.

Watkins, J.C. and Collingridge, G.L. (1994) 'Phenylglycine derivatives as antagonists of metabotropic glutamate receptors', *Trends Pharmacol. Sci.* **15**: 333–342.

Watkins, J.C. and Evans, R.H. (1981) 'Excitatory amino acid transmitters', *Ann. Rev. Pharmacol. Toxicol.* **21**: 165–204.

Wermuth, C.G., Mann, A., Schoenfelder, A., Wright, R.A., Johnson, B.G., Burnett, J.P., Mayne, N.G. and Schoepp, D.D. (1996) '(2*S*,4*S*)-2-amino-4-(4,4-diphenylbut-1-yl)pentane-1,5-dioic acid: A potent and selective antagonist for metabotropic glutamate receptors negatively linked to adenylate cyclase', *J. Med. Chem.* **39**: 814–816.

Wilsch, V.W., Pidoplichko, V.I., Opitz, T., Shinozaki, H. and Reymann, K.G. (1994) 'Metabotropic glutamate receptor agonist DCG-IV as NMDA receptor agonist in immature rat hippocampus', *Eur. J. Pharmacol.* **262**: 287–291.

Wright, R.A., McDonald, J.W. and Schoepp, D.D. (1994) 'Distribution and ontogeny of 1S,3R-1-aminocyclopentane-1,3-dicarboxylic acid-sensitive and quisqualate-insensitive [^{3}H]glutamate binding sites in the rat brain', *J. Neurochem.* **63**: 938–945.

Wu, S., Wright, R.A., Rockey, P.K., Burgett, S.G., Arnold, P.R., Johnson, B.G., Schoepp, D.D. and Belagaje, R. (1998) 'Group III human metabotropic glutamate receptors 4: 7: and 8: molecular cloning, functional expression, and comparison of pharmacological properties in RGT cells', *Mol. Brain Res.* **53**: 88–97.

Yamamoto, C., Sawada, S. and Takada, S. (1983) 'Suppressing action of 2-amino-4-phosphonobutyric acid on mossy fiber-induced excitation in the guinea pig hippocampus', *Exp. Brain Res.* **51**: 128–134.

Part 3

$GABA_A$ receptors

Chapter 8

Insights into $GABA_A$ receptor complexity from the study of cerebellar granule cells

Synaptic and extrasynaptic receptors

William Wisden and Mark Farrant

Introduction

In vertebrate brains, inhibition is provided mainly by γ-aminobutyric acid (GABA) acting on type A ($GABA_A$) receptors. These receptors are anion-permeable channels (Cl^-/HCO_3^-) formed as pentameric assemblies of subunits. A gene family encodes different $GABA_A$ receptor subunits (α1–α6, β1–β3, γ1–γ3, δ, ε, θ, π and ρ1–ρ3). The genes are differentially transcribed (Persohn *et al.*, 1992; Wisden *et al.*, 1992; Fritschy and Möhler, 1995; Hedblom and Kirkness, 1997; Whiting *et al.*, 1997; Bonnert *et al.*, 1999) and the subunits assembled into different combinations depending on the cell-type. Most $GABA_A$ receptors are likely to be αβγ or αβδ forms. Within the αβγ class the subunit ratio is probably 2α/2β/1γ (Tretter *et al.*, 1997; Farrar *et al.*, 1999). The large family of subunit genes results in considerable receptor diversity. These $GABA_A$ receptors differ in properties such as their affinity for neurotransmitter, activation rate, desensitization rate, and channel conductance (Barnard *et al.*, 1998). Important questions are: 'How many receptor subtypes are there?', 'What receptor subtypes are used by which types of neuron and where are they located on the cell?', 'What are the functions of these different receptor subtypes?', and 'How are the expression patterns of the receptor subunit genes controlled?'

In this chapter we review some key properties of $GABA_A$ receptors in one cell type: the cerebellar granule cell. The expression of $GABA_A$ receptors by these cells shares many features seen in more complex cells: receptor heterogeneity and specific subunit assembly; differential targeting of receptors to synaptic and extrasynaptic locations; and developmental changes in receptor properties. By documenting the receptor complexity in granule cells we hope to illustrate how other neuronal types use the range of $GABA_A$ receptor subunits available to them. To place the physiology of granule cell $GABA_A$ receptors in context, we first briefly review the circuitry of the cerebellum.

The cerebellum

The cerebellum consists of the cortex and the deep cerebellar nuclei (DCN) (Llinás and Walton, 1998; Voogd *et al.*, 1996; Voogd and Glickstein, 1998) (Figure 8.1). The cortex, a folded sheet of circuitry reiterated on a vast scale, contains the Purkinje cells and the granule cells. The granule cells receive glutamatergic excitatory input from mossy fibres originating from diverse areas of the brain. The cortical circuitry is the same in different regions of the cortex (lobules and vermis), and so presumably

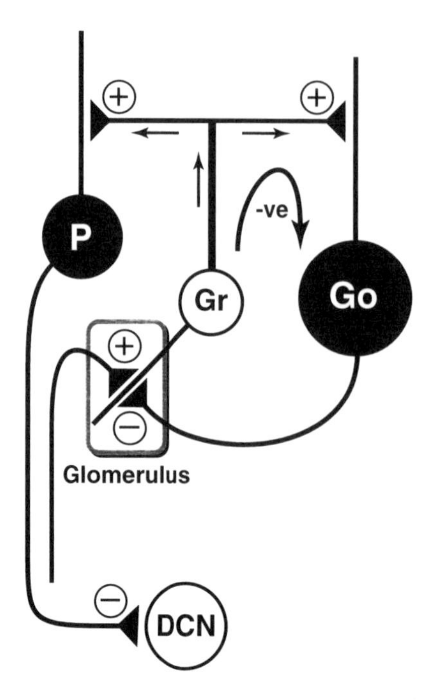

Figure 8.1 The Golgi cell-granule cell loop of the cerebellar cortex.

Key
DCN = deep cerebellar nucleus cell; Go = Golgi cell; Gr = granule cell; P = Purkinje cell; '+' indicates glutamatergic synapse; '−' indicates GABAergic synapse.

performs a similar computation; but different groups of granule cells are activated by mossy fibres carrying information from different brain regions.

The cerebellum controls movement, stores motor memories (Raymond *et al.*, 1996), and does various cognitive processes (reviewed by Bower, 1997; Fiez, 1996; Desmond and Fiez, 1998). The functions of the cortex are expressed solely through the modulation of the firing of cells in the DCN (Llinás and Walton, 1998). The only output of the cerebellar cortex comes from the GABAergic Purkinje cells and is inhibitory onto the DCN cells. The DCN are located in the white matter beneath the cortex and project to many parts of the CNS.

The GABAergic inhibition of granule cells

Within the cerebellar cortex, the circuitry exists to modulate the firing of Purkinje cells (Figure 8.1). Central to this circuit are the granule cells, the most numerous neurons in the cerebellum. They receive excitatory input from many brain areas via mossy fibres and relay this information via their ascending axon and its parallel fibres. The parallel fibres form *en passant* glutamatergic synapses in the molecular layer with the dendrites of Purkinje cells, stellate/basket interneurons and Golgi cells, all of which can be contacted by a single parallel fibre.

Although a few granule cells may be contacted by GABA-positive mossy fibre-like terminals (Hámori and Takács, 1989), the principal GABAergic input to these cells comes from Golgi cells (Llinas and Walton, 1998). The Golgi cell axon forms an extensive plexus contacting thousands of granule cells. Whereas the apical dendrites of Golgi cells extend into the molecular layer and receive excitatory input from parallel fibres, the basolateral dendrites and soma are contacted by mossy fibres and climbing fibres. Golgi cells thus provide both feed-forward and feed-back inhibition onto the granule cells (reviewed in Dieudonne, 1998). The importance of Golgi cells to cerebellar function is clear: their selective ablation causes acute disruption of motor co-ordination (Watanabe *et al.*, 1998).

No synaptic connections are formed on the granule cell soma; instead both the excitatory and inhibitory inputs are confined to the distal ends of the short granule cell dendrites within a specialized structure, the glomerulus (Figure 8.1). The glomeruli form around single mossy fibre terminals, and a single mossy fibre contacts dendrites from up to fifty granule cells. As each granule cell dendrite also receives input from two to three Golgi cell axon varicosities, each glomerulus may contain up to 150 GABA-releasing synapses enclosed within its glial sheath (reviewed by Llinás and Walton, 1998; Voogd *et al.*, 1996).

The Golgi cell-granule cell circuit may aid the filtering of mossy fibre sensory input during its relay to Purkinje cells (Eccles *et al.*, 1967; Marr, 1969; Gabiani *et al.*, 1994; Brickley *et al.*, 1996). The feedback excitation of Golgi cells by granule cell axons could synchronize activity of both cell types, converting the spatial signal of mossy fibre input to a temporal pattern of parallel fibre activity (Maex and Schutter, 1998; Vos *et al.*, 1999).

Assembly rules for $GABA_A$ receptor subunits

Neurons express multiple types of $GABA_A$ receptor subunits, and so can potentially assemble many subunit combinations (Figure 8.2). Cerebellar granule cells are no exception: in the adult they have significant $\alpha1$, $\beta2/3$ and δ expression, moderate $\gamma2$ expression (Fritschy and Möhler, 1995; Laurie *et al.*, 1992a; Somogyi *et al.*, 1996; Nusser *et al.*, 1995, 1996, 1998; Shivers *et al.*, 1989; Nusser and Somogyi, 1997), and weak $\alpha4$, $\beta1$ and $\gamma3$ subunit expression (Laurie *et al.*, 1992a).

Based on expression experiments in non-neuronal cells such as HEK-293 cells or *Xenopus* oocytes, many subunit combinations can form functional $GABA_A$ receptors (Barnard *et al.*, 1998). For example, $\alpha1$, β and δ subunits co-assemble to form functional receptors (Krishek *et al.*, 1996; Saxena and Macdonald, 1994). Granule cells, however, although they express these subunits, do not form receptors with this composition; instead the δ subunit associates specifically with the $\alpha6$ subunit (Figure 8.2). In $\alpha6$ knock-out mice ($\alpha6$ $-/-$), the δ subunit protein is eliminated from the cell surface of cerebellar granule cells, despite normal levels of δ subunit mRNA, and only a 20 per cent reduction in $\alpha1$ protein levels (Jones *et al.*, 1997; Nusser *et al.*, 1999); the $\alpha1$ protein cannot rescue the δ subunit from degradation. Although $\alpha1$ is found in some $\alpha6\delta$-containing complexes, the $\alpha6$ subunit is a cornerstone for the incorporation of the δ. By contrast, in the cerebella of δ $-/-$ mice $\alpha6$ protein is not degraded (Tretter *et al.*, 1999; 2001). The $\alpha6$ subunit can be incorporated into receptors without the δ (as would be expected, since we know from receptor pharmacology that $\alpha6\beta\gamma2$ receptors form *in vivo* – Korpi *et al.*, 1995; Jones *et al.*, 1997; Mäkelä *et al.*, 1997), but the δ protein cannot exist without $\alpha6$.

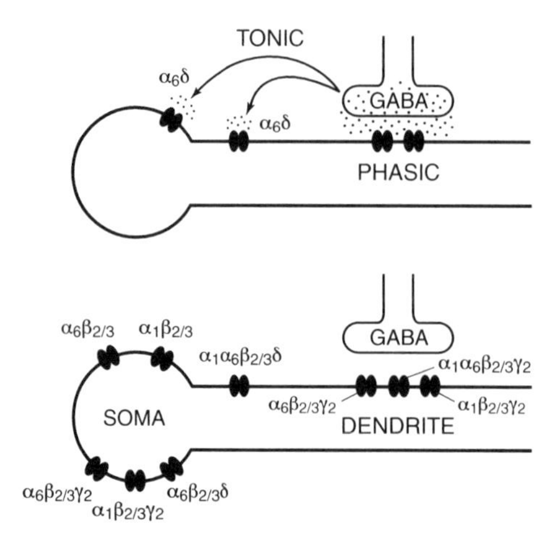

Figure 8.2 Differential distribution of GABA$_A$ receptor subtypes on cerebellar granule cells.

Upper part: GABA released from Golgi cell terminals activates post-synaptic GABA$_A$ receptors to give IPSPs (phasic inhibition), while GABA diffuses from the synapse to tonically activate extrasynaptic receptors. *Lower part:* Possible GABA$_A$ receptor subunit combinations assembled on granule cells. The synapses contain a mixture of α6β2/3γ2, α1β2/3γ2 and possibly α1α6β2/3γ2 receptors. The δ subunit is exclusively extrasynaptic and is always associated with the α6 subunit (Quirk *et al.*, 1994, Jones *et al.*, 1997; Nusser *et al.*, 1998, 1999; Jechlinger *et al.*, 1998). Extrasynaptic αβ combinations may also exist (Brickley *et al.*, 1999a).

Some GABA$_A$ receptor subunit combinations seem highly specific, possibly requiring specific chaperone molecules to guide them together (since there is no specificity when cDNAs are expressed in cell lines). Other subunit combinations might be determined simply by the levels of unassembled subunits in the endoplasmic reticulum. These precursor subunits might compete for assembly. Excess subunits would be degraded. Perturbing this pool of precursors (by knocking out a subunit gene) leads to a redistribution of subunits, indicating a chain of interconnected assembly reactions. For example, in cerebella which have no α6 protein, α1, β2, β3 and γ2 protein levels all decrease (α1, 20 per cent; β2, 50 per cent; β3, 20 per cent; γ2, 40 per cent) (Nusser *et al.*, 1999). Most probably the loss of the α6 and δ proteins results in the loss of those β2/3 and γ2 subunits that would normally have been incorporated into α6βγ2, α6βδ and/or αβγδ receptors (Nusser *et al.*, 1999). As β3 subunit levels decrease less than those of β2, the α6 subunit may associate more with the β2 subunit. The 20 per cent reduction in the α1 subunit in cerebella of α6 −/− mice suggests that α1 and α6 exist partially in the same complex (e.g. as α1α6βδ or α1α6βγ2) (Pollard *et al.*, 1995; Khan *et al.*, 1996; Jechlinger *et al.*, 1998). Mice with a disrupted δ subunit gene illustrate further how the unassembled subunits appear to be processed in ways suggesting strict interdependency. In δ −/− cerebella, γ2 subunit levels increase, and the number of α6βγ2 receptors increase too (Tretter *et al.*, 1999; 2001). Presumably, the α6 protein that would have ordinarily assembled with δ combines with γ2 protein instead (in wild-type cerebella, this excess γ2 protein is degraded).

These data mean that GABA$_A$ receptor subunit assembly is not random. Even so, mature granule cells still make a complex mixture of receptor subtypes, including

$\alpha1\beta2/3\gamma2$, $\alpha6\beta2/3\gamma2$ and $\alpha6\delta$-containing receptors (Figure 8.2). In the following sections we ask whether it is possible to assign different functions to these receptors.

$GABA_A$ receptor expression during granule cell development

In granule cells, as in many other cell types, the pattern of $GABA_A$ receptor subunit expression changes during development (Figure 8.3; reviewed by Carlson *et al.*, 1998; Wisden *et al.*, 1996; Siegel, 1998). In species such as rats and mice the cerebellum develops postnatally. At birth granule cell precursors are on the cerebellum's exterior as a thin layer of cells that have migrated from the rhombic lip – the external germinal layer (or external granule cell layer, EGr in Figure 8.3). In the first postnatal weeks these precursors proliferate extensively. After mitosis the cells extend axonal processes and migrate through the molecular layer and enter the internal granule cell layer (IGr in Figure 8.3); there, the deepest granule cells are the oldest, and the layer develops in an 'inside-out' order. The end product – the adult granule cell layer – is a convoluted sheet of densely packed cells (reviewed by Altman and Bayer, 1997).

Different $GABA_A$ receptors are expressed at different stages of granule cell development (Figure 8.3). Prior to synapse formation, premigratory granule cells in the external germinal layer of the cerebellar cortex express the $\alpha2$, $\alpha3$, $\beta3$, $\gamma1$ and $\gamma2$ subunit genes (Laurie *et al.*, 1992b) and form functional receptors (Farrant *et al.*, 1995). These immature cells assemble at least three distinct types of $GABA_A$ receptor as inferred from their single-channel properties (Brickley *et al.*, 1999a). Although GABA can influence the cell cycle of proliferative neurons (e.g. LoTurco *et al.*, 1995; Fiszman *et al.*, 1999), the source of GABA and its precise role in the external granule cell layer are unknown.

Following migration into the internal layer, granule cells form synaptic contacts with glutamatergic mossy fibres and GABAergic Golgi cell axons (Figure 8.3). Levels of $\alpha2$, $\alpha3$ and $\gamma1$ mRNA decline, whereas levels of $\alpha1$ and $\gamma2$ mRNA increase significantly (Figure 8.3; Laurie *et al.*, 1992b). As in the case of premigratory cells, postmigratory granule cells express multiple $GABA_A$ receptors with distinct single-channel conductances (Brickley *et al.*, 1999a; Kaneda *et al.*, 1995). After the first postnatal week, $\alpha6$

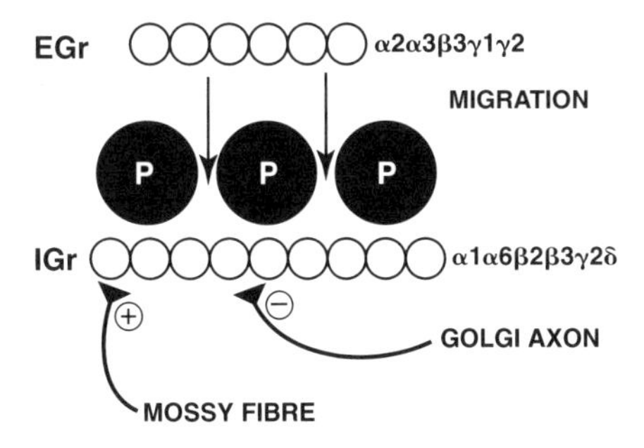

Figure 8.3 The changes in $GABA_A$ receptor subunit gene expression in developing cerebellar granule cells.

Key
Egr = external granule cell layer; Igr = internal granule cell layer; P = Purkinje cells; '+' marks glutamatergic synapses; '−' marks GABAergic synapses; arrows mark the direction of migration.

and δ subunit gene expression turns on (Laurie *et al.*, 1992b; Mellor *et al.*, 1998), leading to the adult pattern of α1, α6, β2, β3, γ2 and δ subunit gene expression (Wisden *et al.*, 1996). The factors regulating the developmental control of subunit gene expression are diverse. In cultured granule cells, neuronal depolarization, cell-density, NMDA receptor activation, cAMP levels, GABA, and growth factors such as BDNF can all affect subunit mRNA levels (e.g. Mellor *et al.*, 1998, Thompson *et al.*, 1996a, 1996b; Carlson *et al.*, 1998). The influence of these factors differs for each subunit gene and can also differ between species (Mellor *et al.*, 1998). Although factors such as cAMP and Ca^{2+} entry may fine-tune the expression levels, or may determine the exact onset of expression, the subunit expression of granule cells is intrinsically specified (Siegel, 1998; Bahn *et al.*, 1999).

In developing granule cells two features of GABAergic synaptic transmission stand out: the appearance of a tonic $GABA_A$ receptor-mediated conductance and a change in kinetics of the inhibitory postsynaptic current (IPSC) (Figure 8.4). At most GABAergic synapses, including those on cerebellar granule cells, the release of GABA causes the rapid activation of postsynaptic $GABA_A$ receptors, resulting in IPSCs usually lasting less than a few tens of milliseconds. In addition to this phasic receptor activation, granule cells exhibit a tonic conductance that results from the persistent activation of receptors by ambient GABA (Kaneda *et al.*, 1995; Tia *et al.*, 1996; Brickley *et al.*, 1996; Wall and Usowicz, 1997). This conductance develops during the second postnatal week (Figure 8.4a) and appears due to the 'overspill' of synaptically released GABA resulting in the

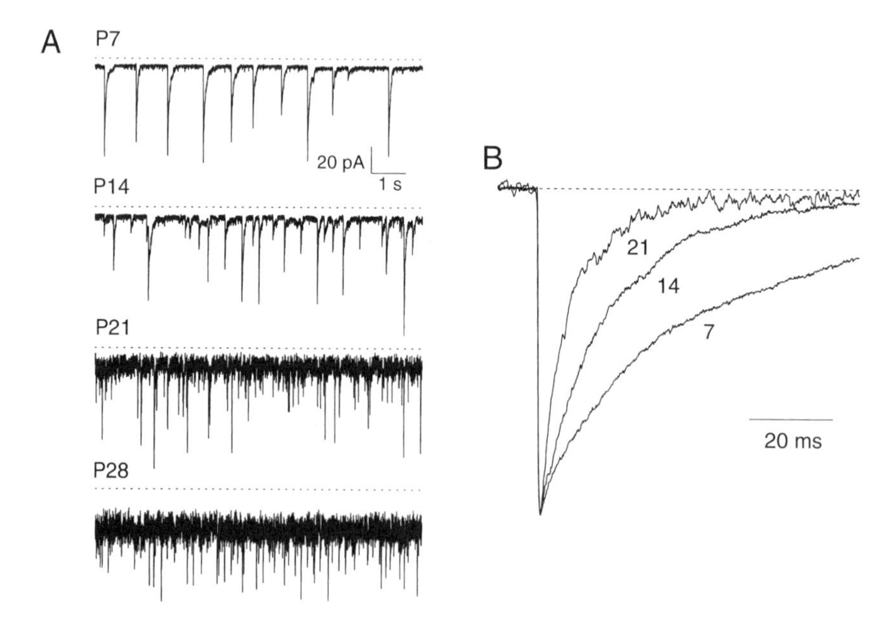

Figure 8.4 Changes in $GABA_A$ receptor properties during granule cell development in the rat.

(a) Spontaneous IPSCs at postnatal (P) days 7, 14, 21 and 28 (symmetrical $[Cl^-]_i$, pipette potential −70 mV) recorded in the presence of CNQX, AP5 and strychnine. Dashed lines indicate zero current level. (b) Superimposed averaged IPSCs at P7, 14 and 21. Traces are normalised to the peak and show the first 80 ms of the decay. The dashed line indicates the pre-event baseline. All recordings are from rat cerebellar slices at room temperature (modified from Brickley *et al.*, 1996).

activation of receptors containing the α6 subunit (Farrant *et al.*, 1999; Brickley *et al.*, 1999b; 2001 – see section 'Extrasynaptic $GABA_A$ receptors'). Over the same period of postnatal development there is a clear speeding of the IPSC decay, leading to a substantial reduction in charge transfer via discrete synaptic events relative to that resulting from the tonic conductance (Figure 8.4b; Brickley *et al.*, 1996). These kinetic changes still take place in mice lacking the α6 and δ subunits (Farrant *et al.*, 1999; Brickley *et al.*, 1999b), indicating that they result from changes in other subunits, or from changes unrelated to subunit composition such as post-translational modification or changes in synaptic architecture.

As in other neurons, $GABA_A$ receptor activation in immature granule cells causes depolarization, and this can be sufficient to evoke action potentials (Brickley *et al.*, 1996). In mature granule cells the GABA reversal potential is shifted to more negative values close to the resting potential (Brickley *et al.*, 1996), reflecting the expression of, for example, the K^+/Cl^- co-transporter KCC2 (Rivera *et al.*, 1999).

Differential subcellular targeting of $GABA_A$ receptors

In young postmigratory granule cells the single-channel conductance of synaptic $GABA_A$ receptors is approximately 28 pS (Brickley *et al.*, 1999a), consistent with the expression of high-conductance (benzodiazepine-sensitive) αβγ subunit combinations. Similar channels are found in the extrasynaptic (somatic) membrane of the same cells, but these are present – together with channels that display lower conductances typical of αβ subunit combinations. Thus channels lacking a γ2 subunit may be excluded from the synapse (Brickley *et al.*, 1999). The differential targeting mechanism of γ-containing receptors could involve $GABA_A$-receptor-associated protein (GABARAP, Wang *et al.*, 1999) or gephyrin (Essrich *et al.*, 1998).

The situation is probably more complicated in mature granule cells; immunohistochemical data supports at least four subunit combinations (see Nusser *et al.*, 1998 and Figure 8.2). In granule cells from adult animals, $GABA_A$ receptor subunits are enriched in the postsynaptic membrane, but are also in the extrasynaptic dendritic and somatic membranes (Nusser *et al.*, 1995; Nusser and Somogyi, 1997). The α1, α6, β2/3 and γ2 subunit proteins are present at both sites; however, the δ protein is exclusively extrasynaptic (Nusser *et al.*, 1998; Figure 8.2), consistent with the idea that a γ subunit is required for synaptic targeting (see above, and Baer *et al.*, 1999).

Synaptic $GABA_A$ receptors

In the cerebella of adult rodents, the α1 and β2/3 subunits are 230- and 180-fold more concentrated, respectively, in the postsynaptic membrane at Golgi cell synapses than on the extrasynaptic membrane of the cell body (Nusser *et al.*, 1995). Whereas β2/3 and γ2 immunoreactivities are enriched at every Golgi–granule cell synapse (Somogyi *et al.*, 1996; Nusser *et al.*, 1995, 1998; Nusser and Somogyi, 1997) (Figure 8.2), the distribution of α subunits is variable. Some synapses contain mainly α1 immunoreactivity; some contain mainly α6; others have equal amounts of both (Nusser *et al.*, 1996; Nusser and Somogyi, 1997). At mature synaptic contacts $GABA_A$ receptors may exist as α1β2/3γ2, α6β2/3γ2 or α1α6β2/3γ2 types (Figure 8.2; Quirk *et al.*, 1994; Pollard *et al.*, 1995; Khan *et al.*, 1996; Jechlinger *et al.*, 1998; Nusser *et al.*, 1998). The α6, β2/3 and $\gamma2$ subunits are also found at mossy fibre synapses (marked '+' in Figure 8.1), but their function at these sites, if any, is unknown (Nusser *et al.*, 1996, 1998).

Extrasynaptic $GABA_A$ receptors

$GABA_A$ receptor subunits are concentrated at Golgi cell synapses, and decrease in density away from the GABA release sites. However, far more subunits overall are found in extrasynaptic (intraglomerular dendritic and somatic) membrane than at points of synaptic specialization (Nusser *et al.*, 1995). The δ protein is found exclusively in extrasynaptic membrane (Nusser *et al.*, 1998). Thus in addition to $\alpha1\beta2/3\gamma2$, $\alpha6\beta2/3\gamma2$ or $\alpha1\alpha6\beta2/3\gamma2$ receptors, which are also found at the synapse, extrasynaptic sites have $\alpha6\beta2/3\delta$, and possibly $\alpha1\alpha6\beta2/3\delta$ receptors (Quirk *et al.*, 1994; Jechlinger *et al.*, 1998; Nusser *et al.*, 1998) as well as $\alpha\beta$ combinations (see above and Figure 8.2).

The special geometry of the glomerulus, with its numerous Golgi cell terminals and glial sheath, may allow synaptically released GABA to persist within the extracellular space. This extrasynaptic GABA is thought responsible for the $GABA_A$ receptor-meditated tonic conductance in granule cells (Brickley *et al.*, 1996;Tia *et al.*, 1996, Wall and Usowicz, 1997), and for the prolonged spill-over currents seen following electrical stimulation of Golgi cell axons (Rossi and Hamann, 1998). Receptors containing $\alpha6$ and δ subunits are well-suited to allow their tonic activation by low concentrations of GABA. For example, recombinant $\alpha6\beta3\delta$ receptors have a high-affinity for GABA ($EC_{50} < 1\,\mu M$), are non-desensitizing and have prolonged channel openings (Saxena and Macdonald, 1996). In support of this idea, the tonic bicuculline-sensitive conductance seen in wild-type animals is absent from granule cells of $\alpha6$ $-/-$ mice (Farrant *et al.*, 1999; Brickley *et al.*, 1999b; 2001).

Other brain regions which use $\alpha6\delta$- and $\alpha4\delta$-containing receptors

Extrasynaptic $GABA_A$ receptors are probably also important beyond the cerebellum. One such region is the cochlear nucleus in the brain stem, the first relay station of the auditory pathway. This nucleus is embryologically related to the cerebellum and has a similar organization. The $\alpha6$ and δ subunit genes are both expressed in granule cells of the ventral cochlear nucleus whereas granule cells of the dorsal cochlear nucleus have $\alpha6$, but no δ expression (Campos *et al*, 2001). $GABA_A$ receptors containing $\alpha6$ and δ subunits probably have a forebrain analogue: receptors containing $\alpha4$ and δ subunits. Both subtypes are selectively marked-out by autoradiography with [^{3}H]muscimol (high-affinity binding) (Jones *et al.*, 1997; Mihalek *et al.*, 1999). The distribution of [^{3}H]muscimol binding sites is almost identical to the combined distribution of the $\alpha4$, $\alpha6$ and δ subunits, i.e. within the cerebellar granule cell layer, thalamus, hippocampus, layers II and III of the neocortex, and the granule cells of the olfactory bulb (Shivers *et al.*, 1989; Olsen *et al.*, 1990; Wisden *et al.*, 1992; Mihalek *et al.*, 1999). The $\alpha4$ and $\alpha6$ subunits are closely related. Uniquely among $GABA_A$ subunits, they share an arginine (R) residue at position 100 (Wisden *et al.*, 1991). R100 may be part of an 'assembly box' which promotes the assembly of the $\alpha4$ and $\alpha6$ subunits with δ. The $\alpha4\delta$-containing receptors may also be located extrasynaptically.

Evolution and redundancy of the α6 system

If conservation of a feature across species reflects the importance of that feature, then α6-containing $GABA_A$ receptors should be important for granule cell physiology. During the 450–500 million year gap separating fish from mammals, the α6 genes' regulatory elements and amino acid sequence (except the TM3–TM4 loop peptide sequence) have stayed constant – α6 expression has stayed selectively associated with cerebellar granule cells, including those of humans (Bahn *et al.*, 1996; Hadingham *et al.*, 1996). Modelling studies of granule cells suggest that a tonic conductance may, under some circumstances, modify granule cell excitability to maintain a reliance on closely timed mossy fibre input (see Gabbiani *et al.*, 1994; Maex and deSchutter, 1998). It is surprising, therefore, that α6 knock-out mice (α6 −/−) lacking any α6δ-containing receptors, are indistinguishable from wild-types in simple motor tasks (Jones *et al.*, 1997; Korpi *et al.*, 1999).

Compensation for the lack of tonic GABA-mediated conductance in the granule cells of these mice is a possibility (Brickley *et al.*, 2001). This view is supported by the recent finding that the GABAergic input to granule cells is partially dispensable for basic cerebellar functions (Watanabe *et al.*, 1998). If the mouse cerebellar cortex is depleted of Golgi cells (by genetic lesion), so that there is no GABA input onto granule cells, there is an initial ataxic phase; subsequently, though, motor function recovers to a large extent. On simple rotorod tasks the lesioned mice perform normally, but on challenging tasks (high-speed of rod rotation) they remain impaired (Watanabe *et al.*, 1998). This partial recovery may involve a down-regulation of granule cell NMDA receptors (Watanabe *et al.*, 1998). If mice with no GABA input onto their granule cells can partially compensate, then it is perhaps unsurprising that they can still function normally when lacking one $GABA_A$ receptor subtype. Indeed, α6 knockout granule cells upregulate a two pore domain potassium channel (TASK-1-like) to compensate for the loss of the $GABA_A$ receptors. This K^+ channel provides a tonic leak conductance (Brickley *et al.*, 2001).

Conclusions

Granule cells are simple cells receiving most of their inhibitory input from one cell type: the Golgi cell. Nevertheless, these cells express multiple $GABA_A$ receptor subtypes in both the synaptic and extrasynaptic membrane. The synaptic receptors are immediately opposite the GABA release sites, enabling fast point-to-point communication (phasic inhibition). The extrasynaptic receptors appear capable of sensing GABA diffusing from the synapse, and generate a persistent background conductance. If present *in vivo*, this conductance would significantly influence the overall level of granule cell excitability (tonic inhibition). Similar mechanisms may operate in other brain regions.

Acknowledgements

William Wisden's research is supported by Human Frontier Science Program Organization grant RG 17/98 and a Deutsche Forschungsgemeinshaft grant – WI1951/1-1. Mark Farrant's research is supported by grants from the Wellcome Trust.

References

Altman, J. and Bayer, S.A. (1997) *Development of the Cerebellar System in Relation to its Evolution, Structure, and Functions*, Boca Raton, New York, London, Tokyo: CRC Press.

Baer, K., Essrich, C., Benson, J.A., Banke, D., Bluethmann, H., Fritschy, J.-M. and Lüscher, B. (1999) 'Postsynaptic clustering of γ-aminobutyric acid type A receptors by the γ3 subunit *in vivo*', *Proceedings of the National Academy of Sciences USA* **96**: 12,860–12,865.

Bahn, S., Harvey, R.J., Darlison, M.G. and Wisden, W. (1996) 'Conservation of γ-aminobutyric acid type A receptor α6 subunit gene expression in cerebellar granule cells', *Journal of Neurochemistry* **66**: 1810–1818.

Bahn, S., Wisden, W., Dunnett, S.B. and Svendsen, C. (1999) 'The intrinsic specification of γ-aminobutyric acid type A receptor α6 subunit gene expression in cerebellar granule cells', *European Journal of Neuroscience* **11**: 2194–2198.

Barnard, E.A., Skolnick, P., Olsen, R.W., Mohler, H., Sieghart, W., Biggio, G., Braestrup, C., Bateson, A.N. and Langer, S.Z. (1998) 'International union of pharmacology. XV. Subtypes of γ-aminobutyric $acid_A$ receptors: classification on the basis of subunit structure and receptor function', *Pharmacological Reviews* **50**: 291–313.

Bonnert, T.P., McKernan, R.M., Farrar, S., Bourdelles, B.l., Heavens, R.P., Smith, D.W., Hewson, L., Rigby, M.R., Sirinathsinghji, D.J.S., Brown, N., Wafford, K.A. and Whiting, P.J. (1999) 'θ, a novel γ-aminobutyric acid type A receptor subunit', *Proceedings of the National Academy of Science USA* **96**: 9891–9896.

Bower, J.M. (1997) 'Is the cerebellum sensory for motor's sake, or motor for sensory's sake: the view from the whiskers of a rat', in C.I. de Zeeuw, P. Strata and J. Voogd (eds) *Progress in Brain Research*, vol. 114: Amsterdam: Elsevier Science B.V., pp. 463–496.

Brickley, S.G., Cull-Candy, S.G. and Farrant, M. (1996) 'Development of a tonic form of synaptic inhibition in rat cerebellar granule cells resulting from persistent activation of $GABA_A$ receptors', *Journal of Physiology* **497**: 753–759.

Brickley, S.G., Cull-Candy, S.G. and Farrant, M. (1999a) 'Single-channel properties of synaptic and extrasynaptic $GABA_A$ receptors suggest differential targeting of receptor subtypes', *Journal of Neuroscience* **19**: 2960–2973.

Brickley, S.G. Farrant, M., Wisden, W. and Cull-Candy, S.G. (1999b) 'The role of α6- and δ-containing $GABA_A$ receptors in the generation of tonic inhibition in cerebellar granule cells', *Journal of Physiology* 518P, 140P.

Brickley, S.G., Revilla, V., Cull-Candy, S.G., Wisden, W. and Farrant, M. (2001) 'Adaptive regulation of neuronal excitability by a voltage-independent potassium conductance, *Nature* **409**: 88–92.

Campos, M.L., De Cabo, C., Wisden, W., Juiz, J.M. and Merlo, D. (2001) 'Expression of $GABA_A$ receptors in rat brainstem auditory pathways: cochlear nuclei, superior olivary complex and nucleus of the lateral lemniscus', *Neuroscience* **102**: 625–638.

Carlson, B.X., Elster, L. and Schousboe, A. (1998) 'Pharmacological and functional implications of developmentally-regulated changes in $GABA_A$ receptor subunit expression in the cerebellum', *European Journal of Pharmacology* **352**: 1–14.

Desmond, J.E. and Fiez, J.A. (1998) 'Neuroimaging studies of the cerebellum: language, learning and memory', *Trends in Cognitive Sciences* **2**: 355–362.

Dieudonne, S. (1998) 'Submillisecond kinetics and low efficacy of parallel fibre–Golgi cell synaptic currents in the rat cerebellum', *Journal of Physiology*, **510**: 845–866.

Eccles, J.C., Ito, M. and Szentágothai, J. (1967) *The Cerebellum as a Neuronal Machine*, Berlin: Springer.

Essrich, C., Lorez, M., Benson, J.A., Fritschy, J.M. and Luscher, B. (1998) 'Postsynaptic clustering of major $GABA_A$ receptor subtypes requires the gamma 2 subunit and gephyrin', *Nature Neuroscience*, **1**: 563–571.

Farrant, M., Kaneda, M. and Cull-Candy, S.G. (1995) 'Benzodiazepine modulation of GABA-activated currents in granule cells of the rat cerebellum', *Journal of Physiology* **489P**: 17.

Farrant, M., Wisden, W., Cull-Candy, S.G. and Brickley, S.G. (1999) 'Absence of tonic $GABA_A$ receptor-mediated conductance in cerebellar granule cells of α6 −/− mice', *Society of Neuroscience* **25**: 1246.

Farrar, S.J., Whiting, P.J., Bonnert, T.P. and McKernan, R.M. (1999) 'Stoichiometry of a ligand-gated ion channel determined by fluorescence energy transfer', *Journal of Biological Chemistry* **274**: 10,100–10,104.

Fiez, J.A. (1996) 'Cerebellar contributions to cognition', *Neuron* **16**: 13–15.

Fiszman, M.L., Borodinsky, L.N. and Neale, J.H. (1999) 'GABA induces proliferation of immature cerebellar granule cells grown in vitro', *Developmental Brain Research* **115**: 1–8.

Fritschy, J.-M. and Möhler, H. (1995) 'GABA$_A$-receptor heterogeneity in the adult rat brain: differential regional and cellular distribution of seven major subunits', *The Journal of Comparative Neurology* **359**: 154–194.

Gabbiani, F., Midtgaard, J. and Knöpfel, T. (1994) 'Synaptic integration in a model of cerebellar granule cells', *Journal of Neurophysiology* **72**: 999–1009.

Hadingham, K.L., Garrett, E.M., Wafford, K.A., Bain, C., Heavens, R.P., Sirinathsinghji, D.J.S. and Whiting, P.J. (1996) 'Cloning of cDNAs encoding the human γ-aminobutyric acid type A receptor α6 subunit and characterization of the pharmacology of α6-containing receptors', *Molecular Pharmacology* **49**: 253–259.

Hámori, J. and Takács, J. (1989) 'Two types of GABA-containing axon terminals in cerebellar glomeruli of cat: an immunogold-EM study', *Experimental Brain Research* **74**: 471–479.

Hedblom, E. and Kirkness, E.F. (1997) 'A novel class of GABA$_A$ receptor subunit in tissues of the reproductive system', *Journal of Biological Chemistry* **272**: 15,346–15,350.

Jechlinger, M., Pelz, R., Tretter, V., Klausberger, T. and Sieghart, W. (1998) 'Subunit composition and quantitative importance of hetero-oligomeric receptors: GABA$_A$ receptors containing α6 subunits', *Journal of Neuroscience* **18**: 2449–2457.

Jones, A., Korpi, E.R., McKernan, R.M., Pelz, R., Nusser, Z., Mäkelä, R., Mellor, J.R., Pollard, S., Bahn, S., Stephenson, F.A., Randall, A.D., Sieghart, W., Somogyi, P., Smith, A.J.H. and Wisden, W. (1997) 'Ligand-gated ion channel subunit partnerships: GABA$_A$ receptor α_6 subunit gene inactivation inhibits δ subunit expression', *Journal of Neuroscience* **17**: 1350–1362.

Kaneda, M., Farrant, M. and Cull-Candy, S.G. (1995) 'Whole-cell and single-channel currents activated by GABA and glycine in granule cells of the rat cerebellum', *Journal of Physiology* **485**: 419–435.

Khan, Z.U., Gutiérrez, A. and DeBlas, A.L. (1996) 'The α1 and α6 subunits can coexist in the same cerebellar GABA$_A$ receptor maintaining their individual benzodiazepine-binding specificities', *Journal of Neurochemistry* **66**: 685–691.

Korpi, E.R., Kuner, T., Seeburg, P.H. and Lüddens, H. (1995) 'Selective antagonist for the cerebellar granule cell-specific GABA$_A$ receptor', *Mol. Pharmacol.* **47**: 283–289.

Korpi, E.R., Koikkalainen, P., Vekovischeva, O.Y., Makela, R., Kleinz, R., Uusi-Oukari, M. and Wisden, W. (1999) 'Cerebellar granule-cell-specific GABA$_A$ receptors attenuate benzodiazepine-induced ataxia: evidence from alpha 6-subunit-deficient mice', *European Journal of Neuroscience* **11**: 233–240.

Krishek, B.J., Amato, A., Connolly, C.N., Moss, S.J. and Smart, T.G. (1996) 'Proton sensitivity of the GABA$_A$ receptor is associated with the receptor subunit composition', *Journal of Physiology* **492**: 431–443.

Laurie, D.J., Seeburg, P.H. and Wisden, W. (1992a) 'The distribution of thirteen GABA$_A$ receptor subunit mRNAs in the rat brain. II. Olfactory bulb and cerebellum', *Journal of Neuroscience* **12**: 1063–1076.

Laurie, D.J., Wisden, W. and Seeburg, P.H. (1992b) 'The distribution of thirteen GABA$_A$ receptor subunit mRNAs in the rat brain. III. Embryonic and postnatal development', *Journal of Neuroscience* **12**: 4151–4172.

Llinás, R.R. and Walton, K.D. (1998) 'Cerebellum', in E.M. Shepherd (ed.) *The Synaptic Organization of the Brain*, New York and Oxford: Oxford University Press, pp. 214–245.

LoTurco, J.J., Owens, D.F., Heath, M.J., Davis, M.B. and Kriegstein, A.R. (1995) 'GABA and glutamate depolarize cortical progenitor cells and inhibit DNA synthesis', *Neuron* **15**: 1287–1298.

Mäkelä, R., Uust-Oukari, M., Homanics, G.E., Wisden, W. and Korpi, E.R. (1997) 'Cerebellar GABA$_A$ receptors: pharmacological subtypes as revealed by mutant mouse lines', *Mol. Pharmacol.* **52**: 380–388.

Maex, R. and Schutter, E.D. (1998) 'Synchronization of Golgi and granule cell firing in a detailed network model of the cerebellar granule cell layer', *Journal of Neurophysiology* **80**: 2521–2537.

Marr, D. (1969) 'A theory of cerebellar cortex', *Journal of Physiology* **202**: 437–470.

Mellor, J.R., Merlo, D., Jones, A., Wisden, W. and Randall, A.D. (1998) 'Mouse cerebellar granule cell differentiation: electrical activity regulates the $GABA_A$ receptor α6 subunit gene', *Journal of Neuroscience* **18**: 2822–2833.

Mihalek, R.M., Banerjee, P.K., Korpi, E.R., Quinlan, J.J., Firestone, L.L., Mi, Z.-P., Lagenaur, C., Tretter, V., Sieghart, W., Anagnostaras, S.L., Sage, J.R., Fanselow, M.S., Guidotti, A., Spigelman, I., Li, Z., DeLorey, T.M., Olsen, R.W. and Homanics, G.E. (1999) 'Attenuated sensitivity to neuroactive steroids in γ-aminobutyrate type A receptor delta subunit knockout mice', *Proceedings of the National Academy of Sciences USA* **96**: 12,905–12,910.

Nusser, Z., Roberts, J.D.B., Baude, A., Richards, J.G. and Somogyi, P. (1995) 'Relative densities of synaptic and extrasynaptic $GABA_A$ receptors on cerebellar granule cells as determined by a quantitative immunogold method', *Journal of Neuroscience* **15**: 2948–2960.

Nusser, S., Sieghart, W., Stephenson, F.A. and Somogyi, P. (1996) 'The α6 subunit of the $GABA_A$ receptor is concentrated in both inhibitory and excitatory synapses on cerebellar granule cells', *Journal of Neuroscience* **16**: 103–114.

Nusser, Z. and Somogyi, P. (1997) 'Compartmentalised distribution of $GABA_A$ and glutamate receptors in relation to transmitter release sites on the surface of cerebellar neurons', in C.I. de Zeeuw, P. Strata and J. Voogd (eds) *Progress in Brain Research* vol. 114: Elsevier Science BV, pp. 109–127.

Nusser, Z., Sieghart, W. and Somogyi, P. (1998) 'Segregation of different $GABA_A$ receptors to synaptic and extrasynaptic membranes of cerebellar granule cells', *Journal of Neuroscience* **18**: 1693–1703.

Nusser, Z., Ahmad, Z., Tretter, V., Fuchs, K., Wisden, W., Sieghart, W. and Somogyi, P. (1999) 'Alterations in the expression of $GABA_A$ receptor subunits in cerebellar granule cells after the disruption of the α6 subunit gene', *European Journal of Neuroscience* **11**: 1685–1697.

Olsen, R.W., McCabe, R.T. and Wamsley, J.K. (1990) '$GABA_A$ receptor subtypes: autoradiographic comparison of GABA, benzodiazepine, and convulsant binding sites in the rat central nervous system', *Journal of Chemical Neuroanatomy* **3**: 59–76.

Persohn, E., Malherbe, P. and Richards, J.G. (1992) 'Comparative molecular neuroanatomy of cloned $GABA_A$ receptor subunits in the rat CNS', *Journal of Comparative Neurology* **326**: 193–216.

Pollard, S., Thompson, C.L. and Stephenson, F.A. (1995) 'Quantitative characterization of α6 and α1α6 subunit-containing native γ-aminobutyric $acid_A$ receptors of adult rat cerebellum demonstrates two α subunits per receptor oligomer', *Journal of Biological Chemistry* **270**: 21,285–21,290.

Quirk, K., Gillard, N.P., Ragan, C.I., Whiting, P.J. and McKernan, R.M. (1994) 'Model of subunit composition of γ-aminobutyric acid A receptor subtypes expressed in rat cerebellum with respect to their α and γ/δ subunits', *Journal of Biological Chemistry* **269**: 16,020–16,028.

Raymond, J.L., Lisberger, S.G. and Mauk, M.D. (1996) 'The cerebellum: a neuronal learning machine?', *Science* **272**: 1126–1131.

Rivera, C., Voipio, J., Payne, J.A., Ruusuvuori, E., Lahtinen, H., Lamsa, K., Pirvola, U., Saarma, M. and Kaila, K. (1999) 'The K^+/Cl^- co-transporter KCC2 renders GABA hyperpolarizing during neuronal maturation', *Nature* **397**: 251–255.

Rossi, D.J. and Hamann, M. (1998) 'Spillover-mediated transmission at inhibitory synapses promoted by high affinity α6 subunit $GABA_A$ receptors and glomerular geometry', *Neuron* **20**: 783–795.

Saxena, N.C. and Macdonald, R.L. (1994) 'Assembly of $GABA_A$ receptor subunits: role of the δ subunit', *Journal of Neuroscience* **14**: 7077–7086.

Saxena, N.C. and Macdonald, R.L. (1996) 'Properties of putative cerebellar γ-aminobutyric $acid_A$ receptor isoforms', *Molecular Pharmacology* **49**: 567–579.

Shivers, B.D., Killisch, I., Sprengel, R., Sontheimer, H., Köhler, M., Schofield, P.R. and Seeburg, P.H. (1989) 'Two novel $GABA_A$ receptor subunits exist in distinct neuronal subpopulations', *Neuron* **3**: 327–337.

Siegel, R.E. (1998) 'Developmental expression of cerebellar $GABA_A$-receptor subunit mRNAs', *Perspectives on Developmental Neurobiology* **5**: 207–217.

Somogyi, P., Fritschy, J.-M., Benke, D., Roberts, J.D.B. and Sieghart, W. (1996) 'The $\gamma 2$ subunit of the $GABA_A$ receptor is concentrated in synaptic junctions containing the $\alpha 1$ and $\beta 2/3$ subunits in hippocampus, cerebellum and globus pallidus', *Neuropharmacology* **35**: 1425–1444.

Thompson, C.L., Pollard, S. and Stephenson, F.A. (1996a) 'Bidirectional regulation of $GABA_A$ receptor $\alpha 1$ and $\alpha 6$ subunit expression by a cyclic AmP-mediated signalling mechanism in cerebellar granule cells in primary culture', *Journal of Neurochemistry* **67**: 434–437.

Thompson, C.L., Pollard, S. and Stephenson, F.A. (1996b) 'Developmental regulation of expression of $GABA_A$ receptor $\alpha 1$ and $\alpha 6$ subunits in cultured rat cerebellar granule cells', *Neuropharmacology* **35**: 1337–1346.

Tia, S., Wang, J.F., Kotchabhakdi, N. and Vicini, S. (1996) 'Developmental changes of inhibitory synaptic currents in cerebellar granule neurons: role of $GABA_A$ receptor $\alpha 6$ subunit', *Journal of Neuroscience* **16**: 3630–3640.

Tretter, V., Ehya, N., Fuchs, K. and Sieghart, W. (1997) 'Stoichiometry and assembly of a recombinant $GABA_A$ receptor subtype', *Journal of Neuroscience* **17**: 2728–2737.

Tretter, V., Hauer, B., Nusser, Z., Somogyi, P., Mihalek, R., Homanics, G. and Sieghart, W. (1999) 'Up-regulation of γ_2 subunits in $GABA_A$ receptor δ subinit knockout mice', *Journal of Neurochemistry* **73**: S140.

Tretter, V., Hauer, B., Nusser, Z., Mihalek, R.M., Hogar, H., Homanics, G.E., Somogyi, P. and Sieghart, W. (2001) 'Targeted disruption of the $GABA_A$ receptor delta subunit gene leads to an upregulation of gamma 2 subunit-containing receptors in cerebellar granule cells', *Journal of Biological Chemistry* **276**: 10532–10538.

Voogd, J. and Glickstein, M. (1998) 'The anatomy of the cerebellum', *Trends in Neuroscience*, **21**: 370–375.

Voogd, J., Jaarsma, D. and Marani, E. (1996) 'The cerebellum: chemoarchitecture and anatomy', in L.W. Swanson, A. Björkland and T. Hökfelt (eds) *Handbook of Chemical Neuroanatomy. Integrated Systems of the CNS, Part III: Cerebellum, Basal Ganglia and Olfactory System*, Amsterdam: Elsevier.

Vos, B.P., Maex, R., Volny-Luraghi, A. and Schutter, E.D. (1999a) 'Parallel fibers synchronize spontaneous activity in cerebellar Golgi cells', *Journal of Neuroscience* **19**: 1–5.

Vos, B.P., Volny-Luraghi, A. and Schutter, E.D. (1999b) 'Cerebellar Golgi cells in the rat: receptive fields and timing of responses to facial stimulation', *European Journal of Neuroscience* **11**: 2621–2634.

Wall, M.J. and Usowicz, M.M. (1997) 'Development of action potential-dependent and independent spontaneous $GABA_A$ receptor-mediated currents in granule cells of postnatal rat cerebellum', *European Journal of Neuroscience* **9**: 533–548.

Wang, H., Bedford, F.K., Brandon, N.J., Moss, S.J. and Olsen, R.W. (1999) '$GABA_A$-receptor-associated protein links $GABA_A$ receptors and the cytoskeleton', *Nature* **397**: 69–72.

Watanabe, D., Inokawa, H., Hashimoto, K., Suzuki, N., Kano, M., Shigemoto, R., Hirano, T., Toyama, K., Kaneko, S., Yokoi, M., Moriyoshi, K., Suzuki, M., Kobayashi, K., Nagatsu, T., Kreitman, R.J., Pastan, I. and Nakanishi, S. (1998) 'Ablation of cerebellar Golgi cells disrupts synaptic integration involving GABA inhibition and NMDA receptor activation in motor coordination', *Cell* **95**: 17–27.

Whiting, P.J., McAllister, G., Vasilatis, D., Bonnert, T.P., Heavens, R.P., Smith, D.W., Hewson, L., O'Donnell, R., Rigby, M.R., Sirinathsinghji, D.J.S., Marshall, G., Thompson, S.A. and Wafford, K.A. (1997) ' Neuronally restricted RNA splicing regulates the expression of a novel $GABA_A$ receptor subunit conferring atypical functional properties', *Journal of Neuroscience* **17**: 5027–5037.

Wisden, W., Herb, A., Wieland, H., Kainänen, K., Lüddens, H. and Seeburg, P.H. (1991) 'Cloning, pharmacological characteristics and expression pattern of the rat $GABA_A$ receptor $\alpha 4$ subunit', *FEBS Letters* **289**: 227–230.

Wisden, W., Korpi, E.R. and Bahn, S. (1996) 'The cerebellum: a model system for studying $GABA_A$ receptor diversity', *Neuropharmacology* **35**: 1139–1160.

Wisden, W., Laurie, D.J., Monyer, H. and Seeburg, P.H. (1992) 'The distribution of 13 $GABA_A$ receptor subunit mRNAs in the rat brain. I. Telencephalon, diencephalon, mesencephalon', *Journal of Neuroscience* **12**: 1040–1062.

Chapter 9

$GABA_A$ receptor complex

Structure and function

Richard W. Olsen and Robert L. Macdonald

Introduction

The $GABA_A$ receptor (GABAR) is a heteropentameric membrane glycoprotein with membrane-spanning, extracellular and intracellular regions (Figure 9.1). All subunit members of the ligand-gated ion channels (LGICs) are about 50 kDa and share a common structural model, based on hydropathy plots. The first N-terminal half of the

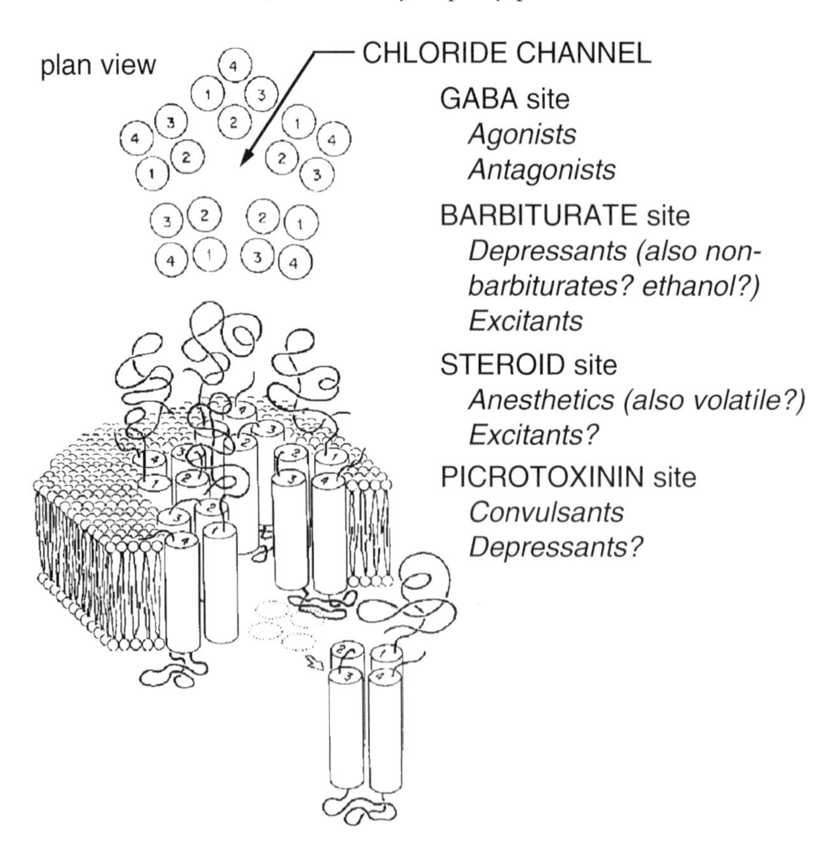

Figure 9.1 Schematic drawing of postulated ligand-gated ion channel receptor structure, including GABAR, with heteropentamer showing membrane topography and various ligand binding sites (modified from Olsen and Tobin, 1990).

polypeptide is a hydrophilic, glycosylated extracellular domain that undoubtedly contains the GABA binding site. Four stretches of hydrophobic residues are thought to fold into membrane-spanning domains, probably α-helices, ending with a short C-terminal extracellular tail. A long intracellular loop is present between the third and fourth membrane-spanning regions M3 and M4. M2 likely forms the wall of the ion channel, and each of the five subunits in the pseudo-symmetric heteropentamer contribute equally to the channel, which resides between them (Figure 9.1: Olsen and Tobin, 1990; Smith and Olsen, 1995; Barnard *et al.*, 1998). The stoichiometry of native receptors is not conclusively known. Most receptors appear to contain at least one copy of an α, a β and a γ, δ or ϵ subunit, although ρ subunits may form homopentamers (Macdonald and Olsen, 1994).

Functional domains within the GABAR protein have been identified for the walls of the ion channel, ligand binding sites, and phosphorylation substrates within the intracellular loop. Considerable homology with other members of the ligand-gated ion channel (LGIC) super-family has afforded models relevant to the structure of GABAR proteins based on observations with these other receptors, including affinity labeling, site-directed mutagenesis, cryo-electron microscopy, and computer modeling.

Native receptors

The nineteen GABAR subunit subtypes (α(1–6), β(1–4 [(β4 aka θ]), γ(1–3), δ, ϵ, π and ρ(1–3) and splice variants can be organized into myriad heteropentameric combinations. Each subtype has its own unique anatomical and developmental expression profile and differences in biological properties, such as channel kinetics, affinity for GABA and endogenous modulators, as well as sensitivity to drugs (Olsen and Tobin, 1990; Macdonald and Olsen, 1994; Burt and Kamatchi, 1994; Sieghart, 1995; McKernan and Whiting, 1996).

The subunit combination of native receptors has been determined primarily by using two approaches. First, one determines which subunit mRNAs and polypeptides are located in a given cell type in a certain brain region and age, and which subunit polypeptides co-purify using subunit subtype-specific antibodies. Second, one attempts to reconstitute the pharmacological and physiological properties of GABARs identified in a given native neuron with recombinant expression of identified subunit cDNAs in heterologous cells. In this way, a developing profile of native receptors has emerged (Table 9.1).

For various reasons, it has proven somewhat difficult to fill in this table for some brain regions, and all native receptor isoforms have not been determined. Nevertheless, several abundant isoforms have been identified (for review, see McKernan and Whiting, 1996) (Table 9.1). The most abundant isoform in the adult rat brain is the α1β2γ2 receptor, which has a widespread distribution in the brain (Benke *et al.*, 1991). α1β2γ2 receptors have Type I benzodiazepine receptor pharmacology and a low sensitivity to inhibition by zinc. Receptors containing α2βγ2, α3βγ2 or α5βγ2 subtypes form Type II benzodiazepine receptors. α2βγ2 receptors are expressed throughout the forebrain, especially in the hippocampus, amygdala and striatum and in the spinal motorneurons (Wisden *et al.*, 1992; McKernan and Whiting, 1996). α3βγ2 receptors are less abundant and have a more restricted distribution to the thalamus and cortex (Wisden *et al.*, 1992; McKernan and Whiting, 1996). Receptors containing α5βγ2

Table 9.1 Identified subunit combinations in native GABARs

GABAR isoform	*Location of high levels of expression*
$\alpha 1\beta 2\gamma 2$[1,2]	In most areas of brain; including cerebellar Purkinje cells
$a2\beta 2/3\gamma 2$[2]	Forebrain, especially hippocampus/striatum; spinal cord motor neurons
$\alpha 3\beta n\gamma 2/\gamma 3$[2]	Primarily in cortex
$\alpha 2\beta n\gamma 1$[2,7]	Bergmann glia
$\alpha 4\beta\delta$[2,6]	Thalamus and hippocampal dentate gyrus
$\alpha 4\beta\gamma 2$[2,6]	Thalamus and hippocampal dentate gyrus
$\alpha 5\beta 3\gamma 2/\gamma 3$[2,3,5]	Hippocampus, cortex and olfactory bulb
$\alpha 6\beta\gamma 2$[2,4]	Cerebellar granule cells
$\alpha 6\beta\delta$[2,4]	Cerebellar granule cells

Notes
1 Benke *et al.* (1991); 2 McKernan and Whiting (1996); 3 Mertens *et al.* (1993); 4 Quirk *et al.* (1994); 5 Sur *et al.* (1998); 6 Sur *et al.* (1999); 7 Wisden *et al.* (1992).

subtypes are almost exclusively expressed in the hippocampus, with minor expression in cortex and olfactory bulb (Wisden *et al.*, 1992; Quirk *et al.*, 1996; Sur *et al.*, 1998) and form GABARs with Type II benzodiazepine receptor sensitivity that are zolpidem-insensitive and that display current rectification (Burgard *et al.*, 1996). Receptors containing α4βγ and α4βδ subtypes are limited primarily to the thalamus, hippocampus, cortex, striatum and olfactory bulb (Benke *et al.*, 1997; Sur *et al.*, 1999) and form benzodiazepine-insensitive GABARs. GABARs containing the α6 subtype are restricted to the cerebellar granule cells. Several subtypes (α2, α3, α5, β3) are even more abundantly expressed in perinatal brain (Laurie *et al.*, 1992; Poulter *et al.*, 1992; Kim *et al.*, 1996), where GABARs might play unique functions in differentiation and establishment of neural circuits (Belhage *et al.*, 1998). GABARs with differing pharmacology and channel properties have been observed in the multiple cell types found in hippocampus (Kapur and Macdonald, 1999; Poisbeau *et al.*, 1999), retina (Pan and Lipton, 1995), cortex (Galarreta and Hestrin, 1997) and cerebellum (Maconochie *et al.*, 1994; Tia *et al.*, 1996a). Variable single channel kinetics were also reported for different regions within a single cell (Banks *et al.*, 1998), possibly representing synaptic and extra-synaptic GABARs (Brickley *et al.*, 1999).

In the cerebellum, the predominant GABAR subunit subtype mRNAs are α1, α6, β2, β3, γ2 and δ mRNAs (Wisden *et al.*, 1992). These mRNAs are also differentially expressed in different cells; α1, β2, β3 and γ2 mRNAs are expressed in Purkinje cells, α1, α6, β2, β3, and γ2 and δ mRNAs are expressed in granule cells and α1, β2 and γ2 mRNAs are expressed in stellate/basket cells. Immunoprecipitation studies have clarified the primary subtype combinations that occur in the cerebellum (Pollard *et al.*, 1993, 1995; Quirk *et al.*, 1994; Kahn *et al.*, 1994, 1996; Jechlinger *et al.*, 1998) (Table 9.1). Initial studies suggested that the primary GABAR isoforms were the α1β2/β3γ2 receptor in Purkinje, stellate/basket and granule cells and the α6β2/β3γ and α6β2/β3δ receptors in granule cells (Quirk *et al.*, 1994). These isoforms were shown to have unique pharmacological and physiological properties (Saxena and Macdonald, 1994; Fisher and Macdonald, 1997a). GABARs containing α1β2/β3γ2 subtypes are benzodiazepine- and zolpidem-sensitive but have low zinc sensitivity, while GABARs containing α6β2/β3γ2 or α6β2/β3δ subtypes are benzodiazepine-insensitive and relatively sensitive to inhibition by zinc. However, several other immunoprecipitation

Table 9.2 Identified subunit combinations containing the α6 subunit in native cerebellar GABARs

GABAR Isoform by α subtype	*% Expression*[1]	*GABAR Isoform by β subtype*	*Expression (%)*[1]
α6βγ2[1,4]	32	α6β1γ/δ[1,4]	10
α1α6βγ2[1,2,3]	37	α6β2γ/δ[1,4]	51
α6βδ[1,4]	14	α6β3γ/δ[1,4]	21
α1α6βδ[1,2,3]	15	α6β1,2or3γ/δ[1,4]	18

Notes
1 Jechlinger *et al.* (1998); 2 Kahn *et al.* (1994, 1996); 3 Pollard *et al.* (1993, 1995); 4 Quirk *et al.* (1994).

studies suggested the presence of GABAR isoforms containing both α1 and α6 subtypes (Pollard *et al.*, 1993, 1995; Kahn *et al.*, 1994, 1996). Furthermore, while studies using immunogold localization demonstrated synapses immunopositive for only one of the α subtypes, the α1 and α6 subtypes were co-localized in many Golgi synapses (Nusser *et al.*, 1996). Coassembly of α1 and α6 subtypes has been confirmed recently (Jechlinger *et al.*, 1998) (Table 9.2), and the predominant α6 subtype-containing isoforms were shown to be the α6βγ2 (32 percent) and α1α6βγ2 (37 percent) isoforms with substantial contributions from α6βδ (14 percent) and α1α6βδ (15 percent) isoforms. In addition, the predominant β subtype coexpressed with the α6 subtype was the β2 subtype (51 percent) with smaller contributions from the β1 (10 percent) and β3 (21 percent) subtypes. In addition, receptors containing two different β subtypes were also present (18 percent). It has also been shown the GABARs containing more than one β subtype have pharmacological properties that differ from receptors containing only one β subtype (Fisher and Macdonald, 1997b).

Knock-out and mutant mice have provided many insights into functions of gene products, including GABARs. Deducing the subunit composition of native GABAR isoforms can be made easier by examining animals lacking certain subunit subtypes – e.g. the α6 subtype (Makela *et al.*, 1997). Mice lacking the α6 subtype show reduced δ subunit expression in the cerebellum, indicating an obligatory partnering of δ with α6 subtypes in the granule cells, although not in other brain regions (Jones *et al.*, 1997). GABAR pharmacology and the subunits expressed in Purkinje cells and the molecular layer were unaltered. However, granule cells showed an overall 50 percent loss in GABAR protein, including 50 percent, 20 percent, and 40 percent reduction in β2, β3, and β1 subtype protein, respectively (Nusser *et al.*, 1999). Mice lacking the α5 subtype show reduced β3 subtype expression (Fritschy *et al.*, 1997) and vice versa (Homanics *et al.*, 1997). Lack of the α5 subtype also leads to reduction of synaptically clustered γ2 subtypes on dendrites in the same cells, but not α2 subtypes in synaptic clusters on axon membranes (Fritschy *et al.*, 1998), indicating that expression, targeting, and synaptic clustering of the different α subtypes are separately regulated.

The differences in properties between subunits can be used to identify the amino acid residues (functional domains) responsible for the functional difference, e.g. benzodiazepine agonist binding associated with the α1, α2, α3, and α5 subunit subtypes can be shown to require the residue H101, missing (replaced with R) in the α4 and α6 subtypes, which does not bind these ligands. Further sections will describe what is currently known about the domains involved in the various GABAR functions.

Functional domains for assembly, targeting, and trafficking

The stoichiometry, as well as nearest-neighbor wheel arrangement of pentamers in native GABARs, has been difficult to determine. Further, cells expressing multiple subunits produce isoforms composed of only certain subunit combinations, and the rules for assembly of GABARs are not understood. Information on subunit-subunit interactions is coded in the N-terminal extracellular domain of ligand-gated ion channels, including nicotinic acetylcholine receptors (review, Hall and Sanes, 1993), glycine receptors (Kuhse *et al.*, 1993), and GABARs (Hackam *et al.*, 1997). A tentative conclusion obtained from recombinant expression studies is that a given complement of subunits will assemble only one type of pentamer. This is based on single-channel kinetics, e.g. few channels with properties of αβ are produced when α, β, and γ or δ subunits are simultaneously transfected in mammalian cells (Angelotti and Macdonald, 1993; Saxena and Macdonald, 1994; Fisher and Macdonald, 1998). Second, expression of α, β, and γ subunits in Sf9 cells yields only one type of receptor based on ligand binding. Both αβ and αβγ subunits form a single oligomeric complex, shown to be pentamers using sucrose gradient centrifugation with correction for detergent binding (Hartnett *et al.*, 1996; Knight *et al.*, 1998; Elster *et al.*, 2000). Thus it seems that even if some subunit subtypes, e.g. the β3 subtype, can form homomeric channels in expression systems when other residues are present, a pentamer containing α, β, and γ (or δ or ε) subunits will be produced (Connolly *et al.*, 1996a, 1999; Gorrie *et al.*, 1997). An exception is the ρ subunit, which apparently makes homomer pentamers in retina (Barnard *et al.*, 1998; Filippova *et al.*, 1999). Residues in the β3 subtype that are needed for homomeric assembly have been identified in the extracellular domain, at G171, K173, E179, R180, between the invariant disulfide loop and TM1 (Taylor *et al.*, 1999).

Subunit-specific antibodies have been used with recombinant receptors in HEK 293 cells to show that the apparent subunit stoichiometry in a pentamer is 2α, 2β, 1γ, with spatial arrangement -αβαβγ- or -βαβαγ-; αβ pentamers were suggested to be 3β2α (Tretter *et al.*, 1997). The stoichiometry of 2α2β1γ is in agreement with conclusions based on an alternative approach using electrophysiology to examine stoichiometry based on engineered mutations in individual subunit channel domains (Chang *et al.*, 1996). Others using this latter approach reached a different conclusion, suggesting 2α1β2γ (Backus *et al.*, 1993). Antibodies and recombinant GABAR in oocytes with α and β subunits (no γ subunit) were employed to measure the ratio of α/β subunits: the value of 1.1 suggested a tetramer 2α2β, or possibly a 3α2β pentamer (Kellenberger *et al.*, 1996). Tandem concatamer constructs of the C-terminus of the α6 subtype attached via a short spacer in frame to the N-terminus of the β2 subtype (Im *et al.*, 1995) were able to form functional GABARs when expressed with either monomeric α6 or γ2 but not β2 subtypes. This is consistent with (a) an even number of membrane-spanning regions per subunit, and (b) the presence of no more than two copies of β subunit per pentamer. Studies on oocytes (M. Gordey and R.W. Olsen, unpublished) suggest that the stoichiometry can be affected by the expression conditions, thus a possible confounding variable in heterologous cell over-expression studies.

Subunits that do not produce surface channels may be trapped in the endoplasmic reticulum or may be degraded (Tehrani and Barnes, 1991, 1993). The γ2 subtype, and perhaps other subunits, is also needed for GABAR interaction with cellular proteins targeting receptors to cellular compartments, including endocytosis and either recycling or degradation, and this is modulated by PKC (Connolly *et al.*, 1999).

The subcellular localization of neurotransmitter receptors is critical for synaptic functioning and has developmental and plasticity implications. Using heteromeric GABARs composed of α and β, or α, β, and γ subunits expressed in the polar MD canine kidney cells, receptors containing the β1 subtype did not polarize, whereas those incorporating the β2 and β3 subtypes were localized to the basolateral domains of the cells. The β3 subtype-containing GABARs were targeted to the basolateral domain via the apical surface, suggesting transcytosis (Connolly *et al.*, 1996b). Interactions of synaptic receptors with cellular proteins, especially the cytoskeleton, have recently been "uncovered" and generated much excitement. Linker proteins have been found that cluster, transport, and target receptor proteins and ion channels (Colledge and Froehner, 1998). Distinct clusters of different subtypes of GABARs or glutamate receptors (Koulen *et al.*, 1996; Allison *et al.*, 1998), or different locations of a single subtype, e.g. dendritic spines versus shafts, require a mechanism to selectively "trap" or immobilize each receptor subtype in its own synaptic cluster and/or to send newly synthesized clusters of a given subtype to a specific membrane target area. In a given cell, the density of GABARs in synapses is constant, but the size of the synaptic area may vary; different cells may vary in both their average synaptic area and in the density of receptors in a synapse (Somogyi *et al.*, 1996; Nusser *et al.*, 1995, 1998). Plasticity of GABAergic synapses may involve changes in the synapse size, composition and density of GABARs, e.g. changes in subunit composition and numbers of receptor proteins have been described in animal models of epilepsy (Brooks-Kayal *et al.*, 1998; Nusser *et al.*, 1998).

GABAR clustering in the subsynaptic membrane is likely to involve cytoskeleton attachment (e.g., Sieghart, 1995). Interestingly, disruption of GABAR clusters can be achieved with microtubule-disrupting drugs in expression cells, and this also reduces GABAR function (Whatley *et al.*, 1994). Actin filament disruption does not perturb clusters of GABARs in brain sections, as it does for certain glutamate receptors (Allison *et al.*, 1998). Cytoplasmic proteins have been found to be associated with other ligand-gated ion channels. The 43K-rapsyn protein links the nicotinic acetylcholine receptor to actin and synaptic proteins (Gautam *et al.*, 1995; Gillespie *et al.*, 1996), and the 93K-gephyrin protein links the glycine receptor/chloride channel to microtubules (Kirsch and Betz, 1995). Rapsyn and gephyrin, or members of a family of such proteins, may also bind to GABAR, and these proteins might have functional effects, acutely, or long term. Rapsyn, co-expressed with GABAR in QT6 cells, led to increased GABAR clustering (Yang *et al.*, 1997). Functional domains for rapsyn/nACh (Ramarao and Cohen, 1998) have been identified.

Co-localization of gephyrin with certain GABAR subunits has been reported (Sassoe-Pognetto *et al.*, 1995). More definitively, gephyrin and GABAR synaptic localization in cultured neurons is disrupted in cells from both gephyrin knock-out mice (Kneussel *et al.*, 1999) and the GABAR γ2 subtype knock-out mouse, strongly suggesting interaction of the two proteins (Essrich *et al.*, 1998). Heterozygotes of the γ2 subtype knock-out mouse have reduced GABAR and gephyrin clustering, and exhibit a hyperanxious phenotype (Crestani *et al.*, 1999). Gephyrin does not bind GABARs *in vitro*, although the intracellular loop of GABAR β, but not γ, subunits have a weak homology with the glycine receptor β subunit intracellular domain that binds gephyrin

(Meyer *et al.*, 1995). Finally, the GABAR ρ subunit has been shown to interact with microtubule-associated protein MAP1B (Hanley *et al.*, 1999).

A recently described protein called GABARAP links GABAR γ subunits to microtubules (Wang *et al.*, 1999) and clusters GABARs (Chen *et al.*, 1999). This new protein is homologous to light chain 3 of microtubule associated protein MAP 1A. GABARAP contains domains for interaction with tubulin (residues 1–35), with GABAR γ subunits (41–52), with gephyrin, and with itself. The domain for binding GABARAP in the intracellular loop of GABAR γ2 subtypes is at residues 394–411. It has been proposed that the interaction of GABARAP with GABAR and microtubules, as well as self-interactions of both GABARAP and the γ2 subtype, provides a mechanism for GABAR clustering at synapses (Figure 9.2). Such protein aggregates may also include signaling proteins such as kinases and phosphatases.

Preliminary results show that GABARAP co-localizes with GABARs in recombinant expression cells, in primary cultured neurons, and in brain sections. It is found inside the cell with the cytoskeleton and at postsynaptic membranes with GABAR. It appears to be present in synaptic clusters on some cell bodies containing some, but not all, γ2 subtype-containing GABAR. By linking GABAR to microtubules, GABARAP may function in GABAR trafficking and/or endocytosis, rather than, or in addition to, anchoring (H. Wang and R.W. Olsen, unpublished). GABARAP is widely expressed in the body and undoubtedly has functions other than GABAR clustering, possibly related to its microtubule association. Gephyrin is also widely expressed, and even has a separate function as part of a molybdoenzyme cofactor activity (Feng *et al.*, 1998). Exciting new results regarding GABAR synaptic structure and plasticity are on the horizon.

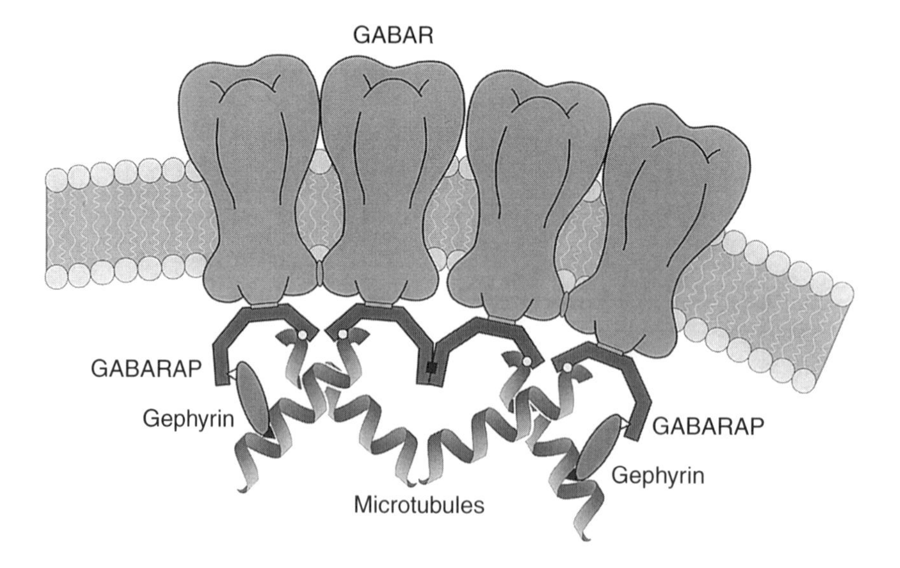

Figure 9.2 Cartoon showing that GABARAP clusters GABAR with six types of protein–protein interaction: GABARAP–GABAR; GABARAP–tubulin; GABARAP–GABARAP; GABARAP–gephyrin; gephyrin–tubulin; and GABAR–GABAR (not to scale).

Phosphorylation substrates and functional correlates

Protein phosphorylation might regulate GABAR function rapidly at the channel level, or slowly at the cell biology level (assembly, targeting, transport, clustering, endocytosis and degradation). Phosphorylation effects are also consistent with "run-down" phenomena, in which whole cell recording, which dialyzes the cellular contents, leads to a gradual reduction over minutes of GABAR responses. Run-down can often be prevented by including ATP in the recording pipette (Gyenes *et al.*, 1988; Chen *et al.*, 1990). This suggests that GABAR function is maintained by phosphorylation of some [unknown] protein, by an unknown kinase, and reduced by calcium-dependent phosphatases like calcineurin (Chen *et al.*, 1990; Jones and Westbrook, 1997), proposed possibly to account also for zinc and calcium modulation of GABAR function (e.g., De Koninck and Mody, 1996). Endogenous kinase activity has also been associated with purified GABARs (Bureau and Laschet, 1995).

In vitro studies suggest that GABARs are functionally modulated by phosphorylation with cyclic AMP-dependent protein kinase (PKA), protein kinase C (PKC), tyrosine kinase, and perhaps CAM kinase II (CamKII) and cyclic GMP-dependent protein kinase (PKG). GABARs are phosphorylated by PKA (Tehrani and Barnes, 1994) on the β subunits (Browning *et al.*, 1990, 1993). Elevation of cyclic AMP decreased GABA-activated chloride ion flux in rat cortical synaptoneurosomes (Heuschneider and Schwartz, 1989) and cultured chick cortical neurons (Tehrani *et al.*, 1989). Injection of the catalytic subunit of PKA (cPKA) into cultured mouse spinal cord neurons (Porter *et al.*, 1990) reduced GABAR current by decreasing channel opening frequency. CPKA also reduced GABAR chloride ion flux into lysed and resealed rat synaptoneurosomes and increased phosphorylation of a polypeptide that was immunoprecipitated with antibodies specific for GABARs (Leidenheimer *et al.*, 1991). These results were confirmed in recombinant α1β1 and α1β1γ2 receptor channels in HEK293 cells where cAMP reduced GABAR current; the effect of cAMP was prevented by site-directed mutagenesis of β1 subtype residue Ser-409, suggesting that phosphorylation of this residue was responsible for the reduction of GABAR current (Moss *et al.*, 1992b). Intracellular cPKA decreased mIPSC amplitude in rat hippocampal CA1 pyramidal cells (Poisbeau *et al.*, 1999). Since GABARs do not have a PKA function, regulation of GABAR currents by PKA must be heterologous; that is, PKA must be regulated by activation of another transmitter system to modify PKA activity and thus to indirectly alter GABAR function by phosphorylation. Such heterologous regulation has been shown in rat retinal amacrine cells (Feigenspan and Bormann, 1994) and in rat olfactory bulb granule cells (Brunig *et al.*, 1999), where a dopamine D1 agonist increased GABAR currents via a PKA pathway. In the amacrine cells the effect of PKA involved G_s, and the enhancement was due to an increase in GABA affinity and an increase in channel open probability.

In contrast, other studies have suggested that PKA acutely increased GABAR current. In cerebellar Purkinje cells (Cheun and Yeh, 1992, 1996) and hippocampal dentate granule cells (Kapur and Macdonald, 1996), cAMP increased GABAR current, mimicking the physiological effect of norepinephrine, known to be mediated by β-adrenergic receptors coupled to PKA. Expression of α1β1γ2 subtype combinations in mammalian cells stably transfected with cPKA cDNA resulted in enhanced GABAR currents compared to expression of α1β1γ2 subtype combinations in the parent L929

Table 9.3 Consensus sequence sites for GABAR phosphorylation

	PKA	*PKC*	*PKG*	*CamII K*
$\alpha 1$	X	x	x	X
$\alpha 2$	X	x	x	x
$\alpha 3$	X	x	S365 (KKIS*)	x
$\alpha 4$	X	S327 (KKKIS*K) S443 (RSAS*AR)	S327 (KKKIS*K)	S423 (RGLS*) S443 (RSAS*)
$\alpha 5$	X	S379 (RAS*EEK) S384 (KTS*ESK)	x	x
$\alpha 6$	x	S331 (KS*K) S373 (KVLS*R)	S356 (KKRIS*)	S357 (RISS*)
$\beta 1$	S409 (RRAS*)[1]	S409 (RAS*QLK)[1,2] S317 (KGAS*K) S384 (KPLS*SR)	S409 (RRRAS*)[3]	S409 (RRAS*)[3] S384 (KPLS*)[3]
$\beta 2$	S410 (RRAS*)	S410 (RAS*QLK) S388 (RHS*FGR) S403 (KKS*R)	S410 (RRRAS*)	S410 (RRAS*)
$\beta 3$	S407 (RRRS*S) S408 (RRRSS*) S381 (RKQS*)	S408 (RSS*QLK) S329 (RS*K) S381 (KQS*MPK)	S407 (RRRS*S) S408 (RRRSS*) S381 (RKQS*)	S407 (RRRS*) S408 (RRSS*) S381 (RKQS*)
$\gamma 1$	x	S338 (KLKS*K) S391 (RTGS*WR)	S327 (RKPS*)	S391 (RTGS*) S327 (RKPS*)
$\gamma 2S$	x	S327 (RKPS*K)[1,2]	x	S348 (RPRS*)[3] T350 (RSAT*)[3]
$\gamma 2L$	x	S327 (RKPS*K)[1,2] S343 (RMFS*FK)[1,3]	x	S343 (RMFS*) S355 (RPRS*)
$\gamma 3$	x	S411 (KSGS*WR)	S336 (KKKTS*)	
δ	x	T332 (KVT*K) S398 (RS*RLK)	x	x

Confirmed phosphorylation sites.

Notes
1 Moss *et al.* (1992a); 2 Krishek *et al.* (1994); 3 McDonald and Moss (1994).

cell line (Angelotti *et al.*, 1993). Enhancement of GABAR current was abolished by mutation of the β subunit PKA phosphorylation site. Recently, PKA has been reported to either enhance or reduce GABAR currents depending on the β subtype. cAMP was reported to enhance $\alpha 1\beta 3\gamma 2S$ currents, not alter $\alpha 1\beta 2\gamma 2S$ currents, and to reduce $\alpha 1\beta 1\gamma 2S$ currents (McDonald *et al.*, 1998). $\beta 1S409A$ or $\beta 3S408A$ mutations abolished the cAMP effects, but a $\beta 3S407A$ mutation converted the cAMP effect from enhancement to inhibition (see Table 9.3).

The functional consequences of tyrosine phosphorylation remain unclear (Valenzuela *et al.*, 1995; Moss *et al.*, 1995; Wan *et al.*, 1997; Huang and Dillon, 1998) but might involve heteropentamer assembly, membrane insertion, subcellular targeting, clustering, or internalization and degradation, possibly including run-down phenomena (e.g. Connolly *et al.*, 1996a, 1996b). The possible stimulation of surface expression by GABAR tyrosine phosphorylation such as that of insulin receptor activation (Wan *et al.*, 1997) is of interest.

PKC phosphorylates GABAR β and $\gamma 2$ subtypes (Browning *et al.*, 1990; Moss *et al.*,

1992a). In oocytes injected with brain or GABAR subunit mRNA, phorbol esters reduced GABAR function via activation of PKC (Sigel and Baur, 1988; Sigel *et al.*, 1991). Site-directed mutagenesis of the β2 subtype S410 and the γ2S subtype S327 demonstrated that phosphorylation of these GABAR residues by phorbol-stimulated kinase was responsible for the inhibition of function (Kellenberger *et al.*, 1992). Phorbol esters also reduced recombinant α1β1 and α1β1γ2S/L currents expressed in oocytes and HEK293 cells and GABAR currents recorded from sympathetic neurons (Krishek *et al.*, 1994). In HEK293 cells no effect of intracellular phorbol esters was obtained without coexpression of SRα-PKCα cDNA. Phorbol esters produced current rundown over 10–50 min. This effect of the phorbol esters was abolished in each of the recombinant receptors when all of the PKC phosphorylation sites were mutated to alanine. The phorbol ester effects on recombinant GABAR currents suggested that they reduce the amplitude of GABAR currents and may reduce the proportion of fast desensitization of α1β1 receptor currents. Phorbol ester inhibition of GABAR function expressed in oocytes also could be shown in brain microsacs, where PKC activation appeared to inhibit selectively the fraction of GABAR flux which was not rapidly desensitized by prolonged (several sec) exposure to agonist (Leidenheimer *et al.*, 1993). Interestingly, it has been recently demonstrated that PKC reduced α1β2γ2L GABAR current in *Xenopus* oocytes (Chappell *et al.*, 1998) and homomeric ρ1 receptors expressed in HEK 293 or COS-7 cells (Filippova *et al.*, 1999) by internalizing the receptors, and that the internalization did not require phosphorylation of any of the known γ, β or ρ subunit phosphorylation sites. The internalization of α1β2γ2L receptors by activation of PKC reduced GABAR current by 80 percent and was maximal by 45 min. Internalization of ρ1 homomeric $GABA_C$ receptors was reversible, recovering over 30–40 min. The inhibitor of actin filament formation cytochalasin B also produced a time-dependent reduction of ρ1 current when added to the cytoplasm, suggesting that the actin cytoskeleton plays a role in phosphorylation-mediated downregulation of ρ1 receptor current. PKC activation was also shown to stimulate internalization of GABARs in primary cultured neurons (Johnston *et al.*, 1998).

In contrast, constitutively active PKC (PKM) has been shown to enhance recombinant α1β1γ2L GABAR currents expressed in L929 cells (Lin *et al.*, 1994). The enhancement was due to an increase in the maximal current evoked by high GABA concentrations and not by an increase in GABA affinity. The effect of PKM was blocked by mutation of β1(S409) and γ2L(S327, S343) subtypes (Lin *et al.*, 1996). Dopamine also enhanced GABAR current in rat olfactory bulb mitral/tufted cells by activating PKC (Brunig *et al.*, 1999) and increased mIPSC amplitude in rat hippocampal dentate granule cells but not CA1 pyramidal cells (Poisbeau *et al.*, 1999).

Additional effects of PKC on GABAR function are abundant, e.g. in the sensitivity to ethanol potentiation of function (Harris *et al.*, 1995; Hodge *et al.*, 1999). The functional relevance of PKC on GABAR function *in vivo* is still in question, but will probably be sorted out in the near future. A protein called Receptor for Activated C Kinase (RACK1) has been described (Ron *et al.*, 1994) that links soluble PKC-β to membrane-bound substrate proteins (Mochly-Rosen and Gordon, 1998). RACK1 interacts with the intracellular loop of α1 subtype, and PKC and RACK1 bind to α1 and β1 subtypes *in vitro*, and PKC phosphorylates β subunits (Brandon *et al.*, 1999).

Functional domains involved in the ion channel

The GABAR conduction pathway

The GABAR is a highly anion-selective channel. When GABA is applied to GABARs, it evokes bursts of single channel chloride currents composed of relatively uniform, square current pulses (Figure 9.4). GABAR single channel conductances are influenced by both subunit and subunit subtype. αβ heterodimeric pentamers have relatively small conductance levels, ranging from 11 to 15 pS (Verdoorn *et al.*, 1990; Moss *et al.*, 1990; Angelotti and Macdonald, 1993; Fisher and Macdonald, 1997a), whereas heterotrimeric pentamers composed of αβε, αβπ, αβγ, and αβδ subunits have larger conductance levels ranging from 24 to 27 pS (Fisher and Macdonald, 1997a; Neelands *et al.*, 1999; Neelands and Macdonald, 1999). Based on the permeability sequence for large polyatomic anions, it has been estimated that the GABAR channel main-conductance level has an effective pore diameter of 5.6 nm (Bormann *et al.*, 1987). As with nicotinic cholinergic receptor channels (Unwin, 1995; Miyazawa *et al.*, 1999), GABAR channels (Nayeem *et al.*, 1994) are presumed to be heteropentamers arranged pseudosymmetrically around a central channel, and the M2 membrane spanning segment likely lines the anion channel. Picrotoxinin is a universal GABAR channel blocker, and residues in the M2 region affect picrotoxin sensitivity (Xu *et al.*, 1995; Gurley *et al.*, 1995; Wang *et al.*, 1995a; Zhang *et al.*, 1995), as first shown in glycine receptor subunits (Pribilla *et al.*, 1992). In fact, genetic resistance to the insecticide dieldrin, a picrotoxin-like GABAR channel blocker, was used to clone the first insect GABAR, identified on the basis of a mutation in the M2 domain (ffrench-Constant *et al.*, 1993). Several structural features are conserved in cation and anion channels, including cytoplasmic rings of negatively and positively charged amino acids, two polar rings of threonines or serines and a hydrophobic ring of leucines (e.g., Galzi and Changeux, 1994).

The amino acids in the M2 segment that line the pore have been inferred using the cysteine-scanning technique (Xu and Akabas, 1993, 1996). The rat α1 subtype M2 segment is composed of 26 amino acids that span the membrane from the cytoplasmic side (E250) to the extracellular side (N275). When the α1 subtype was coexpressed the β1 and γ2 subtypes, nine α1 subtype M2 residues (V257, T261, T262, L264, T265, T268, I271, S272 and N275) were inferred to be exposed to the aqueous portion of the channel. On a helical wheel plot, all of the exposed residues were placed on the same side of the wheel, except T262, suggesting that the M2 segment forms an α helix that is interrupted in the region of T262 (Figure 9.3).

The selectivity of ligand-gated channels for anions or cations does not appear to involve the conserved charge rings. While the M2 segment of cation and anion channels share many structural features (Figure 9.3), the N-terminal ends of anion channel M2 segments contain an additional amino acid in the short segment linking M1 to M2. In the GABAR there is an additional proline (P253) in the α1 subtype and an additional alanine (A248) in the β2 subtype. Substitution of three amino acids from the GABAR M2 segment into the M2 segment of the α7 neuronal nicotinic receptor converted the cation channel into an anion channel (Galzi *et al.*, 1992). The residues included introduction into the α7 M2 segment of the additional P or A at position 236 and two amino acids at positions 237 (E to A) and 251 (V to T). Thus, protein geometrical constraints in M2, rather than charge, appear to determine the polarity of ion selectivity.

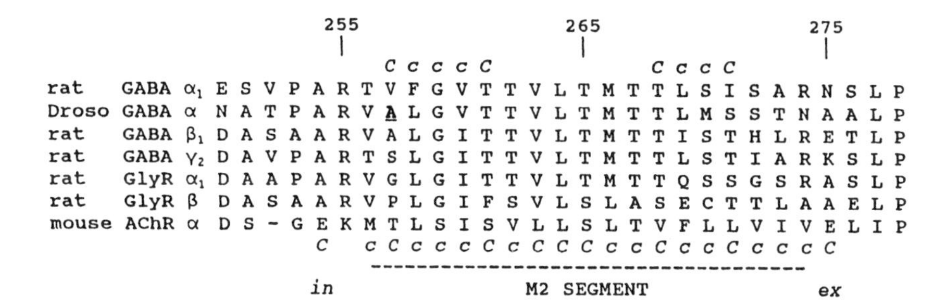

Figure 9.3 The aligned sequences of amino acid residues in and flanking the M2 membrane-spanning segment of the rat GABAR α_1, β_1, and γ_2 subunits, the Drosophila GABAR subunit, the rat glycine receptor α_1 and β subunits, and the mouse muscle acetylcholine receptor α subunit (from Xu et al., 1995).

The upper and lower case italic letters *C* and *c* above and below the aligned sequences indicate residues substituted by cysteine in the GABA$_A$ (Xu and Akabas, 1993; Xu *et al.*, 1995) and acetylcholine receptors (Akabas *et al.*, 1994). Residues exposed in the channel are indicated by an upper case *C*; residues not exposed in the channel are indicated by a lower case *c*. The M2 segment is underlined. "*In*" indicates the intracellular end and "*ex*" indicates the extracellular end of M2. The numbers at the top indicate the residue number of the α_1 GABAR subunit. In *Drosophila* strains resistant to picrotoxin and cyclodiene insecticides, the alanine in bold is mutated to serine (ffrench-Constant *et al.*, 1993).

Portions of the M2 domain also appear to be critical for allosteric modulation of GABARs by several compounds (e.g., Thompson *et al.*, 1999b), including zinc, loreclezole, etomidate, DMCM/benzodiazepines and ethanol.

The GABAR gate

GABA increases the probability of channel opening, and after the channel opens it can close and rapidly reopen to create bursts of openings (Sakmann *et al.*, 1983; Bormann *et al.*, 1987; Macdonald *et al.*, 1989; Twyman *et al.*, 1990) (Figure 9.4). The GABAR channel has been shown to open into three different open states with mean durations of about 0.5 ms, 2.5 ms and 8 ms. The gating pattern of GABARs is dependent on subunit composition. For example, recombinant $\alpha\beta$, $\alpha\beta\gamma$ and $\alpha\beta\delta$ GABAR isoforms exhibited distinct single channel kinetic properties (Figure 9.4) (Angelotti *et al.*, 1993; Fisher and Macdonald, 1997a). Recombinant $\alpha 1\beta 1\gamma 2$ receptor single channel currents have multiple open and closed states, while $\alpha 1\beta 1\delta$ receptor single channel currents are briefer, and the channel does not open in complex bursts (Angelotti and Macdonald, 1993; Fisher and Macdonald, 1997a; Haas and Macdonald, 1999) (Figure 9.4). The location of the channel gate and the bases for these differences in gating among subunits are not clear. The M2 domain of the $\gamma 2$ and δ subunits only differs by four amino acids, and interchanging the M2 domains between $\gamma 2$ and δ subunits did not interchange gating patterns (M.T. Bianchi, K. Haas and R.L. Macdonald, unpublished). Studies on nicotinic acetylcholine receptors suggested that a conserved leucine residue at position 9′ in the middle of M2 is the major barrier to ion flow (Unwin, 1995), and changes in this residue affect GABAR channel opening (Tierney *et al.*, 1996; Pan *et al.*, 1997; Chang and Weiss, 1998) and sensitivity to block by picrotoxin (Xu *et al.*, 1995). However,

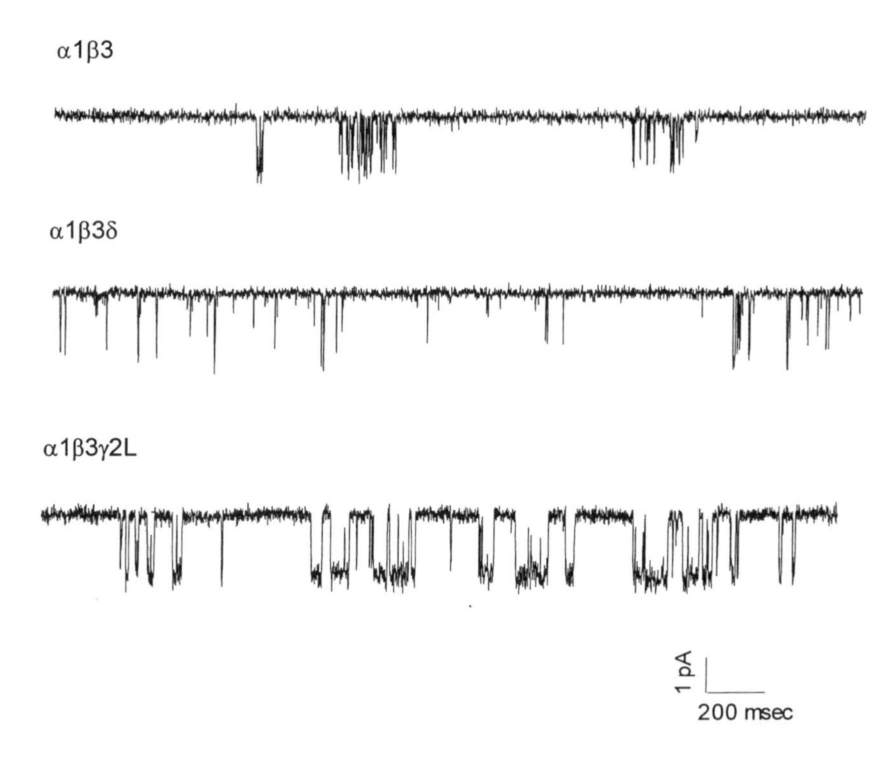

Figure 9.4 Recombinant $GABA_A$ receptor single channel currents. Single channel GABAR currents were obtained from outside-out patches. Traces shown were continuous 2 sec recordings from patches in response to 3 μM (α1β3, α1β3δ) or 10 μM GABA (α1β3γ2L). Channel openings are downward. Currents were obtained from patches held at −70 mV for α1β3δ and α1β3γ2L receptors, and at −100 mV for α1β3 receptors (from Fisher and Macdonald, 1997a).

with the cysteine scanning technique, residues as cytoplasmic as α1V257 at position 2′ were modified in the absence of applied GABA, suggesting that the narrowest portion of the channel (the channel gate?) is at least as cytoplasmic in location as V257 (Xu and Akabas, 1996).

GABAR desensitization

Receptor desensitization occurs when agonist-evoked current declines in the presence of continued application of agonist. Using rapid application of GABA to outside-out patches from acutely dissociated hippocampal neurons, Celentano and Wong (1994) demonstrated rapid activation and at least three phases of desensitization for GABAR channels. Multiple desensitization states of GABAR can even be observed by biochemical assays using rapid kinetic measurements (Cash *et al.*, 1997). Jones and Westbrook (1995) demonstrated that GABAR channels entered long closed states and subsequently reopened, suggesting the presence of desensitized states that result in prolonging currents evoked by brief applications of GABA as seen at synapses. Similar results have been obtained using recombinant α5β3γ2L GABARs (Burgard *et al.*, 1999). Rapid desensitization has been shown to be important in shaping the time course of ligand-gated ion channel currents (Jones and Westbrook, 1996). Multiple studies of native and recombinant GABARs have correlated rapid desensitization with prolonged deactivation of macropatch currents (Jones and Westbrook, 1995; Tia *et al.*, 1996b; Mellor and Randall, 1997; Zhu and Vicini, 1997), suggesting that desensitization prolongs the time course of IPSCs (Jones and Westbrook, 1995.). Desensitization rates also correlate with the fast phase of current deactivation (Jones and Westbrook, 1995; Galarreta and Hestrin, 1997).

GABAR subunit composition influences current desensitization rates (Verdoorn *et al.*, 1990; Saxena and Macdonald, 1994; Verdoorn, 1994; Gingrich *et al.*, 1995; Fisher and Macdonald, 1997a; Dominguez-Perrot *et al.*, 1997), but these studies have reported desensitization rates in the order of hundreds of milliseconds to several seconds, and the slow GABA application rates most likely obscured the most rapid phase of desensitization. Using rapid application techniques, $\alpha1\beta2\gamma2L$ currents desensitized more rapidly than $\alpha6\beta2\gamma2L$ currents in HEK293 cells (Tia *et al.*, 1996b), and $\alpha1\beta_x$ and $\alpha1\beta_x\gamma2L$ currents desensitized more rapidly and completely than $\alpha1\beta_x\delta$ currents (Figure 9.5) (Haas and Macdonald, 1999). While the desensitization time courses were similar for $\alpha1\beta3$ and $\alpha1\beta3\gamma2L$ currents, there were subtle differences. Thus, addition of the δ subunit to $\alpha1$ and $\beta3$ subtypes significantly reduced both the rate and the extent of desensitization, while addition of the $\gamma2L$ subtype to $\alpha1$ and $\beta3$ subtypes changed the pattern of desensitization.

The structural basis for desensitization of anion channel currents is unclear. In the glycine receptor, desensitization is influenced by amino acids at the distal end of M1, leading to the hypothesis that this region and the M1–M2 connector served as a hinge involved in the gating of the channel (Lynch *et al.*, 1997). Mutation of residues at M2

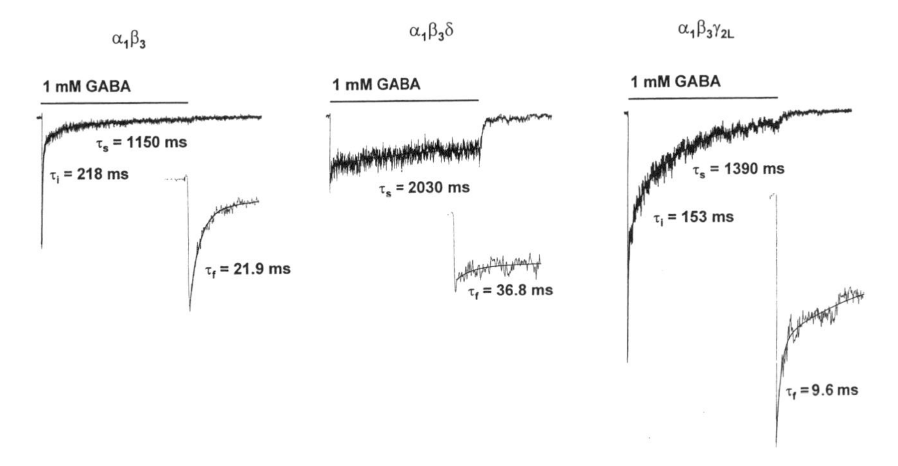

Figure 9.5 Rapid phases of GABAR current desensitization. GABA (1 mM) was rapidly applied for 4000 ms to outside-out membrane patches containing $\alpha1\beta3$, $\alpha1\beta3\delta$, and $\alpha1\beta3\gamma2L$ isoforms. For the representative traces shown, current desensitization was fit with multicomponent exponential equations with time constants of 21.9 ms (inset), 218 ms, and 1150 ms for the $\alpha1\beta3$ current, 36.8 ms (inset) and 2030 ms for $\alpha1\beta3\delta$ current (5), and 9.6 ms (inset), 153 ms, and 1390 ms for $\alpha1\beta3\gamma2L$ current. Time calibrations for *traces* and insets are 1000 ms and 150 ms, respectively (from Haas and Macdonald, 1999).

positions 5′, 9′, and 12′ have been shown to influence macroscopic desensitization rates of recombinant α1β1 GABAR currents (Tierney *et al.*, 1996), but since the amino acids at these positions are conserved between δ and γ2L subunits, they are unlikely to determine their different rates of desensitization. To explore the structural basis of GABAR desensitization, a chimera strategy was employed, focusing upon the different desensitization rates of GABARs containing γ2L and δ subunits. Using GABAR subunit chimeras that were constructed with N-terminal δ sequence spliced to γ2L sequence at various points within the first two transmembrane domains and using rapid agonist application, desensitization was demonstrated to be dependent on the structure of M1 (Bianchi *et al.*, 1999). Mutation in the M1 region of the β2 subunit also altered desensitization (Carlson *et al.*, 2000).

Functional domains for agonist and allosteric drug actions

The GABAR is a unique allosteric protein in that its function, as a membrane chloride ion channel, is regulated by binding of an allosteric ligand, the neurotransmitter, GABA. In addition, it is the target of myriad neuroactive drugs which act as allosteric modulators directly on the GABAR protein. Exactly how the drug modulatory sites affect agonist-regulated channel opening is a major contemporary question in pharmacology and neurobiology. In addition to the binding of allosteric modulators like benzodiazepines, a conformational change must be triggered that involves other domains in the protein, potentially modifying the effect of agonist on channel gating. In the continuing absence of X-ray crystallography information on the structure of these multi-subunit membrane receptor proteins, models will be based on information from various approaches, and these can be applied to any one of the LGIC receptors. For example, models for the glycine receptor strychnine binding site (Gready *et al.*, 1997) may assist in understanding the action of modulators like steroids, barbiturates, alcohols and volatile anesthetics at other domains of the protein related to channel activation.

Functional domains for GABA binding

Amino acid residues likely to participate in GABA binding (Figure 9.6) were identified by photoaffinity labeling with [^{3}H]muscimol, fragmentation and microsequencing (Smith and Olsen, 1994). One of two major-labeled fragments carried covalent label on an amino acid identified as F65 in the bovine α1 subtype (F64 in the rat α1 subtype, also present in α2, α3, and α5 subtypes, but not β subunits). This residue was also shown by mutagenesis to affect GABA site ligand affinities (Sigel *et al.*, 1992). The other major affinity-labeled fragment might be on a β subunit. Another residue affecting GABA site affinities identified in α subunits was I121 (Westh-Hansen *et al.*, 1997); this group also showed that the adjacent residue R120 was important (M. Nielsen, personal communication). The residues within the invariant disulfide loop found in the extracellular domain of all LGIC are not considered essential to the agonist site (Smith and Olsen, 1995), but one residue within this loop of the ρ subunit (ρ1Q189) has been found to affect GABA binding (Kusama *et al.*, 1994); some effect of mutating ρ1H141 (outside the S–S loop) was also noted.

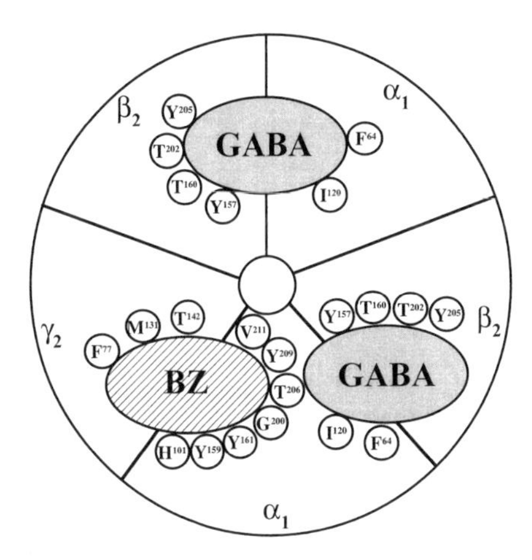

Figure 9.6 Donut model of heteropentameric GABAR showing amino acid residues implicated in ligand binding sites, as discussed and referenced in the text (modified from Smith and Olsen, 1995; Srinivasan et al., 1999b).

Secondary structure predictions for the region around the GABA binding residue F64 suggested a short β-strand from 59–65, then a short connecting loop 66–69, often associated with protein ligand binding pockets, followed by a piece of α-helix at 69–78, including 4 residues invariant in all ligand-gated ion channel complexes W69, D71, R73, and L74 (Smith and Olsen, 1995; Olsen *et al.*, 1996). However, mutation of α1 R66A/Q to eliminate the + charge putatively involved in GABA binding (also, A and Q are the residue present in β and γ subunits) did not alter [^{3}H]muscimol binding, which remained at 30–50 percent of wild type when expressed with β2 and γ2 in Sf9 cells (S. Srinivasan and R. Olsen, unpublished). Also, W69 was mutated to several other amino acids, and the mutations failed to produce membrane-expressed pentameric channels, although they continued to bind [^{3}H]muscimol (Srinivasan *et al.*, 1999b). However, Boileau *et al.* (1999) made several cysteine scanning mutations in the region 60–70 that are consistent with a GABA binding site including F64, R66, and S68 on one side of a β-strand.

Two domains in β subunits (Y157, T160, and T202, Y205) were identified as critical for the GABA binding pocket (Amin and Weiss, 1993; Figure 9.6). The corresponding residues in the ρ subunit (Y198, T244) also affect GABA affinity, as well as channel activation and desensitization kinetics (Amin and Weiss, 1994). These residues are homologous to those identified in the agonist binding sites in other members of the LGIC family (Smith and Olsen, 1995; Galzi and Changeux, 1994; Figure 9.7). Implication of residues in both α and β subunits, on different sections of the polypeptide, suggests that the GABAR, like other members of the LGIC family, probably have ligand binding sites at the interface of two subunits (α/β in Figure 9.6: Smith and Olsen, 1995; Figure 9.7). The corresponding residues in α subunits appear to participate in formation of the benzodiazepine binding site (see below).

Alternatively, there may be two distinct types of GABA binding sites per pentamer, possibly even present in two copies, or even three types of site. For example, mutation of β2 subtype Y64 had a large effect on GABA binding (Newell *et al.*, 1999). It is unlikely that this β residue, corresponding to the F64 on α1 described above as integral to the GABA binding pocket, could be part of the same pocket (Figure 9.6), so if it is in a pocket, it would be a second one. This β residue would likely contribute to a GABA site at either the β/α or β/γ interface, each present in one copy per pentamer, or both, potentially resulting in a total of three sorts of GABA site (Figure 9.6). If homomeric β pentamers exist, and if they bind and respond to GABA (e.g. Sieghart, 1995), they would probably contribute a loop D, including Y64, to each of five identical pockets (Figure 9.7).

Functional domains for benzodiazepine action

Similarly, photoaffinity labeling with [^{3}H]flunitrazepam of α subunits (Sieghart, 1995) was shown to occur in the N-terminal extracellular domain using subtype-specific antibodies (Stephenson and Duggan, 1989). One amino acid was identified (Figure 9.6) as the covalent attachment site for H102 in bovine α1 (H101 in rat, also present in α2, α3, α5: Smith and Olsen, 1995, 2000; Duncalfe *et al.*, 1996). This residue had been

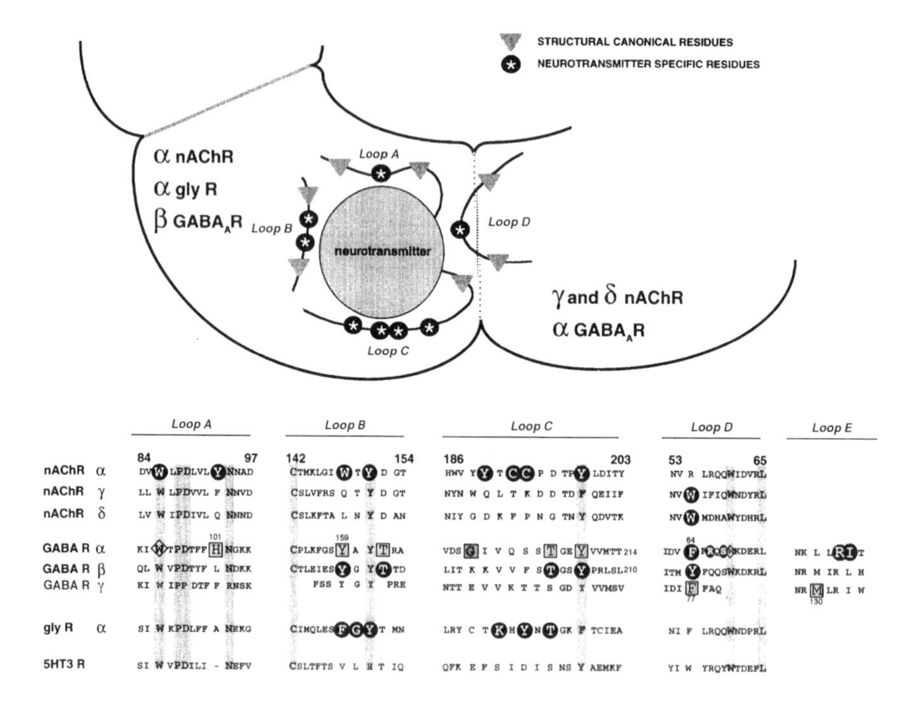

Figure 9.7 Five-loop model of ligand binding sites in ligand-gated ion channels, indicating homologous use of the same residues in nicotinic acetylcholine, GABAR, glycine, and 5HT-3 receptors. GABAR residues (referenced in text) shown for GABA site (black circles), benzodiazepine site (gray squares), and structural canonical residues (gray columns), including diamonds for α1 W69 and W94 demonstrated to be essential to membrane expression by Srinivasan *et al.* (1999b); drawing modified from Galzi and Changeux (1994).

identified earlier as the most important amino acid in determining the poor affinity for benzodiazepine agonists in the α6 subunit, a cerebellar subtype. Replacement of the H in the α1 subtype with R as in the α6 subtype reduced benzodiazepine binding, while replacement of R(100) in the α6 subtype with H as in the α1 subtype resulted in benzodiazepine binding (Wieland *et al.*, 1992). Furthermore, the same residue shows an allelic variant α6Q100 in an inbred rat strain that shows low sensitivity to the behavioral effects of ethanol and benzodiazepines on motor coordination involving the cerebellum. A mouse knock-in for point mutation H101R in the α1 subtype loses sensitivity to the sedative-hypnotic action of benzodiazepines, but not the anxolytic or motor-impairing effects, which must be mediated by α2, α3, α5 subtypes (Rudolph *et al.*, 1999). Comparison of other amino acids with histidine at this position revealed that α1H 101 is critical for benzodiazepine binding and efficacy (Dunn *et al.*, 1999). Additional studies revealed that the α4 subtype, also producing GABAR with low affinity for benzodiazepine agonists, has R, not H in the corresponding position 101. Chimeras and mutagenesis (Im *et al.*, 1997) also implicated the N-terminal area of α in the benzodiazepine binding site, with two additional residues P161 and I211 identified on α6 (Wieland and Lüddens, (1994). In addition, the variable selectivity of α subunits for benzodiazepines allowed determination of a residue involved in this specificity: residue α1G200 = α3E225 was shown to be responsible for differences in α1 vs. α2/3 subtypes in affinity for CL 218,872 and other ligands (Pritchett and Seeburg, 1991); this suggests the possibility that these residues participate in the benzodiazepine binding pocket.

The γ subunit is required for benzodiazepine binding by GABARs, and γ2T142 was shown to affect benzodiazepine binding (Mihic *et al.*, 1994). Other residues in the γ2 subtype (F77, T55, M57) have been implicated in benzodiazepine binding (Buhr *et al.*, 1996, 1997a). Chimeras between γ2 and α1 subtypes showed that γ2 K41-W82 and γ2 R114-D161 are needed for BZ binding in HEK cells, but the chimera of N-terminal 1-161γ2 with the rest of the α1 subtype sequence (expressed with the β2 subtype) gave poor BZ enhancement of GABAR currents in oocytes and poor GABA enhancement of BZ binding in HEK cells (Boileau *et al.*, 1998). Finally, residues at the extracellular mouth of M2 and in the M2/3 loop in α subunits have been shown to be critical for benzodiazepine modulatory action, as opposed to binding *per se* (Boileau and Czajkowski, 1999).

Recently, the residues found in β subunits to affect GABA binding (Amin and Weiss, 1993) were mutagenized in the α1 subtype (α1Y159, Y161, T206, Y209) and found to affect benzodiazepine binding (Figure 9.6; Buhr *et al.*, 1997b; Wingrove *et al.*, 1997; Amin *et al.*, 1997). The latter are situated near the already mentioned G200 residue (Schaerer *et al.*, 1998).

This correspondence of domains for benzodiazepine binding pockets with those for GABA binding in the same pentamer was noted (Galzi and Changeux, 1994; Smith and Olsen, 1995; Buhr *et al.*, 1996; Sigel and Buhr, 1997) as a possible evolutionary modification of an agonist site into one for an exogenous allosteric modulator. A sequence homology surrounding the residues in the α1 subtype implicated in binding muscimol and flunitrazepam (Smith and Olsen, 1995; Olsen *et al.*, 1996), suggested the structural connection, as well as a functional correlate: in whatever manner GABA binding on β/α subunits is conformationally coupled to channel opening, benzodiazepine binding on α/γ subunits might also promote that physical change. The benzodiazepine binding sites in some subunits may thus replace or compete with "accessory" agonist binding sites. The word "accessory" is used because these sites are probably not obligatory for channel activation but, rather, modulatory.

Functional domains for anesthetic modulation of GABAR

Barbiturates and related drugs that enhance GABAR function also allosterically enhance the binding of radioligands to the GABA site and to the benzodiazepine site on the GABAR, and have a biphasic effect on the binding of $[^{35}S]$TBPS to the picrotoxin site in CNS homogenates and cells expressing GABARs (Olsen *et al.*, 1986, 1991). These interactions, especially for neuroactive steroids, showed a heterogeneity with brain region (Olsen *et al.*, 1991; Sapp *et al.*, 1992; Bureau and Olsen, 1993; Nguyen *et al.*, 1995; Carlson *et al.*, 1997). This heterogeneity in anesthetic potency and efficacy extended to different polypeptide subunits of the purified GABAR protein (Bureau and Olsen, 1993) and was consistent with the developing appreciation for multiple GABAR subtypes based on variable subunit composition (Olsen and Sapp, 1995).

Recombinant GABAR expression studies have revealed minor differences in sensitivity to steroids for isoforms differing in their α subunit: the $\alpha 1$ subtype was more sensitive than $\alpha 2$, $\alpha 3$ (Shingai *et al.*, 1991), or $\alpha 4$ (Kim *et al.*, 1997) subtypes, using electrophysiology in *Xenopus* oocytes. In HEK cells, using binding, Lan *et al.* (1991) found $\alpha 3 > \alpha 2 > \alpha 1$, while Puia *et al.* (1993) found $\alpha 5 = \alpha 6 > \alpha 1 = \alpha 3$, and $\gamma 1 > \gamma 2$, using electrophysiology. The δ subunit combines with $\alpha 4$ or $\alpha 6$ subtypes in nature to form isoforms that generally lack γ subunits (Macdonald and Olsen, 1994). These δ-containing isoforms are particularly insensitive to neurosteroids (Zhu *et al.*, 1996). It was also found that the presence of $\gamma 2$ was needed for the enhancement phase of neurosteroid modulation of $[^{35}S]$TBPS binding in Sf9 cells expressing $\alpha 1\beta 2$ GABAR; $\alpha 6$-containing GABAR were not enhanced; in brain homogenates, this enhancement was seen in cortex but not cerebellum (Srinivasan *et al.*, 1999a); the functional correlate of this allosteric binding effect is not known, however.

Barbiturate modulation of $[^3H]$flunitrazepam and $[^3H]$muscimol binding is lower for the $\alpha 5\beta 3$ combination than for the $\alpha 1\beta 2$ combination expressed in Sf9 cells (Carlson *et al.*, 1995). Barbiturate enhancement is considerably greater for $\alpha 6$ than other α (Thompson *et al.*, 1996). Wafford *et al.* (1994) identified a β-subunit specific interaction: while isoforms containing the $\beta 2$ and $\beta 3$ subtypes are sensitive to the anxiolytic non-benzodiazepine drug loreclezole, the $\beta 1$ subtype is not. Further, the same β selectivity applies to the anesthetic etomidate, a structural analog of loreclezole (Belelli *et al.*, 1997; Moody *et al.*, 1997).

The most dramatic subunit selectivity for anesthetics is afforded by the total insensitivity of the ρ variety of GABAR (Kusama *et al.*, 1994; Mihic *et al.*, 1997; Serafini *et al.*, 1997); the ρ subunit, expressed primarily in retina, is also insensitive to classic GABAR drugs like benzodiazepines and bicuculline. A new GABAR subunit ϵ combines with α and β subunits to produce isoforms with poor sensitivity to anesthetics (Davies *et al.*, 1997).

Very little information is available to date on domains within GABAR that participate in the action of anesthetics, although this situation will hopefully change rapidly. Studies comparing anesthetic-sensitive and anesthetic-insensitive subunits are in progress (e.g., Carlson *et al.*, 1995; Koltchine *et al.*, 1996; Kim *et al.*, 1997; Serafini *et al.*, 1997). Mutation of certain amino acid residues within the TM2 region (β1T262, Figure 9.3) reduces pentobarbital sensitivity of GABAR (Birnir *et al.*, 1997). Because channel residues can affect, by allosteric coupling, the affinity of the agonist, which is not thought to bind within the pore, it is not clear that the barbiturate binding site must be within TM2. Likewise residues in TM1 and in the intracellular loop may affect agonist efficacy (Fisher *et al.*, 1997; Kim *et al.*, 1997; Carlson *et al.*, 2000). However, other studies also implicate TM2 for anesthetic actions.

The selectivity of β subunits for modulation by loreclezole (Wafford *et al.*, 1994) led to the discovery that a single residue (β1S265 = β3N290) could change the sensitivity to this drug (Wingrove *et al.*, 1994). The same residue was shown to alter barbiturate activation (Cestari and Yang, 1996). This residue now has been shown to affect sensitivity to the anesthetic etomidate, a structural analog of loreclezole (Moody *et al.*, 1997; Belelli *et al.*, 1997). Serine, as in β1, at this position in β2 or β3 leads to loss of activation by loreclezole and etomidate. In addition, this same residue has now been identified in a comparison of GABAR ρ (insensitive to volatile anesthetic enhancement, and inhibited by high concentrations of ethanol) and glycine receptor α subunit (enhanced by these drugs). Conversion of the residue in ρ to that in glycine receptors gives anesthetic sensitivity; conversely, mutation of the glycine receptor residue to that in ρ eliminates anesthetic sensitivity. Furthermore, the mutation of the corresponding residue in GABAR α subunits (S270 in α1 or α2) eliminates the sensitivity to enflurane.

This residue in TM2 is the most implicated residue in anesthetic action to date; nevertheless, it may not be the anesthetic binding site, but merely conformationally coupled to anesthetic action. However, modeling of the α-helix of TM2 suggests that this residue is not actually within the channel lumen but may be on the opposite side of the helix, thus in contact with lipids or other subunit polypeptides. This therefore may represent a domain for anesthetic binding. In addition, a residue within TM3 is affected in this GABAR ρ-glycine receptor interaction (Mihic *et al.*, 1997): mutation A291W in the GABAR α2 subtype eliminated enflurane enhancement of GABAR currents, but not that of propofol. This may provide additional evidence for functional domains of anesthetic modulation of GABARs. Further characterization of the M2 mutations showed that the cutoff for size of long chain alcohol interactions was altered, consistent with the existence of a binding "pocket" (Wick *et al.*, 1998). These residues in TM2 and TM3 were not found to affect the sensitivity of GABAR to intravenous anesthetics like barbiturates, steroids, as well as propofol (Mihic *et al.*, 1997; Krasowski *et al.*, 1998), implicating the region N-terminal to TM2 (extracellular or TM1) in their action (Koltchine *et al.*, 1996). Other chimeras also implicated the extracellular N-terminus in action of allosteric modulators (e.g., Fisher *et al.*, 1997). However, the same residue W328 in TM3 of the anesthetic-insensitive ρ subunit, when mutated to one of several amino acids present in non-ρ (α,β,γ,δ) subunits, led to pentobarbital sensitivity (Amin, 1999). Carlson *et al.* (2000) replaced the β2 residue G223 with the ρ amino acid F and lost enhancement of binding and GABA currents by pentobarbital, alphaxalone, propofol, and etomidate. This residue is at the mouth of TM1, thus defining a third area of interest for anesthetic modulation, this one joining the other two in TM2 and TM3 in position at the membrane/extracellular water interface of the GABAR protein (Figure 9.8). Further work is needed to determine the precise role of these critical amino acids in drug modulation.

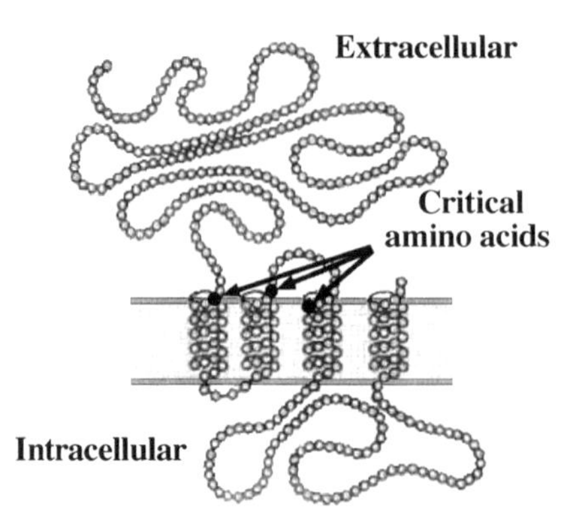

Figure 9.8 Schematic model of individual subunit of GABAR showing membrane topology with extracellular N-terminus, four membrane-spanning helices, and large intracellular loop between M3 and M4. Sites identified as critical for modulation by anesthetics (Mihic *et al.*, 1997; Carlson *et al.*, 2000) are shown by arrows (modified after Franks and Lieb, 1997) (figure modified from Olsen and Tobin, 1990).

Functional domains for other allosteric modulator actions

The M2 region is implicated in the efficacy of agonists as well as allosteric modulators, including volatile and intravenous anesthetics, as well as non-anesthetics like loreclezole. As mentioned, it is unclear if this region represents a binding pocket for any of these agents. Other agents of interest include several naturally occurring as well as pharmacologically active metal ions.

Zinc inhibits GABAR currents (Smart, 1992). The highest zinc sensitivity is found with $\alpha\beta$ GABARs, and β subunits contain a histidine in the extracellular portion of M2 (H267 in the $\beta 1$ subtype) that confers high zinc sensitivity to $\alpha\beta$ GABARs (Wooltorton *et al.*, 1997; Horenstein and Akabas, 1998). A histidine residue in the extracellular domain (not M2) of ρ subunits was shown to be responsible for low zinc sensitivity (Wang *et al.*, 1995b). Zinc sensitivity of GABARs, however, is also affected by the presence and/or nature of the α, δ and γ subunits. $\alpha 6$ subtype-containing GABARs are more sensitive than $\alpha 1$ subtype-containing GABARs; zinc sensitivity is somewhat reduced by the presence of a ρ subunit and strongly reduced by the presence of a γ subunit (Draguhn *et al.*, 1990; Saxena and Macdonald, 1994). The C-terminal domains of $\alpha 1$ and $\alpha 6$ subtypes contain sites for regulation of zinc sensitivity (Fisher *et al.*, 1997). The differential sensitivity is due to the presence of an asparagine residue (N274) in $\alpha 1$ subtypes conferring relatively low zinc sensitivity and a corresponding histidine residue (H273) in $\alpha 6$ subtypes conferring a relatively high sensitivity to zinc (Fisher and Macdonald, 1998). The basis for the differential δ and γ subunit reduction of zinc sensitivity is more complex, involving both N-terminal and C-terminal domains (Nagaya *et al.*, 1998).

Lanthanum sensitivity is also regulated by α subtype, but in a different manner: $\alpha 1$ subtype-containing GABAR currents are enhanced while $\alpha 6$ subtype–containing GABAR currents are reduced by lanthanum (Saxena *et al.*, 1997). In the $\alpha 1$ subtype, the N-terminal domain between the cysteine loop and the beginning of M1 contains a site for enhancement of GABAR current by lanthanum (Fisher *et al.*, 1997; Kim and Macdonald, 1999).

Sensitivity to the membrane chloride pump inhibitor furosemide is also affected by α subtype. GABARs have high furosemide sensitivity with the $\alpha4$ and $\alpha6$ subtypes, but much lower sensitivity with the $\alpha1$, $\alpha2$, $\alpha3$ or $\alpha5$ subtypes (Korpi *et al.*, 1995). An $\alpha6$ subtype domain N-terminal to the middle of M1 contains sites for high affinity reduction of GABAR current by furosemide (Fisher *et al.*, 1997; Jackel *et al.*, 1998), while a low affinity site is present in the remaining C-terminal portions of the subtype (Fisher *et al.*, 1997). The high affinity site was an isoleucine (I228) in the $\alpha6$ subtype, which conferred high affinity inhibition by furosemide, while a threonine (T230) at this position in the $\alpha1$ subtype conferred low affinity inhibition by furosemide (Thompson *et al.*, 1999a).

In summary, considerable progress has been made in identifying functional domains in the GABAR protein that are critical for ion channel function and for ligand binding, as well as for the coupling between the binding of allosteric modulators, including GABA itself, to ion channel gating. Additional physicochemical and functional analysis of the biological and pharmacological transduction activity of this protein can now be made. For example, in the case of benzodiazepines, amino acids in the extracellular N-terminal region have been recently identified to mediate subtype-specific binding of the drugs (Rudolph *et al.*, 1999), while residues in the M2 channel vestibule and M2–M3 extracellular loop have been implicated in the transduction event (Boileau and Czajkowski, 1999).

Acknowledgements

The authors would like to thank the members of their laboratories for contributions to this chapter. Supported by NIH grants NS28772 and NS35985 to Richard W. Olsen and NS33300 to Robert L. Macdonald.

References

Akabas, M.H., Kaufmann, C., Archdeacon, P. and Karlin, A. (1994) 'Identification of acetylcholine receptor channel-lining residues in the entire M2 segment of the α subunit', *Neuron* **13**: 919–927.

Allison, D.W., Gelfand, V.I., Spector, I. and Craig, A.M. (1998) 'Role of actin in anchoring postsynaptic receptors in cultured hippocampal neurons: differential attachment of NMDA versus AMPA receptors', *J. Neurosci.* **18**: 2423–2436.

Amin, J. (1999) 'A single hydrophobic residue confers barbiturate sensitivity to $GABA_C$ receptor', *Mol. Pharmacol.* **55**: 411–423.

Amin, J. and Weiss, D.S. (1993) '$GABA_A$ receptor needs two homologous domains of the β-subunit for activation by GABA but not by pentobarbital', *Nature* **366**: 565–569.

Amin, J. and Weiss, D.S. (1994) 'Homomeric $\rho1$ GABA channels: activation properties and domains', *Recept. Chann.* **2**: 227–236.

Amin, J., Brooks-Kayal, A. and Weiss, D.S. (1997) 'Two tyrosine residues on the α subunit are crucial for benzodiazepine binding and allosteric modulation of $GABA_A$ receptors', *Mol. Pharmacol.* **51**: 833–841.

Angelotti, T.P. and Macdonald, R.L. (1993) 'Assembly of $GABA_A$ receptor subunits $\alpha1\beta1$ and $\alpha1\beta1\gamma2S$ subunits produce unique ion channels with dissimilar single-channel properties', *J. Neurosci.* **13**: 1429–1440.

Angelotti, T.P., Uhler, M.D. and Macdonald, R.L. (1993) 'Enhancement of recombinant $GABA_A$ receptor currents by chronic activation of cAMP-dependent protein kinase', *Mol. Pharmacol.* **44**: 1202–1210.

Backus, K.H., Arigoni, M., Drescher, U., Scheurer, L., Malherbe, P., Mohler, H. and Benson, J.A. (1993) 'Stoichiometry of a recombinant $GABA_A$ receptor deduced from mutation-induced rectification', *Neuroreport* **5**: 285–288.

Banks, M.I, Li, T.B. and Pearce, R.A. (1998) 'The synaptic basis of $GABA_A$, slow', *J. Neurosci.* **18**: 1305–1317.

Barnard, E., Skolnick, P., Olsen, R., Mohler, H., Sieghart, W., Biggio, G., Braestrup, C., Bateson, A. and Langer, S. (1998) 'International Union of Pharmacology. XV. Subtypes of γ-aminobutyric $acid_A$ receptors: classification on the basis of subunit structure and receptor function', *Pharmacol. Rev.* **50**: 291–313.

Belelli, D., Lambert, J.J., Peters, J.A., Wafford, K. and Whiting, P.J. (1997) 'The interaction of the general anesthetic etomidate with the γ-aminobutyric acid type A receptor is influenced by a single amino acid', *Proc. Natl. Acad. Sci. USA* **94**: 11,031–11,036.

Belhage, B., Hansen, G.H., Elster, L. and Schousboe, A. (1998) 'Effects of γ-aminobutyric acid (GABA) on synaptogenesis and synaptic function', *Perspect. Develop. Neurobio.* **5**: 235–246.

Benke, D., Mertens, S., Trzeciak, A. Gillessen, D. and Möhler, H. (1991) '$GABA_A$ receptors display association of γ2-subunit with α1- and β2/3 subunits', *J. Biol. Chem.* **266**: 4478–4483.

Benke, D., Michel, C. and Möhler, H. (1997) '$GABA_A$ receptors containing the α4-subunit: prevalence, distribution, pharmacology and subunit architecture in situ', *J. Neurochem.* **69**: 806–814.

Bianchi, M.T., Haas, K.F., Chung, H., Zhang., J. and Macdonald, R.L. (1999) 'Structural determinants of $GABA_A$ receptor gating and desensitization', *Abstr. Soc. Neurosci.* **25**: 1710.

Birnir, B., Tierney, M.L., Dalziel, J.E., Cox, G.B. and Gage, P.W. (1997) 'A structural determinant of desensitization and allosteric regulation by pentobarbitone of the $GABA_A$ receptor', *J. Memb. Biol.* **155**: 157–166.

Boileau, A.J. and Czajkowski, C. (1999) 'Identification of transduction elements for benzodiazepine modulation of the $GABA_A$ receptor: three residues are required for allosteric coupling', *J. Neurosci.* **19**: 10,213–10,220.

Boileau, A.J., Kucken, A.M., Evers, A. and Czajkowski, C. (1998) 'Molecular dissection of benzodiazepine binding and allosteric coupling using chimeric $GABA_A$ receptor subunits', *Mol. Pharmacol.* **53**: 295–303.

Boileau, A.J., Evers, A.R., Davus, A.F. and Czajkowski, C. (1999) 'Mapping the agonist binding site of the $GABA_A$ receptor', *J. Neurosci.* **19**: 4847–4854.

Bormann, J., Hamill, O.P., Sakmann, B. (1987) 'Mechanism of anion permeation through channels gated by glycine and gamma-aminobutyric acid in mouse cultured spinal neurones', *J. Physiol.* **385**: 243–286.

Brandon, N.J., Uren, J.M., Kittler, J.T., Wang, H., Olsen, R.W., Parker, P.J. and Moss, S.J. (1999) 'Subunit specific association of protein kinase C and the receptor for activated C kinase with γ-aminobutyric acid type A receptors', *J. Neurosci.* **19**: 9228–9234.

Brickley, S.G., Cull-Candy, S.G. and Farrant, M. (1999) 'Single-channel properties of synaptic and extrasynaptic $GABA_A$ receptors suggest differential targeting of receptor subtypes', *J. Neurosci.* **19**: 2960–2973.

Brooks-Kayal, A.P., Shumate, M.D., Jin, H., Rikhter, T.Y. and Coulter, D.A. (1998) 'Selective changes in single cell GABA receptor expression and function in temporal lobe epilepsy', *Nature Med.* **4**: 1166–1172.

Browning, M.D., Bureau, M., Dudek, E.M. and Olsen, R.W. (1990) 'Protein kinase C and cAMP-dependent protein kinase phosphorylate the β-subunit of the purified $GABA_A$ receptor', *Proc. Natl. Acad. Sci. USA* **87**: 1315–1318.

Browning, M.D., Endo, S., Smith, G., Dudek, E.M. and Olsen, R.W. (1993) 'Phosphorylation of the $GABA_A$ receptor by cAMP-dependent protein kinase and by protein kinase C: analysis of the substrate domain', *Neurochem. Res.* **18**: 95–100.

Brunig, I., Sommer, M., Hatt, H. and Bormann J. (1999) 'Dopamine receptor subtypes modulate olfactory bulb γ-aminobutyric acid type A receptors', *Proc. Natl. Acad. Sci. USA* **96**: 2456–2460.

Buhr, A. and Sigel, E. (1997) 'A point mutation in the γ2 subunit of GABA$_A$ receptors results in altered benzodiazepine binding site specificity', *Proc. Natl. Acad. Sci. USA* **94**: 8824–8829.

Buhr, A., Baur, R., Malherbe, P. and Sigel, E. (1996) 'Point mutations of the α1β2γ2 GABA$_A$ receptor affecting modulation of the channel by ligands of the benzodiazepine binding site', *Mol. Pharmacol.* **49**: 1080–1084.

Buhr, A., Baur, R. and Sigel, E. (1997a) 'Subtle changes in residue 77 of the γ subunit of α1β2γ2 GABA$_A$ receptors drastically alter the affinity for ligands of the benzodiazepine binding site', *J. Biol. Chem.* **272**: 11,799–11,804.

Buhr, A., Schaerer, M.T., Baur, R. and Sigel, E. (1997b) 'Residues at positions 206 and 209 of the α1 subunit of GABA$_A$ receptors influence affinities for benzodiazepine binding site ligands', *Mol. Pharmacol.* **52**: 676–682.

Bureau, M.H. and Laschet, J.J. (1995) 'Endogenous phosphorylation of distinct GABA$_A$ receptor polypeptides by Ser/Thr and Tyr kinase activities associated with the purified receptor', *J. Biol. Chem.* **270**: 26,482–26,487.

Bureau, M.H. and Olsen, R.W. (1993) 'GABA$_A$ receptor subtypes: ligand binding heterogeneity demonstrated by photoaffinity labeling and autoradiography', *J. Neurochem.* **61**: 1479–1491.

Burgard, E.C., Haas, K. and Macdonald, R.L. (1999) 'Channel properties determine the transient activation kinetics of recombinant GABA$_A$ receptors', *Mol. Brain. Res.* **73**: 28–36.

Burgard, E.C., Tietz, E.I., Neelands, T.R. and Macdonald, R.L. (1996) 'Properties of recombinant γ-aminobutyric acid$_A$ receptor isoforms containing the α5 subunit subtype', *Mol. Pharmacol.* **50**: 119–127.

Burt, D.R. and Kamatchi, G.L. (1994) 'GABA$_A$ receptor subtypes: from pharmacology to molecular biology', *FASEB J.* **5**: 2916–2923.

Carlson, B.X., Srinivasan, S. and Olsen, R.W. (1995) 'Subunit selectivity for anesthetic modulation of GABA$_A$ receptors expressed in Sf9 cells using baculovirus', *Abstr. Soc. Neurosci.* **21**: 849.

Carlson, B.X., Hales, T.G. and Olsen, R.W. (1997) 'GABA$_A$ receptors and anesthesia', in T.L. Yaksh, C. Lynch, W.M. Zapol, M. Maze, J.F. Biebuyck and L.J. Saidman (eds) *Anesthesia, Biologic Foundations*, New York: Lippincott-Raven Publishers, pp. 259–275.

Carlson, B.X., Engblom, C., Kristiansen, U., Schousboe, A. and Olsen R.W. (2000) 'A single glycine residue at the entrance to the first membrane-spanning domain of the GABA$_A$ receptor β2 subunit affects allosteric sensitivity to GABA and anesthetics', *Mol. Pharmacol.* **57**: 474–484.

Cash, D.J., Serfozo, P. and Allan, A.M. (1997) 'Desensitization of a GABA$_A$ receptor in rat is increased by chronic treatment with chlordiazepoxide: a molecular mechanism of dependence', *J. Pharmacol. Exptl. Ther.* **283**: 704–711.

Celentano, J.J. and Wong, R.K. (1994) 'Multiphasic desensitization of the GABA$_A$ receptor in outside-out patches', *Biophys. J.* **66**: 1039–1050.

Cestari, I. and Yang, J. (1996) 'Pentobarbital opening of a homomeric chloride ion channel requires a specific asparagine residue', *Abstr. Soc. Neurosci.* **22**: 1288.

Chang, Y., Wang, R., Barot, S. and Weiss, D.S. (1996) 'Stoichiometry of a recombinant GABA$_A$ receptor', *J. Neurosci.* **16**: 5415–5424.

Chang, Y. and Weiss, D.S. (1998) 'Substitution of the highly conserved M2 leucine creates spontaneously opening ρ1 GABA receptors', *Mol. Pharmacol.* **53**: 511–523.

Chappell, R., Bueno, O.F., Alvarez-Hernandez, X., Robinson, L.C. and Leidenheimer, N.J. (1998) 'Activation of protein kinase C induces GABA$_A$ receptor internalization in Xenopus oocytes', *J. Biol. Chem.* **273**: 32,595–32,601.

Chen, L., Wang, H., Vicini, S. and Olsen, R.W. (1999) 'The GABA$_A$ receptor associated protein (GABARAP) promotes GABA$_A$ receptor clustering and modulates the channel kinetics', *Abstr. Soc. Neurosci.* **25**: 1226.

Chen, Q.X., Stelzer, A., Kay, A.R. and Wong, R.K.S. (1990) 'GABA$_A$ receptor function is regulated by phosphorylation in acutely dissociated guinea-pig hippocampal neurones', *J. Physiol.* **420**: 207–221.

Cheun, J.E. and Yeh, H.H. (1992) 'Modulation of $GABA_A$ receptor-activated current by norepinephrine in cerebellar Purkinje cells', *Neuroscience* **51**: 951–960.

Cheun, J.E. and Yeh, H.H. (1996) 'Noradrenergic potentiation of cerebellar Purkinje cell responses to GABA: cyclic AMP as intracellular intermediary', *Neuroscience* **74**: 835–844.

Colledge, M. and Froehner, S.C. (1998) 'To muster a cluster: anchoring neurotransmitter receptors at synapses (Commentary)', *Proc. Natl. Acad. Sci. USA* **95**: 3341–3343.

Connolly, C.N., Krishek, B.J., McDonald, B.J., Smart, T.G. and Moss, S.J. (1996a) 'Assembly and cell surface expression of heteromeric and homomeric $GABA_A$ receptors', *J. Biol. Chem.* **271**: 89–96.

Connolly, C.N., Wooltorton, J.R.A., Smart, T.G. and Moss, S.J. (1996b) 'Subcellular localization of $GABA_A$ receptors is determined by receptor β subunits', *Proc. Natl. Acad. Sci. USA* **93**: 9899–9904.

Connolly, C.N., Uren, J.M., Thomas, P., Gorrie, G.H., Gibson, A., Smart, T.G. and Moss, S.J. (1999) 'Subcellular localisation and preferential assembly of γ2 subunit splice variants of $GABA_A$ receptors', *Mol. Cell. Neurobiol.* **13**: 259–271.

Crestani, F., Lorez, M., Baer, K., Essrich, C., Benke, D., Laurent, J.P., Belzung, C., Fritschy, J.M., Lüscher, B. and Möhler, H. (1999) 'Decreased $GABA_A$ receptor clustering results in enhanced anxiety and a bias for threat cues', *Nature Neurosci.* **2**: 833–839.

Davies, P.A., Hanna, M.C., Hales, T.G. and Kirkness, E.F. (1997) 'Insensitivity to anaesthetic agents conferred by a class of $GABA_A$ receptor subunit', *Nature* **385**: 820–823.

De Koninck, Y. and Mody, I. (1996) 'The effects of raising intracellular calcium on synaptic $GABA_A$ receptor-channels', *Neuropharmacology* **35**: 1365–1374.

DeLorey, T.M. and Olsen, R.W. (1992) 'γ-aminobutyric $acid_A$ receptor structure and function', *J. Biol. Chem.* **267**: 16,747–16,750.

Dominguez-Perrot, C., Feltz, P. and Poulter, M.O. (1997) 'Recombinant $GABA_A$ receptor desensitization: the role of the gamma2 subunit and its physiological significance', *J. Physiol.* **497.1**: 145–159.

Draguhn, A., Verdoorn, T., Ewer, M., Seeburg, P. and Sakmann, B. (1990) 'Functional and molecular distinction between recombinant rat $GABA_A$ receptor subtypes by Zn^{2+}', *Neuron* **5**: 781–788.

Duncalfe, L.L., Carpenter, M.R., Smillie, L.B., Martin, I.L. and Dunn, S.M.J. (1996) 'The major site of photoaffinity labeling of the γ-aminobutyric acid type A receptor by [^{3}H] flunitrazepam is histidine 102 of the α subunit', *J. Biol. Chem.* **271**: 9209–9214.

Dunn, S.M.J., Davies, M., Munton, A.L. and Lambert, J.J. (1999) 'Mutagenesis of the rat α1 subunit of the $GABA_A$ receptor reveals the importance of residue 101 in determining the allosteric effects of benzodiazepine site ligands', *Mol. Pharmacol.* **56**: 768–774.

Elster, L., Schousboe, A. and Olsen, R.W. (2000) 'Stable $GABA_A$ receptor intermediates in Sf9 cells expressing α1: β2: and γ2L subunits', *J. Neurosci. Res.* **61**: 193–205.

Essrich, C., Fritschy, J.M., Lorez, M., Benson, J. and Luscher, B. (1998) 'Essential roles in postsynaptic clustering of $GABA_A$ receptors for the γ2 subunit and gephyrin', *Nature Neurosci.* **1**: 563–571.

Feigenspan, A. and Bormann, J. (1994) 'Facilitation of GABAergic signaling in the retina by receptors stimulating adenylate cyclase', *Proc. Natl. Acad. Sci. USA* **91**: 10,893–10,897.

Feng, G., Tintrup, H., Kirsch, J., Nichol, M.C., Kuhse, J., Betz, H. and Sanes, J.R. (1998) 'Dual requirement for gephyrin in glycine receptor clustering and molybdoenzyme activity', *Science* **282**: 1321–1324.

ffrench-Constant, R.H., Rocheleau, T.A., Steichen, J.C. and Chalmers, A.E. (1993) 'A point mutation in a *Drosophila* GABA receptor confers insecticide resistance', *Nature* **363**: 449–451.

Filippova, N., Dudley, R. and Weiss, D.S. (1999) 'Evidence for phosphorylation-dependent internalization of recombinant human ρ1 $GABA_C$ receptors', *J. Physiol.* 518.**2**: 385–399.

Fisher, J.L. and Macdonald, R.L. (1997a) 'Single channel properties of recombinant $GABA_A$ receptors containing γ2 or δ subtypes expressed with α1 and β3 subtypes in L929 cells', *J. Physiol.* **505**: 283–297.

Fisher, J.L. and Macdonald, R.L. (1997b) 'Functional properties of recombinant $GABA_A$ receptors composed of single or multiple β subunit subtypes', *Neuropharmacology* **36**: 1601–1610.

Fisher, J.L. and Macdonald, R.L. (1998) 'The role of an α subtype M_2–M_3 His in regulating inhibition of $GABA_A$ receptor current by zinc and other divalent cations', *J. Neurosci.* **18**: 2944–2953.

Fisher, J.L., Zhang, J. and Macdonald, R.L. (1997) 'The role of α1 and α6 subtype amino-terminal domains in allosteric regulation of $GABA_A$ receptors', *Mol. Pharmacol.* **52**: 714–724.

Franks, N. and Lieb, W. (1997) 'Anaesthetics set their sites on ion channels', *Nature* **389**: 334–335.

Fritschy, J.M., Benke, D., Johnson, D.K., Möhler, H. and Rudolph, U. (1997) '$GABA_A$ receptor α-subunit is an essential prerequisite for receptor formation in vivo', *Neuroscience* **81**: 1043–1053.

Fritschy, J.M., Johnson, D.K., Möhler, H. and Rudolph, U. (1998) 'Independent assembly and subcellular targeting of $GABA_A$-receptor subtypes demonstrated in mouse hippocampal and olfactory neurons in vivo', *Neurosci. Lett.* **249**: 99–102.

Galarreta, M. and Hestrin S. (1997) 'Properties of $GABA_A$ receptors underlying inhibitory synaptic currents in neocortical pyramidal neurons', *J.Neurosci.* **17**: 7220–7227.

Galzi, J.L. and Changeux, J.P. (1994) 'Ligand-gated ion channels as unconventional allosteric proteins', *Curr. Opin. Struct. Biol.* **4**: 554–565.

Galzi, J.-L., Devillers-Thiery, A., Hussy, N., Bertrand, S., Changeux, J.-P. and Bertrand, D. (1992) 'Mutations in the channel domain of a neuronal nicotinic receptor convert ion selectivity from cationic to anionic', *Nature* **359**: 500–505.

Gautam, M., Noakes, P., Mudd, J., Nichol, M., Chu, G., Sanes, J. and Merlie, J. (1995) 'Failure of postsynaptic specialization to develop at neuromuscular junctions of rapsyn-deficient mice', *Nature* **377**: 232–236.

Gillespie, S.K.H., Balasubramanian, S., Fung, E.T. and Huganir, R.L. (1996) 'Rapsyn clusters and activates the synapse-specific receptor tyrosine kinase MuSK', *Neuron* **16**: 953–962.

Gingrich, K.J., Roberts, W.A and Kass, R.S. (1995) 'Dependence of the $GABA_A$ receptor gating kinetics on the α-subunit isoform: implications for structure–function relations and synaptic transmission', *J. Physiol.* **489.2**: 529–543.

Gorrie, G.H., Vallis, Y., Stephenson, A., Whitfield, J., Browning, G., Smart, T.G. and Moss, S.J. (1997) 'Assembly of $GABA_A$ receptors composed of α1 and β2 subunits in both cultured neurons and fibroblasts', *J. Neurosci.* **17**: 6587–6596.

Gready, J.E., Ranganathan, S., Schofield, P.R., Matsuo, Y. and Nishikawa, K. (1997) 'Predicted structure of the extracellular region of ligand-gated ion-channel receptors shows SH2-like and SH3-like domains forming the ligand-binding site', *Protein Sci.* **6**: 983–998.

Gurley, D., Amin, J., Ross, P.C., Weiss, D.S. and White, G. (1995) 'Point mutations in the M2 regions of the α, β, or γ subunit of the $GABA_A$ channel that abolish block by picrotoxin', *Recept. Chann.* **3**: 13–20.

Gyenes, M., Farrant, M. and Farb, D.H. (1988) 'Run-down of $GABA_A$ function during whole-cell recording: a possible role for phosphorylation', *Mol. Pharmacol.* **34**: 719–723.

Haas, K. and Macdonald, R.L. (1999) '$GABA_A$ receptor subunit γ2 and δ subtypes confer unique kinetic properties on recombinant $GABA_A$ receptor currents in mouse fibroblasts', *J. Physiol.* **514.1**: 27–45.

Hackam, A.S., Wang, T.L., Guggino, W.B. and Cutting, G.R. (1997) 'The N-terminal domain of human GABA receptor ρ1 subunits contains signals for homooligomeric and heterooligomeric interaction', *J. Biol. Chem.* **272**: 13,750–13,757.

Hall, Z. and Sanes, J.R. (1993) 'Synaptic structure and development of the neuromuscular junction', *Cell* **72** (suppl.): 99–121.

Hanley, J.G., Koulen, P., Bedford, F., Gordon-Weeks, P.R. and Moss, S.J. (1999) 'The protein MAP 1B links $GABA_C$ receptors to the cytoskeleton at retinal synapses', *Nature* **397**: 66–69.

Harris, R.A., McQuilken, S.J., Paylor, R., Abeliovitch, A., Tonegawa, S. and Wehner, J. (1995) 'Mutant mice lacking the γ-isoform of protein kinase C show decreased behavioral actions of ethanol and altered function of $GABA_A$ receptors', *Proc. Natl. Acad. Sci. USA* **92**: 3658–3662.

Hartnett, C., Brown, M.S., Yu, J., Primus, R.J., Meyyappan, M., White, G., Stirling, V.B., Tallman, J.F., Ramabhadran, T.V. and Gallager, D.W. (1996) 'Effect of subunit composition on $GABA_A$ receptor complex characteristics in a Baculovirus expression system', *Recept. Chann.* **4**: 179–195.

Heuschneider, G. and Schwartz, R.D. (1989) 'cAMP and forskolin decrease gamma-aminobutyric acid-gated chloride flux in rat brain synaptoneurosomes', *Proc. Natl. Acad. Sci. USA* **86**: 2938–2942.

Hodge, C.W., Mehmert, K.K, Kelley, S.P., McMahon, T., Haywood, A., Olive, M.F., Wang, D., Sanchez-Perez, A.M. and Messing, R.O. (1999) 'Supersensitivity to allosteric $GABA_A$ receptor modulators and alcohol in mice lacking PKCϵ', *Nature Neurosci.* **2**: 997–1002.

Homanics, G.E., DeLorey, T.M., Firestone, L.L., Quinlan, J.J., Handforth, A., Harrison, N.L., Krasowski, M.D., Rick, C.E.M., Korpi, E.R., Makela, R., Brilliant, M.H., Hagiwara, N., Ferguson, C., Snyder, K. and Olsen, R.W. (1997) 'Mice devoid of $GABA_A$ receptor $\beta 3$ subunit have epilepsy, cleft palate, and hypersensitive behavior', *Proc. Natl. Acad. Sci. USA* **94**: 4143–4148.

Horenstein, J. and Akabas, M.H. (1998) 'Location of a high affinity Zn^{2+} binding site in the channel of $\alpha 1\beta 1$ γ-aminobutyric $acid_A$ receptors', *Mol. Pharmacol.* **53**: 870–877.

Huang, R.Q. and Dillon, G.H. (1998) 'Maintenance of recombinant $GABA_A$ receptor function: role of protein tyrosine phosphorylation and calcineurin', *J. Pharmacol. Exp. Ther.* **286**: 243–255.

Im, W.B., Pregenzer, J.F., Binder, J.A., Dillon, G.H. and Alberts, G.L. (1995) 'Chloride channel expression with the tandem construct of $\alpha 6$-$\beta 2$ $GABA_A$ receptor subunit requires a monomeric $\alpha 6$ or $\gamma 2$', *J. Biol. Chem.* **270**: 26,063–26,066.

Im, W.B., Pregenzer, J.F., Binder, J.A., Alberts, G.L. and Im, H.K. (1997) 'Alterations in the benzodiazepine site of rat $\alpha 6\beta 2\gamma 2$-$GABA_A$ receptors by replacement of several divergent amino-terminal regions with the a1 counterparts', *Br. J. Pharmacol.* **120**: 559–564.

Jackel, C., Kleinz, R., Makela, R., Hevers, W., Jezequel, S., Korpi, E.R. and Lüddens, H. (1998) 'The main determinant of furosemide inhibition on $GABA_A$ receptors is located close to the first transmembrane domain', *Eur. J. Pharmacol.* **357**: 251–256.

Jechlinger, M., Pelz, R., Tretter, V., Klausberger, T. and Sieghart, W. (1998) 'Subunit composition and quantitative importance of hetero-oligomeric receptors: $GABA_A$ receptors containing $\alpha 6$ subunits', *J. Neurosci.* **18**: 2449–2457.

Johnston, J.D., Price, S.A. and Bristow, D.R. (1998) 'Benzodiazepine chronic exposure leads to protein kinase C mediated internalization of $GABA_A$ receptors in primary cultured cerebellar granule cells', *Br. J. Pharmacol.* **124**: 1328–1340.

Jones, M.V. and Westbrook, G.L. (1995) 'Desensitized states prolong $GABA_A$ channel responses to brief agonist pulses', *Neuron* **15**: 181–191.

Jones, M.V. and Westbrook, G.L. (1996) 'The impact of receptor desensitization on fast synaptic transmission', *Trends Neurosci.* **19**: 96–101.

Jones, M.V. and Westbrook, G.L. (1997) 'Shaping of IPSCs by endogenous calcineurin activity', *J. Neurosci.* **17**: 7626–7633.

Jones, A., Korpi, E.R., McKernan, R.M., Pelz, R., Nusser, Z., Makela, R., Mellor, J.R., Pollard, S., Bahn, S., Stephenson, F.A., Randall, A.D., Sieghart, W., Somogyi, P., Smith, A.J.H. and Wisden, W. (1997) 'Ligand-gated ion channel subunit partnerships: $GABA_A$ receptor $\alpha 6$ subunit gene inactivation inhibits δ subunit expression', *J. Neurosci.* **17**: 1350–1362.

Kahn, Z., Gutierrez, A. and De Blas, A. (1994) 'The subunit composition of a $GABA_A$/benzodiazepine receptor from rat cerebellum', *J. Neurochem.* **63**: 371–374.

Kahn, Z., Gutierrez, A. and De Blas, A., (1996) 'The $\alpha 1$ and $\alpha 6$ subunits can coexist in the same cerebellar $GABA_A$ receptor maintaining their individual benzodiazepine-binding specificities', *J. Neurochem.* **66**: 685–691.

Kapur, J. and Macdonald, R.L. (1996) 'Cyclic AMP-dependent protein kinase enhances hippocampal dentate granule cell $GABA_A$ receptor currents', *J. Neurophysiol.* **76**: 2626–2634.

Kapur, J. and Macdonald, R.L. (1999) 'Postnatal development of hippocampal dentate granule cell $GABA_A$ receptor pharmacological properties', *Mol. Pharmacol.* **55**: 444–452.

Kellenberger, S., Eckerstein, S., Baur, R., Malherbe, P., Buhr, A. and Sigel, E. (1996) 'Subunit stoichiometry of oligomeric membrane proteins: $GABA_A$ receptors isolated by selective immunoprecipitation from the cell surface', *Neuropharmacology* **35**: 1403–1411.

Kellenberger, S., Malherbe, P. and Sigel, E. (1992) 'Function of the $\alpha1\beta2\gamma2S$ $GABA_A$ receptor is modulated by protein kinase C via multiple phosphorylation sites', *J. Biol. Chem.* **267**: 25,660–25,663.

Kim, H. and Macdonald, R.L. (1999) 'The role of the amino terminal domain of α subunits in regulation of γ-aminobutyric $acid_A$ receptors by La', *Abstr. Soc. Neurosci.* **25**: 1710.

Kim, H.Y., Gordey, M. and Olsen, R.W. (1997) '$\alpha1$ and $\alpha4$ chimeric $GABA_A$ receptor polypeptides show differential sensitivity to GABA and the neuroactive steroid alphaxalone', *Abstr. Soc. Neurosci.* **23**: 959.

Kim, H.Y., Olsen, R.W. and Tobin, A.J. (1996) 'GABA and $GABA_A$ receptors: development and regulation', in C.A. Shaw (ed.) *Receptor Dynamics in Neural Development*, Boca Raton, Fla.: CRC Press, pp. 59–72.

Kirsch, J. and Betz, H. (1995) 'The postsynaptic localization of the glycine receptor-associated protein gephyrin clusters is regulated by the cytoskeleton', *J. Neurosci.* **15**: 4148–4156.

Kneussel, M., Brandstatter, J.H., Laube, B., Stahl, S., Muller, U. and Betz, H. (1999) 'Loss of postsynaptic $GABA_A$ receptor clustering in gephyrin-deficient mice', *J. Neurosci.* **19**: 9289–9297.

Knight, A.R., Hartnett, C., Marks, C., Brown, M., Gallager, D., Tallman, J. and Ramabhadran, T.V. (1998) 'Molecular size of recombinant $\alpha1\beta1$ and $\alpha1\beta1\gamma2$ $GABA_A$ receptors expressed in Sf9 cells', *Recept. Chann.* **6**: 1–18.

Koltchine, V.V., Ye, Q., Finn, S.E. and Harrison, N.L. (1996) 'Chimeric $GABA_A$/glycine receptors: expression and barbiturate pharmacology', *Neuropharmacology* **35**: 1445–1456.

Korpi, E.R., Kuner, T., Seeburg, P.H. and Lüddens, H. (1995) 'Selective antagonist for the cerebellar granule cell-specific γ-aminobutyric acid type A receptor', *Mol. Pharmacol.* **47**: 283–289.

Koulen, P., Sassoe-Pognetto, M., Grunert, U. and Wassle, H. (1996) 'Selective clustering of $GABA_A$ and glycine receptors in the mammalian retina', *J. Neurosci.* **16**: 2127–2140.

Krasowski, M.D., Koltchine, V.V., Rick, C.E., Ye, Q., Finn, S.E. and Harrison, N.L. (1998) 'Propofol and other intravenous anesthetics have sites of action on the $GABA_A$ receptor distinct from isoflurane', *Mol. Pharmacol.* **53**: 530–538.

Krishek, B.J., Xie, X., Blackstone, C., Huganir, R.L., Moss, S.J. and Smart, T.G. (1994) 'Regulation of $GABA_A$ receptor function by protein kinase C phosphorylation', *Neuron* **12**: 1081–1095.

Kuhse, J., Laube, B., Magalei, D. and Betz, H. (1993) 'Assembly of the inhibitory glycine receptor: identification of amino acid sequence motifs governing subunit stoichiometry', *Neuron* **11**: 1049–1056.

Kusama, T., Wang, J.B., Spivak, C.E. and Uhl, G.R. (1994) 'Mutagenesis of the GABA $\rho1$ receptor alters agonist affinity and channel gating', *Neuroreport* **5**: 1209–1212.

Lan, N.C., Gee, K.W., Bolger, M.B. and Chen, J.S. (1991) 'Differential responses of expressed recombinant human $GABA_A$ receptors to neurosteroids', *J. Neurochem.* **57**: 1818–1821.

Laurie, D.J., Wisden, W. and Seeburg, P.H. (1992) 'The distribution of thirteen $GABA_A$ receptor subunit mRNAs in the rat brain. III. Embryonic and postnatal development', *J. Neurosci.* **12**: 4151–4172.

Leidenheimer, N.J., Machu, T.K., Endo, S., Olsen, R.W., Harris, R.A. and Browning, M.D. (1991) 'Cyclic AMP-dependent protein kinase decreases gamma-aminobutyric $acid_A$ receptor-mediated 36Cl− uptake by brain microsacs', *J. Neurochem.* **57**: 722–725.

Leidenheimer, N.J., Whiting, P.J. and Harris, R.A. (1993) 'Activation of calcium-phospholipid-dependent protein kinase enhances benzodiazepine and barbiturate potentiation of the $GABA_A$ receptor', *J. Neurochem.* **60**: 1972–1975.

Lin, Y.F., Angelotti, T.P., Dudek, E.M., Browning, M.D. and Macdonald, R.L. (1996) 'Enhancement of recombinant $\alpha1\beta1\gamma2L$ $GABA_A$ receptor whole-cell currents by protein kinase C is mediated through phosphorylation of both $\beta1$ and $\gamma2L$ subunits', *Neuron* **13**: 1421–1431.

Lin, Y.F., Browning, M.D., Dudek, E.M. and Macdonald, R.L. (1994) 'Protein kinase C enhances recombinant bovine $\alpha 1\beta 1\gamma 2L$ $GABA_A$ receptor whole-cell currents expressed in L929 fibroblasts', *Neuron* **13**: 1421–1431.

Lynch, J.W., Rajendra, S., Pierce, K.D., Handford, C.A., Barry, P.H. and Schofield, P.R. (1997) 'Identification of intracellular and extracellular domains mediating signal transduction in the inhibitory glycine receptor chloride channel', *EMBO J.* **16**: 110–120.

Macdonald, R.L. and Olsen, R.W. (1994) '$GABA_A$ receptor channels', *Ann. Rev. Neurosci.* **17**: 569–602.

Macdonald, R.L, Rogers, C.J. and Twyman, R.E. (1989) 'Kinetic properties of the $GABA_A$ receptor main conductance state of mouse spinal cord neurons in culture', *J. Physiol.* **410**: 479–499.

Maconochie, D.J., Zempel, J.M. and Steinbach, J.H. (1994) 'How quickly can $GABA_A$ receptors open?', *Neuron* **12**: 61–71.

Makela, R., Uusi-Oukari, M., Homanics, G.E., Quinlan, J.J., Firestone, L.L., Wisden, W. and Korpi, E.R. (1997) 'Cerebellar $GABA_A$ receptors: pharmacological subtypes revealed by mutant mouse lines', *Mol. Pharmacol.* **52**: 380–388.

McDonald, B.J., Amato, A., Connolly, C.N., Benke, D., Moss, S.J. and Smart, T.G. (1998) 'Adjacent phosphorylation of $GABA_A$ receptor β subunits determine regulation by cAMP-dependent protein kinase', *Nature Neurosci.* **1**: 23–28.

McDonald, B.J. and Moss, S.J. (1997) 'Conserved phosphorylation of the intracellular domains of $GABA_A$ receptor $\beta 2$ and $\beta 3$ subunits by cAMP-dependent protein kinase, cGMP-dependent protein kinase, protein kinase C and Ca^{2+}/calmodulin Type II-dependent protein kinase', *Neuropharmacology* **36**: 1377–1385.

McKernan, R.M. and Whiting, P.J. (1996) 'Which $GABA_A$-receptor subtypes really occur in the brain?', *Trends Neurosci.* **19**: 139–143.

Mellor, J.R. and Randall, A.D. (1997) 'Frequency-dependent actions of benzodiazepines on $GABA_A$ receptors in cultured murine cerebellar granule cells', *J. Physiol.* **503.2**: 353–369.

Mertens, S., Benke, D. and Möhler, H. (1993) '$GABA_A$ receptor populations with novel subunit combinations and drug binding profiles identified in brain by $\alpha 5$- and δ-subunit-specific immunopurification', *J. Biol. Chem.* **268**: 5965–5973.

Meyer, G., Kirsch, J., Betz, H. and Langosch, D. (1995) 'Identification of a gephyrin binding motif on the glycine receptor β subunit', *Neuron* **15**: 563–572.

Mihic, S.J., Whiting, P.J., Klein, R.L., Wafford, K.A. and Harris, R.A. (1994) 'A single amino acid of the human $GABA_A$ receptor $\gamma 2$ subunit determines benzodiazepine efficacy', *J. Biol. Chem.* **269**: 32,768–32,783.

Mihic, S.J., Ye, Q., Wick, M.J., Koltchine, V.V., Krasowski, M.D., Finn, S.E., Mascia, M.P., Valenzuela, C.F., Hanson, K.K., Greenblatt, E.P., Harris, R.A. and Harrison, N.L. (1997) 'Sites of alcohol and volatile anesthetic action on $GABA_A$ and glycine receptors', *Nature* **389**: 385–389.

Miyazawa, A., Fujiyoshi, Y., Stowell, M. and Unwin, N. (1999) 'Nicotinic acetylcholine receptor at 4.6 A resolution: Transverse tunnels in the channel wall', *J. Mol. Biol.* **14**: 765–786.

Mochly-Rosen, D. and Gordon, A.S. (1998) 'Anchoring proteins for protein kinase C: a means for isozyme selectivity', *FASEB J.* **12**: 35–42.

Moody, E.J., Knauer, C., Granja, R., Strakhova, M. and Skolnick, P. (1997) 'Distinct loci mediate the direct and indirect actions of the anesthetic etomidate at $GABA_A$ receptors', *J. Neurochem.* **69**: 1310–1313.

Moss S.J., Smart T.A., Porter N.M., Nayeem, N., Devine, J., Stephenson, F.A., Macdonald, R.L. and Barnard, E.A. (1990) 'Cloned GABA receptors are maintained in a stable cell line: Allosteric and channel properties', *Eur. J. Pharmacol.* **189**: 77–88.

Moss, S.J., Doherty, C.A. and Huganir, R.L. (1992a) 'Identification of the cAMP-dependent protein kinase and protein kinase C phosphorylation sites within the major intracellular domains of the $\beta 1$: $\gamma 2S$, and $\gamma 2L$ subunits of the $GABA_A$ receptor', *J. Biol. Chem.* **267**: 14,470–14,476.

Moss, S.J., Smart, T.G., Blackstone, C.D. and Huganir, R.L. (1992b) 'Functional modulation of $GABA_A$ receptors by cAMP-dependent protein phosphorylation', *Science* **257**: 661–665.

Moss, S.J., Gorrie, G.H., Amato, A. and Smart, T.G. (1995) 'Modulation of $GABA_A$ receptors by tyrosine phosphorylation', *Nature* **377**: 344–348.

Nagaya, N., Sun, F. and Macdonald, R.L. (1998) 'Low zinc sensitivity of γ-containing $GABA_A$ receptors is conferred by at least two γ subunit domains', *Abstr. Soc. Neurosci.* **24**: 1990.

Nayeem N., Green, T.P., Martin, I.L. and Barnard E.A. (1994) 'Quaternary structure of the native $GABA_A$ receptor determined by electron microscopic image analysis', *J. Neurochem.* **62**: 815–818.

Neelands, T.R., Fisher, J., Bianchi, M. and Macdonald, R.L. (1999) 'ϵ subunit-containing $GABA_a$ receptor channels produce spontaneous and GABA-evoked currents', *Mol. Pharmacol.* **55**: 168–178.

Neelands, T.R. and Macdonald RL (1999) 'Incorporation of the π subunit into functional $GABA_A$ receptors', *Mol. Pharmacol.* **56**: 598–610.

Newell, J.G., Bateson, A.N. and Dunn, S.M.J. (1999) 'Y64 of the $\beta 2$ subunit of the $GABA_A$ receptor is an important determinant for high affinity muscimol recognition', *Abstr. Soc. Neurosci.* **25**: 1711.

Nguyen, Q., Sapp, D.W., Van Ness, P.C. and Olsen, R.W. (1995) 'Modulation of $GABA_A$ receptor binding in human brain by neuroactive steroids: Species and brain regional differences', *Synapse* **19**: 77–87.

Nusser, Z., Roberts, J.D., Baude, A., Richards, J.G. and Somogyi, P. (1995) 'Relative densities of synaptic and extrasynaptic $GABA_A$ receptors on cerebellar granule cells as determined by quantitative immuno-gold method', *J. Neurosci.* **15**: 2948–2960.

Nusser, Z., Hajos, N., Somogyi, P. and Mody, I. (1998) 'Increase in the number of synaptic $GABA_A$ receptors underlies change in synaptic efficacy at hippocampal inhibitory synapses', *Nature* **395**: 172–177.

Nusser, Z., Sieghart, W., Stephenson, F.A. and Somogyi, P. (1996) 'The $\alpha 6$ subunit of the $GABA_A$ receptor is concentrated in both inhibitory and excitatory synapses on cerebellar granule cells', *J. Neurosci.* **16**: 103–144.

Nusser, Z., Ahmad, Z., Tretter, V., Fuchs, K., Wisden, W. Sieghart, W. and Somogyi, P. (1999) 'Alterations in the expression of $GABA_A$ receptor subunits in cerebellar granule cells after the disruption of the $\alpha 6$ subunit gene', *Eur. J. Neurosci.* **11**: 1685–1697.

Olsen, R.W. (1998) 'The molecular mechanism of action of general anesthetics: structural aspects of interactions with $GABA_A$ receptors', *Toxicol. Lett.* **101**: 193–201.

Olsen, R.W. and Sapp, D.W. (1995) 'Neuroactive steroid modulation of $GABA_A$ receptors', in G. Biggio, E. Sanna, M. Serra and E. Costa (eds) *$GABA_A$ Receptors and Anxiety: From Neurobiology to Treatment* [*Adv. Biochem. Psychopharm.* **48**], New York: Raven Press, pp. 57–74.

Olsen, R.W. and Tobin, A.J. (1990) 'Molecular biology of $GABA_A$ receptors', *FASEB J.* **4**: 1469–1480.

Olsen R.W., Fischer J.B. and Dunwiddie T.V. (1986) 'Barbiturate enhancement of GABA receptor binding and function as a mechanism of anesthesia', in S. Roth, and K.W. Miller (eds) *Molecular and Cellular Mechanisms of Anaesthetics*, New York: Plenum Publishing, pp. 165–177.

Olsen, R.W., Sapp, D.W, Bureau, M.H., Turner, D.M. and Kokka, N. (1991) 'Allosteric actions of CNS depressants including anesthetics on subtypes of the inhibitory $GABA_A$ receptor-chloride channel complex', in E. Rubin, K.W. Miller and S.H. Roth (eds), *Molecular and Cellular Mechanisms of Alcohol and Anesthetics* [*Ann. N.Y. Acad. Sci.* **625**]: 145–154.

Olsen, R.W., Smith, G.B. and Srinivasan, S. (1996) 'Modelling functional domains of the $GABA_A$ receptor chloride channel protein', in C. Tanaka and N. Bowery (eds) *GABA: Receptors, Transporters and Metabolism*, Basel: Birkhauser, pp. 145–155.

Pan, Z.H. and Lipton, S.A. (1995) 'Multiple GABA receptor subtypes mediate inhibition of calcium influx at rat retinal bipolar cell terminals', *J. Neurosci.* **15**: 2668–2679.

Pan, Z.H., Zhang, D., Zhang, X. and Lipton, S.A. (1997) 'Agonist-induced closure of constitutively open GABA channels with mutated domains', *Proc. Natl. Acad. Sci. USA* **94**: 6490–6495.

Poisbeau, P., Cheney, M.C., Browning, M.D. and Mody, I. (1999) 'Modulation of synaptic $GABA_A$ receptor function by PKA and PKC in adult hippocampal neurons', *J. Neurosci.* **19**: 674–683.

Pollard, S., Duggan, M. and Stephenson, F. (1993) 'Further evidence for the existence of α subunit heterogeneity within discrete γ-aminobutyric $acid_A$ receptor subpopulations', *J. Biol. Chem.* **268**: 3753–3757.

Pollard, S., Thompson, C., Stephenson, F. (1995) 'Quantitative characterization of α6 and α1α6 subunit-containing native γ-aminobutyric $acid_A$ receptors of adult rat cerebellum demonstrates two α subunits per receptor oligomer', *J. Biol. Chem.* **270**: 21,285–21,290.

Porter, N.M., Twyman, R.E., Uhler, M.D. and Macdonald, R.L. (1990) 'Cyclic AMP-dependent protein kinase decreases $GABA_A$ receptor current in mouse spinal neurons', *Neuron* **5**: 789–796.

Poulter, M.O., Barker, J.L., O'Carroll, A.M., Lolait, S.J. and Mahan, L.C. (1992) 'Differential and transient expression of the $GABA_A$ receptor α subunit mRNA in the developing rat CNS', *J. Neurosci.* **12**: 2888–2900.

Pribilla, I., Takagi, T., Langosch, D., Bormann, J. and Betz, H. (1992) 'The atypical M2 segment of the β subunit confers picrotoxinin resistance to inhibitory glycine receptor channels', *EMBO J.* **11**: 4305–4311.

Pritchett, D.B. and Seeburg, P.H. (1991) '$GABA_A$ receptor point mutations increases the affinity of compounds for the benzodiazepine site', *Proc. Natl. Acad. Sci. USA* **88**: 1421–1425.

Puia, G., Ducic, I., Vicini, S. and Costa, E. (1993) 'Does neurosteroid modulatory efficacy depend on $GABA_A$ receptor subunit composition?', *Recept. Chann.* **1**: 135–142.

Quirk, K., Gillard, N.P., Ragan, I., Whiting, P.J. and McKernan, R.M. (1994) 'Model of subunit composition of γ-aminobutyric acid A receptor subtypes expressed in rat cerebellum with respect to their α and γ/δ subunits', *J. Biol. Chem.* **266**: 16,020–16,028.

Quirk, K., Whiting, P.J., Ragan, I. and McKernan, R.M. (1995) 'Characterization of δ-subunit containing $GABA_A$ receptors from rat brain', *Eur. J. Pharmacol.* **290**: 175–181.

Quirk, K., Blurton, P., Fletcher, S., Leeson, P., Tang, F., Mellilo, D., Ragan, C.I. and McKernan, R.M. (1996) '[^{3}H] L-655,708: a novel ligand selective for the benzodiazepine site of $GABA_A$ receptors which contain the alpha 5 subunit', *Neuropharmacology* **35**: 1331–1335.

Ramarao, M.K. and Cohen, J.B. (1998) 'Mechanism of nicotinic acetylcholine receptor cluster formation by rapsyn', *Proc. Natl. Acad. Sci. USA* **95**: 4007–4012.

Ron, D., Chen, C.H., Caldwell, J., Jamieson, L., Orr, E. and Mochly-Rosen, D. (1994) 'Cloning of an intracellular receptor for protein kinase C: a homolog of the β subunit of G proteins', *Proc. Natl. Acad. Sci. USA* **91**: 839–843.

Rudolph, U. Crestani, F., Benke, D., Brunig, I., Benson, J.A., Fritschy, J.M., Martin, J.R., Bluethmann, H. and Möhler, H. (1999) 'Benzodiazepine actions mediated by specific $GABA_A$ receptor subtypes', *Nature* **401**: 796–800.

Sakmann, B., Hamill, O.P. and Bormann, J. (1983) 'Patch-clamp measurements of elementary chloride currents activated by the putative inhibitory transmitters GABA and glycine in mammalian spinal neurons', *J. Neural. Trans.* **18** (suppl.): 83–95.

Sapp, D.W., Witte, U., Turner, D.M., Longoni, B., Kokka, N. and Olsen, R.W. (1992) 'Regional variation in steroid anesthetic modulation of [35S] TBPS binding to γ-aminobutyric acidA receptors in rat brain', *J. Pharmacol. Exp. Ther.* **262**: 801–808.

Sassoe-Pognetto, M., Kirsch, J., Grunert, U., Gregerath, U., Fritschy, J.M., Möhler, H., Betz, H. and Wassle, H. (1995) 'Co-localization of gephyrin and $GABA_A$ receptor subunits', *J. Comp. Neurol.* **357**: 1–14.

Saxena N.C. and Macdonald, R.L. (1994) 'Assembly of $GABA_A$ receptor subunits: role of the δ subunit', *J. Neurosci.* **14**: 7077–7086.

Saxena, N.C., Neelands, T.R. and Macdonald, R.L. (1997) 'Contrasting actions of lanthanum on different recombinant γ-aminobutyric $acid_A$ receptor isoforms expressed in L929 fibroblasts', *Mol. Pharmacol.* **51**: 328–335.

Schaerer, M.T., Buhr, A., Baur, R. and Sigel, E. (1998) 'Amino acid residue 200 on the α1 subunit of $GABA_A$ receptors affects the interaction with selected benzodiazepine binding site ligands', *Eur. J. Pharmacol.* **354**: 283–287.

Serafini, R., Bracamontes, J. and Steinbach, J.H. (1997) 'Structural domains involved in the effects of general anesthetics on recombinant $GABA_A$ receptors', *Abstr. Soc. Neurosci.* **23**: 110.

Shingai, R., Sutherland, M.L. and Barnard, E.A. (1991) 'Effects of subunit types of the cloned $GABA_A$ receptor on the response to a neurosteroid', *Eur. J. Pharmacol.* **206**: 77–80.

Sieghart, W. (1995) 'Structure and pharmacology of $GABA_A$ receptor subtypes', *Pharmacol. Rev.* **47**: 181–234.

Sigel, E. and Baur, R. (1988) 'Activation of protein kinase C differentially modulates neuronal Na^+: Ca^{2+}: and γ-aminobutyrate type A channels', *Proc. Natl. Acad. Sci. USA* **85**: 6192–6196.

Sigel, E., Baur, R. and Malherbe, P. (1991) 'Activation of protein kinase C results in down-modulation of different recombinant $GABA_A$-channels', *FEBS Lett.* **291**: 150–152.

Sigel, E., Baur, R., Kellenberger, S. and Malherbe, P. (1992) 'Point mutations affecting antagonist affinity and agonist dependent gating of $GABA_A$ receptor channels', *EMBO J.* **11**: 2017–2023.

Sigel, E. and Buhr, A. (1997) 'The benzodiazepine binding site of $GABA_A$ receptors', *Trends Pharmacol. Sci.* **18**: 425–429.

Smart, T.G. (1992) 'A novel modulatory binding site for zinc on the $GABA_A$ receptor complex in cultured rat neurones', *J. Physiol.* **447**: 587–625.

Smith, G.B. and Olsen, R.W. (1994) 'Identification of a $[^3H]$muscimol photoaffinity substrate in the bovine γ-aminobutyric acidA receptor α subunit', *J. Biol. Chem.* **269**: 20,380–20,387.

Smith, G.B. and Olsen, R.W. (1995) 'Functional domains of $GABA_A$ receptors', *Trends Pharmacol. Sci.* **16**: 162–168.

Smith, G.B. and Olsen, R.W. (2000) 'Deduction of amino acid residues in the GABA(A) receptor alpha subunits photoaffinity labeled with the benzodiazepine flunitrazepam', *Neuropharmacology* **39**: 55–64.

Somogyi, P., Fritschy, J.M., Benke, D., Roberts, J.D. and Sieghart, W. (1996) 'The γ2 subunit of the $GABA_A$ receptor is concentrated in synaptic junctions containing the α1 and β2/3 subunits in hippocampus, cerebellum, and globus pallidus', *Neuropharmacology* **35**: 1425–1444.

Srinivasan, S., Sapp, D.W., Tobin, A.J. and Olsen, R.W. (1999a) 'Biphasic modulation of $GABA_A$ receptor binding by steroids suggests functional correlates', *Neurochem. Res.*, **24**: 1363–1372.

Srinivasan, S., Nichols, C.J., Olsen, R.W. and Tobin, A.J. (1999b) 'Two invariant tryptophans define domains necessary for $GABA_A$ receptor assembly', *J. Biol. Chem.*, **274**: 26,633–26,638.

Stephenson, F.A. and Duggan, M.J. (1989) 'Mapping the benzodiazepine-affinity labelling site with sequence-specific $GABA_A$ receptor antibodies', *Biochem. J.* **264**: 199–206.

Sur, C., Quirk, K., Dewar, D., Atack, J. and McKernan, R.M. (1998) 'Rat and human hippocampal α5 subunit-containing $GABA_A$ receptors have α5β3γ2 pharmacological characteristics', *Mol. Pharmacol.* **54**: 928–933.

Sur, C., Farrar, S.J., Kerby, J., Whiting, P.J., Atack, J. and McKernan, R.M. (1999) 'Preferential coassembly of α4 and δ subunits of the γ-aminobutyric $acid_A$ receptor in rat thalamus', *Mol. Pharmacol.* **56**: 110–115.

Taylor, P.M., Thomas, P., Gorrie, G.H., Gibson, A., Smart, T.G. and Moss, S.J. (1999) 'Identification of amino acids residues within $GABA_A$ receptor β subunits that mediate both homomeric and heteromeric receptor expression', *J. Neurosci.* **19**: 6360–6371.

Tehrani, M.H.J. and Barnes, E.M. (1991) 'Agonist-dependent internalization of $GABA_A$/benzodiazepine receptors in chick cortical neurons', *J. Neurochem.* **57**: 1307–1312.

Tehrani, M.H.J. and Barnes, E.M. (1993) 'Identification of $GABA_A$/benzodiazepine receptors on clathrin-coated vesicles from rat brain', *J. Neurochem.* **60**: 1755–1761.

Tehrani, M.H.J. and Barnes, E.M. (1994) '$GABA_A$ receptors in mouse cortical homogenates are phosphorylated by endogenous protein kinase A', *Mol. Brain Res.* **24**: 55–64.

Tehrani, M.H.J., Hablitz, J.J. and Barnes, E.M. (1989) 'cAMP increases the rate of $GABA_A$ receptor desensitization in chick cortical neurons', *Synapse* **4**: 126–131.

Thompson, S.A., Arden, S.A., Marshall, G., Wingrove, P.B., Whiting, P.J. and Wafford, K.A.

(1999a) 'Residues in transmembrane domains I and II determine $GABA_A$ receptor subtype-selective antagonism by furosemide', *Mol. Pharmacol.* **55**: 993–999.

Thompson, S.A., Smith, M.Z., Wingrove, P.B., Whiting, P.J. and Wafford, K.A. (1999b) 'Mutation of the putative $GABA_A$ ion-channel gate reveals changes in allosteric modulation', *Br. J. Pharmacol.* **127**: 1349–1358.

Thompson, S.A., Whiting, P.J. and Wafford, K.A. (1996) 'Barbiturate interactions at the human $GABA_A$ receptor: dependence on receptor subunit composition', *Br. J. Pharmacol.* **117**: 521–527.

Tia, S., Wang, J.F., Kotchabhakdi, N. and Vicini, S. (1996a) 'Developmental changes of inhibitory synaptic currents in cerebellar granule neurons: role of $GABA_A$ receptor α6 subunit', *J. Neurosci.* **16**: 3630–3640.

Tia, S., Wang, J.F., Kotchabhakdi, N. and Vicini, S. (1996b) 'Distinct deactivation and desensitization kinetics of recombinant $GABA_A$ receptors', *Neuropharmacology* **35**: 1375–1382.

Tierney, M.L., Birnir, B., Pillai, N.P., Clements, J.D., Howitt, S.M., Cox, G.B. and Gage, P.W. (1996) 'Effects of mutating leucine to threonine in the M2 segment of α1 and β1 subunits of $GABA_A$ α1β1 receptors', *J. Membr. Biol.* **154**: 11–21.

Tretter, V., Ehya, N., Fuchs, K. and Sieghart, W. (1997) 'Stoichiometry and assembly of a recombinant $GABA_A$ receptor subtype', *J. Neurosci.* **17**: 2728–2737.

Twyman, R.E., Rogers, C.J. and Macdonald, R.L. (1990) 'Intraburst kinetic properties of the $GABA_A$ receptor main conductance state of mouse spinal cord neurones in culture', *J. Physiol.* **423**: 193–220.

Unwin, N. (1995) 'Acetylcholine receptor channel imaged in the open state', *Nature* **373**: 37–43.

Valenzuela, C.F., Machu, T.K., McKernan, R.M., Whiting, P., Van Renterghem, B.B., McManaman, J.L., Brozowski, S.J., Smith, G.B., Olsen, R.W. and Harris, R.A. (1995) 'Tyrosine kinase phosphorylation of $GABA_A$ receptors', *Mol. Brain. Res.* **31**: 165–172.

Verdoorn, T.A. (1994) 'Formation of heteromeric γ-aminobutyric acid type A receptors containing two different alpha subunits', *Mol. Pharmacol.* **45**: 475–480.

Verdoorn, T.A., Draguhn, A., Ymer, S., Seeburg, P.H. and Sakmann, B. (1990) 'Functional properties of recombinant rat $GABA_A$ receptors depend upon subunit composition', *Neuron* **4**: 919–928.

Wafford, K.A., Bain, C.J., Quirk, K., McKernan, R.M., Wingrove, P.B., Whiting, P.J. and Kemp, J.A. (1994) 'A novel allosteric modulatory site on the $GABA_A$ receptor β subunit', *Neuron* **12**: 775–782.

Wan, Q., Man, H.Y., Braunton, J., Wang, W., Salter, M.W., Becker, L. and Wang, Y.T. (1997) 'Modulation of $GABA_A$ receptor function by tyrosine phosphorylation of β subunits', *J. Neurosci.* **17**: 5062–5069.

Wang, H., Bedford, F.K., Brandon, N.J., Moss, S.J. and Olsen, R.W. (1999) '$GABA_A$-receptor-associated protein links $GABA_A$ receptors and the cytoskeleton', *Nature* **397**: 69–72.

Wang, T., Hackam, A.S., Guggino, W.B. and Cutting, G.R. (1995a) 'A single amino acid in GABA ρ1 receptors affects competitive and noncompetitive components of picrotoxin inhibition', *Proc. Natl. Acad. Sci. USA* **92**: 11,751–11,755.

Wang, T., Hackam, A.S., Guggino, W.B. and Cutting, G.R. (1995b) 'A single histidine residue is essential for zinc inhibition of GABA ρ1 receptors', *J. Neurosci.* **15**: 7684–7691.

Weiss, D.S. and Magleby, K.L. (1989) 'Gating scheme for single GABA-activated Cl− channels determined from stability plots, dwell-time distributions, and adjacent-interval durations', *J. Neurosci.* **9**: 1314–1324.

Westh-Hansen, D.E., Rasmussen, P.B., Hastrup, S., Nabekura, J., Noguchi, K., Akaike, N., Witt, M.R. and Nielsen, M. (1997) 'Decreased agonist sensitivity of human $GABA_A$ receptors by an amino acid variant, isoleucine to valine, in the α1 subunit', *Eur. J. Pharmacol.* **329**: 253–257.

Whatley, V.J., Mihic, S.J., Allan, A.M., McQuilkin, S.J. and Harris, R.A. (1994) '$GABA_A$ receptor function is inhibited by microtubule depolymerization', *J. Biol. Chem.* **269**: 19,546–19,552.

Wick, M.J., Mihic, S.J., Ueno, S., Mascia, M.P., Trudell, J.R., Brozowski, S.J., Ye, Q., Harrison, N.L. and Harris R.A. (1998) 'Mutations of GABA and glycine receptors change alcohol cutoff: evidence for an alcohol receptor?', *Proc. Natl. Acad. Sci. USA* **95**: 6504–6509.

Wieland, H., Lüddens, H. and Seeburg, P.H. (1992) 'A single histidine in $GABA_A$ receptors is essential for benzodiazepine agonist binding', *J. Biol. Chem.* **257**: 1426–1429.

Wieland, H.A. and Lüddens, H. (1994) 'Four amino acid exchanges convert a diazepam-insensitive inverse agonist-preferring $GABA_A$ receptor into a diazepam-preferring $GABA_A$ receptor', *J. Med. Chem.* **37**: 4576–4580.

Wingrove, P.B., Thompson, S.A., Wafford, K.A. and Whiting, P.J. (1997) 'Key amino acids in the γ subunit of the $GABA_A$ receptor that determine ligand binding and modulation at the benzodiazepine site', *Mol. Pharmacol.* **52**: 874–881.

Wingrove, P.B., Wafford, K.A., Bain, C. and Whiting, P.J. (1994) 'The modulatory action of loreclezole at the $GABA_A$ receptor is determined by a single amino acid in the β2 and β3 subunit', *Proc. Natl. Acad. Sci. USA* **91**: 4569–4573.

Wisden, W., Laurie, D.J., Monyer, H.M. and Seeburg, P.H. (1992) 'The distribution of 13 $GABA_A$ receptor subunit mRNAs in rat brain. I. Telencephalon, diencephalon, mesencephalon', *J. Neurosci.* **12**: 1040–1062.

Wooltorton, J.R., McDonald, B.J., Moss, S.J. and Smart, T.G. (1997) 'Identification of a Zn^{2+} binding site on the murine $GABA_A$ receptor complex: dependence on the second transmembrane domain of β subunits', *J. Physiol.* **505.3**: 633–640.

Xu, M. and Akabas, M. (1993) 'Amino acids lining the channel of the γ-aminobutyric acid type A receptor identified by cysteine substitution', *J. Biol. Chem.* **268**: 21,505–21,508.

Xu, M. and Akabas, M. (1996) 'Identification of channel-lining residues in the M2 membrane-spanning segment of the $GABA_A$ receptor α1 subunit', *J. Gen. Physiol.* **107**: 195–205.

Xu, M., Covey, D.F. and Akabas, M.H. (1995) 'Interaction of picrotoxin with $GABA_A$ receptor channel-lining residues probed in cysteine mutants', *Biophys. J.* **69**: 1858–1867.

Yang, S.H., Armson, P.F., Cha, J. and Phillips, W.D. (1997) 'Clustering of $GABA_A$ receptors by rapsyn/43 kD protein in vitro', *Mol. Cell. Neurosci.* **8**: 430–438.

Zhang, D., Pan, Z.H., Brideau, A.D. and Lipton, S.A. (1995) 'Cloning of a $GABA_C$ receptor subunit in rat retina with a methionine residue critical for picrotoxinin channel block', *Proc. Natl. Acad. Sci. USA* **92**: 11,756–11,760.

Zhu, W.J. and Vicini, S. (1997) 'Neurosteroid prolongs $GABA_A$ channel deactivation by altering kinetics of desensitized states', *J. Neurosci.* **17**: 4022–4031.

Zhu, W.J., Wang, J.F., Krueger, K.E. and Vicini, S. (1996) 'δ subunit inhibits neurosteroid modulation of $GABA_A$ receptors', *J. Neurosci.* **16**: 6648–6656.

Chapter 10

Ligands for the $GABA_A$ receptor complex

Povl Krogsgaard-Larsen, Bente Frølund, Uffe Kristiansen and Bjarke Ebert

GABA inhibition and disinhibition

The neutral amino acid, 4-aminobutanoic acid (GABA), is an inhibitory transmitter in the central nervous system (CNS). Furthermore, GABA is involved as a neurotransmitter and/or a paracrine effector in the regulation of a variety of physiological mechanisms in the periphery. Some of these latter functions may be under central GABA control, whereas others are managed by local GABA neurons. A large percentage, perhaps the majority, of central neurons are under GABA control. The complex mechanisms underlying the GABA-mediated neurotransmission have been extensively studied using a broad spectrum of electrophysiological, neurochemical, pharmacological, and, in recent years, molecular biological techniques (Curtis and Johnston, 1974; Krnjevic, 1974; Olsen and Venter, 1986; Redburn and Schousboe, 1987; Bowery and Nistico, 1989; Biggio and Costa, 1990; Bowery *et al.*, 1990; Schousboe *et al.*, 1992a; Doble and Martin, 1996; Krogsgaard-Larsen *et al.*, 1997).

The overall activity of the brain is basically determined by two superior functions: (1) excitation by the major excitatory amino acid transmitter, glutamic acid (Glu), which depolarizes neurons through a large number of receptor subtypes (Lodge, 1988; Krogsgaard-Larsen and Hansen, 1992; Wheal and Thomson, 1995; Monaghan and Wenthold, 1997); and (2) inhibition by GABA, which hyperpolarizes neurons, likewise through multiple receptors. It may be mentioned, however, that depolarizing actions of GABA may occur, particularly during early postnatal development of the brain (Walton *et al.*, 1993).

It has been proposed that a third mechanism may play a fundamental role in the function of the brain – namely, disinhibition (Roberts, 1976, 1991). This indirect neuronal excitation implies synaptic contact between two inhibitory neurons. The operation of this indirect excitatory mechanism in the CNS has never been unequivocally proved or disproved, but many apparently paradoxical observations have been explained on the basis of disinhibition (Kardos *et al.*, 1994).

GABA in the peripheral nervous system: transmitter and paracrine effector

The role of GABA as a central transmitter is fully established, and there is growing evidence that GABA also has a broad spectrum of physiological functions in the peripheral nervous system (PNS) (Erdö, 1985; Erdö and Bowery, 1986; Bowery and Nistico,

1989; Bowery *et al.*, 1990; Ong and Kerr, 1990; Tanaka and Bowery, 1996). In a wide range of peripheral tissues, notably parts of the PNS, endocrine glands, smooth muscles, and the female reproductive systems, GABA receptors have been detected. In all tissues so far analysed, $GABA_A$ as well as $GABA_B$ receptors (Figure 10.1) have been identified.

There are many unsolved questions regarding the peripheral actions of GABA and its interactions with other physiological mechanisms, using acetylcholine (ACh), norepinephrine, serotonin, and various peptides as transmitter or paracrine mediators (Ong and Kerr, 1990). It is possible that disinhibitory mechanisms between GABA neurons or between $GABA_A$ and $GABA_B$ receptors at the cellular level (Kardos *et al.*, 1994), as described above, also contribute to the apparently very complex functions of GABA in the periphery.

The peripheral GABA receptors, or other GABAergic synaptic mechanisms, obviously have interest as drug targets, but so far, GABA drug design projects have been focused on sites at central GABA-operated synapses. It should, however, be emphasized that even for GABAergic drugs, which easily penetrate the blood–brain barrier (BBB), most of the drug administered is found in the periphery, where it may cause adverse effects. On the other hand, most GABA analogues of pharmacological interest do not easily penetrate the BBB, making it possible to develop GABAergic drugs specifically targeted at peripheral GABA receptors.

In light of the identification of GABA receptors different from $GABA_A$ or $GABA_B$

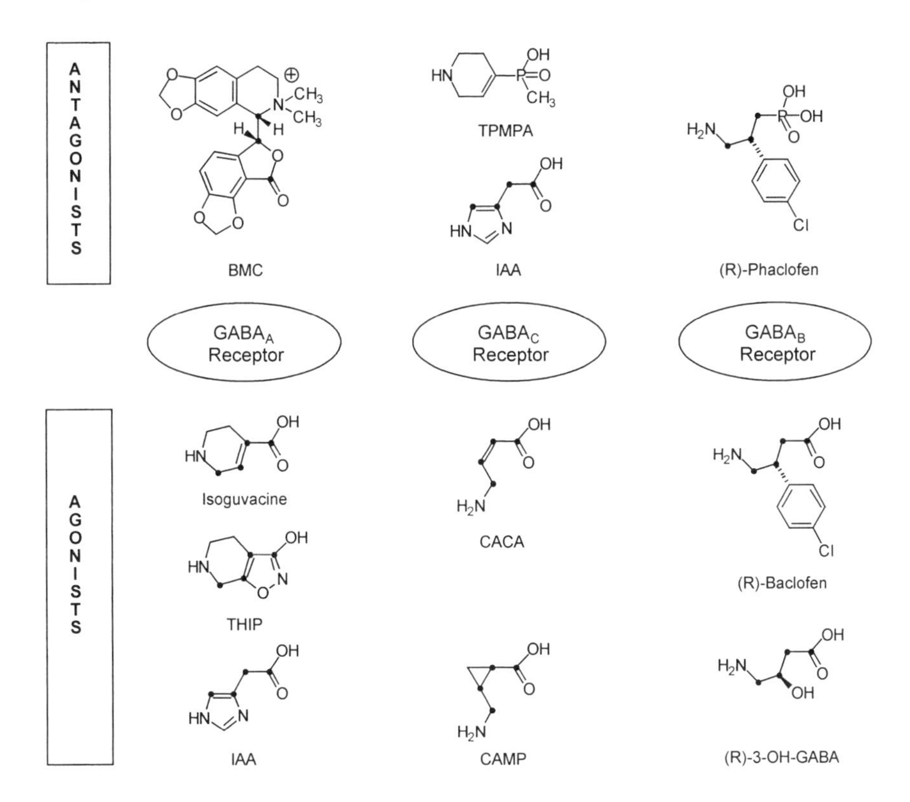

Figure 10.1 Schematic illustration of the different classes of GABA receptors and the structures of some key agonists and antagonists.

receptors, probably of the proposed $GABA_C$ type (Figure 10.1), in the retina, it seems likely that other types of GABA receptors may be identified in peripheral tissues. There is, for example, circumstantial evidence of an involvement of non-classical $GABA_A$ receptors in the regulation of the release of luteinizing hormone from rat pituitary cells (Virmani *et al.*, 1990). Such atypical GABA receptors in tissues showing specific physiological responsiveness to GABA may be particularly interesting targets for drug design.

The function of GABA, mediated by $GABA_A$ as well as $GABA_B$ receptors, in the enteric nervous system has been quite extensively studied (Kerr and Ong, 1986). Whilst it seems unlikely that GABA is playing a major role in intestinal secretory activity, $GABA_A$ and, perhaps in particular, $GABA_B$ receptors may play important roles in the control of gut motility. Thus, peristalsis can be markedly reduced and ultimately stopped after administration of $GABA_A$ and $GABA_B$ antagonists, separately or in combination (Ong and Kerr, 1990).

GABA receptors in the gut, as well as GABA receptors in the gall bladder, the lung and the urinary bladder (Erdö and Bowery, 1986), may give rise to adverse effects in GABA drug therapies or may be important new therapeutic targets in drug-induced or pathological dysfunctions of these organs.

The involvement of GABA in the regulation of blood pressure has been the object of numerous studies (Dannhardt *et al.*, 1993). The relative importance of $GABA_A$ and $GABA_B$ receptors in these very complex mechanisms is still unclear, and the effects of GABA mimetics are, to some extent, species-dependent. In this area, the $GABA_A$ receptor-mediated dilatation by GABA of cerebral blood vessels (Edvinsson *et al.*, 1980) is of particular interest and may have major therapeutic implications. It is possible that this dilatation actually is mediated by ACh released via a $GABA_A$ receptor-regulated mechanism (Saito *et al.*, 1985). The effect of the $GABA_A$ agonist THIP on cortical blood flow in humans has been used for diagnostic purposes as a new test for hemispheric dominance (Roland and Friberg, 1988). Thus, THIP decreases in a dose-dependent manner blood flow in non-activated brain areas. This sub-maximal depression can be counteracted physiologically by the patient (Roland and Friberg, 1988).

The involvement of GABA in the endocrine pancreatic functions (Erdö, 1985; Solimena and De Camilli, 1993) is an area of growing therapeutic interest. Autoantibodies to glutamic acid decarboxylase (GAD) appear to play an important role for the initiation of insulin dependent diabetes (Tirsch *et al.*, 1993), underlining the importance of the GABA system in pancreatic function. GABA is present in the endocrine part of the pancreas at concentrations comparable to those encountered in the CNS, and co-localizes with insulin in the pancreatic b-cells. GABA seems to mediate part of the inhibitory action of glucose on glucagon secretion by activating $GABA_A$ receptors in a_2 cells (Rorsman *et al.*, 1989). Thus, $GABA_A$ receptors probably are playing a key role in the feedback regulation of glucagon release, which seems to be an important mechanism in the hypersecretion of glucagon, frequently associated with diabetes. These $GABA_A$ receptors of as yet undisclosed subunit composition are potential targets for therapeutic GABA intervention.

There is a growing interest in the role of GABA as a transmitter in hearing mechanisms. This interest has been stimulated by the demonstration of a substantial, selective, and age-related loss of GABA in the central nucleus of the inferior colliculus (CIC) in rat (Caspary *et al.*, 1990). There is immunocytochemical evidence for the existence of a

GABAergic system in the guinea pig vestibule and of a role of GABA as a vestibular afferent transmitter (Ryan and Schwartz, 1986; Usami *et al.*, 1989; Lopez *et al.*, 1992). Impairment of inhibitory GABAergic transmission in the CIC may contribute to abnormal auditory perception and processing in neural presbycusis (Caspary *et al.*, 1990). These observations may lead to the identification of novel targets for GABAergic therapeutic intervention in different age-related diseases and conditions involving defective hearing.

In the PNS, $GABA_A$ agonists as well as antagonists are potential therapeutic agents. Whereas the latter class of GABAergic drugs may be rather difficult to use therapeutically in CNS diseases due to seizure potential, $GABA_A$ antagonists being unable to penetrate the BBB may be of great therapeutic value in the periphery.

GABA receptors

Multiplicity

The discovery of GABA in the early 1950s and the identification of the alkaloid bicuculline (Curtis *et al.*, 1970; for review see Roberts, 1986) and its quaternized analogue bicuculline methochloride (BMC) (Johnston *et al.*, 1972) as competitive GABA antagonists in CNS tissues initiated the pharmacological characterization of GABA receptors. The subsequent design of isoguvacine, 4,5,6,7-tetrahydroisoxazolo[5,4-*c*]pyridin-3-ol (THIP) (Krogsgaard-Larsen *et al.*, 1977) (Figure 10.1) and piperidine-4-sulphonic acid (P4S) (Krogsgaard-Larsen *et al.*, 1981) as a novel class of specific GABA agonists further stimulated studies of the pharmacology of the GABA receptors.

The GABA analogue baclofen did, however, disturb the picture of a uniform class of GABA receptors. Baclofen, which was designed as a lipophilic analogue of GABA capable of penetrating the BBB, is an antispastic agent (Burke *et al.*, 1971), but its GABA agonistic effect could not be antagonized by BMC (Curtis *et al.*, 1974). In the early 1980s Bowery and co-workers demonstrated that baclofen, or rather (*R*)-(–)-baclofen, was selectively recognized as an agonist by a distinct sub-population of GABA receptors, named $GABA_B$ receptors (Figure 10.1) (Bowery *et al.*, 1980; Bowery, 1983). The 'classical' GABA receptors were designated $GABA_A$ receptors. This receptor classification represents an important step in the development of the pharmacology of GABA.

During this period the exploration of the GABA receptors was dramatically intensified by the observation that the binding site for the benzodiazepines (BZDs) (Haefely and Polc, 1986; Möhler and Okada, 1977; Squires and Braestrup, 1977) was associated with the $GABA_A$ receptors (Tallman *et al.*, 1980; Barnard and Costa, 1989; Biggio and Costa, 1990; Stephenson and Turner, 1998). After the cloning of a large number of $GABA_A$ receptor subunits, this area of the pharmacology of GABA continues to be in a state of almost explosive development.

Substitution of a phosphono group for the carboxyl group of baclofen gives a $GABA_B$ antagonist, phaclofen (Kerr *et al.*, 1987), and in agreement with the competitive nature of this antagonism the $GABA_B$ receptor affinity of phaclofen resides in the (*R*)-enantiomer (Frydenvang *et al.*, 1994) (Figure 10.1). On the other hand, replacement of the aromatic group of baclofen by a hydroxy group to give 3-hydroxy-4-aminobutanoic acid (3–OH–GABA) gives a $GABA_B$ agonist. It is the (*R*)-form of

3–OH–GABA, which interacts with the $GABA_B$ receptors, and since the aromatic substituent of baclofen and the hydroxy group of (3–OH–GABA) have opposite orientations these groups probably bind to different sub-structures of the $GABA_B$ receptor site (Falch *et al.*, 1986). This observation has been exploited in the $GABA_B$ antagonist field and has led to the development of new effective antagonists (Kristiansen *et al.*, 1992; Froestl *et al.*, 1995a, 1995b). Cloning of the $GABA_BR1$ and $GABA_BR2$ receptors and the identification of functional heterodimeric $GABA_B$ receptors have greatly stimulated $GABA_B$ receptor research (Martin *et al.*, 1999; Möhler and Fritschy, 1999).

In connection with the design of conformationally restricted analogues of GABA another 'disturber of the peace' appeared on the GABA scene; namely, *cis*-4-aminobut-2-enoic acid (CACA) (Figure 10.1). This compound and the structurally related GABA analogue, *cis*-2-aminomethylcyclopropanecarboxylic acid (CAMP) are GABA-like neuronal depressants, which are not sensitive to BMC (Johnston *et al.*, 1975a), and they bind to a class of GABA receptor sites which do not recognize isoguvacine or (*R*)-baclofen (Johnston, 1986, 1997). The phosphinic acid analogue of isoguvacine, TPMPA, has been shown to be a potent and selective antagonist at $GABA_C$ receptors (Murata *et al.*, 1996). These receptors have been named $GABA_C$ receptors (Johnston, 1997) or non-$GABA_A$, non-$GABA_B$ (NANB) receptors for GABA (Drew and Johnston, 1992; Johnston, 1996). It is possible that this not very well understood class of GABA receptors is heterogeneous (Johnston, 1997). It has been proposed that a cloned GABA receptor subunit (r_1) showing some homology with the a and b subunits of $GABA_A$ receptors (Cutting *et al.*, 1991) may confer BMC-resistant properties of ionotropic GABA receptors structurally related to $GABA_A$ receptors (Drew and Johnston, 1992). $GABA_A$-like receptors containing this subunit seem to be identical with/similar to the proposed $GABA_C$ receptors (Stephenson, 1998). A NANB receptor sensitive to CACA (Figure 10.1) has been identified in the retina, and this ionotropic receptor probably comprises the r_1 subunit (Feigenspan *et al.*, 1993; Qian and Dowling, 1993b; Woodward *et al.*, 1993). Three r subunits have now been cloned, and r1–3 are expressed in the retina as well as in various brain regions (Boue-Grabot *et al.*, 1998).

The physiology and pharmacology of NANB GABA receptors are still very incompletely elucidated, but these receptors, which seem to exist in the PNS as well as the CNS (Drew and Johnston, 1992), may be interesting novel targets for drug development. Imidazole-4-acetic acid (IAA) has been shown to be an antagonist at the retinal GABA receptors, probably of the $GABA_C$ type (Figure 10.1) (Qian and Dowling, 1993a, 1994). At least some of the $GABA_C$ receptors are homomeric r subunit-containing ligand-gated chloride channels (Figure 10.2) (Djamgoz, 1995; Stephenson, 1998).

Structure, function and modulation

The introduction of molecular biological techniques has revolutionized receptor research, and during the past few years the number of papers describing structure and function of G protein-coupled receptors as well as ligand-gated ion channels has virtually exploded (Biggio and Costa, 1990; Olsen and Tobin, 1990; Verdoorn *et al.*, 1990; Olsen *et al.*, 1991; Barnard, 1992; DeLorey and Olsen, 1992; Sieghart, 1992; Macdonald and Olsen, 1994; Tanaka and Bowery, 1996; Doble and Martin, 1996; Stephenson and Turner, 1998). A detailed review of this research area is beyond the scope of this

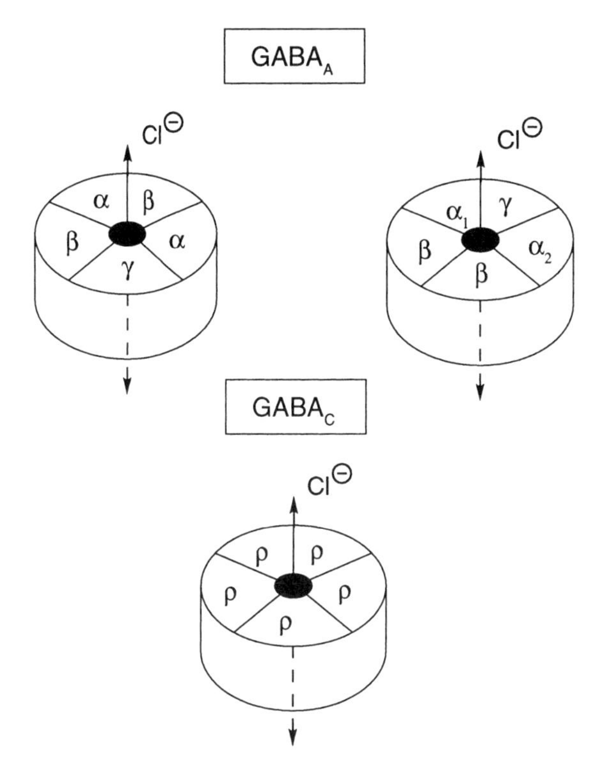

Figure 10.2 Schematic illustration of some proposed subunit combinations of $GABA_A$ and $GABA_C$ receptors.

chapter, and only a few aspects of particular relevance to the design and development of $GABA_A$ receptor ligands shall be mentioned.

$GABA_A$ receptors are known from cDNA cloning and expression studies to contain a combination of homologous subunits primarily from three classes, α, β and γ, whereas additional types, notably d and r, have been identified in certain types of neurons (Sieghart, 1995; Whiting, 1999). Each subunit is present in the brain in several independently expressed isoforms, and so far six α-subunits (α_1–α_6), four b-subunits (β_1–β_4), three g-subunits (γ_1–γ_3), and one δ-subunit have been identified (Stephenson, 1998; Sieghart *et al.*, 1999). The $GABA_A$ receptor is probably assembled as a pentameric structure (Nayeem *et al.*, 1994) (Figure 10.2) from different subunit families, making it possible that a very large number of such heteromeric $GABA_A$ receptors exist in the mammalian CNS and PNS. The number of physiologically relevant $GABA_A$ receptors, their subunit stoichiometry, and their regional distributions are, however, far from being fully elucidated, but studies of these aspects are in rapid progress (Barnard, 1992; Möhler *et al.*, 1995; Sieghart, 1995; Sieghart *et al.*, 1999).

The gating properties of recombinant $GABA_A$ receptors vary markedly with subunit subtype combinations (Macdonald and Olsen, 1994; Stephenson, 1998). Subunits assemble with different efficiencies, and, when expressed in fibroblasts, $\alpha_1\beta_1$ for example, but not $\alpha_1\gamma_2$ or $\beta_1\gamma_2$, subtypes assemble to produce BZD-insensitive $GABA_A$

receptor channels. In contrast to $\alpha_1\beta_1$ channels, which have only two open states, $\alpha_1\beta_1\gamma_2$ $GABA_A$ channels (Figure 10.2) have gating properties similar to those of neuronal $GABA_A$ receptors. The importance of the α subunit is emphasized by the observation that $\alpha_6\beta_1\gamma_2$ channels show different properties.

Recombinant techniques have made it possible to determine the primary structure of receptor glycoproteins and to disclose a degree of heterogeneity of all classes of receptors, which was beyond imagination only a decade ago. A broad spectrum of problems regarding structure and function of receptors remains to be elucidated (Macdonald and Olsen, 1994; Sieghart, 1995; Stephenson, 1998), and molecular biologists and pharmacologists are faced with a number of unanswered questions: Which type of cells express the individual receptor protein mRNAs? Do all subunit subtypes assemble into $GABA_A$ pentameric receptors? How many different heteromeric $GABA_A$ receptors do actually become inserted into the cell membrane? Do all $GABA_A$ receptors assembled form functional and physiologically relevant $GABA_A$ and, perhaps, $GABA_C$ receptors? These (Sieghart *et al.*, 1999) and many other problems regarding structure and function of receptors will be intensively studied during the next decade (Kardos, 1999).

A major goal of such studies is to uncover the mechanisms underlying the extremely complex operational and regulatory mechanisms of the $GABA_A$ receptor complex (Stephenson and Turner, 1998). Such future studies undoubtedly will uncover the molecular mechanisms of key importance for receptor activation and desensitization. Elucidation of these aspects of $GABA_A$ receptors may make rational design of non-desensitizing partial agonists and novel types of $GABA_A$ receptor modulating drugs possible.

There is an urgent need for $GABA_A$ agonists, partial agonists, and antagonists with specific effects at physiologically relevant and pharmacologically important $GABA_A$ receptors of different subunit composition. The observations that THIP binds selectively to a β-subunit of such receptors (Bureau and Olsen, 1990) and that potency as well as relative efficacy of partial $GABA_A$ agonists, such as THIP and P4S, is dependent on the receptor subunit composition (Ebert *et al.*, 1994, 1997) (Figure 10.3) suggest that this is not an unrealistic objective.

In the $GABA_A$ receptor field there are many examples of design of specific receptor ligands following systematic stereochemical and bioisosteric approaches. Identification and topographical analysis of the $GABA_A$ recognition site(s) using molecular modelling and X-ray crystallography may allow rational design of new specific drugs in the future. Identification and localization of the $GABA_A$ recognition site(s) may be facilitated by the availability of agents capable of binding irreversibly to different amino acid residues at these sites. m-sulphonic acid benzene diazonium chloride (MSBD) (Figure 10.4) has been introduced as a compound capable of alkylating $GABA_A$ binding sites (Bouchet *et al.*, 1992), whereas the potent and specific $GABA_A$ agonist, isoguvacine oxide, which contains a chemically reactive epoxy group, has been shown not to bind irreversibly to $GABA_A$ receptors *in vitro* or *in vivo* (Krogsgaard-Larsen *et al.*, 1980b). Muscimol has been used with varying success as a photoaffinity label of $GABA_A$ receptor sites. It has been proposed that photochemical cleavage of the N–O bond converts muscimol into chemically reactive species at the receptor sites (Cavalla and Neff, 1985) (Figure 10.4). More recently, thiomuscimol has been shown to be an effective photolabel for the $GABA_A$ receptor (Frølund *et al.*, 1995a; Nielsen *et al.*, 1995). Thiomuscimol, which is a specific $GABA_A$ agonist approximately equipotent with muscimol (Krogsgaard-Larsen

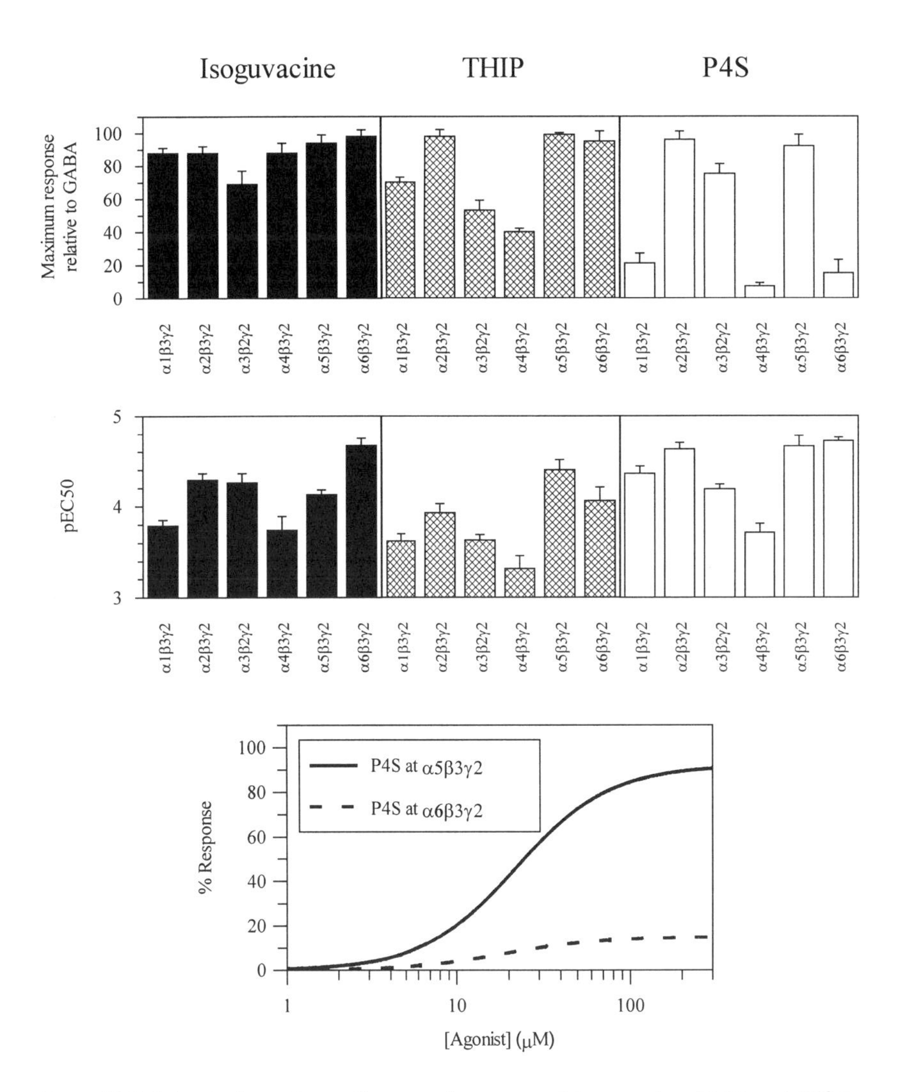

Figure 10.3 Subunit dependent efficacy and potency of isoguvacine, THIP and P4S in oocytes injected with different α-subunits in the presence of β3γ2 or β2γ2.

Upper panel: efficacy expressed as the maximal plateau response obtained relative to that of GABA; *middle panel:* potency expressed as the pEC50 value. *Lower panel:* concentration-response curves for P4S at GABA$_A$ receptors containing α5β3γ2 (filled line) or α6β3γ2 (dotted line). The curves are normalized relative to the maximum response to GABA at the same oocytes.

et al., 1979; Ebert *et al.*, 1997), contains a 3-isothiazolol heterocyclic unit. Since the pK_aI value (6.1) of this acidic group of thiomuscimol is higher than the pK_aI value (4.8) of muscimol, the fraction of thiomuscimol molecules, containing an un-ionized acidic group, is markedly higher than that of monoionized muscimol molecules (Figure 10.4).

Since it is the un-ionized forms of these heterocyclic units which appear to be sensitive to irradiation by UV light (see Figure 10.4), and since thiomuscimol absorbs UV

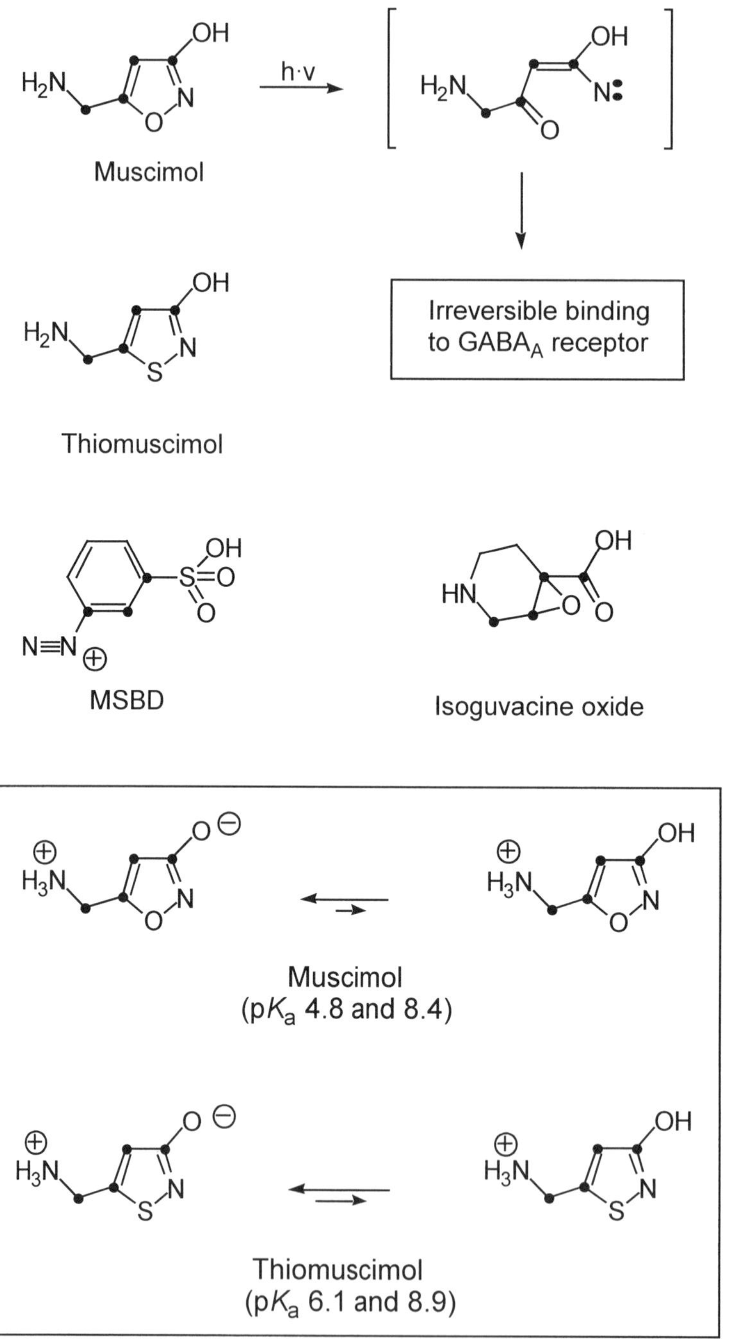

Muscimol

h·v

Irreversible binding to $GABA_A$ receptor

Thiomuscimol

MSBD

Isoguvacine oxide

Muscimol (pK_a 4.8 and 8.4)

Thiomuscimol (pK_a 6.1 and 8.9)

Figure 10.4 Structures of the $GABA_A$ receptor photoaffinity labels, muscimol and thiomuscimol, the irreversible $GABA_A$ receptor ligand, *m*-sulphonic acid benzene diazonium chloride (MSBD), and the reversible $GABA_A$ receptor agonist, isoguvacine oxide. Structures, p*K*a values, and equilibrium between mono- and diionized species of the photosensitive $GABA_A$ agonists, muscimol and thiomuscimol are indicated.

light of higher wavelength than muscimol, thiomuscimol was predicted and subsequently shown to be more effective than muscimol as a $GABA_A$ receptor photoaffinity ligand (Frølund *et al.*, 1995a; Nielsen *et al.*, 1995). Thiomuscimol has been tritiated (Frølund *et al.*, 1995a), and tritiated thiomuscimol may be a useful tool for the localization of the $GABA_A$ receptor recognition site (Ebert *et al.*, 1999).

Molecular cloning studies have disclosed a very high degree of heterogeneity of the $GABA_A$ receptors. The challenge for medicinal chemists is to further develop these observations into rational drug design projects and to develop receptor ligands, which show specificity at the level of receptor subtypes.

Benzodiazepines

The $GABA_A$ receptor complex (Figure 10.5) comprises a large number of binding sites for drugs, notably BZDs, steroids, barbiturates, and the compounds shown in Figure 10.6 (Tallman *et al.*, 1980; Haefely and Polc, 1986; Olsen and Venter, 1986; Barnard and Costa, 1989; Sieghart, 1995; Stephenson, 1998), and virtually all steps in the recognition and gating processes of the $GABA_A$ receptor have been shown to be subject to modulation by such drugs (Macdonald and Olsen, 1994). The binding step(s) appear to be modified by BZDs, β-carbolines acting at the BZD site, and possibly by steroids including endogenous as well as synthetic steroids. The gating process is apparently regulated by steroids and barbiturates. The sensitivity of the desensitization mechanism(s) and state(s) to such drugs has, so far, not been studied in detail. This degree of complexity of the $GABA_A$ receptor function is comparable to that of the NMDA subtype of Glu receptor channels (Lodge, 1988; Wheal and Thomson, 1995; Krogsgaard-Larsen and Hansen, 1992; Monaghan and Wenthold, 1997).

The relationship between $GABA_A$ receptor subunit composition and molecular pharmacology of the $GABA_A$ receptor modulating BZDs has been extensively studied (Gammill and Carter, 1993), and on the basis of mutation studies it has been possible to identify a single amino acid residue of key importance for the binding of BZD ligands

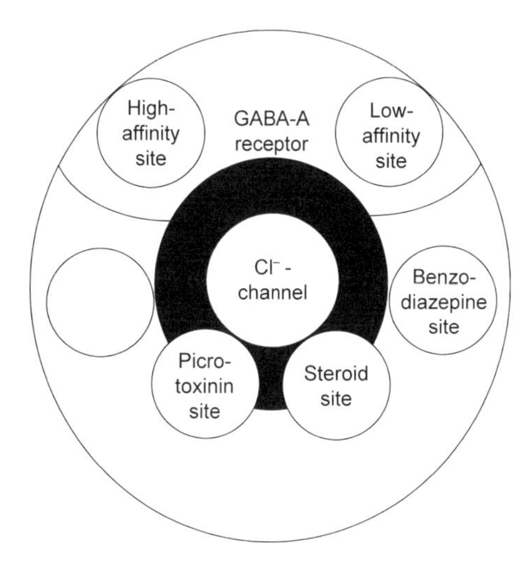

Figure 10.5 Binding sites at $GABA_A$ receptor complex.

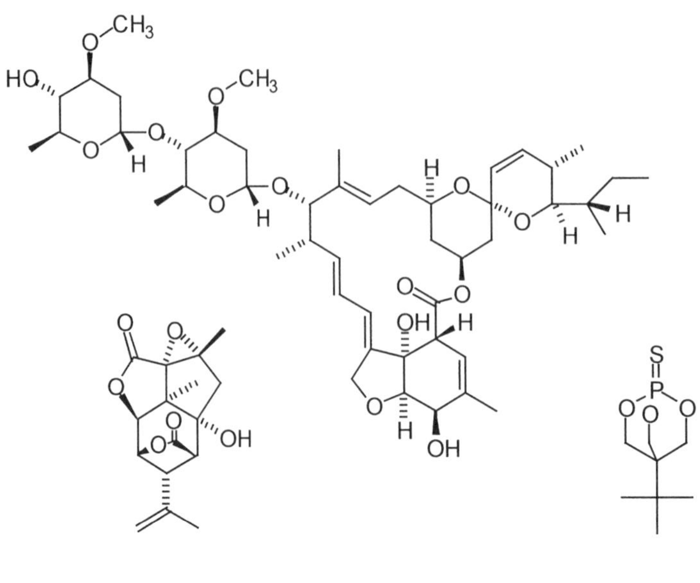

Picrotoxinin Avermectin B_{1a} TBPS

Figure 10.6 Structures of the noncompetitive $GABA_A$ receptor antagonists TBPS and picrotoxinin, and the $GABA_A$ receptor modulatory agent avermectin B_{1a}.

(Figure 10.7) (Wieland *et al.*, 1992; Smith and Olsen, 1995). It may be possible to identify and localize distinct subtypes of $GABA_A$ receptors associated with different physiological and pathophysiological functions. The aim of such studies in the BZD field is to design compounds with appropriately balanced agonist/antagonist or inverse agonist/antagonist properties capable of interacting selectively with $GABA_A$ receptors of

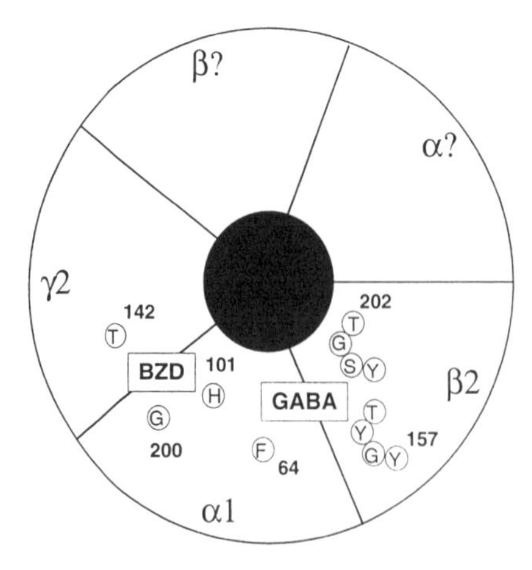

Figure 10.7 An illustration of the binding domains of GABA and BZDs at the $GABA_A$ receptor complex and the amino acid residues implicated in these ligand bindings (redrawn after figure in Smith and Olsen, 1995).

particular relevance to anxiety, epilepsy and sleep disorders (Gammill and Carter, 1993; Stephenson and Turner, 1998).

The pharmacological profile of ligands binding to the BZD sites spans the entire continuum from full and partial agonists through antagonists to partial and full inverse agonists. The BZD agonists act by increasing the channel function in response to GABA or a $GABA_A$ agonist, whereas the inverse agonists exert the opposite effect by decreasing the channel function (Hunkeler *et al.*, 1981). Antagonists do not influence GABA-induced chloride flux but antagonize the action of agonists or inverse agonists (Hunkeler *et al.*, 1981). The full BZD agonists show anxiolytic, anticonvulsant, sedative, and muscle relaxant effects, whereas the inverse agonists produce anxiety and convulsions. Of particular therapeutic interest are the reports of partial agonists at the BZD site displaying potent anxiolytic and anticonvulsive effects with markedly less sedation and muscle relaxation (Guidotti, 1992; Serra *et al.*, 1992). The BZD sites do not only recognize and react to BZDs having different efficacies, but they also accept ligands of very different structures; e.g. β-carbolines, imidazobenzodiazepines, pyrazoloquinolinones, triazolopyridazines, and imidazoquinoxalines (e.g. PNU-97035) (Gammill and Carter, 1993; Jacobsen *et al.*, 1999) (Figure 10.8).

It is generally accepted that heterogeneity exists in BZD binding sites. This has been demonstrated by differences in binding properties of several BZD ligands from different chemical classes. Ro15–4513, a partial inverse agonist, was found to bind to sites in the cerebellum where BZDs (e.g. diazepam, flunitrazepam), some β-carbolines (e.g. 3-carbomethoxy-β-carboline), and triazolopyridazines (e.g. CL 218872) that are high-affinity ligands at other diazepam-sensitive BZD binding sites, show only low-affinity (Sieghart *et al.*, 1987; Malminiemi and Korpi, 1989). The α_6 subunit, which is only expressed in cerebellar granule cells, appears to be responsible for this diazepam-insensitive receptor isoform (Lüddens *et al.*, 1990; Wieland *et al.*, 1992).

A number of structural classes including β-carbolines (e.g. FG 7142 and DMCM) (Korpi *et al.*, 1992) and pyrazoloquinolinones (e.g. I) (Zhang *et al.*, 1995), depicted in

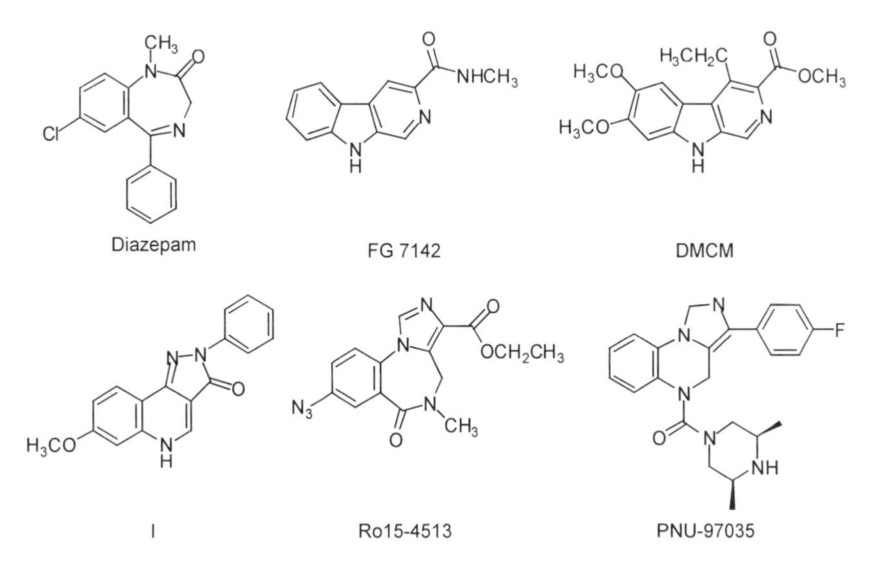

Figure 10.8 Structures of some compounds interacting with different types of BZD receptor sites.

Figure 10.8, bind to the diazepam-insensitive sites with high to low affinity and low selectivity, compared to diazepam-sensitive BZD binding sites, which makes the characterization of the physiological and pharmacological role of the diazepam-insensitive binding site difficult.

Neurosteroids and neuroactive steroids

There is a considerable interest in the steroid binding site of the $GABA_A$ receptor complex (Figure 10.5) as a target for pharmacological and therapeutic intervention (McNeil *et al.*, 1992; Johnston, 1996). The physiological role of this receptor site is still not fully understood (Sieghart, 1995), and steroids interacting with this site should be classified as neurosteroids or neuroactive steroids (Johnston, 1996). Whereas the former class of steroids are synthesized in the brain, neuroactive steroids show pharmacological effects in the CNS but are not necessarily biosynthesized in CNS tissue. Pregnenolone and its metabolites, such as 3α-hydroxy-5β-pregnan-20-one (3α–OH–DHP), are biosynthesized from cholesterol in brain tissue and these compounds belong to the class of neurosteroids (Baulieu, 1991) (Figure 10.9). 3α,21-dihydroxy-5α-pregnan-20-one (5α–THDOC) and 3β-hydroxy-5β-pregnan-20-one (epipregnanolone) are neuroactive steroids, and the results of analyses of the antagonism by epipregnanolone of the effects of neuroactive steroids are consistent with the existence of more than one class of steroid binding sites at the $GABA_A$ receptor complex (Prince and Simmonds, 1993).

Alphaxalone (Figure 10.9) is a potent steroid anaesthetic agent, which has been shown electrophysiologically to enhance the activation of $GABA_A$ receptors by GABA (Harrison and Simmonds, 1984). As a result of this observation of fundamental importance, a large number of industrial and academic drug research groups initiated projects with the object of identifying or developing drugs acting specifically at the steroid site of the $GABA_A$ receptor complex (McNeil *et al.*, 1992), as exemplified in Figure 10.10 (Gasior *et al.*, 1999). Ganaxolone recently failed clinical trials for the treatment of acute migraine (Williams *et al.*, 1999).

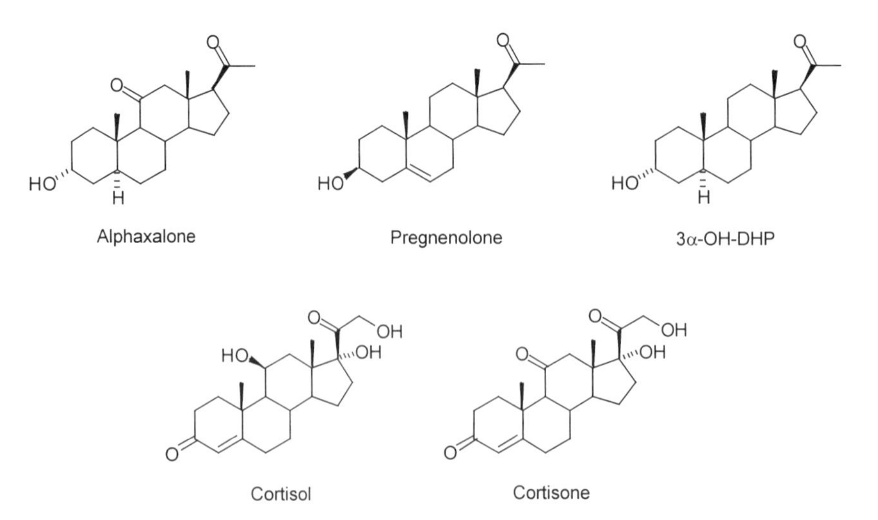

Figure 10.9 Structures of some neurosteroids and neuroactive steroids.

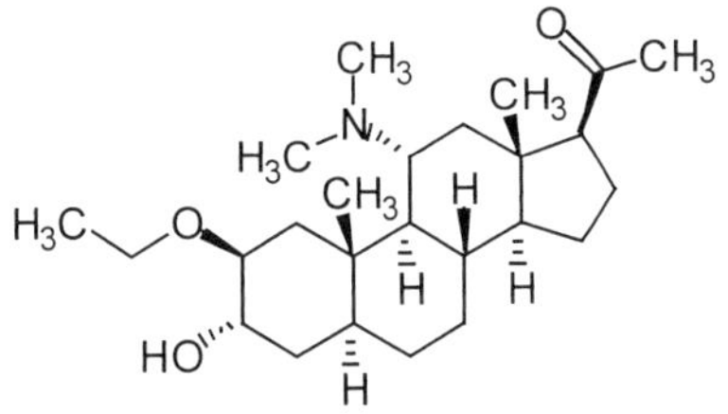

Minaxolone

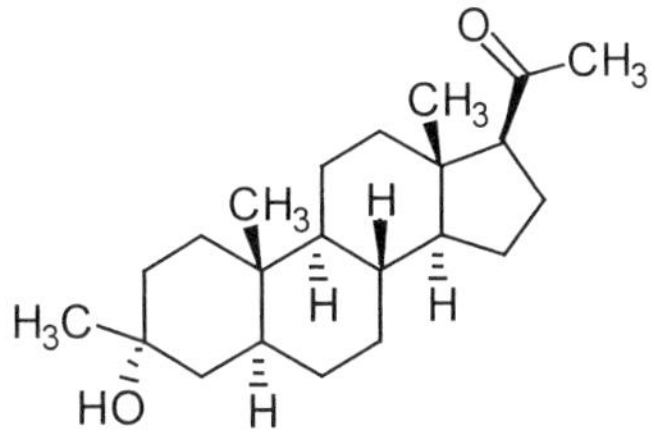

Ganaxolone

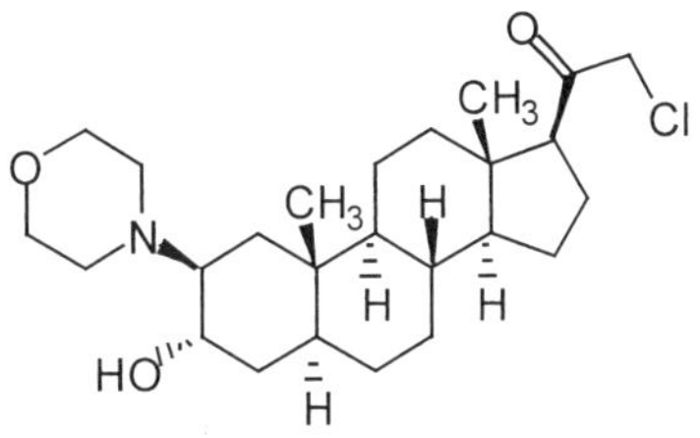

ORG20599

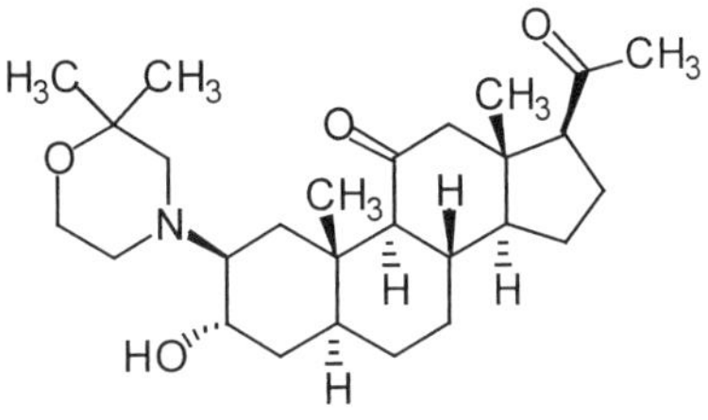

ORG21465

Figure 10.10 Structures of some recently synthesized neuroactive steroids.

The studies of neurosteroids were stimulated by the observation that cortisol is a potent bidirectional modulator of $GABA_A$ receptors, being an enhancer at low concentrations, but an inhibitor at higher concentrations (Ong *et al.*, 1987). This corticosteroid and cortisone (Figure 10.9), which is a noncompetitive antagonist at the steroid site of the $GABA_A$ receptor, are among the most potent agents acting at this site (Ong *et al.*, 1990), which may have a physiological modulatory function (Johnston, 1996; Sieghart, 1995).

Interaction between $GABA_A$ and $GABA_B$ receptors

Neurons intrinsic to cerebellum utilize either Glu or GABA as the neurotransmitter (Palay and Chan-Palay, 1982). The glutamatergic excitatory innervation of the Purkinje neurons by granule cell parallel fibres is fine-tuned by GABAergic interneurons. In other words, GABA exerts a modulatory action inhibiting glutamatergic activity, a process involving activation of both $GABA_A$ and $GABA_B$ receptors. The molecular mechanisms for this fine-tuning of the excitatory glutamatergic activity are as yet not fully clarified, but are likely to involve an interaction between $GABA_A$ and $GABA_B$ receptors leading to disinhibitory phenomena at the level of single neurons, i.e. the granule cells (Kardos *et al.*, 1994).

Cerebellar granule neurons have been shown to have specific binding sites for baclofen (Huston *et al.*, 1990; Travagli *et al.*, 1991; DeErasquin *et al.*, 1992; Kardos *et al.*, 1994), and using cultured cells it has recently been demonstrated that the number of binding sites can be increased by exposure of the neurons to THIP, during the culture period (Kardos *et al.*, 1994). This appears to be analogous to the ability of THIP to induce low-affinity $GABA_A$ receptors on these neurons. The $GABA_B$ receptors on the granule cells are functionally involved in regulation of transmitter release since (*R*)-baclofen has been shown to inhibit K^+-stimulated Glu release from these neurons (Belhage *et al.*, 1990; Huston *et al.*, 1990, 1993; Travagli *et al.*, 1991; Kardos *et al.*, 1994). The inhibitory actions of $GABA_A$ and $GABA_B$ receptors on evoked Glu release in cerebellar granule neurons were recently characterized. It was shown that the inhibitory actions of baclofen and isoguvacine were not additive, which strongly indicates that the two receptors are functionally coupled to each other (Kardos *et al.*, 1994; Schousboe, 1999). An inhibitory action of $GABA_B$ receptors on $GABA_A$ receptors, as also previously suggested (Hahner *et al.*, 1991), will result in a disinhibitory action of the $GABA_B$ receptors at the cellular level, which could possibly explain numerous reports on excitatory actions of GABA or baclofen in multicellular systems or in the intact brain (Mitchell, 1980; Levi and Gallo, 1981; Nielsen *et al.*, 1989; Cherubini *et al.*, 1991). Such a disinhibitory interaction between $GABA_A$ and $GABA_B$ receptors at the cellular level may be functionally indistinguishable from the originally described disinhibitory organization of neuronal networks (Roberts, 1976, 1991).

Cerebellar granule neurons are rich in $GABA_A$ receptors which, based on analysis of GABA binding, can be divided into high- and low-affinity receptors with affinity constants of 5–10 and 500 nM, respectively (Meier *et al.*, 1984). Using a monotypic cerebellar culture system, it has been shown that expression of the low-affinity receptors is dependent upon whether or not the neurons are exposed to GABA or THIP during early development (Meier *et al.*, 1984). Therefore, since cerebellar granule neurons in culture can be grown under conditions where either high-affinity GABA receptors are expressed alone, or the two distinct receptors are expressed together, it is possible to obtain information about the functional properties of these receptors.

As mentioned above, the granule neurons in cerebellum are excitatory in nature using Glu as the neurotransmitter (Stone, 1979), and transmitter release can be inhibited by GABA, dependent upon the expression of the two types of $GABA_A$ receptors as well as the depolarizing signal. In cells expressing only high-affinity GABA receptors, GABA is able to inhibit transmitter release evoked by a moderately depolarizing signal (30 mM KCl), whereas that evoked by a strong depolarizing pulse (55 mM KCl) cannot be inhibited by GABA. On the contrary, in neurons expressing both high- and low-affinity GABA receptors, GABA is able to inhibit transmitter release regardless of the depolarizing condition. This action of GABA can be mimicked by THIP or muscimol (Figure 10.11) and blocked by bicuculline in keeping with the notion that the low-affinity GABA receptors are $GABA_A$ receptors. Interestingly, the action of GABA mediated by the low-affinity GABA receptor has been shown to be insensitive to the chloride channel blocker, picrotoxinin (Figure 10.6), indicating that these receptors may be mechanistically different from the high-affinity receptors which are clearly coupled to a chloride channel and blocked by picrotoxinin (Belhage *et al.*, 1991).

The inhibition of evoked Glu release mediated by high-affinity $GABA_A$ receptors is clearly dependent upon the $GABA_A$ receptor gated chloride channels, since this action of GABA can be blocked by picrotoxinin (Belhage *et al.*, 1991). It is not clear how the inhibitory action of GABA mediated by the inducible low-affinity receptors may be mediated. Since this action of GABA cannot be inhibited by picrotoxinin it is unlikely that the classical mechanism involving the chloride channel is playing a major role

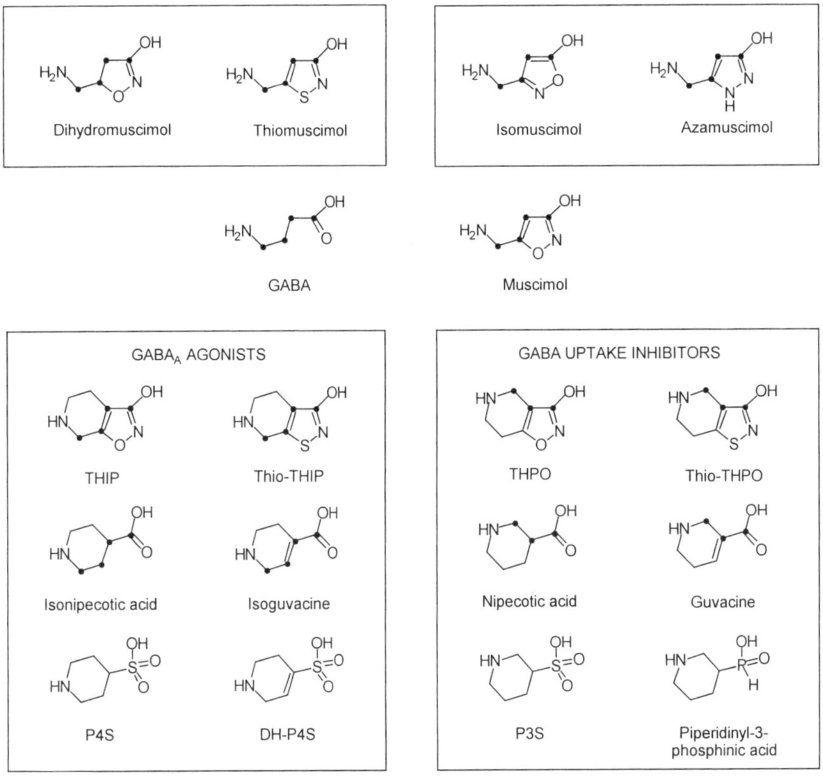

Figure 10.11 Comparison of the structures of some $GABA_A$ agonists and GABA uptake inhibitors.

(Belhage *et al.*, 1991). It has, however, been shown that the induction of the low-affinity $GABA_A$ receptors is closely associated with a similar increase in the number of voltage-gated calcium channels (Hansen *et al.*, 1992). More importantly, it was observed that in nerve processes, but not in cell bodies, there was a tight spatial coupling between $GABA_A$ receptors and calcium channels in neurons expressing the low-affinity GABA receptors – but not in cells expressing high-affinity receptors alone (Hansen *et al.*, 1992). This led to the suggestion that the inhibitory action of GABA mediated by the low-affinity receptors could involve a regulation of the activity of voltage-gated calcium channels (Belhage *et al.*, 1991; Schousboe *et al.*, 1992b). $GABA_B$ receptors in cerebellar granule neurons regulate the intracellular calcium level via a G-protein-dependent mechanism (Sivilotti and Nistri, 1991; Kardos *et al.*, 1994; Schousboe, 1999). If a coupling between $GABA_A$ and $GABA_B$ receptors in these neurons is of functional importance this could explain how $GABA_A$ receptors may modulate transmitter release in a manner involving calcium channels.

$GABA_A$ agonists

The basically inhibitory nature of the central GABA neurotransmission prompted the design and development of different structural types of GABA agonists. Conformational restriction of various parts of the molecule of GABA and bioisosteric replacements of the functional groups of this amino acid have led to a broad spectrum of specific $GABA_A$ agonists. Some of these molecules have played a key role in the development of the pharmacology of the $GABA_A$ receptor, or rather receptor family (Krogsgaard-Larsen *et al.*, 1994, 1997).

The histamine metabolite, IAA, is a relatively potent $GABA_A$ agonist and a $GABA_C$ antagonist (Figure 10.1), which may play a role as a central and/or peripheral endogenous GABA receptor ligand.

Muscimol, a constituent of the mushroom *Amanita muscaria*, has been extensively used as a lead for the design of different classes of GABA analogues (Figure 10.11). The 3-isoxazolol carboxyl group bioisostere of muscimol can be replaced by a 3-isothiazolol or 3-hydroxyisoxazoline group to give thiomuscimol and dihydromuscimol, respectively, without significant loss of $GABA_A$ receptor agonism (Krogsgaard-Larsen *et al.*, 1979). (*S*)-dihydromuscimol is the most potent $GABA_A$ agonist so far described (Krogsgaard-Larsen *et al.*, 1986). The structurally related muscimol analogues, isomuscimol and azamuscimol (Figure 10.11), on the other hand, are virtually inactive, emphasizing the very strict structural constraints imposed on agonist molecules by the $GABA_A$ receptors (Krogsgaard-Larsen *et al.*, 1979).

Conversion of muscimol into THIP (Krogsgaard-Larsen *et al.*, 1977) and the isomeric compound 4,5,6,7-tetrahydroisoxazole[4,5-*c*]pyridin-3-ol (THPO) effectively separated $GABA_A$ receptor and GABA uptake affinity, THIP being a specific $GABA_A$ agonist and THPO a GABA uptake inhibitor (Figure 10.11) (Krogsgaard-Larsen and Johnston, 1975).

Using THIP as a lead, a series of specific monoheterocyclic $GABA_A$ agonists, including isoguvacine and isonipecotic acid, was developed (Krogsgaard-Larsen *et al.*, 1985, 1988). Thio-THIP is weaker than THIP as a $GABA_A$ agonist, but recent studies have disclosed a unique pharmacological profile of this compound. Whereas Thio-THIP shows distinct $GABA_A$ agonist effects on cat spinal neurons (Krogsgaard-Larsen *et al.*, 1983), recent studies using human brain recombinant $GABA_A$ receptors have disclosed that at such receptors Thio-THIP does express very low-efficacy partial agonism (Brehm *et al.*, 1997).

In light of the structural similarity of THIP and Thio-THIP (Figure 10.11) the markedly different pharmacology of these compounds is noteworthy and emphasizes the very strict structural requirements of $GABA_A$ receptors. The pK_a values of THIP (4.4, 8.5) and Thio-THIP (6.1, 8.5) (Krogsgaard-Larsen *et al.*, 1983) are different, and a significant fraction of the molecules of the latter compound must contain a nonionized 3-isothiazolol group at physiological pH. Furthermore, the different degree of charge delocalization of the zwitterionic forms of THIP and Thio-THIP and other structural parameters of these two compounds, as well as the bioisosteric 3-isoxazolol and 3-isothiazolol groups, may have to be considered in order to explain their different potency and efficacy at $GABA_A$ receptors.

A series of cyclic amino acids derived from THPO, including nipecotic acid (Krogsgaard-Larsen and Johnston, 1975) and guvacine (Johnston *et al.*, 1975b), was developed as GABA uptake inhibitors. Whereas nipecotic acid and guvacine potently inhibit neuronal as well as glial GABA uptake (Schousboe *et al.*, 1979), THPO interacts selectively with the latter uptake system (Schousboe *et al.*, 1981; Krogsgaard-Larsen *et al.*, 1987). Thio-THPO is slightly weaker than THPO as an inhibitor of GABA uptake (Krogsgaard-Larsen *et al.*, 1983).

It is interesting to note that although isoguvacine is an order of magnitude more potent than the corresponding saturated cyclic amino acid isonipecotic acid as a $GABA_A$ agonist (Krogsgaard-Larsen *et al.*, 1977), the unsaturated analogue of P4S, DH-P4S is an order of magnitude weaker than P4S as an agonist at $GABA_A$ receptors (Krogsgaard-Larsen *et al.*, 1981, 2000; Falch *et al.*, 1985) (Figure 10.11). The sulphonic acid analogue of nipecotic acid, P3S does not significantly affect GABA uptake (Krogsgaard-Larsen *et al.*, 1980a), whereas the corresponding phosphinic acid, piperidinyl-3-phosphinic acid (Figure 10.11) is a GABA uptake inhibitor, though somewhat weaker than nipecotic acid (Kehler *et al.*, 1999).

Stereostructure–activity relationships for $GABA_A$ agonists

The molecule of GABA is highly flexible and prochiral (Krogsgaard-Larsen, 1988). The degree of stereoselectivity of chiral GABA analogues of known absolute stereochemistry (Figure 10.12) depends on the structure of the compounds and is a function of the conformational flexibility of the molecules (Krogsgaard-Larsen *et al.*, 1985, 1988; Krogsgaard-Larsen, 1988). Thus, the (*S*)- and (*R*)-forms of the flexible GABA analogue 4-aminopentanoic acid (4–Me–GABA) are equally effective at $GABA_A$ receptor sites, and both enantiomers interact with neuronal as well as glial GABA uptake systems *in vitro* (Schousboe *et al.*, 1979). Introduction of a double bond into these molecules to give the (*S*)- and (*R*)-forms of *trans*-4-aminopent-2-enoic acid (4–Me–TACA) has quite dramatic consequences. Thus, whilst (*S*)-4–Me–TACA interacts specifically with $GABA_A$ receptors *in vivo* and *in vitro*, (*R*)-4–Me–TACA interacts with the GABA uptake systems and does not affect $GABA_A$ receptor binding detectably (Schousboe *et al.*, 1979; Krogsgaard-Larsen, 1988).

Homo-β-proline is a cyclic but still rather flexible analogue of GABA (Figure 10.12). Whereas (*R*)homo-β-proline is about an order of magnitude more potent than (*S*)-homo-β-proline as an inhibitor of $GABA_A$ receptor binding, the latter enantiomer selectively binds to $GABA_B$ receptor sites, and both enantiomers bind with equal

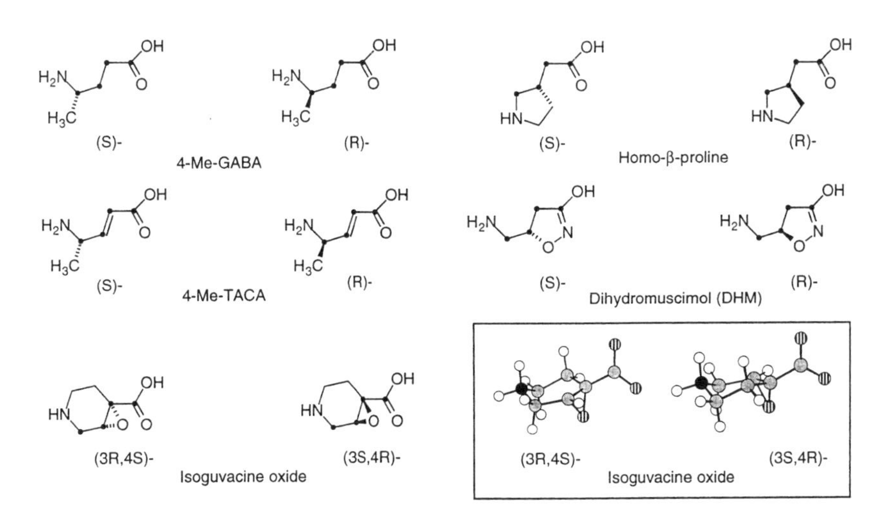

Figure 10.12 Structures of the enantiomers of some chiral GABA analogues.

affinity to the synaptosomal GABA uptake system (Nielsen *et al.*, 1990). Dihydromuscimol is a more rigid analogue of GABA, and whilst (*S*)-dihydromuscimol is a selective and extremely potent $GABA_A$ agonist, (*R*)-dihydromuscimol is a weak $GABA_A$ agonist and an inhibitor of GABA uptake (Krogsgaard-Larsen *et al.*, 1986).

Isoguvacine oxide, which contains a chemically reactive epoxy group (Figure 10.12), was designed as a $GABA_A$ agonist capable of interacting irreversibly with the $GABA_A$ receptor (Krogsgaard-Larsen *et al.*, 1980b). Although isoguvacine oxide is a potent and high-efficacy $GABA_A$ agonist, it shows no sign of irreversible receptor interaction. Isoguvacine oxide has been resolved, and, quite surprisingly, the (3*R*,4*S*)- and the (3*S*,4*R*)-enantiomers were shown to be equally active as $GABA_A$ agonists (Frølund *et al.*, 1995b). The equipotency of the enantiomers of this $GABA_A$ agonist, containing a rather bulky epoxy group (Figure 10.12), indicates that a relatively spacious cavity is present at the $GABA_A$ recognition site, and that this proposed cavity does not contain an appropriately positioned nucleophilic group capable of reacting with the epoxy group of isoguvacine oxide.

Behavioural and clinical effects of $GABA_A$ agonists

GABA in analgesia, anxiety and insomnia

The involvement of central $GABA_A$ receptors in pain mechanisms and analgesia has been thoroughly studied, and the results have been discussed and reviewed (Krogsgaard-Larsen *et al.*, 1984, 1985, 1988; DeFeudis, 1989; Sawynok, 1989). The demonstration of potent antinociceptive effects of the specific and metabolically stable $GABA_A$ agonist THIP in different animal models and the potent analgesic effects of THIP in man greatly stimulated studies in this area of pain research. THIP-induced analgesic effects were shown to be insensitive to the opiate antagonist naloxone, indicating that these effects are not mediated by the opiate receptors (Kendall *et al.*, 1982). Quite surprisingly, THIP analgesia could not be reversed by bicuculline, which may reflect the involvement of a distinct class of $GABA_A$ receptors or, perhaps, a NANB-type of GABA receptors. On the other hand, THIP-induced analgesia could be reduced by atropine and potentiated by cholinergics such as physostigmine, reflecting as yet unclarified functional interactions between GABA and ACh neurons and, possibly, the central opiate systems rather than a direct action of THIP on muscarinic ACh receptors (Zorn and Enna, 1987).

THIP and morphine are approximately equipotent as analgesics, although their relative potencies are dependent on the animal species and experimental models used. Acute injection of THIP potentiates morphine-induced analgesia, and chronic administration of THIP produces a certain degree of functional tolerance to its analgesic effects. In contrast to morphine, THIP does not cause respiratory depression. Clinical studies on post-operation patients, and patients with chronic pain of malignant origin, have disclosed potent analgesic effects of THIP, in the latter group of patients at total doses of 5–30 mg (intramuscular injection) of THIP.

In these cancer patients, and also in patients with chronic anxiety (Hoehn-Saric, 1983), the desired effects of THIP were accompanied by side effects, notably sedation, nausea, and in a few cases euphoria (Kjaer and Nielsen, 1983; Williams *et al.*, 1999). The side effects of THIP have been described as mild and similar in quality to those of other GABA-mimetics (Hoehn-Saric, 1983).

It is assumed that the postsynaptic $GABA_A$ receptor complex is mediating the anxiolytic effects of the BZDs, and, consequently, it is of interest to see whether $GABA_A$ agonists have anxiolytic effects. Muscimol has proved effective in conflict tests, though with a pharmacological profile different from that of diazepam, and in humans muscimol in low doses was found to sedate and calm schizophrenic patients (Tamminga *et al.*, 1979). In a number of patients with chronic anxiety the effects of THIP were assessed on several measures of anxiety (Hoehn-Saric, 1983). Although these effects were accompanied by side effects, the combination of analgesic and anxiolytic effects of THIP would seem to have therapeutic prospects.

The neuronal and synaptic mechanisms underlying THIP- and, in general, GABA-induced analgesia are still only incompletely understood. The insensitivity of THIP-induced analgesia to naloxone has been consistently demonstrated. Sensitivity of THIP analgesia to a serotonin agonist seems to indicate an interaction between central GABA and serotonin systems. GABA-induced analgesia does not seem to be mediated primarily by spinal $GABA_A$ receptors but rather by GABA mechanisms in the forebrain, and it appears also to involve neurons in the midbrain. The naloxone-insensitivity and apparent lack of dependence liability of $GABA_A$ agonist-mediated analgesia suggest that GABAergic drugs may play a role in future treatment of pain. Furthermore, it has been suggested that pharmacological manipulation of GABA mechanisms may have some relevance for future treatment of opiate drug addicts.

Other observations further emphasize the complexity of the role of GABA in pain mechanisms. Thus, THIP has been shown to inhibit its own analgesic action at higher doses producing a bell-shaped dose–response curve (Zorn and Enna, 1987). In addition, subconvulsant doses of bicuculline were shown to increase the latency of licking in the hot plate test in mice, an effect which was not modified by naloxone or atropine but was antagonized by the $GABA_B$ antagonist, CGP 35348 (Malcangio *et al.*, 1992). These latter observations suggest that more than one type of $GABA_A$ receptor, perhaps including autoreceptors, are involved in pain mechanisms and, furthermore, that interactions between $GABA_A$ and $GABA_B$ receptors are playing a role.

Curiously, the full $GABA_A$ agonist muscimol, the very efficacious partial $GABA_A$ agonist THIP, as well as the $GABA_A$ antagonist bicuculline show potent antinociceptive effects. It is possible that the side effects of THIP in patients somehow are associated with its high efficacy at $GABA_A$ receptors. If so, partial $GABA_A$ agonists may have particular interest as analgesics.

THIP has been shown to increase non-REM sleep and to enhance delta activity in the rat (Lancel and Faulhaber, 1996). These effects on sleep pattern, which are markedly different from those of benzodiazepines, have prompted an interest in THIP and related $GABA_A$ agonists as potential hypnotics in man.

GABA in neurological disorders

There is an overwhelming amount of indirect evidence, derived from experimental models of epilepsy, supporting the view that pharmacological stimulation of the GABA neurotransmission may have therapeutic interest in epilepsy (Morselli *et al.*, 1981; Fariello *et al.*, 1984; Nistico *et al.*, 1986; Sutch *et al.*, 1999). The anticonvulsant effects of THIP and muscimol have been compared in a variety of animal models. THIP typically is 2–5 times weaker than muscimol in suppressing seizure activities. In mice and in gerbils with genetically determined epilepsy systemically administered THIP has proved very effective in suppressing seizure activity, and THIP is capable of reducing audiogenic seizures in DBA/2 mice. However, THIP failed to protect baboons with photosensitive epilepsy against photically induced myoclonic responses.

THIP has been subjected to a single-blind controlled trial in patients with epilepsy, in which THIP was added to the concomitant antiepileptic treatment. Under these conditions no significant effects of THIP were detected, although a trend was observed for lower seizure frequency during a period of submaximal doses of THIP (Petersen *et al.*, 1983).

In light of these quite surprising effects of THIP in photosensitive baboons and the lack of clinical antiepileptic effects of this specific $GABA_A$ agonist, its effects on human brain glucose metabolism has been studied using positron emission tomography (PET) scanning techniques (Peyron *et al.*, 1994a, 1994b). Due to the sedative effects of THIP observed in animals and patients (Krogsgaard-Larsen *et al.*, 1984), the sleepiness and decrease of alpha rhythms observed in the patients and normal volunteers involved in these PET studies were not unexpected. Accordingly, brain glucose hypometabolism was anticipated in these volunteers and epileptic patients receiving clinically relevant doses of THIP (0.2 mg/kg). Surprisingly and paradoxically, brain glucose metabolism increased globally, showing an average increase in grey matter of 17 per cent (Peyron *et al.*, 1994b).

Dysfunctions of the central GABA neurotransmitter system(s) have also been associated with other neurological disorders such as Huntington's chorea (DiChiara and Gessa, 1981) and tardive dyskinesia (Thaker *et al.*, 1983, 1989). In Huntington's chorea there is a marked loss of GABA neurons, whereas no significant changes in numbers and binding characteristics of $GABA_A$ receptors could be detected (Van Ness *et al.*, 1982). Nevertheless, replacement therapies using the specific $GABA_A$ agonists muscimol or THIP did not significantly ameliorate the symptoms of choreic patients (Foster *et al.*, 1983). Similarly, THIP only marginally improved the symptoms of patients suffering from tardive dyskinesia (Korsgaard *et al.*, 1982).

There are several possible explanations of these largely negative effects of THIP in the neurological disorders mentioned: (1) Disinhibitory neuronal mechanisms, converting inhibitory effects into functional excitation, may play a key role in the brain areas affected in these disorders; (2) $GABA_A$ autoreceptors regulating GABA release may be more sensitive to the $GABA_A$ agonists studied than the hyperpolarizing postsynaptic $GABA_A$ receptors; (3) The $GABA_A$ agonists studied may cause rapid and effective receptor desensitization after prolonged activation by systemically administered agonists (Mathers, 1987; Löscher, 1989); (4) The ρ-like ($GABA_C$) receptors found in retina (Feigenspan *et al.*, 1993; Woodward *et al.*, 1993), where THIP shows very weak antagonistic effects (Woodward *et al.*, 1993) may play a role in certain parts of the human brain.

These aspects open up the prospects of designing new types of therapeutic $GABA_A$ receptor ligands. Antagonists at postsynaptic $GABA_A$ receptors, notably those involved in disinhibitory mechanisms, may in principle have therapeutic interest, but selective antagonists at $GABA_A$ autoreceptors seem to have major interest. There is, however, an obvious need for partial $GABA_A$ agonists showing different levels of efficacy and, in addition, showing selectivity for $GABA_A$ receptors in brain areas of particular relevance for the disorders under discussion.

GABA in Alzheimer's disease

There is a well-documented loss of central ACh nerve terminals in certain brain areas of patients suffering from Alzheimer's disease (Wurtman *et al.*, 1990). Consequently, most efforts for a therapeutic treatment of this neurodegenerative disease have hitherto been focused on the processes and mechanisms at cholinergic synapses.

Central cholinergic neurons appear to be under inhibitory GABAergic control (Supavilai and Karobath, 1985; Friedman and Redburn, 1990), and consequently the function of such neurons may be indirectly stimulated by blockade of the $GABA_A$ receptors involved in this regulation. These $GABA_A$ receptors may be located pre- or postsynaptically on ACh neurons. Therapies based on agents with antagonist actions at $GABA_A$ receptors, or at one of the modulatory sites of the $GABA_A$ receptor complex, should, at least theoretically, be applicable in Alzheimer's disease. $GABA_A$ receptor antagonists, which in addition show low-efficacy $GABA_A$ agonist effects, might stimulate ACh release, and, thus, improve learning and memory in Alzheimer patients without causing convulsions.

The results of studies on different $GABA_A$ receptor ligands in animal models relevant to learning and memory seem to support such GABAergic therapeutic approaches in Alzheimer's disease. Thus, whereas administration of $GABA_A$ agonists impairs learning and memory in animals (Brioni *et al.*, 1989, 1990) via modulation of cholinergic pathways (Brioni *et al.*, 1990), memory enhancement was observed after injection of the $GABA_A$ antagonist BMC (Brioni *et al.*, 1989). Similarly, agonists and inverse agonists at the BZD site of the $GABA_A$ receptor complex impair and enhance, respectively, performance in learning and memory tasks (Venault *et al.*, 1986). Administration of THIP to Alzheimer patients failed to improve cognitive performance significantly (Bruno *et al.*, 1984).

The lack of positive and, in particular, negative effects of THIP in Alzheimer patients is very interesting and may reflect that THIP, as mentioned earlier, is an efficacious partial $GABA_A$ agonist rather than a full $GABA_A$ agonist. These observations seem to support the view that low-efficacy partial $GABA_A$ agonists may be of clinical interest in Alzheimer's disease.

GABA in schizophrenia

Evidence supporting the hypothesis of GABA receptor mediated hyperactivity as an important component of schizophrenic symptoms is, as in the case of the Glu hypothesis, indirect. Increased GABAergic activity either via direct stimulation with $GABA_A$ agonists such as muscimol or THIP (Theobald *et al.*, 1968; Tamminga *et al.*, 1978), or indirect stimulation via inhibition of enzymes essential for the metabolism of GABA (Meldrum, 1982; Brodie and McKee, 1990; Ring and Reynolds, 1990; Robinson *et al.*, 1990; Sander and Hart, 1990), produces psychotomimetic effects. Likewise, BZD agonists may produce schizophrenia-like symptoms in a minority of the treated patients (Hall and Zisool, 1981; Bixler *et al.*, 1987). Partial inverse agonists at the BZD site of the $GABA_A$ receptor complex reduce schizophrenic symptoms (Haefely, 1984; Merz *et al.*, 1988), which supports the view that hyperactivity of GABAergic pathways may be responsible for symptoms associated with schizophrenia (Squires and Saedrup, 1991). Receptor binding studies have revealed an increased muscimol binding and increased flunitrazepam binding (Hanada *et al.*, 1987; Kiuchi *et al.*, 1989; Benes *et al.*, 1992; Dean *et al.*, 1999) in brains from schizophrenic patients. However, flunitrazepam binding also has been reported to be significantly reduced in post mortem schizophrenic brains (Squires *et al.*, 1993).

A possible role of hyperactivity at $GABA_A$ receptors in schizophrenic symptoms suggests that blockade of $GABA_A$ receptor mediated synaptic transmission may be a relevant approach to the treatment of schizophrenia. A GABAergic strategy in the treatment of schizophrenia is therefore analogous to a GABAergic strategy for the treatment of Alzheimer's disease.

$GABA_A$ antagonists

Specific receptor antagonists are essential tools for studies of the physiological role and pharmacological importance of the particular receptors. The classical $GABA_A$ antagonists bicuculline (Figure 10.13) and BMC (Figure 10.1) have played a key role in such studies on $GABA_A$ receptors, although BMC has recently been shown to lack the selectivity of bicuculline as a $GABA_A$ antagonist (Seutin and Johnson, 1999).

Subsequently, new structural classes of $GABA_A$ antagonists have been developed, and this line of GABA drug research has been stimulated by the growing interest in such compounds as potential therapeutic agents, at least in theory. Whereas the bicyclic 5-isoxazolol compound, Iso-THAZ (Figure 10.13), derived from THIP, is a moderately potent $GABA_A$ antagonist (Arnt and Krogsgaard-Larsen, 1979; Rognan *et al.*, 1992), a series of arylaminopyridazine analogues of GABA, notably Gabazine, show very potent and selective $GABA_A$ antagonist effects (Chambon *et al.*, 1985; Wermuth and Biziére, 1986; Wermuth *et al.*, 1987; Rognan *et al.*, 1992; Seutin and Johnson,

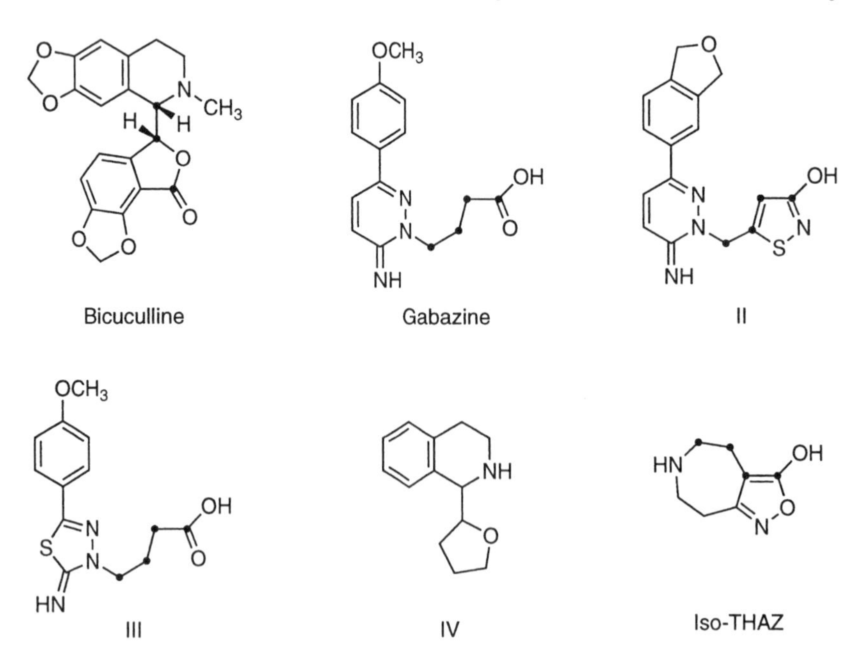

Figure 10.13 Structures of some $GABA_A$ antagonists.

1999). These compounds bind tightly to $GABA_A$ receptor sites, and tritiated Gabazine is now used as a standard receptor ligand. Although Gabazine and related compounds containing a GABA structure element show convulsant effects after systemic administration (Melikian *et al.*, 1992), these zwitterionic compounds do not easily penetrate the BBB. Compound II, in which the GABA structure element has been replaced by a thiomuscimol unit, is the most potent $GABA_A$ antagonist in the arylaminopyridazine series (Melikian *et al.*, 1992). This increased potency has been explained by the more pronounced lipophilic character of compound II as compared with the corresponding analogues of GABA (Melikian *et al.*, 1992). Bioisosteric substitution of a 2-amino-1,3,4-thiadiazole unit for the 3-aminopyridazine part of Gabazine gives compound III (Figure 10.13), which also shows $GABA_A$ antagonistic properties, though markedly weaker than those of Gabazine (Allan *et al.*, 1990).

$GABA_A$ autoreceptors, which regulate GABA release via a negative feed-back mechanism, are interesting novel targets for GABAergic drug design. Although such autoreceptors basically are $GABA_A$ receptors, they have pharmacological characteristics markedly different from those of postsynaptic $GABA_A$ receptors (Minchin *et al.*, 1992a). Selective $GABA_A$ autoreceptor antagonists may function as positive modulators of GABA neurotransmission processes. Interestingly, compound IV, which is a 'peeled' analogue of bicuculline, and a number of other related compounds are two orders of magnitude more potent as $GABA_A$ autoreceptor antagonists than as antagonists at post-synaptic $GABA_A$ receptors (Minchin *et al.*, 1992a, 1992b). This particular class of GABA receptor antagonists may have therapeutic potential.

Partial $GABA_A$ agonists

Full $GABA_A$ agonists or antagonists may be rather difficult to introduce as drugs of practical clinical applicability, at least in diseases in the CNS. Whilst the former class of compounds may induce rapid desensitization of the target receptors after constant activation by systemically administered agonist drugs, $GABA_A$ antagonists are potential anxiogenics, proconvulsants, or frank convulsants.

In clinical conditions where $GABA_A$ agonist therapies may be relevant, relatively efficacious agonists may be appropriate drugs, the levels of efficacy probably being dependent on the particular disease. The very potent analgesic effects of THIP seem to indicate that the relatively high level of efficacy of this partial $GABA_A$ agonist (Krogsgaard-Larsen *et al.*, 1988; Maksay, 1994) is close to optimal with respect to pain treatment, although it may be postulated that a slightly less efficacious $GABA_A$ agonist may show fewer side effects than does THIP.

Analogously, very low-efficacy $GABA_A$ agonists showing predominant antagonist profiles may have clinical usefulness in certain diseases. Such compounds showing sufficient $GABA_A$ receptor agonism to avoid seizure side effects may, theoretically, be useful therapeutic agents in, for example, Alzheimer's disease or, quite paradoxically, in epileptic disorders.

The heterocyclic GABA bioisosteres IAA (Braestrup *et al.*, 1979; Maksay, 1994) (Figure 10.1) and P4S (Braestrup *et al.*, 1979) (Figure 10.11) show the characteristics of partial $GABA_A$ agonists. The nonfused THIP analogue, 5-(4-piperidyl)isoxazol-3-ol (4-PIOL) (Figure 10.14) is a moderately potent agonist at $GABA_A$ receptors in the cat spinal cord (Byberg *et al.*, 1987, 1993). However, 4-PIOL did not show significant

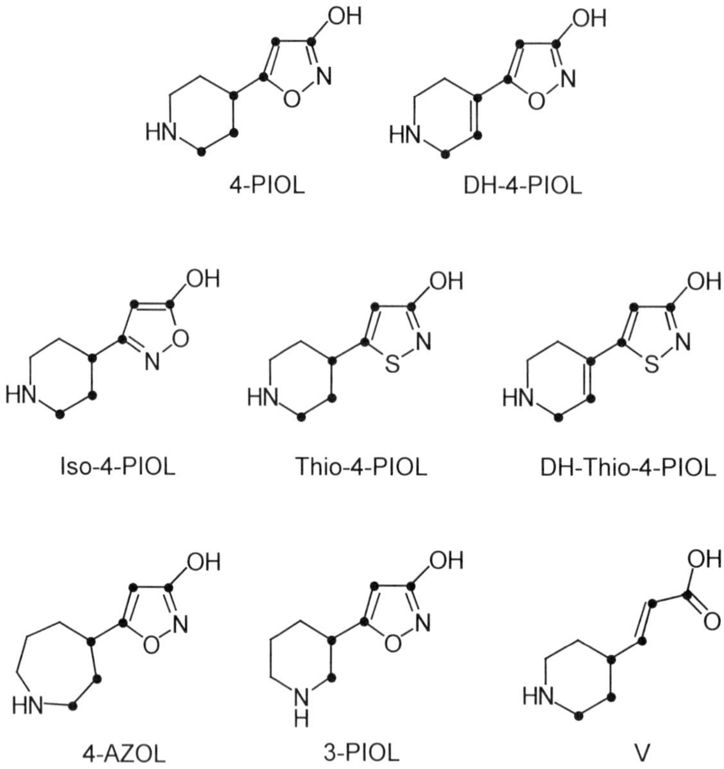

4-PIOL DH-4-PIOL

Iso-4-PIOL Thio-4-PIOL DH-Thio-4-PIOL

4-AZOL 3-PIOL V

Figure 10.14 Structures of the low-efficacy partial $GABA_A$ agonist, 4-PIOL, and some active and inactive (*bottom row*) analogues.

stimulatory effects on BZD binding, but it antagonized dose-dependently, muscimol-induced stimulation of BZD binding in a manner similar to that of the $GABA_A$ antagonist BMC (Falch *et al.*, 1990).

Whole-cell patch-clamp recordings from cultured hippocampal (Kristiansen *et al.*, 1991) or cerebral cortical (Frølund *et al.*, 1995c) neurons (Figure 10.15) have been used to further characterize the action of 4-PIOL. The action of 4-PIOL was compared with those of the full $GABA_A$ agonist isoguvacine (Kristiansen *et al.*, 1991; Frølund *et al.*, 1995c) (Figure 10.15) and the $GABA_A$ antagonist BMC (Kristiansen *et al.*, 1991). The response to 4-PIOL was competitively antagonized by BMC. 4-PIOL was about 200 times less potent as an agonist than isoguvacine. The maximum response to 4-PIOL was only a small fraction of that to submaximal concentrations of isoguvacine, and 4-PIOL antagonized the response to isoguvacine (Frølund *et al.*, 1995c). On the basis of these studies it is concluded that 4-PIOL is a low-efficacy partial $GABA_A$ agonist showing a predominant $GABA_A$ antagonist profile, being about 30 times weaker than BMC as a $GABA_A$ antagonist. Importantly, repeated administration of 4-PIOL did not cause significant desensitization of the $GABA_A$ receptors studied (Kristiansen *et al.*, 1991). Unfortunately, 4-PIOL does not show pharmacological effects after systemic administration (Falch *et al.*, 1990). In contrast to THIP, which penetrates the BBB very

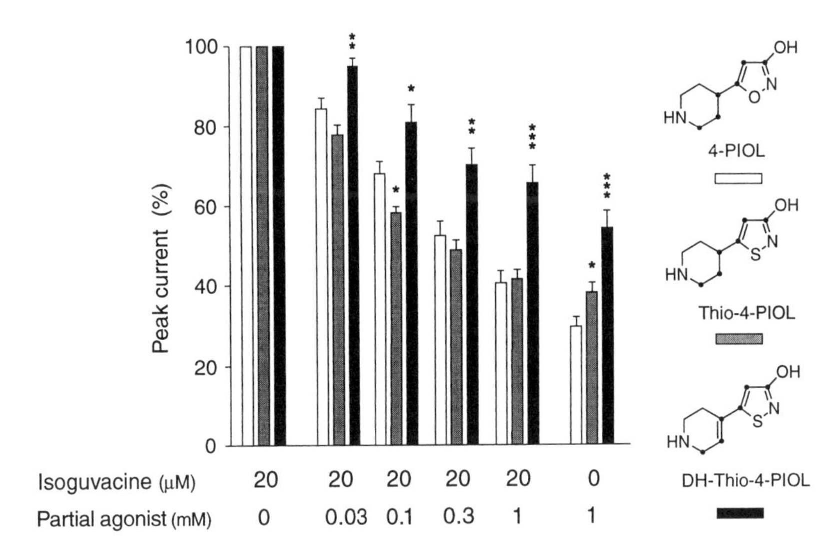

Figure 10.15 The dose-dependent reduction by the partial $GABA_A$ agonists, 4-PIOL, Thio-4-PIOL and DH-Thio-4-PIOL, of the effects of the full $GABA_A$ agonist, isoguvacine, on cultured cortical neurons using whole-cell patch-clamp techniques (for details, see Frølund *et al.*, 1995c).

easily (Krogsgaard-Larsen *et al.*, 1984), the protolytic properties of 4-PIOL do not allow this compound to pass the BBB (Falch *et al.*, 1990).

In an attempt to overcome this pharmacokinetic obstacle and to shed further light on the relationship between structure and $GABA_A$ agonist efficacy of this class of heterocyclic GABAergic compounds, a number of 4-PIOL analogues have been synthesized and tested. Some of these 4-PIOL analogues, notably 4-AZOL (Frølund *et al.*, 1992), 3-PIOL (Byberg *et al.*, 1987, 1993), and the monocyclic analogue V (Frølund *et al.*, 1995c) (Figure 10.14) were completely inactive, emphasizing the very strict structural requirements for binding to and activation of $GABA_A$ receptors (Krogsgaard-Larsen *et al.*, 1985, 1988).

On the other hand, the 4-PIOL analogues, DH-4-PIOL, Iso-4-PIOL, Thio-4-PIOL, and DH-Thio-4-PIOL (Figure 10.14), showed qualitatively similar effects on cultured cortical neurons (Frølund *et al.*, 1995c). Interestingly, however, the relative efficacies of these compounds as partial $GABA_A$ agonists ranged from levels markedly below that of 4-PIOL to significantly higher levels (Figure 10.15). It should be stressed that neither potency nor efficacy of the compounds under study can be determined precisely under the present experimental conditions. On the assumption that each compound acts as a true partial agonist capable of completely displacing isoguvacine from the $GABA_A$ receptors at high concentration, its maximal agonist effect must lie between the agonist levels produced by a high concentration (1 mM) of the compound in the absence or presence of isoguvacine (20 μM).

Figure 10.16 illustrates the relative efficacies of the full $GABA_A$ agonist isoguvacine and the partial agonists 4-PIOL and 4-PIOL analogues. The 3-isothiazolol analogue of 4-PIOL, Thio-4-PIOL, which binds more tightly to $GABA_A$ receptor sites than 4-PIOL, was approximately equieffective with 4-PIOL, whereas the unsaturated analogue of Thio-4-PIOL, DH-Thio-4-PIOL, was significantly more efficacious. In

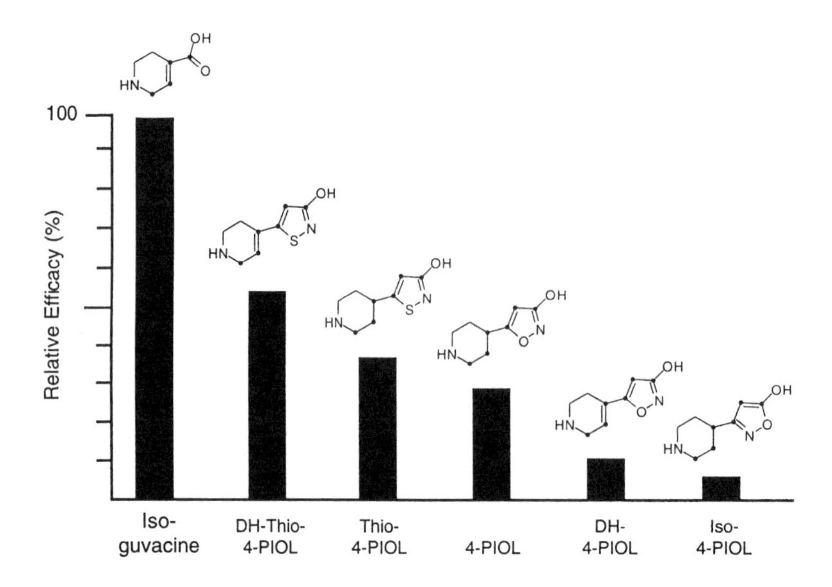

Figure 10.16 An illustration of the relative efficacies of the full $GABA_A$ agonist isoguvacine and the partial $GABA_A$ agonists 4-PIOL and analogues (based on data from Frølund *et al.*, 1995c).

light of these relative efficacies it was quite surprising to observe the reverse relative efficacies of 4-PIOL and the corresponding unsaturated analogue, DH-4-PIOL (Figure 10.16) (Frølund *et al.*, 1995c).

Introduction of alkyl groups into the 4 positions of molecules of muscimol or THIP severely inhibits interaction with the $GABA_A$ receptor recognition site. Thus, 4-Me-muscimol is 3–4 orders of magnitude weaker than muscimol as an inhibitor of $GABA_A$ receptor agonist binding, whereas 4-Me-THIP is inactive (Krogsgaard-Larsen *et al.*, 1988) (Figure 10.17). In contrast, introduction of alkyl groups into the 4-position of

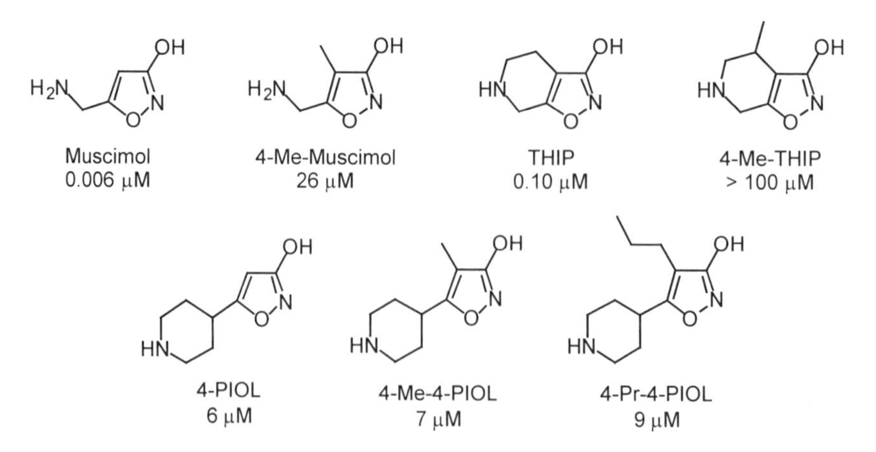

Figure 10.17 $GABA_A$ agonist binding data (IC_{50}, μM) for muscimol, THIP, 4-PIOL and some alkylated analogues of these $GABA_A$ agonist ligands.

the isoxazole ring of 4-PIOL is tolerated by the $GABA_A$ recognition site (Figure 10.17). These structure–activity relationships indicate that the binding modes of the $GABA_A$ agonists muscimol and THIP, showing full or high-efficacy agonist effects, respectively, and of the low-efficacy partial agonists 4-PIOL and alkylated 4-PIOL analogues, are different.

Future challenges

Molecular biology studies have disclosed a high degree of heterogeneity of the $GABA_A$ receptors. A necessary condition for elucidation of the physiological relevance, pharmacological importance, and potential as therapeutic targets of subtypes of this receptor is the availability of subtype-selective or subtype-specific ligands of different types. The design of such $GABA_A$ receptor ligands, using systematic stereochemical and bioisosteric approaches, represents a major challenge for medicinal and computational chemists.

Using affinity labelling techniques, site-directed mutagenesis, and domain-specific antibodies, the functional domains of the $GABA_A$ receptors are being mapped out (Smith and Olsen, 1995). One of the prerequisites for design of $GABA_A$ receptor ligands on a strictly rational basis is very detailed information about the topography of the recognition site(s) derived from X-ray crystallographic structure determination of crystals of $GABA_A$ receptors or complexes of $GABA_A$ receptor domains and specific ligands. Using conventional or novel crystallization techniques, studies along these lines are likely to be possible in the not too distant future.

For different reasons, neither full $GABA_A$ agonists nor antagonists may be useful therapeutic agents for the treatment of psychiatric and neurological diseases. Whereas the high-efficacy partial $GABA_A$ agonist THIP shows very potent analgesic effects in the human clinic, it seems likely that low-efficacy partial $GABA_A$ agonists, such as Thio-4-PIOL or related 4-PIOL analogues, may have considerable therapeutic interest in certain CNS disorders.

It is beyond doubt that $GABA_A$ receptor modulatory agents, notably different types of BZDs (Gammill and Carter, 1993) and neurosteroids or neuroactive steroids (Lambert *et al.*, 1995; Gasior *et al.*, 1999), will have growing therapeutic interest and utility in the future. It seems likely that novel types of modulatory sites at the $GABA_A$ receptor complex may be disclosed and shown to be interesting targets for therapeutic intervention.

Acknowledgements

This work was supported by grants from the Danish Technical Research Council, the Lundbeck Foundation, the Alfred Benzon Foundation and the Danish State Biotechnology Programme (1991–1995). The secretarial assistance of Mrs Anne Nordly is gratefully acknowledged.

References

Allan, R.D., Apostopoulos, C. and Richardson, J.A. (1990) '2-amino-1,3,4-thiadiazole derivatives of GABA as $GABA_A$ antagonists', *Aust. J. Chem.* **43**: 1767–1772.

Arnt, J. and Krogsgaard-Larsen, P. (1979) 'GABA agonists and potential antagonists related to muscimol', *Brain Res.* **177**: 395–400.

Barnard, E.A. (1992) 'Receptor classes and the transmitter-gated ion channels', *Trends Biochem. Sci.* **17**: 368–374.

Barnard, E.A. and Costa, E. (eds) (1989) *Allosteric Modulation of Amino Acid Receptors: Therapeutic Implications*, New York: Raven Press.

Baulieu, E.E. (1991) 'Neurosteroids: a new function in the brain', *Biol. Cell.* **71**: 3–10.

Belhage, B., Damgaard, I., Saederup, E., Squires, R.F. and Schousboe, A. (1991) 'High- and low-affinity GABA-receptors in cultured cerebellar granule cells regulate transmitter release by different mechanisms', *Neurochem. Intl.* **19**: 475–482.

Belhage, B., Hansen, G.H., Meier, E. and Schousboe, A. (1990) 'Effects of inhibitors of protein synthesis and intracellular transport on the GABA-agonist induced functional differentiation of cultured cerebellar granule cells', *J. Neurochem.* **55**: 1107–1113.

Benes, F.M., Vincent, S.L., Alsterberg, G., Bird, E.D. and SanGiovanni, J.P. (1992) 'Increased GABAA receptor binding in superficial layers of cingulate cortex in schizophrenics', *J. Neurosci.* **12**: 924–929.

Biggio, G. and Costa, E. (eds) (1990) *GABA and Benzodiazepine Receptor Subtypes*, New York: Raven Press.

Bixler, E.O., Kales, A., Brubaker, B.H. and Kales, J.D. (1987) 'Adverse reactions to benzodiazepine hypnotics: spontaneous reporting system', *Pharmacology* **35**: 286–300.

Bouchet, M.-J., Jacques, P., Ilien, B., Goeldner, M. and Hirth, C. (1992) 'm-sulfonate benzene diazonium chloride: a powerful affinity label for the γ-aminobutyric acid binding site from rat brain', *J. Neurochem.* **59**: 1405–1413.

Boue-Grabot, E., Roudbaraki, M., Tramu, G., Block, B. and Garret, M. (1998) 'Expression of GABA receptor ρ subunits in rat brain', *J. Neurochem.* **70**: 899–907.

Bowery, N.G. (1983) 'Classification of GABA receptors', in S.J. Enna (ed.) *The GABA Receptors*, Clifton: The Humana Press, pp. 177–213.

Bowery, N.G., Bittiger, H. and Olpe, H.-R. (eds) (1990) *$GABA_B$ Receptors in Mammalian Function*, Chichester: John Wiley.

Bowery, N.G., Hill, D.R., Hudson, A.L., Doble, A., Middlemiss, D.N., Shaw, J. and Turnbull, M. (1980) '(−) baclofen decreases neurotransmitter release in the mammalian CNS by an action at a novel GABA receptor', *Nature* **283**: 92–94.

Bowery, N.G. and Nistico, G. (eds) (1989) *GABA: Basic Research and Clinical Applications*, Rome: Pythagora Press.

Braestrup, C., Nielsen, M., Krogsgaard-Larsen, P. and Falch, E. (1979) 'Partial agonists for brain GABA/benzodiazepine receptor complex', *Nature* **280**: 331–333.

Brehm, L., Ebert, B., Kristiansen, U., Wafford, K.A., Kemp, J.A. and Krogsgaard-Larsen, P. (1997) 'Structure and pharmacology of 4,5,6,7-tetrahydroisothiazolo[5,4-*c*]pyridin-3-ol (Thio-THIP), an agonist/antagonist at $GABA_A$ receptors', *Eur. J. Med. Chem.* **32**: 357–363.

Brioni, J.D., Decker, M.W., Gamboa, L.P., Izquierdo, I. and McGaugh, J.L. (1990) 'Muscimol injections in the medial septum impair spatial learning', *Brain Res.* **522**: 227–234.

Brioni, J.D., Nagahara, A.H. and McGaugh, J.L. (1989) 'Involvement of the amygdala GABAergic system in the modulation of memory storage', *Brain Res.* **487**: 105–112.

Brodie, M.J. and McKee, P.J.W. (1990) 'Vigabatrin and psychosis', *Lancet* **335**: 1279.

Bruno, G., Foster, N.L., Fedio, P., Mohr, E., Cox, C., Gillespie, M.M. and Chase, T.N. (1984) 'THIP therapy of Alzheimer's disease', *Neurology Suppl.* **34**: 225.

Bureau, M. and Olsen, R.W. (1990) 'Multiple distinct subunits of the γ-aminobutyric acid-A receptor protein show different ligand-binding affinities', *Mol. Pharmacol.* **37**: 497–502.

Burke, D., Andrews, C.J. and Knowles, L. (1971) 'The action of a GABA derivative in human spasticity', *J. Neurol.* **14**: 199–208.

Byberg, J.R., Hjeds, H., Krogsgaard-Larsen, P. and Jørgensen, F.S. (1993) 'Conformational analysis and molecular modelling of a partial $GABA_A$ agonist and a glycine antagonist related to the $GABA_A$ agonist, THIP', *Drug Des. Delivery* **10**: 213–229.

Byberg, J.R., Labouta, I.M., Falch, E., Hjeds, H., Krogsgaard-Larsen, P., Curtis, D.R. and Gynther, B.D. (1987) 'Synthesis and biological activity of a GABA-A agonist which has no effect on benzodiazepine binding and structurally related glycine antagonists', *Drug Des. Delivery* **1**: 261–274.

Caspary, D.M., Raza, A., Lawhorn Armour, B.A., Pippin, J. and Arneri, S.P. (1990) 'Immunocytochemical and neurochemical evidence for age-related loss of GABA in the inferior colliculus: implications for neural presbycusis', *J. Neurosci.* **10**: 2363–2372.

Cavalla, D. and Neff, N.H. (1985) 'Photoaffinity labeling of the $GABA_A$ receptor with [^{3}H]muscimol', *J. Neurochem.* **44**: 916–921.

Chambon, J.-P., Feltz, P., Heaulme, M., Restlé, S., Schlichter, R., Biziére, K. and Wermuth, C.G. (1985) 'An arylaminopyridazine derivative of γ-aminobutyric acid (GABA) is a selective and competitive antagonist of the receptor sites', *Proc. Natl. Acad. Sci. USA* **82**: 1832–1836.

Cherubini, E., Gaiarsa, J.L. and Ben-Ari, Y. (1991) 'GABA: an excitatory transmitter in early postnatal life', *Trends Neurosci.* **14**: 515–519.

Curtis, D.R., Duggan, A.W., Felix, D. and Johnston, G.A.R. (1970) 'GABA, bicuculline and central inhibition', *Nature* **226**: 1222–1224.

Curtis, D.R., Game, C.J.A., Johnston, G.A.R. and McCulloch, R.M. (1974) 'Central effects of β-(p-chlorophenyl)-γ-aminobutyric acid', *Brain Res.* **70**: 493–499.

Curtis, D.R. and Johnston, G.A.R. (1974) 'Amino acid transmitters in the mammalian central nervous system', *Ergebn. Physiol.* **69**: 97–188.

Cutting, G.R., Lu, L., O'Hara, B.F., Kasch, L.M., Montrose-Rafizaheh, C., Donovan, D.M., Shimada, S., Antonorakis, S.E., Guggino, W.B., Uhl, G.R. and Kazazian, H.H. (1991) 'Cloning of the γ-aminobutyric acid (GABA) ρ_1 cDNA: a GABA receptor subunit highly expressed in retina', *Proc. Natl. Acad. Sci. USA* **88**: 2673–2677.

Dannhardt, G., Dominiak, P. and Laufer, S. (1993) 'Hypertensive effects and structure–activity relationships of 5-ω-aminoalkyl isoxazoles', *Drug Res.* **43**: 441–444.

Dean, B., Hussain, T., Hayes, W., Scarr, E., Kitsoulis, S., Hill, C., Opeskin, K. and Copolov, D.L. (1999) 'Changes in $serotonin_{2A}$ and $GABA_A$ receptors in schizophrenia: studies on the human dorsolateral prefrontal cortex', *J. Neurochem.* **72**: 1593–1599.

DeErasquin, G., Grooker, G., Costa, E. and Woscik, W.J. (1992) 'Stimulation of high affinity γ-aminobutyric $acid_B$ receptors potentiates the depolarization induced increase of intraneuronal ionised calcium content in cerebellar granule neurons', *Mol. Pharmacol.* **42**: 407–414.

DeFeudis, F.V. (1989) 'GABA agonists and analgesia', *Drug News Perspect.* **2**: 172–173.

DeLorey, T.M. and Olsen, R.W. (1992) 'γ-aminobutyric $acid_A$ receptor structure and function', *J. Biol. Chem.* **267**: 16,747–16,750.

DiChiara, G. and Gessa, G.L. (eds) (1981) *GABA and the Basal Ganglia*, New York: Raven Press.

Djamgoz, M.B.A. (1995) 'Diversity of GABA receptors in the vertebrate outer retina', *Trends Neurosci.* **18**: 118–120.

Doble, A. and Martin, I.L. (1996) *The $GABA_A$/Benzodiazepine Receptor as a Target for Psychoactive Drugs*, Heidelberg: Springer.

Drew, C.A. and Johnston, G.A.R. (1992) 'Bicuculline- and baclofen-insensitive γ-aminobutyric acid binding to rat cerebellar membranes', *J. Neurochem.* **58**: 1087–1092.

Ebert, B., Frølund, B., Diemer, N.H. and Krogsgaard-Larsen, P. (1999) 'Equilibrium binding characteristics of [^{3}H]thiomuscimol', *Neurochem. Intl.* **34**: 427–434.

Ebert, B., Thompson, S.A., Saounatsou, K., McKernan, R., Krogsgaard-Larsen, P. and Wafford, K.A. (1997) 'Differences in agonist/antagonist binding affinity and receptor transduction using recombinant human γ-aminobutyric acid type A receptors', *Mol. Pharmacol.* **52**: 1150–1156.

Ebert, B., Wafford, K.A., Whiting, P.J., Krogsgaard-Larsen, P. and Kemp, J.A. (1994) 'Molecular pharmacology of γ-aminobutyric acid type A receptor agonists and partial agonists in oocytes injected with different α, β and γ receptor subunit combinations', *Mol. Pharmacol.* **46**: 957–963.

Edvinsson, L., Larsson, B. and Skärby, T. (1980) 'Effect of the GABA receptor agonist muscimol on regional cerebral blood flow in the rat', *Brain Res.* **185**: 445–448.

Erdö, S.L. (1985) 'Peripheral GABAergic mechanisms', *Trends Pharmacol. Sci.* **6**: 205–208.

Erdö, S.L. and Bowery, N.G. (eds) (1986) *GABAergic Mechanisms in Mammalian Periphery*, New York: Raven Press.

Falch, E., Hedegaard, A., Nielsen, L., Jensen, B.R., Hjeds, H. and Krogsgaard-Larsen, P. (1986) 'Comparative stereostructure–activity studies on $GABA_A$ and $GABA_B$ receptor sites and GABA uptake using rat brain membrane preparations', *J. Neurochem.* **47**: 898–903.

Falch, E., Jacobsen, P., Krogsgaard-Larsen, P., Curtis, D.R. (1985) 'GABA-mimetic activity and effects on diazepam binding of aminosulphonic acids structurally related to piperidine-4-sulphonic acid', *J. Neurochem.* **44**: 68–75.

Falch, E., Larsson, O.M., Schousboe, A. and Krogsgaard-Larsen, P. (1990) 'GABA-A agonists and GABA uptake inhibitors: structure–activity relationships', *Drug Dev. Res.* **21**: 169–188.

Fariello, R.G., Morselli, P.L., Lloyd, K.G., Quesney, L.F. and Engel, J. (eds) (1984) *Neurotransmitters, Seizures, and Epilepsy II*, New York: Raven Press.

Feigenspan, A., Wässle, H. and Bormann, J. (1993) 'Pharmacology of GABA receptor Cl^- channels in rat retinal bipolar cells', *Nature* **361**: 159–162.

Foster, N.L., Chase, T.N., Denaro, A., Hare, T.A. and Tamminga, C.A. (1983) 'THIP treatment and Huntington's chorea', *Neurology* **33**: 637–639.

Friedman, D.E. and Redburn, D.A. (1990) 'Evidence for functionally distinct subclasses of γ-aminobutyric acid receptors in rabbit retina', *J. Neurochem.* **55**: 1189–1199.

Froestl, W., Mickel, S.J., Hall, R.G., von Sprecher, G., Strub, D., Baumann, P.A., Brugger, F., Gentsch, C., Jaekel, J., Olpe, H.-R., Rihs, G., Vassout, A., Waldmeier, P.C. and Bittiger, H. (1995a) 'Phosphinic acid analogues of GABA. 1. New potent and selective $GABA_B$ agonists', *J. Med. Chem.* **38**: 3297–3312.

Froestl, W., Mickel, S.J., von Sprecher, G., Diel, P.J., Hall, R.G., Maier, L., Strub, D., Melillo, V., Baumann, P.A., Bernasconi, R., Gentsch, C., Hauser, K., Jaekel, J., Karlsson, G., Klebs, K., Maître, L., Marescaux, C., Pozza, M.F., Schmutz, M., Steinmann, M.W., van Riezen, H., Vassout, A., Mondadori, C., Olpe, H.-R., Waldmeier, P.C. and Bittiger, H. (1995b) 'Phosphinic acid analogues of GABA. 2. Selective, orally active $GABA_B$ antagonists', *J. Med. Chem.* **38**: 3313–3331.

Frydenvang, K., Krogsgaard-Larsen, P., Hansen, J.J., Mitrovic, A., Tran, H., Drew, C.A. and Johnston, G.A.R. (1994) '$GABA_B$ antagonists: resolution, absolute stereochemistry and pharmacology of (R)- and (S)-phaclofen', *Chirality* **6**: 583–589.

Frølund, B., Ebert, B., Lawrence, L.W., Hurt, S.D. and Krogsgaard-Larsen, P. (1995a) 'Synthesis and receptor binding of 5-amino[^{3}H]$_2$methyl-3-isothiazolol ([^{3}H]thiomuscimol), a specific $GABA_A$ agonist photoaffinity label', *J. Labelled Compd. Radiopharm.* **36**: 877–889.

Frølund, B., Jeppesen, L., Krogsgaard-Larsen, P. and Hansen, J.J. (1995b) '$GABA_A$ agonists: resolution and pharmacology of (+)- and (−)-isoguvacine oxide', *Chirality* **7**: 434–438.

Frølund, B., Kristiansen, U., Brehm, L., Hansen, A.B., Krogsgaard-Larsen, P. and Falch, E. (1995c) 'Partial $GABA_A$ receptor agonists. Synthesis and in vitro pharmacology of a series of nonannulated analogs of 4,5,6,7-tetrahydroisoxazolo[5,4-c]pyridin-3-ol', *J. Med. Chem.* **38**: 3287–3296.

Frølund, B.F., Kristiansen, U., Nathan, T., Falch, E., Lambert, J.D.C. and Krogsgaard-Larsen, P. (1992) '4-PIOL, a low-efficacy partial $GABA_A$ agonist', in A. Schousboe, N.H. Diemer and H. Kofod (eds) *Drug Research Related to Neuroactive Amino Acids, Alfred Benzon Symposium 32*: Copenhagen: Munksgaard, pp. 449–460.

Gammill, R.B. and Carter, D.B. (1993) 'Neuronal BZD receptors: new ligands, clones and pharmacology', *Annu. Rep. Med. Chem.* **28**: 19–27.

Gasior, M., Carter, R.B. and Witkin, J.M. (1999) 'Neuroactive steroids: potential therapeutic use in neurological and psychiatric disorders', *Trends Pharmacol. Sci.* **20**: 107–112.

Guidotti, A. (1992) 'Imidazenil: a new partial positive allosteric modulator of the $GABA_A$ receptor', *Neurosci. Facts*, **3**: 71–72.

Haefely, W. (1984) 'Pharmacological profile of two benzodiazepine partial agonists: Ro 16–6028 and Ro 17–1812', *Clin. Neuropharmacol.* **7** (suppl. 1): 670–671.

Haefely, W. and Polc, P. (1986) 'Physiology of GABA enhancement by benzodiazepines and barbiturates', in R.W. Olsen and J.C. Venter (eds) *Benzodiazepine/GABA Receptors and Chloride Channels: Structural and Functional Properties*, New York: Alan R. Liss, pp. 97–133.

Hahner, L., McQuilkin, S. and Harris, R.A. (1991) 'Cerebellar $GABA_B$ receptors modulate function of $GABA_A$ receptors', *FASEB J.* **5**: 2466–2472.

Hall, R.C. and Zisool, S. (1981) 'Paradoxical reactions to benzodiazepines', *Br. J. Clin. Pharmacol.* **11**: 99S–104S.

Hanada, S., Mita, S., Nishino, N. and Tanaka, C. (1987) '[^{3}H]muscimol binding sites increased in autopsied brains of chronic schizophrenics', *Life Sci.* **40**: 259–266.

Hansen, G.H., Belhage, B. and Schousboe, A. (1992) 'First direct electron microscopic visualization of a tight spatial coupling between $GABA_A$-receptors and voltage sensitive calcium channels', *Neurosci. Lett.* **137**: 14–18.

Harrison, N.L. and Simmonds, M.A. (1984) 'Modulation of the GABA receptor complex by a steroid anaesthetic', *Brain Res.* **323**: 287–292.

Hoehn-Saric, R. (1983) 'Effects of THIP on chronic anxiety', *Psychopharmacology* **80**: 338–341.

Hunkeler, W., Möhler, H., Pieri, L., Polc, P., Bonetti, E.P., Cumin, R., Schaffner, R. and Haefely, W. (1981) 'Selective antagonists of benzodiazepines', *Nature* **290**: 514–516.

Huston, E., Gullen, G., Sweeney, M.I., Pearson, H., Fazeli, M.S. and Dolphin, A.C. (1993) 'Pertussis toxin treatment increases glutamate release and dihydropyridine binding sites in cultured rat cerebellar granule neurons', *Neuroscience* **52**: 787–798.

Huston, E., Scott, R.H. and Dolphin, A.C. (1990) 'A comparison of the effect of calcium channel ligands and $GABA_B$ agonists and antagonists in transmitter release and somatic calcium currents in cultured neurons', *Neuroscience* **38**: 721–729.

Jacobsen, E.J., Stelzer, L.S., TenBrink, R.E., Belonga, K.L., Carter, D.B., Im, H.K., Im, W.B., Sethy, V.H., Tang, A.H., von Voigtlander, P.F., Petke, J.D., Zhong, W.-Z. and Mickelson, J.W. (1999) 'Piperazine imidazo[1,5-*a*]quinoxaline ureas as high-affinity $GABA_A$ ligands of dual functionality', *J. Med. Chem.* **42**: 1123–1144.

Johnston, G.A.R. (1986) 'Multiplicity of GABA receptors', in R.W. Olsen and J.C. Venter (eds) *Benzodiazepine/GABA Receptors and Chloride Channels: Structural and Functional Properties*, New York: Alan R. Liss, pp. 57–71.

Johnston, G.A.R. (1996) '$GABA_A$ receptor pharmacology', *Pharmacol. Ther.* **69**: 173–198.

Johnston, G.A.R. (1997) '$GABA_C$ receptors: molecular biology, pharmacology and physiology', in S.J. Enna and N.G. Bowery (eds) *The GABA Receptors*, Totowa: The Humana Press, pp. 297–323.

Johnston, G.A.R., Beart, P.M., Curtis, D.R., Game, C.J.A., McCulloch, R.M. and MacLachlan, R.M. (1972) 'Bicuculline methochloride as a GABA antagonist', *Nature (New Biol.)* **240**: 219–220.

Johnston, G.A.R., Curtis, D.R., Beart, P.M., Game, C.J.A., McCulloch, R.M. and Twitchin, B. (1975a) 'Cis- and trans-4-aminocrotonic acid as GABA analogues of restricted conformation', *J. Neurochem.* **24**: 157–160.

Johnston, G.A.R., Krogsgaard-Larsen, P. and Stephanson, A. (1975b) 'Betel nut constituents as inhibitors of γ-aminobutyric acid uptake', *Nature* **258**: 627–628.

Kardos, J. (1999) 'Recent advances in GABA research', *Neurochem. Intl.* **34**: 353–358.

Kardos, J., Elster, L., Damgaard, I., Krogsgaard-Larsen, P. and Schousboe, A. (1994) 'Role of $GABA_B$ receptors in intracellular Ca^{2+} homeostasis and possible interaction between $GABA_A$ and $GABA_B$ receptors in regulation of transmitter release in cerebellar granule neurons', *J. Neurosci. Res.* **39**: 646–655.

Kehler, J., Stensbøl, T.B. and Krogsgaard-Larsen, P. (1999) 'Piperidinyl-3-phosphinic acids as novel uptake inhibitors of the neurotransmitter 4-aminobutyric acid (GABA)', *Bioorg. Med. Chem. Lett.* **9**: 811–814.

Kendall, D.A., Browner, M. and Enna, S.J. (1982) 'Comparison of the antinociceptive effect of GABA agonists: evidence for a cholinergic involvement', *J. Pharmacol. Exp. Ther.* **220**: 482–487.

Kerr, D.I.B. and Ong, J. (1986) '$GABA_B$-receptors in peripheral function', in S.L. Erdö and N.G. Bowery (eds) *GABAergic Mechanisms in Mammalian Periphery*, New York: Raven Press, pp. 239–259.

Kerr, D.I.B., Ong, J., Prager, R.H., Gynther, B.D. and Curtis, D.R. (1987) 'Phaclofen: a peripheral and central baclofen antagonist', *Brain Res.* **405**: 150–154.

Kiuchi, Y., Kobayashi, T., Takeuchi, J., Shimuzu, H., Ogata, H. and Toru, M. (1989) 'Benzodiazepine receptors increase in post-mortem brain of chronic schizophrenics', *Eur. Arch. Psychiat. Neurol. Sci.* **239**: 71–78.

Kjaer, M. and Nielsen, H. (1983) 'The analgesic action of the GABA agonist THIP in patients with chronic pain of malignant origin', *Br. J. Clin. Pharmacol.* **16**: 477–485.

Korpi, E.R., Uusi-Oukari, M. and Wegelius, K. (1992) 'Substrate specificity of diazepam-insensitive cerebellar [^{3}H]Ro 15–4513 binding sites', *Eur. J. Pharmacol.* **213**: 323–329.

Korsgaard, S., Casey, D.E., Gerlach, J., Hetmar, O., Kaldan, B. and Mikkelsen, L.B. (1982) 'The effect of tetrahydroisoxazolopyridinol (THIP) in tardive dyskinesia', *Arch. Gen. Psychiatry* **39**: 1017–1021.

Kristiansen, U., Hedegaard, A., Herdeis, C., Lund, T.M., Nielsen, B., Hansen, J.J., Falch, E., Hjeds, H. and Krogsgaard-Larsen, P. (1992) 'Hydroxylated analogues of 5-aminovaleric acid as 4-aminobutyric $acid_B$ receptor antagonists: stereostructure – activity relationships', *J. Neurochem.* **58**: 1150–1159.

Kristiansen, U., Lambert, J.D.C., Falch, E. and Krogsgaard-Larsen, P. (1991) 'Electrophysiological studies of the $GABA_A$ receptor ligand, 4-PIOL, on cultured hippocampal neurons', *Br. J. Pharmacol.* **104**: 85–90.

Krnjevic, K. (1974) 'Chemical nature of synaptic transmission in vertebrates', *Physiol. Rev.* **54**: 418–540.

Krogsgaard-Larsen, P. (1988) 'GABA synaptic mechanisms: stereochemical and conformational requirements', *Med. Res. Rev.* **8**: 27–56.

Krogsgaard-Larsen, P., Falch, E. and Christensen, A.V. (1984) 'Chemistry and pharmacology of the GABA agonists THIP (gaboxadol) and isoguvacine', *Drugs Fut.* **9**: 597–618.

Krogsgaard-Larsen, P., Falch, E. and Hjeds, H. (1985) 'Heterocyclic analogues of GABA: chemistry, molecular pharmacology and therapeutic aspects', *Prog. Med. Chem.* **22**: 67–120.

Krogsgaard-Larsen, P., Falch, E., Larsson, O.M. and Schousboe, A. (1987) 'GABA uptake inhibitors: relevans to antiepileptic drug research', *Epilepsy Res.* **1**: 77–93.

Krogsgaard-Larsen, P., Falch, E., Schousboe, A., Curtis, D.R. and Lodge, D. (1980a) 'Piperidine-4-sulphonic acid, a new specific GABA agonist', *J. Neurochem.* **34**: 756–759.

Krogsgaard-Larsen, P., Frølund, B. and Frydenvang, K. (2000) 'GABA uptake inhibitors. Design, molecular pharmacology and therapeutic aspects', *Curr. Pharm. Des.* **6**: 1193–1209.

Krogsgaard-Larsen, P., Frølund, B., Jørgensen, F.S. and Schousboe, A. (1994) '$GABA_A$ receptor agonists, partial agonists, and antagonists. Design and therapeutic prospects', *J. Med. Chem.* **37**: 2489–2505.

Krogsgaard-Larsen, P., Frølund, B., Kristiansen, U., Frydenvang, K. and Ebert, B. (1997) '$GABA_A$ and $GABA_B$ receptor agonists, partial agonists, antagonists and modulators: design and therapeutic prospects', *Eur. J. Pharm. Sci.* **5**: 355–384.

Krogsgaard-Larsen, P. and Hansen, J.J. (eds) (1992) *Excitatory Amino Acid Receptors: Design of Agonists and Antagonists*, Chichester: Ellis Horwood.

Krogsgaard-Larsen, P., Hjeds, H., Curtis, D.R., Lodge, D. and Johnston, G.A.R. (1979) 'Dihydromuscimol, thiomuscimol and related heterocyclic compounds as GABA analogues', *J. Neurochem.* **32**: 1717–1724.

Krogsgaard-Larsen, P., Hjeds, H., Falch, E., Jørgensen, F.S. and Nielsen, L. (1988) 'Recent advances in GABA agonists, antagonists and uptake inhibitors: structure–activity relationships and therapeutic potential', *Adv. Drug Res.* **17**: 381–456.

Krogsgaard-Larsen, P., Jacobsen, P., Brehm, L., Larsen, J.-J. and Schaumburg, K. (1980b) 'GABA agonists and uptake inhibitors designed as agents with irreversible actions', *Eur. J. Med. Chem.* **15**: 529–535.

Krogsgaard-Larsen, P. and Johnston, G.A.R. (1975) 'Inhibition of GABA uptake in rat brain slices by nipecotic acid, various isoxazoles and related compounds', *J. Neurochem.* **25**: 797–802.

Krogsgaard-Larsen, P., Johnston, G.A.R., Lodge, D. and Curtis, D.R. (1977) 'A new class of GABA agonist', *Nature* **268**: 53–55.

Krogsgaard-Larsen, P., Mikkelsen, H., Jacobsen, P., Falch, E., Curtis, D.R., Peet, M.J. and Leah, J.D. (1983) '4,5,6,7-tetrahydroisothiazolo[5,4-c]pyridin-3-ol and related analogues of THIP. Synthesis and biological activity', *J. Med. Chem.* **26**: 895–900.

Krogsgaard-Larsen, P., Nielsen, L., Falch, E. and Curtis, D.R. (1986) 'GABA agonists. Resolution, absolute stereochemistry, and enantioselectivity of (S)-(+)- and (R)-(−)-dihydromuscimol', *J. Med. Chem.* **28**: 1612–1617.

Krogsgaard-Larsen, P., Snowman, A., Lummis, S.C. and Olsen, R.W. (1981) 'Characterization of the binding of the GABA agonist [^{3}H]piperidine-4-sulfonic acid to bovine brain synaptic membranes', *J. Neurochem.* **37**: 401–409.

Lambert, J.J., Belelli, D., Hill-Venning, C. and Peters, J.A. (1995) 'Neurosteroids and $GABA_A$ receptor function', *Trends Pharmacol. Sci.* **16**: 295–303.

Lancel, M. and Faulhaber, J. (1996) 'The $GABA_A$ agonist THIP (gaboxadol) increases non-REM sleep and enhances delta activity in the rat', *NeuroReport* **7**: 2241–2245.

Levi, G. and Gallo, V. (1981) 'Glutamate as a putative transmitter in the cerebellum: stimulation by GABA of glutamic acid release from specific pools', *J. Neurochem.* **37**: 22–31.

Lodge, D. (ed.) (1988) *Excitatory Amino Acids in Health and Disease*, Chichester: John Wiley.

Lopez, I., Wu, J.Y. and Meza, G. (1992) 'Immunocytochemical evidence for an afferent GABAergic neurotransmission in the guinea pig vestibular system', *Brain Res.* **589**: 341–348.

Löscher, W. (1989) 'GABA and the epilepsies. Experimental and clinical conditions', in N.G. Bowery and G. Nistico (eds) *GABA: Basic Research and Clinical Applications*, Rome: Pythagora Press, pp. 260–300.

Lüddens, H., Pritchett, D.B., Köhler, M., Killisch, I., Keinänen, K., Moneyer, H., Sprengel, R. and Seeburg, P.H. (1990) 'Cerebellar $GABA_A$ receptor selective for a behavioural alcohol antagonist', *Nature* **346**: 648–651.

Macdonald, R.L. and Olsen, R.W. (1994) '$GABA_A$ receptor channels', *Annu. Rev. Neurosci.* **17**: 569–602.

Maksay, G. (1994) 'Thermodynamics of γ-aminobutyric acid type A receptor binding differentiate agonists from antagonists', *Mol. Pharmacol.* **46**: 386–390.

Malcangio, M., Malmberg-Aiello, P., Giotti, A., Ghelardini, C. and Bartolini, A. (1992) 'Desensitization of $GABA_B$ receptors and antagonism by CGP 35348 prevent bicuculline- and picrotoxin-induced antinociception', *Neuropharmacology* **31**: 783–791.

Malminiemi, O. and Korpi, E.S. (1989) 'Diazepam-insensitive [^{3}H]Ro 15–4513 binding in intact cultured cerebellar granule cells', *Eur. J. Pharmacol.* **169**: 53–60.

Martin, S.C., Russek, S.J. and Farb, D.H. (1999) 'Molecular identification of the human $GABA_BR2$: cell surface expression and coupling to adenylyl cyclase in the absence of $GABA_BR1$', *Mol. Cell Neurosci.* **13**: 180–191.

Mathers, D.A. (1987) 'The $GABA_A$ receptor: new insights from single-channel recording', *Synapse* **1**: 96–101.

McNeil, R.G., Gee, K.W., Bolger, M.B., Lan, N.C., Wieland, S., Belelli, D., Purdy, R.H. and Paul, S.M. (1992) 'Neuroactive steroids that act at $GABA_A$ receptors', *Drug News & Perspect.* **5**: 145–152.

Meier, E., Drejer, J. and Schousboe, A. (1984) 'GABA induces functionally active low-affinity GABA receptors on cultured cerebellar granule cells', *J. Neurochem.* **43**: 1737–1744.

Meldrum, B. (1982) 'GABA and acute psychoses', *Psychol. Med.* **12**: 1–5.

Melikian, A., Schlewer, G., Chambon, J.-P. and Wermuth, C.G. (1992) 'Condensation of muscimol or thiomuscimol with aminopyridazines yields GABA-A antagonists', *J. Med. Chem.* **35**: 4092–4097.

Merz, W.A., Alterwain, P., Ballmer, U., Bechelli, L., Capponi, R., Munoz, J.G., Marquez, C., Nestoros, J., Almanzor, L.R., Udabe, R.U. and Versiani, M. (1988) 'Treatment of paranoid schizophrenia with the partial benzodiazepine agonist, Ro 16–6028', *Psychopharmacol. Suppl.* **237**: 95–96.

Minchin, M.C.W., Ennis, C., Lattimer, N., White, J.F., White, A.C. and Lloyd, K.G. (1992a) 'The $GABA_A$-like autoreceptor is a pharmacologically novel GABA receptor', in G. Biggio, A. Concas and E. Costa (eds) *GABAergic Synaptic Transmission*, New York: Raven Press, pp. 199–203.

Minchin, M.C.W., White, A.C. and White, J.F. (1992b) 'Novel GABA autoreceptor antagonists', *Current Drugs* **2**: 1878–1880.

Mitchell, R. (1980) 'A novel GABA receptor modulates stimulus induced glutamate release from cortico-striatal terminals', *Eur. J. Pharmacol.* **67**: 119–122.

Möhler, H. and Fritschy, J.-M. (1999) '$GABA_B$ receptors make it to the top – as dimers', *Trends. Pharmacol. Sci.* **20**: 87–89.

Möhler, H., Knoflach, F., Paysan, J., Motejlek, K., Benke, D., Lüscher, B. and Fritschy, J.M. (1995) 'Heterogeneity of $GABA_A$-receptors: cell-specific expression, pharmacology, and regulation', *Neurochem. Res.* **20**: 631–636.

Möhler, H. and Okada, T. (1977) 'Benzodiazepine receptors: demonstration in the central nervous system', *Science* **198**: 849–851.

Monaghan, D.T. and Wenthold, R.J. (1997) *The Ionotropic Glutamate Receptors*, Totowa, N.J.: Humana Press.

Morselli, P.L., Löscher, W., Lloyd, K.G., Meldrum, B. and Reynolds, E.H. (eds) (1981) *Neurotransmitters, Seizures, and Epilepsy*, New York: Raven Press.

Murata, Y., Woodward, R.M., Miledi, R. and Overman, L.E. (1996) 'The first selective antagonist for a $GABA_C$ receptor', *Bioorg. Med. Chem. Lett.* **6**: 2073–2076.

Nayeem, N., Green, T.P., Martin, I.L. and Barnard, E.A. (1994) 'Quaternary structure of the native $GABA_A$ receptor determined by electron microscopic image analysis', *J. Neurochem.* **62**: 815–818.

Nielsen, E.Ø., Aarslew-Jensen, M., Diemer, N.H., Krogsgaard-Larsen, P. and Schousboe, A. (1989) 'Baclofen-induced, calcium-dependent stimulation of in vivo release of D-[^{3}H]aspartate from rat hippocampus monitored by intracerebral microdialysis', *Neurochem. Res.* **14**: 321–326.

Nielsen, L., Brehm, L. and Krogsgaard-Larsen, P. (1990) 'GABA agonists and uptake inhibitors. Synthesis, absolute stereochemistry, and enantioselectivity of (R)-(−)- and (S)-(+)-homo-β-proline', *J. Med. Chem.* **33**: 71–77.

Nielsen, M., Witt, M.-R., Ebert, B. and Krogsgaard-Larsen, P. (1995) 'Thiomuscimol, a new photoaffinity label for the $GABA_A$ receptor', *Eur. J. Pharmacol.* **289**: 109–112.

Nistico, G., Morselli, P.L., Lloyd, K.G., Fariello, R.G. and Engel, J. (eds) (1986) *Neurotransmitters, Seizures, and Epilepsy III*, New York: Raven Press.

Olsen, R.W., Bureau, M.H., Edno, S. and Smith, G. (1991) 'The $GABA_A$ receptor family in the mammalian brain', *Neurochem. Res.* **16**: 317–325.

Olsen, R.W. and Tobin, A.J. (1990) 'Molecular biology of $GABA_A$ receptors', *FASEB J.* **4**: 1469–1480.

Olsen, R.W. and Venter, J.C. (eds) (1986) *Benzodiazepine/GABA Receptors and Chloride Channels: Structural and Functional Properties*, New York: Alan R. Liss.

Ong, J. and Kerr, D.I.B. (1990) 'GABA-receptors in peripheral tissues', *Life Sci.* **46**: 1489–1501.

Ong, J., Kerr, D.I.B., Capper, H.R. and Johnston, G.A.R. (1990) 'Cortisone, a potent $GABA_A$ antagonist in the guinea-pig isolated ileum', *J. Pharm. Pharmacol.* **42**: 662–664.

Ong, J., Kerr, D.I.B. and Johnston, G.A.R. (1987) 'Cortisol: a potent biphasic modulator at $GABA_A$-receptor-complexes in the guinea-pig isolated ileum', *Neurosci. Lett.* **82**: 101–106.

Palay, S. and Chan-Palay, V. (1982) 'The cerebellum – new vistas', *Exp. Brain Res.*, **6** (suppl.): 1–620.

Petersen, H.R., Jensen, I. and Dam, M. (1983) 'THIP: a single-blind controlled trial in patients with epilepsy', *Acta Neurol. Scand.* **67**: 114–117.

Peyron, R., Le Bars, D., Cinotti, L., Garcia-Larrea, L., Galy, G., Landais, P., Millet, P., Lavenne, F., Froment, J.C., Krogsgaard-Larsen, P. and Mauguière, F. (1994a) 'Effects of $GABA_A$ receptor activation on brain glucose metabolism in normal subjects and temporal lope epilepsy (TLE) patients. A positron emission tomography (PET) study. I. Brain glucose metabolism is increased after $GABA_A$ receptors activation', *Epilepsy Res.* **19**: 45–54.

Peyron, R., Cinotti, L., Le Bars, D., Garcia-Larrea, L., Galy, G., Landais, P., Millet, P., Lavenne, F., Froment, J.C., Krogsgaard-Larsen, P. and Mauguière, F. (1994b) 'Effects of $GABA_A$ receptor activation on brain glucose metabolism in normal subjects and temporal lobe epilepsy (TLE) patients. A positron emission tomography (PET) study. II. The focal hypometabolism is reactive to $GABA_A$ agonist administration in TLE', *Epilepsy Res.* **19**: 55–62.

Prince, R.J. and Simmonds, M.A. (1993) 'Differential antagonism by epipregnanolone of alphaxalone and pregnanolone potentiation of [^{3}H]flunitrazepam binding suggests more than one class of binding site for steroids at $GABA_A$ receptors', *Neuropharmacology* **32**: 59–63.

Qian, H. and Dowling, J.E. (1993a) 'GABA responses on retinal bipolar cells', *Biol. Bull.*, **185**: 312.

Qian, H. and Dowling, J.E. (1993b) 'Novel GABA responses from roddriven retinal horizontal cells', *Nature* **361**: 162–164.

Qian, H. and Dowling, J.E. (1994) 'Pharmacology of novel GABA receptors found on rod horizontal cells of the white perch retina', *J. Neurosci.* **14**: 4299–4307.

Redburn, D.A. and Schousboe, A. (eds) (1987) *Neurotrophic Activity of GABA during Development*, New York: Alan R. Liss.

Ring, H.A. and Reynolds, E.H. (1990) 'Vigabatrin and behaviour disturbance', *Lancet* **335**: 970.

Roberts, E. (1976) 'Disinhibition as an organizing principle in the nervous system – the role of the GABA system. Application to neurologic and psychiatric disorders', in E. Roberts, T.N. Chase and D.B. Tower, (eds) *GABA in Nervous System Function*, New York: Raven Press, pp. 515–539.

Roberts, E. (1986) 'GABA: The road to neurotransmitter status', in R.W. Olsen. and J.C. Venter (eds) *Benzodiazepine/GABA Receptors and Chloride Channels: Structural and Functional Properties*, New York: Alan R. Liss, pp. 1–39.

Roberts, E. (1991) 'Living systems are tonically inhibited, autonomous optimizers, and disinhibition coupled to a variability generation is their major organizing principle: inhibitory command-control at levels of membrane, genome, metabolism, brain, and society', *Neurochem. Res.* **16**: 409–421.

Robinson, M.K., Richens, A. and Oxley, R. (1990) 'Vigabatrin and behaviour disturbances', *Lancet* **336**: 504.

Rognan, D., Boulanger, T., Hoffmann, R., Vercauteren, D.P., Andre, J.-M., Durant, F. and Wermuth, C.G. (1992) 'Structure and molecular modeling of $GABA_A$ receptor antagonists', *J. Med. Chem.* **35**: 1969–1977.

Roland, P.E. and Friberg, L. (1988) 'The effect of the $GABA_A$ agonist THIP on regional cortical blood flow in humans. A new test of hemispheric dominance', *J. Cereb. Blood Flow Metab.* **8**: 314–323.

Rorsman, P., Berggren, P.-O., Bokvist, K., Ericson, H., Möhler, H., Östenson, C.-G. and Smith, P.A. (1989) 'Glucose-inhibition of glucagon secretion involves activation of $GABA_A$-receptor chloride channels', *Nature* **341**: 233–236.

Ryan, A.F. and Schwartz, I.R. (1986) 'Nipecotic acid: preferential accumulation in the cochlea by GABA uptake systems and selective retrograde transport to brainstem', *Brain Res.* **399**: 399–403.

Saito, A., Wu, J.Y. and Lee, T.J. (1985) 'Evidence for the presence of cholinergic nerves in cerebral arteries: an immunohistochemical demonstration of choline acetyltransferase', *J. Cereb. Blood Flow Metab.* **5**: 327–334.

Sander, J.W. and Hart, Y.M. (1990) 'Vigabatrin and behaviour disturbances', *Lancet* **335**: 57.

Sawynok, J. (1989) 'GABAergic agents as analgesics', in N.G. Bowery and G. Nistico (eds) *GABA: Basic Research and Clinical Applications*, Rome: Pythagora Press, pp. 383–399.

Schousboe, A. (1999) 'Pharmacologic and therapeutic aspects of the developmentally regulated expression of $GABA_A$ and $GABA_B$ receptors: cerebellar granule cells as a model system', *Neurochem. Intl.* **34**: 373–377.

Schousboe, A., Diemer, N.H. and Kofod, H. (eds) (1992a) *Drug Research Related to Neuroactive Amino Acids*, Copenhagen: Munksgaard.

Schousboe, A., Hansen, G.H. and Belhage, B. (1992b) 'Regulation of neurotransmitter release by $GABA_A$ receptors in glutamatergic neurons', in P. Krogsgaard-Larsen, S.B. Christensen and H. Kofod (eds) *New Leads and Targets in Drug Research*, Copenhagen: Munksgaard, pp. 176–186.

Schousboe, A., Larsson, O.M., Hertz, L. and Krogsgaard-Larsen, P. (1981) 'Heterocyclic GABA analogues as new selective inhibitors of astroglial GABA transport', *Drug Dev. Res.* **1**: 115–127.

Schousboe, A., Thorbek, P., Hertz, L. and Krogsgaard-Larsen, P. (1979) 'Effects of GABA analogues of restricted conformation on GABA transport in astrocytes and brain cortex slices and on GABA receptor binding', *J. Neurochem.* **33**: 181–189.

Serra, M., Foddi, M.C., Ghiani, C.A., Melis, M.A., Motzo, C., Concas, A., Sanna, E. and Biggio, G. (1992) 'Pharmacology of γ-aminobutyric $acid_A$ receptor complex after the in vivo administration of the anxioselective and anticonvulsant β-carboline derivative abecarnil', *J. Pharmacol. Exp. Ther.* **263**: 1360–1368.

Seutin, V. and Johnson, S.W. (1999) 'Recent advances in the pharmacology of quaternary salts of bicuculline', *Trends. Pharmacol. Sci.* **20**: 268–270.

Sieghart, W. (1992) '$GABA_A$ receptors: ligand-gated Cl^- ion channels modulated by multiple drug-binding sites', *Trends Pharmacol. Sci.* **13**: 446–450.

Sieghart, W. (1995) 'Structure and pharmacology of γ-aminobutyric $acid_A$ receptor subtypes', *Pharmacol. Rev.* **47**: 181–234.

Sieghart, W., Eichinger, A., Richards, J.G. and Möhler, H. (1987) 'Photoaffinity labeling of benzodiazepine receptor proteins with the partial inverse agonist [^{3}H]Ro 15–4513: a biochemical and autoradiographic study', *J. Neurochem.* **48**: 46–52.

Sieghart, W., Fuchs, K., Tretter, V., Ebert, V., Jechlinger, M., Höger, H. and Adamiker, D. (1999) 'Structure and subunit composition of $GABA_A$ receptors', *Neurochem. Intl.* **34**: 379–385.

Sivilotti, L. and Nistri, A. (1991) 'GABA inhibits neuronal activity by activating $GABA_B$ receptors coupled to K^+ channels', *Prog. Neurobiol.* **36**: 35–92.

Smith, G.B. and Olsen, R.W. (1995) 'Functional domains of $GABA_A$ receptors', *Trends Pharmacol. Sci.* **16**: 162–169.

Solimena, M. and De Camilli, P. (1993) 'Spotlight on a neuronal enzyme', *Nature*, **366**: 15–17.

Squires, R.F. and Braestrup, C. (1977) 'Benzodiazepine receptors in rat brain', *Nature* **266**: 732–734.

Squires, R.F., Lajtha, A., Saedrup, E. and Palkovits, M. (1993) 'Reduced [^{3}H]flunitrazepam binding in cingulate cortex and hippocampus of post morten schizophrenic brains: is selective loss of glutamatergic neurons associated with major psychoses?', *Neurochem. Res.* **18**: 219–223.

Squires, R.F. and Saedrup, E. (1991) 'A review of evidence for GABAergic predominance/glutamatergic deficit as a common etiological factor in both schizophrenia and affective psychoses: more support for a continuum hypothesis of "functional" psychosis', *Neurochem. Res.* **16**: 1099–1111.

Stephenson, F.A. (1998) 'Molecular structure of $GABA_A$ receptors', in F.A. Stephenson and A.J. Turner (eds) *Amino Acid Neurotransmission*, London: Portland Press.

Stephenson, F.A. and Turner, A.J. (eds) (1998) *Amino Acid Neurotransmission*, London: Portland Press.

Stone, T.W. (1979) 'Glutamate as the neurotransmitter of cerebellar granule cells in the rat: electrophysiological evidence', *Br. J. Pharmacol.* **66**: 291–296.

Supavilai, P. and Karobath, M. (1985) 'Modulation of acetylcholine release from rat striatal slices by the GABA/benzodiazepine receptor complex', *Life Sci.* **36**: 417–426.

Sutch, R.J., Davies, C.C. and Bowery, N.G. (1999) 'GABA release and uptake measured in crude synaptosomes from Genetic Absence Epilepsy Rats from Strasbourg (GAERS)', *Neurochem. Intl.* **34**: 415–425.

Tallman, J.F., Paul, S.M., Skolnick, P. and Gallager, D.W. (1980) 'Receptors for the age of anxiety: pharmacology of the benzodiazepines', *Science* **207**: 274–281.

Tamminga, C.A., Crayton, J.W. and Chase, T.N. (1978) 'Muscimol: GABA agonist therapy in schizophrenia', *Am. J. Psychiat.* **135**: 746–747.

Tamminga, C.A., Crayton, J.W. and Chase, T.N. (1979) 'Improvement of tardive dyskinesia after muscimol therapy', *Arch. Gen. Psychiat.* **36**: 595–598.

Tanaka, C. and Bowery, N.G. (1996) *GABA: Receptors, Transporters and Metabolism*, Basel: Birkhäuser Verlag.

Thaker, G.K., Hare, T.A. and Tamminga, C.A. (1983) 'GABA system: clinical research and treatment of tardive dyskinesia', *Mod. Probl. Pharmacopsychiat.* **21**: 155–167.

Thaker, G.K., Nguyen, J.A. and Tamminga, C.A. (1989) 'Increased saccadic distractability in tardive dyskinesia: functional evidence for subcortical GABA dysfunction', *Biol. Psychiatry* **25**: 49–59.

Theobald, W., Buch, O., Kunz, H.A., Krupp, P., Stenger, E.G. and Heimann, H. (1968) 'Pharmakologische und experimentalpsychologische untersuchungen mit 2 inhaltsstoffen des fliegenpilzes (Amanita Muscaria)', *Arzneim. Forsch.* **18**: 311–315.

Tirsch, R., Yang, X.-D., Singer, S.M., Liblau, R.S., Fugger, L. and McDevitt, H.O. (1993) 'Immune response to glutamic acid decarboxylase correlates with insulitis in non-obese diabetic mice', *Nature* **366**: 72–75.

Travagli, R.A., Ulivi, M. and Wojcik, W.J. (1991) 'γ-aminobutyric acid-B receptors inhibit glutamate release from cerebellar granule cells: consequences of inhibiting cyclic AMP formation and calcium influx', *J. Pharm. Exp. Ther.* **258**: 903–909.

Usami, S., Hozawa, J., Tazawa, M., Igarashi, M., Thompson, G.C., Wu, J.Y. and Wenthold, R.J. (1989) 'Immunocytochemical study of the GABA system in chicken vestibular endorgans and the vestibular ganglion', *Brain Res.* **503**: 214–218.

Van Ness, P.C., Watkins, A.E., Bergman, M.O., Tourtelotte, W.W. and Olsen, R.W. (1982) 'γ-aminobutyric acid receptors in normal human brain and Huntington's disease', *Neurology* **32**: 63–68.

Venault, P., Chapouthier, G., Prado de Carvalho, L., Simiand, J., Morre, M., Dodd, R.H. and Rossier, J. (1986) 'Benzodiazepine impairs and β-carboline enhances performance in learning and memory tasks', *Nature* **321**: 864–866.

Verdoorn, T.A., Draguhn, A., Ymer, S., Seeburg, P.H. and Sakmann, B. (1990) 'Functional properties of recombinant rat $GABA_A$ receptors depend upon subunit composition', *Neuron* **4**: 919–928.

Virmani, M.A., Stojilkovic, S.S. and Catt, K.J. (1990) 'Stimulation of luteinizing hormone release by γ-aminobutyric acid (GABA) agonists: mediation by $GABA_A$-type receptors and activation of chloride and voltage-sensitive calcium channels', *Endocrinology* **126**: 2499–2505.

Walton, M.K., Schaffner, A.E. and Barker, J.L. (1993) 'Sodium channels, $GABA_A$ receptors, and glutamate receptors develop sequentially on embryonic rat spinal cord cells', *J. Neurosci.* **13**: 2068–2084.

Wermuth, C.G. and Biziére, K. (1986) 'Pyridazinyl-GABA derivatives: a new class of synthetic $GABA_A$ antagonists', *Trends Pharmacol. Sci.* **7**: 421–424.

Wermuth, C.G., Bourguignon, J.-J., Schlewer, G., Gies, J.-P., Schoenfelder, A., Melikian, A., Bouchet, M.-J., Chantreux, D., Molimard, J.-C., Heaulme, M., Chambon, J.-P. and Biziére, K. (1987) 'Synthesis and structure–activity relationships of a series of aminopyridazine derivatives of γ-aminobutyric acid acting as selective GABA-A antagonists', *J. Med. Chem.* **30**: 239–249.

Wheal, H. and Thomson, A. (eds) (1995) *Excitatory Amino Acids and Synaptic Transmission*, London: Academic Press.

Whiting, P.J. (1999) 'The $GABA_A$ receptor gene family: new targets for therapeutic intervention', *Neurochem. Intl.* **34**: 387–390.

Wieland, H.A., Lüddens, H. and Seeburg, P.H. (1992) 'A single histidine in $GABA_A$ receptors is essential for benzodiazepine agonist binding', *J. Biol. Chem.* **267**: 1426–1429.

Williams, M., Kowaluk, E.A. and Arneric, S.P. (1999) 'Emerging molecular approaches to pain therapy', *J. Med. Chem.*, **42**: 1481–1500.

Woodward, R.M., Polenzani, L. and Miledi, R. (1993) 'Characterization of bicuculline/baclofen-in-sensitive (ρ-like) γ-aminobutyric acid receptors expressed in Xenopus oocytes. II. Pharmacology of γ-aminobutyric $acid_A$ and γ-aminobutyric $acid_B$ receptor agonists and antagonists', *Mol. Pharmacol.* **43**: 609–625.

Wurtman, R.J., Corkin, S., Growdon, J.H. and Ritter-Walker, E. (eds) (1990) *Advances in Neurology*. Vol. 51. *Alzheimer's Disease*, New York: Raven Press.

Zhang, P., Zhang, W., Liu, R., Harris, B., Skolnick, P. and Cook, J.M. (1995) 'Synthesis of novel imidazobenzodiazepines as probes of the pharmacophore for "diazepam-insensitive" $GABA_A$ receptors', *J. Med. Chem.* **38**: 1679–1688.

Zorn, S.H. and Enna, S.J. (1987) 'The GABA agonist THIP attenuates antinociception in the mouse by modifying central cholinergic transmission', *Neuropharmacology* **26**: 433–437.

Part 4

Metabotropic GABA receptors

Chapter 11

Molecular structure of the $GABA_B$ receptors

N. Klix and B. Bettler

Introduction

The neurotransmitter γ-amino-butyric acid (GABA) is the major inhibitory neurotransmitter in the mammalian central nervous system. Diverse actions of GABA neurotransmission are mediated by three types of receptors. On the basis of pharmacological, physiological and molecular criteria they can be classified into two distinct groups: GABA-gated chloride channels ($GABA_{A/C}$) and metabotropic G-protein coupled receptors ($GABA_B$; for reviews see Bowery, 1993, Borman and Feigenspan, 1995; Kerr and Ong, 1995; Smith and Olson, 1995). It was only in 1981 that Norman Bowery and his colleagues were able to define the $GABA_B$ receptor pharmacologically by the selective agonist L-baclofen and its insensitivity to bicuculline, a $GABA_A$ receptor antagonist (Hill and Bowery, 1981). Whereas binding of GABA to $GABA_{A/C}$ leads to fast synaptic inhibition, the effect of activating $GABA_B$ receptors results in activating multistep pathways via guanine-nucleotide-binding proteins. Hence slow and long-lasting inhibitory signals are produced. To date, the physiological role of $GABA_B$ receptors is only poorly understood. Neurophysiological and pharmacological studies suggest, however, that malfunctioning of the receptor could be involved in the development of chronic pain, spasticity and epileptogenesis of absence seizures.

The $GABA_B$ receptor

Cloning of the $GABA_B$ receptors

Over the last 20 years or so many attempts had been made to isolate the $GABA_B$ receptor protein. Due to the lack of ligands which bind the receptor irreversibly or with high affinity this proved to be difficult. In addition, the $GABA_B$ receptor did not efficiently couple to signalling pathways in *Xenopus* oocytes, which rendered functional cloning strategies unsuccessful. It was not until 1997 that the development of the high-affinity antagonist [^{125}I]CGP64213, combined with an expression cloning approach, allowed the isolation of the rat $GABA_B$R1a cDNA (BR1a) which encodes a protein of 960 amino acids (Kaupmann *et al.*, 1997). Subsequently, by using homology screening, the $GABA_B$R1b (BR1b) cDNA was isolated and identified as a shorter N-terminal splice variant: the mature BR1a protein differs from BR1b in that 147 different residues replace the N-terminal 18 ones. BR1a and BR1b derive from the same gene by alternative splicing (Peters *et al.*, 1998). The BR1a-specific region contains two copies of short consensus repeats (SCRs), about 60 amino acid residues each, also known as complement control protein (CCP) or sushi repeats (Bettler *et al.*, 1998; Hawrot *et al.*, 1998). As these repeats exist in a wide variety of complement and adhesion proteins (e.g. the selectins) it is conceivable that sushi domains may direct protein–protein interactions and, for example, serve as an extracellular targeting signal for BR1a. Additional splice variants, designated $GABA_BR1c$ (BR1c) and 1d (BR1d), generate isoforms with sequence differences in presumed extracellular and intracellular domains (Isomoto *et al.*, 1998; Pfaff *et al.*, 1999). By using $GABA_BR1$ sequence information, database searches led to the discovery of a new $GABA_B$ receptor subtype called $GABA_BR2$ (BR2; Jones *et al.*, 1998; Kaupmann *et al.*, 1998a; White *et al.*, 1998; Kuner *et al.*, 1999). At the protein level the receptors are 35 per cent identical and 54 per cent similar. Two C-terminal splice variants have been reported for the human BR2 receptors so far (Figure 11.1) (Ng *et al.*, 1999).

The topological organization of $GABA_B$ receptors is typical for G-protein coupled receptors (GPCR): a seven-transmembrane-domain-protein, a signal peptide followed by an extracellular N-terminal domain and a C-terminal cytoplasmic domain (which in the case of $GABA_BR2$ is unusually long; for review see Bockaert and Pin, 1999). Based on sequence and structural similarities, $GABA_B$ receptors belong to a gene family comprising mGluRs, the Ca^{2+}-sensing receptor, a family of vomeronasal receptors, taste receptors and periplasmic bacterial amino acid binding proteins, such as the leucine isoleucine valine binding protein (LIVBP) and the leucine binding protein (LBP; Bargmann, 1997; Bettler *et al.*, 1998; Galvez *et al.*, 1999; Hoon *et al.*, 1999).

The genomic localization, tissue expression and function of the human $GABA_BR1$ gene identifies it as a potential candidate for neurobehavioural disorders with a genetic locus on 6p21.3 (mouse chromosome 17B3), such as schizophrenia, juvenile myoclonic epilepsy, multiple sclerosis and dyslexia (Goei *et al.*, 1998; Grifa *et al.*, 1998; Kaupmann *et al.*, 1998a). So far association analysis of exonic variants of the $GABA_BR1$ gene and families with idiopathic generalized epilepsy have not revealed any amino acid substitutions that might be crucial for disease development (Sander *et al.*, 1999). The $GABA_BR2$ gene maps to human and mouse chromosome 9q22.2–22.3 and 4B, respectively. This chromosomal localization, however, does not point towards a link between any neurologic disorders and a possible involvement of $GABA_B$ receptors so far.

Heteromerization of $GABA_B$ receptors

Although the cloned $GABA_BR1$ subtypes showed many of the expected properties in terms of structure and pharmacology, the characterization of their functional activation was impeded by poor coupling when being expressed either in mammalian cells or oocytes. However, biochemical studies indicated that activation of BR1a receptors in HEK293 cells leads to an inhibition of adenylyl cyclase activity (Kaupmann *et al.*, 1997). The measured effect on forskolin-stimulated cAMP production was weak (30 per cent) but clearly inhibited by $GABA_B$ antagonists. BR2 couples to adenylyl cyclase slightly more efficiently (approximately 40 per cent inhibition), demonstrating that BR2 is a bona fide $GABA_B$ receptor (Kuner *et al.*, 1999; Ng *et al.*, 1999). The coupling of the cloned receptors to potential effector Ca^{2+}/K^+ occurred even at a lower success rate. Like BR1 (Kaupmann *et al.*, 1998b), BR2 failed to activate Kir3 channels in oocytes (Jones *et al.*, 1998; Kaupmann *et al.*, 1998a; White *et al.*, 1998) and showed weak coupling in HEK293 cells (Jones *et al.*, 1998; Kaupmann *et al.*, 1998a). The failure of BR1a/b receptors to recruit signalling pathways could have been explained by poor cell surface expression (Couve *et al.*, 1998; White *et al.*, 1998). In contrast, however, BR2 efficiently translocates to the cell membrane (White *et al.*, 1998), and therefore the low rate of Kir3 coupling was unexpected because the assay is known to be a sensitive read-out for many cloned G-protein coupled receptors. Lack of robust coupling indicated that each of the subtypes do not form fully functional $GABA_B$ receptors when expressed in heterologous systems. Therefore the involvement of auxiliary factors that are limiting or missing in non-neuronal expression systems was postulated. Interestingly, similar problems had been reported for the calcitonin-receptor-like-receptor and an odorant receptor which, however, were overcome by the identification of associated proteins called RAMP and ODR-4, respectively (McLatchie *et al.*, 1998; Dwyer *et al.*, 1998).

The observed overlapped expression patterns of BR1 and BR2 *in situ* hybridization studies (see colour plate Figure 11.2) in many neuronal populations indicated that co-expression was possibly needed for robust functional activity. Furthermore, analysis of hybridization signals on adjacent brain sections provided direct evidence for co-expression of BR1 and BR2 transcripts within individual neurons, e.g. in Purkinje cells (Jones *et al.*, 1998; Kaupmann *et al.*, 1998a). Was heterodimerization of both receptors a prerequisite for the formation of a functional $GABA_B$ receptor? Indeed, this is the case. While neither BR1a/b nor BR2 alone efficiently activated Kir3 channels, their co-expression in HEK293 cells and *Xenopus* oocytes yielded robust GABA evoked currents (Jones *et al.*, 1998; Kaupmann *et al.*, 1998a; White *et al.*, 1998; Kuner *et al.*, 1999). Co-expression of BR1 and BR2 in heterologous cells allowed robust stimulation of GTP [^{35}S] binding as well (White *et al.*, 1998). These functional responses exhibited pharmacological characteristics reminiscent of those reported for abundant native $GABA_B$ receptors. Independent evidence for a required heterodimerization was derived from a search for putative BR1 trafficking factors (White *et al.*, 1998; Kuner *et al.*, 1999). In this case, BR2 was identified in a yeast two-hybrid screen using the C-terminal domain of BR1 as a bait. BR1 and BR2 receptors seem to specifically interact via coiled–coil structures in their intracellular C-terminal tails via a dimerization signal similar to the one used by leucine zipper transcription factors. Immunoelectron microscopy analysis showed an extensive co-localization of BR1 and BR2 proteins at selected neuronal sites like Purkinje cell dendritic spines. Furthermore, the existence of heteromeric $GABA_B$ receptor *in vivo* was substantiated by co-immunoprecipitation of native and recombinant $GABA_B$R2 protein with $GABA_B$R1a/b proteins (Kaupmann *et al.*, 1998a). To sum up, the amount of data clearly supports the idea that the assembly of a heteromeric receptor represents the predominant form of native $GABA_B$ receptors. This, however, does not rule out the presence of homomeric receptors.

In binding assays, BR1a and BR1b show similar high affinities to known $GABA_B$ receptors agonists and antagonists so that these splice variants probably do not represent pharmacological subtypes. Interestingly, $GABA_B$R2 does not bind the $GABA_B$ receptor antagonist CGP54626A, which indicates that $GABA_B$R2 is a $GABA_B$-receptor subtype with pharmacological properties distinct from $GABA_B$R1a/b. When BR1a or BR1b are expressed together with BR2 an up to tenfold increase in agonist and partial agonist binding potency is observed in the inhibition of [^{125}I]CGP64213 antagonist binding (Kaupmann *et al.*, 1998a). Based on the experiments discussed above it is expected that BR2 enables BR1a/b to be translocated to the cell surface (White *et al.*, 1998).

Therefore the observed increase in agonist binding potency could arise from a more efficient coupling of the heteromeric receptor to G-proteins. Nevertheless, there is a remaining hundredfold discrepancy in apparent agonist binding potency between heteromeric recombinant and native receptors. This might reflect limitations of a heterologous expression system; for example, differences in the relative expression levels of G-proteins and receptors (Kenakin, 1997), lack of specific receptor modification (e.g. phosphorylation) or the need for associated factors.

Molecular determinants of ligand binding at $GABA_B$ receptors

The importance of the extracellular N-terminal domain for ligand binding has been demonstrated to be critical in the metabotropic receptors for L-glutamate, the mGluRs (O'Hara *et al.*, 1993; Tones *et al.*, 1995, Parmentier *et al.*, 1998). Moreover, the N-terminal portion of $GABA_B$ receptors shares structural similarity with bacterial amino acid binding proteins (Kaupmann *et al.*, 1997; Galvez *et al.*, 1999) which shuttle ions and nutrients from the periplasm via transporter proteins across the plasma membrane. Several lines of evidence support the idea that the ligand binding site of $GABA_B$ receptors has evolved from these ancestral bacterial amino acid binding proteins. For example, a soluble protein encompassing the extracellular N-terminal domain of BR1b closely reproduces the binding pharmacology wild-type $GABA_B$ receptors (Malitschek *et al.*, 1999). This demonstrates that the N-terminal extracellular domain can correctly fold when dissociated from the transmembrane domains and contains all the structural information that is necessary and sufficient for agonist/antagonist binding.

The crystal structure of the bacterial proteins indicates that two globular lobes that are connected through a hinge region form the amino acid binding pocket. Based on the known structure of LBP and LIVBP, a three-dimensional model of the ligand-binding site of $GABA_B$ receptors was constructed (Galvez *et al.*, 1999) and substantiated by extended mutagenesis studies in the vicinity of the presumed ligand-binding pocket. These experiments have highlighted the importance of several residues for binding. Serine 246, a residue homologous to Serine 79 in LBP that forms a hydrogen bond with the ligand, is critical for antagonist binding. Similarly the mutation of Serine 269 was found to affect the affinity of various GABA analogs differentially. Finally, the mutation of Serine 247 and Glutamine 312 was found to increase the affinity of agonists and to decrease the affinity of antagonists, respectively. The effects of these point mutations clearly support not only an evolutionary relationship between the ligand binding sites of the LBP/LIVBP and $GABA_B$ receptors but might, as well, give us a hint about how receptor activation could operate. One could envisage a model in which the two lobes of the binding domain close upon ligand binding similar to a Venus flytrap mechanism (Galvez *et al.*, 1999). Subsequently, a conformational change would be transduced to cytoplasmatic receptor regions at which G-proteins will become activated.

Neurophysiology of $GABA_B$ receptor subtypes

Differences in the efficacies of antagonists in the modulation of neurotransmitter release led to the postulation that presynaptic $GABA_B$ receptors are heterogeneous and distinct from postsynaptic receptors (Cunningham and Enna, 1996; Bonanno *et al.*, 1997; Deisz *et al.*, 1997). Furthermore there is evidence that, depending on their synaptic localization, $GABA_B$ receptors differ in their coupling preferences. It was shown, for example, that pertussis toxin (PTX) is unable to uncouple presynaptic $GABA_B$ receptors from their effectors, while the action of postsynaptic $GABA_B$ receptors can be blocked by treatment with PTX. This not only indicates coupling to G_i/G_o-type G-proteins (Dutar and Nicoll, 1988; Harrison, 1990) but supports the idea that pre- and postsynaptic $GABA_B$ receptors are coupled to different G-proteins. The general picture, however, seems to be much more complex and the claim of presynaptic receptor subtypes has been challenged by synaptosomal release experiments (Waldmeier *et al.*, 1994).

The physiological roles of $GABA_B$ receptors can be mainly attributed to the regulation of G-protein gated Ca^{2+} and K^+ channels (Lüscher *et al.*, 1997; Poncer *et al.*, 1997; Slesinger *et al.*, 1997; Wu and Saggau, 1997). Presynaptic $GABA_B$ receptors influence neurotransmission by suppression of neurotransmitter and neuropeptide release on excitatory and inhibitory terminals, presumably by diminution of Ca^{2+} conductance. $GABA_B$ heteroreceptors are involved in the release control of several other neurotransmitters (e.g. glutamate, dopamine, noradrenaline, substance P, cholecystokinin and somatostatin), whereas $GABA_B$ autoreceptors inhibit the release of GABA thus providing a negative feedback mechanism. In addition, a Ca^{2+} independent interaction of $GABA_B$ receptors with the presynaptic secretion machinery has been suggested (Capogna *et al.*, 1996). The rapid time course of $GABA_B$ receptor-mediated inhibition of N and P/Q-type Ca^{2+} channels indicates a membrane-delimited pathway through the G-protein $\beta\gamma$-subunits (Mintz and Bean, 1993; Thompson *et al.*, 1993) similar to other G-protein coupled receptors.

In contrast to presynaptic $GABA_B$ receptors, postsynaptic subtypes hyperpolarize neurons by activating an outward K^+ current that underlies the late inhibitory postsynaptic potentials (IPSPs). The activation of K^+ channels is sensitive to pertussis toxin and blocked by Ba^{2+} and Cs^{2+}. Characteristically the late IPSP is slower in onset and has a prolonged duration as compared to the fast IPSP, which derives from the Cl^- permeable $GABA_A$ receptors. Recent studies indicate that inwardly rectifying K^+ channels of the Kir3 type (formerly GIRK) are prominent effectors of postsynaptic $GABA_B$ receptors (Lüscher *et al.*, 1997). For example the late IPSP evoked by L-baclofen, a selective $GABA_B$ receptor agonist, is largely absent in Kir3.2 knockout mice, whereas presynaptic $GABA_B$ receptor responses are unaltered. Moreover the amplitude of the $GABA_B$ receptor-activated K^+ current is significantly attenuated in weaver mutant mice, which carry a point mutation in the pore-forming region of the Kir3.2 subunit (Slesinger *et al.*, 1997). These results support the idea that postsynaptic $GABA_B$ receptors couple to K^+ channels assembled with Kir3.2. Apart from ion channel modulation, $GABA_B$ receptors were shown to couple negatively to adenylyl cyclase and to inhibit forskolin-stimulated cAMP levels (Wojcik and Neff, 1984). So far no direct coupling to phospholipase C, and the release of Ca^{2+} from internal stores, has been demonstrated.

Spatial and temporal distribution of cloned $GABA_B$ receptors

Since heteromerization seems to be a prerequisite for robust functional coupling, at least in heterologous cells, it is important to find a largely overlapping pattern of BR1 and BR2 transcripts in the brain as studied by *in situ* hybridization. Interestingly, the *in situ* hybridization pattern qualitatively reflects those of $GABA_B$ agonist (e.g. Wilkin *et al.*, 1981; Gehlert *et al.*, 1985; Chu *et al.*, 1990; Turgeon and Albin, 1993) and antagonist binding sites (Towers *et al.*, 1997; Kaupmann *et al.*, 1998a; Bischoff *et al.*, 1999). This suggests that BR1 and BR2 constitute the majority of native $GABA_B$ binding sites. Possibly, $GABA_BR1a/b$ subtypes are targeted to specific cellular and subcellular sites. Indeed, their distribution can differ quite drastically. In the cerebellum BR1a transcripts are confined to the granule cell layer that comprises the cell bodies of the parallel fibres, which are excitatory to the Purkinje cell dendrites in the molecular layer. In contrast, BR1b transcripts are mostly found in Purkinje cells (dendrites of which possess $GABA_B$ receptors that would be postsynaptic to GABAergic basket and stellate cells or glutamatergic parallel fibres). Similarly in dorsal root ganglia the density of BR1a, but not BR1b, transcripts is high and confined to the neuronal cell bodies. This supports the idea that $GABA_BR1a$ mRNA might encode for presynaptic receptors, whereas $GABA_BR1b$ corresponds to postsynaptic sides – at least at selected synapses.

In immunohistological studies the BR1a/b and BR2 protein levels appear to be differentially regulated during postnatal development, and the relative ratios vary between tissues over time (Malitschek *et al.*, 1998; Fritschy *et al.*, 1999). This is accompanied by increasing affinity of BR1a/b for L-baclofen. In the cerebellum BR1b and BR2 protein expression is mostly restricted to the Purkinje cell dendrites and spines (Kaupmann *et al.*, 1998a; Fritschy *et al.*, 1999). It is worth mentioning that, at GABAergic synapses in the rat retina, BR1 is localized at pre-, post- and extrasynaptic sites (Koulen *et al.*, 1998). In Purkinje cells the BR1b and BR2 proteins are localized in the vicinity of excitatory synapses, whereas the BR1 protein seems to be largely absent at GABAergic inputs. Taking all this into consideration, current data suggest that $GABA_B$ receptors are present at a variety of synaptic sites, at both inhibitory and excitatory synapses.

Conclusion

It is evident that the extent of genetic diversity in the $GABA_B$ receptor gene family is less than that of the mGluR family. Possibly the targeting of receptor splice variants to distinct subcellular sites allows the coupling to various signalling pathways required in specific cellular contexts – a strategy to compensate for a rather limited diversity. Heterologous coupling of $GABA_B$ receptors to Kir3 and adenylyl cyclase, together with the demonstration that BR1a/b containing receptors inhibit high voltage-activated Ca^{2+} channels (Morris *et al.*, 1998), indicate that all major actions of native $GABA_B$ receptors could relate to the cloned receptors. While there is increasing evidence for homodimerization of mGluRs (Romano *et al.*, 1996), calcium-sensing receptors (Bai *et al.*, 1998) and muscarinic/adrenergic receptors (Maggio *et al.*, 1993; Hebert *et al.*, 1996), $GABA_B$ receptors provide the first example for the existence of GPCR heterodimers *in vivo* and their importance for receptor function – a new principle which gained further support by demonstrating that the opioid receptors κ and δ acquire properties different to those of homomeric receptors (Jordan and Devi, 1999). Thus, heterodimerization of GPCRs represents a novel regulatory level at which inputs can initiate and integrate distinct sets of signalling pathways.

References

Bai, M., Trivedi, S. and Brown, E.M. (1998) 'Dimerization of the extracellular calcium-sensing receptor (CaR) on the cell surface of CaR-transfected HEK293 cells', *J. Biol. Chem.* **273**: 23,605–23,610.

Bargmann, C. (1997) 'Olfactory receptors, vomeronasal receptors and the organization of olfactory information', *Cell* **90**: 585–587.

Bettler, B., Kaupmann, K. and Bowery, N.G. (1998) 'GABA$_B$ receptors: drugs meet clones', *Curr. Opin. Neurobiol.* **8**: 345–350.

Bischoff, S., Leonhard, S., Reymann, N., Schuler, V., Felner, A., Bittiger, H., Shigemoto, R., Kaupmann, K. and Bettler, B. (1999) 'Spatial distribution of GABA$_B$R1 Receptor mRNA and binding sites in the rat brain', *J. Comp. Neurol.* **142**: 1–16.

Bockaert, J. and Pin, J.-P. (1999) 'Molecular tinkering of G protein-coupled receptors: an evolutionary success', *EMBO J.* **18**: 1723–1729.

Bonanno, G., Fassio, A., Schmid, G., Severi, P., Sala, R. and Raiteri, M. (1997) 'Pharmacologically distinct GABA$_B$ receptors that mediate inhibition of GABA and glutamate release in human neocortex', *Br. J. Pharmacol.* **120**: 60–64.

Borman, J. and Feigenspan, A. (1995) 'GABA$_C$ receptors', *Trends Neurosci.* **18**: 515–519.

Bowery, N.G. (1993) 'GABA$_B$ receptor pharmacology', *Annu. Rev. Pharmacol. Toxicol.* **33**: 109–117.

Capogna, M., Gahwiler, B.H. and Thompson, S.M. (1996) 'Presynaptic inhibition of calcium-dependent and -independent release elicited with ionomycin, gadolinium, and alpha-latrotoxin in the hippocampus', *J. Neurophysiol.* **75**: 2017–2028.

Chu, D.C., Albin, R.L., Young, A.B. and Penney, J.B. (1990) 'Distribution and kinetics of GABA$_B$ binding sites in rat central nervous system: a quantitative autoradiographic study', *Neuroscience* **34**: 341–357.

Couve, A., Filippov, A.K., Connolly, C.N., Bettler, B., Brown, D.A. and Moss, S.J. (1998) 'Intracellular retention of recombinant GABA$_B$ receptors', *J. Biol. Chem.* **273**: 26,361–26,367.

Cunningham, M.D. and Enna, S.J. (1996) 'Evidence for pharmacologically distinct GABA$_B$ receptors associated with cAMP production in rat brain', *Brain Res.* **720**: 220–224.

Deisz, R.A., Billard, J.M. and Zieglgansberger, W. (1997) 'Presynaptic and postsynaptic GABA$_B$ receptors of neocortical neurons of the rat in vitro: differences in pharmacology and ionic mechanisms', *Synapse* **25**: 62–72.

Dutar, P. and Nicoll, R.A. (1988) 'Pre- and postsynaptic GABA$_B$ receptors in the hippocampus have different pharmacological properties', *Neuron* **1**: 585–591.

Dwyer, N.D., Troemel, E.R., Sengupta, P. and Bargmann, C.I. (1998) 'Odorant receptor localization to olfactory cilia is mediated by ODR-4, a novel membrane-associated protein', *Cell* **93**: 455–466.

Fritschy, J.M., Meskenaite, V., Weinmann, O., Honer, M., Benke, D. and Möhler, H. (1999) 'GABA$_B$-receptor splice variants GB1a and GB1b in rat brain: developmental regulation, cellular distribution and extrasynaptic localization', *Eur. J. Neurosci.* **11**: 761–768.

Galvez, T., Parmentier, M.L., Joly, C., Malitschek, B., Kaupmann, K., Kuhn, R., Bittiger, H., Froestl, W., Bettler, B. and Pin, J.-P. (1999) 'Mutagenesis and modeling of the GABA$_B$ receptor extracellular domain support a 'Venus Flytrap' mechanism of ligand binding', *J. Biol. Chem.* **274**: 13,362–13,369.

Gehlert, D.R., Yamamura, H.I. and Wamsley, J.K. (1985) 'Gamma-aminobutyric acid B receptors in the rat brain: quantitative autoradiographic localization using [^{3}H]-baclofen', *Neurosci. Lett.* **56**: 183–188.

Goei, V.L., Choi, J., Ahn, J., Bowlus, C.L., Raha-Chowdhury, R. and Gruen, J.R. (1998) 'Human gamma-aminobutyric acid B receptor gene: Complementary DNA cloning, expression, chromosomal location, and genomic organization', *Biol. Psychiatry* **44**: 659–666.

Grifa, A., Totaro, A., Rommens, J.M., Carella, M., Roetto, A., Borgato, L., Zelante, L. and Gasparini, P. (1998) 'GABA (gamma-amino-butyric-acid) neurotransmission: Identification and fine mapping of the human GABA$_B$ receptor gene', *Biochem. Biophys. Res. Commun.* **250**: 240–245.

Harrison, N.L. (1990) 'On the presynaptic action of baclofen at inhibitory synapses between cultured rat hippocampal neurones', *J. Physiol. (Lond.)* **422**: 433–446.

Hawrot, E., Yuanyuan, X., Shi, Q.-L., Norman, D., Kirkitadze, M. and Barlow, P.N. (1998) 'Demonstration of a tandem pair of complement protein modules in $GABA_B$ receptor 1a', *FEBS Lett.* **432**: 103–108.

Hebert, T.E., Moffett, S., Morello, J.P., Loisel T.P., Bichet, D.G., Barret, C. and Bouvier, M. (1996) 'A peptide derived from a beta2-adrenergic receptor transmembrane domain inhibits both receptor dimerization and activation', *J. Biol. Chem.* **271**: 16,384–16,392.

Hill, D.R. and Bowery, N.G. (1981) '^{3}H-baclofen and ^{3}H-GABA bind to bicuculline-insensitive $GABA_B$ sites in rat brain', *Nature* **290**: 149–152.

Hoon, M.A., Adler, E., Lindemeier, J., Battey, J.F., Ryba, N.J.P. and Zuker, C.S. (1999) 'Putative mammalian taste receptors: a class of taste-specific GPCRs with distinct topographic selectivity', *Cell* **96**: 541–551.

Isomoto, S., Kaibara, M., Sakurai-Yamashita, Y., Nagayama, Y., Uezono, Y., Yano, K. and Taniyama, K. (1998) 'Cloning and tissue distribution of novel splice variants of the $GABA_B$ receptor', *Biochem. Biophys. Res. Commun.* **253**: 10–15.

Jones, K.A., Borowsky, B., Tamm, J.A., Craig, D.A., Durkin, M.M., Dai, M., Yao, W.-J., Johnson, M., Gunwaldsen, C., Huang, L.-Y., Tang, C., Shen, O., Salon, J.A., Morse, K., Laz, T., Smith, K.E., Nagarathnam, D., Noble, S.A., Branchek, T.A. and Gerald, C. (1998) '$GABA_B$ receptors function as a heteromeric assembly of the subunits $GABA_BR1$ and $GABA_BR2$', *Nature* **396**: 674–679.

Jordan, B.A. and Devi, L.A. (1999) 'G-protein-coupled receptor heterodimerization modulates receptor function', *Nature* **399**: 697–700.

Kaupmann, K., Huggel, K., Heid, J., Flor, P.J., Bischoff, S., Mickel, S.J., McMaster, G., Angst, C., Bittiger, H., Froestl, W. and Bettler, B. (1997) 'Expression cloning of $GABA_B$ receptors uncovers similarity to metabotropic glutamate receptors', *Nature* **386**: 239–246.

Kaupmann, K., Malitschek, B., Schuler, V., Heid, J., Froestl, W., Beck, P., Mosbacher, J., Bischoff, S., Kulik, A., Shigemoto, R., Karschin, A. and Bettler, B. (1998a) '$GABA_B$-receptor subtypes assemble into functional heteromeric complexes', *Nature* **396**: 683–687.

Kaupmann, K., Schuler, V., Mosbacher, J., Bischoff, S., Bittiger, H., Heid, J., Fröstl, W., Leonhardt, T., Pfaff, T., Karschin, A. and Bettler, B. (1998b) 'Human $GABA_B$ receptors are differentially expressed and regulate inwardly rectifying K^+ channels', *Proc. Natl. Acad. Sci. USA* **95**: 14,991–14,996.

Kenakin, T. (1997) 'Differences between natural and recombinant G-protein coupled receptor systems with varying receptor/G-protein stoichiometry', *Trends Pharmacol. Sci.* **18**: 456–464.

Kerr, D.I.B. and Ong, J. (1995) '$GABA_B$ receptors', *Pharmacol. Ther.* **76**: 187–246.

Koulen, P., Malitschek, B., Kuhn, R., Bettler, B., Wässle, H. and Brandstätter, J.H. (1998) 'Presynaptic and postsynaptic localization of $GABA_B$ receptors in neurons of the rat retina', *Eur. J. Neurosci.* **10**: 1446–1456.

Kuner, R., Kohr, G., Grunewald, S., Eisenhardt, G., Bach, A. and Kornau, H.C. (1999) 'Role of heteromer formation in $GABA_B$ receptor function', *Science* **283**: 74–77.

Lüscher, C., Jan, L.Y., Stoffel, M., Malenka, R.C. and Nicoll, R.A. (1997) 'G protein-coupled inwardly rectifying K^+ channels (GIRKs) mediate postsynaptic but not presynaptic transmitter actions in hippocampal neurons', *Neuron* **19**: 687–695.

Maggio, R., Vogel, Z. and Wess, J. (1993) 'Coexpression studies with mutant muscarinic/ adrenergic receptors provide evidence for intermolecular "cross-talk" between G-protein-linked receptors', *Proc. Natl. Acad. Sci. USA* **90**: 3103–3107.

Malitschek, B., Rüegg, D., Heid, J., Kaupmann, K., Bittiger, H., Fröstl, W., Bettler, B. and Kuhn, R. (1998) 'Developmental changes in agonist affinity at $GABA_BR1$ receptor variants in rat brain', *Mol. & Cell Neurosci.* **12**: 56–64.

Malitschek, B., Schweizer, C., Keir, M., Heid, J., Froestl, W., Mosbacher, J., Kuhn, R., Henley, J.,

Joly, C., Pin, J.-P., Kaupmann, K. and Bettler, B. (1999) 'The N-terminal domain of $GABA_B$ receptors is sufficient to specify agonist and antagonist binding', *Mol. Pharmacol.* **56**: 448–454.

McLatchie, L.M., Fraser, N.J., Main, M.J., Wise, A., Brown, J., Thompson, N., Solari, R., Lee, M.G. and Foord, S.M. (1998) 'RAMPs regulate the transport and ligand specificity of the calcitonin-receptor-like receptor', *Nature* **393**: 333–339.

Mintz, I.M. and Bean, B.P. (1993) '$GABA_B$ receptor inhibition of P-type Ca^{2+} channels in central neurons', *Neuron* **10**: 889–898.

Morris, S.J., Beatty, D.M. and Chronwall, B.M. (1998) '$GABA_BR1a/R1b$-type receptor anti-sense deoxynucleotide treatment of melanotropes blocks chronic $GABA_B$ receptor inhibition of high-voltage-activated Ca^{2+} channels', *J. Neurochem.* **71**: 1329–1332.

Ng, G.Y.K., Clark, J., Coulombe, N., Ethier, N., Hebert, T.E., Sullivan, R., Kargman, S., Chateauneuf, A., Tsukamoto, N., McDonald, T., Whiting, P., Mezey, E., Johnson, M.P., Liu, Q., Kolakowski, L.F., Evans, J.F., Bonner, T.I. and O'Neill, G.P. (1999) 'Identification of a $GABA_B$ receptor subunit, gb2, required for functional $GABA_B$ receptor activity', *J. Biol. Chem.* **274**: 7607–7610.

O'Hara, P.J., Sheppard, P.O., Thogersen, H., Venezia, D., Haldeman, B.A., McGrane, V., Houamed, K.M., Thomsen, C., Gilbert, T.L. and Mulvihill, E.R. (1993) 'The ligand-binding domain in metabotropic glutamate receptors is related to bacterial periplasmatic binding proteins', *Neuron* **11**: 41–52.

Parmentier, M.L., Joly, C., Restituito, S., Bockaert, J., Grau, Y. and Pin, J.-P. (1998) 'The G protein-coupling profile of metabotropic glutamate receptors, as determined with exogenous G proteins, is independent of their ligand recognition domain', *Mol. Pharmacol.* **53**: 778–786.

Peters, H.C., Kämmer, G., Volz, A., Kaupmann, K., Ziegler, A., Bettler, B., Epplen, J.T., Sander, T. and Riess, O. (1998) 'Mapping, genomic structure, and polymorphisms of the human $GABA_BR1$ receptor gene: evaluation of its involvement in idiopathic generalized epilepsy', *Neurogenetics* **2**: 47–54.

Pfaff, T., Malitschek, B., Kaupmann, K., Prézeau, L., Pin, J.-P., Bettler, B. and Karschin, A. (1999) 'Alternative splicing generates a novel isoform of the rat metabotropic $GABA_BR1$ receptor', *Eur. J. Neurosci.* **11**: 2874–2882.

Poncer, J.C., McKinney, R.A., Gahwiler, B.H. and Thompson, S.M. (1997) 'Either N- or P-type calcium channels mediate GABA release at distinct hippocampal inhibitory synapses', *Neuron* **18**: 463–472.

Romano, C., Yang, W.L. and O'Malley, K.L. (1996) 'Metabotropic glutamate receptor 5 is a disulfide-linked dimer', *J. Biol. Chem.* **271**: 2861–2866.

Sander, T., Peters, H.C., Kämmer, G., Samochowiec, J., Zirra, M., Mischke, A., Ziegler, A., Kaupmann, K., Bettler, B., Epplen, J.T. and Riess, O. (1999) 'Association analysis of exonic variants of the gene encoding the $GABA_B$ receptor and idiopathic generalized epilepsy', *Am. J. Med. Genet. (Neuropsychiatric Genetics)* **88**: 305–310.

Slesinger, P.A., Stoffel, M., Jan, Y.N. and Jan, L.Y. (1997) 'Defective γ-amino butyric acid type B receptor-activated inwardly rectifying K^+ currents in cerebellar granule cells isolated from weaver and GIRK2 null mutant mice', *Proc. Natl. Acad. Sci. USA* **94**: 12,210–12,217.

Smith, G.B. and Olsen, R.W. (1995) 'Functional domains of $GABA_A$ receptors', *Trends Pharmacol. Sci.* **16**: 162–168.

Thompson, S.M., Capogna, M. and Scanziani, M. (1993) 'Presynaptic inhibition in the hippocampus', *Trends Neurosci.* **16**: 222–227.

Tones, M.A., Bendali, N., Flor, P.J., Knöpfel, T. and Kuhn, R. (1995) 'The agonist selectivity of a class III metabotropic glutamate receptor, human mGluR4a, is determined by the N-terminal extracellular domain', *Neuroreport* **7**: 117–120.

Towers, S., Meoni, P., Billinton, A., Kaupmann, K., Bettler, B., Urban, L., Bowery, N.G. and Spruce, A. (1997) '$GABA_B$ receptor expression in spinal cord and dorsal root ganglia of neuropathic rats', *Society for Neuroscience Abstract* **23**: 955.

Turgeon, S.M. and Albin, R.L. (1993) 'Pharmacology, distribution, cellular localization, and development of $GABA_B$ binding in rodent cerebellum', *Neuroscience* **55**: 311–323.

Waldmeier, P.C., Wicki, P., Feldtrauer, J.J., Mickel, S.J., Bittiger, H. and Baumann, P.A. (1994) 'GABA and glutamate release affected by $GABA_B$ receptor antagonist with similar potency: no evidence for pharmacologically different presynaptic receptors', *Br. J. Pharmacol.* **113**: 1515–1521.

White, J.H., Wise, A., Main, M.J., Green, A., Fraser, N.J., Disney, G.H., Barnes, A.A., Emson, P., Foord, S.M. and Marshall, F.H. (1998) 'Heteromerization is required for the formation of a functional $GABA_B$ receptor', *Nature* **396**: 679–682.

Wilkin, G.P., Hudson, A.L., Hill, D.R. and Bowery, N.G. (1981) 'Autoradiographic localization of $GABA_B$ receptors in the cerebellum', *Nature* **294**: 584–587.

Wojcik, W.J. and Neff, N.H. (1984) 'γ-aminobutyric acid B receptors are negatively coupled to adenylate cyclase in brain and in the cerebellum these receptors may be associated with granule cells', *Mol. Pharmacol.* **25**: 24–28.

Wu, L.G. and Saggau, P. (1997) 'Presynaptic inhibition of elicited neurotransmitter release', *Trends Neurosci.* **20**: 204–212.

Chapter 12

Pharmacology of $GABA_B$ receptors

N.G. Bowery

Introduction

Metabotropic, or indirectly coupled, receptors are activated by a variety of neurotransmitters or paracrine hormones in the body and are responsible for mediating the actions of more than 70 per cent of the therapeutic agents which are currently in clinical use. These receptors depend on relatively few types of G-protein to link them to their effector mechanisms. Thus, it is the unique structure and conformation of individual receptors that provides the specific and individual characteristics of ligand action. In many cases the receptor is expressed in cell membranes simply as a monomer after transcription of the protein in the endoplasmic reticulum. In some instances the receptor is expressed as a functional unit by combination of two molecules of the entire receptor linked to form a homodimer. Up to 3 years ago there was no evidence for any functional receptors that comprised two forms of the complete receptor with distinct molecular structures coupled together. But then the unique conformation of the functional $GABA_B$ receptor(s) was revealed showing that it exists as a heterodimer (White *et al.*, 1998; Kaupmann *et al.*, 1998a; Jones *et al.*, 1998; Kuner *et al.*, 1999; Ng *et al.*, 1999). This followed the original cloning of the receptor in 1997 by Kaupmann and colleagues. However, soon after the structure was announced it became apparent that the single individual receptor subunit is not normally expressed in the plasma membrane but remains in the endoplasmic reticulum (see Couve *et al.*, 1998). Initial thoughts were focused on the possibility of an unknown trafficking protein 'RAMP' (receptor activity modifying protein) being required to transport the receptor protein to the external membrane (cf. McLatchie *et al.*, 1998). For detailed information about the structure of the $GABA_B$ receptor see Chapter 11.

The $GABA_B$ receptor was first described much earlier based on pharmacological observations. It emerged during studies aimed at determining whether the long-established Cl^- dependent GABA receptor exists on peripheral autonomic nerve terminals and whether its activation would suppress transmitter release. Evidence was obtained and the predicted reduction in evoked release of neurotransmitter from the sympathetic fibres did arise (Bowery *et al.*, 1981). The GABA receptor responsible for this effect appeared to have different characteristics, and the pharmacological profile in particular was strikingly different from that of the receptor responsible for the Cl^- dependent action of GABA. The effect could not be blocked by bicuculline, was not mimicked by isoguvacine and was only activated by high concentrations of the normally potent agonist muscimol. Most striking of all was that the therapeutic agent, baclofen (β-chlorophenyl GABA), which had been introduced into therapeutics a few years earlier as an antispastic agent on the basis that it would mimic the effect of GABA (Keberle and Faigle, 1972; Bein, 1972), was stereospecifically active at this new receptor. It soon became clear that the action of GABA on sympathetic terminals was not Cl^- dependent, as originally predicted, but instead was dependent on the presence of external $[Ca^{2+}]$.

Subsequent experiments showed that this receptor was also present, more importantly, in the mammalian CNS. Membrane receptor binding studies using tritiated baclofen and tritiated GABA finally provided the crucial evidence and it was then that the term '$GABA_B$ receptor' was coined to distinguish the receptor from the bicuculline-sensitive receptor, which was in turn designated '$GABA_A$' (Hill and Bowery, 1981).

Functional role of $GABA_B$ receptors in mammals

Both pre- and postsynaptically located $GABA_B$ receptors have been described in the mammalian brain, where they have each been suggested to play a physiological role. Whilst activation at the latter site derives from GABAergic innervation of neuronal $GABA_B$ sites, the former probably stems from the action of GABA released from the same (autoreceptors) or an adjacent synapse (heteroreceptors) (Isaacson *et al.*, 1993).

$GABA_B$ receptor stimulation at postsynaptic sites normally produces a long-lasting neuronal hyperpolarization mediated by an increase in membrane conductance to K^+. The presynaptic action manifests as a reduction in postsynaptic potential resulting from a decrease in transmitter output. This is presumed to result from a decrease in Ca^{2+} conductance in the presynaptic membrane following activation of the $GABA_B$ receptor, although other mechanisms may contribute. This effect can be readily demonstrated in brain slices, e.g. hippocampal (Isaacson *et al.*, 1993), where there is no evidence for any axo-axonic nerve terminal junctions. Instead the evidence would suggest that GABA 'washes-over' from adjacent synapses to activate the heteroreceptors (Isaacson *et al.*, 1993). This does not seem unreasonable, as the estimated concentration of GABA in GABAergic synapses is in the millimolar range whilst the affinity of GABA for $GABA_B$ sites is in the submicromolar range.

The only site where there exists evidence for innervation of presynaptic $GABA_B$ receptors is in the spinal cord, where transmitter release from primary afferent fibres is attenuated by GABAergic interneurons that synapse on to the afferent fibre terminals (e.g. Barber *et al.*, 1978). This can be mimicked by $GABA_B$ agonists *in vitro* such that baclofen and GABA will suppress the evoked release of substance P(SP) and glutamate (see Malcangio and Bowery, 1996) which are believed to be sensory transmitters and are co-localised in primary afferent terminals of the dorsal horn. A physiological role in the spinal cord is supported by evidence showing that $GABA_B$-mediated inhibition of substance P release, which leads to substance P (NK1) receptor internalization, is a tonic process (Marvizon *et al.*, 1999).

Recent studies in which the transcript for the $GABA_{B1}$ subunit has been detected in a variety of peripheral tissues including the heart, lung, intestine, kidney and urinary bladder using an RT-PCR technique (Castelli *et al.*, 1999) would support the possibility that the receptor may be functional outside the CNS. However, this would, of course, depend on the presence of a natural receptor agonist and receptor protein in any of the organs. The enteric nervous system of the intestine may be a particularly important focus in this regard. GABA neurons as well as an abundance of $GABA_B$ receptors are present and the action of $GABA_B$ agonists has been well documented in this system (Ong and Kerr, 1990). Other effects on peripheral organs are probably of more pharmacological significance, although central $GABA_B$ mechanisms do appear to influence peripheral cardiovascular and respiratory function as well as hormone release (see Bowery, 1993; Ferreira *et al.*, 1996; Rey-Roldan *et al.*, 1996).

$GABA_B$ receptor distribution in CNS

Native $GABA_B$ as well as the $GABA_{B1}$ and $GABA_{B2}$ subunits are widely distributed throughout the mammalian brain (Table 12.1). Brain regions with the highest density of $GABA_B$ binding sites are the thalamic nuclei, the molecular layer of the cerebellum, the cerebral cortex and interpeduncular nucleus in higher centres (Bowery *et al.*, 1987; Chu *et al.*, 1990). However, high densities are also present in laminae II and III of the spinal cord (Bowery *et al.*, 1987).

Distribution of mRNA for the splice variants of subunit $GABA_{B1}$, $GABA_{B(1a)}$ and $GABA_{B(1b)}$ using *in situ* hybridisation techniques has revealed that $GABA_{B(1a)}$ may be more associated with presynaptic receptors whereas $GABA_{B(1b)}$ may be responsible for postsynaptic receptor formation in certain brain regions. Thus, in the dorsal horn of the rat spinal cord the density of $GABA_{B(1a)}$ mRNA is low, whereas in the dorsal root ganglia, which contain the cell bodies of the primary afferent fibres, >90 per cent of the $GABA_B$ subunit mRNA is $GABA_{B(1a)}$ with $GABA_{B(1b)}$ mRNA comprising less than 10 per cent of the total $GABA_{B1}$ mRNA (Towers *et al.*, submitted). In support of this conclusion we have recently observed, using immunocytochemistry, that the levels of $GABA_{B(1a)}$ protein are higher than $GABA_{B(1b)}$ in the dorsal horn of the spinal cord. A similar conclusion has arisen from observations in rat and human cerebellum where $GABA_{B(1a)}$ mRNA was detected over the granule cells which send their axons into the molecular layer to innervate the Purkinje cell dendrites (Kaupmann *et al.*, 1998b; Billinton *et al.*, 1999). In contrast, $GABA_{B(1b)}$ mRNA was observed to be associated with the Purkinje cells, which would be expected to express $GABA_B$ receptors on their dendrites in the molecular layer, which would then most likely be postsynaptic to GABAergic stellate cells. These observations with the subunit transcripts may only be pertinent to the spinal cord and cerebellum as supporting data from other regions have yet to emerge.

The patterns of distribution of $GABA_{B1}$ and $GABA_{B2}$ protein subunits are in broad alignment with the native receptor although in some brain regions, such as the caudate putamen, $GABA_{B1}$ and the native receptor are present whereas $GABA_{B2}$ appears to be absent (Clark *et al.*, 1998). In addition, it has been noted that there is only a low level of $GABA_{B2}$ mRNA relative to $GABA_{B1}$ mRNA in the hypothalamus (Jones *et al.*, 1998). Does this mean that another unknown subunit or association protein can couple with $GABA_{B1}$ to produce a functional receptor, since the expressed subunit does not appear to exist as a monomer (Benke *et al.*, 1999)? A variety of association proteins which couple readily with $GABA_{B1}$ have recently been described by White and colleagues (1999), but whether any of these would allow full expression of the receptor in

Table 12.1 Distribution of $GABA_B$ receptor protein and mRNA in rat brain sections

	$GABA_B$ binding (^{3}H-GABA autoradiography)	$GABA_{B1}$-LI (immunocytochemistry)	$GABA_{B1}$ mRNA (in situ hybridisation)	$GABA_{B2}$ mRNA (in situ hybridisation)
Olfactory bulb				
Glomerular layer	2	5		1.5
Internal granule layer	1	2		1.5
External plexiform layer	1.5	1		
Anterior olfactory nucleus	2.5	2		3
Cerebral cortex				
Frontal	3	3	3	3
Temporal	2	2.5	3	3
Cingulate		2.5	3	3
Basal ganglia				
Caudate putamen	0.5	2.5	3	1
Nucleus accumbens	0.5	2.5	3	1
Medial septum	0.5	2	3	2
Globus pallidus	1	1	3	1
Hippocampal formation				
Dentate gyrus mol. layer	1.5	3		3
Dentate gyrus gran. layer	1	3	3	3
CA1	1	3	3	2
CA2	1	2	3	4
Subiculum		3.5		1
Thalamic nuclei				
Medial geniculate	2.5	4	3	4
Dorsal lateral geniculate	2.5	5		3
Ventral lateral geniculate	1	2.5		
Lateral dorsal	1.5	4		3
Lateral posterior	3	2.5		2
Interpeduncular nucleus	3.5	5	3	2
Medial habenular	1.5	5	3	4
Lateral habenular	1	4	3	1
Amygdaloid nucleus	2.5	2		3
Fascicular retroflexus tract	1.5	4		
Superior colliculus	2.5	3		2
Substantia nigra compacta	0.5	5		
Substantia nigra reticulata	1	2		
Dorsal raphé	1	4		2
Pontine nucleus	1.5	3.5		3
Cerebellar cortex				
Granule cell layer	1.5	2	2	1
Molecular/Purkinje layer	5	4	3	3
Prepositus Hypoglossal Nucleus	1	3		1
Spinal trigeminal nucleus	1	2		1
Spinal cord				
Substantia gelatinosa	1.5	3.5		1
Laminae II/III	3	4		1
Lamina X	0.5	3.5		
Laminae VII/VIII	<0.5	2.5		1

Note
Values reflect density (higher value = greater density) of labelling in autoradiography, *in situ* hybridisation or immunocytochemistry. Each value has been normalised from the range of densities detected across the brain section. Gaps in the table indicate that a value was not determined.

the absence of $GABA_{B2}$ is unknown. What seems more likely is that they somehow provide a structural base for receptor expression.

$GABA_B$ receptor effector mechanisms

Neuronal $GABA_B$ receptors are coupled indirectly via G-proteins to their effector mechanisms; namely, membrane Ca^{2+} and K^+ channels (Inoue *et al.*, 1985; Andrade *et al.*, 1986; Dolphin *et al.*, 1990; Bindokas and Ishida, 1991; Gage, 1992) and adenylate cyclase (Karbon *et al.*, 1984; Hill *et al.*, 1984; Hill, 1985; Xu and Wojcik, 1986).

Adenylate cyclase

It is well established that $GABA_B$ agonists inhibit basal and forskolin-stimulated neuronal adenylate cyclase activity in brain slices (e.g. Xu and Wocjik, 1986). Whilst this effect is unrelated to the channel events it is similarly mediated via G-proteins (G_i/G_o) and produces a reduced level of intracellular cAMP. Enhancement of cAMP formation produced by G_s coupled receptor agonists such as isoprenaline, is also a well-documented response to $GABA_B$ receptor agonists in brain slice preparations, but not in neuronal membranes. Both of these actions have also been observed *in vivo*. Using a microdialysis technique in freely moving rats, Hashimoto and Kuriyama (1997) were able to show that baclofen and GABA could reduce the increase in cAMP generated by infusion of forskolin in the cerebral cortex. This effect was blocked by the $GABA_B$ antagonist CGP 54626. Baclofen also potentiated the generation of cAMP by isoprenaline in this preparation.

Whilst it appears that both $GABA_{B1}$ and $GABA_{B2}$ can individually modulate adenylate cyclase activity in cell lines (Kaupmann *et al.*, 1998a; Martin *et al.*, 1999) there is currently no information about the nature of the G-protein coupling to the heterodimer.

CHANNEL CONDUCTANCES

Receptor activation decreases Ca^{2+} conductance but increases membrane conductance to K^+ ions. The decrease in Ca^{2+} conductance appears to be primarily associated with presynaptic sites (e.g. Chen and Van der Pol, 1998; Takahashi *et al.*, 1998) suppressing 'P/Q' and 'N' type channels (e.g. Santos *et al.*, 1995; Lambert and Wilson, 1996) although facilitation of an 'L' type channel in non-mammalian retina has been described (Shen and Slaughter, 1999). Modulation of K^+ conductances appears to be primarily linked with postsynaptic $GABA_B$ sites (e.g. Lüscher *et al.*, 1997) and possibly more than one type of K^+ channel (Wagner and Dekin, 1993) and even Ca^{2+} channel events may be involved in certain postsynaptic responses (Harayama *et al.*, 1998). Conversely a K^+(A) current appears to be coupled to $GABA_B$ receptors on presynaptic terminals in hippocampal cultures (Saint *et al.*, 1990). But overall, changes in membrane K^+ flux appear to make the major contribution to postsynaptic $GABA_B$ receptor activation.

Whilst suppression of Ca^{2+} influx is probably the most frequently observed mechanism associated with presynaptic $GABA_B$ sites (e.g. Doze *et al.*, 1995; Wu and Saggau, 1995; Isaacson, 1998), a process independent of Ca^{2+} or K^+ channels has been noted in rodent CA1 hippocampal pyramidal cells by Jarolimek and Misgeld (1997) who suggest that activation of protein kinase C(PKC) may be responsible. $GABA_B$ receptor activation has previously been shown to induce a rapid increase in PKC activity in rat hippocampal slices, but this was only apparent in the early postnatal period of life (Tremblay *et al.*, 1995).

Low threshold Ca^{2+} T-currents, which are inactivated at normal resting membrane potentials, may also be involved in the response to $GABA_B$ receptor activation, at least within the thalamus (Scott *et al.*, 1990). $GABA_B$ receptor activation produces a post-synaptic hyperpolarisation of long duration, which initiates Ca^{2+} spiking activity in thalamocortical cells and may be implicated in the generation of spike and wave discharges associated with absence seizures (Crunelli and Leresche, 1991).

PHARMACOLOGICAL AND THERAPEUTIC EFFECTS OF $GABA_B$ RECEPTOR AGONISTS

β-[4-chlorophenyl]GABA(baclofen), the prototypical selective agonist, was shown not only to have efficacy at the $GABA_B$ receptor but also to be stereospecifically active (Bowery *et al.*, 1980, 1981). Unfortunately, relatively few compounds have subsequently emerged with selectivity for $GABA_B$ sites and with greater efficacy or affinity than baclofen.

3-aminopropyl phosphinic acid (2APPA) and its methyl homologue(AMPPA,SKF 97541) were reported to be 3–7 times more potent at $GABA_B$ receptors than the active isomer of baclofen. A variety of other phosphinic based agonist ligands have been produced (Froestl *et al.*, 1995a) which have varying potencies but which have so far not been compared clinically with baclofen.

A wide variety of effects have been attributed to the action of $GABA_B$ receptor agonists and $GABA_B$-mediated synaptic events. Some of these include central muscle relaxant action, antitussive action, bronchiolar relaxation, urinary bladder relaxation, gastric motility increase, epileptogenesis, antinociception, yawning, hypotension, brown fat thermogenesis, cognitive impairment, reduction in release of hormones such as corticotrophin releasing hormone, prolactin releasing hormone, luteinizing hormone and melanocyte stimulating hormone, and reduced gastric acid secretion. This is not an exhaustive list but illustrates the variety of effects that may arise.

First on any list of $GABA_B$-mediated effects is surely the centrally mediated muscle relaxant action for which baclofen has been used clinically for more than 25 years. The mechanism underlying this action of the $GABA_B$ agonist appears to derive from its ability to reduce the release of excitatory neurotransmitter on to motoneurons in the ventral horn of the spinal cord. Its effectiveness has made it the drug of choice in the treatment of spasticity irrespective of the cause. Unfortunately, baclofen is not without significant side effects in some patients, making it poorly tolerated after systemic administration. This can be avoided by intrathecal infusion of very low amounts of the drug. This mode of treatment has proved to be very successful in spasticity associated with tardive dystonia, brain and spinal cord injury, cerebral palsy, tetanus, multiple sclerosis and stiff-man syndrome (e.g. Penn and Mangieri, 1993; Ochs *et al.*, 1989; Dressler *et al.*, 1997; Meythaler *et al.*, 1997; Armstrong *et al.*, 1997; Francois *et al.*, 1997; Becker *et al.*, 1995; Albright *et al.*, 1996; Paret *et al.*, 1996; Dressnandt and Conrad, 1996; Azouvi *et al.*, 1996; Seitz *et al.*, 1995).

Baclofen, has been shown to be very effective in the clinical treatment of otherwise intractable hiccups (e.g. Guelaud *et al.*, 1995; Marino, 1998; Nickerson *et al.*, 1997; Kumar and Dromerick, 1998). This effect is believed to stem from an inhibition of the hiccup reflex arc and possibly involves GABAergic inputs from the nucleus raphe magnus (Oshima *et al.*, 1998). Baclofen also has an antitussive action in low oral doses in man (Dicpinigaitis and Dobkin, 1997), which confirms earlier reports of an antitussive action in the cat and guinea pig (Bolser *et al.*, 1994).

Although pain relief has been noted with baclofen in trigeminal neuralgia in man (Fromm, 1994) as well as in a rodent model (Idänpäänheikklä and Guilbaud, 1999), its usefulness as an analgesic has always been questioned (see Hansson and Kinnman, 1996). Nevertheless, more recent clinical observations have indicated that baclofen can reduce pain due to stroke or spinal cord injury and musculoskeletal pain when administered by intrathecal infusion (Taira *et al.*, 1995; Loubser and Akman, 1996). But, despite these reports, $GABA_B$-mediated clinical analgesia has still to be established – possibly because of receptor desensitisation or receptor inactivation following systemic administration of baclofen.

In animal acute pain models baclofen has long been known to have an antinociceptive action. These tests include the tail flick, acetic acid writhing, formalin and hot plate tests in rodents (e.g. Cutting and Jordan, 1975; Levy and Proudfit, 1979; Serrano *et al.*, 1992; Przesmycki *et al.*, 1998). Even in chronic neuropathic pain models in rats, baclofen clearly exhibits an antinociceptive or anti-allodynic response (Smith *et al.*, 1994; Wiesenfeld Hallin *et al.*, 1997; Cui *et al.*, 1998). The locus of this action is probably, in part, within higher centres of the brain (Liebman and Pastor, 1980; Thomas *et al.*, 1995), but an action within the spinal cord also makes an important contribution (Sawynok and Dickson, 1985; Hammond and Washington, 1993; Dirig and Yaksh, 1995; Thomas *et al.*, 1996). Even a single intrathecal injection of GABA given within one week of nerve injury in the rat could permanently reverse neuropathic pain (Eaton *et al.*, 1999). This led the authors to suggest that altered spinal GABA levels may contribute to the induction phase of the chronic pain. A recent report supporting the analgesic potential of $GABA_B$ agonists stems from examining the effect of the GABA transport inhibitor tiagabine as an antinociceptive compound in rodents (Ipponi *et al.*, 1999). These authors reported that tiagabine was effective in a variety of paradigms, and its action was associated with an increase in the extracellular concentration of GABA within the thalamus. Most importantly the antinociceptive effect was blocked by the $GABA_B$ antagonist, CGP 35348.

The majority of $GABA_B$ receptors in the rat dorsal horn of the spinal cord appear to be located on small diameter afferent fibre terminals (Price *et al.*, 1987) where their activation decreases the evoked release of sensory transmitters such as substance P and glutamate (see Malcangio and Bowery, 1996). This suppression of transmitter release appears to contribute to the antinociceptive action of baclofen after systemic or intrathecal administration. Baclofen is also able to suppress the output from mechanosensitive primary afferents (Page and Blackshaw, 1999), which, arguably, might contribute to its apparent analgesic action.

In a recent preliminary report it has been shown that baclofen can also be beneficial in the prophylactic treatment of migraine to suppress headache pain (Heringhanit, 1999).

One important and recent observation with baclofen indicates that it may be a very effective treatment for cocaine addiction by reducing the craving for the drug. In rats, baclofen, administered at doses of 1–5 mg/kg, suppressed the self-administration of cocaine without affecting responding for food reinforcement (Roberts and Andrews, 1997; Shoaib *et al.*, 1998; Campbell *et al.*, 1999). Moreover the selective $GABA_B$ receptor agonist, CGP 44532(0.063–0.5 mg/kg) mimicked the action of baclofen and failed to disrupt responding for food (Brebner *et al.*, 1999). Preliminary clinical observations indicate that the effect may well extend to man (Ling *et al.*, 1998). This is an important observation, which could have major consequences in the future therapy of drug addiction, as similar results have been obtained in rats self-administering heroin. Baclofen was reported to suppress significantly the intake of heroin (Xi and Stein, 1999).

The effects of $GABA_B$ agonist administration are not confined to the CNS in mammals but may derive from actions on peripheral organs as well. For example, in asthma it has been suggested that there is a dysfunction of presynaptic $GABA_B$ systems, which might normally attenuate cholinergic contraction of airway smooth muscle (Tohda *et al.*, 1998).

PHARMACOLOGICAL AND POTENTIAL THERAPEUTIC EFFECTS OF $GABA_B$ RECEPTOR ANTAGONISTS

Thus far there are no reports on the pharmacological actions of $GABA_B$ receptor antagonists in man. Nevertheless, we can make a number of predictions on the basis of results obtained in animals and knowledge of the effects of the $GABA_B$ receptor agonist, baclofen (Bowery, 1993).

$GABA_B$ receptor antagonists improve cognitive performance in a variety of animal paradigms (Mondadori *et al.*, 1993; Carletti *et al.*, 1993; Getova *et al.*, 1997, 1998; Yu *et al.*, 1997; Nakagawa and Takashima, 1997; but see Brucato *et al.*, 1996). Perhaps not surprisingly, therefore, $GABA_B$ agonists impair learning behaviour in animal models (Tong and Hasselmo, 1996; Arolfo *et al.*, 1998; McNamara and Skelton, 1996; Nakagawa *et al.*, 1995), and this induced amnesia appears to be mediated via G-protein linked receptors as the impairment produced by baclofen in mice can be blocked by pertussis toxin administered intracerebroventricularly (Galeotti *et al.*, 1998). Extending these studies to man is not a simple task, but if the eventual outcome is positive this could provide the basis for providing important therapeutic agents in the treatment of cognitive deficits.

Another important effect of the antagonists is their ability to suppress absence seizures in a variety of animal models. Marescaux *et al.* (1992) first showed that $GABA_B$ antagonists administered systemically or directly into the thalamus can prevent the spike and wave discharges manifest in the EEG of genetic absence rats (GAERS). Similar observations have been made in the lethargic mouse (Hosford *et al.*, 1992) and also in rats injected with gamma-hydroxybutyric acid, which produces seizure activity reminiscent of absence epilepsy (Snead, 1992). In all cases $GABA_B$ antagonists dose-dependently reduced the seizure activity. These and other data have prompted the suggestion that $GABA_B$ mechanisms may be involved in the generation of the absence syndrome within the thalamus, possibly through Ca^{2+} spike generation (Crunelli and Leresche, 1991).

At much higher doses certain $GABA_B$ antagonists can, conversely, produce convulsant seizures in rats (Vergnes *et al.*, 1997), but why and how this occurs is unknown – although $GABA_B$ receptor agonists will 'block' the effect. Not every antagonist appears to produce the same effect. For example, we have failed to observe any convulsant activity with SCH 50911 at doses 10–100 times higher than the dose which completely blocks absence seizures in the genetic absence rat (Richards and Bowery, 1996).

The production of absence-like seizures by γ-hydroxybutyric acid in rats appears to be due, at least in part, to a weak partial agonist action at $GABA_B$ receptors (Bernasconi *et al.*, 1999; Lingenhoehl *et al.*, 1999).

Other potential areas in which $GABA_B$ antagonist intervention may prove to be beneficial are anxiety, depression and neurodegeneration, but the evidence for these indications is currently very limited.

$GABA_B$ antagonists and agonists could both have the potential to produce neuroprotection. Lal *et al.*(1995) suggest that the agonist, baclofen, could be cytoprotective in a cerebral ischaemia model in gerbils; but very large doses, well in excess of that producing muscle relaxation, were required. By contrast Heese *et al.* (2000) have described the effect of low doses of $GABA_B$ antagonists on NGF and BDNF levels in rats. A single dose of CGP 36742, CGP 56433A or CGP 56999A increased growth factor levels in hippocampus, neocortex and spinal cord by twofold to threefold, which could have a major influence on the neurodegenerative process. Other studies on mouse cultured striatal neurons have shown that $GABA_B$ receptor activation can enhance the neurotoxic effects of NMDA (Lafoncazal *et al.*, 1999). This would also support the view that $GABA_B$ antagonists are more likely to be neuroprotective. However, Beskid *et al.* (1999) have recently reported that baclofen attenuates the neurotoxic effect of quinolinic acid on cells in the CA1 region of the rat hippocampus.

A role for $GABA_B$ mechanisms in depression was originally suggested some time ago by Lloyd and colleagues (1985), but this was subsequently challenged by other groups. More recently, however, further suggestions that $GABA_B$ antagonists (e.g. CGP 36742) are effective in animal models of depression have emerged (Nakagawa *et al.*, 1999). This might be supported in due course by the observations of Heese *et al.* (2000) who showed that $GABA_B$ antagonists produce a rapid increase in nerve growth factors. As antidepressants produce the same response, but only after 2–3 weeks, this might be linked.

The design and development of selective $GABA_B$ receptor antagonists with increasing receptor affinity and potency has been an important process in establishing the significance and structure of the $GABA_B$ receptor. The original selective antagonists, phaclofen and 2-hydroxy saclofen, were discovered by Kerr and colleagues (1987, 1988). These represented a major breakthrough, even though they have low potency with affinities of 100 μM and 12 μM respectively for ^{3}H-GABA binding sites in cortical membranes.

Subsequent discoveries stemmed largely from a second group led by W. Froestl and S. Mickle. This group developed the first antagonist that was able to cross the blood–brain barrier after intraperitoneal injection, CGP35348 (Olpe *et al.*, 1990), and subsequently CGP36742, which was shown to be centrally active after oral administration in rats (Olpe *et al.*, 1993). However, both of these compounds, and others in the same series, have low potency even though they are selective for the $GABA_B$ receptor. The only other low affinity compound of note, which exhibits significant CNS activity after peripheral administration, is SCH50911, which, unlike all of the CGP antagonist series, does not contain the phosphorus moiety (Bolser *et al.*, 1995).

The most crucial breakthrough in the discovery of antagonists came with the development, by Froestl and colleagues, of compounds with affinities about 10,000 times higher than any previous antagonist. This major advance stemmed from the substitution of a dichlorobenzene moiety into the existing molecules. This produced a profusion of compounds with affinities in the nanomolar or even subnanomolar range (Froestl *et al.*, 1995b). Perhaps the most notable compounds among these are CGP55845, CGP54626 and CGP62349, although many more were produced. This series eventually led to the development of the iodinated high affinity antagonist ^{125}I-CGP64213, which was used in the elucidation of the structure of $GABA_{B1}$ (Kaupmann *et al.*, 1997).

MULTIPLE $GABA_B$ RECEPTORS?

It seems unlikely that $GABA_B$ receptors are homogeneous, but at present definitive evidence for functionally distinct receptor subtypes is lacking. However, if heterodimers comprising different isomers of $GABA_{B1}$ and $GABA_{B2}$ exist this could give rise to pharmacologically distinct subtypes. Transmitter release studies suggest differences between receptors on different nerve terminals and between heteroreceptors and autoreceptors (Gemignani *et al.*, 1994; Ong *et al.*, 1998; Bonanno *et al.*, 1998; Phelan, 1999), and the dual action of $GABA_B$ agonists on adenylate cyclase in brain slices would support the concept of receptor subtypes (Cunningham and Enna, 1996). Electrophysiological studies in mammalian brain also indicate that there are subtle distinctions between pre- and postsynaptic receptors (Dutar and Nicoll, 1988; Harrison *et al.*, 1990; Colmers and Williams, 1988; Thompson and Gahwiler, 1992; Deisz *et al.*, 1997; Chan *et al.*, 1998).

Unfortunately, the only receptor ligands currently available do not appear to distinguish reliably between any receptor subtypes. Although certain antagonists appear to select for the four subtypes described by Gemignani *et al.* (1994) on synaptosomes, these same compounds have not been reported to produce the same separation in other neuronal systems. Equally, the suggested distinctions in other systems such as cAMP generation in brain slices (Cunningham and Enna, 1996) are not necessarily supported in other systems.

References

Albright, A.L., Barry, M.J., Fasick, P., Barron, W. and Shultz, B. (1996) 'Continuous intrathecal baclofen infusion for symptomatic generalized dystonia', *Neurosurgery* **38**: 934–938.

Andrade, R., Malenka, R.C. and Nicoll, R.A. (1986) 'A G protein couples serotonin and $GABA_B$ receptors to the same channels in hippocampus', *Science* **234**: 1261–1265.

Armstrong, R.W., Steinbrok, P., Cochrane, D.D., Kube, S.D. and Fife, S.E. (1997) 'Intrathecally administered baclofen for treatment of children with spasticity of cerebral origin', *Journal of Neurosurgery* **87**: 409–414.

Arolfo, M.P., Zanudio, M.A. and Ramirez, O.A. (1998) 'Baclofen infused in rat hippocampal formation impairs spatial learning', *Hippocampus* **8**: 109–113.

Azouvi, P., Mane, M., Thiebaut, J.B., Denys, P., Remyneris, O. and Bussel, B. (1996) 'Intrathecal baclofen administration for control of severe spinal spasticity: Functional improvement and long-term follow-up', *Archives of Physical Medicine and Rehabilitation* **77**: 35–38.

Barber, R.P., Vaughn, J.E., Saito, K., McLaughlin, B.J. and Roberts, E. (1978) 'GABAergic terminals are presynaptic to primary afferent terminals; in the substantia gelatinosa of the rat spinal cord', *Brain Research* **141**: 35–55.

Becker, W.J., Harris, C.J., Long, M.L., Ablett, D.P., Klein, G.M. and DeForge, D.A. (1995) 'Long-term intrathecal baclofen therapy in patients with intractable spasticity', *Canadian Journal of Neurological Sciences* **22**: 208–217.

Bein, H.J. (1972) 'Pharmacological differentiations of muscle relaxants', in W. Birkmayer, *Spasticity: A Topical Survey*, Vienna: Hans Huber.

Benke, D., Honer, M., Michel, C., Bettler, B. and Möhler, H. (1999) 'Gamma-aminobutyric acid type B receptor splice variant proteins GBR1a and GBR1b are both associated with GBR2 in situ and display differential regional and subcellular distribution', *J. Biol. Chem.* **274**: 27,323–27,330

Bernasconi, R., Mathivet, P., Bischoff, S. and Marescaux, C. (1999) 'Gamma-hydroxybutyric acid: an endogenous neuromodulator with abuse potential?' *Trends in Pharmacological Sciences* **20**: 135–141.

Beskid, M., Rozycka, Z. and Taraszewska, A. (1999) 'Quinolinic acid and $GABA_B$ receptor ligand: effect on pyramidal neurons of the CA1 sector of rat's dorsal hippocampus following peripheral administration', *Folia Neuropathologica* **37**: 99–106

Billinton, A., Upton, N. and Bowery, N.G. (1999) '$GABA_B$ receptor isoforms GBR1a and GBR1b, appear to be associated with pre- and post-synaptic elements respectively in rat and human cerebellum', *British Journal of Pharmacology* **126**: 1387–1392.

Bindokas, V.P. and Ishida, A.T. (1991) '(−)-baclofen and gamma-aminobutyric acid inhibit calcium currents in isolated retinal ganglion cells', *Proc. Natl. Acad. Sci. USA* **88**: 10,759–10,763.

Bolser, D.C., Blythin, D.J., Chapman, R.W., Egan, R.W., Hey, J.A., Rizzo, C., Kuo, S.C. and Kreutner, W. (1995) 'The pharmacology of SCH 50911: a novel, orally-active GABA-beta receptor antagonist', *J. Pharmacol. Exp. Ther.* **274**: 1393–1398.

Bolser, D.C., DeGennaro, F.C., O'Reilly, S., Chapman, R.W., Kreutner, W., Egan, R.W. and Hey, J.A. (1994) 'Peripheral and central sites of action of GABA-B agonists to inhibit the cough reflex in the cat and guinea pig', *British Journal of Pharmacology* **113**: 1344–1348.

Bonanno, G., Fassio, A., Sala, R., Schmid, G. and Raiteri, M. (1998) '$GABA_B$ receptors as potential targets for drugs able to prevent excessive excitatory amino acid transmission in the spinal cord', *Eur. J Pharmacol.* **362**: 143–148.

Bowery N.G. (1993) '$GABA_B$ receptor pharmacology', *Annu. Rev. Pharmacol. Toxicol.* **33**: 109–147.

Bowery, N.G., Doble, A., Hill, D.R., Hudson, A.L., Shaw, J.S., Turnbull, M.J. and Warrington, R. (1981) 'Bicuculline-insensitive GABA receptors on peripheral autonomic nerve terminals', *Eur. J. Pharmacol.* **71**: .53–70.

Bowery, N.G., Hill, D.R., Hudson, A.L., Doble, A., Middlemiss, D.N., Shaw, J.S. and Turnbull, M.J. (1980) '(−)baclofen decreases neurotransmitter release in the mammalian CNS by an action at a novel GABA receptor', *Nature* **283**: 92–94.

Bowery, N.G., Hudson, A.L. and Price, G.W. (1987) '$GABA_A$ and $GABA_B$ receptor site distribution in the rat central nervous system', *Neuroscience* **20**: 365–383.

Brebner, K., Froestl, W., Andrews, M., Phelan, R. and Roberts, D.C.S. (1999) 'The $GABA_B$ agonist CGP44532 decreases cocaine self-administration in rats: demonstration using a progressive ratio and a discrete trials procedure', *Neuropharmacology* **38**: 1797–1804

Brucato, F.H., Levin, E.D., Mott, D.D., Lewis, D.V., Wilson, W.A. and Swartzwelder, H.S. (1996) 'Hippocampal long-term potentiation and spatial learning in the rat: Effects of $GABA_B$ receptor blockade', *Neuroscience* **74**: 331–339.

Campbell, U.C., Lac, S.T. and Carroll, M.E. (1999) 'Effects of baclofen on maintenance and reinstatement of intravenous cocaine self-administration in rats', *Psychopharmacology* **143**: 209–214

Carletti, R., Libri, V. and Bowery, N.G. (1993) 'The $GABA_B$ antagonist CGP 36742 enhances spatial learning performance and antagonises baclofen-induced amnesia in mice', *British Journal of Pharmacology* **109**: 74P.

Castelli, M.P., Ingianni, A., Stefanini, E. and Gessa, G.L. (1999) 'Distribution of $GABA_B$ receptor mRNAs in the rat brain and peripheral organs', *Life Sciences* **64**: 1321–1328.

Chan, P.K.Y., Leung, C.K.S. and Yung, W.H. (1998) 'Differential expression of pre- and postsynaptic $GABA_B$ receptors in rat substantia nigra pars reticulata neurones', *Eur. J. Pharmacol.* **349**: 187–197.

Chen, G. and Van der Pol, A.N. (1998) 'Presynaptic $GABA_B$ autoreceptor modulation of P/Q-type calcium channels and GABA release in rat suprachiasmatic nucleus neurons', *J. Neurosci.* **18**: 1913–1922.

Chu, D.C.M., Albin, R.L., Young, A.B. and Penney, J.B. (1990) 'Distribution and kinetics of $GABA_B$ binding sites in rat central nervous system: a quantitative autoradiographic study', *Neuroscience* **34**: 341–357.

Clark, J.A, Mezey, E., Lam, A.S. and Bonner, T.I. (1998) 'Functional expression and distribution of GB2 a second $GABA_B$ receptor', *Soc. Neurosci. Abstr.* **24**: 795.8.

Colmers, W.F. and Williams, J.T. (1988) 'Pertussis toxin pretreatment discriminates between pre- and postsynaptic actions of baclofen in rat dorsal raphe nucleus in vitro', *Neurosci. Lett.* **93**: 300–306.

Couve, A., Filippov, A.K., Connolly, C.N., Bettler, B., Brown, D.A. and Moss, S.J. (1998) 'Intracellular retention of recombinant $GABA_B$ receptors', *J. Biol. Chem.* **273**: 26,361–26,367.

Crunelli, V. and Leresche, N. (1991) 'A role for the $GABA_B$ receptors in excitation and inhibition of thalamocortical cells', *Trends in Neurosciences* **14**: 16–21.

Cui, J.G., Meyerson, B.A., Sollevi, A. and Linderoth, B. (1998) 'Effect of spinal cord stimulation on tactile hypersensitivity in mononeuropathic rats is potentiated by simultaneous $GABA_B$ and adenosine receptor activation', *Neurosci. Lett.* **247**: 183–186.

Cunningham, M.D. and Enna, S.J. (1996) 'Evidence for pharmacologically distinct $GABA_B$ receptors associated with cAMP production in rat brain', *Brain Research* **720**: 220–224.

Cutting, D.A. and Jordan, C.C. (1975) 'Alternative approaches to analgesia: baclofen as a model compound', *Br. J. Pharmacol.* **54**: 171–179.

Deisz, R.A., Billard, J.M. and Zieglgänsberger, W. (1997) 'Presynaptic and postsynaptic $GABA_B$ receptors of neocortical neurons of the rat in vitro: Differences in pharmacology and ionic mechanisms', *Synapse* **25**: 62–72.

Dicpinigaitis, P.V. and Dobkin, J.B. (1997) 'Antitussive effect of the GABA-agonist baclofen', *Chest* **111**: 996–999.

Dirig, D.M. and Yaksh, T.L. (1995) 'Intrathecal baclofen and muscimol, but not midazolam, are antinociceptive using the rat-formalin model', *J. Pharmacol. Exp. Ther.* **275**: 219–227.

Dolphin, A.C., Huston, E. and Scott, R.H. (1990) '$GABA_B$-mediated inhibition of calcium currents: a possible role in presynaptic inhibition', in N.G. Bowery, H. Bittiger, and H.-R. Olpe (eds) *$GABA_B$ Receptors in Mammalian Function*, Chichester: J. Wiley, pp.259–271.

Doze, V.A., Cohen, G.A. and Madison, D.V. (1995) 'Calcium channel involvement in $GABA_B$ receptor-mediated inhibition of GABA release in area CA1 of the rat hippocampus', *Journal of Neurophysiology* **74**: 43–53.

Dressler, D., Oeljeschlager, R.O. and Ruther, E. (1997) 'Severe tardive dystonia: Treatment with continuous intrathecal baclofen administration', *Movement Disorders* **12**: 585–587.

Dressnandt, J. and Conrad, B. (1996) 'Lasting reduction of severe spasticity after ending chronic treatment with intrathecal baclofen', *Journal of Neurology, Neurosurgery and Psychiatry* **2**: 168–173.

Dutar, P. and Nicoll, R.A. (1988) 'Pre- and postsynaptic $GABA_B$ receptors in the hippocampus have different pharmacological properties', *Neuron* **1**: 585–591.

Eaton, M.J., Martinez, M.A. and Karmally, S. (1999) 'A single intrathecal injection of GABA permanently reverses neuropathic pain after nerve injury', *Brain Research* **835**: 334–339

Ferreira, S.A, Scott, C.J., Kuehl, D.E. and Jackson, G.L. (1996) 'Differential regulation of luteinizing hormone release by gamma-aminobutyric acid receptor subtypes in the arcuate-ventromedial region of the castrated ram', *Endocrinology* **137**: 3453–3460.

Francois, B., Clavel, M., Desachy, A., Vignon, P., Salle, J.Y. and Gastinne, H. (1997) 'Continuous intrathecal baclofen in tetanus – An alternative management', *Presse Medicale* **26**: 1045–1047.

Froestl, W., Mickel, S.J., Hall, R.G., Von Sprecher, G., Strub, D., Baumann, P.A., Brugger, F., Gentsch, C., Jaekel, J., Olpe, H.R., Rihs, G., Vassout, A., Waldmeier, P.C. and Bittiger, H.

(1995a) 'Phosphinic acid analogues of GABA. 1. New potent and selective $GABA_B$ agonists', *J. Med. Chem.* **38**: 3297–3312.

Froestl, W., Mickel, S.J., Von Sprecher, G., Diel, P.J., Hall, R.G., Maier, L., Strub, D., Melillo, V., Baumann, P.A., Bernasconi, R., Gentsch, C., Hauser, K., Jaekel, J., Karlsson, G., Klebs, K., Maltre, L., Marescaux, C., Pozza, M.F., Schmutz, M., Steinmann, M. W., van Reizen, H., Vassout, A., Mondadori, C., Olpe, H.-R., Walmeier, P.C. and Bittiger, H. (1995b) 'Phosphinic acid analogues of GABA. 2. Selective, orally active $GABA_B$ antagonists', *J. Med. Chem.* **38**: 3313–3331.

Fromm, G.H. (1994) 'Baclofen as an adjuvant analgesic', *Journal of Pain Symptom Management* **9**: 500–509.

Gage, P.W. (1992) 'Activation and modulation of neuronal K^+ channels by GABA', *Trends Neurosci.* **15**: 46–51.

Galeotti, N., Ghelardini, C. and Bartolini, A. (1998) 'Effect of pertussis toxin on baclofen- and diphenhydramine-induced amnesia', *Psychopharmacology* **136**: 328–334.

Gemignani, A., Paudice, P., Bonanno, G. and Raiteri, M. (1994) 'Pharmacological discrimination between gamma-aminobutyric acid type B receptors regulating cholecystokinin and somatostatin release from rat neocortex synaptosomes', *Mol. Pharmacol.* **46**: 558–562.

Getova, D. and Bowery, N.G. (1998) 'The modulatory effects of high affinity $GABA_B$ receptor antagonists in an active avoidance learning paradigm in rats', *Psychopharmacology* **137**: 369–373

Getova, D., Bowery, N.G. and Spassov, V. (1997) 'Effects of $GABA_B$ receptor antagonists on learning and memory retention in a rat model of absence epilepsy', *Eur. J. Pharmacol.* **320**: 9–13.

Guelaud, C., Similowski, T., Bizec, J.L., Cabane, J., Whitelaw, W.A. and Derenne, J.P. (1995) 'Baclofen therapy for chronic hiccup', *European Respiratory Journal* **8**: 235–237.

Hammond, D.L. and Washington, J.D. (1993) 'Antagonism of L-baclofen-induced antinociception by CGP 35348 in the spinal cord of the rat', *Eur. J. Pharmacol.* **234**: 255–262.

Hansson, P. and Kinnman, E. (1996) 'Unmasking mechanisms of peripheral neuropathic pain in a clinical perspective', *Pain Review* **3**: 272–292.

Harayama, N., Shibuya, I., Tanaka, K., Kabashima, N., Ueta, Y. and Yamashita, H. (1998) 'Inhibition of N- and P/Q-type calcium channels by postsynaptic $GABA_B$ receptor activation in rat supraoptic neurones', *Journal of Physiology* **509**: 371–383.

Harrison, N.L., Lambert, N.A. and Lovinger, D.M. (1990) 'Presynaptic $GABA_B$ receptors on rat hippocampal neurons', in N.G. Bowery, H. Bittiger, and H.-R. Olpe (eds) *$GABA_B$ Receptors in Mammalian Function*, Chichester: Wiley, pp. 208–221.

Hashimoto, T. and Kuriyama, K. (1997) 'In vivo evidence that $GABA_B$ receptors are negatively coupled to adenylate cyclase in rat striatum', *J. Neurochem.* **69**: 365–370.

Heese, K., Otten, U., Mathivet, P., Raiteri, M., Marescaux, C. and Bernasconi, R. (2000) '$GABA_B$ receptor antagonists elevate both mRNA and protein levels of the neurotrophins nerve growth factor (NGF) and brain-derived neurotrophic factor (BDNF) but not neurotrophin-3 (NT-3) in brain and spinal cord of rats', *Neuropharmacology* **39**: 449–462.

Heringhanit, R. (1999) 'Baclofen for prevention of migraine', *Cephalalgia* **19**: 589–591

Hill, D.R. (1985) 'GABAB receptor modulation of adenylate cyclase activity in rat brain slices', *British Journal of Pharmacology* **84**: 249–257.

Hill, D.R and Bowery, N.G. (1981) '^{3}H-baclofen and ^{3}H-GABA bind to bicuculine-insensitive $GABA_B$ sites in rat brain', *Nature* **290**: 149–152.

Hill, D.R., Bowery, N.G. and Hudson, A.L. (1984) 'Inhibition of GABAB receptor binding by guanyl nucleotides', *J. Neurochem.* **42**: 652–657.

Hosford, D.A., Clark, S., Cao, Z., Wilson, W.A., Jr., Lin, F., Morrisett, R.A. and Huin, A. (1992) 'The role of $GABA_B$ receptor activation in absence seizures of lethargic (*lh/lh*) mice', *Science* **257**: 398–401.

Idänpään Heikkilä, J.J. and Guilbaud, G. (1999) 'Pharmacological studies on a rat model of trigeminal neuropathic pain: baclofen, but not carbamazepine, morphine or tricyclic antidepressants, attenuate the allodynia-like behaviours', *Pain* **79**: 281–290.

Inoue, M., Matsuo, T. and Ogata, N. (1985) 'Possible involvement of K^+ conductance in the action of gamma-aminobutyric acid in the guinea-pig hippocampus', *British Journal of Pharmacology* **86**: 515–524.

Ipponi, A., Lamberti, C., Medica, A., Bartolini, A. and Malmberg Aiello, P. (1999) 'Tiagabine antinociception in rodents depends on $GABA_B$ receptor activation: parallel antinociception testing and medial thalamus GABA microdialysis', *Eur. J. Pharmacol.* **368**: 205–211

Isaacson, J.S. (1998) '$GABA_B$ receptor-mediated modulation of presynaptic currents and excitatory transmission at a fast central synapse', *Journal of Neurophysiology* **80**: 1571–1576.

Isaacson, J.S., Solís, J.M. and Nicoll, R.A. (1993) 'Local and diffuse synaptic actions of GABA in the hippocampus', *Neuron* **10**: 165–175.

Jarolimek, W. and Misgeld, U. (1997) '$GABA_B$ receptor-mediated inhibition of tetrodotoxin-resistant GABA release in rodent hippocampal CA1 pyramidal cells', *J. Neuroscience* **17**: 1025–1032.

Jones, K.A, Borowsky, B., Tamm, J.A., Craig, D.A., Durkin, M.M., Dai, M., Yao, W.J., Johnson, M., Gunwaldsen, C., Huang, L.Y., Tang, C., Shen, Q.R., Salon, J.A., Morse, K., Laz, T., Smith, K.E., Nagarathnam, D., Noble, S.A., Branchek, T.A. and Gerald, C. (1998) '$GABA_B$ receptors function as a heteromeric assembly of the subunits $GABA_BR1$ and $GABA_BR2$', *Nature* **396**: 674–679.

Karbon, E.W., Duman, R.S. and Enna, S.J. (1984) '$GABA_B$ receptors and norepinephrine-stimulated cAMP production in rat brain cortex', *Brain Research* **306**: 327–332.

Kaupmann, K., Huggel, K., Heid, J., Flor, P.J., Bischoff, S., Mickel, S.J., McMaster, G., Angst, C., Bittiger, H., Froestl, W. and Bettler, B. (1997) 'Expression cloning of $GABA_B$ receptors uncovers similarity to metabotropic glutamate receptors', *Nature* **386**: 239–246.

Kaupmann, K., Malitschek, B., Schuler, V., Heid, J., Froestl, W., Beck, P., Mosbacher, J., Bischoff, S., Kulik, A., Shigemoto, R., Karschin, A. and Bettler, B. (1998a) '$GABA_B$-receptor subtypes assemble into functional heteromeric complexes', *Nature* **396**: 683–687.

Kaupmann, K., Schuler, V., Mosbacher, J., Bischoff, S., Bittiger, H., Heid, J., Froestl, W., Leonhard, S., Pfaff, T., Karschin, A. and Bettler, B. (1998b) 'Human gamma-aminobutyric acid type B receptors are differentially expressed and regulate inwardly rectifying K+ channels', *Proc. Natl. Acad. Sci. USA* **95**: 14,991–14,996.

Keberle, H. and Faigle, J.W. (1972) 'Synthesis and structure–activity relationship of the gamma-aminobutyric acid derivatives', *in* W. Birkmayer *Spasticity: A Topical Survey*, Vienna: Hans Huber.

Kerr, D.I.B., Ong, J., Johnston, G.A.R., Abbenante, J. and Prager, R.H. (1988) '2-hydroxy-saclofen: an improved antagonist at central and peripheral $GABA_B$ receptors', *Neurosci. Lett.* **92**: 92–96.

Kerr, D.I.B., Ong, J., Prager, R.H., Gynther, B.D. and Curtis, D.R. (1987) 'Phaclofen: a peripheral and central baclofen antagonist', *Brain Research* **405**: 150–154.

Kumar, A. and Dromerick, A.W. (1998) 'Intractable hiccups during stroke rehabilitation', *Archives of Physical Medicine and Rehabilitation* **79**: 697–699.

Kuner, R., Köhr, G., Grünewald, S., Eisenhardt, G., Bach, A. and Kornau, H.C. (1999) 'Role of heteromer formation in $GABA_B$ receptor function', *Science* **283**: 74–77.

Lafoncazal, M., Viennois, G., Kuhn, R., Malitschek, B., Pin, J.P., Shigemoto, R. and Bockaert, J. (1999) 'mGluR7-like receptor and $GABA_B$ receptor activation enhance neurotoxic effects of N-methyl-D-aspartate in cultured mouse striatal GABAergic neurones', *Neuropharmacology* **38**: 1631–1640.

Lal, S., Shuaib, A. and Ijaz, S. (1995) 'Baclofen is cytoprotective to cerebral ischemia in gerbils', *Neurochem. Res.* **20**: 115–119.

Lambert, N.A. and Wilson, W.A. (1996) 'High-threshold Ca^{2+} currents in rat hippocampal interneurones and their selective inhibition by activation of $GABA_B$ receptors', *Journal of Physiology* **492**: 115–127.

Levy, R.A. and Proudfit, H.K. (1979) 'Analgesia produced by microinjection of baclofen and morphine at brain stem sites', *Eur. J. Pharmacol.* **57**: 43–55.

Liebman, J.M. and Pastor, G. (1980) 'Antinociceptive effects of baclofen and muscimol upon intraventricular administration', *Eur. J. Pharmacol.* **61**: 225–230.

Ling, W., Shoptaw, S. and Majewska, D. (1998) 'Baclofen as a cocaine anti-craving medication: a preliminary clinical study', *Neuropsychopharmacology* **18**: 403–404.

Lingenhoehl, K., Brom, R., Heid, J., Beck, P., Froestl, W., Kaupmann, K., Bettler, B. and Mosbacher, J. (1999) 'Gamma-hydroxybutyrate is a weak agonist at recombinant GABA(B) receptors', *Neuropharmacology* **38**: 1667–1673.

Lloyd, K.G., Thuret, F. and Pilc, A.(1985) 'Upregulation of gamma-aminobutyric acid ($GABA_B$) binding sites in rat frontal cortex: a common action of repeated administration of different classes of antidepressants and electroshock', *J. Pharmacol. Exp. Ther.* **235**: 191–199.

Loubser, P.G. and Akman, N.M. (1996) 'Effects of intrathecal baclofen on chronic spinal cord injury pain', *Journal of Pain and Symptom Management* **12**: 241–247.

Lüscher, C., Jan, L.Y., Stoffel, M., Malenka, R.C. and Nicoll, R.A. (1997) 'G protein-coupled inwardly rectifying K+ channels (GIRKs) mediate postsynaptic but not presynaptic transmitter actions in hippocampal neurons', *Neuron* **19**: 687–695.

Malcangio, M. and Bowery, N.G. (1996) 'GABA and its receptors in the spinal cord', *Trends in Pharmacological Sciences* **17**: 457–462.

Marescaux, C., Vergnes, M. and Bernasconi, R. (1992) '$GABA_B$ receptor antagonists: potential new anti-absence drugs', *J. Neural Transm.* **35**: 179–188.

Marino, R.A. (1998) 'Baclofen therapy for intractable hiccups in pancreatic carcinoma', *American Journal of Gastroenterology* **93**: 2000.

Martin, S.C., Russek, S.J and Farb, D.H. (1999) 'Molecular identification of the human $GABA_B$ R2: cell surface expression and coupling to adenylyl cyclase in the absence of $GABA_B$ R1', *Molecular and Cellular Neuroscience* **13**: 180–191.

Marvizon, J.C.G., Grady, E.F., Stefani, E., Bunnett, N.W. and Mayer, E.A. (1999) 'Substance P release in the dorsal horn assessed by receptor internalization: NMDA receptors counteract a tonic inhibition by $GABA_B$ receptors', *Eur. J. Neurosci.* **11**: 417–426.

McLatchie, L.M., Fraser, N.J., Main, M.J., Wise, A., Brown, J., Thompson, N., Solari, R., Lee, M.G. and Foord, S.M. (1998) 'RAMPs regulate the transport and ligand specificity of the calcitonin-receptor-like receptor', *Nature* **393**: 333–339.

McNamara, R.K. and Skelton, R.W. (1996) 'Baclofen, a selective $GABA_B$ receptor agonist, dose-dependently impairs spatial learning in rats', *Pharmacology, Biochemistry and Behavior* **53**: 303–308.

Meythaler, J.M., McCary, A. and Hadley, M.N. (1997) 'Prospective assessment of continuous intrathecal infusion of baclofen for spasticity caused by acquired brain injury: a preliminary report', *Journal of Neurosurgery* **87**: 415–419.

Mondadori, C., Jaekel, J. and Preiswerk, G. (1993) 'CGP 36742: The first orally active GABAB blocker improves the cognitive performance of mice, rats, and rhesus monkeys', *Behavioral and Neural Biology* **60**: 62–68.

Nakagawa, Y. and Takashima, T. (1997) 'The $GABA_B$ receptor antagonist CGP36742 attenuates the baclofen- and scopolamine-induced deficit in Morris water maze task in rats', *Brain Research* **766**: 101–106.

Nakagawa, Y., Ishibashi, Y., Yoshii, T. and Tagashira, E. (1995) 'Involvement of cholinergic systems in the deficit of place learning in Morris water maze task induced by baclofen in rats', *Brain Research* **683**: 209–214.

Nakagawa, Y., Sasaki, A. and Takashima, T. (1999) 'The $GABA_B$ receptor antagonist CGP36742 improves learned helplessness in rats', *Eur. J. Pharmacol.* **381**: 1–7.

Ng, G.Y.K., Clark, J., Coulombe, N., Ethier, N., Hebert, T.E., Sullivan, R., Kargman, S., Chateauneuf, A., Tsukamoto, N., McDonald, T., Whiting, P., Mezey, E., Johnson, M.P., Liu, Q., Kolakowski, L.F., Evans, J.F., Bonner, T.I. and O'Neill, G.P. (1999) 'Identification of a $GABA_B$ receptor subunit, gb2, required for functional $GABA_B$ receptor activity', *J. Biol. Chem.* **274**: 7607–7610.

Nickerson, R.B., Atchison, J.W., Van Hoose, J.D. and Hayes, D. (1997) 'Hiccups associated with lateral medullary syndrome. A case report', *American Journal of Physical Medicine and Rehabilitation* **76**: 144–146.

Ochs, G., Struppler, A., Meyerson, B.A., Linderoth, G. and Gybels, J. (1989) 'Intrathecal baclofen for long term treatment of spasticity: a multicentre study', *Journal of Neurology, Neurosurgery & Psychiatry* **52**: 933–939.

Olpe, H.-R., Karlsson, G., Pozza, M.F., Brugger, F., Steinmann, M., Van Riezen, H., Fagg, G., Hall, R.G., Froestl, W. and Bittiger, H. (1990) 'CGP 35348: a centrally active blocker of GABAB receptors', *Eur. J. Pharmacol.* **187**: 27–38.

Olpe, H.-R., Steinmann, M.W., Ferrat, T., Pozza, M.F., Greiner, K., Brugger, F., Froestl, W., Mickel, S.J., and Bittiger, H. (1993) 'The actions of orally active $GABA_B$ receptor antagonists on GABAergic transmission in vivo and in vitro', *Eur. J. Pharmacol.* **233**: 179–186.

Ong, J and Kerr, D.I. (1990) 'GABA-receptors in peripheral tissues', *Life Sci.* **46**: 1489–1501.

Ong, J., Marino, V., Parker, D.A.S. and Kerr, D.I.B. (1998) 'Differential effects of phosphonic analogues of GABA on $GABA_B$ autoreceptors in rat neocortical slices', *Naunyn-Schmiedebergs Archives of Pharmacology* **357**: 408–412.

Oshima, T., Sakamoto, M., Tatsuta, H. and Arita, H. (1998) 'GABAergic inhibition of hiccup-like reflex induced by electrical stimulation in medulla of cats', *Neuroscience Research* **30**: 287–293.

Page, A.J. and Blackshaw, L.A. (1999) '$GABA_B$ receptors inhibit mechanosensitivity of primary afferent endings', *J. Neuroscience* **19**: 8597–8602.

Paret, G., Tirosh, R., Benzeev, B., Vardi, A., Brandt, N. and Barzilay, Z. (1996) 'Intrathecal baclofen for severe torsion dystonia in a child', *Acta Paediatrica* **85**: 635–637.

Penn, R.D. and Mangieri, E.A. (1993) 'Stiff-man syndrome treated with intrathecal baclofen', *Neurology* **43**: 2412

Phelan, K.D. (1999) 'N-ethylmaleimide selectively blocks presynaptic $GABA_B$ autoreceptor but not heteroreceptor-mediated inhibition in adult rat striatal slices', *Brain Research* **847**: 308–313.

Price, G.W., Kelly, J.S. and Bowery, N.G. (1987) 'The location of GABAB receptor binding sites in mammalian spinal cord', *Synapse* **1**: 530–538.

Przesmycki, K., Dzieciuch, J.A., Czuczwar, S.J. and Kleinrok, Z. (1998) 'An isobolographic analysis of drug interaction between intrathecal clonidine and baclofen in the formalin test in rats', *Neuropharmacology* **37**: 207–214.

Rey-Roldan, E.B., Lux-Lantos, A.R., Gonzalez-Iglesias, A.E., Becu-Villalobos, D. and Libertun, C. (1996) 'Baclofen, a gamma-aminobutyric acid B agonist, modifies hormonal secretion in pituitary cells from infantile female rats', *Life Sci.* **58**: 1059–1065.

Richards, D.A. and Bowery, N.G. (1996) 'Anti-seizure effects of the $GABA_B$ antagonist, SCH-50911, in the genetic absence epilepsy rat from Strasbourg (GAERS)', *Pharmacol. Comm.* **8**: 227–230.

Roberts, D.C.S. and Andrews, M.M. (1997) 'Baclofen suppression of cocaine self-administration: Demonstration using a discrete trials procedure', *Psychopharmacology* **131**: 271–277.

Saint, D.A., Thomas, T. and Gage, P.W. (1990) 'GABAB agonists modulate a transient potassium current in cultured mammalian hippocampal neurons', *Neurosci Lett.* **118**: 9–13.

Santos, A.E, Carvalho, C.M., Macedo, T.A. and Carvalho, A.P. (1995) 'Regulation of intracellular $[Ca^{2+}]$ and GABA release by presynaptic $GABA_B$ receptors in rat cerebrocortical synaptosomes', *Neurochem. Intl.* **27**: 397–406.

Sawynok, J. and Dickson, C. (1985) 'D-baclofen is an antagonist at baclofen receptors mediating antinociception in the spinal cord', *Pharmacology* **31**: 248–259.

Scott, R.H., Wootton, J.F. and Dolphin, A.C. (1990) 'Modulation of neuronal T-type calcium channel currents by photoactivation of intracellular guanosine 5′-O(3-thio) triphosphate', *Neuroscience* **38**: 285–294.

Seitz, R.J., Blank, B., Kiwit, J.C.W. and Benecke, R. (1995) 'Stiff-person syndrome with anti-glutamic acid decarboxylase autoantibodies – complete remission of symptoms after intrathecal baclofen administration', *Journal of Neurology* **242**: 618–622.

Serrano, I., Ruiz, R.M., Serrano, J.S. and Fernandez, A. (1992) 'GABAergic and cholinergic mediation in the antinociceptive action of homotaurine', *General Pharmacology* **23**: 421–426.

Shen, W. and Slaughter, M.M. (1999) 'Metabotropic GABA receptors facilitate *L*-type and inhibit N-type calcium channels in single salamander retinal neurons', *J. Physiol.* **516**: 711–718

Shoaib, M., Swanner, L.S., Beyer, C.E., Goldberg, S.R. and Schindler, C.W. (1998) 'The $GABA_B$ agonist baclofen modifies cocaine self-administration in rats', *Behavioural Pharmacology* **9**: 195–206.

Smith, G.D., Harrison, S.M., Birch, P.J., Elliott, P.J., Malcangio, M. and Bowery, N.G. (1994) 'Increased sensitivity to the antinociceptive activity of (+/−)-baclofen in an animal model of chronic neuropathic, but not chronic inflammatory hyperalgesia', *Neuropharmacology* **33**: 1103–1108.

Snead, O.C, III (1992) 'Evidence for $GABA_B$-mediated mechanisms in experimental generalized absence seizures', *Eur. J. Pharmacol.* **213**: 343–349.

Taira, T., Kawamura, H., Tanikawa, T., Iseki, H., Kawabatake, H. and Takakura, K. (1995) 'A new approach to control central deafferentation pain: spinal intrathecal baclofen', *Stereotactic and Functional Neurosurgery* **65**: 101–105.

Takahashi, T., Kajikawa, Y. and Tsujimoto, T. (1998) 'G-protein-coupled modulation of presynaptic calcium currents and transmitter release by a GABA(B) receptor', *J. Neuroscience* **18**: 3138–3146.

Thomas, D.A., McGowan, M.K. and Hammond, D.L. (1995) 'Microinjection of baclofen in the ventromedial medulla of rats: antinociception at low doses and hyperalgesia at high doses', *J. Pharmacol. Exp. Ther.* **275**: 274–284.

Thomas, D.A., Navarrete, I.M., Graham, B.A., McGowan, M.K. and Hammond, D.L. (1996) 'Antinociception produced by systemic R(+)-baclofen hydrochloride is attenuated by CGP 35348 administered to the spinal cord or ventromedial medulla of rats', *Brain Research* **718**: 129–137.

Thompson, S.M. and Gahwiler, B.W. (1992) 'Comparison of the actions of baclofen at pre- and postsynaptic receptors in the rat hippocampus in vitro', *J. Physiol.* **451**: 329–345.

Tohda, Y., Ohkawa, K., Kubo, H., Muraki, M., Fukuoka, M. and Nakajima, S. (1998) 'Role of GABA receptors in the bronchial response: studies in sensitized guinea-pigs', *Clinical and Experimental Allergy* **28**: 772–777.

Tong, A.C. and Hasselmo, M.E. (1996) 'Effects of long term baclofen treatment on recognition memory and novelty detection', *Behavioural Brain Research* **74**: 145–152.

Towers, S., Princivalle, A., Billinton, A., Edmunds, M., Bettler, B., Urban, L., Castro-Lopes, J. and Bowery, N.G. (2000) 'GABAB receptor protein and mRNA distribution in rat spinal cord and dorsal root ganglia', *European Journal of Neuroscience* **12**: 3201–3210.

Tremblay, E., Ben-Ari, Y. and Roisin, M.P. (1995) 'Different GABAB-mediated effects on protein kinase C activity and immunoreactivity in neonatal and adult rat hippocampal slices', *J. Neurochem.* **65**: 863–870.

Vergnes, M., Boehrer, A., Simler, S., Bernasconi, R. and Marescaux, C. (1997) 'Opposite effects of $GABA_B$ receptor antagonists on absences and convulsive seizures', *Eur. J. Pharmacol.* **332**: 245–255.

Wagner, P.G. and Dekin, M.S. (1993) '$GABA_B$ receptors are coupled to a barium-insensitive outward rectifying potassium conductance in premotor respiratory neurons', *J. Neurophysiol.* **69**: 286–289.

White, J.H., Wise, A., Main, M.J., Green, A., Fraser, N.J., Disney, G.H., Barnes, A.A., Emson, P., Foord, S.M. and Marshall, F.H. (1998) 'Heterodimerization is required for the formation of a functional $GABA_B$ receptor', *Nature* **396**: 679–682.

White, J.H., Wise, A. and Marshall, F.H. (1999) 'Analysis of the $GABA_B$ heterodimer: interacting protein partners as revealed by two hybrid studies', *Soc. Neurosciences*, Abstract 681.27.

Wiesenfeld Hallin, Z., Aldskogius, H., Grant, G., Hao, J.X., Hokfelt, T. and Xu, X.J. (1997) 'Central inhibitory dysfunctions: Mechanisms and clinical implications', *Behavioral and Brain Sciences* **20**: 420–430.

Wu, L.G. and Saggau, P. (1995) 'GABAB receptor-mediated presynaptic inhibition in guinea-pig hippocampus is caused by reduction of presynaptic Ca2+ influx', *Journal of Physiology* **485**: 649–657.

Xi, Z.X. and Stein, E.A. (1999) 'Baclofen inhibits heroin self-administration behavior and mesolimbic dopamine release', *J. Pharmacol. Exptl. Ther.* **290**: 1369–1374.

Xu, J. and Wojcik, W.J. (1986) 'Gamma aminobutyric acid B receptor-mediated inhibition of adenylate cyclase in cultured cerebellar granule cells: blockade by islet-activating protein', *J. Pharmacol. Exp. Ther.* **239**: 568–573.

Yu, Z.F., Cheng, G.J. and Hu, B.R. (1997) 'Mechanism of colchicine impairment on learning and memory, and protective effect of CGP36742 in mice', *Brain Research* **750**: 53–58.

Part 5

Glutamate and GABA transporters

Chapter 13

Structure, function and regulation of glutamate transporters

Line M. Levy

Introduction

The amino acid glutamate is the major excitatory neurotransmitter in the central nervous system (CNS) (Collingridge and Lester, 1989; Fonnum, 1984; Ottersen and Storm-Mathisen, 1984). It is released Ca^{2+}-dependently by exocytosis from both neurons and astrocytes (Bezzi *et al.*, 1998). In addition, both neurons and glial cells can release glutamate in Ca^{2+}-independent ways (Nicholls and Attwell, 1990; Szatkowski *et al.*, 1990; Jabaudon *et al.*, 1999). Glutamate in the extracellular fluid acts on glutamate receptors in the plasma membrane of neurons (Ozawa *et al.*, 1998) and glial cells (Steinhäuser and Gallo, 1996). Glutamate must be removed rapidly from the extracellular fluid so that the stimulation of glutamate receptors reflects release. In addition, prolonged exposure of neurons to high concentrations of glutamate is harmful (Choi *et al.*, 1987). Since there are no extracellular enzymes that can metabolise glutamate, removal of glutamate from the extracellular fluid must be mediated by uptake into surrounding cells. Glutamate transporters are membrane proteins localised in both neurons and glial cells (for review, see Danbolt, 1994; Danbolt *et al.*, 1998). They use the transmembrane electrochemical gradients of Na^+, K^+ and H^+ to drive the uptake of glutamate into the cell (Billups *et al.*, 1998a), thereby helping to terminate excitatory synaptic transmission and preventing the extracellular glutamate concentration from rising to toxic levels.

Transmembrane glutamate transport in the CNS

Introduction to Na^+-dependent glutamate transport across plasma membranes

Fifty years ago Krebs and co-workers incubated brain slices in phosphate buffered saline and demonstrated that the slices could concentrate glutamate from the medium by an energy-dependent process (Stern *et al.*, 1949). After the excitatory action of glutamate was discovered (Curtis *et al.*, 1959, 1960; Hayashi, 1954), it was suggested that the Na^+-dependent high-affinity uptake of glutamate (Balcar and Johnston, 1972; Logan and Snyder, 1971) was involved in the termination of the putative transmitter activity of glutamate (Johnston, 1981). Perfusion dialysis revealed that the extracellular concentration of glutamate ($[glu]_o$) is in the low micromolar range (Hamberger *et al.*, 1983). Combined with the knowledge that the glutamate content in adult brain tissue is high (about 10 mmol/kg wet weight) (Agrawal *et al.*, 1966), it could be deduced that the intracellular concentration of glutamate ($[glu]_i$) is more than a thousand times higher than $[glu]_o$ (Hamberger *et al.*, 1983). The pharmacological profile of Na^+-dependent glutamate uptake in different brain regions revealed heterogeneity of the uptake systems (Balcar and Li, 1992; Ferkany and Coyle, 1986; Fletcher and Johnston, 1991; Rauen *et al.*, 1992; Robinson *et al.*, 1991, 1993). From 1992 onwards, several subtypes of Na^+-dependent glutamate transporters were identified by DNA cloning (see below). The proteins could now be expressed in eukaryotic cells, and peptides corresponding to parts of the deduced sequences were synthesised and used to raise antibodies selectively recognising the different transporter subtypes.

The topic of this review is Na^+-dependent glutamate transporters in the plasma membrane. However, some other glutamate transporters in the CNS will be briefly mentioned in the following sections.

Cystine-glutamate exchange across plasma membranes is Na^+-independent

Glutamate can be transported across the plasma membrane in exchange for cystine (Cho and Bannai, 1990). The high transmembrane gradient of glutamate (established by Na^+-dependent glutamate uptake) favours net uptake of cystine and release of glutamate under physiological conditions. Cystine can in turn be reduced to cysteine, which is rate-limiting in the synthesis of the antioxidant glutathione. Although cysteine can be taken up into the cell by a glutamate-insensitive transport mechanism, the cystine–glutamate exchange seems to play a crucial role for the availability of cysteine and thereby for glutathione synthesis: it has been shown that the glutathione level of retinal Müller glial cells is dependent on the high-affinity Na^+-dependent uptake of glutamate (Reichelt *et al.*, 1997), in accordance with glutamate transporters being responsible for maintaining the plasma membrane gradient of glutamate that drives uptake of cystine into the cell. A plasma membrane cystine/glutamate exchange transporter has been cloned from activated macrophages (Sato *et al.*, 1999).

Mitochondrial glutamate transporters

Most of the enzymes metabolising glutamate are located in mitochondria. In agreement with this, mitochondria possess at least two different mechanisms for taking up glutamate: the first is a proton-glutamate symporter and the second is a glutamate/aspartate antiporter (Dennis *et al.*, 1976; Kramer and Heberger, 1986; Kramer and Palmieri, 1989; Minn and Gayet, 1977). For review, see Sluse, 1996.

Synaptic vesicle glutamate transporter

Glutamate is transported from cytosol into synaptic vesicles and the electrochemical proton gradient driving this uptake is established by an H^+-ATPase localised in the membrane of synaptic vesicles. Glutamate uptake into synaptic vesicles is chloride dependent (Fykse *et al.*, 1989; Naito and Ueda, 1985), but the exchange stoichiometry of the uptake process is unknown. The K_m value for glutamate was determined to be 1.6 mM (Naito and Ueda, 1985). The uptake of glutamate into synaptic vesicles is regulated by endogenous protein factors (for review, see Fonnum *et al.*, 1998; Özkan and

Ueda, 1998). A cloned transporter (Ni *et al.*, 1994) displays all the characteristics of vesicular glutamate uptake (Bellocchio *et al.*, 2000; Takamori *et al.*, 2000).

Na^+-dependent glutamate transporters

Cloning of Na^+-dependent glutamate transporters

Five different Na^+-dependent glutamate transporters have so far been cloned: GLAST (EAAT1) (Storck *et al.*, 1992; Tanaka, 1993), GLT (EAAT2) (Pines *et al.*, 1992), EAAC (EAAT3) (Kanai and Hediger, 1992), EAAT4 (Fairman *et al.*, 1995) and EAAT5 (Arriza *et al.*, 1997). GLAST and GLT were originally cloned from rat while EAAC1 was cloned from rabbit. EAAT4 and EAAT5 were both cloned from man. Corresponding cDNAs have later been cloned and sequenced from different species, including man (Arriza *et al.*, 1994; Kanai *et al.*, 1994; Kawakami *et al.*, 1994; Manfras *et al.*, 1994; Shashidharan *et al.*, 1994a, 1994b; Shashidharan and Plaitakis, 1993), mouse (Kirschner *et al.*, 1994a, 1994b; Maeno-Hikichi *et al.*, 1997; Mukainaka *et al.*, 1995; Sutherland *et al.*, 1995), rat (Bjørås *et al.*, 1996; Kanai *et al.*, 1995; Velaz-Faircloth *et al.*, 1996), cow (Inoue *et al.*, 1995), nematode (Kawano *et al.*, 1996) and insect (Besson *et al.*, 1999; Donly *et al.*, 1997; Seal *et al.*, 1998). Within a single species, the amino acid sequence homology of GLT, GLAST, EAAC and EAAT4 is about 50 per cent, whereas EAAT5 shares only 30–40 per cent of the amino acids with the four other transporters. When each of the five subtypes is compared to the equivalent protein of another species, there is about 90 per cent amino acid sequence homology.

Localisation of glutamate transporters

The distribution of the subtypes of glutamate transporters in the CNS is highly differentiated and most of them are also present in tissues outside the CNS (Danbolt, 2000; Danbolt *et al.*, 1998).

GLAST is expressed in all parts of the mammalian CNS; it is exclusively in astroglia (Lehre *et al.*, 1995; Schmitt *et al.*, 1997) and in highest concentrations where glial cells face neuropil (Chaudhry *et al.*, 1995). GLAST is the most abundant glutamate transporter in the molecular layer of cerebellum (Lehre and Danbolt, 1998; Lehre *et al.*, 1995) and in the retina, where it is expressed both in Müller cells and astrocytes (Derouiche and Rauen, 1995; Lehre *et al.*, 1997; Rauen *et al.*, 1996). An initial paper (Rothstein *et al.*, 1994) reported that GLAST was present in neurons as well as in astroglia, but this has later been corrected by the same authors (Ginsberg *et al.*, 1995; Rothstein *et al.*, 1995).

GLT is the most abundant glutamate transporter in the forebrain (Haugeto *et al.*, 1996; Lehre and Danbolt, 1998). Immunocytochemically it has only been detected in astrocytes throughout the normal postnatal and adult CNS, at highest concentrations in hippocampus and cerebral cortex (Danbolt *et al.*, 1992; Lehre *et al.*, 1995; Levy *et al.*, 1993; Ullensvang *et al.*, 1997). GLT and GLAST are expressed in the same astrocytes, but in different proportions in different parts of the brain (Chaudhry *et al.*, 1995; Lehre *et al.*, 1995). Also the density of GLT is highest in the parts of the astrocytic plasma membrane facing the neuropil (Chaudhry *et al.*, 1995). In contrast to the GLT protein, GLT mRNA is present in both astrocytes and several populations of neurons (Berger and Hediger, 1998; Schmitt *et al.*, 1996; Torp *et al.*, 1994, 1997), but the rationale for this is not known. Although neurons do not normally express the GLT protein in the mature brain, neonatal neurons may express GLT after hypoxia-ischaemia (Martin *et al.*, 1997). Further, postnatal neurons in culture may express GLT protein (Brooks-Kayal *et al.*, 1998) and GLT protein is found in embryonic neurons, both in culture (Meaney *et al.*, 1998; Mennerick *et al.*, 1998; Plachez *et al.*, 2000) and in vivo (Northington *et al.*, 1999). In the retina, GLT is not expressed in glial cells but in some bipolar cells and amacrine cells (Euler and Wassle, 1995; Rauen and Kanner, 1994; Rauen *et al.*, 1996).

EAAC is widely distributed in the brain (Rothstein *et al.*, 1994; Conti *et al.*, 1998), but is present at low concentration (Haugeto *et al.*, 1996). The EAAC protein (Kugler and Schmitt, 1999; Rothstein *et al.*, 1994) and its mRNA (Kanai *et al.*, 1995; Kugler and Schmitt, 1999; Torp *et al.*, 1997; Velaz-Faircloth *et al.*, 1996) have been detected, respectively by immunocytochemistry and *in situ* hybridisation, in most glutamatergic neurons and some GABAergic neurons. Ultrastructural studies showed that EAAC immunoreactivity was restricted to postsynaptic elements of neurons and to some astrocytes (Conti *et al.*, 1998). In the retina, EAAC is found in horizontal cells, amacrine cells and ganglion cells (Rauen *et al.*, 1996; Schultz and Stell, 1996).

The EAAT4 protein is predominantly expressed in the molecular layer of the cerebellum, with a much lower expression in the forebrain. More precisely EAAT4 is localised in plasma membranes of Purkinje cell dendrites and spines (Dehnes *et al.*, 1998; Furuta *et al.*, 1997a; Itoh *et al.*, 1997; Nagao *et al.*, 1997; Tanaka *et al.*, 1997a; Yamada *et al.*, 1996).

EAAT5 is selectively expressed in the retina (Arriza *et al.*, 1997). Its main localisation is to Müller cells, but some may also be expressed in neurons (Eliasof *et al.*, 1998).

Pharmacology of glutamate transporters

The physiological substrates for glutamate transporters are L-glutamate, L-aspartate, D-aspartate (Arriza *et al.*, 1994, 1997; Balcar and Johnston, 1972; Danbolt, 1994; Fairman *et al.*, 1995; Hashimoto and Oka, 1997), as well as some sulphur containing amino acids (Bouvier *et al.*, 1991; Wilson and Pastuszko, 1986). The transport is stereospecific as the affinity for D-glutamate is much lower than for L-glutamate (Arriza *et al.*, 1994; Balcar and Johnston, 1972). In contrast both L-aspartate and D-aspartate are transported with similar affinities and with K_m values within the range of K_m values reported for L-glutamate (1–100 μM) (Arriza *et al.*, 1994; Balcar and Johnston, 1972; Fairman *et al.*, 1995).

Glutamate transporter substrates share certain features: They are α-amino-dicarboxylic acids where the carboxyl groups are separated by 2–3 methylene groups. One exception is β-glutamate, which can be transported although the amino group is at carbon 3 (Balcar and Johnston, 1972). The amino group can be part of a cyclic structure, as in L-trans-pyrrolidine-2,4-dicarboxylate (L-trans-PDC) (Bridges *et al.*, 1991), but no other modifications of this group are tolerated (Bridges *et al.*, 1993, 1999).

Some sulphur-containing analogues of glutamate (L-homocysteic acid (HCA) and L-homocysteinesulphinic acid (HCSA)) and aspartate (L-cysteic acid (CA), L-cysteinesulphinic acid (CSA)) have been proposed as neurotransmitter candidates (for review, see Thompson and Kilpatrick, 1996). Uptake of these amino acids has been demonstrated in synaptosomes and/or in retinal Müller cells (Bouvier *et al.*, 1991; Wilson and Pastuszko, 1986), where also the putative neurotoxin S-sulpho-L-cysteine (SC) is transported (Bouvier *et al.*, 1991).

Kainate (KA) and the derivative dihydrokainate (DHK) are not transported, but competitively inhibit glutamate uptake (Johnston *et al.*, 1979). KA and DHK showed regional selectivity with respect to the inhibitory effect on glutamate uptake (Ferkany and Coyle, 1986; Killinger *et al.*, 1996). This has partly been explained by the differential distribution of glutamate transporters (see above) and the observation that DHK and KA are much more potent inhibitors of glutamate uptake mediated by GLT than uptake mediated by other cloned glutamate transporters (Arriza *et al.*, 1994). DHK has therefore been used to study the properties of GLT in isolation. It should, however, be noted that there is an as yet molecularly unidentified glutamate transporter in glutamatergic nerve endings (Gundersen *et al.*, 1993), and this transporter may also be sensitive to KA and DHK (for review, see Danbolt, 2001).

Another non-transported competitive inhibitor has recently been synthesised (Lebrun *et al.*, 1997): DL-threo-beta-benzyloxyaspartate (DL-TBOA) is a novel derivative of DL-threo-beta-hydroxyaspartate, and it is a potent inhibitor of EAAT1 and EAAT2 (Shimamoto *et al.*, 1998).

Electrophysiological characteristics of glutamate transporters

Uptake of glutamate into the cell is driven by the transmembrane electrochemical gradients of Na^+, K^+ and H^+. Using electrophysiological techniques, the uptake process can be detected as an inward current (Brew and Attwell, 1987), partly because of the thermodynamically coupled transport of the ions mentioned above and partly due to the activation of an anion conductance.

Stoichiometry

The stoichiometry of glutamate uptake, here defined as the number of different ions transported across the membrane per glutamate, determines the steady state glutamate gradient maintainable by glutamate transporters under different ionic conditions. It follows that knowing the stoichiometry of glutamate transporters makes it possible to predict under which ionic conditions $[glu]_o$ will rise to neurotoxic levels. The stoichiometry of glutamate uptake also determines the energy consumption of this process.

Most of the energy is gained from the co-transport of Na^+, which is the main extracellular cation. Therefore it has been crucial to determine the number of sodium ions translocated with each glutamate. Studying the dependence of the reversal potential on various ions established that three Na^+ ions and two net charges are moved into the cell with each glutamate transported by human EAAC (Zerangue and Kavanaugh, 1996a). The same results were obtained for rat GLT-1 (Levy *et al.*, 1998).

Using membrane vesicles it was demonstrated that glutamate uptake is dependent on internal K^+ (Kanner and Sharon, 1978). Whole-cell clamp experiments on retinal Müller cells showed that one K^+ ion is countertransported per glutamate molecule (Amato *et al.*, 1994b; Barbour *et al.*, 1988).

As glutamate is taken up, a fall in pH inside the cell and a rise in pH outside the cell can be detected (Amato *et al.*, 1994a; Bouvier *et al.*, 1992; Erecinska *et al.*, 1983). Based on anion substitution experiments this was first attributed to the countertransport of OH^- (Bouvier *et al.*, 1992). However, a glutamate gated anion conductance was discovered in the same cell type (see below), leading to reinterpretation of previous data to be equally compatible with cotransport of an H^+ into the cell (Billups *et al.*, 1996; Eliasof and Jahr, 1996). Zerangue and Kavanaugh found support to the idea of H^+ being co-transported as they observed that human EAAC can transport both anionic and neutral/protonated cysteine (Zerangue and Kavanaugh, 1996b).

In conclusion, the transport of one glutamate anion is coupled to the co-transport of three Na^+ and one H^+ and to the countertransport of one K^+, both in the case of human EAAC (Zerangue and Kavanaugh, 1996a) and rat GLT-1 (Levy *et al.*, 1998).

Anion conductance

Substrate binding activates anion currents in glutamate transporters. A glutamate-activated chloride current with the pharmacology of glutamate transporters was observed in cone photoreceptors (Eliasof and Werblin, 1993; Picaud *et al.*, 1995; Sarantis *et al.*, 1988; Tachibana and Kaneko, 1988) and in bipolar cells (Grant and Dowling, 1995). Soon after, it turned out that a significant component of the uptake currents of human GLAST, human GLT and human EAAC expressed in oocytes is due to activation of a reversible anion flux that is not thermodynamically coupled to amino acid transport (Wadiche *et al.*, 1995a). In *Xenopus* oocytes expressing EAAT4 (Fairman *et al.*, 1995) or EAAT5 (Arriza *et al.*, 1997), glutamate elicited a current predominantly carried by chloride ions. The selectivity sequence of the ligand-activated conductance was found to be $NO_3^- > I^- > Br^- > Cl^- > F^-$ (Wadiche *et al.*, 1995a). In agreement with this, a retinal glia cell glutamate transporter is coupled to an anion conductance (Billups *et al.*, 1996; Eliasof and Jahr, 1996) with the selectivity sequence $SCN^- > ClO_4^- > NO_3^- > Cl^-$ (Billups *et al.*, 1996) and a postsynaptic excitatory amino acid transporter in Purkinje cells to an anion conductance with a sequence of relative current amplitude $ClO_4^- > Cl^-$ (Takahashi *et al.*, 1996) and $SCN^- > NO_3^- > I^- > Cl^-$ (Kataoka *et al.*, 1997). A kinetic scheme including one anion conducting state per transport cycle was proposed for transport in Müller cells (Billups *et al.*, 1996, 1998b), whereas a recently constructed cyclic kinetic model of glutamate transport in Purkinje neurons proposes two separate open-channel states (Otis and Jahr, 1998). In addition to glutamate-activated anion conductances, glutamate transporters exhibit constitutive anion leak conductances in the absence of substrates (Bergles and Jahr, 1997; Levy *et al.*, 1998; Otis and Jahr, 1998). Plasma membrane glutamate transporter proteins may thus mediate both transporter- and channel-like modes of permeation, providing a potential mechanism for dampening cell excitability, in addition to removal of transmitter. Further, the anion conductance can be modulated, for instance by the presence of zinc (Spiridon *et al.*, 1998; Vandenberg *et al.*, 1998).

Topology of glutamate transporters

The cloned glutamate transporters show about 50 per cent amino acid homology, and the hydropathy patterns are almost identical. Hydrophobicity analysis (Kanai *et al.*, 1993) clearly predicts six α-helical transmembrane (TM) segments in the amino-terminal half of the glutamate transporters (Kanai and Hediger, 1992; Pines *et al.*, 1992;

Storck *et al.*, 1992). However, hydrophobicity analysis of the more C-terminal portions of the proteins has been inconclusive, and several very different models have been proposed recently (Grunewald *et al.*, 1998; Grunewald and Kanner, 2000; Seal *et al.*, 2000; Slotboom *et al.*, 1996; Wahle and Stoffel, 1996). There is as yet no consensus, but the two more favoured models (Grunewald and Kanner, 2000; Seal *et al.*, 2000) have several features in common. Both models contain eight TMs and both suggest an unusual topology of the highly hydrophobic stretch between TMs 6 and 8. The unusual features include superficial membrane-associated linker regions and re-entrant loops (loops that go into the membrane without going through it). The models disagree on the location of these elements and on the location of TM7.

Intracellular localisation of the amino- and carboxy termini was suggested by hydropathy analysis (Eliasof *et al.*, 1998; Kanai and Hediger, 1992; Pines *et al.*, 1992; Storck *et al.*, 1992) and demonstrated by immunocytochemistry for GLAST and GLT (Lehre *et al.*, 1995) as well as EAAT4 (Dehnes *et al.*, 1998).

Identification of functionally important amino acid residues

A central question to answer is which functions are associated with which parts of the glutamate transporters. The main approaches used to unravel this are sequence comparisons, site-directed mutagenesis and generation of chimeras (combining parts of different transporters).

The highly conserved amino acid sequence AA(I/V)FIAQ (amino acids 409–415 in EAAT1 and 407–413 in GLT-1), which is located in one of the hydrophobic parts of the glutamate transporters, probably has an important function. A 76-residues-long segment corresponding to residues 364–439 of GLT-1 (and thereby comprising AA(I/V)FIAQ) contains the site responsible for the unique sensitivity of GLT type transporters to kainate and dihydrokainate (Vandenberg *et al.*, 1995). A marked reduction in sensitivity to dihydrokainate is observed when serine residues 440 and 443 of GLT-1 are mutated to glycine and glutamine, which, respectively, occupy these positions in the other homologous glutamate transporters (Zhang and Kanner, 1999).

Serine-440 seems to have a function in sodium binding, as mutation of serine-440 to glycine enables not only sodium but also lithium ions to drive a net influx of acidic amino acids (Zhang and Kanner, 1999).

Site directed mutagenesis has shown that histidine residue 326 (located in TM6) is required for the intrinsic activity of GLT-1 (Zhang *et al.*, 1994). This positively charged TM residue is shared by all known members of the family. As histidine residues have been implicated in proton translocation, the authors proposed that this residue could be involved in the pH-changing effect of the transporter.

Out of five conserved, negatively charged TM residues, aspartate 398, glutamate 404 and aspartate 470 are critical for glutamate transport (Pines *et al.*, 1995). Negative charge is not enough, since the mutated forms D398E (aspartate 398 mutated to glutamate), E404D and D470E are all unable to transport glutamate. However, binding of glutamate is not impaired in E404D, and this modified protein can still transport D- and L-aspartate and bind them at increased affinity. The site thus appears involved in substrate recognition as well as translocation. Subsequently, glutamate 404 has been shown (Kavanaugh *et al.*, 1997) to be essential for the building up of a transmembrane glutamate gradient, this important function being dependent on a single methylene group: the mutant E404D cannot sustain sodium/potassium coupled net transport, but does catalyse sodium-dependent exchange of excitatory amino acids and shows sodium- and amino- acid-dependent chloride conductance.

Tyrosine 403 is also important for potassium coupling, and mutation to phenylalanine (Y403F) resulted in an electroneutral obligate exchange mode of glutamate transport. This mutation also caused an approximately eightfold increase in the apparent Na^+ affinity, with no change in the apparent affinity for L-glutamate or D-aspartate (Zhang *et al.*, 1998). Consistent with its role as a structural determinant of the potassium binding site, tyrosine 403 is alternately accessible from either side of the membrane (Zarbiv *et al.*, 1998).

GLAST residues tyrosine 405 and arginine 479, conserved in glutamate and dicarboxylate transporters, are essential for transport, Y405F and R479T being inactive (Conradt and Stoffel, 1995). On the other hand, arginine residues 122 and 280 of GLAST, situated in the first and second putative cytosolic loops close to TM3 and TM5, respectively, appear to be subtly involved in substrate recognition: the mutants R122I and R280V transport glutamate normally, but show increased affinity for aspartate and D,L-threo-3-hydroxy aspartate.

The region from amino acid 442 to 499 of GLT (corresponding to the linker and the C-terminal TM in recent topology models) seems to play an important role in defining functional differences, as the differences in chloride permeability, between EAAT1 and EAAT2 (Mitrovic *et al.*, 1998). However, it is not clear whether the anion conductance is conferred by the cloned glutamate transporters themselves or by ion channels associated with them. A recent finding is in favour of the latter: The EAAT5 sequence contains a carboxy-terminal motif, ETNV, identified previously in *N*-methyl-D-aspartate receptors and potassium channels. The motif E(S/T)XV is shown to confer interactions with a family of synaptic proteins that promote ion channel clustering (Arriza *et al.*, 1997).

Role of glutamate transporters in normal synaptic transmission

Glutamate transporters represent the only mechanism for removal of excitatory amino acids from the extracellular fluid in the brain, and their role for long-term maintenance of a low extracellular concentration of glutamate is well documented (for review see Danbolt, 1994; Robinson and Dowd, 1997). However, it is not yet clear to what extent glutamate transporters contribute to shaping the fall in the free extracellular concentration of glutamate occurring in the synaptic cleft the first few milliseconds after glutamate release. According to mathematical models, passive diffusion alone can reduce the glutamate concentration in the synaptic cleft rapidly (Barbour and Hausser, 1997; Clements, 1996; Holmes, 1995; Kleinle *et al.*, 1996). Since the cycling time of glutamate transporters is estimated to be 50–100 msec (Wadiche *et al.*, 1995b), uptake of glutamate may be too slow to contribute to the rapid fall in glutamate concentration. However, if many copies of glutamate transporters are located close to the release sites, binding (followed by uptake) of glutamate could effectively reduce the concentration of free glutamate in the cleft on a 100 μs timescale (Tong and Jahr, 1994).

GLT and GLAST are intermingled in the astrocytic membranes, the concentrations being highest in the parts of the membranes facing neuropil (Chaudhry *et al.*, 1995). Lehre and Danbolt (1998) conclude that these glutamate transporters are present at sufficiently high average densities to support the notion (Diamond and Jahr, 1997; Tong and Jahr, 1994) that they can contribute to glutamate inactivation on the short timescale by binding rather than by transport. However, their importance in the control of extrasynaptic and intersynaptic glutamate diffusion is likely to vary considerably between different synapses because the transporters are predominantly associated with astrocytes and thereby not evenly distributed in the extracellular space (Lehre and Danbolt, 1998).

The role of EAAC is not clear at the moment, since the concentration of EAAC may be one or two orders of magnitude lower than the concentrations of GLT and GLAST (Haugeto *et al.*, 1996; Lehre and Danbolt, 1998). In addition, EAAC does not appear to be concentrated at synapses (Coco *et al.*, 1997). EAAC is present in cerebellar Purkinje cells (Rothstein *et al.*, 1994; Conti *et al.*, 1998) together with EAAT4, which is concentrated in the glia-covered parts of the membranes of dendrites, with the highest densities at the spines and thinner dendrites close to synapses (Danbolt *et al.*, 1998; Dehnes *et al.*, 1998). It has been established that functional postsynaptic glutamate transporters in Purkinje cells contribute to terminating synaptic transmission in the immature brain (Takahashi *et al.*, 1996).

The protein mediating the uptake of excitatory amino acids into excitatory nerve terminals (Gundersen *et al.*, 1993, 1996) remains to be identified molecularly. The possibility exists that this may be localised closer to the release sites than the so far identified glial and postsynaptic glutamate transporters.

Role of glutamate transporters in pathological conditions

In view of the important functions of glutamate transporters in normal synaptic transmission one would expect malfunctioning of these proteins to play a role in neuropathology. In animal models, knocking out glutamate transporters leads to disease (see below). Lack of a glutamate transporter protein has not yet been demonstrated in any human disease. However, malfunctioning of glutamate transporters seems to be involved in common conditions like neurotoxicity and ischaemia.

The important role of glutamate transporters in limiting glutamate neurotoxicity has been known for several decades: glutamate is toxic to nerve cells in high concentrations (Choi *et al.*, 1987) and glutamate transporters represent the only mechanism for its removal from the extracellular fluid. The challenge over the last few years has been to identify the role of each subtype of glutamate transporters. One approach to this is to study the physiological and anatomical abnormalities that result when their expression is reduced or eliminated. In agreement with GLT being the most abundant glutamate transporter in the mammalian forebrain (Haugeto *et al.*, 1996; Lehre and Danbolt, 1998), homozygous mice deficient in GLT show lethal spontaneous seizures and increased susceptibility to acute cortical injury (Tanaka *et al.*, 1997b). Mice deficient in EAAC develop dicarboxylic aminoaciduria and behavioural abnormalities, but no neurodegeneration (Peghini *et al.*, 1997). GLAST is required for normal signal transmission between photoreceptors and bipolar cells, and both GLAST and GLT play neuroprotective roles during ischaemia in the retina as retinal damage after ischaemia is exacerbated in GLAST deficient mice, and a similar but milder effect is observed in GLT-deficient mice (Harada *et al.*, 1998). Inactivating the mouse GLAST gene showed that GLAST plays active roles both in the cerebellar climbing fibre synapse formation and in preventing excitotoxic cerebellar damage after acute brain injury (Watase *et al.*, 1998).

Although the glutamate transporters protect against toxic extracellular levels of excitatory amino acids they can also do the opposite; namely, release the same amino acids. This reversed operation may occur during periods of impaired blood supply (ischaemia) or oxygen supply (hypoxia) leading to lack of ATP and consequently to rundown of ion gradients. It was early suggested that the ischaemia-mediated release, rather than being exocytotic, results mainly from a reversal of the Na^+-dependent high-affinity glutamate transporter in the plasma membrane (Kauppinen *et al.*, 1988; Nicholls and Attwell, 1990; Torgner and Kvamme, 1990). With more knowledge about the stoichiometry of several glutamate transporters (Levy *et al.*, 1998; Zerangue and Kavanaugh, 1996a), it is possible to estimate under which ionic conditions they will be capable of maintaining the extracellular concentration of glutamate below neurotoxic levels (see discussion in Levy *et al.*, 1998). If all neuronal and glial glutamate transporters have the same ion stoichiometry, neurons are better candidates than glial cells for releasing glutamate via glutamate transporters early in ischaemia, partly because the pre-ischaemic intracellular concentration of glutamate is higher in neurons than in glial cells (Storm-Mathisen *et al.*, 1992). This idea is supported by the observation that ATP-depletion decreases glutamate-like immunoreactivity primarily in axonal terminal-like structures, suggesting a release primarily from terminals (Madl and Burgesser, 1993). Further endogenous glutamate is reduced in neuronal perikarya but increased in glia during ischaemia (Torp *et al.*, 1991, 1993). Finally, a recent study of complete energy failure in hippocampal slices demonstrates that glutamate release is largely by reversed operation of neuronal glutamate transporters, and that it plays a key role in generating the anoxic depolarisation that abolishes information processing in the central nervous system a few minutes after the start of ischaemia (Rossi *et al.*, 2000). During reperfusion after a period of ischaemia, the oxidative stress is high. The effects of oxidation on glutamate transporters (see below) are therefore relevant for understanding the role of glutamate transporters during and after reperfusion, when the extracellular glutamate concentration approaches normal values, but remains elevated despite normalisation of ATP-levels and ion gradients.

Ischaemia in turn leads to changes in glutamate transporter expression. The expression of GLT protein in the rat hippocampus is decreased by 20 per cent 6 hours after transient forebrain ischaemia (Torp *et al.*, 1995). In the striata of newborn pigs total levels of GLT were reduced by 15 per cent 24 hours after hypoxia-ischaemia, and while GLT disappeared in astrocytes it appeared in neurons (Martin *et al.*, 1997). Bilateral occlusion of the common carotid arteries for 10 minutes results in a 36–46 per cent reduction in GLT expression between days 1–3 and a 42–68 per cent decrease in EAAC expression between days 1–7 in the gerbil hippocampus (Raghavendra Rao *et al.*, 2000). Similarly the levels of EATT4 and GLAST decrease in the human neonatal cerebellum after hypoxia-ischaemia (Inage *et al.*, 1998). In the cerebral cortex expression of GLAST mRNA in the penumbra is reduced 12 hours after ischaemia (Yin *et al.*, 1998) but increased 72 hours after ischaemia (Yan *et al.*, 1998; Yin *et al.*, 1998). Moreover, experimental occlusion of the central retinal artery followed by reperfusion for 48 hours resulted in degeneration of neurons and a marked increase in GLAST mRNA expression (Otori *et al.*, 1994). The latter four studies are in agreement with studies of uptake activity in mixed neuronal–glial cell cultures after ischaemia

(Sher and Hu, 1990). The reported effects of ischaemia on expression and activity of different glutamate transporters are diverse, and although some of this diversity may reflect actual differences in regulation of the transporter subtypes it should be kept in mind that different models for ischaemia have been used so that the results cannot be compared directly.

Regulation of glutamate transport

Glutamate transport can be regulated at many different levels: transporter mRNA synthesis, stability and alternative splicing as well as transporter protein synthesis, degradation, cell surface expression and activity. In addition, modulation of the transmembrane electrochemical gradients driving the transport process will strongly influence glutamate uptake (Billups *et al.*, 1998a; Levy *et al.*, 1998), as will the presence of competing substrates, both endogenous and exogenous.

Alternative splicing of glutamate transporter proteins

Some primary gene transcripts can be spliced in different ways to produce distinct RNA molecules, each encoding a different protein. Alternative splicing of the 5′ end of the EAAT2 cDNA has been observed, but the potential functional meaning of this remains to be demonstrated (Munch *et al.*, 1998). Both brain and liver GLT-1 possess two types of 3′-ends. Although functional properties are not changed by the alteration of N-termini and C-termini when expressed in *Xenopus laevis* oocytes, co-expression of two liver type GLT-1 with different C-termini (mGLT-1A and mGLT-1B) has been found to result in the increase in V_{max} of transport without changing K_m. These results suggest the tissue-specific alternative splicing at 5′-ends of GLT-1 messages and the interesting association of spliced variants with different C-termini (Utsunomiya-Tate *et al.*, 1997).

Soluble factors from neurons regulate glial glutamate transporter expression

Lesioning of glutamatergic pathways in the CNS leads to reduced expression of GLT and GLAST in the target area. By measuring immunoreactivity it was demonstrated that expression of GLT and GLAST in striatum was reduced 7 days after lesions of the rat cerebral cortex (Ginsberg *et al.*, 1995; Levy *et al.*, 1995), whereas EAAC expression was unaltered (Ginsberg *et al.*, 1995). Transection of another glutamatergic pathway, fimbria-fornix, likewise leads to reduced expression of GLT and GLAST but not EAAC (Ginsberg *et al.*, 1995, 1996). A likely interpretation of these findings is that neurons influence transporter expression in glial cells. To shed light on this, cultures of astrocytes have been used as model systems for the regulation of the expression of GLT and GLAST (for review, see Gegelashvili *et al.*, 2000; Gegelashvili and Schousboe, 1998; Sims and Robinson, 1999). Cultured in the absence of neurons astrocytes express GLAST, but very little GLT (Gegelashvili *et al.*, 1996; Kondo *et al.*, 1995). Co-culturing astrocytes with neurons increases the levels of GLAST and induces the expression of GLT (Gegelashvili *et al.*, 1997; Schlag *et al.*, 1998; Swanson *et al.*, 1997). The neuron-dependent upregulation of GLT expression is also observed when pure astrocyte cultures are cultured with conditioned media from neuron cultures (Gegelashvili *et al.*, 1997; Schlag *et al.*, 1998; Swanson *et al.*, 1997), therefore it does not seem to require direct contact between neurons and astrocytes. In contrast, neuron-conditioned media have little effect on the expression of GLAST (Gegelashvili *et al.*, 1997) but induce supporting machinery for differential regulation of GLAST expression via astroglial metabotropic receptors, mGluR3 and mGluR5 (Gegelashvili *et al.*, 2000). The stimulatory factors in neuron-conditioned media are not yet identified. One candidate is glutamate, and modulation of glutamate transporters by glutamate (e.g. via glutamate receptors) might represent a feedback regulatory mechanism for glutamate uptake in the brain. In agreement with this, the astrocytic expression of GLAST is up-regulated by stimulation of the kainate-preferring type of glutamate receptors (Gegelashvili *et al.*, 1996). This stimulation does not up-regulate GLT protein expression (Gegelashvili *et al.*, 1997), although low concentrations of glutamate (1 nM to 10 μM for 48 hours) increased the levels of GLT-1 mRNA in astroglial cells in culture (Thorlin *et al.*, 1998). Thus the neuronal soluble factors do not seem to regulate GLT expression via glutamate receptors, but possibly via the tyrphostin-sensitive receptor tyrosine kinase (Gegelashvili *et al.*, 2000).

Developmental changes in glutamate transporter expression

Glutamate uptake in the rat CNS is low at birth and increases to adult levels during the first few postnatal weeks (Schousboe *et al.*, 1976; Schmidt and Wolf, 1988; Erdö and Wolff, 1990; Collard *et al.*, 1993. This correlates with a strong increase in the uptake of glutamate into synaptic vesicles and with the active period of synaptogenesis (Kish *et al.*, 1989; Christensen and Fonnum, 1991, 1992). In mouse, GLAST mRNA and GLT mRNA increase dramatically the first two postnatal weeks, whereas the increase in EAAC mRNA is more modest (Shibata *et al.*, 1996). Similar results were obtained for murine GLAST and GLT mRNA, and distinct but overlapping embryonic and postnatal patterns of localisation were observed for these two transporter transcripts (Sutherland *et al.*, 1996). Using immunocytochemistry it has been demonstrated that glutamate transporter subtypes are expressed differentially during rat CNS development (Furuta *et al.*, 1997b; Ullensvang *et al.*, 1997). At birth GLT is undetectable whereas GLAST is present at significant concentrations, both in the forebrain and in the cerebellum. GLT is first detected in the forebrain and cerebellum in the second and third week, respectively (Ullensvang *et al.*, 1997). The postnatal patterns of expression resemble those in the adult brain, but in the prenatal brain there are notable exceptions: in the ovine foetal brain, GLT is transiently expressed in various neuronal populations at midgestation (Northington *et al.*, 1998), and in the mouse spinal cord GLT is transiently localised on growing axons before establishing astrocytic expression (Yamada *et al.*, 1998).

Phosphorylation modulates glutamate transporter expression and activity

Phosphorylation can modulate glutamate transport by at least three different mechanisms: it can regulate the total expression of glutamate transporters, their activity and their relative surface expression (by trafficking of already synthesised glutamate transporters between intracellular membranes and the plasma membrane, see below).

Phosphorylation is mediated by protein kinases, and both protein kinase A (PKA) and protein kinase C (PKC) are candidates for phosphorylating glutamate transporters. PKA is activated by cyclic AMP (cAMP), and it is thus of interest to clarify whether the observed effects of cAMP analogues on glutamate transporters (see below) are mediated by PKA. The natural activators of PKC are arachidonic acid (for review, see Khan *et al.*, 1995) and diacylglycerol (DAG). However, for *in vitro* studies it is common to use the potent PKC activator phorbol 12-tetradecanoyl-13-acetate (TPA) [the abbreviation PMA is also frequently encountered].

Dibutyryl cyclic AMP (dbcAMP) is a membrane-permeable, poorly metabolised cyclic AMP (cAMP) analog that activates cAMP dependent protein kinases. Culturing mouse astrocytes in the presence of dbcAMP up-regulates the glutamate uptake activity (Hertz *et al.*, 1978). In agreement with this, long-term treatment with dbcAMP of rat astrocytes in primary culture increases levels of GLAST and GLT-1 mRNA (Eng *et al.*, 1997) as well as GLAST protein expression (Gegelashvili *et al.*, 1996; Schlag *et al.*, 1998; Swanson *et al.*, 1997). In addition, induction of GLT expression by dbcAMP was found in some of these studies (Schlag *et al.*, 1998; Swanson *et al.*, 1997). These effects mimic the effects of neuronal soluble factors on the expression of glial glutamate transporters (see above), but it remains to be demonstrated whether the effect of neuronal soluble factors is actually mediated by an increase in intracellular cAMP.

Regulation of GLT by direct phosphorylation has been suggested but not confirmed: GLT-1 expressed in HeLa cells could be stimulated by TPA, and mutation of serine 113 to asparagine abolished the stimulatory effect of TPA without affecting the levels of expression (Casado *et al.*, 1993). Strikingly different effects were found in Y-79 human retinoblastoma cells as a DHK-sensitive glutamate transporter was strongly inhibited by TPA. TPA increased the K_m of this transporter from 1.7 to 9.2 mM (Ganel and Crosson, 1998). In the most recent studies, modulation of PKC or phosphatase activity had no effect on GLT-1-mediated activity in GLT-1 transfected cell lines. To determine if GLT-1 regulation by PKC is cell-specific, HeLa cells, which endogenously express the EAAC1 subtype of transporter, were stably transfected with GLT-1. Although EAAC1-mediated activity was increased by activation of PKC, there was no evidence for regulation of GLT-1. The authors suggest (Tan *et al.*, 1999) that the PKC-mediated increase in glutamate transport is either due to increased cell surface expression of endogenous EAAC or to the mutant vaccinia virus (Blakely *et al.*, 1991; Fuerst *et al.*, 1986) used to express GLT.

In contrast, regulation of GLAST by direct phosphorylation has been demonstrated: GLAST was expressed in *Xenopus* oocytes and human embryonic kidney cells (HEK293), and in both systems TPA decreased glutamate transport activity to 25 per cent of the initial transport activity. Removal of all predicted PKC sites of wild-type GLAST by site-directed mutagenesis did not abolish inhibition of glutamate transport. Neither the stability nor the subcellular distribution of GLAST seemed to be affected by phosphorylation (Conradt and Stoffel, 1997). TPA produced a significant decrease in D-[^{3}H]aspartate uptake in chick Bergmann glial cell cultures. This effect was dose and time dependent and sensitive to staurosporine, a Ca^{2+}/DAG-dependent PKC inhibitor. Long-term exposure of the culture to TPA resulted in a dramatic fall of the transporter activity and a decrease in the amount of GLAST protein. These findings suggest that PKC is involved in transport modulation and possibly in the regulation of the transporter gene expression (Gonzalez and Ortega, 1997). In cultured Müller glia cells from chick retina, treatment with TPA produced a decrease in D-[^{3}H]aspartate uptake which was reversed by staurosporine and partially by two PKC inhibitors. Long-term treatment with TPA resulted in a drastic decrease in the uptake activity, correlated with a substantial fall in the expression GLAST. These findings suggest that PKC is involved in transport modulation at two different levels: phosphorylation and transporter expression in retinal Müller glial cells (Gonzalez *et al.*, 1999).

There are several plausible physiological links between glutamatergic transmission and the activation of PKC. One example is as follows: in astrocytes, activation of the inositol cycle was observed after exposure to excitatory amino acids (Pearce *et al.*, 1986). Among the glutamate receptors expressed in glial cells are the metabotropic glutamate receptors coupled to phospholipase C (PLC) (Gallo and Russell, 1995). When PLC acts on phosphatidylinositol it cleaves off DAG, which is a natural activator of PKC (Berridge, 1984). DAG also results from the hydrolysis of phosphatidylcholine by PLC or phospholipase D (Exton, 1990). Another example is glutamate stimulating NMDA receptors, thereby causing release of AA (see below), which in turn activates PKC.

Glutamate transporter trafficking is regulated by several mechanisms

Translocation of glutamate transporters between intracellular storage compartments and the plasma membrane can regulate the surface expression of these proteins in minutes, much faster than protein synthesis. Regulated trafficking has not been demonstrated for GLT, but trafficking of EAAC seems to be regulated by several pathways and trafficking of GLAST and EAAT4 by yet other mechanisms.

EAAC, but not GLAST or GLT, is expressed in C6 glioma (Dowd *et al.*, 1996; Palos *et al.*, 1996). By incubating C6 glioma cells with the PKC activator TPA, the effect of phosphorylation on the surface expression of EAAC has been studied (Davis *et al.*, 1998; Dowd and Robinson, 1996; Ganel and Crosson, 1998). Preincubation with TPA caused a time-dependent and concentration-dependent increase in EAAC-mediated L-[^{3}H]glutamate transport activity. Kinetic analyses demonstrated that the increase in transport activity was due to a 2.5-fold increase in V_{max} with no change in K_m (Dowd and Robinson, 1996). A two-minute preincubation with TPA was sufficient to cause more than a twofold increase in transport activity, and the protein synthesis inhibitor cycloheximide had no effect on the increase. These data suggest that this increase is independent of protein synthesis (Dowd and Robinson, 1996). The increase in activity correlated with an increase in cell surface expression of EAAC. Both effects of TPA were blocked by the PKC inhibitor bisindolylmaleimide II (Davis *et al.*, 1998). The trafficking of EAAC may be regulated by two independent signalling pathways since wortmannin, which inhibits phosphatidylinositol 3-kinase-mediated trafficking, decreased L-[^{3}H]-glutamate uptake activity by more than 50 per cent within minutes (Davis *et al.*, 1998). The platelet-derived growth factor (PDGF) increases cell surface expression of EAAC in C6 glioma cells through activation of phosphatidylinositol 3-kinase (Sims *et al.*, 2000). This increase and the increase induced by TPA are not additive, suggesting that the cell surface expression is controlled by two independent but converging pathways (Sims *et al.*, 2000).

Glutamate incubation inducing rapid up-regulation of astrocyte glutamate transport also produces an increase in GLAST expression at the astrocyte cell surface, demonstrated by biotinylation labelling of membrane surface proteins (Duan *et al.*, 1999). This effect is likely to be triggered by increased activity of glutamate transporters as it is mimicked by D-aspartate (which can be transported, or can stimulate NMDA glutamate receptors), blocked in sodium-free medium (inhibiting transport) but not blocked by glutamate receptor antagonists that are not transported.

The cell surface expression of EAAT4 in BT4 glioma cells can also undergo rapid changes (Gegelashvili *et al.*, 2000), and as for GLAST, the transporter substrate (e.g. aspartate or glutamate) seems to trigger these changes.

Arachidonic acid regulates glutamate transport activity

When glutamate activates NMDA receptors, the resulting calcium influx activates phospholipase A_2 which releases arachidonic acid (AA) from membrane phospholipids (Dumuis *et al.*, 1988). The polyunsaturated fatty acid AA is a diffusable signalling molecule. AA was found to reduce glutamate uptake (Chan *et al.*, 1983; Yu *et al.*, 1986; Barbour *et al.*, 1989; Volterra *et al.*, 1992b). This could either be a direct effect of AA or be mediated by AA derivatives (e.g. leukotrienes, prostaglandins or thromboxans). The finding that AA inhibits purified and liposome-reconstituted rat GLT directly from the water phase, and not via incorporation into the phospholipid bilayer (Trotti *et al.*, 1995), shows that this effect neither depends on AA derivatives nor on other proteins. AA inhibition can be mimicked by *cis*-polyunsaturated fatty acids, but not by *trans*-unsaturated and saturated ones (Barbour *et al.*, 1989; Rhoads *et al.*, 1983; Volterra *et al.*, 1992a), suggesting that the relevant molecular feature is the folded carbon chain of AA and its analogues. Further, the carboxyl group must be free (Trotti *et al.*, 1995).

A recent study (Zerangue *et al.*, 1995) revealed differential modulation by AA of human glutamate transporters expressed in oocytes. Micromolar levels of AA inhibited glutamate uptake mediated by human GLAST by reducing the V_{max} by about 30 per cent. In contrast to previously described results, AA increased transport mediated by human GLT by causing the affinity to increase more than twofold. The mechanisms behind the different effects of AA on purified GLT in liposomes and human GLT expressed in oocytes have not been demonstrated, but it should be kept in mind that AA activates PKC (see above) which is present in oocytes.

Application of AA to oocytes expressing rat EAAT4 increased glutamate-induced currents threefold. However, AA did not cause an increase in the rate of glutamate transport or in the chloride current associated with glutamate transport; rather, it activated a proton-selective conductance, revealing a novel action of AA on a glutamate transporter and suggesting a mechanism by which synaptic activity may decrease intracellular pH in neurons where this transporter is localised (Tzingounis *et al.*, 1998).

A postsynaptic excitatory amino acid transporter with chloride conductance is functionally regulated by neuronal activity in cultured cerebellar Purkinje cells, and the effect may be mediated by Ca^{2+}-dependent activation of phospholipase A_2 releasing AA (Kataoka *et al.*, 1997).

Redox modulation of glutamate transporter activity

Redox modulation of glutamate transporters may contribute to physiological regulation of these proteins but also to irreversible damage. Reactive oxygen species include free radicals and non-radicals centred on O_2. They can inhibit glutamate uptake (Berman and Hastings, 1997; Piani *et al.*, 1993; Volterra *et al.*, 1994), both directly and indirectly (for instance via lipid peroxidation).

Several different mechanisms seem to underlie the direct effect of oxidative stress on glutamate transporters. One mechanism is oxidation of cysteine sulphydryl groups by H_2O_2 resulting in reduced transport activity (Muller *et al.*, 1998; Trotti *et al.*, 1997b; Volterra *et al.*, 1994). This effect can be reversed by reducing agents (Muller *et al.*, 1998; Trotti *et al.*, 1997b), *in vivo* candidates for this being (among others) glutathione, thioredoxins and α-lipoic acid (Trotti *et al.*, 1998). These results suggest a transporter modulation by a redox-related event. In contrast, oxidation, and thereby inhibition of glutamate transporters by the strong oxidant peroxynitrite ($ONOO^-$) (Volterra *et al.*, 1994), can only be partially reversed, suggesting other mechanisms in addition to cysteine modulation (Trotti *et al.*, 1997a, 1997b). For review see (Trotti *et al.*, 1998).

Indirect effects of oxidation on glutamate transporters include effects via lipid peroxidation. Recent reports demonstrate that glutamate transport is impaired by 5-hydroxynonenal (HNE), which is an aldehyde generated by peroxidation of membrane lipids (Blanc *et al.*, 1998; Keller *et al.*, 1997a, 1997b; Springer *et al.*, 1997).

The effect of oxidation on glutamate transporters may start vicious circles: reduced uptake of glutamate may lead to increased $[glu]_o$, and high $[glu]_o$ competitively inhibits cystine uptake (see above), leading to further oxidative stress by depletion of intracellular glutathione (Murphy *et al.*, 1989, 1990).

In general, oxidized proteins undergo enhanced degradation (Grune *et al.*, 1997), but effects of oxidation on glutamate transporter protein degradation have not been published.

Zinc modulates glutamate transporter activity

In certain regions of the brain, notably in the hippocampal mossy fibres, zinc ions (Zn^{2+}) are stored in synaptic vesicles with glutamate. When released into the synapse, Zn^{2+} may modulate the activity of various membrane proteins, among these glutamate receptors (Peters *et al.*, 1987; Westbrook and Mayer, 1987) and glutamate transporters (Balcar and Johnston, 1972; Gabrielsson *et al.*, 1986; Spiridon *et al.*, 1998). Zn^{2+} is a noncompetitive, partial inhibitor of glutamate transport by EAAT1 with an IC_{50} value of 9.9 +/− 2.3 μM, but has no effect on glutamate transport by EAAT2 at concentrations up to 300 μM. Zn^{2+} selectively inhibits transport and increases the relative chloride flux through EAAT1, and site-directed mutagenesis revealed that the histidine residues at positions 146 and 156 form part of the Zn^{2+}-binding site (Vandenberg *et al.*, 1998). This modulation is likely to be of physiological relevance as the concentration of Zn^{2+} in the synaptic cleft may be higher than 100 μM (Assaf and Chung, 1984; Spiridon *et al.*, 1998).

pH modulates glutamate transporter activity

Since H^+ is co-transported with glutamate it can be predicted that a decrease in intracellular pH, as well as an increase in extracellular pH, will reduce glutamate uptake. Recent findings indicate that the effect of pH on glutamate transporters is more complex: in retinal glial cells, lowering the extracellular pH inhibits both glutamate uptake and efflux (Billups and Attwell, 1996). In principle, this mechanism may protect the brain during ischaemia, as the low pH occurring during ischaemia (Mutch and Hansen, 1984; Silver and Erecinska, 1992) may inhibit release of glutamate mediated by glutamate transporters (Madl and Burgesser, 1993; Szatkowski *et al.*, 1990). However, this is only valid if the subtype of glutamate transporters that mediate the glutamate release is inhibited by lowering the extracellular pH. At present it is believed that glutamate is released mainly from nerve terminals when energy supply is compromised (Madl and Burgesser, 1993; Torp *et al.*, 1993), whereas the inhibition described above affects a glial glutamate transporter, presumably GLAST.

Concluding remarks

Glutamate transporters play a key role in normal synaptic transmission as well as in pathological conditions like hypoxia and ischaemia. The present knowledge about the regulation of glutamate transporters demonstrates that their function can be regulated at many different levels and by a multitude of factors. However, it is not yet clear to which extent the effects described above contribute to the functional regulation of glutamate transporters under physiological and pathophysiological conditions.

Acknowledgements

I would like to thank Niels Christian Danbolt for carefully reading the manuscript.

References

Agrawal, H.C., Davis, J.M. and Himwich, W.A. (1966) 'Postnatal changes in free amino acid pool of rat brain', *J. Neurochem.* **13**: 607–615.

Amato, A., Ballerini, L. and Attwell, D. (1994a) 'Intracellular pH changes produced by glutamate uptake in rat hippocampal slices', *J Neurophysiol* **72**: 1686–1696.

Amato, A., Barbour, B., Szatkowski, M. and Attwell, D. (1994b) 'Counter-transport of potassium by the glutamate uptake carrier in glial cells isolated from the tiger salamander retina', *J. Physiol. (Lond.)* **479**: 371–380.

Arriza, J.L., Eliasof, S., Kavanaugh, M.P. and Amara, S.G. (1997) 'Excitatory amino acid transporter 5, a retinal glutamate transporter coupled to a chloride conductance', *Proc. Natl. Acad. Sci. USA* **94**: 4155–4160.

Arriza, J.L., Fairman, W.A., Wadiche, J.I., Murdoch, G.H., Kavanaugh, M.P. and Amara, S.G. (1994) 'Functional comparisons of three glutamate transporter subtypes cloned from human motor cortex', *J. Neurosci.* **14**: 5559–5569.

Assaf, S.Y. and Chung, S.H. (1984) 'Release of endogenous Zn2+ from brain tissue during activity', *Nature* **308**: 734–736.

Balcar, V.J. and Johnston, G.A. (1972) 'The structural specificity of the high affinity uptake of L-glutamate and L-aspartate by rat brain slices', *J. Neurochem.* **19**: 2657–2666.

Balcar, V.J. and Li, Y. (1992) 'Heterogeneity of high affinity uptake of L-glutamate and L-aspartate in the mammalian central nervous system', *Life Sci.* **51**: 1467–1478.

Barbour, B., Brew, H. and Attwell, D. (1988) 'Electrogenic glutamate uptake in glial cells is activated by intracellular potassium', *Nature* **335**: 433–435.

Barbour, B. and Hausser, M. (1997) 'Intersynaptic diffusion of neurotransmitter', *Trends Neurosci.* **20**: 377–384.

Barbour, B., Szatkowski, M., Ingledew, N. and Attwell, D. (1989) 'Arachidonic acid induces a prolonged inhibition of glutamate uptake into glial cells', *Nature* **342**: 918–920.

Bellocchio, E.E., Reimer, R.J., Fremeau, R.T. Jr, and Edwards, R.H. (2000) 'Uptake of glutamate into synaptic vesicles by an inorganic phosphate transporter', *Science* **289**: 957–960.

Berger, U.V. and Hediger, M.A. (1998) 'Comparative analysis of glutamate transporter expression in rat brain using differential double in situ hybridization', *Anat. Embryol. (Berl.)* **198**: 13–30.

Bergles, D.E. and Jahr, C.E. (1997) 'Synaptic activation of glutamate transporters in hippocampal astrocytes', *Neuron* **19**: 1297–1308.

Berman, S.B. and Hastings, T.G. (1997) 'Inhibition of glutamate transport in synaptosomes by dopamine oxidation and reactive oxygen species', *J. Neurochem.* **69**: 1185–1195.

Berridge, M.J. (1984) 'Inositol trisphosphate and diacylglycerol as second messengers', *Biochem. J.* **220**: 345–360.

Besson, M.T., Soustelle, L. and Birman, S. (1999) 'Identification and structural characterization of two genes encoding glutamate transporter homologues differently expressed in the nervous system of Drosophila melanogaster', *FEBS Lett.* **443**: 97–104.

Bezzi, P., Carmignoto, G., Pasti, L., Vesce, S., Rossi, D., Rizzini, B.L., Pozzan, T. and Volterra, A. (1998) 'Prostaglandins stimulate calcium-dependent glutamate release in astrocytes', *Nature* **391**: 281–285.

Billups, B. and Attwell, D. (1996) 'Modulation of non-vesicular glutamate release by pH', *Nature* **379**: 171–174.

Billups, B., Rossi, D. and Attwell, D. (1996) 'Anion conductance behavior of the glutamate uptake carrier in salamander retinal glial cells', *J. Neurosci.* **16**: 6722–6731.

Billups, B., Rossi, D., Oshima, T., Warr, O., Takahashi, M., Sarantis, M., Szatkowski, M. and Attwell, D. (1998a) 'Physiological and pathological operation of glutamate transporters', *Progress in Brain Research* **116**: 45–57.

Billups, B., Szatkowski, M., Rossi, D. and Attwell, D. (1998b) 'Patch-clamp, ion-sensing, and glutamate-sensing techniques to study glutamate transport in isolated retinal glial cells [In Process Citation]', *Methods Enzymol.* **296**: 617–632.

Bjørås, M., Gjesdal, O., Erickson, J.D., Torp, R., Levy, L.M., Ottersen, O.P., Degree, M., Storm-Mathisen, J., Seeberg, E. and Danbolt, N.C. (1996) 'Cloning and expression of a neuronal rat brain glutamate transporter', *Brain Res. Mol. Brain Res.* **36**: 163–168.

Blakely, R.D., Clark, J.A., Rudnick, G. and Amara, S.G. (1991) 'Vaccinia-T7 RNA polymerase expression system: evaluation for the expression cloning of plasma membrane transporters', *Anal. Biochem* **194**: 302–308.

Blanc, E.M., Keller, J.N., Fernandez, S. and Mattson, M.P. (1998) '4-hydroxynonenal, a lipid peroxidation product, impairs glutamate transport in cortical astrocytes', *Glia* **22**: 149–160.

Bouvier, M., Miller, B.A., Szatkowski, M. and Attwell, D. (1991) 'Electrogenic uptake of sulphur-containing analogues of glutamate and aspartate by Muller cells from the salamander retina', *J. Physiol. (Lond.)* **444**: 441–457.

Bouvier, M., Szatkowski, M., Amato, A. and Attwell, D. (1992) 'The glial cell glutamate uptake carrier countertransports pH-changing anions', *Nature* **360**: 471–474.

Brew, H. and Attwell, D. (1987) 'Electrogenic glutamate uptake is a major current carrier in the membrane of axolotl retinal glial cells' [published erratum appears in *Nature*, 1987, 20–26 Aug.; 328 (6132): 742], *Nature* **327**: 707–709.

Bridges, R.J., Kavanaugh, M.P. and Chamberlin, A.R. (1999) 'A pharmacological review of competitive inhibitors and substrates of high-affinity, sodium-dependent glutamate transport in the central nervous system', *Curr. Pharm. Des.* **5**: 363–379.

Bridges, R.J., Lovering, F.E., Humphrey, J.M., Stanley, M.S., Blakely, T.N., Cristofaro, M.F. and Chamberlin, A.R. (1993) 'Conformationally restricted inhibitors of the high affinity L-glutamate transporter', *Bioorganic & Medicinal Chemistry Letters* **3**: 115–120.

Bridges, R.J., Stanley, M.S., Anderson, M.W., Cotman, C.W. and Chamberlin, A.R. (1991) 'Conformationally defined neurotransmitter analogues. Selective inhibition of glutamate uptake by one pyrrolidine-2,4-dicarboxylate diastereomer', *J. Med. Chem.* **34**: 717–725.

Brooks-Kayal, A.R., Munir, M., Jin, H. and Robinson, M.B. (1998) 'The glutamate transporter, GLT-1, is expressed in cultured hippocampal neurons', *Neurochem. Intl.* **33**: 95–100.

Casado, M., Bendahan, A., Zafra, F., Danbolt, N.C., Aragon, C., Gimenez, C. and Kanner, B.I. (1993) 'Phosphorylation and modulation of brain glutamate transporters by protein kinase C', *J. Biol. Chem.* **268**: 27,313–27,317.

Chan, P.H., Kerlan, R. and Fishman, R.A. (1983) 'Reductions of gamma-aminobutyric acid and glutamate uptake and (Na+ + K+)-ATPase activity in brain slices and synaptosomes by arachidonic acid', *J. Neurochem.* **40**: 309–316.

Chaudhry, F.A., Lehre, K.P., van Lookeren Campagne, M., Ottersen, O.P., Danbolt, N.C. and Storm-Mathisen, J. (1995) 'Glutamate transporters in glial plasma membranes: highly differentiated localizations revealed by quantitative ultrastructural immunocytochemistry', *Neuron* **15**: 711–720.

Cho, Y. and Bannai, S. (1990) 'Uptake of glutamate and cysteine in C-6 glioma cells and in cultured astrocytes', *J. Neurochem.* **55**: 2091–2097.

Choi, D.W., Maulucci-Gedde, M. and Kriegstein, A.R. (1987) 'Glutamate neurotoxicity in cortical cell culture', *J. Neurosci.* **7**: 357–368.

Christensen, H. and Fonnum, F. (1991) 'The ontogeny of the uptake systems for glycine, GABA and glutamate in synaptic vesicles isolated from rat spinal cord-medulla', *Brain Res. Dev. Brain Res.* **64**: 155–159.

Christensen, H. and Fonnum, F. (1992) 'The ontogeny of the uptake systems for glutamate, GABA, and glycine in synaptic vesicles isolated from rat brain', *Neurochem. Res.* **17**: 457–462.

Clements, J.D. (1996) 'Transmitter timecourse in the synaptic cleft: its role in central synaptic function', *Trends Neurosci.* **19**: 163–171.

Coco, S., Verderio, C., Trotti, D., Rothstein, J.D., Volterra, A. and Matteoli, M. (1997) 'Non-synaptic localization of the glutamate transporter EAAC1 in cultured hippocampal neurons', *Eur. J. Neurosci.* **9**: 1902–1910.

Collard, K.J., Edwards, R. and Liu, Y. (1993) 'Changes in synaptosomal glutamate release during postnatal development in the rat hippocampus and cortex', *Brain Res. Dev. Brain Res.* **71**: 37–43.

Collingridge, G.L. and Lester, R.A. (1989) 'Excitatory amino acid receptors in the vertebrate central nervous system', *Pharmacol. Rev.* **41**: 143–210.

Conradt, M. and Stoffel, W. (1995) 'Functional analysis of the high affinity, Na(+)-dependent glutamate transporter GLAST-1 by site-directed mutagenesis', *J. Biol. Chem.* **270**: 25,207–25,212.

Conradt, M. and Stoffel, W. (1997) 'Inhibition of the high-affinity brain glutamate transporter GLAST-1 via direct phosphorylation', *J. Neurochem.* **68**: 1244–1251.

Conti, F., DeBiasi, S., Minelli, A., Rothstein, J.D. and Melone, M. (1998) 'EAAC1, a high-affinity glutamate transporter, is localized to astrocytes and gabaergic neurons besides pyramidal cells in the rat cerebral cortex', *Cereb. Cortex* **8**: 108–116.

Curtis, D.R., Phillis, J.W. and Watkins, J.C. (1959) 'Chemical excitation of spinal neurones', *Nature* **183**: 611.

Curtis, D.R., Phillis, J.W. and Watkins, J.C. (1960) 'The chemical excitation of spinal neurones by certain acidic amino acids', *J. Physiol.* **150**: 656–682.

Danbolt, N.C. (1994) 'The high affinity uptake system for excitatory amino acids in the brain', *Prog. Neurobiol.* **44**: 377–396.

Danbolt, N.C. (2000) 'Sodium- and potassium-dependent excitatory amino acid transporters in brain plasma membranes', in O.P. Ottersen and S.-M. Ottersen (eds) *Handbook of Chemical Neuroanatomy.* Vol. 18: *Glutamate*, J. Amsterdam: Elsevier Science B.V., pp. 213–254.

Danbolt, N.C. (2001) 'Glutamate uptake', *Prog. Neurobiol.* **65**: 1–105.

Danbolt, N.C., Chaudhry, F.A., Dehnes, Y., Lehre, K.P., Levy, L.M., Ullensvang, K. and Storm-Mathisen, J. (1998) 'Properties and localization of glutamate transporters', *Progress in Brain Research* **116**: 23–43.

Danbolt, N.C., Storm-Mathisen, J. and Kanner, B.I. (1992) 'An [Na++K+]coupled L-glutamate transporter purified from rat brain is located in glial cell processes', *Neuroscience* **51**: 295–310.

Davis, K.E., Straff, D.J., Weinstein, E.A., Bannerman, P.G., Correale, D.M., Rothstein, J.D. and Robinson, M.B. (1998) 'Multiple signaling pathways regulate cell surface expression and activity of the excitatory amino acid carrier 1 subtype of Glu transporter in C6 glioma', *J. Neurosci.* **18**: 2475–2485.

Dehnes, Y., Chaudhry, F.A., Ullensvang, K., Lehre, K.P., Storm-Mathisen, J. and Danbolt, N.C. (1998) 'The glutamate transporter EAAT4 in rat cerebellar Purkinje cells: a glutamate-gated chloride channel concentrated near the synapse in parts of the dendritic membrane facing astroglia', *J. Neurosci.* **18**: 3606–3619.

Dennis, S.C., Land, J.M. and Clark, J.B. (1976) 'Glutamate metabolism and transport in rat brain mitochondria', *Biochem. J.* **156**: 323–331.

Derouiche, A. and Rauen, T. (1995) 'Coincidence of L-glutamate/L-aspartate transporter (GLAST) and glutamine synthetase (GS) immunoreactions in retinal glia: evidence for coupling of GLAST and GS in transmitter clearance', *J. Neurosci. Res.* **42**: 131–143.

Diamond, J.S. and Jahr, C.E. (1997) 'Transporters buffer synaptically released glutamate on a submillisecond time scale', *J. Neurosci.* **17**: 4672–4687.

Donly, B.C., Richman, A., Hawkins, E., McLean, H. and Caveney, S. (1997) 'Molecular cloning and functional expression of an insect high-affinity Na+-dependent glutamate transporter', *Eur. J. Biochem.* **248**: 535–542.

Dowd, L.A., Coyle, A.J., Rothstein, J.D., Pritchett, D.B. and Robinson, M.B. (1996) 'Comparison of Na+-dependent glutamate transport activity in synaptosomes, C6 glioma, and Xenopus oocytes expressing excitatory amino acid carrier 1 (EAAC1)', *Mol. Pharmacol.* **49**: 465–473.

Dowd, L.A. and Robinson, M.B. (1996) 'Rapid stimulation of EAAC1-mediated Na+-dependent L-glutamate transport activity in C6 glioma cells by phorbol ester', *J. Neurochem.* **67**: 508–516.

Duan, S., Anderson, C.M., Stein, B.A. and Swanson, R.A. (1999) 'Glutamate induces rapid upregulation of astrocyte glutamate transport and cell-surface expression of GLAST', *J. Neurosci.* **19**: 10,193–10,200.

Dumuis, A., Sebben, M., Haynes, L., Pin, J.P. and Bockaert, J. (1988) 'NMDA receptors activate the arachidonic acid cascade system in striatal neurons', *Nature* **336**: 68–70.

Eliasof, S., Arriza, J.L., Leighton, B.H., Kavanaugh, M.P. and Amara, S.G. (1998) 'Excitatory amino acid transporters of the salamander retina: identification, localization, and function', *J. Neurosci.* **18**: 698–712.

Eliasof, S. and Jahr, C.E. (1996) 'Retinal glial cell glutamate transporter is coupled to an anionic conductance', *Proc. Natl. Acad. Sci. USA* **93**: 4153–4158.

Eliasof, S. and Werblin, F. (1993) 'Characterization of the glutamate transporter in retinal cones of the tiger salamander', *J. Neurosci.* **13**: 402–411.

Eng, D.L., Lee, Y.L. and Lal, P.G. (1997) 'Expression of glutamate uptake transporters after dibutyryl cyclic AMP differentiation and traumatic injury in cultured astrocytes', *Brain Res.* **778**: 215–221.

Erdö, S.L. and Wolff, J.R. (1990) 'Postnatal development of the excitatory amino acid system in visual cortex of the rat. Changes in uptake and levels of aspartate and glutamate', *Intl. J. Dev. Neurosci.* **8**: 205–208.

Erecinska, M., Wantorsky, D. and Wilson, D.F. (1983) 'Aspartate transport in synaptosomes from rat brain', *J. Biol. Chem.* **258**: 9069–9077.

Euler, T. and Wassle, H. (1995) 'Immunocytochemical identification of cone bipolar cells in the rat retina', *J. Comp. Neurol.* **361**: 461–478.

Exton, J.H. (1990) 'Signaling through phosphatidylcholine breakdown', *J. Biol. Chem.* **265**: 1–4.

Fairman, W.A., Vandenberg, R.J., Arriza, J.L., Kavanaugh, M.P. and Amara, S.G. (1995) 'An excitatory amino-acid transporter with properties of a ligand-gated chloride channel', *Nature* **375**: 599–603.

Ferkany, J. and Coyle, J.T. (1986) 'Heterogeneity of sodium-dependent excitatory amino acid uptake mechanisms in rat brain', *J. Neurosci. Res.* **16**: 491–503.

Fletcher, E.J. and Johnston, G.A. (1991) 'Regional heterogeneity of L-glutamate and L-aspartate high-affinity uptake systems in the rat CNS', *J. Neurochem.* **57**: 911–914.

Fonnum, F. (1984) 'Glutamate: a neurotransmitter in mammalian brain', *J. Neurochem.* **42**: 1–11.

Fonnum, F., Fykse, E.M. and Roseth, S. (1998) 'Uptake of glutamate into synaptic vesicles', *Progress in Brain Research* **116**: 87–101.

Fuerst, T.R., Niles, e.g. Studier, F.W. and Moss, B. (1986) 'Eukaryotic transient-expression system based on recombinant vaccinia virus that synthesizes bacteriophage T7 RNA polymerase', *Proc. Natl. Acad. Sci. USA* **83**: 8122–8126.

Furuta, A., Martin, L.J., Lin, C.L., Dykes-Hoberg, M. and Rothstein, J.D. (1997a) 'Cellular and synaptic localization of the neuronal glutamate transporters excitatory amino acid transporter 3 and 4', *Neuroscience* **81**: 1031–1042.

Furuta, A., Rothstein, J.D. and Martin, L.J. (1997b) 'Glutamate transporter protein subtypes are expressed differentially during rat CNS development', *J. Neurosci.* **17**: 8363–8375.

Fykse, E.M., Christensen, H. and Fonnum, F. (1989) 'Comparison of the properties of gamma-aminobutyric acid and L-glutamate uptake into synaptic vesicles isolated from rat brain', *J. Neurochem.* **52**: 946–951.

Gabrielsson, B., Robson, T., Norris, D. and Chung, S.H. (1986) 'Effects of divalent metal ions on the uptake of glutamate and GABA from synaptosomal fractions', *Brain Res.* **384**: 218–223.

Gallo, V. and Russell, J.T. (1995) 'Excitatory amino acid receptors in glia: different subtypes for distinct functions?' *J. Neurosci. Res.* **42**: 1–8.

Ganel, R. and Crosson, C.E. (1998) 'Modulation of human glutamate transporter activity by phorbol ester', *J. Neurochem.* **70**: 993–1000.

Gegelashvili, G., Civenni, G., Racagni, G., Danbolt, N.C., Schousboe, I. and Schousboe, A. (1996) 'Glutamate receptor agonists up-regulate glutamate transporter GLAST in astrocytes', *Neuroreport* **8**: 261–265.

Gegelashvili, G., Danbolt, N.C. and Schousboe, A. (1997) 'Neuronal soluble factors differentially regulate the expression of the GLT1 and GLAST glutamate transporters in cultured astroglia', *J. Neurochem.* **69**: 2612–2615.

Gegelashvili, G., Dehnes, Y., Danbolt, N.C. and Schousboe, A. (2000) 'The high-affinity glutamate transporters GLT1, GLAST, and EAAT4 are regulated via different signalling mechanisms', *Neurochem. Intl.* **37**: 163–170.

Gegelashvili, G. and Schousboe, A. (1998) 'Cellular distribution and kinetic properties of high-affinity glutamate transporters', *Brain Res. Bull.* **45**: 233–238.

Ginsberg, S.D., Martin, L.J. and Rothstein, J.D. (1995) 'Regional deafferentation down-regulates subtypes of glutamate transporter proteins', *J. Neurochem.* **65**: 2800–2803.

Ginsberg, S.D., Rothstein, J.D., Price, D.L. and Martin, L.J. (1996) 'Fimbria-fornix transections selectively down-regulate subtypes of glutamate transporter and glutamate receptor proteins in septum and hippocampus', *J. Neurochem.* **67**: 1208–1216.

Gonzalez, M.I., Lopez-Colom, A.M. and Ortega, A. (1999) 'Sodium-dependent glutamate transport in Muller glial cells: regulation by phorbol esters', *Brain Res.* **831**: 140–145.

Gonzalez, M.I. and Ortega, A. (1997) 'Regulation of the Na+-dependent high affinity glutamate/aspartate transporter in cultured Bergmann glia by phorbol esters', *J. Neurosci. Res.* **50**: 585–590.

Grant, G.B. and Dowling, J.E. (1995) 'A glutamate-activated chloride current in cone-driven ON bipolar cells of the white perch retina', *J. Neurosci.* **15**: 3852–3862.

Grune, T., Reinheckel, T. and Davies, K.J. (1997) 'Degradation of oxidized proteins in mammalian cells', *Faseb. J.* **11**: 526–534.

Grunewald, M., Bendahan, A. and Kanner, B.I. (1998) 'Biotinylation of single cysteine mutants of the glutamate transporter GLT-1 from rat brain reveals its unusual topology [In Process Citation]', *Neuron* **21**: 623–632.

Grunewald, M. and Kanner, B.I. (2000) 'The accessibility of a novel reentrant loop of the glutamate transporter GLT-1 is restricted by its substrate', *J. Biol. Chem.* **275**: 9684–9689.

Gundersen, V., Danbolt, N.C., Ottersen, O.P. and Storm-Mathisen, J. (1993) 'Demonstration of glutamate/aspartate uptake activity in nerve endings by use of antibodies recognizing exogenous D-aspartate', *Neuroscience* **57**: 97–111.

Gundersen, V., Ottersen, O.P. and Storm-Mathisen, J. (1996) 'Selective excitatory amino acid uptake in glutamatergic nerve terminals and in glia in the rat striatum: quantitative electron microscopic immunocytochemistry of exogenous (D)-aspartate and endogenous glutamate and GABA', *Eur. J. Neurosci.* **8**: 758–765.

Hamberger, A., Berthold, C.-L., Karlsson, B., Lehmann, A. and Nyström, B. (1983) 'Extracellular GABA, glutamate and glutamine in vivo – perfusion-dialysis of the rabbit hippocampus', *Neurology and Neurobiology* **7**: 473–492.

Harada, T., Harada, C., Watanabe, M., Inoue, Y., Sakagawa, T., Nakayama, N., Sasaki, S., Okuyama, S., Watase, K., Wada, K. and Tanaka, K. (1998) 'Functions of the two glutamate transporters GLAST and GLT-1 in the retina', *Proc. Natl. Acad. Sci. USA* **95**: 4663–4666.

Hashimoto, A. and Oka, T. (1997) 'Free D-aspartate and D-serine in the mammalian brain and periphery', *Prog. Neurobiol.* **52**: 325–353.

Haugeto, Ø., Ullensvang, K., Levy, L.M., Chaudhry, F.A., Honoré, T., Nielsen, M., Lehre, K.P. and Danbolt, N.C. (1996) 'Brain glutamate transporter proteins form homomultimers', *J. Biol. Chem.* **271**: 27,715–27,722.

Hayashi, T. (1954) 'Effects of sodium glutamate on the nervous system', *Keio J. Med.* **3**: 183–192.

Hertz, L., Bock, E. and Schousboe, A. (1978) 'GFA content, glutamate uptake and activity of glutamate metabolizing enzymes in differentiating mouse astrocytes', *Dev. Neurosci.* **1**: 226–238.

Holmes, W.R. (1995) 'Modeling the effect of glutamate diffusion and uptake on NMDA and non-NMDA receptor saturation', *Biophys. J.* **69**: 1734–1747.

Inage, Y.W., Itoh, M., Wada, K. and Takashima, S. (1998) 'Expression of two glutamate transporters, GLAST and EAAT4, in the human cerebellum: their correlation in development and neonatal hypoxic- ischemic damage', *J. Neuropathol. Exp. Neurol.* **57**: 554–562.

Inoue, K., Sakaitani, M., Shimada, S. and Tohyama, M. (1995) 'Cloning and expression of a bovine glutamate transporter', *Brain Res. Mol. Brain Res.* **28**: 343–348.

Itoh, M., Watanabe, Y., Watanabe, M., Tanaka, K., Wada, K. and Takashima, S. (1997) 'Expression of a glutamate transporter subtype, EAAT4, in the developing human cerebellum', *Brain Res.* **767**: 265–271.

Jabaudon, D., Shimamoto, K., Yasuda-Kamatani, Y., Scanziani, M., Gahwiler, B.H. and Gerber, U. (1999) 'Inhibition of uptake unmasks rapid extracellular turnover of glutamate of nonvesicular origin', *Proc. Natl. Acad. Sci. USA* **96**: 8733–8738.

Johnston, G.A., Kennedy, S.M. and Twitchin, B. (1979) 'Action of the neurotoxin kainic acid on high affinity uptake of L- glutamic acid in rat brain slices', *J. Neurochem.* **32**: 121–127.

Johnston, G.A.R. (1981) 'Glutamate uptake and its possible role in neurotransmitter inactivation', in P.J. Roberts, J. Storm-Mathisen and G.A.R. Johnston (eds) *Glutamate: Transmitter in the Central Nervous System*, Chichester, UK: John Wiley & Sons, pp. 77–87.

Kanai, Y., Bhide, P.G., DiFiglia, M. and Hediger, M.A. (1995) 'Neuronal high-affinity glutamate transport in the rat central nervous system', *Neuroreport* **6**: 2357–2362.

Kanai, Y. and Hediger, M.A. (1992) 'Primary structure and functional characterization of a high-affinity glutamate transporter', *Nature* **360**: 467–471.

Kanai, Y., Smith, C.P. and Hediger, M.A. (1993) 'The elusive transporters with a high affinity for glutamate', *Trends Neurosci.* **16**: 365–370.

Kanai, Y., Stelzner, M., Nussberger, S., Khawaja, S., Hebert, S.C., Smith, C.P. and Hediger, M.A. (1994) 'The neuronal and epithelial human high affinity glutamate transporter. Insights into structure and mechanism of transport', *J. Biol. Chem.* **269**: 20,599–20,606.

Kanner, B.I. and Sharon, I. (1978) 'Active transport of L-glutamate by membrane vesicles isolated from rat brain', *Biochemistry* **17**: 3949–3953.

Kataoka, Y., Morii, H., Watanabe, Y. and Ohmori, H. (1997) 'A postsynaptic excitatory amino acid transporter with chloride conductance functionally regulated by neuronal activity in cerebellar Purkinje cells', *J. Neurosci.* **17**: 7017–7024.

Kauppinen, R.A., McMahon, H.T. and Nicholls, D.G. (1988) 'Ca2+-dependent and Ca2+-independent glutamate release, energy status and cytosolic free Ca2+ concentration in isolated nerve terminals following metabolic inhibition: possible relevance to hypoglycaemia and anoxia', *Neuroscience* **27**: 175–182.

Kavanaugh, M.P., Bendahan, A., Zerangue, N., Zhang, Y. and Kanner, B.I. (1997) 'Mutation of an amino acid residue influencing potassium coupling in the glutamate transporter GLT-1 induces obligate exchange', *J. Biol. Chem.* **272**: 1703–1708.

Kawakami, H., Tanaka, K., Nakayama, T., Inoue, K. and Nakamura, S. (1994) 'Cloning and expression of a human glutamate transporter', *Biochem. Biophys. Res. Commun.* **199**: 171–176.

Kawano, T., Takuwa, K. and Nakajima, T. (1996) 'Molecular cloning of a cDNA for the glutamate transporter of the nematode Caenorhabditis elegans', *Biochem. Biophys. Res. Commun.* **228**: 415–420.

Keller, J.N., Mark, R.J., Bruce, A.J., Blanc, E., Rothstein, J.D., Uchida, K., Waeg, G. and Mattson, M.P. (1997a) '4-hydroxynonenal, an aldehydic product of membrane lipid peroxidation, impairs glutamate transport and mitochondrial function in synaptosomes', *Neuroscience* **80**: 685–696.

Keller, J.N., Pang, Z., Geddes, J.W., Begley, J.G., Germeyer, A., Waeg, G. and Mattson, M.P. (1997b) 'Impairment of glucose and glutamate transport and induction of mitochondrial oxidative stress and dysfunction in synaptosomes by amyloid beta-peptide: role of the lipid peroxidation product 4– hydroxynonenal', *J. Neurochem.* **69**: 273–284.

Khan, W.A., Blobe, G.C. and Hannun, Y.A. (1995) 'Arachidonic acid and free fatty acids as second messengers and the role of protein kinase C', *Cell Signal* **7**: 171–184.

Killinger, S., Blume, G.L., Bohart, L., Bested, A., Dias, L.S., Cooper, B., Allan, R.D. and Balcar, V.J. (1996) 'Autoradiographic studies indicate regional variations in the characteristics of L-glutamate transporters in the rat brain', *Neurosci. Lett.* **216**: 101–104.

Kirschner, M.A., Arriza, J.L., Copeland, N.G., Gilbert, D.J., Jenkins, N.A., Magenis, E. and Amara, S.G. (1994a) 'The mouse and human excitatory amino acid transporter gene (EAAT1) maps to mouse chromosome 15 and a region of syntenic homology on human chromosome 5', *Genomics* **22**: 631–633.

Kirschner, M.A., Copeland, N.G., Gilbert, D.J., Jenkins, N.A. and Amara, S. G. (1994b) 'Mouse excitatory amino acid transporter EAAT2: isolation, characterization, and proximity to neuroexcitability loci on mouse chromosome 2', *Genomics* **24**: 218–224.

Kish, P.E., Kim, S.Y. and Ueda, T. (1989) 'Ontogeny of glutamate accumulating activity in rat brain synaptic vesicles', *Neurosci. Lett.* **97**: 185–190.

Kleinle, J., Vogt, K., Luscher, H.R., Muller, L., Senn, W., Wyler, K. and Streit, J. (1996) 'Transmitter concentration profiles in the synaptic cleft: an analytical model of release and diffusion', *Biophys. J.* **71**: 2413–2426.

Kondo, K., Hashimoto, H., Kitanaka, J., Sawada, M., Suzumura, A., Marunouchi, T. and Baba, A. (1995) 'Expression of glutamate transporters in cultured glial cells', *Neurosci. Lett.* **188**: 140–142.

Kramer, R. and Heberger, C. (1986) 'Functional reconstitution of carrier proteins by removal of detergent with a hydrophobic ion exchange column', *Biochim. Biophys. Acta* **863**: 289–296.

Kramer, R. and Palmieri, F. (1989) 'Molecular aspects of isolated and reconstituted carrier proteins from animal mitochondria', *Biochim. Biophys. Acta* **974**: 1–23.

Kugler, P. and Schmitt, A. (1999) 'Glutamate transporter EAAC1 is expressed in neurons and glial cells in the rat nervous system', *Glia* **27**: 129–142.

Lebrun, B., Sakaitani, M., Shimamoto, K., Yasuda-Kamatani, Y. and Nakajima, T. (1997) 'New beta-hydroxyaspartate derivatives are competitive blockers for the bovine glutamate/aspartate transporter', *J. Biol. Chem.* **272**: 20,336–20,339.

Lehre, K.P. and Danbolt, N.C. (1998) 'The number of glutamate transporter subtype molecules at glutamatergic synapses: chemical and stereological quantification in young adult rat brain', *J. Neurosci.* **18**: 8751–8757.

Lehre, K.P., Davanger, S. and Danbolt, N.C. (1997) 'Localization of the glutamate transporter protein GLAST in rat retina', *Brain Res.* **744**: 129–137.

Lehre, K.P., Levy, L.M., Ottersen, O.P., Storm-Mathisen, J. and Danbolt, N.C. (1995) 'Differential expression of two glial glutamate transporters in the rat brain: quantitative and immunocytochemical observations', *J. Neurosci.* **15**: 1835–1853.

Levy, L.M., Lehre, K.P., Rolstad, B. and Danbolt, N.C. (1993) 'A monoclonal antibody raised against an [Na+ +K+]coupled L-glutamate transporter purified from rat brain confirms glial cell localization', *FEBS Lett.* **317**: 79–84.

Levy, L.M., Lehre, K.P., Walaas, S.I., Storm-Mathisen, J. and Danbolt, N.C. (1995) 'Down-regulation of glial glutamate transporters after glutamatergic denervation in the rat brain', *Eur. J. Neurosci.* **7**: 2036–2041.

Levy, L.M., Warr, O. and Attwell, D. (1998) 'Stoichiometry of the glial glutamate transporter GLT-1 expressed inducibly in a Chinese hamster ovary cell line selected for low endogenous Na+-dependent glutamate uptake', *J. Neurosci.* **18**: 9620–9628.

Logan, W.J. and Snyder, S.H. (1971) 'Unique high affinity uptake systems for glycine, glutamic and aspartic acids in central nervous tissue of the rat', *Nature* **234**: 297–299.

Madl, J.E. and Burgesser, K. (1993) 'Adenosine triphosphate depletion reverses sodium-dependent, neuronal uptake of glutamate in rat hippocampal slices', *J. Neurosci.* **13**: 4429–4444.

Maeno-Hikichi, Y., Tanaka, K., Shibata, T., Watanabe, M., Inoue, Y., Mukainaka, Y. and Wada, K. (1997) 'Structure and functional expression of the cloned mouse neuronal high-affinity glutamate transporter', *Brain Res. Mol. Brain Res.* **48**: 176–180.

Manfras, B.J., Rudert, W.A., Trucco, M. and Boehm, B.O. (1994) 'Cloning and characterization of a glutamate transporter cDNA from human brain and pancreas', *Biochim. Biophys. Acta* **1195**: 185–188.

Martin, L.J., Brambrink, A.M., Lehmann, C., Portera-Cailliau, C., Koehler, R., Rothstein, J. and Traystman, R.J. (1997) 'Hypoxia-ischemia causes abnormalities in glutamate transporters and death of astroglia and neurons in newborn striatum', *Ann. Neurol.* **42**: 335–348.

Meaney, J.A., Balcar, V.J., Rothstein, J.D. and Jeffrey, P.L. (1998) 'Glutamate transport in cultures from developing avian cerebellum: presence of GLT-1 immunoreactivity in Purkinje neurons', *J. Neurosci. Res.* **54**: 595–603.

Mennerick, S., Dhond, R.P., Benz, A., Xu, W., Rothstein, J.D., Danbolt, N.C., Isenberg, K.E. and Zorumski, C.F'. (1998) 'Neuronal expression of the glutamate transporter GLT-1 in hippocampal microcultures', *J. Neurosci.* **18**: 4490–4499.

Minn, A. and Gayet, J. (1977) 'Kinetic study of glutamate transport in rat brain mitochondria', *J. Neurochem.* **29**: 873–881.

Mitrovic, A.D., Amara, S.G., Johnston, G.A. and Vandenberg, R.J. (1998) 'Identification of functional domains of the human glutamate transporters EAAT1 and EAAT2', *J. Biol. Chem.* **273**: 14,698–14,706.

Mukainaka, Y., Tanaka, K., Hagiwara, T. and Wada, K. (1995) 'Molecular cloning of two glutamate transporter subtypes from mouse brain', *Biochim. Biophys. Acta* **1244**: 233–237.

Muller, A., Maurin, L. and Bonne, C. (1998) 'Free radicals and glutamate uptake in the retina', *Gen. Pharmacol.* **30**: 315–318.

Munch, C., Schwalenstocker, B., Knappenberger, B., Liebau, S., Volkel, H., Ludolph, A.C. and Meyer, T. (1998) '5′-heterogeneity of the human excitatory amino acid transporter cDNA EAAT2 (GLT-1)', *Neuroreport* **9**: 1295–1297.

Murphy, T.H., Miyamoto, M., Sastre, A., Schnaar, R.L. and Coyle, J.T. (1989) 'Glutamate toxicity in a neuronal cell line involves inhibition of cystine transport leading to oxidative stress', *Neuron* **2**: 1547–1558.

Murphy, T.H., Schnaar, R.L. and Coyle, J.T. (1990) 'Immature cortical neurons are uniquely sensitive to glutamate toxicity by inhibition of cystine uptake', *Faseb. J.* **4**: 1624–1633.

Mutch, W.A. and Hansen, A.J. (1984) 'Extracellular pH changes during spreading depression and cerebral ischemia: mechanisms of brain pH regulation', *J. Cereb. Blood Flow Metab.* **4**: 17–27.

Nagao, S., Kwak, S. and Kanazawa, I. (1997) 'EAAT4, a glutamate transporter with properties of a chloride channel, is predominantly localized in Purkinje cell dendrites, and forms parasagittal compartments in rat cerebellum', *Neuroscience* **78**: 929–933.

Naito, S. and Ueda, T. (1985) 'Characterization of glutamate uptake into synaptic vesicles', *J. Neurochem.* **44**: 99–109.

Ni, B., Rosteck, P.R. Jr, Nadi, N.S. and Paul, S.M. (1995) 'Cloning and expression of a cDNA encoding a brain-specific Na^+-dependent inorganic phosphate cotransporter', *Proc. Natl. Acad. Sci. USA* **91**: 5607–5611.

Nicholls, D. and Attwell, D. (1990) 'The release and uptake of excitatory amino acids', *Trends Pharmacol. Sci.* **11**: 462–468.

Northington, F.J., Traystman, R.J., Koehler, R.C. and Martin, L.J. (1999) 'GLT1, glial glutamate transporter, is transiently expressed in neurons and develops astrocyte specificity only after midgestation in the ovine fetal brain', *J. Neurobiol.* **39**: 515–526.

Northington, F.J., Traystman, R.J., Koehler, R.C., Rothstein, J.D. and Martin, L.J. (1998) 'Regional and cellular expression of glial (GLT1) and neuronal (EAAC1) glutamate transporter proteins in ovine fetal brain', *Neuroscience* **85**: 1183–1194.

Otis, T.S. and Jahr, C.E. (1998) 'Anion currents and predicted glutamate flux through a neuronal glutamate transporter', *J. Neurosci.* **18**: 7099–7110.

Otori, Y., Shimada, S., Tanaka, K., Ishimoto, I., Tano, Y. and Tohyama, M. (1994) 'Marked increase in glutamate-aspartate transporter (GLAST/GluT-1) mRNA following transient retinal ischemia', *Brain Res. Mol. Brain Res.* **27**: 310–314.

Ottersen, O.P. and Storm-Mathisen, J. (1984) 'Neurons containing or accumulating transmitter amino acids', in A. Björklund, T. Hökfelt and M.J. Kuhar (eds) *Classical Transmitters and Transmitter Receptors in the CNS*, Part II Vol. 3, Amsterdam: Elsevier, pp. 141–246.

Ozawa, S., Kamiya, H. and Tsuzuki, K. (1998) 'Glutamate receptors in the mammalian central nervous system', *Prog. Neurobiol.* **54**: 581–618.

Özkan, E.D. and Ueda, T. (1998) 'Glutamate transport and storage in synaptic vesicles', *Jpn. J. Pharmacol.* **77**: 1–10.

Palos, T.P., Ramachandran, B., Boado, R. and Howard, B.D. (1996) 'Rat C6 and human astrocytic tumor cells express a neuronal type of glutamate transporter', *Brain Res. Mol. Brain Res.* **37**: 297–303.

Pearce, B., Albrecht, J., Morrow, C. and Murphy, S. (1986) 'Astrocyte glutamate receptor activation promotes inositol phospholipid turnover and calcium flux', *Neurosci. Lett.* **72**: 335–340.

Peghini, P., Janzen, J. and Stoffel, W. (1997) 'Glutamate transporter EAAC-1-deficient mice develop dicarboxylic aminoaciduria and behavioral abnormalities but no neurodegeneration', *Embo. J.* **16**: 3822–3832.

Peters, S., Koh, J. and Choi, D.W. (1987) 'Zinc selectively blocks the action of N-methyl-D-aspartate on cortical neurons', *Science* **236**: 589–593.

Piani, D., Frei, K., Pfister, H.W. and Fontana, A. (1993) 'Glutamate uptake by astrocytes is inhibited by reactive oxygen intermediates but not by other macrophage-derived molecules including cytokines, leukotrienes or platelet-activating factor', *J. Neuroimmunol.* **48**: 99–104.

Picaud, S.A., Larsson, H.P., Grant, G.B., Lecar, H. and Werblin, F.S. (1995) 'Glutamate-gated chloride channel with glutamate-transporter-like properties in cone photoreceptors of the tiger salamander', *J. Neurophysiol.* **74**: 1760–1771.

Pines, G., Danbolt, N.C., Bjoras, M., Zhang, Y., Bendahan, A., Eide, L., Koepsell, H., Storm-Mathisen, J., Seeberg, E. and Kanner, B.I. (1992) 'Cloning and expression of a rat brain L-glutamate transporter' [published erratum appears in *Nature*, 1992 24–31 Dec.; 360 (6406): 768], *Nature* **360**: 464–467.

Pines, G., Zhang, Y. and Kanner, B.I. (1995) 'Glutamate 404 is involved in the substrate discrimination of GLT-1, a (Na+ +K+)-coupled glutamate transporter from rat brain', *J. Biol. Chem.* **270**: 17,093–17,097.

Plachez, C., Danbolt, N.C. and Recasens, M. (2000) 'Transient expression of the glial glutamate transporters GLAST and GLT in hippocampal neurons in primary culture', *J. Neurosci. Res.* **59**: 587–593.

Raghavendra Rao, V.L., Rao, A.M., Dogan, A., Bowen, K.K., Hatcher, J., Rothstein, J.D. and Dempsey, R.J. (2000) 'Glial glutamate transporter GLT-1 down-regulation precedes delayed neuronal death in gerbil hippocampus following transient global cerebral ischemia', *Neurochem. Intl.* **36**: 531–537.

Rauen, T., Jeserich, G., Danbolt, N.C. and Kanner, B.I. (1992) 'Comparative analysis of sodium-dependent L-glutamate transport of synaptosomal and astroglial membrane vesicles from mouse cortex', *FEBS Lett.* **312**: 15–20.

Rauen, T. and Kanner, B.I. (1994) 'Localization of the glutamate transporter GLT-1 in rat and macaque monkey retinae', *Neurosci. Lett.* **169**: 137–140.

Rauen, T., Rothstein, J.D. and Wassle, H. (1996) 'Differential expression of three glutamate transporter subtypes in the rat retina', *Cell Tissue Res.* **286**: 325–336.

Reichelt, W., Stabel-Burow, J., Pannicke, T., Weichert, H. and Heinemann, U. (1997) 'The glutathione level of retinal Müller glial cells is dependent on the high-affinity sodium-dependent uptake of glutamate', *Neuroscience* **77**: 1213–1224.

Rhoads, D.E., Ockner, R.K., Peterson, N.A. and Raghupathy, E. (1983) 'Modulation of membrane transport by free fatty acids: inhibition of synaptosomal sodium-dependent amino acid uptake', *Biochemistry* **22**: 1965–1970.

Robinson, M.B. and Dowd, L.A. (1997) 'Heterogeneity and functional properties of subtypes of sodium-dependent glutamate transporters in the mammalian central nervous system', *Adv. Pharmacol.* **37**: 69–115.

Robinson, M.B., Hunter-Ensor, M. and Sinor, J. (1991) 'Pharmacologically distinct sodium-dependent L-[3H]glutamate transport processes in rat brain', *Brain Res.* **544**: 196–202.

Robinson, M.B., Sinor, J.D., Dowd, L.A. and Kerwin, J.F., Jr. (1993) 'Subtypes of sodium-dependent high-affinity L-[3H]glutamate transport activity: pharmacologic specificity and regulation by sodium and potassium', *J. Neurochem.* **60**: 167–179.

Rossi, D.J., Oshima, T. and Attwell, D. (2000) 'Glutamate release in severe brain ischaemia is mainly by reversed uptake', *Nature* **403**: 316–321.

Rothstein, J.D., Martin, L., Levey, A.I., Dykes-Hoberg, M., Jin, L., Wu, D., Nash, N. and Kuncl, R.W. (1994) 'Localization of neuronal and glial glutamate transporters', *Neuron* **13**: 713–725.

Rothstein, J.D., Van Kammen, M., Levey, A.I., Martin, L.J. and Kuncl, R.W. (1995) 'Selective loss of glial glutamate transporter GLT-1 in amyotrophic lateral sclerosis', *Ann. Neurol.* **38**: 73–84.

Sarantis, M., Everett, K. and Attwell, D. (1988) 'A presynaptic action of glutamate at the cone output synapse', *Nature* **332**: 451–453.

Sato, H., Tamba, M., Ishii, T. and Bannai, S. (1999) 'Cloning and expression of a plasma membrane cystine/glutamate exchange transporter composed of two distinct proteins', *J. Biol. Chem.* **274**: 11,455–11,458.

Schlag, B.D., Vondrasek, J.R., Munir, M., Kalandadze, A., Zelenaia, O.A., Rothstein, J.D. and Robinson, M.B. (1998) 'Regulation of the glial Na+-dependent glutamate transporters by cyclic AMP analogs and neurons', *Mol. Pharmacol.* **53**: 355–369.

Schmidt, W. and Wolf, G. (1988) 'High-affinity uptake of L-[3H]glutamate and D-[3H]aspartate during postnatal development of the hippocampal formation: a quantitative autoradiographic study', *Exp. Brain Res.* **70**: 50–54.

Schmitt, A., Asan, E., Puschel, B., Jons, T. and Kugler, P. (1996) 'Expression of the glutamate transporter GLT1 in neural cells of the rat central nervous system: non-radioactive in situ hybridization and comparative immunocytochemistry', *Neuroscience* **71**: 989–1004.

Schmitt, A., Asan, E., Puschel, B. and Kugler, P. (1997) 'Cellular and regional distribution of the glutamate transporter GLAST in the CNS of rats: nonradioactive in situ hybridization and comparative immunocytochemistry', *J. Neurosci.* **17**: 1–10.

Schousboe, A., Lisy, V. and Hertz, L. (1976) 'Postnatal alterations in effects of potassium on uptake and release of glutamate and GABA in rat brain cortex slices', *J. Neurochem.* **26**: 1023–1027.

Schultz, K. and Stell, W.K. (1996) 'Immunocytochemical localization of the high-affinity glutamate transporter, EAAC1, in the retina of representative vertebrate species', *Neurosci. Lett.* **211**: 191–194.

Seal, R.P., Daniels, G.M., Wolfgang, W.J., Forte, M.A. and Amara, S.G. (1998) 'Identification and characterization of a cDNA encoding a neuronal glutamate transporter from Drosophila melanogaster [In Process Citation]', *Receptors Channels* **6**: 51–64.

Seal, R.P., Leighton, B.H. and Amara, S.G. (2000) 'A model for the topology of excitatory amino acid transporters determined by the extracellular accessibility of substituted cysteines', *Neuron* **25**: 695–706.

Shashidharan, P. and Plaitakis, A. (1993) 'Cloning and characterization of a glutamate transporter cDNA from human cerebellum', *Biochim. Biophys. Acta* **1216**: 161–164.

Shashidharan, P., Huntley, G.W., Meyer, T., Morrison, J.H. and Plaitakis, A. (1994a) 'Neuron-specific human glutamate transporter: molecular cloning, characterization and expression in human brain', *Brain Res.* **662**: 245–250.

Shashidharan, P., Wittenberg, I. and Plaitakis, A. (1994b) 'Molecular cloning of human brain glutamate/aspartate transporter II', *Biochim. Biophys. Acta* **1191**: 393–396.

Sher, P.K. and Hu, S.X. (1990) 'Increased glutamate uptake and glutamine synthetase activity in neuronal cell cultures surviving chronic hypoxia', *Glia* **3**: 350–357.

Shibata, T., Watanabe, M., Tanaka, K., Wada, K. and Inoue, Y. (1996) 'Dynamic changes in expression of glutamate transporter mRNAs in developing brain', *Neuroreport* **7**: 705–709.

Shimamoto, K., Lebrun, B., Yasuda-Kamatani, Y., Sakaitani, M., Shigeri, Y., Yumoto, N. and Nakajima, T. (1998) 'DL-threo-beta-benzyloxyaspartate, a potent blocker of excitatory amino acid transporters', *Mol. Pharmacol.* **53**: 195–201.

Silver, I.A. and Erecinska, M. (1992) 'Ion homeostasis in rat brain in vivo: intra- and extracellular [Ca2+] and [H+] in the hippocampus during recovery from short-term, transient ischemia', *J. Cereb. Blood Flow Metab.* **12**: 759–772.

Sims, K.D. and Robinson, M.B. (1999) 'Expression patterns and regulation of glutamate transporters in the developing and adult nervous system', *Crit. Rev. Neurobiol.* **13**: 169–197.

Sims, K.D., Straff, D.J. and Robinson, M.B. (2000) 'Platelet-derived growth factor rapidly increases activity and cell surface expression of the EAAC1 subtype of glutamate transporter through activation of phosphatidylinositol 3-kinase', *J. Biol. Chem.* **275**: 5228–5237.

Slotboom, D.J., Lolkema, J.S. and Konings, W.N. (1996) 'Membrane topology of the C-terminal half of the neuronal, glial, and bacterial glutamate transporter family', *J. Biol. Chem.* **271**: 31,317–31,321.

Sluse, F.E. (1996) 'Mitochondrial metabolite carrier family, topology, structure and functional properties: an overview', *Acta Biochim. Pol.* **43**: 349–360.

Spiridon, M., Kamm, D., Billups, B., Mobbs, P. and Attwell, D. (1998) 'Modulation by zinc of the glutamate transporters in glial cells and cones isolated from the tiger salamander retina', *J. Physiol. (Lond.)* **506**: 363–376.

Springer, J.E., Azbill, R.D., Mark, R.J., Begley, J.G., Waeg, G. and Mattson, M.P. (1997) '4-hydroxynonenal, a lipid peroxidation product, rapidly accumulates following traumatic spinal cord injury and inhibits glutamate uptake', *J. Neurochem.* **68**: 2469–2476.

Steinhäuser, C. and Gallo, V. (1996) 'News on glutamate receptors in glial cells', *Trends Neurosci.* **19**: 339–345.

Stern, J.R., Eggleston, L.V., Hems, R. and Krebs, H.A. (1949) 'Accumulation of glutamic acid in isolated brain tissue', *Biochem. J.* **44**: 410–418.

Storck, T., Schulte, S., Hofmann, K. and Stoffel, W. (1992) 'Structure, expression, and functional analysis of a Na(+)-dependent glutamate/aspartate transporter from rat brain', *Proc. Natl. Acad. Sci. USA* **89**: 10,955–10,959.

Storm-Mathisen, J., Danbolt, N.C., Rothe, F., Torp, R., Zhang, N., Aas, J.E., Kanner, B.I., Langmoen, I. and Ottersen, O.P. (1992) 'Ultrastructural immunocytochemical observations on the localization, metabolism and transport of glutamate in normal and ischemic brain tissue', *Prog. Brain Res.* **94**: 225–241.

Sutherland, M.L., Delaney, T.A. and Noebels, J.L. (1995) 'Molecular characterization of a high-affinity mouse glutamate transporter', *Gene* **162**: 271–274.

Sutherland, M.L., Delaney, T.A. and Noebels, J.L. (1996) 'Glutamate transporter mRNA expression in proliferative zones of the developing and adult murine CNS', *J. Neurosci.* **16**: 2191–2207.

Swanson, R.A., Liu, J., Miller, J.W., Rothstein, J.D., Farrell, K., Stein, B.A. and Longuemare, M.C. (1997) 'Neuronal regulation of glutamate transporter subtype expression in astrocytes', *J. Neurosci.* **17**: 932–940.

Szatkowski, M., Barbour, B. and Attwell, D. (1990) 'Non-vesicular release of glutamate from glial cells by reversed electrogenic glutamate uptake', *Nature* **348**: 443–446.

Tachibana, M. and Kaneko, A. (1988) 'L-glutamate-induced depolarization in solitary photoreceptors: a process that may contribute to the interaction between photoreceptors in situ', *Proc. Natl. Acad. Sci. USA* **85**: 5315–5319.

Takahashi, M., Sarantis, M. and Attwell, D. (1996) 'Postsynaptic glutamate uptake in rat cerebellar Purkinje cells', *J. Physiol. (Lond.)* **497**: 523–530.

Takamori, S., Rhee, J.S., Rosenmund, C. and Jahn, R. (2000) 'Identification of a vesicular glutamate transporter that defines a glutamatergic phenotype in neurons', *Nature* **407**: 189–194.

Tan, J., Zelenaia, O., Correale, D., Rothstein, J.D. and Robinson, M.B. (1999) 'Expression of the GLT-1 subtype of Na+-dependent glutamate transporter: pharmacological characterization and lack of regulation by protein kinase C', *J. Pharmacol. Exp. Ther.* **289**: 1600–10.

Tanaka, J., Ichikawa, R., Watanabe, M., Tanaka, K. and Inoue, Y. (1997a) 'Extra-junctional localization of glutamate transporter EAAT4 at excitatory Purkinje cell synapses', *Neuroreport* **8**: 2461–2464.

Tanaka, K. (1993) 'Expression cloning of a rat glutamate transporter. *Neurosci. Res.* **16**: 149–153.

Tanaka, K., Watase, K., Manabe, T., Yamada, K., Watanabe, M., Takahashi, K., Iwama, H., Nishikawa, T., Ichihara, N., Kikuchi, T., Okuyama, S., Kawashima, N., Hori, S., Takimoto, M. and Wada, K. (1997b) 'Epilepsy and exacerbation of brain injury in mice lacking the glutamate transporter GLT-1', *Science* **276**: 1699–1702.

Thompson, G.A. and Kilpatrick, I.C. (1996) 'The neurotransmitter candidature of sulphur-containing excitatory amino acids in the mammalian central nervous system', *Pharmacol. Ther.* **72**: 25–36.

Thorlin, T., Roginski, R.S., Choudhury, K., Nilsson, M., Ronnback, L., Hansson, E. and Eriksson, P.S. (1998) 'Regulation of the glial glutamate transporter GLT-1 by glutamate and delta-opioid receptor stimulation', *FEBS Lett.* **425**: 453–459.

Tong, G. and Jahr, C.E. (1994) 'Block of glutamate transporters potentiates postsynaptic excitation', *Neuron* **13**: 1195–1203.

Torgner, I. and Kvamme, E. (1990) 'Interrelationship between glutamate and membrane-bound ATPases in nerve cells', *Mol. Chem. Neuropathol.* **12**: 19–25.

Torp, R., Andine, P., Hagberg, H., Karagulle, T., Blackstad, T.W. and Ottersen, O.P. (1991) 'Cellular and subcellular redistribution of glutamate-, glutamine- and taurine-like immunoreactivities during forebrain ischemia: a semiquantitative electron microscopic study in rat hippocampus', *Neuroscience* **41**: 433–447.

Torp, R., Arvin, B., Le Peillet, E., Chapman, A.G., Ottersen, O.P. and Meldrum, B.S. (1993) 'Effect of ischaemia and reperfusion on the extra- and intracellular distribution of glutamate, glutamine, aspartate and GABA in the rat hippocampus, with a note on the effect of the sodium channel blocker BW1003C87', *Exp. Brain Res.* **96**: 365–376.

Torp, R., Danbolt, N.C., Babaie, E., Bjoras, M., Seeberg, E., Storm-Mathisen, J. and Ottersen, O.P. (1994) 'Differential expression of two glial glutamate transporters in the rat brain: an in situ hybridization study', *Eur. J. Neurosci.* **6**: 936–942.

Torp, R., Hoover, F., Danbolt, N.C., Storm-Mathisen, J. and Ottersen, O.P. (1997) 'Differential distribution of the glutamate transporters GLT1 and rEAAC1 in rat cerebral cortex and thalamus: an in situ hybridization analysis', *Anat. Embryol. (Berl.)* **195**: 317–326.

Torp, R., Lekieffre, D., Levy, L.M., Haug, F.M., Danbolt, N.C., Meldrum, B.S. and Ottersen, O.P. (1995) 'Reduced postischemic expression of a glial glutamate transporter, GLT1, in the rat hippocampus', *Exp. Brain Res.* **103**: 51–58.

Trotti, D., Danbolt, N.C. and Volterra, A. (1998) 'Glutamate transporters are oxidant-vulnerable: a molecular link between oxidative and excitotoxic neurodegeneration?' *Trends Pharmacol. Sci.* **19**: 328–334.

Trotti, D., Nussberger, S., Volterra, A. and Hediger, M.A. (1997a) 'Differential modulation of the uptake currents by redox interconversion of cysteine residues in the human neuronal glutamate transporter EAAC1', *Eur. J. Neurosci.* **9**: 2207–2212.

Trotti, D., Rizzini, B.L., Rossi, D., Haugeto, O., Racagni, G., Danbolt, N.C. and Volterra, A. (1997b) 'Neuronal and glial glutamate transporters possess an SH-based redox regulatory mechanism', *Eur. J. Neurosci.* **9**: 1236–1243.

Trotti, D., Volterra, A., Lehre, K.P., Rossi, D., Gjesdal, O., Racagni, G. and Danbolt, N.C. (1995) 'Arachidonic acid inhibits a purified and reconstituted glutamate transporter directly from the water phase and not via the phospholipid membrane', *J. Biol. Chem.* **270**: 9890–8985.

Tzingounis, A.V., Lin, C.L., Rothstein, J.D. and Kavanaugh, M.P. (1998) 'Arachidonic acid activates a proton current in the rat glutamate transporter EAAT4', *J. Biol. Chem.* **273**: 17,315–17,317.

Ullensvang, K., Lehre, K.P., Storm-Mathisen, J. and Danbolt, N.C. (1997) 'Differential developmental expression of the two rat brain glutamate transporter proteins GLAST and GLT', *Eur. J. Neurosci.* **9**: 1646–1655.

Utsunomiya-Tate, N., Endou, H. and Kanai, Y. (1997) 'Tissue specific variants of glutamate transporter GLT-1', *FEBS Lett.* **416**: 312–316.

Vandenberg, R.J., Arriza, J.L., Amara, S.G. and Kavanaugh, M.P. (1995) 'Constitutive ion fluxes and substrate binding domains of human glutamate transporters', *J. Biol. Chem.* **270**: 17,668–17,671.

Vandenberg, R.J., Mitrovic, A.D. and Johnston, G.A. (1998) 'Molecular basis for differential inhibition of glutamate transporter subtypes by zinc ions', *Mol. Pharmacol.* **54**: 189–196.

Velaz-Faircloth, M., McGraw, T.S., Malandro, M.S., Fremeau, R.T., Jr., Kilberg, M.S. and Anderson, K.J. (1996) 'Characterization and distribution of the neuronal glutamate transporter EAAC1 in rat brain', *Am. J. Physiol.* **270**: C67–C75.

Volterra, A., Trotti, D., Cassutti, P., Tromba, C., Galimberti, R., Lecchi, P. and Racagni, G. (1992a) 'A role for the arachidonic acid cascade in fast synaptic modulation: ion channels and transmitter uptake systems as target proteins', *Adv. Exp. Med. Biol.* **318**: 147–158.

Volterra, A., Trotti, D., Cassutti, P., Tromba, C., Salvaggio, A., Melcangi, R.C. and Racagni, G. (1992b) 'High sensitivity of glutamate uptake to extracellular free arachidonic acid levels in rat cortical synaptosomes and astrocytes', *J. Neurochem.* **59**: 600–606.

Volterra, A., Trotti, D., Tromba, C., Floridi, S. and Racagni, G. (1994) 'Glutamate uptake inhibition by oxygen free radicals in rat cortical astrocytes', *J. Neurosci.* **14**: 2924–2932.

Wadiche, J.I., Amara, S.G. and Kavanaugh, M.P. (1995a) 'Ion fluxes associated with excitatory amino acid transport', *Neuron* **15**: 721–728.

Wadiche, J.I., Arriza, J.L., Amara, S.G. and Kavanaugh, M.P. (1995b) 'Kinetics of a human glutamate transporter', *Neuron* **14**: 1019–1027.

Wahle, S. and Stoffel, W. (1996) 'Membrane topology of the high-affinity L-glutamate transporter (GLAST-1) of the central nervous system', *J. Cell Biol.* **135**: 1867–1877.

Watase, K., Hashimoto, K., Kano, M., Yamada, K., Watanabe, M., Inoue, Y., Okuyama, S., Sakagawa, T., Ogawa, S., Kawashima, N., Hori, S., Takimoto, M., Wada, K. and Tanaka, K. (1998) 'Motor discoordination and increased susceptibility to cerebellar injury in GLAST mutant mice', *Eur. J. Neurosci.* **10**: 976–988.

Westbrook, G.L. and Mayer, M.L. (1987) 'Micromolar concentrations of Zn2+ antagonize NMDA and GABA responses of hippocampal neurons', *Nature* **328**: 640–643.

Wilson, D.F. and Pastuszko, A. (1986) 'Transport of cysteate by synaptosomes isolated from rat brain: evidence that it utilizes the same transporter as aspartate, glutamate, and cysteine sulfinate', *J. Neurochem.* **47**: 1091–1097.

Yamada, K., Watanabe, M., Shibata, T., Nagashima, M., Tanaka, K. and Inoue, Y. (1998) 'Glutamate transporter GLT-1 is transiently localized on growing axons of the mouse spinal cord before establishing astrocytic expression', *J. Neurosci.* **18**: 5706–5713.

Yamada, K., Watanabe, M., Shibata, T., Tanaka, K., Wada, K. and Inoue, Y. (1996) 'EAAT4 is a post-synaptic glutamate transporter at Purkinje cell synapses', *Neuroreport* **7**: 2013–2017.

Yan, Y.P., Yin, K.J. and Sun, F.Y. (1998) 'Effect of glutamate transporter on neuronal damage induced by photochemical thrombotic brain ischemia', *Neuroreport* **9**: 441–446.

Yin, K.J., Yan, Y.P. and Sun, F.Y. (1998) 'Altered expression of glutamate transporter GLAST mRNA in rat brain after photochemically induced focal ischemia', *The Anatomical Record* **251**: 9–14.

Yu, A.C., Chan, P.H. and Fishman, R.A. (1986) 'Effects of arachidonic acid on glutamate and gamma-aminobutyric acid uptake in primary cultures of rat cerebral cortical astrocytes and neurons', *J. Neurochem.* **47**: 1181–1189.

Zarbiv, R., Grunewald, M., Kavanaugh, M.P. and Kanner, B.I. (1998) 'Cysteine scanning of the surroundings of an alkali-ion binding site of the glutamate transporter GLT-1 reveals a conformationally sensitive residue', *J. Biol. Chem.* **273**: 14,231–14,237.

Zerangue, N., Arriza, J.L., Amara, S.G. and Kavanaugh, M.P. (1995) 'Differential modulation of human glutamate transporter subtypes by arachidonic acid', *J. Biol. Chem.* **270**: 6433–6435.

Zerangue, N. and Kavanaugh, M.P. (1996a) 'Flux coupling in a neuronal glutamate transporter', *Nature* **383**: 634–637.

Zerangue, N. and Kavanaugh, M.P. (1996b) 'Interaction of L-cysteine with a human excitatory amino acid transporter', *J. Physiol. (Lond.)* **493**: 419–423.

Zhang, Y., Bendahan, A., Zarbiv, R., Kavanaugh, M.P. and Kanner, B.I. (1998) 'Molecular determinant of ion selectivity of a (Na+ +K+)-coupled rat brain glutamate transporter', *Proc. Natl. Acad. Sci. USA* **95**: 751–755.

Zhang, Y. and Kanner, B.I. (1999) 'Two serine residues of the glutamate transporter GLT-1 are crucial for coupling the fluxes of sodium and the neurotransmitter', *Proc. Natl. Acad. Sci. USA* **96**: 1710–1715.

Zhang, Y., Pines, G. and Kanner, B.I. (1994) 'Histidine 326 is critical for the function of GLT-1, a (Na+ + K+)-coupled glutamate transporter from rat brain', *J. Biol. Chem.* **269**: 19,573–19,577.

Chapter 14

GABA transporters

Functional and pharmacological properties

Arne Schousboe and Baruch Kanner

Introduction

The inactivation of GABA, the major inhibitory neurotransmitter in the CNS (Roberts, 1991), is brought about by diffusion away from the receptors followed by high affinity transport into the presynaptic GABAergic neuron as well as surrounding astroglial cells (Schousboe, 1981, 1990). Analysis of the kinetic characteristics of GABA transport in a variety of preparations of neurons and astrocytes has led to the proposal that at GABAergic synapsis by far the majority of the released GABA is taken up back into the GABAergic nerve endings allowing GABA to be recycled as neurotransmitter (Hertz and Schousboe, 1987). Thus only about 20 per cent of the released GABA will be lost as a neurotransmitter by uptake into surrounding astrocytes, where it is metabolized via GABA-transaminase, succinate semialdehyde dehydrogenase and the TCA cycle to CO_2 (Schousboe, 1981; Hertz and Schousboe, 1987). On the basis of these considerations it has been proposed that the activity of the astrocytic GABA transporters may be of critical importance for the efficacy of the GABA neurotransmission; hence the elucidation of the pharmacological properties of neuronal and glial GABA transport systems has become of interest (Schousboe, 1990). The advent of the cloning of a number of GABA transporters (Guastella *et al.*, 1990; Lopéz-Corcuera *et al.*, 1992; Liu *et al.*, 1992, 1993; Clark *et al.*, 1992; Borden *et al.*, 1992, 1994, 1995; Borden, 1996) has provided the tools to investigate in closer detail the distribution, molecular properties and function of the GABA transporters. Moreover, the availability of a variety of GABA analogs of restricted conformation has further facilitated the study of the pharmacological properties of these transporters (Krogsgaard-Larsen *et al.*, 1987; Schousboe *et al.*, 1991; Falch *et al.*, 1999). This review will concentrate on providing an update on these aspects.

Basic properties of GABA-transporters

Stoichiometry

High-affinity GABA transport in CNS preparations is well known to be dependent on an intact sodium gradient across the plasma membrane (Schousboe, 1981). Using a variety of systems, such as reconstitution of the transporters in liposomes or expression in for example, *Xenopus* oocytes, it has been demonstrated that the stoichiometry between Na^+, Cl^- and GABA is 2.5:1:1, which is in perfect agreement with the electrogenicity of the transport process (Radian and Kanner, 1983; Keynan and Kanner, 1988; Kavanaugh *et al.*, 1992; Mager *et al.*, 1993; Kanner, 1997).

Table 14.1 Nomenclature of GABA transporters cloned from rat, human and mouse.

Species		*Nomenclature*		
Rat	GAT-1	BGT-1	GAT-2	GAT-3
Human	GAT-1	BGT-1	NC	GAT-3
Mouse	GAT1	GAT2	GAT3	GAT4

For references, see text.

Notes
NC: Not cloned.

Nomenclature of cloned transporters

The fact that GABA transporters have been cloned from different species by different groups of investigators over several years has led to the existence of somewhat conflicting numbering systems, giving rise to some confusion. The first GABA-transporter was cloned from rat brain (Guastella *et al.*, 1990) and it was called GAT-1 (GABA transporter no. 1). Highly homologous transporters were subsequently or simultaneously cloned from the mouse (Lopéz-Corcuera *et al.*, 1992) and humans (Nelson *et al.*, 1990) and these are also termed GAT1. (Note in the mouse nomenclature no hyphenation is used, hence GAT1.) As additional transporters were subsequently cloned, of which one turned out to transport not only GABA but also betaine, the numbering was no longer synchronized among the species and unfortunately the mouse GAT3 and GAT4 correspond to the rat and human GAT-2 and GAT-3, respectively (Liu *et al.*, 1993; Borden *et al.*, 1992, 1994). The betaine/GABA transporter from humans (BGT-1) corresponds to the GAT2 cloned from mice (Lopéz-Corcuera *et al.*, 1992; Borden *et al.*, 1995). In order to facilitate the understanding of this apparently somewhat confusing situation the current nomenclature has been summarized in Table 14.1.

Basic pharmacology of cloned transporters

Kinetic analysis of the four cloned mouse GABA transporters GAT1–4 have yielded K_m values for GABA of about 10–20 μM for GAT1, 3 and 4 and a K_m of 50–100 μM for GAT2 (Liu *et al.*, 1993; Thomsen *et al.*, 1997; Bolvig *et al.*, 1999). Using a number of classical and novel GABA related compounds (see Figure 14.1) as inhibitors of GABA uptake by these transporters (Table 14.2) it has been clear that they differ from each other with regard to pharmacological properties. Among the classical GABA transport inhibitors, nipecotic acid and guvacine inhibit GAT1, 3 and 4 but have no effect on GAT2. It should also be noted that diaminobutyric acid and 3-aminocyclohexanecarboxylic acid (ACHC) inhibit GAT1 more potently than the other three transporters, while β-alanine exhibits the opposite selectivity (Borden, 1996). This may be of interest in relation to the cellular distribution of the transporters (see p. 343). It may also be noted that THPO (4,5,6,7-tetrahydroisoxazolo[4,5-*c*]pyridin-3-ol) and THAO (4,5,6,7-tetrahydroisoxazolo[4,5-*c*]azepin-3-ol) have no effect on the cloned transporters except for a modest action of THPO on GAT3 and THAO on GAT1 (Bolvig *et al.*, 1999). Among lipophilic derivatives of these inhibitors, as well as a series of novel GABA transport inhibitors introduced by Thomsen *et al.* (1997), some interesting

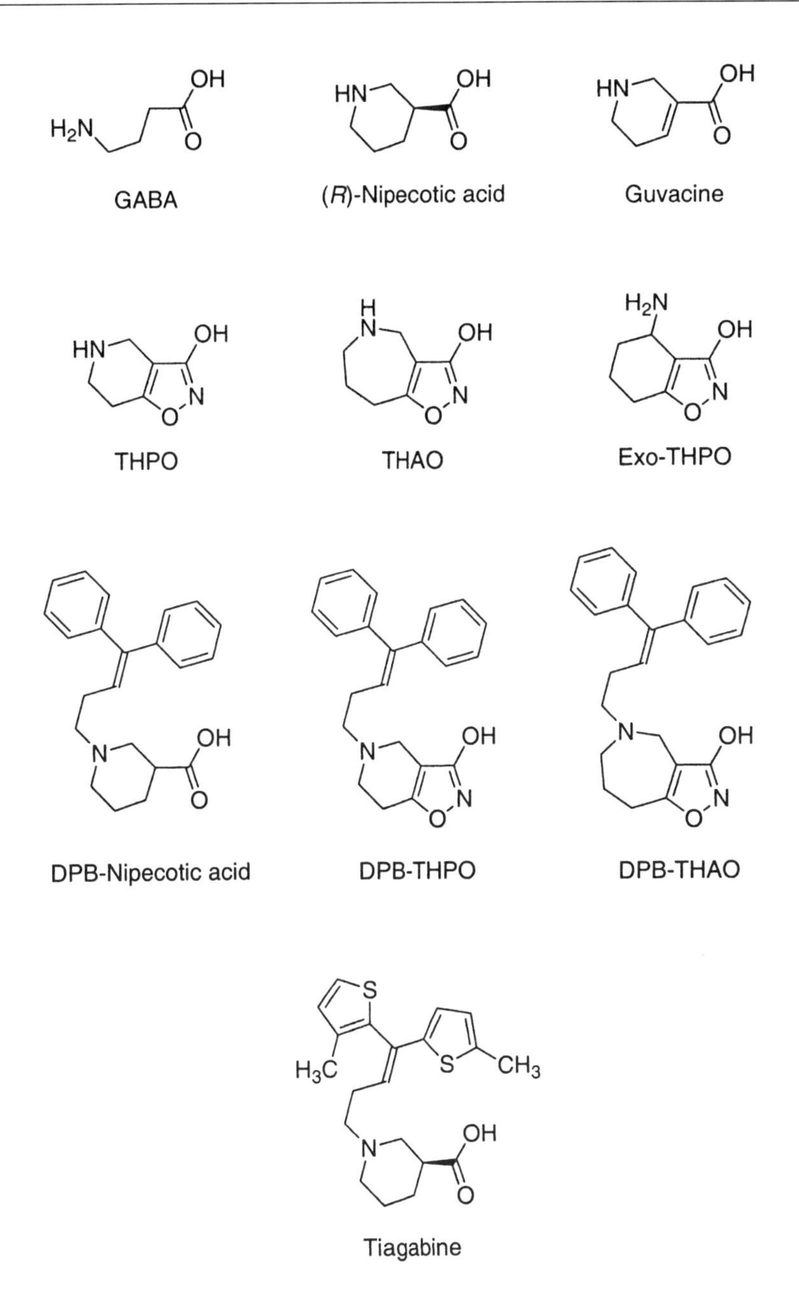

Figure 14.1 Chemical structures of key GABA analogs inhibiting GABA transporters.

results have been obtained (Table 14.2). Tiagabine and DPB-THPO were found to be selective inhibitors of GAT1 while SNAP-5114 preferentially inhibits GAT2–4. Interestingly, the NNC compounds were found to inhibit selectively either GAT2 or GAT4.

Table 14.2 Basic pharmacology of cloned GABA transporters from rat and mouse.

	IC_{50} (μM)			
GABA analogs	*GAT-1/GAT1*	*GAT2*	*GAT-2/GAT3*	*GAT-3/GAT4*
GABA	30/17*	51*	7/15*	33/17*
Nipecotic acid	24	>1000	113	159
Guvacine	39	>1000	228	378
ACHC	132	070†	>1000	>10,000
DABA	128	528†	300	710
β-alanine	2920	1100†	66	110
THPO	1000	3000	800	5000
THAO	1000	4500	>3000	>10,000
DPB-nipecotic acid	0.64	7210†	550	4390
DPB-THPO	30	200	>300	>1000
Tiagabine	0.11	>100	>100	100
SNAP-5114	>30	22	20	6.6
NNC-05 0341	20	4.1	7.4	2.8
NNC-05 1965	23	2.6	10	2.8
NNC-05 1973	43	2.4	34	5.2
NNC-05 2045	27	1.6	14	6.1
NNC-05 2090	19	1.4	41	15

Sources: Data from Borden (1996), Thomsen *et al.* (1997), Bolvig *et al.* (1999) or T. Bolvig, O.M. Larsson and A. Schousboe (unpublished).

Notes
*K_m value; †Human BGT-1.

Pharmacology of neuronal and glial GABA transport

Using cell cultures prepared from mouse cerebral cortex taken from prenatal or neonatal animals, GABA transport in neurons and astrocytes has been investigated with regard to the inhibitory effect of a number of GABA analogs of restricted conformation (Schousboe *et al.*, 1978, 1979, 1981; Larsson *et al.*, 1981, 1983, 1985; Falch *et al.*, 1999). Recently, a large number of diaryloxime and diarylvinyl ether derivatives of nipecotic acid and guvacine have been synthesized and assayed for inhibition of GABA uptake in synaptosomes. Some of these GABA analogs turned out to be more potent inhibitors than tiagabine and the original DPB-derivatives of nipecotic acid and guvacine (Knutsen *et al.*, 1999). No information is at this moment available concerning the effect of these compounds on cloned transporters. Table 14.3 summarizes results from some of the above-mentioned investigations, and it seems clear that the pharmacological profiles of neuronal and glial GABA transport are different from each other. This is particularly obvious for the newly developed second generation isoxazole derivatives exo-THPO and *N*-methyl-exo-THPO, of which the latter exhibits more than a tenfold higher inhibitory potency for glial transport than for neuronal transport of GABA (Falch *et al.*, 1999). This is interesting in the light of the observation that these compounds have equal affinity for GAT1 and no action on any of the other cloned GABA transporters (T. Bolvig, O.M. Larsson and A. Schousboe, unpublished). This therefore raises the question that GABA transporters may exist, the exact nature of which has yet to be elucidated. Obviously, one cannot explain the pharmacological differences between neuronal and glial GABA uptake based on the cellular distribution of

Table 14.3 Basic pharmacology of neuronal and astroglial GABA transport

	IC_{50} values or K_i values (μM)	
GABA analog	*Neurons*	*Astrocytes*
GABA*	8	32
Nipecotic acid	12	16
Guvacine	32	29
ACHC	200	700
DABA	1000	>5000
β-alanine	1666†	843†
THPO	501†	262†
THAO	487†	258†
Exo-THPO	883	208
N-methyl-exo-THPO	423	28
N-ethyl-exo-THPO	391	278
DPB-nip.	1.3†	2.0†
DPB-guv.	4.9	4.2†
DPB-THPO	38†	26†
DPB-THAO	9†	3†
Tiagabine	0.45	0.18

Sources: Data from Schousboe *et al.* (1979), Larsson *et al.* (1981, 1983, 1986, 1988), Braestrup *et al.* (1990) and Falch *et al.* (1999).

Notes
*K_m; †K_i value.

the hitherto cloned GABA transporters (see p. 343) unless it is assumed that the simultaneous expression of several of these transporters in one cell will alter the basic properties of the resulting GABA transport notably. While this may be perfectly possible it has so far not been directly demonstrated.

Molecular studies of ligand binding and transport in GAT-1

According to the original hydropathy analysis of the protein amino acid sequences in GAT-1 by Guastella *et al.* (1990) the GABA transporter contains twelve putative transmembrane α-helices and, moreover, it was assumed that the amino and carboxyl termini were located intracellularly (Figure 14.2). Positioning of consensus phosphorylation sites near the N and C termini, as well as glycosylation sites in the loop between transmembrane domains 3 and 4, is in keeping with this model. Due to conflicting results concerning a possible glycosylation site in the loop between α-helices 2 and 3 the proposed structure of the first three transmembrane domains has been debated (Bennett and Kanner, 1997; Olivares *et al.*, 1997; Yu *et al.*, 1998). However, based on recent studies of the serotonin transporter which belongs to the same superfamily of Na^+ dependent transporters (Chen *et al.*, 1998), the originally proposed model (Guastella *et al.*, 1990) appears to be correct.

Further studies using truncated and mutated GAT-1 have provided information about the parts of the protein which are directly involved in GABA binding. Thus, deletions in both the N- and C-terminal parts of the transporter (Mabjeesh and Kanner, 1992; Bendahan and Kanner, 1993) showed that these parts of the protein are not

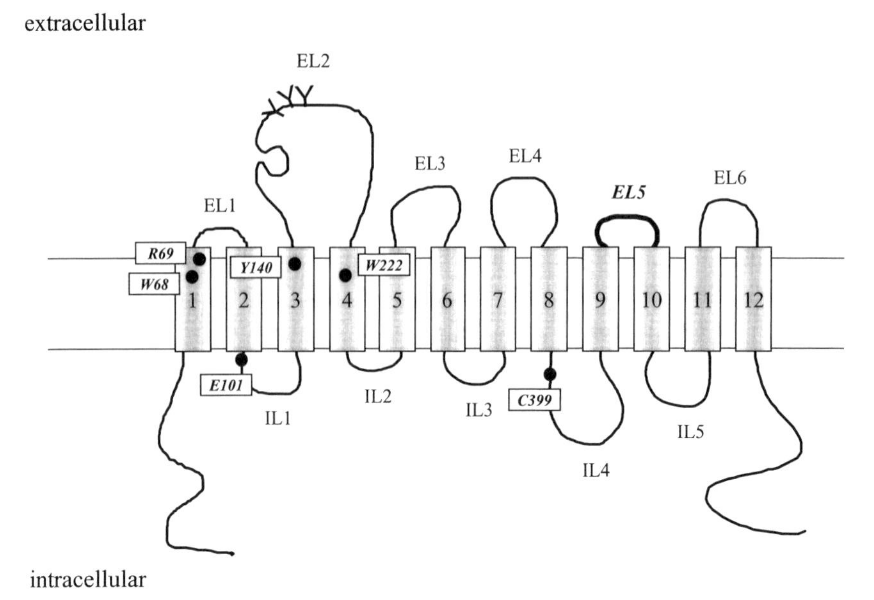

Figure 14.2 Schematic drawing of the cloned GABA transporter GAT-1 showing the transmembrane domains as well as amino acid positions thought to be involved in glycosylation, phosphorylation or interaction with the substrate. For further details, see text.

essential for the ability of the transporter to mediate sodium- and chloride-dependent GABA uptake. Since transport of the ligand by this superfamily of transporters is always sodium- and chloride-dependent it is likely that the membrane domains of the transporters directly involved in this function will contain charged amino acids. Mutation studies aimed at this question have shown that arginine-69 in the first transmembrane α-helix in the GABA transporter is essential for activity (Pantanowitz *et al.*, 1993). In addition to this, aromatic amino acids could be essential for function (Sussman and Silman, 1992). Thus, mutations of tryptophan 68 and 222 in α-helix 1 and 4, respectively, to serine or leucine (Kleinberger-Doron and Kanner, 1994) led to almost complete loss of the GABA transport capacity. Additionally, tyrosine-140 has been shown by mutation to either phenylalanine or tryptophan to be essential for GABA binding and transport (Bismuth *et al.*, 1997). It has also been shown by deletions in the loops between helices 7 and 8 or 8 and 9 that a minimal length of these loops is necessary for activity (Kanner *et al.*, 1994). In this context it is interesting that replacement of glutamate-101 located in the first intracellular loop of GAT-1 by an aspartate residue almost completely abolishes the transport capacity (Keshet *et al.*, 1995). As mentioned above, GAT-1 is inhibited by ACHC but not by β-alanine, while GAT-2 and GAT-3 exhibit the opposite sensitivity to these compounds. It has been demonstrated that the fifth extracellular loop between α-helices 9 and 10 plays a crucial role in this selectivity. Thus, replacement of the amino acid residues of this loop in GAT-1, with the corresponding residues of GAT-3, confers β-alanine sensitivity to GAT-1 (Tamura *et al.*, 1995).

Recently the role of cysteine residues for GABA transport by GAT-1 has been investigated (Golovanevsky and Kanner, 1999). Replacement of cysteine residues by other amino acids by site-directed mutagenesis has revealed an essential role of cysteine-399 in the sensitivity of GAT-1 to the polar -SH reagent (2-aminoethyl) methanethiosulfonate. As this cysteine residue is located intracellularly in the loop between α-helices 8 and 9 it will not normally be accessible to this -SH reagent, but if GABA is bound, the carrier is more sensitive to this reagent. Conversely, binding of the non-transportable (Larsson *et al.*, 1985) GABA analogue *N*-diphenylbutenyl-guvacine (SKF-100 330A) protects the carrier against the -SH reagent, an effect not seen in GAT-1 where cysteine-399 had been replaced with serine (Golovanevsky and Kanner, 1999). It thus appears that the accessibility of this cysteine residue is dependent upon the conformation of the GABA carrier.

Regional and cellular localization of GABA carriers

Among the GABA transporters cloned from rat (GAT-1, GAT-2 and GAT-3), GAT-1 is by far the most abundant and it is found in all brain regions (Ikegaki *et al.*, 1994; Swan *et al.*, 1994; Durkin *et al.*, 1995; Ribak *et al.*, 1996a, 1996b). These studies, together with investigations in neuronal and glial cell cultures, strongly suggest a predominant neuronal localization of GAT-1. Recently, it was reported (Norenberg *et al.*, 2000) that cultured neurons and astrocytes from rat brain express all known GABA transporters, and several studies have identified GAT-1 in glial cells both in the brain and in retina (Radian *et al.*, 1990; Rattray and Priestley, 1993; Brecha and Weigmann, 1994; Durkin *et al.*, 1995; Johnson *et al.*, 1996; Ribak *et al.*, 1996a, 1996b; Yang *et al.*, 1997; De Biasi *et al.*, 1998). Rat GAT-2 is probably not present in brain cells but only in the parenchyma and leptomeninges (Ikegaki *et al.*, 1994). However, GAT3, the mouse homolog of GAT-2, is found in the neonatal brain (Liu *et al.*, 1993). Rat GAT-3 and the mouse homolog GAT4 are found in the olfactory bulb, the midbrain and in cerebellum but are less expressed in the hippocampus and cerebral cortex (Ikegaki *et al.*, 1994; Durkin *et al.*, 1995). At the ultrastructural level astrocytes appear generally to have a higher expression than neurons (Minelli *et al.*, 1996; Ribak *et al.*, 1996a, 1996b; De Biasi *et al.*, 1998). The betain carrier which is denoted GAT2 in the mouse has been found in several areas (Borden, 1996), and it appears to be present both in neurons and glia (Borden, 1996; Norenberg *et al.*, 2000).

It is interesting that both GAT-1 and GAT-3 are present apically (i.e. in axons and cell bodies), whereas GAT-2 (mouse GAT3) and mouse GAT2 are located basolaterally (i.e. in cell bodies and dendrites) (Borden, 1996; Ahn *et al.*, 1996; Muth *et al.*, 1998). This strongly suggests different functional roles of these different carriers. However, at the present time too little is known about these aspects firmly to associate particular functions to specific GABA carriers. In this context it seems imperative that subtype-specific inhibitors are developed; so far only very few drugs are available with such properties (Thomsen *et al.*, 1997; Falch *et al.*, 1999).

Functional and behavioral correlates of GABA transport

GABAergic neurotransmission is likely to be terminated by the action of high affinity GABA transporters (Martin, 1976; Schousboe, 1981), and the cellular and subcellular localization of GAT-1 indicates that this particular transporter may be of special importance. In this context it is interesting that CI-966, which is a potent inhibitor of GAT-1, enhances the action of GABA in the hippocampal CA1 retion (Ebert and Krnjevic, 1990). Moreover, external GABA taken up into presynaptic nerve endings via the GABA carriers is extensively utilized in vesicular release of synaptically active GABA (Belhage *et al.*, 1993; Waagepetersen *et al.*, 1999). In keeping with this, inhibitors of GABA uptake such as THPO and lipophilic derivatives of nipecotic acid, guvacine and THPO increase the extracellular concentration of GABA in the brain (Fink-Jensen *et al.*, 1992; Richards and Bowery, 1996; Juhász *et al.*, 1997). That this effect is indeed of functional importance is substantiated by the repeated observation that these GABA transport inhibitors are anticonvulsants in different seizure models (Schousboe *et al.*, 1983; Gonsalves *et al.*, 1989a, 1989b; Taylor *et al.*, 1990; Swinyard *et al.*, 1991; White *et al.*, 1993; Smith *et al.*, 1995; Suszdak and Jansen, 1995; Dalby *et al.*, 1997, Falch *et al.*, 1999). One of these compounds, tiagabine, is now utilized as an antiepileptic drug (Richens *et al.*, 1995). This clearly shows that epilepsy which is associated with a deficient GABAergic neurotransmission activity (Schousboe, 1990) can be ameliorated by inhibition of GABA transport, leading to increased efficacy of GABAergic synapses. So far it is believed that GAT-1 inhibitors are the most likely candidates to produce efficient antiepileptic drugs. It should, however, be kept in mind that GABA transport inhibitors which not only inhibit GAT-1 but also have a similar or even more potent inhibitory action on the other GABA transporters have proven efficient anticonvulsants (Dalby *et al.*, 1997). More detailed studies of the individual GABA carriers are thus warranted to develop subtype selective inhibitors which may turn out to have as yet undiscovered pharmacological properties. In this context it should be kept in mind that THPO and THAO, which show some selectivity for glial GABA transport but which only marginally inhibit the cloned transporters, have pronounced anticonvulsant activity in several seizure models (Gonsalves *et al.*, 1989a, 1989b). This, together with the recent demonstration of pronounced selectivity for inhibition of glial GABA uptake by exo-THPO and its *N*-methyl derivative, combined with anticonvulsant activity (Falch *et al.*, 1999), underlines the potential importance of control of glial GABA transport with regard to fine tuning of GABAergic activity.

Regulation of expression and activity of GABA transporters

The capacity for cellular GABA transport appears to be regulated at several different levels. Thus, it has been reported that GABA transport in astroglial cells is enhanced by the action of neuronally produced and released proteins (Drejer *et al.*, 1983; Nissen *et al.*, 1992), and an activity dependent increase in GAT-1 activity in rat brain has recently been reported (Bernstein and Quick, 1999). It also appears that phosphorylation of the carrier mediated by protein kinase A and C plays an important role in regulating GABA transport (Corey *et al.*, 1994; Gomeza *et al.*, 1994; Osawa *et al.*, 1994; Tian *et al.*, 1994). Some of these mechanisms apparently involve translocation of transporters from an intracellular site to the cell surface, a phenomenon seen also for glutamate transporters (Davis *et al.*, 1998; Gegelashvili *et al.*, 2000).

Acknowledgements

The expert secretarial assistance by Ms Hanne Danø is cordially acknowledged. Professor P. Krogsgaard-Larsen, Dept. Med. Chem., Royal Danish School of Pharmacy is thanked for drawing the chemical structures shown in Figure 14.1. The work has been supported from the Danish Medical Research Council (grant no. 9700761) and the Lundbeck Foundation.

References

Ahn, J., Mundigl, O., Muth, T.R., Rudnick, G. and Caplan, M.J. (1996) 'Polarized expression of gaba transporters in madin-darby canine kidney-cells and cultured hippocampal-neurons', *J. Biol. Chem.* **271**: 6917–6924.

Belhage, B., Hansen, G.H. and Schousboe, A. (1993) 'Depolarization by K^+ and glutamate activates different neurotransmitter release mechanisms in GABAergic neurons: Vesicular versus non-vesicular release of GABA', *Neuroscience* **54**: 1019–1034.

Bendahan, A. and Kanner, B.I. (1993) 'Identification of domains of a cloned rat brain GABA transporter which are not required for its functional expression', *FEBS Lett.* **318**: 41–44.

Bennett, E.R. and Kanner, B.I. (1997) 'The membrane topology of GAT-1, a (Na^+-Cl^-)-coupled γ-aminobutyric acid transporter from brain', *J. Biol. Chem.* **272**: 1203–1210.

Bernstein, E.M. and Quick, M.W. (1999) 'Regulation of gamma-aminobutyric acid' (GABA) transporters by extracellular GABA', *J. Biol. Chem.* **274**: 889–895.

Bismuth, Y., Kavanaugh, M.P. and Kanner, B.I. (1997) 'Tyrosine 140 of the γ-aminobutyric acid transporter GAT-1 plays a critical role in neurotransmitter recognition', *J. Biol. Chem.* **272**: 16,046–16,102.

Bolvig, T., Larsson, O.M., Pickering, D.S., Nelson, N., Falch, E., Krogsgaard-Larsen, P. and Schousboe, A. (1999) 'Action of bicyclic isoxazole GABA analogues on GABA transporters and its relation to anticonvulsant activity', *Eur. J. Pharmacol.* **375**: 367–374.

Borden, L.A. (1996) 'GABA transporter heterogeneity: pharmacology and cellular localization', *Neurochem. Intl.* **29**: 335–356.

Borden, L.A., Smith, K.E., Hartig, P.R., Branchek, T.A. and Weinshank, R.L. (1992) 'Molecular heterogeneity of the γ-aminobutyric acid (GABA) transport system', *J. Biol. Chem.* **267**: 21,098–21,104.

Borden, L.A., Dhar, T.G.M., Smith, K.E., Weinshank, R.L., Branchek, T.A. and Gluchowski, C. (1994) 'Tiagabine, S.K. and F 89976-A, CI-966, and NNC-711 are selective for the cloned GABA transporter GAT-1', *Eur. J. Pharmacol.* **269**: 219–224.

Borden, L.A., Smith, K.E., Gustafson, E.L., Branchek, T.A. and Weinshank, R.L. (1995) 'Cloning and expression of a betaine/GABA transporter from human brain'. *J. Neurochem.* **64**: 977–984.

Brecha, N.C. and Weigmann, C. (1994) 'Expression of GAT-1, a high-affinity gamma-aminobutyric aid plasma membrane transporter in the rat retina', *J. Neurochem.* **345**: 602–611.

Braestrup, C., Nielsen, E.B., Sonnewald, U., Knutsen, L.J.S., Andersen, K.E., Jansen, J. A., Frederiksen, K., Andersen, P.H., Mortensen, A. and Suzdak, P.D. (1990) '(*R*)-*N*-[4,4-bis(3-methyl-2-thienyl)but-3-en-1-yl]nipecotic acid binds with high affinity to the brain γ-aminobutyric acid uptake carrier', *J. Neurochem.* **54**: 639–647.

Chen, J.G., Chen, S.L. and Rudnick, G. (1998) 'Determination of external loop topology in the serotonin transporter by site-directed chemical labeling', *J. Biol. Chem.* **273**: 12,675–12,681.

Clark, J.A., Deutch, A.Y., Gallipoli, P.Z. and Amara, S.G. (1992) 'Functional expression and CNS distribution of a β-alanine sensitive neuronal GABA transporter', *Neuron* **9**: 337–348.

Corey, J.L., Davison, N., Lester, H.A., Brecha, N. and Quick, M.W. (1994) 'Protein kinase C modulates the activity of a cloned γ-aminobutyric acid transporter expressed in *Xenopus* oocytes via regulated subcellular distribution of the transporter', *J. Biol. Chem.* **269**: 14,759–14,767.

Dalby, N.O., Thomsen, C., Fink-Jensen, A., Lundbeck, J., Sokilde, B., Man, C.M., Sorensen, P.O. and Meldrum, B. (1997) 'Anticonvulsant properties of two GABA uptake inhibitors NNC 05–2045 and NNC 05–2090, not acting preferentially on GAT-1', *Epilepsy Res.* **28**: 51–61.

Davis, K.E., Sraff, D.J., Weinstein, E.A., Bannermann, P.G., Correale, D.M., Rothstein, J.D. and Robinson, M.B. (1998) 'Multiple signaling pathways regulate cell surface expression and activity of the excitatory amino acid carrier 1 subtype of Glu transporter in C6 glioma', *J. Neurosci.* **18**: 2475–2485.

De Biasi, S., Vitellaro-Zuccarello, L. and Brecha, N.C. (1998) 'Immunoreactivity for the GABA transporter-1 and GABA transporter-3 is restricted to astrocytes in the rat thalamus. A light and electron-microscopic immunolocalization', *Neuroscience* **83**: 815–828.

Drejer, J., Meier, E. and Schousboe, A. (1983) 'Novel neuron-related regulatory mechanisms for astrocytic glutamate and GABA high affinity uptake', *Neurosci. Lett.* **37**: 301–306.

Durkin, M.M., Smith, K.e., Borden, L.A., Weinshank, R.L., Branchek, T.A. and Gustafson, E.L. (1995) 'Localization of messenger RNAs encoding three GABA transporters in rat brain: an in situ hybridization study', *Mol. Brain Res.* **33**: 7–21.

Ebert, U. and Krnjevic, K. (1990) 'Systemic CI-966, a new gamma-aminobutyric acid uptake blocker, enhances gamma-aminobutyric acid action in CA1 pyramidal layer in situ', *Can. J. Physiol. Pharmacol.* **68**: 1194–1199.

Falch, E., Perregaard, J., Frølund, B., Søkilde, B., Buur, A., Hansen, L.M., Frydenvang, K., Brehm, L., Bolvig, T., Larsson, O.M., Sanchez, C., White, H.S., Schousboe, A. and Krogsgaard-Larsen, P. (1999) 'Selective inhibitors of glial GABA uptake: Synthesis, absolute stereochemistry and pharmacology of the enantiomers of 3-hydroxy-4-amino-4,5,6,7-tetrahydro-1,2-benzisoxazole (Exo-THPO) and analogues', *J. Med. Chem.* **42**: 5402–5414.

Fink-Jensen, A., Suzdak, P.D., Swedberg, M.D.B., Judge, M.E., Hansen, L., and Nielsen, P.G. (1992) 'The γ-aminobutyric acid (GABA) uptake inhibitor, tiagabine, increases extracllular brain levels of GABA in awake rats', *Eur. J. Pharmacol.* **220**: 197–201.

Gegelashvili, G., Dehnes, Y., Danbolt, N.C. and Schousboe, A. (2000) 'The high-affinity glutamate transporters GLT1, GLAST, and EAAT4 are regulated via different signalling mechanism', *Neurochem. Intl.* **37**: 163–170.

Golovanevsky, V. and Kanner, B.I. (1999) 'The reactivity of the γ-aminobutyric acid transporter GAT-1 toward sulf-hydryl reagents is conformationally sensitive. Identification of a major target residue', *J. Biol. Chem.* **274**: 23,020–23,026.

Gomeza, J., Gimenez, C. and Zafra, F. (1994) 'Cellular distribution and regulation by cAMP of the GABA transporter (GAT-1) mRNA', *Mol. Brain Res.* **21**: 150–156.

Gonsalves, S.F., Twitchell, B., Harbaugh, R.E., Krogsgaard-Larsen, P. and Schousboe, A. (1989a) 'Anticonvulsant activity of intracerebroventricularly administered glial GABA uptake inhibitors and other GABA mimetics in chemical seizure models', *Epilepsy Res.* **4**: 34–41.

Gonsalves, S.F., Twitchell, B., Harbaugh, R.E., Krogsgaard-Larsen, P. and Schousboe, A. (1989b) 'Anticonvulsant activity of the glial GABA uptake inhibitor, THAO, in chemical seizures', *Eur. J. Pharmacol.* **168**: 265–268.

Guastella, J., Nelson, N., Nelson, H., Czyzyk, L., Keynan, S., Miedel, M.C., Davidson, N., Lester, H.A. and Kanner, B.I. (1990) 'Cloning and expression of a rat brain GABA transporter', *Science* **249**: 1303–1306.

Hertz, L. and Schousboe, A. (1987) 'Primary cultures of GABAergic and glutamatergic neurons as model systems to study neurotransmitter functions. I. Differentiated cells', in A. Vernadakis, A. Privat, J.M. Lauder, P.S. Timiras, and E. Giacobini (eds.) *Model Systems of Development and Aging of the Nervous System*, Boston: M. Nijhoff, pp. 19–31.

Ikegaki, N., Saito, N., Hashima, M. and Tanaka, C. (1994) 'Production of specific antibodies against GABA transporter subtypes (GAT1, GAT2, GAT3) and their application to immunocytochemistry, *Mol. Brain Res.* **26**: 47–54.

Johnson, J., Chen, T.K., Rickman, D.W., Evans, C. and Brecha, N.C. (1996) 'Multiple gamma-aminobutyric acid plasma membrane transporters (GAT-1, GAT-2, GAT-3) in the rat retina', *J. Comparative Neurol.* **375**: 212–224.

Juhász, G., Kékesi, K.A., Nyitrai, G., Dobolyi, A., Krogsgaard-Larsen, P. and Schousboe, A. (1997) 'Differential effects of nipecotic acid and 4,5,6,7-tetrahydroisoxazolo[4,5-*c*]pyridin-3-ol on extracellular γ-aminobutyrate levels in rat thalamus', *Eur. J. Pharmacol.* **331**: 139–144.

Kanner, B.I. (1997) 'Sodium-coupled GABA and glutamate transporters: structure and function', in M.E.A. Reith (ed.) *Neurotransmitter Transporters – Structure, Function, and Regulation*, Totowa, New Jersey: Humana Press, pp. 151–169.

Kanner, B.I., Bendahan, A., Pantanowitz, S. and Su, H. (1994) 'The number of amino acid residues in hydrophilic loops connecting transmembrane domains of the GABA transporter GAT-1 is critical for its function', *FEBS Lett.* **356**: 191–194.

Kavanaugh, M.P., Arriza, J.L., North, R.A. and Amara, S.G. (1992) 'Electrogenic uptake of γ-aminobutyric acid by a cloned transporter expressed in oocytes', *J. Biol. Chem.* **267**: 22,007–22,009.

Keshet, G.I., Bendahan, A., Su, H., Mager, S., Lester, H.A. and Kanner, B.I. (1995) 'Glutamate-101 is critical for the function of the sodium and chloride-coupled GABA transporter GAT-1', *FEBS Lett.* **371**: 39–42.

Keynan, S. and Kanner, B.I. (1988) 'γ-Aminobutyric acid transport in reconstituted preparations from rat brain: coupled sodium and chloride fluxes', *Biochemistry* **27**: 12–17.

Kleinberger-Doron, N. and Kanner, B.I. (1994) 'Identification of tryptophan residues critical for the function and targeting of the γ-aminobutyric acid transporter (subtype A)', *J. Biol. Chem.* **269**: 3063–3067.

Knutsen, L.J.S., Andersen, K.E., Lau, J., Lundt, B.F., Henry, R.F., Morton, H.E., Nærum, L., Petersen, H., Stephensen, H., Suzdak, P.D., Swedberg, M.D.B., Thomsen, C. and Sørensen, P.O. (1999) 'Synthesis of novel GABA uptake inhibitors. 3. Diaryloxime and diarylvinyl ether derivatives of nipecotic acid and guvacine as anticonvulsant agents', *J. Med. Chem.* **42**: 3447–3462.

Krogsgaard-Larsen, P., Falch, E., Larsson, O.M. and Schousboe, A. (1987) 'GABA uptake inhibitors: Relevance to antiepileptic drug research', *Epilepsy Res.* **1**: 77–93.

Larsson, O.M., Thorbek, P., Krogsgaard-Larsen, P. and Schousboe, A. (1981) 'Effect of homo-β-proline and other heterocyclic GABA analogues on GABA uptake in neurons and astroglial cells and on GABA receptor binding', *J. Neurochem.* **37**: 1509–1516.

Larsson, O.M., Johnston, G.A.R. and Schousboe, A. (1983) 'Differences in uptake kinetics of cis-3-aminocyclohexane carboxylic acid into neurons and astrocytes in primary cultures', *Brain Res.* **260**: 279–285.

Larsson, O.M., Krogsgaard-Larsen, P. and Schousboe, A. (1985) 'Characterization of the uptake of GABA, nipecotic acid and *cis*-4-OH-nipecotic acid in cultured neurons and astrocytes', *Neurochem. Intl.* **7**: 853–860.

Larsson, O.M., Griffiths, R., Allen, I.C. and Schousboe, A. (1986) 'Mutual inhibition kinetic analysis of γ-aminobutyric acid, taurine, taurine and β-alanine high affinity transport into neurons and astrocytes: Evidence for similarity between the taurine and β-alanine carriers in both cell types' *J. Neurochem.* **47**: 426–432.

Larsson, O.M., Falch, E., Krogsgaard-Larsen, P. and Schousboe, A. (1988) 'Kinetic characterization of inhibition of τ-aminobutyric acid uptake into cultured neurons and astrocytes by 4,4-diphenyl-3-butenyl derivatives of nipecotic acid and guvacine', *J. Neurochem.* **50**: 818–823.

Liu, Q.R., Lopéz-Corcuera, B., Nelson, H., Mandiyan, S. and Nelson, N. (1992) 'Cloning and expression of a cDNA encoding the transporter of taurine and β-alanine in mouse brain', *Proc. Natl. Acad. Sci. USA* **89**:12,145–12,149.

Liu, Q.R., Lopéz-Corcuera, B., Mandiyan, S., Nelson, H. and Nelson, N. (1993) 'Molecular characterization of four pharmacologically distinct γ-aminobutyric acid transporters in mouse brain', *J. Biol. Chem.* **268**: 2104–2112.

Lopéz-Corcuera, B., Liu, Q.R., Mandiyan, S., Nelson, H. and Nelson, N. (1992) 'Expression of a mouse brain cDNA encoding novel γ-aminobutyric acid transporter', *J. Biol. Chem.* **267**: 17,491–17,493.

Mabjeesh, N.J. and Kanner, B.I. (1992) 'The substrates of a sodium- and chloride-coupled γ-aminobutyric acid transporter from rat brain', *J. Biol. Chem.* **267**: 2563–2568.

Mager, S., Naeve, J., Quick, M., Labarca, C., Davidson, N. and Lester, H.A. (1993) 'Steady states, charge movements and rates for a cloned GABA transporter expressed in *Xenopus* oocytes', *Neuron* **10**: 177–188.

Martin, D.L. (1976) 'Carrier-mediated transport and removal of GABA from synaptic regions', in E. Roberts, T.N. Chase, and D.B. Tower, (eds) *GABA in Nervous System Function*, New York: Raven Press, pp. 347–386.

Minelli, A., DeBiasi, S., Brecha, N.C., Zuccarello, L.V. and Conti, F. (1996) 'GAT-3, a high-affinity GABA plasma membrane transporter, is localized to astrocytic processes, and is not confined to the vicinity of GABAergic synapses in the cerebral cortex', *J. Neurosci.* **16**: 6255–6264.

Muth, T.R., Ahn, J. and Caplan, M.J. (1998) 'Identification of sorting determinants in the c-terminal cytoplasmic tails of the gamma-aminobutyric-acid transporters gat-2 and gat-3', *J. Biol. Chem.* **273**: 25,616–25,627.

Nelson, H., Mandiyan, S. and Nelson, N. (1990) 'Cloning of the human brain GABA transporter', *FEBS Lett.* **269**: 181–184.

Nissen, J., Schousboe, A., Halkier, T. and Schousboe, I. (1992) 'Purification and characterization of an astrocyte GABA-carrier inducing protein (GABA-CIP) released from cerebellar granule cells', *Glia* **6**: 236–243.

Norenberg, M.D., Vastag, M. and Zhou, B.-G. (2000) 'GABA transporters in cultured rat astrocytes and neurons', *J. Neurochem.* **74** (suppl.): S80.

Olivares, L., Aragon, C., Gimenez, C. and Zafra, F. (1997) 'Analysis of the transmembrane topology of the glycine transporter GlyT1', *J. Biol. Chem.* **272**: 1211–1217.

Osawa, I., Saito, N., Koga, T. and Tanaka, C. (1994) 'Phorbol ester-induced inhibition of GABA uptake by synaptosomes and by *Xenopus* oocytes expressing GABA transporter (GAT-1)', *Neurosci. Res.* **19**: 287–293.

Pantanowitz, S., Bendahan, A. and Kanner, B.I. (1993) 'Only one of the charged amino acids located in the transmembrane α-helices of the γ-aminobutyric acid transporter (subtype A) is essential for its activity', *J. Biol. Chem.* **268**: 3222–3225.

Radian, R. and Kanner, B.I. (1983) 'Stoichiometry of sodium- and chloride-coupled γ-aminobutyric acid transport by synaptic plasma membrane vesicles isolated from rat brain', *Biochemistry* **22**: 1236–1241.

Radian, R., Ottersen, O.P., Storm-Mathisen, J., Castel, M. and Kanner, B.I. (1990) 'Immunocytochemical localization of the GABA transporter in rat brain', *J. Neurosci.* **10**: 1319–1330.

Rattray, M. and Priestley, J.V. (1993) 'Differential expression of GABA transporter-1 messenger RNA in subpopulations of GABA neurons', *Neurosci. Lett.* **156**: 163–166.

Ribak, C.E., Tong, W.M. and Brecha, N.C. (1996a) 'GABA plasma membrane transporters, GAT-1 and GAT-3, display different distributions in the rat hippocampus', *J. Comp. Neurol.* **367**: 595–606.

Ribak, C.E., Tong, W.M. and Brecha, N.C. (1996b) 'Astrocytic processes compensate for the apparent lack of GABA transporters in the axon terminals of cerebellar Purkinje cells', *Anat. Embryol.* **193**: 379–390.

Richards, D.A. and Bowery, N.G. (1996) 'Comparative effects of the GABA uptake inhibitors, tiagabine and NNC-711, on extracellular GABA levels in the rat ventrolateral thalamus', *Neurochem. Res.* **21**: 135–140.

Richens, A., Chadwick, D.W., Duncan, J.S., Dam, M., Gram, L., Mikkelsen, M., Morrow, J., Mengel, H., Shu, V., Mckelvy, J.F. and Pierce, M.W. (1995) 'Adjunctive treatment of partial seizures with tiagabine: A placebo trial', *Epilepsy Res.* **21**: 37–42.

Roberts, E. (1991) 'Living systems are tonically inhibited, autonomous optimizers, and disinhibition coupled to variability generation is their major organizing principle: Inhibitory command-control at levels of membrane, genome, metabolism, brain, and society', *Neurochem. Res.* **16**: 409–421.

Schousboe, A. (1981) 'Transport and metabolism of glutamate and GABA in neurons and glial cells', *Intl. Rev. Neurobiol.* **22**: 1–45.

Schousboe, A. (1990) 'Neurochemical alterations associated with epilepsy or seizure activity', in M. Dam, and L. Gram, (eds) *Comprehensive Epileptology*, New York: Raven Press, pp. 1–16.

Schousboe, A., Krogsgaard-Larsen, P., Svenneby, G. and Hertz, L. (1978) 'Inhibition of the high affinity net uptake of GABA into cultured astrocytes by β-proline, nipecotic acid and other compounds', *Brain Res.* **153**: 623–626.

Schousboe, A., Thorbek, P., Hertz, L. and Krogsgaard-Larsen, P. (1979) 'Effects of GABA analogues of restricted conformation on GABA transport in astrocytes and brain cortex slices and on GABA receptor binding', *J. Neurochem.* **33**: 181–189.

Schousboe, A., Larsson, O.M., Hertz, L. and Krogsgaard-Larsen, P. (1981) 'Heterocyclic GABA analogues as new selective inhibitors of astroglial GABA transport', *Drug Dev. Res.* **1**: 115–127.

Schousboe, A., Larsson, O.M., Wood, J.D. and Krogsgaard-Larsen, P. (1983) 'Transport and metabolism of GABA in neurons and glia: Implications for epilepsy', *Epilepsia* **24**: 531–538.

Schousboe, A., Larsson, O.M. and Krogsgaard-Larsen, P. (1991) 'GABA uptake inhibitors as anticonvulsants', in G. Tunnicliff, and B.U. Raess, (eds) *GABA Mechanisms in Epilepsy*, New York: Wiley Liss, pp. 165–187.

Smith, K.E., Fried, S.G., Durkin, M.M., Gustafson, E.L., Borden, L.A., Branchek, T.A. and Weinshank, R.L. (1995) 'Molecular cloning of an orphan transporter: a new member of the neurotransmitter transporter family', *FEBS Lett.* **357**: 86–92.

Sussman, J.L. and Silman, I. (1992) 'Acetylcholinesterase: structure and use as a model for specific cation–protein interactions', *Curr. Opin. Struc. Biol.* **2**: 721–729.

Suszdak, P.D. and Jansen, J.A. (1995) 'A review of the preclinical pharmacology of tiagabine – a potent and selective anticonvulsant GABA uptake inhibitor', *Epilepsia* **36**: 612–626.

Swan, M., Najlerahim, A., Watson, R.E. and Bennett, J.P. (1994) 'Distribution of mRNA for the GABA transporter GAT-1 in the rat brain: evidence that GABA uptake is not limited to presynaptic neurons', *J. Anat.* **185** (Pt 2): 315–323.

Swinyard, E.A., White, H.S., Wolf, H.H. and Bondilli, W.E. (1991) 'Anticonvulsant profiles of the potent and orally active GABA uptake inhibitors SK&F 89976-A and SK&F 100330-A and four prototype antiepileptic drugs in mice and rats', *Epilepsia* **32**: 569–577.

Tamura, S., Nelson, H., Tamura, A. and Nelson, N. (1995) 'Short external loops as potential substrate-binding site of gamma-aminobutyric-acid transporters', *J. Biol. Chem.* **270**: 28,712–28,715.

Taylor, C.P., Vartanian, M.G., Schwarz, R.D., Rock, D.M., Callahan, M.J. and Davis, M.D. (1990) 'Pharmacology of Cl-966: A potent GABA uptake inhibitor, *in vitro* and in experimental animals', *Drug Dev. Res.* **21**: 195–215.

Thomsen, C., Sørensen, P.O. and Egebjerg, J. (1997) '1-(3-(9H-carbazol-9-yl)-1-propyl)-4-(2-methoxyphenyl)-4-piperidinol, a novel subtype selective inhibitor of the mouse type II GABA transporter', *Br. J. Pharmacol.* **120**: 983–985.

Tian, Y., Kapatos, G., Granneman, J.G. and Bannon, M.J. (1994) 'Dopamine and γ-aminobutyric acid transporters: differential regulation by agents that promote phosphorylation', *Neurosci. Lett.* **173**: 143–146.

Waagepetersen, H.S., Sonnewald, U. and Schousboe, A. (1999) 'The GABA paradox: Multiple roles as metabolite, neurotransmitter, and neurodifferentiative agent', *J. Neurochem.* **73**: 1335–1342.

White, H.S., Hunt, J., Wolf, H.H., Swinyard, E.A., Falch, E., Krogsgaard-Larsen, P. and Schousboe, A. (1993) 'Anticonvulsant activity of the γ-aminobutyric acid uptake inhibitor *N*-4,4-diphenyl-3-butenyl-4,5,6,7-tetrahydroisoxazolo[4,5-*c*]pyridin-3-ol', *Eur. J. Pharmacol.* **236**: 147–149.

Yang, C.Y., Brecha, N.C. and Tsao, E. (1997) 'Immunocytochemical localization of γ-aminobutyric acid plasma membrane transporters in the tiger salamander retina', *J. Comp. Neurol.* **389**: 117–126.

Yu, N., Cao, Y., Mager, S. and Lester, H.A. (1998) 'Topological localization of cysteine 74 in the GABA transporter, GAT-1, and its importance in ion binding and permeation', *FEBS Lett.* **426**: 174–178.

Part 6

Pathophysiology and therapeutic prospects

Chapter 15

Transgenic models for glutamate receptor function

A.J. Doherty and G.L. Collingridge

Introduction

Glutamate is the major excitatory neurotransmitter in the mammalian central nervous system. The effects of glutamate are mediated by a diverse array of ligand gated ion channels (ionotropic receptors) and G-protein coupled (metabotropic) receptors, revealed over recent years by molecular cloning techniques (reviewed in Hollmann and Heinemann, 1994; Mori and Mishina, 1995; Bettler and Mulle, 1995; Pin and Duvoisin, 1995; Conn and Pin, 1997). The ionotropic receptors are multi-subunit proteins that are subdivided into the NMDA, AMPA and kainate receptors, based on their pharmacological properties. These pharmacological distinctions are reflected in the molecular organisation of these receptors. Thus NMDA receptors are built up from complexes formed by NR1 and NR2A–D subunits while AMPA receptors are built from subunits GluR1–4 and kainate receptors are composed of GluR5–7 and KA1–2. In contrast, the metabotropic glutamate (mGlu) receptors are each composed of a single polypeptide chain and couple to trimeric G-proteins. To date, eight subtypes have been cloned, which can be subdivided into three groups, based on pharmacology, structural similarity and signal transduction mechanisms.

Glutamate plays central roles in neuronal functioning. Excitatory synaptic transmission in the mammalian CNS is almost exclusively glutamatergic. Long-term modulation of synaptic transmission via long-term potentiation (LTP) and long-term depression (LTD) may be the molecular basis of learning and memory (Bliss and Collingridge, 1993; Bear and Abraham, 1996), and alterations in such plasticity can be seen in transgenic models for Alzheimer's disease (reviewed in Seabrook and Rosahl, 1999). Given the complexity of the potential glutamate receptor population in even a single neuron, it is an enormous challenge to delineate the functional responses seen following stimulation to a single receptor sub-type. Although great strides have been made in recent years towards the development of sub-type selective pharmacological agents (detailed in Chapter 4), for the majority of glutamate receptors, very few specific agonists or antagonists exist. The use of transgenic animals in which specific receptors or subunits are knocked out or mutated are a complementary tool for investigating the roles these proteins play in neuronal functioning. In this review, we shall describe transgenic mouse models for glutamate receptors, focusing principally on the role of glutamate receptor sub-types in synaptic plasticity and behaviour. The functional and behavioural responses of these mice have been summarised in Tables 15.1 and 15.2. Work describing the generation of the mice and the major findings are summarised in Figures 15.1 and 15.2.

Updated information regarding these models can found at http://www.bris.ac.uk/synaptic/infor/tools.html

Table 15.1 Summary of the effects of altered NMDA receptor subunit expression in mice. Both cellular and behavioural responses to altered gene expression are indicated. CA1, CA3 and D.G. refer to the CA1, CA3 and dentate gyrus regions of the hippocampal formation; CB refers to the cerebellum, and ? refers to effects that have been inconsistently observed at present

Mouse strain		*Whisker-related map*	*LTP*	*LTD*	*Transmission*	*Spatial learning*	*Contextual learning*	*Motor responses*	*Seizures*
NR1	Lethal	Absent							
NR1 (CA1 k/o)			Not inducible in CA1 Normal in D.G.	Not inducible in CA1		Impaired			
NR2A			Impaired in CA1 and CA3			Impaired	Impaired		
NR2A (C-term)							Impaired	?	
NR2B (over expressed)			Enhanced in CA1	Normal	Normal	Enhanced	Enhanced	Normal	
NR2B	Lethal	Absent		Not inducible in CA1					
NR2B (C-term)	Lethal								
NR2B (2/3 C-term)	Lethal 2–3 days	Absent		Impaired in CA1					
NR2B$^{-/+}$			Impaired in CA3						
NR2C					Impaired in CA3/CB			?	
NR2D (over expressed)			Normal in juvenile in CA1; impaired in adult in CA1	Impaired in juvenile in CA1; normal in adult in CA1		Normal			Retardation of motor seizure development
NR3	Increased dendritic spine density				Increased NMDA-induced current density				
δ – Grid2								Impaired	

Table 15.2 Summary of the effects of altered AMPA/kainate receptor subunit and mGlu receptor expression in mice. Both cellular and behavioural responses to altered gene expression are indicated. CA1, CA3 and D.G. refer to the CA1, CA3 and dentate gyrus regions of the hippocampal formation, CB refers to the cerebellum

Mouse strain		*LTP*	*LTD*	*PPF*	*Transmission*	*Spatial learning*	*Contextual learning*	*Motor responses*	*Seizures*
GluR1		Not inducible in CA1			Normal	Normal	Normal		
GluR2		Increased in CA1	Normal in CA1					Impaired	
GluR2 (edits)	Lethal by P20	Increased in CA1							Mice are epileptic
GluR6				Normal in CA3	Normal in CA3	Normal		Normal	Protected against kainate-induced seizure
mGlu1		Impaired in CA3; normal in CA1	Not inducible in CB	Normal		Impaired	Impaired	Impaired	
mGlu5		Impaired in CA1; normal in CA3	DHPG-induced LTD absent			Impaired	Impaired		
mGlu2		Normal in CA3	Not inducible in CA3	Normal in CA3	Normal in CA3	Normal			
mGlu4				Impaired in CB				Impaired learning of motor skills	
mGlu6	Delayed ON response								
mGlu7		Normal in CA1		Normal in CA1	Lower synaptic response during tetanus		Impaired fear response; impaired CTA		

Ionotropic glutamate receptors

NMDA receptors

A critical role for NMDA receptors in synaptic transmission (Biscoe *et al.*, 1977) and in many forms of synaptic plasticity, including long-term potentiation (Collingridge *et al.*, 1983), long-term depression (Dudek and Bear, 1992; Mulkey and Malenka, 1992), ocular dominance plasticity (Kleinschmidt *et al.*, 1987), formation of somatosensory maps (Simon *et al.*, 1992), epileptogenesis (Anderson *et al.*, 1987) and types of learning and memory (Morris *et al.*, 1986), were established using pharmacological approaches – in particular their selective and reversible blockade using the highly specific NMDA receptor antagonist D-AP5 (originally called D-APV) (Davies *et al.*, 1981). Their key role in synaptic plasticity has made the NMDA receptor a prime target for homologous recombination gene targeting to further understand the roles of this receptor in brain function.

NMDAR1

The first reported NMDA receptor subunit knockout mice were those in which the NR1 allele was disrupted (Li *et al.*, 1994). This resulted in the total loss of the NMDA-induced increase in intracellular free calcium and membrane currents (Forrest *et al.*, 1994), showing that NR1 expression is essential to NMDA receptor function *in vivo*, as had been proposed previously *in vitro* (Ishii *et al.*, 1993). Although global disruption of NR1 expression was lethal within a day of birth, within this time these mice were found to lack whisker-related patterns in the trigeminal nuclei of the brainstem (Li *et al.*, 1994; Forrest *et al.*, 1994), suggesting a role for NMDA receptor activation in somatosensory map formation. Ectopic expression of a NR1-1a transgene has since been used to rescue NR1 knockout mice (Iwasato *et al.*, 1997). The viability of these mice was dependent on the level of expression of the NR1-1a splice variant, as was the formation of somatosensory patterns in the cortex, thalamus, principal nucleus and sub-nucleus interpolaris. Mice expressing high levels of the NR1-1a transgene developed normal maps, while patterning in those mice with low levels of NR1-1a expression was generally poorly segmented or absent.

A second approach to producing viable NR1 knockout mice has been to localise the loss of expression to a particular region of the brain. This has been achieved using a *cre-loxP* recombination technique (Tsien *et al.*, 1996a; see also Wilson and Tonegawa, 1997), to produce a mouse in which the deletion of NR1 expression is restricted to the CA1 region of the hippocampus (Tsien *et al.*, 1996b). These mice have been used to study the effects of NR1 deletion on synaptic plasticity in the CA1 region of the hippocampus. Neither short-term potentiation (STP) nor long-term potentiation (LTP) could be induced in this region in the mutant mice, although LTP was induced in the dentate gyrus (Tsien *et al.*, 1996b). In addition, long-term depression could not be induced by low frequency stimulation, whereas all three forms of synaptic plasticity were readily induced in CA1 neurons in control mice. Mutant mice were also found to be selectively deficient in spatial memory in a Morris water maze test (Tsien *et al.*, 1996b). In addition, they showed a decrease in the specificity of individual CA1 neuronal place fields and a deficit in the co-ordinated firing of pairs of neurons tuned to similar place fields (McHugh *et al.*, 1996). These results have demonstrated the import-

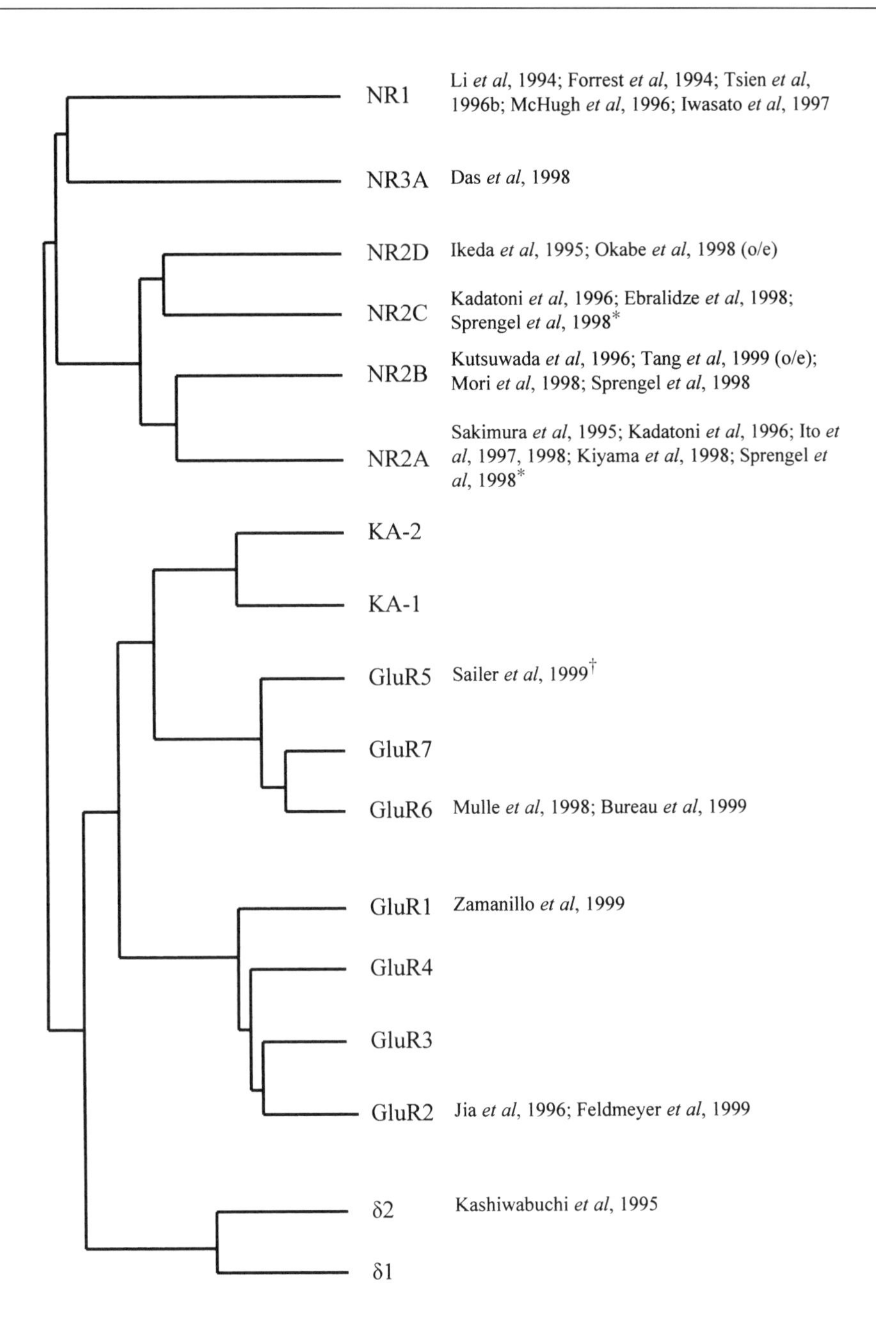

Figure 15.1 Studies using ionotropic glutamate receptor knockout mice. Phylogenic tree of the ionotropic glutamate receptor superfamily based on amino-acid sequence homology (modified from Das *et al.*, 1998). Studies describing the generation and characterisation of mice with altered expression of particular receptor subunits are detailed.

Key
* = C-terminal deletion; † = point mutation in Q/R site; o/e = over-expression; all other mice have a functional deletion of the subunit.

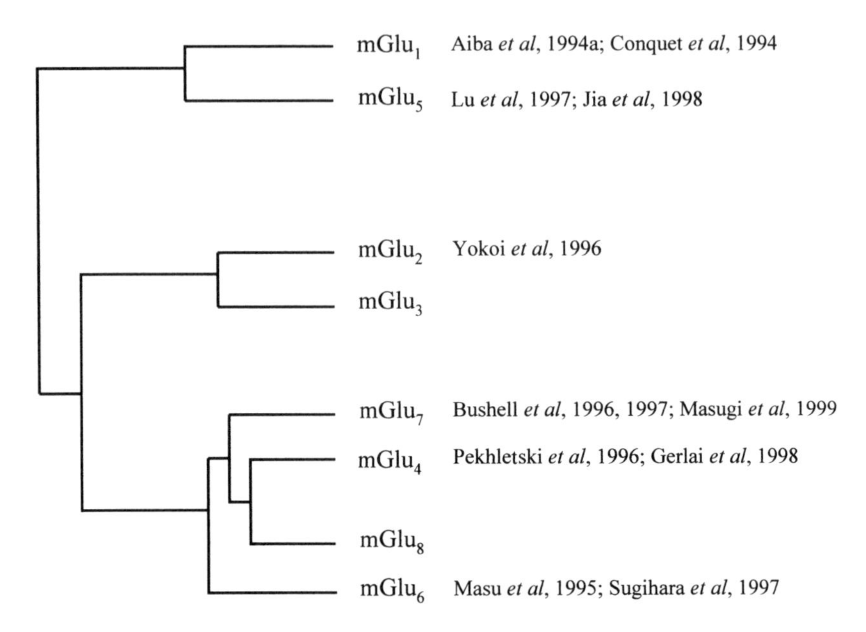

Figure 15.2 Studies using metabotropic glutamate (mGlu) receptor knockout mice. Phylogenic tree of the metabotropic glutamate receptor superfamily based on amino-acid sequence homology (modified from Bortolotto *et al.*, 1999). Studies describing the generation and characterisation of mice with loss of expression of particular receptor subtypes are detailed.

ance of NMDA-mediated synaptic plasticity in CA1 hippocampal neurons to the aquisition and representation of spatial information.

NR2A

Knockout of the NR2A (ϵ1) subunit was shown to reduce, but not eliminate, the NMDA receptor-mediated component of synaptic transmission and the induction of LTP in the CA1 region of the hippocampus (Sakimura *et al.*, 1995) and cerebellum (Kadotani *et al.*, 1996). There was a similar impairment in NMDA receptor-mediated synaptic transmission and LTP in the associational input to CA3 neurons, but interestingly not the fimbrial input to the same cells (Ito *et al.*, 1997). Deletion of the NR2B (ϵ2) subunit had the converse effect, reducing LTP in the fimbrial but not associational inputs to these cells, suggesting that NMDA receptors with different subunit compositions are targeted to CA3 neurons in a synapse-specific manner. Similar effects on LTP in CA1 neurons were seen when the C-terminus of the NR2A subunit was deleted (Sprengel *et al.*, 1998). This deletion did not interfere with the formation of gateable, synaptically activated channels, suggesting that the C-terminus has a specific role in synaptic plasticity, possibly by enabling the appropriate localisation of downstream signalling molecules. Further analysis of the NR1 knockout showed that the LTP deficit could be overcome by giving a stronger tetanic stimulation and that saturating levels were the same as in wild-type mice (Kiyama *et al.*, 1998). Thus the effect of the NR2A knockout is to increase the threshold for LTP rather than to reduce the maximum level of plasticity that can be achieved.

Deletion of the NR2A subunit leads to deficits in Morris water maze learning (Sakimura *et al.*, 1995), aquisition of the conditioned eyeblink response (Kishimoto *et al.*, 1997), and in associative learning immediately after contextual fear conditioning (Kiyama *et al.*, 1998). In this case, mutant mice required a longer period of exposure to a conditioning chamber prior to a footshock to induce a maximum behavioural response (freezing), although the maximum responses in both mutant and wild-type mice were indistinguishable. Deletion of the C-terminus alone induced impairments in contextual memory and motor co-ordination (Sprengel *et al.*, 1998), although no effects on motor co-ordination were seen following ablation of the NR2A subunit itself (Kadatoni *et al.*, 1996).

NR2B

Like the NR1 subunit, functional NR2B (ϵ2) expression appears to be essential for neonatal survival. Mice lacking NR2B were unable to feed due to a defective suckling response and died within 24 hours of birth (Kutsuwada *et al.*, 1996). The life span of the mutant mice could be extended to 2–3 days by hand feeding, and overall brain anatomy and development during this time were indistinguishable from wild-type mice. However, no whisker-related barrels were detected in the trigeminal nucleus of the brainstem at P2, although such structures were easily located in wild-type mice. In addition, LTD could not be induced in hippocampal CA1 neurons from mutant mice, although a robust LTD was induced in age matched controls. These results are remarkably similar to those found for the NR1 global knockout (Li *et al.*, 1994; Forrest *et al.*, 1994). It is interesting to note that loss of the NR1 subunit also resulted in a reduced level of NR2B, although the expression of this gene was not completely lost (Forrest *et al.*, 1994). In addition to the results described above, LTP induction in hippocampal CA3 neurons receiving inputs from fimbrial fibres was severely impaired in heterozygotic mutant mice (Ito *et al.*, 1997).

Similar results have been found for global C-terminal truncation mutants of NR2B. Truncation of the whole of the C-terminus is lethal prenatally (Sprengel *et al.*, 1998), while deletion of about two-thirds results in mice that die within three postnatal days (Mori *et al.*, 1998). These latter mice (termed *GluRϵ2dC*) could be rescued by hand feeding as described for the NR2B knockout (Kutsuwada *et al.*, 1996). *GluRϵ2dC* mice showed disrupted barrel formation in the trigeminal nucleus of the brainstem, but unlike the NR2B null mutant, LTD was demonstrated in hippocampal CA1 neurons in some slices (Mori *et al.*, 1998). C-terminal deletion of the NR2B subunit also affected the distribution and clustering of NR2B containing receptors. Fewer puncta of NR2B immunoreactivity were detected in the dendritic layers of the hippocampus from *GluRϵ2dC* mice compared to wild-type controls, although the number of synapses, measured using an anti-synaptophysin antibody, appeared normal. Similar results were seen in cultured cerebellar neurons prepared from either wild-type or *GluRϵ2dC* mice. These results suggest that the effect of the C-terminal deletion is to disrupt the synaptic targeting and clustering of NMDA receptors. This changed synaptic localisation of NMDA receptors could thus affect the probability of LTD induction, leading to the measured deficit in synaptic plasticity.

Mice have been produced in which the NR2B subunit was over-expressed by linkage to the CaMKII promoter (Tang *et al.*, 1999). These mice showed normal development, brain anatomy and open-field behaviour. No differences in NMDA receptor-mediated synaptic currents were detected within the first 14 post-natal days. However, after 18 days, transgenic mice displayed a significantly longer decay phase of the current and a larger peak amplitude, reminiscent of currents seen in juvenile mice. This is interesting because the NR2B subunit undergoes a down-regulation during the transition to adulthood (Sheng *et al.*, 1994; Okabe *et al.*, 1998), correlating with a shortening of the NMDA receptor-mediated EPSP over this same period.

The effect of over-expression of NR2B is to prolong the opening of the NMDA receptor. This might be expected to enhance the ability of neurons to display NMDA receptor-dependent plasticity in response to synaptic activation. This is indeed the case. Mice over-expressing the NR2B subunit show greatly enhanced LTP in CA1 hippocampal neurons, evoked by either a 10 Hz or 100 Hz tetanus. LTP could not be demonstrated in either transgenic or wild-type mice using a 5 Hz tetanus. LTD expression in both transgenic and wild-type mice was also indistinguishable, as was paired-pulse facilitation. Hence, over-expression of NR2B leads to a selective enhancement of LTP evoked by stimulation in the 10–100 Hz range, the same frequency range that many neurons fire at during behavioural experience.

The ability of mice over-expressing NR2B to retain learned information is enhanced over that displayed by wild-type mice in a number of behavioural tests. Transgenic mice showed enhanced preference in exploring novel objects up to three days following training but not when left for only one hour, demonstrating enhanced long-term visual recognition memory. They also showed stronger freezing responses than wild-type mice in contextual and cued fear conditioning tests when tested between one hour and ten days after training, although no difference could be detected in immediate freezing responses. In addition, transgenic mice showed an enhanced ability to disassociate the fear response from either contextual or tonal cues (fear-extinction) over that displayed by wild-type mice. Spatial learning was also faster in mice over-expressing NR2B, as tested using a hidden platform in a Morris water maze. These tests demonstrate that although both wild-type and transgenic mice are generally able to learn equally well, mice over-expressing NR2B are able to retain information over a much longer time and thus show enhanced long-term memory.

NR2C

Loss of NR2C expression has also been shown to reduce the NMDA receptor-mediated component of EPSCs recorded from cerebellar granule cells (Kadotani *et al.*, 1996). In a second NR2C knockout strain, the stimulation of the mossy fibre-granule cell synapse resulted in an EPSC with a higher peak amplitude but faster decay time (Ebralidze *et al.*, 1996). This study also showed that the effect of NR2C deletion is to remove low-conductance channels (<37 pS) from the NMDA receptor population.

There are conflicting reports on the effects of the disruption of NR2C expression on motor responses. NR2C C-terminal deletion mutant mice have been reported to display deficits in motor co-ordination (Sprengel *et al.*, 1998). However, other single NR2 knockout mice have shown no impairment of motor control (Kadotani *et al.*,

1996; Ebralidze *et al.*, 1996), although the combined NR2A/NR2C knockout mouse could only manage simple co-ordination tasks, such as staying on a rotating rod (Kadotani *et al.*, 1996). The reason for this discrepancy is unclear, but may be due to differences in the motor tests used in different studies.

NR2D

The effect of the loss of NR2D in adult mice appears to be minimal. Homozygous NR2D knockout mice appear normal, have no obvious histological abnormalities and form whisker-related neuronal patterns (e.g. barrels) (Ikeda *et al.*, 1995). No differences were found between heterozygotes and mutant mice in either motor activity or anxiety tests, although mutants displayed reduced spontaneous activity in the open-field test. This may not be unexpected as NR2D is expressed at very low levels in adult brain.

More recently, a transgenic mouse line was generated in which NR2D was over-expressed (Okabe *et al.*, 1998). In this case, evoked currents had lower amplitudes and slower kinetics than in control mice. LTD in hippocampal CA1 neurons was impaired in juvenile mice but with no significant effect on LTP. In mature mice there was a marked impairment, but not complete loss, of NMDA receptor-dependent LTP, while LTD could not be induced in either wild-type or mutant animals. Spatial behaviour in a Morris water maze was normal. Adult mice have also recently been shown to exhibit a marked suppression of kindling epileptogenesis (Bengzon *et al.*, 1999). This was limited to a retardation of motor seizure development and a suppression of the afterdischarge duration during the development of kindling. No differences were observed in the behavioural or electroencephalographic expression of the established epileptic seizures. Thus, NR2D over-expression affects the development of epilepsy, not the already established condition.

NR3A

The NR3A subunit (also known as χ-1 and NMDAR-L) of the NMDA receptor is a recently characterised subunit expressed primarily during brain development (Ciabarra *et al.*, 1995; Sucher *et al.*, 1995) and has since been shown to exist in receptor complexes with NR1 and NR2 subunits and to modify single channel conductances (Das *et al.*, 1998). Mice lacking this subunit have been reported to show a threefold increase in NMDA-induced current density in acutely dissociated cerebrocortical neurons, compared to wild-type mice (Das *et al.*, 1998). This study also showed that knockout of NR3A subunit expression resulted in a dramatic increase in dendritic spine density, larger spine heads and longer spine necks in regions where NR3A is normally expressed; cells in which NR3A is not normally expressed, such as cerebellar Purkinje cells, showed normal spine development. This increased spine density was evident in both juvenile and adult animals, although overall brain morphology was normal and no behavioural abnormalities were observed. These results suggest that the NR3A subunit of the NMDA receptor functions as a regulatory subunit and is involved in dendritic spine development.

AMPA receptors

AMPA (formerly quisqualate) receptors are the principal mediators of fast excitatory synaptic transmission in the brain (Watkins and Evans, 1981; Collingridge and Lester,

1989). AMPA receptors have been the focus of considerable interest, partly because they mediate a synaptic response that is modifiable during LTP (Collingridge *et al.*, 1983). Thus, it is believed that an understanding of the regulation of AMPA receptors will provide major insights into the molecular basis of learning and memory.

A mouse strain in which the AMPA receptor subunit GluR1 has been deleted was recently reported (Zamanillo *et al.*, 1999). This mouse showed normal brain development and neural architecture. Synapse formation and spine development was indistinguishable from wild-type mice, as was AMPA receptor-mediated synaptic transmission, paired-pulse facilitation and Ca^{2+} transients in spines of CA1 hippocampal neurons. However, the hippocampal distribution of GluR2 in GluR1$^{-/-}$ mice was radically altered, with increased expression in the cell body layer and little expression in the dendritic layers. This may indicate the importance of GluR1 in the assembly or dendritic targeting of GluR2. The presence of normal synaptic currents in GluR1$^{-/-}$ mice is curious, particularly since the size of AMPA receptor-mediated currents recorded from the soma of CA1 neurons was greatly reduced, indicating a severely restricted number of AMPA receptors. It thus seems likely that loss of GluR1 expression results in the reduction of the number of functional AMPA receptors, but that those that remain are preferentially trafficked to synapses. Interestingly, there was a complete absence of LTP at CA3 to CA1 synapses in hippocampal slices obtained from adult GluR1$^{-/-}$ mice. Two explanations were offered for this deficit; the lack of a reserve of AMPA receptors required for their rapid insertion at synapses, or the absence of the critical subunit that is required to enable phosphorylation. A plausible mechanism, surprisingly not proposed by the authors, is that LTP in adult mice involves an increase in AMPA receptor single channel conductance that is mediated by CaMKII-dependent phosphorylation of ser831 on GluR1 (Benke *et al.*, 1998; Derkach *et al.*, 1999). The lack of LTP in GluR1$^{-/-}$ mice did not affect their ability to perform hidden platform tests in a Morris water maze, indicating that spatial learning was unaffected by the loss of GluR1 and LTP in the CA1 of the hippocampus.

All other data published to date regarding AMPA receptor subunit knockouts has concerned the GluR2 subunit. GluR2 is a particularly interesting subunit in the AMPA receptor complex since it controls the Ca^{2+} permeability of the channel via RNA-editing of the Q/R site (Sommer *et al.*, 1991). It is therefore of importance to determine the role of the GluR2 subunit in synaptic function.

The first GluR2 knockout mouse was reported by Jia *et al.* (1996). This line showed multiple effects on both synaptic plasticity and behaviour. Ca^{2+} permeability of AMPA receptors in mutant mice was increased ninefold over that in wild-type mice, consistent with the loss of the Ca^{2+}-impermeable subunit. Interestingly, LTP in CA1 hippocampal neurons was enhanced twofold in these mice, compared to wild-type littermates, whereas LTD was unaffected. Furthermore, whereas the induction of LTP in wild-type mice was eliminated by D-AP5, LTP could be elicited in the GluR2$^{-/-}$ mouse in the presence of D-AP5 (and nifedipine). This suggests that Ca^{2+} permeation through AMPA receptors can induce a form of LTP, and that this is additional to NMDA receptor-dependent LTP. The removal of GluR2 from the AMPA receptor population also had significant effects on the behaviour of mutant mice. Adult mutant mice were normal in appearance, although young mice were smaller than control littermates and had a higher mortality rate. However, behavioural analysis of adult mutant mice showed that they had impaired exploratory capacities (both novelty-induced exploration in the open field and object exploration), reduced grooming, disrupted motor co-ordination (rotating rod) and impaired eye closure responses to approaching objects. GluR2$^{-/-}$ mice are also proving useful in the study of synaptic plasticity. For instance, data from Mainen *et al.* (1998) suggest a postsynaptic location for LTP expression in hippocampal CA1 neurons.

Another approach to the question of the role of Ca^{2+} permeability of AMPA receptors on neuronal function and behaviour has been to generate mutant mice with varying levels of GluR2 expression and varying amounts of editing of the Q/R site in GluR2 (Brusa *et al.*, 1995; Feldmeyer *et al.*, 1999). A series of mice were generated with up to 100 per cent of unedited GluR2 transcripts. A corresponding rise in the Ca^{2+} permeability of the AMPA receptors expressed in these mice was also reported (up to 29-fold), which was associated with the ability to induce AMPA receptor-mediated LTP. These findings support those from the GluR2$^{-/-}$ mice (Jia *et al.*, 1996). However, NMDA receptor-mediated and AMPA receptor-mediated LTP were not additive phenomena, although this *was* the case in the absence of the GluR2 subunit (Jia *et al.*, 1996). This may suggest a modulatory role for GluR2 in the expression of LTP unrelated to global Ca^{2+} permeability, possibly via interactions with intracellular proteins.

Phenotypic differences in the effects of GluR2 knockout and simply altering the Ca^{2+} permeability of the AMPA receptor population are also evident. In contrast to GluR2 knockout mice, which are mostly viable (Jia *et al.*, 1996; Lu *et al.*, 1997), mice carrying mutations of GluR2 that raise the Ca^{2+} permeability are epileptic, hyperexcitable and die between P20–P25 (Brusa *et al.*, 1995; Feldmeyer *et al.*, 1999). Mice that express completely unedited GluR2 receptors are small and display extreme hypotrophy and lethargy; death occurs between P4 and P20 (Feldmeyer *et al.*, 1999). These mice also have architectural defects in individual CA3 neurons, imaged by biocitin labelling, but there was no effect on the formation of the whisker-related barrel-field somatosensory map, suggesting that AMPA receptors play little or no part in the activity dependent structuring of synaptic connections in the neocortex.

Kainate receptors

All five of the kainate receptor genes (GluR5, 6, 7, KA-1, KA-2) have been knocked out, but at the present time only information on the GluR6 knockout has been published (Mulle *et al.*, 1998). Given the high concentration of kainate receptors in the hippocampus, this area was studied in most detail. High affinity [^{3}H]-kainate binding in the CA3 region and dentate gyrus was absent in the GluR6 knockout mouse, and kainate receptor-mediated responses in CA3 neurons were absent (residual affects of high concentrations of kainate were due to activation of AMPA receptors). The synaptic activation of kainate receptors, which can be evoked by high frequency stimulation of the mossy fibre pathway and revealed following blockade of AMPA receptors using GYKI53655 (Castillo *et al.*, 1997; Vignes and Collingridge, 1997), was absent in the GluR6 knockout. It should be noted that this synaptic response is also sensitive to GluR5 selective antagonists, such as LY294486 (Vignes *et al.*, 1997). The simplest explanation for this finding is that the kainate receptor responsible for the mediation of mossy fibre synaptic transmission exists as a heteromer comprising GluR5 and GluR6 (and possibly other) subunits.

Behavioural analysis of the GluR6$^{-/-}$ mice revealed no significant effects on motor learning (latency to fall from rotating rod), or spatial learning (finding a hidden platform in Morris water maze), although GluR6$^{-/-}$ mice were less active that wild-type mice (Mulle *et al.*, 1998). However, GluR6$^{-/-}$ mice were almost all protected against a seizure syndrome induced by intraperitoneally administered kainate ($20\,mg\,kg^{-1}$). GluR6$^{+/+}$ and GluR6$^{+/-}$ mice all suffered seizures under the same conditions (Mulle *et al.*, 1998). Similar results were found for kainate-induced *c-fos* expression, which precedes kainate-induced neuronal damage and neuronal degeneration. GluR6$^{+/+}$ and GluR6$^{+/-}$ mice showed a marked increase in c-Fos immunoreactivity and strong induction of GFAP (an indicator of neuronal damage) in dentate gyrus and CA3 neurons following kainate administration. No detectable increases of either protein were seen in GluR6$^{-/-}$ mice. In summary, analysis of the GluR6 knockout has shown that this subunit is an essential component of postsynaptic kainate receptors in the CA3 region of the hippocampus, and is critically involved in kainate-induced seizures and neuronal death in the hippocampus.

In hippocampal CA1 pyramidal neurons (Bureau *et al.*, 1999), the activation of kainate receptors, by kainate or domoate, resulted in small currents (compared to CA3 neurons) that were also eliminated in the GluR6 knockout. In contrast, the excitation of interneurons and the associated alterations in synaptic inhibition, manifest as a decrease in evoked monosynaptic GABAergic IPSCs and an increase in spontaneous IPSCs, was similar in GluR6 knockout and wild-type mice. These results suggest that GluR6 kainate receptor subunits are not involved in the kainate receptor(s) responsible for these actions. Pharmacological studies have presented evidence that GluR5 subunits are involved in these actions of kainate (Clarke *et al.*, 1997; Cossart *et al.*, 1998). Therefore this suggests the existence of GluR6 lacking kainate receptors that regulate synaptic inhibition.

Recently, GluR5 function has been altered genetically by generating two mouse strains with mutations at amino acid residue 636, which is the Q/R site in the second membrane domain and controls Ca^{2+} permeability and single channel conductance in wild type receptors (Sailer *et al.*, 1999). GluR5(R^{loxP}/R^{loxP}) encodes an arginine at this site whereas GluR5(wt^{loxP}/wt^{loxP}) encodes a glutamine. DRG neurons isolated from the GluR5(R^{loxP}/R^{loxP}) mutant showed fewer and smaller responses to domoate, compared with GluR5(wt^{loxP}/wt^{loxP}) or wild-type mice. However, no differences in the behaviour of these mice could be discerned. It was particularly surprising that there was no difference between the two types of mutants in their responses to noxious thermal or chemical stimuli, given the recent pharmacological evidence implicating GluR5 subunits in nociception (Simmons *et al.*, 1998).

Delta

The δ receptor performs a largely unknown function in the mammalian CNS. Knockout of δ2 (Grid2) expression resulted in the disruption of synaptogenesis at Purkinje cell synapses and LTD of parallel fibre-Purkinje cell synaptic transmission (Kashiwabuchi *et al.*, 1995). Motor co-ordination was also disrupted. Loss of Grid2 expression is also associated with the Hotfoot mutations in mice (allelic to Lurcher), characterised by cerebellar ataxia associated with relatively mild cerebellar abnormalities (Lalouette *et al.*, 1998). Thus a role in motor co-ordination appears likely for this receptor subunit.

Metabotropic glutamate receptors

The effects of a number of mGlu receptor knockout mouse strains were recently reviewed (Baskys, 1998). We shall review the current state of knowledge regarding each knockout mouse within the context of each group of receptors.

Group I mGlu receptors

mGlu$_1$ receptor

Two simultaneous reports described the effects of the loss of the mGlu$_1$ receptor on neuronal activity in mice, generated either by deletion of the start codon and glutamate binding domain (Aiba *et al.*, 1994a, 1994b) or by creation of a non-functional, mGlu$_1$-βGal fusion through insertion of the *lacZ* reporter gene and stop codon into the mGlu$_1$ receptor coding sequence (Conquet *et al.*, 1994; Conquet, 1995). Both mouse strains showed normal neural development, with the exception of a population of cerebellar Purkinje cells that were innervated by multiple climbing fibres in adulthood (Kano *et al.*, 1997; Levenes *et al.*, 1997). Synaptic transmission and hippocampal LTD were also normal, but cerebellar LTD was greatly impaired (Aiba *et al.*, 1994a, 1994b; Conquet *et al.*, 1994). However, different effects on hippocampal LTP were reported. Mice from Aiba *et al.* (1994a) were found to display reduced LTP in CA1 hippocampal neurons, while LTP was found to be normal in CA3 neurons (Hsai *et al.*, 1995). Mice from Conquet *et al.* (1994) were found to be deficient in mossy fibre LTP (CA3 neurons), while LTP in CA1 neurons was normal. The reason for this discrepancy is still unclear, but the results of Conquet *et al.* (1994) are supported by work using the mGlu$_1$ receptor selective antagonists 4-CPG, AIDA and EtCCCP, which were unable to block LTP in hippocampal CA1 neurons (McCaffery *et al.*, 1998). In addition, the mGlu$_1$ receptor has been localised only to a specific type of interneuron in the CA1 region of the hippocampus, whereas it is widely distributed in area CA3 (Luján *et al.*, 1996).

Despite the differences found in hippocampal LTP, similar motor deficits were apparent in the two mouse strains. Both mice developed normally for two or three weeks, but then developed impaired balance, a staggering gait and whole body tremors upon initiation of movement, although exploratory interest was maintained (Conquet *et al.*, 1994). Mice were also found to have deficits in a contextual (but not tone conditioned) fear response (Aiba *et al.*, 1994a), in the conditioned eyeblink response (Aiba *et al.*, 1994b) and in spatial learning (Conquet *et al.*, 1994).

mGlu$_5$ receptor

The other mGlu receptor in group I is the mGlu$_5$ receptor, and a mouse lacking this receptor has been generated (Lu *et al.*, 1997). In contrast to mGlu$_1$ knockout mice, these animals show no developmental or motor co-ordination deficits. Synaptic transmission and paired-pulse facilitation of the Schaffer-collateral pathway in area CA1 of the hippocampus was essentially normal, although there was a slightly reduced NMDA receptor-mediated component of the synaptic response. In addition, mGlu$_5$ receptors were shown to mediate the postsynaptic excitatory effects of (1*S*,3*R*)-ACPD in area CA1, thus supporting immunochemical data that mGlu$_5$ is the principal group I receptor in this region (Luján *et al.*, 1996). The ability of low concentrations of (1*S*,3*R*)-ACPD to induce a reversible depression in synaptic transmission in this region was also absent in mutant mice.

The effect of $mGlu_5$ receptor knockout on LTP in the hippocampus has been investigated in these mice. In the first report (Lu *et al.*, 1997), a small reduction in hippocampal LTP was demonstrated. However, it was subsequently shown that disruption of $mGlu_5$ receptor expression specifically abolished the NMDA receptor-mediated component of LTP (LTP_{NMDA}) in CA1 neurons while AMPA receptor-mediated LTP in the same neurons was unaffected (Jia *et al.*, 1998). The loss of LTP_{NMDA} could be reversed by activation of PKC (Jia *et al.*, 1998). Loss of $mGlu_5$ receptors has also been noted to abolish DHPG-induced potentiation of NMDA responses and DHPG-induced LTD in area CA1 of the hippocampus (Schnabel *et al.*, 1999).

The behavioural response of the $mGlu_5$ mutant mice is mixed. No developmental behavioural deficits were seen and no difference could be detected in their locomotor and exploratory behaviour to wild-type mice in open field tests (Lu *et al.*, 1997). However, like $mGlu_1$ mutant mice (Conquet *et al.*, 1994), significant deficits were seen in the ability of $mGlu_5$ mutant mice to perform the hidden platform test in a Morris water maze, indicating that these mice have impaired spatial learning. Non-spatial learning (e.g. a visible platform test) was comparable to that of control mice. In addition, and again similar to $mGlu_1$ receptor mutant mice (Aiba *et al.*, 1994a), $mGlu_5$ receptor mutant mice were impaired in contextual, but not tone conditioned, fear response. Hence, although these two receptors may play different modulatory roles in LTP, the effects of their loss on spatial and associational learning appear to be very similar.

Group II mGlu receptors

Group II mGlu receptors consist of $mGlu_2$ and $mGlu_3$ receptors. To date the only transgenic mouse reported within this group is one in which the $mGlu_2$ receptor is deleted (Yokoi *et al.*, 1996). The $mGlu_2$ receptor is found in the presynaptic elements of the mossy fibre-CA3 synapse. Loss of this receptor had no effect on basal transmission or paired-pulse facilitation. In addition, no histological changes were seen in mutant mice. Furthermore, $mGlu_2$ receptor mutant mice showed normal LTP at the mossy fibre-CA3 synapse and displayed no deficits in spatial learning, assessed by Morris water maze tests. Interestingly, however, LTD at the mossy fibre synapse was absent in the mutants.

Group III mGlu receptors

Group III receptors consist of $mGlu_4$, $mGlu_6$, $mGlu_7$ and $mGlu_8$. To date, no knockout mouse has been reported for $mGlu_8$.

$mGlu_4$ receptor

A $mGlu_4$ receptor knockout mouse was generated by insertion of a pgk-1/*neo* cassette into an exon in the first third of the N-terminal domain (Pekhletski *et al.*, 1996). Homozygous mutant mice were indistinguishable from their wild-type littermates and immunoblots showed that $mGlu_4$ receptor immunoreactivity was absent from the cerebellum, a region of high $mGlu_4$ receptor expression (Tanabe *et al.*, 1993). In addition, later work demonstrated that levels of [^{3}H]-L-AP4 binding was also greatly reduced in the molecular layer of the cerebellar cortex (Thompsen and Hampson, 1999). This study also showed reduced [^{3}H]-L-AP4 binding in the molecular layer of the dentate gyrus in the hippocampus, again in agreement with the distribution of the $mGlu_4$ receptor (Shigemoto *et al.*, 1997). Expression of other glutamate receptor subunits was unaffected (Pekhletski *et al.*, 1996).

Synaptic plasticity in the parallel fibre–Purkinje cell excitatory synapse was investigated in cerebellar slices. The suppression of synaptic response following application of L-AP4 seen in wild-type mice was absent in mutant mice (Pekhletski *et al.*, 1996). Paired-pulse facilitation was impaired in $mGlu_4$ receptor mutant mice. There was with a smaller facilitation than in control mice seen at all inter-pulse intervals tested, and a small paired-pulse depression at inter-pulse intervals of over 300 msec. In addition, LTD was enhanced in mutant mice compared to wild-type controls.

Further experiments were performed in which the size of a synaptic response was compared to a control response after tetanic stimulation (Pekhletski *et al.*, 1996). In wild-type mice, a small post-tetanic potentiation (PTP) was observed, but in mutant mice the effect was a post-tetanic depression (PTD). It was suggested that the $mGlu_4$ receptor acts as an inhibitory autoreceptor, present on the presynaptic terminals of the parallel fibres, acting to reduce glutamate release.

The behavioural responses of the $mGlu_4$ receptor knockout mice indicate that they are deficient in the ability to learn motor skills, but are not deficient in motor co-ordination *per se*. No differences were seen in either open field, bar cross or rotating rod tests between mutant and wild-type mice (Pekhletski *et al.*, 1996). However, mutant mice learned this latter task at a slower rate and had a shorter latency to fall than wild-type mice. In contrast, mutant mice learned a new platform position faster that wild-type mice in the hidden platform test in a Morris water maze, with reduced escape latencies in the early stages of training (Gerlai *et al.*, 1998). However, when mice were then left for six weeks, mutants displayed impaired accuracy in swimming over the position of the platform, indicating that these mice were unable to retain the learned information as well as wild-type mice.

$mGlu_6$ receptor

The $mGlu_6$ receptor is restricted in its distribution to ON bipolar cells in the retina (Nakajima *et al.*, 1993). Loss of the $mGlu_6$ receptor does not affect retinal morphology or the distribution and connectivity of ON bipolar cells (Tagawa *et al.*, 1999). However, the ON response of these cells is affected. Mutant mice were originally reported to have lost the ON response to light stimulus in retinal bipolar cells and ON-type cone cells, but to have retained the OFF response to the dark (Masu *et al.*, 1995). However, the mice were still able to respond to light and this was interpreted as revealing that the OFF response is important in transmitting visual information. A later study, however, demonstrated that the ON response was in fact delayed (Sugihara *et al.*, 1997). In wild-type mice, the ON response appeared with a latency of 50–100 ms, whereas in the mutant mice the latency for the ON response was 200–300 ms. In addition, the amplitude of this response decreased with increasing light intensity. Thus, loss of the $mGlu_6$ receptor, although altering the ON response from the wild-type condition, is not necessary for the generation of that response. The $mGlu_6$ receptor may therefore be a modulatory receptor.

$mGlu_7$ receptor

Targeted disruption of the $mGlu_7$ receptor gene has been reported to result in impaired synaptic plasticity (Bushell *et al.*, 1996, 1997). Gene disruption resulted in complete loss of $mGlu_7$ receptor expression, including splice variants (Flor *et al.*, 1997). LTP and paired-pulse facilitation in hippocampal CA1 neurons were not statistically different in wild-type and mutant mice, but STP was reduced for over 15 minutes in $mGlu_7^{-/-}$ mice compared to controls. In addition, during tetanic stimulation, synaptic responses were lower in mutant mice than in wild-type littermates, as was the response to a single stimulus delivered either 50 ms or 500 ms following the tetanus (Bushell *et al.*, 1997). These results suggest that $mGlu_7$ receptors expressed in the hippocampus undergo synaptic activation during high frequency transmission and may act to facilitate glutamate release for many minutes following stimulation.

Further work with this mouse has indicated that $mGlu_7$ receptor activation is critical in amygdyla function (Masugi *et al.*, 1999). In this study, the fear response of mice to footshock was assessed, a response in which the amygdyla plays a central role (Maren and Fanselow, 1996). The $mGlu_7^{-/-}$ mice showed a reduced freezing response to footshock, with a lower frequency of response during footshocks and a lower plateau level of response than littermate controls with identical genetic backgrounds. When mice were returned to the shock chamber 24 hours after shock treatment, both wild-type and mutant mice displayed freezing responses at the plateau level reached previously. No differences were seen in locomotor activity in open field tests or in pain threshold tests (e.g. hot-plate avoidance) between the two mice. Thus the loss of freezing response was not due to a non-specific deficit in sensory or motor processing capabilities.

The ability of $mGlu_7^{-/-}$ mice to associate taste stimuli (saccharin) with a toxic effect (malaise-evoking LiCl injection) was tested (conditioned taste aversion, CTA; Masugi *et al.*, 1999). Both wild-type and mutant mice displayed the same taste preferences and sensitivity to LiCl toxicity, but $mGlu_7^{-/-}$ mice were unable to associate the taste of saccharin with the effects of LiCl, whereas this was readily achieved by wild-type animals. Thus CTA, another amygdyla-dependent task (Yamamoto *et al.*, 1994), was impaired in $mGlu_7$ receptor deficient mice.

Concluding remarks

The use of glutamate receptor knockout and other transgenic mice has contributed to our understanding of the roles of these receptors in health and disease. They have been particular useful to address the roles of individual subtypes (e.g., $mGlu_5$ receptors), or subunits (e.g., NR2A), since most available glutamate receptor antagonists are not subtype specific. Of course, the interpretation of all studies to date has been complicated by the absence of the gene throughout development. However, the introduction of new technologies to allow temporal, as well as spatial, control over the genetic manipulation should increase the power of approach enormously. Other compounding issues have been the problem of genetic background and the methods used to analyse experimental data. In particular, it is imperative that all experiments are performed blind; this has not always been the case. Furthermore, since the experimental variable is genotype the n value should, but often is not, the number of animals analysed to enable valid statistics to be carried out. Despite these caveats, knockouts and other transgenics have made a big impact in the field of glutamate receptors and are likely to continue to do so.

References

Aiba, A., Chen, C., Herrup, K., Rosenmund, C., Stevens, C.F. and Tonegawa, S. (1994a) 'Reduced hippocampal long-term potentiation and context-specific deficit in associative learning in mGluR1 mutant mice', *Cell* **79**: 365–375.

Aiba, A., Kano, M., Chen, C., Stanton M.E., Fox, G.D. and Herrup, K. (1994b) 'Deficient cerebellar long-term depression and impaired motor learning in mGluR1 mutant mice', *Cell* **79**: 377–388.

Anderson, W.W., Swartzwelder, H.S. and Wilson, W.A. (1987) 'The NMDA receptor antagonist 2-amino-5-phosphonovalerate blocks stimulus train-induced epileptogenesis but not epileptoform bursting in the rat hippocampal slice', *J. Neurophysiol.* **57**: 1–21.

Baskys, A. (1998) 'Metabotropic glutamate receptor gene knockout studies', in F. Moroni, F. Nicoletti and D.E. Pellegrini-Gampietro (eds), *Metabotropic Glutamate Receptors and Brain Function,* London: Portland Press, pp. 29–36.

Bear, M.F. and Abraham, W.C. (1996) 'Long-term depression in the hippocampus', *Ann. Rev. Neuroscience* **19**: 437–462.

Bengzon, J., Okabe, S., Lindvall, O. and McKay, R.D.G. (1999) 'Suppression of epileptogenesis by modification of N-methyl-D-aspartate receptor subunit composition', *Eur. J. Neurosci.* **11**: 916–922.

Benke, T.A., Lüthi, A., Isaac, J.T.R. and Collingridge, G.L. (1998) 'Modulation of AMPA receptor unitary conductance by synaptic activity', *Nature* **393**: 793–797.

Bettler, B. and Mulle, C. (1995) 'Neurotransmitters II. AMPA and kainate receptors', *Neuropharmacology* **34**: 123–139.

Biscoe, T.J., Evans, R.H., Francis, A.A., Martin, M.R., Watkins, J.C., Davies, J. and Dray, A. (1977) 'D-alpha-Aminoadipate as a selective antagonist of amino acid-induced and synaptic excitation of mammalian spinal neurones', *Nature* **270**: 743–745.

Bliss, T. and Collingridge, G.L. (1993) 'A synaptic model for memory: long-term potentiation in the hippocampus', *Nature* **136**: 31–39.

Bortolotto, Z.A., Fitzjohn, S.M. and Collingridge, G.L. (1999) 'Roles of metabotropic glutamate receptors in LTP and LTD in the hippocampus', *Curr. Opinion Neurobiol.* **9**: 299–304.

Brusa, R., Zimmermann, F., Koh, D.S., Feldmeyer, D., Gass, P., Seeburg, P.H. and Sprengel, R. (1995) 'Early-onset epilepsy and post-natal lethality associated with an editing-deficient GluR-B allele in mice', *Science* **270**: 1677–1680.

Bureau, I., Bischoff, S., Heinemann, S.F. and Mulle, C. (1999) 'Kainate receptor-mediated responses in the CA1 field of wild-type and GluR6 deficient mice', *J. Neurosci.* **19**: 653–663.

Bushell, T.J., Sansig, G., Shigemoto, R., Flor, P., Kuhn, R., Knoepfel, T., Schroeder, M., Collett, V.J., Collingridge, G.L. and van der Putten, H. (1996) 'An impairment of hippocampal synaptic plasticity in mice lacking $mGlu_7$ receptors', *Neuropharmacology* **35**: A6.

Bushell, T.J., Collett, V.J., van der Putten, H. and Collingridge, G.L. (1997) 'Altered high frequency synaptic transmission and short-term potentiation in mice lacking $mGlu_7$ receptors: studies in the hippocampus *in vitro*', *J. Physiol. (Proceedings)* **501.P**: 8P.

Castillo, P.E., Malenka, R.C. and Nicoll, R.A. (1997) 'Kainate receptors mediate a slow postsynaptic current in hippocampal CA3 neurones', *Nature* **388**: 182–188.

Ciabarra, A.M., Sullivan, J.M., Gahn, L.G., Pecht, G., Heinemann, S. and Sevarino, K.A. (1995)

'Cloning and characterisation of χ-1: a developmentally regulated member of a novel class of the ionotropic glutamate receptor family', *J. Neurosci.* **15**: 6498–6508.

Clarke, V.J.R., Ballyk, B.A., Hoo, K.H., Mandelzys, A., Pellizari, A., Bath, C.P., Thomas, J., Sharpe, E.F., Davies, C.H., Ornstein, P.L., Schoepp, D.D., Kamboj, R.K., Collingridge, G.L., Lodge, D. and Bleakman, D. (1997) 'A hippocampal GluR5 kainate receptor regulating inhinbitory synaptic transmission', *Nature* **389**: 599–603.

Collingridge, G.L., Kehl, S.J. and McLennan, H. (1983) 'Excitatory amino acids in synaptic transmission in the Schaffer collateral-commisural pathway of the rat hippocampus', *J. Physiol.* **334**: 33–46.

Collingridge, G.L. and Lester, R.A. (1989) 'Excitatory amino acid receptors in the vertebrate nervous system', *Pharmacological Reviews* **41**: 143–210.

Conn, P. J. and Pin, J.-P. (1997) 'Pharmacology and functions of metabotropic glutamate receptors', *Annu. Rev. Pharmacol. Toxicol.* **37**: 205–237.

Conquet, F. (1995) 'Inactivation *in vivo* of metabotropic glutamate receptor 1 by specific chromosomal insertion of reporter gene *lacZ*', *Neuropharmacology* **34**: 865–870.

Conquet, F., Bashir, Z.I., Davies, C.H., Daniel, H., Ferraguti, F., Bordi, F., Franz-Bacon, K., Reggiani, A., Matarese, V., Conde, F., Collingridge, G.L. and Crepel, F. (1994) 'Motor deficit and impairment of synaptic plasticity in mice lacking mGluR1', *Nature* **372**: 237–243.

Cossart, R., Esclapez, M., Hirsch, J.C., Bernard, C. and Ben-Ari, Y. (1998) 'GluR5 kainate receptor activation in interneurones increases tonic inhibition of pyramidal cells', *Nature Neuroscience* **1**: 470–478.

Das, S., Sasaki, Y.F., Rothe, T., Premkumar, L.S., Takasu, M., Crandall, J.E., Dikkes, P., Conner, D.A., Rayudu, P.V., Cheung, W., Cheung, H.-S.V., Lipton, S.A. and Nakanishi, N. (1998) 'Increased NMDA current and spine density in mice lacking the NMDA receptor subunit NR3A', *Nature* **393**: 377–381.

Davies, J., Francis, A.A., Jones, A.W. and Watkins, J.C. (1981) '2-amino-5-phosphonovalerate (2-APV), a potent and selective antagonist of amino acid-induced and synaptic excitation', *Neurosci. Lett.* **21**: 77–81.

Derkach, V., Barria, A. and Soderling, T.R. (1999) 'Ca^{2+}/calmodulin-kinase II enhances channel conductance of α-amino-3-hydroxy-5-methyl-4-isoxaolproprionate type glutamate receptors', *Proc. Natl. Acad. Sci.* **96**: 3269–3274.

Dudek, S.M. and Bear, M.F. (1992) 'Homosynaptic long-term depression in area CA1 of the hippocampus and effects of N-methyl-D-aspartate receptor blockade', *Proc. Natl. Acad. Sci. USA* **89**: 4363–4367.

Ebralidze, A.K., Rossi, D.J., Tonegawa, S. and Slater, N.T. (1996) 'Modification of NMDA receptor channels and synaptic transmission by targeted disruption of the NR2C gene', *J. Neurosci.* **16**: 5014–5025.

Feldmeyer, D., Kask, K., Brusa, R., Kornau, H.-C., Kolhekar, R., Rozov, A., Burnashev, N., Jensen, V., Hvalby, O., Sprengel, R. and Seeburg, P.H. (1999) 'Neurological dysfunctions in mice expressing different levels of the Q/R site-unedited AMAPR subunit GluR-B', *Nature Neuroscience* **2**: 57–64.

Flor, P.J., van der Putten, H., Ruegg, D., Likic, S., Leonhardt, T., Bence, M., Sansig, G., Knopfel, T. and Kuhn, R. (1997) 'A novel splice variant of a metabotropic glutamate receptor, human mGluR7b', *Neuropharmacology* **36**: 153–159.

Forrest, D., Yuzaki, M., Soares, H.D., Ng, L., Luk, D.C., Sheng, M., Stewart, C.L., Morgan, J.I., Connor, J.A. and Curran, T. (1994) 'Targeted disruption of NMDA receptor 1 gene abolishes NMDA response and results in neonatal death', *Neuron* **13**: 325–338.

Gerlai, R., Roder, J.C. and Hampson, D.R. (1998) 'Altered spatial learning and memory in mice lacking the mGluR4 subtype of metabotropic glutamate receptor', *Behavioural Neuroscience* **112**: 525–532.

Hollmann, M. and Heinemann, S. (1994) 'Clones glutamate receptors', *Ann. Rev. Neurosci.* **17**: 31–108.

Hsai, A.Y., Salin, P.A., Castillo, P.E., Aiba, A., Abeliovich, A., Tonegawa, S. and Nicoll, R.A.

(1995) 'Evidence against a role for metabotropic glutamate receptors in mossy fibre LTP: the use of mutant mice and pharmacological antagonists', *Neuropharmacology* **34**: 1567–1572.

Ikeda, K., Araki, K., Takayama, C., Inoue, Y., Yagi, T., Aizawa, S. and Mishina, M. (1995) 'Reduced spontaneous activity of mice defective in the ϵ4 subunit of the NMDA receptor channel', *Mol. Brain Res.* **33**: 61–75.

Ishii, T., Moriyoshi, K., Sugihara, H., Sakurada, K., Kadotani, H., Yokoi, M., Akazawa, C., Shigemoto, R., Mizuno, N., Masu, M. and Nakanishi, S. (1993) 'Molecular characterisation of the family of the N-methyl-D-aspartate receptor subunits', *J. Biol. Chem.* **268**: 2836–2843.

Ito, I., Futai, K., Katagiri, H., Watanabe, M., Sakimura, K., Mishina, M. and Sugiyama, H. (1997) 'Synapse-selective impairment of NMDA receptor functions in mice lacking NMDA receptor ϵ1 or ϵ2 subunits', *J. Physiol. (Lond)* **500.2**: 401–408.

Ito, I., Akashi, K., Sakimura, K., Mishina, M. and Sugiyama, H. (1998) 'Distribution and development of NMDA receptor activities at hippocampal synapses using mice lacking the ϵ1 subunit gene', *Neurosci. Res.* **30**: 119–123.

Iwasato, T., Erzurumlu, R.S., Huerta, P.T., Chen, D.F. Sasoka, T., Ulupinar, E. and Tonegawa, S. (1997) 'NMDA receptor-dependent refinement of somatotopic maps', *Neuron* **19**: 1201–1210.

Jia, Z., Agopyan, N., Miu, P., Xiong, Z., Henderson, J., Gerlai, R., Taverna, F.A., Velumian, A., MacDonald, J., Carlen, P., Abramow-Nerwerly, W. and Roder, J. (1996) 'Enhanced LTP in mice deficient in the AMPA receptor GluR2 *Neuron* **17**: 945–956.

Jia, Z., Lu, Y., Henderson, J., Taverna, F., Romano, C., Abramow-Newerly, W., Wojtowicz, J.M. and Roder, J. (1998) 'Selective abolition of the NMDA component of long-term potentiation in mice lacking mGluR5', *Learning & Memory* **5**: 331–343.

Kadotani, H., Hirano, T., Masugi, M., Nakamura, K., Nakao, K., Katsuki, M. and Nakanishi, S. (1996) 'Motor discoordination results from the combined gene disruption of the NMDA receptor NR2A and NR2C subunits, but not from single disruption of the NR2A or NR2C subunit', *J. Neurosci.* **16**: 7859–7867.

Kano, M., Hashimoto, K., Kurihara, H., Watanabe, M., Inoue, Y., Aiba, A. and Tonegawa, S. (1997) 'Persistent multiple climbing fibre innervation of cerebellar Purkinje cells in mice lacking mGluR1', *Neuron* **18**: 71–79.

Kashiwabuchi, N., Ikeda, K., Araki, K., Hirano, T., Shibuki, K., Takayama, C., Inoue, Y., Kutsuwada, T., Yagi, T., Kang, Y., Aizawa, S. and Mishina, M. (1995) 'Impairment of motor coordination, Purkinje cell synapse formation, and cerebellar long-term depression in GluRδ2 mutant mice', *Cell* **81**: 245–252.

Kask, K., Zamanillo, D., Rozov, A., Burnashev, N. and Sprengel, R. (1998) 'The AMPA receptor GluR-B in its Q/R site unedited form is not essential for brain development and function', *Proc. Natl. Acad. Sci. USA* **95**: 13,777–13,782.

Kishimoto, Y., Kawahara, S., Kirino, Y., Kadotani, H., Nakamura, Y., Ikeda, M. and Yoshioka, T. (1997) 'Conditioned eyeblink response is impaired in mutant mice lacking NMDA receptor subunit NR2A', *Neuroreport* **8**: 3717–3721.

Kiyama, Y., Manabe, T., Sakimura, K., Kawakami, F., Mori, H. and Mishina, M. (1998) 'Increased thresholds for long-term potentiation and contextual learning in mice lacking the NMDA-type glutamate receptor ϵ1 subunit', *J. Neurosci.* **18**: 6704–6712.

Kleinschmidt, A., Bear, M.F. and Singer, W. (1987) 'Blockade of "NMDA" receptors disrupts experience-dependent plasticity of kitten striate cortex', *Science* **238**: 355–358.

Kutsuwada, T., Sakimura, K., Manabe, T., Takayama, C., Katakura, N., Kushiya, E., Natsume, R., Watanabe, M., Inoue, Y., Yagi, T., Aizawa, S., Arakawa, M., Takahashi, T., Nakamura, Y., Mori, H. and Mishina, M. (1996) 'Impairment of suckling response, trigeminal neuron pattern formation, and hippocampal LTD in NMDA receptor ϵ2 subunit mutant mice', *Neuron* **16**: 333–344.

Lalouette, A., Guenet, J.L. and Vriz, S. (1998) 'Hotfoot mouse mutations affect the δ2glutamate receptor gene and are allelic to lurcher', *Genomics* **50**: 9–13.

Levenes, C., Daniel, H., Jaillard, D., Conquet, F. and Crepel, F. (1997) 'Incomplete regression of multiple climbing fibre innervation of cerebellar Purkinje cells in mGluR1 mutant mice', *Neuroreport* **8**: 571–574.

Li, Y., Erzurumlu, R.S., Chen, C., Jhaveri, S. and Tonegawa, S. (1994) 'Whisker-related neuronal patterns fail to develop in the trigeminal brainstem nuclei of NMDAR1 knockout mice', *Cell* **76**: 427–437.

Lu, Y.-M., Jia, Z., Janus, C., Henderson, J.T., Gerlai, R., Wojtowicz, J.M. and Roder, J.C. (1997) 'Mice lacking metabotropic glutamate receptor 5 show impaired learning and reduced CA1 long-term potentiation (LTP) but normal CA3 LTP', *J. Neurosci.* **17**: 5196–5205.

Luján, R., Nusser, Z., Roberts, J.B., Shigemoto, R. and Somogyi, P. (1996) 'Perisynaptic location of metabotropic glutamate receptors mGluR1 and mGluR5 on dendrites and dendritic spines in the rat hippocampus', *Eur. J. Neuroscience* **8**: 1488–1500.

Mainen, Z.F., Jia, Z., Roder, J. and Malinow, R. (1998) 'Use-dependent AMPA receptor block in mice lacking GluR2 suggests postsynaptic site for LTP expression', *Nature Neuroscience* **1**: 579–586.

Maren, S. and Fanselow, M.S. (1996) 'The amygdyla and fear conditioning: has the nut been cracked?', *Neuron* **16**: 237–240.

Masu, M., Iwakabe, H., Tagawa, Y., Miyoshi, T., Yamahita, M., Fukuda, Y., Sasaki, H., Hiroi, K., Nakamura, Y., Shigemoto, R., Takada, M., Nakamura, K., Nakao, K., Katsuki, M. and Nakanishi, S. (1995) 'Specific deficit of the ON response in visual transmission by targeted disruption of the mGluR6 gene', *Cell* **80**: 757–765.

Masugi, M., Yokoi, M., Shigemoto, R., Muguruma, K., Watanabe, Y., Sansig, G., van der Putten, H. and Nakanishi, S. (1999) 'Metabotropic, glutamate receptor subtype 7 ablation causes deficit in fear response and conditional taste aversion', *J. Neuroscience* **19**: 955–963.

McCaffery, B., Bortolotto, Z.A., Bashir, Z.I., Collett, V.J., Gasparini, F., Flor, P.J., Jane, D.E., Conquet, F. and Collingridge, G.L. (1998) 'Effects of metabotropic glutamate receptor mGluR1 antagonists on long-term potentiation in area CA1 of the hippocampus', in F. Moroni, F. Nicoletti and D.E. Pellegrini-Gampietro (eds) *Metabotropic Glutamate Receptors and Brain Function*, London: Portland Press, pp. 139–145.

McHugh, T.J., Blum, K.I., Tsien, J.Z., Tonegawa, S. and Wilson, M.A. (1996) 'Impaired hippocampal representation of space in CA1-specific NMDAR1 knockout mouse', *Cell* **87**: 1339–1349.

Mori, H., Manabe, T., Watanabe, M., Satoh, Y., Suzuki, N., Toki, S., Nakamura, K., Yagi, T., Kushiya, E., Takahashi, T., Inoue, Y., Sakimura, K. and Mishina, M. (1998) 'Role of the carboxy-terminal region of the GluRε2 subunit in synaptic localization of the NMDA receptor channel', *Neuron* **21**: 571–580.

Mori, H. and Mishina, M. (1995) 'Structure and function of the NMDA receptor channel', *Neuropharmacology* **34**: 1219–1237.

Morris, R.G., Anderson, E., Lynch, G.S. and Baudry, M. (1986) 'Selective impairment of learning and blockade of long-term potentiation by an N-methyl-D-aspartate receptor antagonist, AP5', *Nature* **319**: 774–776.

Mulkey, R.M. and Malenka, R.C. (1992) 'Mechanisms underlying induction of homosynaptic long-term depression in area CA1 of the hippocampus', *Neuron* **9**: 967–975.

Mulle, C., Sailer, A., Perez-Ontano, I., Dickinson-Anson, H., Castillo, P.E., Bureau, I., Maron, C., Gage, F.H., Mann, J.R., Bettler, B. and Heinemann, S.F. (1998) 'Altered synaptic physiology and reduced susceptibility to kainate-induced seizures in GluR6-deficient mice', *Nature* **392**: 601–605.

Nakajima, Y., Iwakabe, H., Akazawa, C., Nawa, H., Sihemoto, R., Mizuno, N. and Nakanishi, S. (1993) 'Molecular characterisation of a novel retinal metabotropic glutamate receptor mGluR6 with a high agonist selectivity for L-2-amino-4-phosphonobutyrate', *J. Biol. Chem.* **268**: 11,868–11,873.

Okabe, S., Collin, C., Auerbach, J.M., Meiri, N., Bengzon, J., Kennedy, M.B., Segal, M. and McKay, R.D. (1998) 'Hippocampal synaptic plasticity in mice overexpressing an embryonic subunit of the NMDA receptor', *J. Neurosci.* **18**: 4177–4188.

Pekhletski, R., Gerlai, R., Overstreet, L.S., Huang, X.-P., Agopyan, N., Slater, N.T., Abramow-

Newerly, W., Roder, J.C. and Hampson, D.R. (1996) 'Impaired cerebellar synaptic plasticity and motor performance in mice lacking the mGluR4 subtype of metabotropic glutamate receptor', *J. Neurosci.* **15**: 6364–6373.

Pin, J.-P. and Duvoisin, R. (1995) 'The metabotropic glutamate receptors: structure and functions', *Neuropharmacology* **34**: 1–26.

Sailer, A., Swanson, G.T., Perez-Otano, I., O'Leary, L., Malkmus, S.A., Dyck, R.H., Dickinson-Anson, H., Schiffer, H.H., Maron, C., Yaksh, T.L., Gage, F.H., O'Gorman, S. and Heinemann, S.F. (1999) 'Generation and analysis of GluR5(Q636R) kainate receptor mutant mice', *J. Neurosci.* **19**: 8757–8764.

Sakimura, K., Kutsuwada, T., Ito, I., Manabe, T., Takayama, C., Kushiya, E., Yagi, T., Aizawa, S., Inoue, Y., Sugiyama, H. and Mishina, M. (1995) 'Reduced hippocampal LTP and spatial learning in mice lacking NMDA receptor ϵ1 subunit', *Nature* **373**: 151–155.

Schnabel, R., Kilpatrick, I.C., Jia, Z., Roder, J.C., Conquet, F. and Collingridge, G.L. (1999) 'DHPG-induced LTD and DHPG-induced potentiation of NMDA actions are absent in mice lacking $mGlu_5$ receptors', Paper presented at the Neuroscience Conference, Miami.

Seabrook, G.R. and Rosahl, T.W. (1999) 'Transgenic animals relevant to Alzheimer's disease', *Neuropharmacology* **38**: 1–17.

Sekiyama, N., Hayashi, Y., Nakanishi, S., Jane, D.E., Tse, H.-W., Birse, E.F. and Watkins, J.C. (1996) 'Structure–activity relationships of new agonists and antagonists of different metabotropic glutamate receptor subtypes', *Br. J. Pharmacol.* **117**: 1493–1503.

Sheng, M., Cummings, J., Roldan, L.A., Yan, Y.N. and Yan, L.Y. (1994) 'Changing subunit composition of hetereomeric NMDA receptors during development of rat cortex', *Nature* **368**: 144–147.

Shigemoto, R., Kinoshita, A., Wada, E., Nomura, S., Ohishi, H., Takada, M., Flor, P., Neki, A., Abe, T., Nakanishi, S. and Mizuno, N. (1997) 'Differential presynaptic localisation of metabotropic glutamate receptor subtypes in the rat hippocampus', *J. Neurosci.* **17**: 7503–7522.

Simmons, R.M., Li, D.L., Hoo, K.H., Deverill, M., Ornsein, P.L. and Iyengar, S. (1998) 'Kainate GluR5 receptor subtype mediates the nociceptive response to formalin', *Neuropharmacology* **37**: 25–36.

Simon, D.K., Prusky, G.T., O'Leary, D.D.M. and Constantine-Paron, M. (1992) 'N-methyl-D-aspartate receptor antagonists disrupt the formation of a neural map, *Proc. Natl. Acad. Sci. USA* **89**: 10,593–10,597.

Sommer, B., Kohler, M., Sprengel, R. and Seeburg, P.H. (1991) 'RNA editing in brain controls a determinant of ion flow in glutamate-gated channels', *Cell* **67**: 11–19.

Sprengel, R., Suchanek, B., Amico, C., Brusa, R., Burnashev, N., Rozov, A., Hvalby, O., Jensen, V., Paulsen, O., Andersen, P., Kim, J.J., Thompson, R.F., Sun, W., Webster, L.C., Grant, S.G., Eilers, J., Konnerth, A., Li, J., McNamara, J.O. and Seeburg, P.H. (1998) 'Importance of the intracellular domain of NR2 subunits for NMDA receptor function *in vivo*', *Cell* **92**: 279–289.

Sucher, N.J., Akbarian, S., Chi, C.L., Leclerc, C.L., Awobuluyi, M., Deitcher, D.L., Wu, M.K., Yuan, J.P., Jones, E.G. and Lipton, S.A. (1995) 'Development and regional expression pattern of a novel NMDA receptor-like subunit (NMDAR-L) in the rodent brain', *J. Neurosci.* **15**: 6509–6520.

Sugihara, H., Inoue, T., Nakanishi, S. and Fuduka, Y. (1997) 'A late ON response remains in visual response of the mGluR6-deficient mouse', *Neurosci. Lett.* **233**: 137–140.

Tagawa, Y., Sawai, H., Ueda, Y., Tauchi, M. and Nakanishi, S. (1999) 'Immunological studies of metabotropic glutamate receptor subtype 6-deficient mice show no abnormality of retinal cell organization and ganglion cell maturation', *J. Neurosci.* **19**: 2568–2579.

Tanabe, Y., Nomura, A., Masu, M., Shigemoto, R. and Nakanishi, S. (1993) 'Signal transduction, pharmacological properties, and expression patterns of two rat metabotropic glutamate receptors, mGluR3 and mGluR4a', *J. Neurosci.* **13**: 1372–1378.

Tang, Y.-P., Shimizu, E., Dube, G.R., Rampon, C., Kerchner, G.A., Zhuo, M., Liu, G., Tsien, J.Z. (1999) 'Genetic enhancement of learning and memory in mice', *Nature* **401**: 63–69.

Thompson, C. and Hampson, D.R. (1999) 'Contribution of metabotropic glutamate receptor mGluR4 to L-2-[^{3}H]amino-4-phosphonobutyrate binding in mouse brain', *J. Neurochem.* **72**: 835–840.

Tsien, J.Z., Chen, D.F., Gerber, D., Tom, C., Mercer, E.H., Anderson, D.J., Mayford, M., Kandel, E.R. and Tonegawa, S. (1996a) 'Subregion- and cell type-restricted gene knockout in mouse brain', *Cell* **87**: 1317–1326.

Tsien, J.Z., Huerta, P.T. and Tonegawa, S. (1996b) 'The essential role of hippocampal CA1 NMDA receptor-dependent synaptic plasticity in spatial memory', *Cell* **87**: 1327–1338.

Vignes, M., Bleakman, D., Lodge, D. and Collingridge, G.L. (1997) 'The synaptic activation of the GluR5 subtype of kainate receptor in area CA3 of the rat hippocampus', *Neuropharmacology* **36**: 1477–1481.

Vignes, M. and Collingridge, G.L. (1997) 'The synaptic activation of kainate receptors', *Nature* **388**: 179–182.

Watkins, J.C. and Evans, R.H. (1981) 'Excitatory amino acid transmitters', *Ann. Rev. Pharmacol.Toxicol.* **21**: 165–204.

Wenthold, R.J., Petralia, R.S., Blahos, J. and Niedzielski, A.S. (1996) 'Evidence for multiple AMPA receptor complexes in hippocampal CA1/CA2 neurons', *J. Neurosci.* **16**: 945–956.

Wilson, M.A. and Tonegawa, S. (1997) 'Synaptic plasticity and spatial memory: study with second generation knockouts', *TINS* **20**: 102–106.

Yamamoto, T., Shimura, T., Sako, N., Yasoshima, Y. and Sakai, N. (1994) 'Neural substrates for conditioned taste aversion in the rat', *Behavioural Neuroscience* **100**: 455–465.

Yokoi, M., Kobayashi, K., Manabe, T., Takahashi, T., Sakaguchi, I., Katsuura, G., Shigemoto, R., Ohishi, H., Nomura, S., Nakamura, K., Nakao, K., Katsuki, M. and Nakanishi, S. (1996) 'Impairment of hippocampal mossy fibre LTD in mice lacking mGluR2', *Science* **273**: 645–647.

Zamanillo, D., Sprengel, R., Hvalby, Ø., Jensen, V., Burnashev, N., Rozov, A., Kaiser, K.M.M., Köster, H.J., Borchardt, T., Worley, P., Lübke, J., Frotscher, M., Kelly, P.H., Sommer, B., Anderson, P., Seeburg, P.H. and Sakmann, B. (1999) 'Importance of AMPA receptors for hippocampal synaptic plasticity but not for spatial learning', *Science* **284**: 1805–1811.

Chapter 16

Transgenic models for $GABA_A$-receptor function

H. Möhler

Introduction

$GABA_A$-receptors are formed by the assembly of various subunits which, in mammalian brain, are encoded by a total of 17 different genes (α1–6, β1–3, γ1-3, δ, ε, ρ1–3) (Möhler *et al.*, 1997; Möhler, 2001; Sieghart, 1995; MacDonald *et al.*, 1994). In heterologous expression systems, no major restrictions for the combinatorial assembly of the subunits were apparent. However, *in vivo*, the expression of $GABA_A$-receptor subunits is regulated in an ontogenetic and cell-specific manner providing particular populations of neurons with distinct sets of $GABA_A$-receptor subtypes (Fritschy and Möhler, 1995). The process of receptor assembly is expected to include specific interactions among the constituent subunits. These interactions may govern not only the subunit stoichiometry but also the formation and membrane targeting of multiple $GABA_A$-receptors in a single cell. The distinct targeting of receptor subtypes is considered to serve specific functions in regulating complex behavior and particular drug actions. Thus, the genetic dissection of $GABA_A$-receptors is expected to provide new insights into the function of the $GABA_A$-receptor system.

In recent years, mutant mouse lines were generated in which the genes for the $GABA_A$-receptor subunits γ2, β3 and α6 were inactivated (Günther *et al.*, 1995; Homanics *et al.*, 1997a, 1997b; Jones *et al.*, 1997; Crestani *et al.*, 1999). In addition, mouse strains with radiation-induced chromosomal deletions, which include the α5, β3 and γ3 subunit genes, are available (Gardner *et al.*, 1992; Lyon *et al.*, 1992; Culiat *et al.*, 1993, 1994; Nakatsu *et al.*, 1993). These animal models provide insight into the role of specific subunits in the assembly and targeting of particular $GABA_A$-receptors and their pathophysiology (Rudolph and Möhler, 1999).

Independent regulation of subunit gene transcription

Gene transcription

The inactivation of a particular $GABA_A$-receptor subunit gene has so far not been found to induce a compensatory up- or down-regulation of the transcription of other $GABA_A$-receptor subunit genes. For instance, in the γ2 subunit null mutants the protein level of the subunits α1, α2, α3, β2,3, γ1, γ3 remained unaltered as shown by Western blotting (Günther *et al.*, 1995). Similarly, in the β3-subunit mutant, the β2 subunit failed to be a substitute, as deduced from radioligand binding experiments performed in newborn and adult animals (Homanics *et al.*, 1997a). In addition, the inactivation of the α6 subunit gene, which selectively affected $GABA_A$-receptors in cerebellar granule cells, left the receptors containing the α1, β2,3 and γ2 subunits in this brain region unaltered as shown by radioligand binding and by α1 subunit immunohistochemistry, although a slight and variable downward trend in α1 protein levels was apparent in Western blotting (Jones *et al.*, 1997). Similarly, in a further α6 null mutant, diazepam-insensitive 3H–Ro 15–4513 binding in cerebellar granule cells, corresponding to α6 receptors, was reduced while the mRNA for α1, α3, β2, γ2 and δ remained unchanged (Homanics *et al.*, 1997b). These results demonstrate that the lack of a particular subunit of $GABA_A$-receptors does not result in a molecular rescue by an enhancement of the transcription of a related $GABA_A$-receptor subunit. However, neuronal elements other than $GABA_A$-receptor subunits can be altered as demonstrated by the induction of a potassium channel in α_6 subunit mutant mice (Brickley *et al.*, 2001).

Post-transcriptional modifications

In mice containing a targeted mutation of the α6 subunit gene, a post-transcriptional influence on a related subunit was discovered (Jones *et al.*, 1997). In these mutants, the lack of the α6 subunit protein was accompanied by a concomitant and selective loss of the δ subunit as demonstrated by immunoprecipitation, immunocytochemistry and immunoblot analysis with δ subunit-specific antibodies (Jones *et al.*, 1997). The δ subunit mRNA was present at wild-type levels in the mutant granule cells, indicating a post-transcriptional loss of the δ subunit. These results provide genetic evidence for a specific association between the α6 and δ subunits in receptor assembly (Jones *et al.*, 1997). In the absence of the α6 subunit, the δ subunit may either be degraded or fails to be translated. An influence of subunit assembly on protein stability of a subunit has so far been observed only for the δ subunit.

Receptor assembly and targeting

Subunit requirement

Most $GABA_A$-receptors contain an α subunit variant, a β subunit variant and the γ2 subunit. However, it is unknown whether three types of subunits are necessarily required for the formation of a receptor, or whether, in mutant animals, two types of subunits are sufficient to form functional receptors. When the β3 subunit gene was inactivated, about half of all $GABA_A$-receptors in the brain were lost (Homanics *et al.*, 1997a). This result indicated that the β3 subunit was essential for the assembly of the corresponding receptors and could not be substituted by the corresponding α and γ2 subunits (Homanics *et al.*, 1997a). Corresponding results were found in those cases in which an α subunit gene was deleted or inactivated. In mice with a deletion chromosome 7, which includes the α5 subunit gene, the corresponding α5 receptor subtype was not formed (Fritschy *et al.*, 1997). Furthermore, the lack of the α6 subunit resulted in the loss of the corresponding α6 receptor subtype in cerebellar granule cells (Homanics *et al.*, 1997b). Thus, it appears that both the α and the β subunit are essential components for the assembly and membrane targeting of $GABA_A$-receptors.

In contrast, in mice devoid of the γ2 subunit, $GABA_A$-receptors were formed in practically unaltered numbers from the remaining α and β subunits. This was evident from radioligand binding studies in newborn mice, in which the maximum number of GABA binding sites (^{3}H-99531 binding) was unaltered (Günther *et al.*, 1995). Only the number of benzodiazepine binding sites (^{3}H-flumazenil binding) was reduced by 90 per cent, concomitant with the absence of the γ2 subunit (Günther *et al.*, 1995). The remaining α and β subunits were targeted to the cell surface as shown immunohistochemically in dorsal root ganglion cells. The receptors displayed a normal dose response curve for GABA, were potentiated by a barbiturate, but lacked the benzodiazepine response. In addition, the single channel conductance was reduced (Günther *et al.*, 1995). Nevertheless, these experiments in newborn mice show that the γ2 subunit is not a necessary prerequisite to permit the assembly and targeting of $GABA_A$-receptors. The γ2 subunit has recently been shown to be rather required for receptor clustering (Essrich *et al.*, 1998).

Independent subcellular targeting of receptor subtypes

Various neurons express more than a single type of $GABA_A$-receptor. This can be most clearly visualised in hippocampal pyramidal cells, where receptors containing the α2 subunit are largely concentrated in the axon-initial segment, while receptors containing the α5 subunit are distributed on the cell soma and dendrites (Nusser *et al.*, 1996; Fritschy *et al.*, 1997, 1998a, 1998b). The mechanisms governing the differential assembly and subcellular targeting of receptor subtypes are unknown. In mice lacking the α5 subunit due to a chromosomal deletion, the α5 receptors are absent in hippocampal pyramidal cells, as shown by a reduction in radioligand binding and immunohistochemical staining of the β2,3 and γ2 subunits. The α2 receptors, however, remained targeted to the axon-initial segment and did not display a compensatory distribution on the soma and dendrites (Fritschy *et al.*, 1998a, 1998b). A selective extrasynaptic localization was shown for receptors containing the δ subunit in cerebellar granule cells (Nusser *et al.*, 1998). These results show that each receptor subtype has its own target identity presumably determined by domains of the distinct α and β subunit variants. Highly selective recognition processes appear to be operative to ensure the differential assembly and subcellular distribution of the receptor subtypes within a single cell.

Anchoring and clustering of $GABA_A$-receptors

The postsynaptic membrane and the subsynaptic cell compartment are specialised for inter- and intracellular signalling, including the long-term regulation of synaptic efficacy. Several subsynaptic proteins have recently been found to be associated – directly or indirectly – with $GABA_A$-receptors. In particular a coordinated indirect interaction of the clustering protein gephyrin and a majority of synaptic $GABA_A$-receptors was recently demonstrated (Essrich *et al.*, 1998). The clustering of major postsynaptic $GABA_A$-receptor subtypes required the presence of both the γ2 subunit and gephyrin. In cultured cortical neurons from γ2 null mutant mice synaptic $GABA_A$-receptor function was reduced in parallel with a reduced amount of gephyrin, as shown immunohistochemically. Conversely, inhibiting gephyrin expression caused a loss of $GABA_A$-receptor clustering (Essrich *et al.*, 1998). Gephyrin has previously been associated with the clustering of glycine receptors, where activity induced aggregation of gephyrin is thought to 'trap' glycine receptors and thus concentrate them opposite to active presynaptic terminals (Kirsch, 1999). However, in the case of $GABA_A$-receptors clustering is less likely to be due to synaptic activity than to interactions with proteins underlying the postsynaptic specialisation. A direct link between the γ2 subunit and an associated protein emerged recently from an analysis using the yeast two hybrid system. The $GABA_A$-receptor associated protein GABARAP interacts selectively with the large cytoplasmic loop of the γ2 subunit–subunit and co-localises with $GABA_A$-receptors in cultured cortical neurons (Wang *et al.*, 1999). It is presently not clear whether GABARAP is the physical link to gephyrin in situ. Sequence analyses of GABARAP revealed similarities to the light chain 3 (LC-3) of the microtubule associated proteins MAP-1A and MAP-1B. Its N-terminus is highly charged and may mediate tubuline binding. Beside its interaction with $GABA_A$-receptors, GABARAP is likely to serve additional functions since its expression is not confined to the nervous system. GABARAP, like gephyrin, is expressed in all tissues and organs examined so far. Gephyrin was recently shown to interact with RAFTA, a member of cell cycle regulators (Sabatini *et al.*, 1999).

Interestingly, the heavy chain of MAP-1B (also known as MAP-5) also interacts with $GABA_A$-receptors, although with a particular subtype, the $GABA_C$-receptor (Hanley *et al.*, 1999), which is expressed almost exclusively in retinal bipolar cells (Feigenspan *et al.*, 1993; Koulen *et al.*, 1998). MAP-1B interacts with the large cytoplasmic loop between TM3 and TM4 of the ρ1 subunit. The ρ and MAP-1B immunoreactivities are co-localised in axon terminals.

Mechanism of action of drugs acting on $GABA_A$-receptors

Studies of the pharmacology of $GABA_A$-receptor subtypes in heterologous expression systems identified specific structural requirements for certain drug actions. This information has subsequently been exploited to identify the subunit composition of particular receptors in mutant mice through their drug response. Conversely, studies in mutant mice shed new light on the types of $GABA_A$-receptors which mediate particular drug actions. This is outlined below, for benzodiazepines, loreclezole, ethanol, anaesthetics and pentobarbital.

Benzodiazepines and loreclezole

In the γ2 null mutants, $GABA_A$-receptors were devoid of a benzodiazepine response, as tested in dorsal root ganglion cells (Günther *et al.*, 1995). These results confirmed that the remaining GABA-induced currents were due to receptors made up of α and β subunits devoid of the γ2 subunit. When a minor mutation was introduced into the γ2L subunit variant by converting it to the γ2S variant, midazolam and zolpidem induced a longer sleep time (+20 per cent), but not pentobarbital and etomidate (Homanics *et al.*, 1999). Somewhat more sensitive to the motor impairing effect of diazepam in the rotarod test were the α6 null mutants, although only in a very limited dose range and in the accelerod version of the test (Korpi *et al.*, 1998), which is surprising since α6 receptors are insensitive to classical benzodiazepines. However, it cannot be excluded that a deficit in motor learning of the mutants (Korpi *et al.*, 1998) or compensatory homeostatic mechanisms may be operative in the α6 null mutant mice (Homanics *et al.*, 1997b; Brickley *et al.*, 2001) and thereby blunt or exacerbate drug responses. In the β3 null mutant mice the small GABA currents remaining were identified in dorsal root ganglion cells to be due to receptors containing the β2 and not the β1 subunit. This conclusion was derived from the fact that the remaining currents were potentiated by loreclezole (Homanics *et al.*, 1997a) which is known to interact with a site on the β2 but not the β1 subunits. Using a point mutation strategy specific benzodiazepine actions were recently attributed to particular GABA$_A$-receptor subtypes. For instance α_1 GABA$_A$-receptors mediate sedative activity, while α_2 receptors mediate anxiolysis (Rudloph *et al.*, 1999; Löw *et al.*, 2000; Rudolph *et al.*, 2001; Möhler *et al.*, 2001).

Ethanol

Mutant GABA$_A$-receptors were used to test the mechanism of action of ethanol. It had been shown earlier that mice lacking the γ isoform of protein-kinase C show a reduced response to ethanol (Harris *et al.*, 1995). This result supported the view that the phosphorylation of GABA$_A$-receptors at sites of the large cytoplasmic loop of the γ2L subunit may be critical for mediating the effect of ethanol. In order to test this hypothesis mice were generated in which a 24 bp exon was deleted which distinguishes the γ2L splice variant from by the γ2S variant (Homanics *et al.*, 1999). However, these animals showed the same sensitivity to ethanol as control mice. There was no difference in the potentiation of GABA currents by ethanol observed in neurons from wild-type or $\gamma 2L^{-/-}$ mice. Furthermore, several behavioral effects of ethanol were likewise unchanged, such as the ethanol-induced sleeptime, anxiolysis, acute tolerance, chronic withdrawal hyperexcitability and hyperlocomotor activity (Homanics *et al.*, 1999). Thus, γ2L does not appear to be required for the ethanol-induced modulation of GABA$_A$-receptors or whole animal behavior. The mechanism of action of ethanol was also analyzed in animals with mutations affecting the α6 subunit. A naturally occurring point mutation in the α6 subunit gene was earlier shown to co-segretate with a phenotype which was more sensitive than controls to the motor impairing effect of alcohol (Korpi *et al.*, 1993; Hellevno *et al.*, 1989). However, α6 null mutant mice failed to display altered responses to ethanol (Homanics *et al.*, 1997b). In particular, ethanol-induced motor impairment, tolerance and withdrawal hyperexcitability were not different between genotypes ($\alpha 6^{+/+}$, $\alpha 6^{-/-}$) (Homanics *et al.*, 1998; Korpi *et al.*, 1998). Thus, the receptors containing α6 subunit do not appear to be critically involved in the behavioral response to this drug (Homanics *et al.*, 1997b).

Anaesthetics and pentobarbital

The role of GABA$_A$-receptors in the mechanism of action of anaesthetics was also assessed by gene targeting. Although mice lacking the β3 subunit gene generally die as neonates, some survive with abnormal behavior (hyperactivity, incoordination, epilepsy). In these animals, the effectiveness of pentobarbital, enflurane and halothane to induce a loss of righting reflex remained unaltered, while midazolam and etomidate were less effective (Quinlan *et al.*, 1998). The latter agents were therefore postulated to produce hypnosis by different molecular mechanisms. However, in contrast to the unaltered effectiveness of the volatile anaesthetics enflurane and halothane in inducing a loss of the righting reflex, their immobilising effect (tail clamp test) was impaired in the β3 null mutant mice. Absence of the $\alpha 6$ subunit did not change the response to pentobarbital and general anaesthetics (Homanics *et al.*, 1997b), a result which is somewhat surprising since at least pentobarbital can directly activate $\alpha 6$ but not $\alpha 1$ receptors at $100\,\mu M$ (Hadingham *et al.*, 1996; Thompson *et al.*, 1996). However, a naturally occurring point mutation in the $\alpha 6$ subunit gene enhanced the ataxic effects of volatile anaesthetics and the loss of righting reflex by pentobarbital (Korpi *et al.*, 1993; Hellevno *et al.*, 1989).

GABA$_A$-receptor mutations and phenotypic pathophysiology

Since GABA$_A$-receptors are expressed early in development, an overt behavioral phenotype may be the result of multiple changes, including developmental aberrations, functional deficits in adulthood as well as compensatory adaptations. In general, the severity of the pathophysiological deficit paralleled the extent of the functional impairment of GABA$_A$-receptors, as evidenced by the neonatal mortality of $\gamma 2$ and $\beta 3$ null mutants. In the following, the contributions of GABA$_A$-receptor mutants to molecular, cellular, developmental and pathophysiological aspects are summarised.

Single channel conductance

Mice which lacked the ubiquitous $\gamma 2$ subunit showed neonatal lethality with only some homozygous mutants surviving up to two weeks. Although the remaining α and β subunits form the normal complement of receptors in neonatal brain, their function is drastically reduced, as shown by their single channel conductance which was reduced from 38 pS in wild-type to 13 pS in the homozygous mutants (Günther *et al.*, 1995). Thus, the $\gamma 2$ subunit contributes not only to the pharmacology and clustering of GABA$_A$-receptors (Günther *et al.*, 1995; Essrich *et al.*, 1998) but also to their single channel properties.

Neuronal oscillations

Mice which were devoid of the $\beta 3$ subunit (Homanics *et al.*, 1997a) mostly die as neonates, displaying only half the normal density of GABA$_A$-receptors in brain. The few $\beta 3$-deficient mice that survive eventually reach normal body size, although with reduced life span. They display many neurological impairments, including deficits in neuronal inhibition in spinal cord and higher cortical centers, as shown by their hyperresponsiveness to sensory stimuli, their strong motor impairment and frequent myoclonus and occasional epileptic seizures (Homanics *et al.*, 1997a). In the reticular nucleus of the thalamus, which normally acts as 'desynchroniser', recurrent GABA mediated inhibitions were abolished in brain slices of $\beta 3$ null mutants. Since $\beta 3$ receptors are present in the reticular nucleus but not in principal neurons of thalamic relay cells, oscillatory synchrony was dramatically intensified (Huntsman *et al.*, 1999).

Cranio-facial development

Some of the $\beta 3$-deficient neonatal mortality, but not all, is accompanied by cleft palate. The role of GABA$_A$-receptors in cranio-facial development is supported by the emergence of the neonatally lethal cleft palate in mice homozygous for the p4THO-II deletion, which includes the α5, γ3 and the β3 subunit gene (Culiat *et al.*, 1993, 1994). Since the cleft palate phenotype could be rescued by introducing a β3 subunit transgene into the p4THO-II homozygous mutants the β3 $GABA_A$-receptors appear to play an essential role in cranio-facial development (Culiat *et al.*, 1995). Finally, β3 subunit null mutants are considered to be a model of Angelman's syndrome in humans (Homanics *et al.*, 1997a; DeLorey *et al.*, 1998).

Anxiety

Most remarkably, mice with a reduced gene dosage for the γ2 subunit displayed enhanced anxiety and a bias for threat cues as tested in mutants on two genetic backgrounds (see Figure 16.1 in colour section; Crestani *et al.*, 1999). The $\gamma2^{+/0}$ mice showed anxiety-related behavior as demonstrated by enhanced behavioral inhibition in response to natural aversive stimuli such as novelty, exposed space and bright light. Remarkably, the $\gamma2^{+/0}$ mice also displayed an enhanced responsiveness in the assessment of negative associations in fear conditioning paradigms. In trace conditioning the perception of the contingency of aversive stimuli was enhanced in $\gamma2^{+/0}$ compared to wild-type. Similarly, in ambiguous cue discrimination learning, a partial stimulus was perceived by $\gamma2^{+/0}$ mice to be as threatening as a fully conditioned stimulus. In contrast to these forms of explicit memory, classical delay conditioning was unaltered, indicating that acquisition and retention of implicit forms of memory were unchanged, which was in line with an intact spatial memory and long-term potentiation in hippocampus. On the molecular level, synaptic $GABA_A$-receptor clustering was reduced in $\gamma2^{+/0}$ mice. This was associated with the emergence of extrasynaptic receptors with a low single conductance (13 pS). These molecular changes were most pronounced in the hippocampus and part of the cerebral cortex, areas in which a deficit of $GABA_A$-receptors is found in patients with panic anxiety. Thus, the $\gamma2^{+/-}$ mice represent a valid animal model of anxiety. This model qualifies $GABA_A$-receptor dysfunction in patients as a causal predisposition for anxiety disorders. An increased level of anxiety was also noted in mice in which the γ2L variant was mutated into the γ2S variant, although this was evident only in a single test (Homanics *et al.*, 1999b).

From genotype to phenotype

Inactivation or alteration of a $GABA_A$-receptor subunit gene is likely to result in a functional impairment of the receptor. The respective phenotype provides information on the mechanism of particular neuroanatomical circuits and human disease. However, the road from the genotype to the phenotype can be circuitous and the phenotype may result from multiple changes. Evidence has arisen for adaptive mechanisms (Homanics *et al.*, 1996, 1997b; Brickley *et al.*, 2001), for an influence of the selectable markers used for the generation of the mutant animals (Homanics *et al.*, 1999), and for strain differences affecting behavioral drug responses (Korpi *et al.*, 1998). To minimise the influence of the genetic background particular breeding schemes should be followed (Banbury Conference, 1997), as demonstrated in the study of the $\gamma2^{+/-}$ mice which displayed an anxiety phenotype (Crestani *et al.*, 1999). In addition, knock-in point mutations have been shown to be advantageous compared to the knock-out approach in identifying specific phenotypic determinants (Rudolph *et al.*, 1999; Löw *et al.*, 2000; Rudolph *et al.*, 2001; Möhler *et al.*, 2001).

References

Banbury Conference on the Genetic Background in Mice (1997) 'Mutant mice and neuroscience: recommendations concerning genetic background', *Neuron* **19**: 755–759.

Brickley, S.G. *et al.* (2001) 'Adaptive regulation of neuronal excitability by a voltage independent potassium conductance' *Nature* **409**: 89–92.

Crestani, F., Lorez, M., Baer, K., Essrich, C., Benke, D., Laurent, J.P., Belzung, C., Fritschy, J.M., Luscher, B. and Möhler, H. (1999) 'Impairment of $GABA_A$-receptor clustering results in enhanced anxiety responses and a bias for threat cues', *Nature Neurosci.* **2**: 833–839.

Culiat, C.T., Stubbs, L., Nicholls, R.D., Montgomery, C.S., Russel, L.B., Johnson, D.K. and Rinchik, E.M. (1993) 'Concordance between isolated cleft palate in mice and alterations within a region including the gene encoding the beta3 subunit of the type A gamma-aminobutyric acid receptor', *Proc. Natl. Acad. Sci. USA* **90**: 5105–5109.

Culiat, C.T., Stubbs, L.J., Montgomery, C.S., Russell, L.B. and Rinchik, E.M. (1994) 'Phenotypic consequences of deletion of the $\gamma3$, $\alpha5$, or $\beta3$ subunit of the type A γ-aminobutyric acid receptor in mice', *Proc. Natl. Acad. Sci. USA* **91**: 2815–2818.

Culiat, C.T., Stubbs, L.J., Woychik, R.P., Russell, L.B., Johnson, D.K. and Rinchik, E.M. (1995) 'Deficiency of the $\beta3$ subunit of the type A γ-aminobutyric acid receptor causes cleft palate in mice', *Nature Genetics* **11**: 344–346.

DeLorey, T.M., Handforth, A., Anagnostaras, S.G., Homanics, G.E., Minassian, B.A., Asatourian, A., Fanselow, M.S., Delgado-Escueta, A., Ellison, G.D. and Olsen, R.W. (1998) 'Mice lacking the $\beta3$ subunit of the $GABA_A$ receptor have the epilepsy phenotype and many of the behavioral characteristics of Angelman syndrome', *J. Neurosci.* **18**: 8505–8514.

Essrich, C., Lorez, M., Benson, J.A., Fritschy, J.M. and Luscher, B. (1998) 'Postsynaptic clustering of major $GABA_A$ receptor subtypes requires the $\gamma2$ subunit and gephyrin', *Nature Neuroscience* **1**: 563–571.

Feigenspan, A., Wassle, H. and Bormann, J. (1993) 'Pharmacology of GABA receptor C1-channels in rat retinal bipolar cells', *Nature* **361**: 159–162.

Fritschy, J.M. and Mohler, H. (1995) 'GABAA-receptor heterogeneity in the adult rat brain: differential regional and cellular distribution of seven major subunits', *J. Comp. Neurol.* **359**: 154–194.

Fritschy, J.M., Benke, D., Johnson, D.K., Mohler, H. and Rudolph, U. (1997) 'GABAA-receptor alpha-subunit is an essential prerequisite for receptor formation in vivo', *Neurosci.* **81**: 1043–1053.

Fritschy, J.M., Weinmann, O., Wenzel, A. and Benke, D. (1998a) 'Synapse-specific localization of NMDA and GABA(A) receptor subunits revealed by antigen-retrieval immunohistochemistry', *J. Comp. Neurol.* **390**: 194–210.

Fritschy, J.M., Johnson, D.K., Möhler, H. and Rudolph, U. (1998b) 'Independent assembly and subcellular targeting of $GABA_A$-receptor subtypes demonstrated in mouse hippocampal and olfactory neurons', *Neurosci. Lett.* **249**: 99–102.

Gardner, J.M., Nakatsu, Y., Gondo, Y., Lee, S., Lyon, M.F., King, R.A. and Brilliant, M.H. (1992) 'The mouse pink-eyed dilution gene: association with human Prader-Willi and Angelman syndromes', *Science* **257**: 1121–1124.

Günther, U., Benson, J., Benke, D., Fritschy, J.M., Reyes, G.H., Knoflach, F., Crestani, F., Aguzzi, A., Arigoni, M., Lang, Y., Bluthmann, H., Möhler, H. and Luscher, B. (1995) 'Benzodiazepine-insensitive mice generated by targeted disruption of the $\gamma2$ subunit gene of γ-aminobutyric acid type A receptors', *Proc. Natl. Acad. Sci. USA* **92**: 7749–7753.

Hadingham, K.L., Garrett, E.M., Wafford, K.A., Bain, C., Heavens, R.P., Sirinathsinghji, D.J. and Whiting, P.J. (1996) 'Cloning of cDNAs encoding the human gamma-aminobutyric acid type A receptor alpha 6 subunit and characterization of the pharmacology of alpha 6-containing receptors', *Mol. Pharmacol.* **49**: 253–259.

Hanley, J.G., Koulen, P., Bedford, F., Gordon-Weeks, P.R. and Moss, S.J. (1999) 'The protein MAP-1B links GABA(C) receptors to the cytoskeleton at retinal synapses', *Nature* **397**: 66–69.

Harris, R.A., McQuilkin, S.J., Paylor, R., Abeliovich, A., Tonegawa, S. and Wehner, J.M. (1995) 'Mutant mice lacking the γ isoform of protein kinase C show decreased behavioral actions of ethanol and altered function of γ-aminobutyrate type A receptors', *Proc. Natl. Acad. Sci. USA* **92**: 3658–3662.

Hellevno, K., Kiianmaa, K. and Korpi, E.R. (1989) 'Effect of GABAergic drugs on motor impairment from ethanol, barbital and lorazepam in rat lines selected for differential sensitivity to ethanol', *Pharm. Biochem. Behav.* **34**: 399–404.

Homanics, G.E., Delorey, T.M., Firestone, L.L., Quinlan, J.J., Handforth, A., Harrison, N.L., Krasowski, M.D., Rick, C.E.M., Korpi, E.R., Makela, R., Brilliant, M.H., Hagiwara, N., Ferguson, C., Snyder, K. and Olsen, R.W. (1997a) 'Mice devoid of γ-aminobutyrate type A receptor β3 subunit have epilepsy, cleft palate, and hypersensitive behavior', *Proc. Natl. Acad. Sci. USA* **94**: 4143–4148.

Homanics, G.E., Ferguson, C., Quinlan, J.J., Daggett, J., Snyder, K., Lagenaur, C., Mi, Z.P., Wang, X.H., Grayson, D.R. and Firestone, L.L. (1997b) 'Gene knockout of the α6 subunit of the γ-aminobutyric acid type A receptor: lack of effect on responses to ethanol, pentobarbital, and general anesthetics', *Mol. Pharmacol.* **51**: 588–596.

Homanics, G.E., Le, N.Q., Kist, F., Mihalek, R., Hart, A.R. and Quinlan, J.J. (1998) 'Ethanol tolerance and withdrawal responses in $GABA_A$ receptor alpha 6 subunit null allele mice and in inbred C57BL/6J and strain 129/SvJ mice', *Alcohol. Clin. Exp. Res.* **22**: 259–265.

Homanics, G.E., Harrison, N.L., Quinlan, J.J., Krasowski, M.D., Rick, C.E.M., de Blas, A.L., Mehta, A.K., Kist, F., Mihalek, R.M., Aul, J.J. and Firestone, L.L. (1999a) 'Normal electrophysiological and behavioral responses to ethanol in mice lacking the long splice variant of the γ2 subunit of the γ-aminobutyrate type A receptor', *Neuropharmacol.* **38**: 253–265.

Huntsman, M.M., Porcello, D.M., Homanics, G.E., DeLorey, T.M. and Huguenard, J.R. (1999) 'Reciprocal inhibitory connections and network synchrony in the mammalian thalamus', *Science* **283**: 541–543.

Jones, A., Korpi, E.R., McKernan, R.M., Pelz, R., Nusser, Z., Makela, R., Mellor, J.R., Pollard, S., Bahn, S., Stephenson, F.A., Randall, A.D., Sieghart, W., Somogyi, P., Smith, A.J.H. and Wisden, W. (1997) 'Ligand-gated ion channel subunit partnerships: $GABA_A$ receptor α6 subunit gene inactivation inhibits δ subunit expression', *J. Neurosci.* **17**: 1350–1362.

Kirsch, J. (1999) 'Assembly of signaling machinery at the postsynaptic membrane', *Curr. Opin. Neurobiol.* **9**: 329–335.

Korpi, E.R., Kleingoor, C., Kettenmann, H. and Seeburg, P.H. (1993) 'Benzodiazepine-induced motor impairment linked to point mutation in cerebellar $GABA_A$ receptor', *Nature* **361**: 356–359.

Korpi, E.R., Koikkalainen, P., Vekovischeva, O.Y., Makela, R., Kleinz, R., Uusi-Oukari, M. and Wisden, W. (1998) 'Cerebellar granule-cell-specific GABAA receptors attenuate benzodiazepine-induced ataxia: evidence from α6-subunit-deficient mice', *Eur. J. Neurosci.* **11**: 233–240.

Koulen, P., Brandstatter, J.H., Enz, R., Bormann, J. and Wässle, H. (1998) 'Synaptic clustering of GABA(C) receptor rho-subunits in the rat retina', *Eur. J. Neurosci.* **10**: 115–127.

Löw K., *et al.* (2000) *'Molecular and neuronal substrates from the selective attenuation of anxiety science* **290**: 131–134.

Lyon, M.F., King, T.R., Gondo, Y., Gardner, J.M., Nakatsu, Y., Eicher, E.M. and Brilliant, M.J. (1992) 'Genetic and molecular analysis of recessive alleles at the pink-eyed dilution (p) locus of the mouse', *Proc. Natl. Acad. Sci. USA* **89**: 6968–6972.

Macdonald, R.L. and Olsen, R.W. (1994) 'GABAA receptor channels', *Ann. Rev. Neurosci.* **17**: 569–602.

Möhler, H. (2001) 'Functions of $GABA_A$-receptors: pharmacology and pathophysiology', in H. Möhler (ed.) *Pharmacology of GABA and glycine neurotransmission*, New York: Springer, pp. 101–116.

Möhler, H., Fritschy, J.M., Lüscher, B., Rudolph, U., Benson, J. and Benke, D. (1996) 'The $GABA_A$-receptors: from subunits to diverse functions', in T. Narahashi (ed.) *Ion Channels*, New York: Plenum Press, pp. 89–113.

Möhler, H., Benke, D., Benson, J., Lüscher, B., Rudolph, U. and Fritschy, J.M. (1997) 'Diversity in structure, pharmacology, and regulation of $GABA_A$-receptors', in S.J. Enna, and N.G. Bowery (eds) *GABA Receptors*, Totowa: Humana Press, pp. 11–36.

Möhler, H., Crestani, F. and Rudolph, U. (2001) '$GABA_A$-receptor subtypes – a new pharmacology', *Current Opinion in Pharmacology* **1**: 22–25.

Nakatsu, Y., Tyndale, R.F., DeLorey, T.M., Durham-Pierre, D., Gardner, J.M., McDanel, H.J., Nguyen, Q., Wagstaff, J., Lalande, M., Sikela, J.M., Olsen, R.W., Tobin, A.J. and Brilliant, M.H. (1993) 'A cluster of three GABAA receptor subunit genes is deleted in a neurological mutant of the mouse p locus', *Nature* **364**: 448–451.

Nusser, Z., Sieghart, W., Benke, D., Fritschy, J.M. and Somogyi, P. (1996) 'Differential synaptic localization of two major gamma-aminobutyric acid type A receptor alpha subunits on hippocampal pyramidal cells', *Proc. Natl. Acad. Sci. USA* **93**: 11,939–11,944.

Nusser, Z., Sieghart, W. and Somogyi, P. (1998) 'Segregation of different GABAA receptors to synaptic and extrasynaptic membranes of cerebellar granule cells', *J. Neurosci.* **18**: 1693–1703.

Quinlan, J.J., Homanics, G.E. and Firestone, L.L. (1998) 'Anesthesia sensitivity in mice that lack the $\beta 3$ subunit of the γ-aminobutyric acid type A receptor', *Anaesthesiology* **88**: 775–780.

Rudolph, U., Crestani, F. and Möhler, H. (2001) '$GABA_A$-receptor subtypes: dissecting their pharmacological functions', *Trends in Pharmacol. Sci.* **22**: 188–194.

Rudolph, U. *et al.* (1999) 'Benzodiazipine actions mediated by specific γ-aminobutyric acid, receptor subtypes', *Nature* **401**: 796–800.

Rudolph, U. and Möhler, H. (1999) 'Transgenic and targeted mutant mice in drug discovery', in *Current Opinion in Drug Discovery and Development*, Vol. 2, pp. 134–141.

Sabatini, D.M., Barrow, R.K., Blackshaw, S., Burnett, P.E., Lai, M.M., Field, M.E., Bahr, B.A., Kirsch, J., Betz, H. and Snyder, S.H. (1999) 'Interaction of RAFT1 with gephyrin required for rapamycin-sensitive signaling', *Science* **284**: 1161–1164.

Sieghart, W. (1995) 'Structure and pharmacology of gamma-aminobutyric acid A receptor subtypes', *Pharmacol. Rev.* **47**: 181–234.

Thompson, S.A., Whiting, P.J. and Wafford, K.A. (1996) 'Barbiturate interactions at the human GABAA receptor: dependence on receptor subunit combination', *Br. J. Pharmacol.* **117**: 521–527.

Wang, H., Bedford, F.K., Brandon, N.J., Moss, S.J. and Olsen, R.W. (1999) '$GABA_A$-receptor-associated protein links $GABA_A$ receptors and the cytoskeleton', *Nature* **397**: 69–72.

Chapter 17

Glutamatergic transmission

Therapeutic prospects for schizophrenia and Alzheimer's disease

Nuri B. Farber, John W. Newcomer and John W. Olney

Introduction

Extensive research focusing on the amino acid, glutamate (Glu), has documented the central role played by this compound in both the normal and abnormal functioning of the central nervous system (CNS). Glu is now recognized to be the main excitatory neurotransmitter in the CNS, estimated to be released at up to half of the synapses in the brain. In addition, Glu is also an excitotoxin that can destroy CNS neurons by excessive activation of excitatory receptors on dendritic and somal surfaces.

Two major classes of Glu receptors – ionotropic and metabotropic – have been identified. Glu exerts excitotoxic activity through three receptor subtypes which belong to the ionotropic family. These three receptors are named after agonists to which they are differentially sensitive – *N*-methyl-D-aspartate (NMDA), amino-3-hydroxy-5-methyl-4-isoxazole propionic acid (AMPA), and kainic acid (KA). Of these three, the NMDA receptor (depicted schematically in Figure 17.1) has been the most extensively studied and the most frequently implicated in CNS diseases (Olney, 1990). Excessive activation of NMDA receptors (NMDA receptor hyperfunction [NRHyper]) plays an important role in the pathophysiology of acute CNS injury syndromes such as hypoxia-ischemia, trauma and status epilepticus (Olney, 1990; Choi, 1992). Recently, hyper-stimulation of AMPA/KA receptors and consequent excitotoxicity has been proposed to underlie neurodegeneration in amyotrophic lateral sclerosis (ALS, Lou Gerhig's Disease (Rothstein *et al.*, 1995; Ikonomidou *et al.*, 1996)). The role of Glu excitotoxicity in the pathology seen in several other neuropsychiatric disorders has been extensively reviewed elsewhere (Choi, 1988; Olney, 1990) and will not be the focus of this chapter. Instead, we have chosen to focus on a recently discovered unconventional form of neurotoxicity that arises from the *under*excitation of NMDA receptors. We postulate that this underexcitation of NMDA receptors is operative in two major neuropsychiatric disorders – schizophrenia and Alzheimer's disease (AD). In the following sections we will describe initially the type of neuronal damage produced by hypoactivation of the NMDA receptor and the complex neural circuitry that is postulated to be perturbed as a consequence of NMDA receptor hypofunction (NRHypo). Then we will discuss the proposal that NRHypo may be involved in both schizophrenia and AD. This will be followed by a discussion about the implications of this model for treatment.

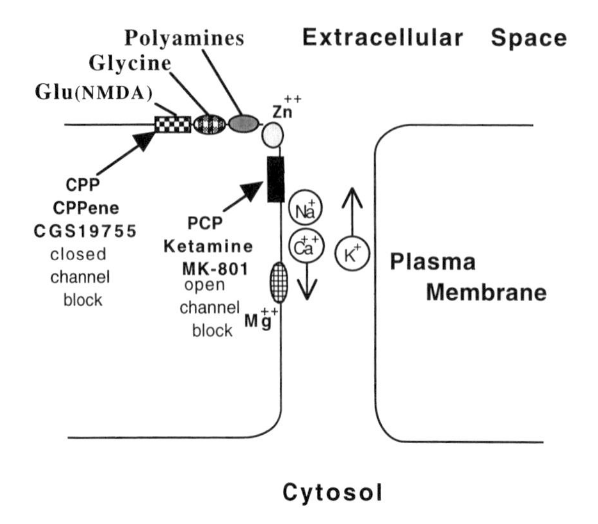

Figure 17.1 Associated with the NMDA receptor are multiple recognition sites through which receptor channel function is modulated. Currently it is recognized that NMDA receptor hypofunction (NRHypo) can be induced by agents (PCP, ketamine, MK-801) acting at the PCP site to perform an open channel block or agents (CPP, CPPene, CGS 19755) acting at the NMDA site to perform a closed channel block.

NRHypo produces brain damage in the rat

Descriptive characterization

Low doses of NMDA receptor antagonists such as ketamine, MK-801, nitrous oxide (laughing gas), phencyclidine (PCP), tiletamine, CPP, and CPP-ene reliably injure certain cerebrocortical neurons (Olney *et al.*, 1989, 1991; Hargreaves *et al.*, 1993; Jevtovic-Todorovic *et al.*, 1998b). At these doses the injury is confined to the posterior cingulate and retrosplenial (PC/RS) cortex and consists of the formation of intracytoplasmic vacuoles in layer III–IV pyramidal neurons. These changes are transient and resolve by 24 hours (Olney *et al.*, 1989). While these neurons will continue to express the 72kDa form of heat shock protein (HSP-72) for up to two weeks (Sharp *et al.*, 1991; Olney *et al.*, 1991), they do not become argyrophilic (de Olmos cupric silver method) or die.

Administering these agents in high dosage or by continuous infusion for several days induces a prolonged NRHypo state which causes irreversible injury involving the death of neurons in many cerebrocortical and limbic brain regions (Allen and Iversen, 1990; Ellison, 1994; Corso *et al.*, 1997; Horvath *et al.*, 1997; Wozniak *et al.*, 1998). Large- to medium-sized pyramidal and multipolar neurons are preferentially affected, although smaller neurons are also involved. The full pattern of damage (Figure 17.2) includes the PC/RS, frontal, temporal, entorhinal, perirhinal, piriform and prefrontal cortices, the amygdala and hippocampus (Corso *et al.*, 1997; Wozniak *et al.*, 1998). At four hours, the reaction in PC/RS cortex consists of intracytoplasmic vacuole formation, but in other brain regions a spongiform reaction featuring edematous swelling of spines on proximal dendrites is the most prominent cytopathological change. At 24 to 48 hours, the affected neurons become argyrophilic and immunopositive for HSP-72 and begin

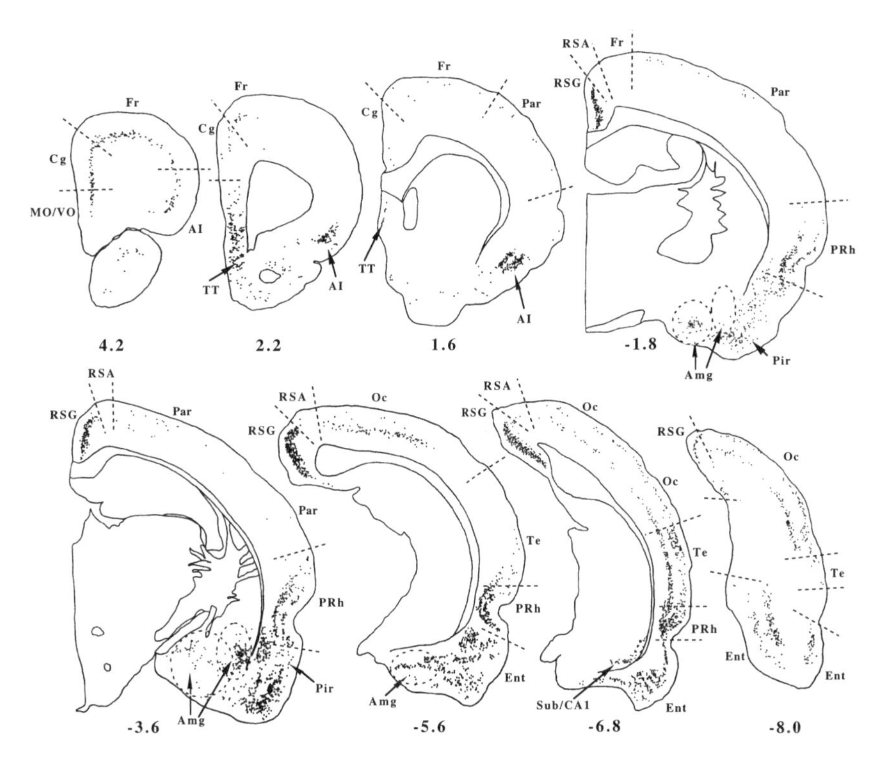

Figure 17.2 Schematic representation of the distribution of degenerating neurons at eight rostrocaudal levels of the adult rat brain four days following treatment with a high dose of an NMDA antagonist. These cell plots illustrate the neuronal populations that are primarily at risk to degenerate when the NMDA receptor system is rendered profoundly hypofunctional. At-risk populations include pyramidal and multipolar neurons in the prefrontal, PC/RS, occipital, temporal, parietal, entorhinal, perirhinal and piriform cortices and in the anterior olfactory nucleus, taenia tecta, amygdala and hippocampus. Rostro-caudal level and anatomical nomenclature are from Paxinos and Watson (1986).

The number under each cell plot is the distance in mm from Bregma. AI = agranular insular cortex; Amg = amygdala; Cg = anterior cingulate; Ent = entorhinal cortex; Fr = frontal cortex; MO/VO = medial orbital/ventral orbital cortex; Oc = occipital cortex; Par = parietal cortex; Pir = piriform cortex; PRh = perirhinal cortex; RSA = retrosplenial agranular cortex; RSG = retrosplenial granular cortex; Sub/CA1 = subiculum/CA1 field of hippocampus; Te = temporal cortex; TT = taenia tecta. Note: The area designated as RSG is the equivalent to the region referred to herein as PC/RS. See Corso *et al.* (1997) for histological documentation of this degeneration pattern.

to display cytoskeletal abnormalities, including a conspicuous corkscrew deformity of their apical dendrites. In the 72 to 96 hour interval many of the degenerating neurons display conspicuous fragmentation, but cytoplasmic organelles and cytoskeletal elements within the cell body and mainstem dendrites of some cells continue to show mixed signs of viability and degeneration for at least ten days. Over this period the degenerative reaction does not elicit a robust glial or phagocytic response and the overall appearance is one of a subacute protracted neurodegenerative process.

Receptor mechanisms and neural circuitry involved

In a series of recent studies, our group (Olney *et al.*, 1991, 1997; Farber *et al.*, 1993; Price *et al.*, 1994; Farber and Olney, 1995; Jevtovic-Todorovic *et al.*, 1998a) and others (Sharp *et al.*, 1992;, 1994, 1995) have found that several different classes of drugs (Table 17.1) effectively block the PC/RS neurotoxic action of NMDA antagonists. These findings, in conjunction with other work (Price *et al.*, 1994; Farber *et al.*, 1997; Kim *et al.*, 1998, 1999; Moghaddam *et al.*, 1997; Adams and Moghaddam, 1998; Giovannini *et al.*, 1994; Hasegawa *et al.*, 1993), indicate that a surprisingly complex neural circuitry is involved in NRHypo neurotoxicity (Figure 17.3).

Main excitatory limbs

The key feature of this neural circuitry is that Glu, acting at NMDA receptors, functions in this circuit as a regulator of inhibitory tone. This regulation is accomplished through tonic stimulation of NMDA receptor-bearing GABAergic interneurons. Each of these interneurons, in turn, inhibits excitatory projections that converge onto the vulnerable PC/RS neuron. One of these excitatory projections consists of a cholinergic neuron that resides in the basal forebrain and projects to the PC/RS cortex where it synapses on a m_3 muscarinic receptor located on the vulnerable pyramidal neuron

Table 17.1 Efficacy of Various Agents in Blocking NRHypo Neurotoxicity

Test Compound	*ED50 (mg/kg ip)*	*Confidence limits (25th & 75th percentile)*
Novel Antipsychotics		
Olanzapine	0.6	(0.2–1.9)
Clozapine	2.8	(1.5–5.1)
α_2 Adrenergic Agonists		
Clonidine	0.044	(0.0088–022)
Guanabenz	0.21	(0.13–0.33)
$5HT_{2A}$ Serotonergic Agonist		
Lisuride	0.17	(0.03–0.85)
DOI	0.98	(0.37–3.57)
GABAergic Agents		
Pentobarbitol	13.1	(9.5–16.6)
Thiopental	27.4	(16.1–38.6)
Diazepam	1.0	*
Antimuscarinics		
Scopolamine	0.13	(0.08–0.18)
Benztropine	1.29	(0.92–1.65)
Trihexyphenidyl	2.41	(1.26–3.57)
Sigma Agents		
Trihexyphenidyl	2.41	(1.26-3.57)
Haloperidol	5.1	(2.4–10.9)
Rimcazole	17.8	(8.9–35.6)
Novel Anticonvulsants		
Lamotrigine	8.6	(4.7–15.5)

*Diazepam did not show protection greater than 50% and thus confidence limits could not be calculated.

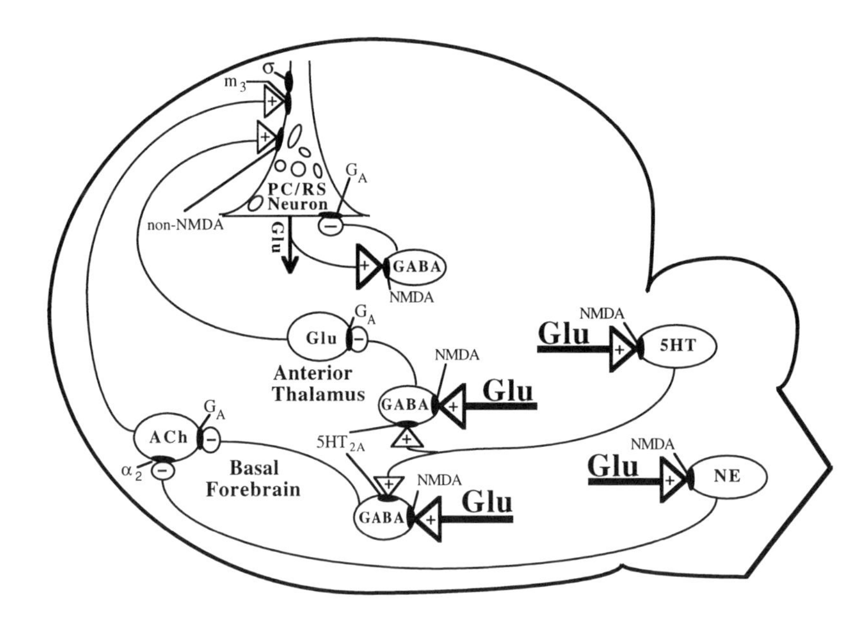

Figure 17.3 To explain NRHypo-induced neurotoxicity of PC/RS neurons, we propose that Glu acting through NMDA receptors on GABAergic, serotonergic and noradrenergic neurons maintains tonic inhibitory control over multiple excitatory pathways that convergently innervate PC/RS neurons. Systemic administration of an NMDA receptor antagonist (or NRHypo produced by any mechanism) would simultaneously abolish inhibitory control over multiple excitatory inputs to PC/RS neurons. This would create chaotic disruption among multiple intracellular second messenger systems, thereby causing derangement of cognitive functions subserved by the afflicted neurons, as well as eventual degeneration of these neurons. This circuit diagram focuses exclusively on PC/RS neurons. We hypothesize that a similar disinhibition mechanism and similar but not necessarily identical neural circuits and receptor mechanisms mediate damage induced in other corticolimbic brain regions by sustained NRHypo.

(+) = excitatory input; (−) = inhibitory input; ACh = acetylcholine; NE = norepinephrine; σ = sigma receptor; 5HT = serotonin; $\alpha_2 = \alpha_2$ subtype of adrenergic receptor; $G_A = GABA_A$ subtype of GABA receptor; $m_3 = m_3$ subtype of muscarinic cholinergic receptor; non-NMDA = non-NMDA subtype of Glu receptor; NMDA = NMDA subtype of Glu receptor; $5HT_{2A} = 5HT_{2A}$ subtype of 5HT receptor.

(Olney *et al.*, 1991; Price *et al.*, 1994; Farber *et al.*, 1997; Kim *et al.*, 1999). A second limb consists of a glutamatergic neuron that originates in the anterior thalamus and synapses on a non-NMDA glutamate receptor on the PC/RS pyramidal neuron (Price *et al.*, 1994; Sharp *et al.*, 1995; Farber *et al.*, 1997). By blocking Glu's tonic stimulation of GABAergic interneurons, NMDA antagonists remove inhibitory control from the two excitatory neurons. The resulting disinhibition of these two excitatory projections, which converge on the vulnerable PC/RS pyramidal neuron, leads to excessive release of neurotransmitter (Giovannini *et al.*, 1994; Hasegawa *et al.*, 1993; Adams and Moghaddam, 1998; Giovannini *et al.*, 1994; Kim *et al.*, 1999) and consequent excessive stimulation of the pyramidal neuron. Through this excessive stimulation the vulnerable pyramidal neuron is injured and develops histological changes (Price *et al.*, 1994; Farber *et al.*, 1997). Thus, blockade of one of the most common excitatory receptors in the CNS paradoxically leads to a hyperexcitatory state that can cause widespread cortico-limbic neurodegeneration.

Accessory control elements

In addition to the primary control provided by the NMDA receptor-bearing GABAergic neurons, there are at least two secondary control elements, which we have termed "accessory" because they provide additional control over the main excitatory projections. One accessory control element consists of an NMDA receptor-bearing noradrenergic (NE) neuron that, via stimulation of an α_2 adrenergic receptor, controls the release of ACh by the basal forebrain cholinergic neuron (Farber and Olney, 1995; Kim *et al.*, 1998, 1999). The second accessory control element consists of an NMDA receptor-bearing serotonergic (5HT) neuron that exerts agonist activity at a $5HT_{2A}$ receptor (Olney *et al.*, 1997) that we postulate is located on a GABAergic neuron. Stimulation of either the α_2 receptor or the $5HT_{2A}$ receptor restores inhibitory control over one or more of the excitatory projections and thus prevents the neurotoxic reaction from occurring (Farber and Olney, 1995; Olney *et al.*, 1997). These two control elements can be thought of as back-up or accessory braking systems that help to maintain inhibition within the circuit. When generalized NRHypo is produced by the systemic administration of a NMDA antagonist, the two excitatory limbs become disinhibited while at the same time Glu is blocked from activating these accessory control elements. Control over the disinhibited excitatory projections is lost, resulting in excessive excitation of the pyramidal neuron and consequent histological damage.

Output control element

In addition to the primary and accessory control elements described above, the circuit also contains an inhibitory feedback loop that directly controls output of the pyramidal neuron. The loop consists of a recurrent collateral given off by the long projecting axon of the PC/RS pyramidal neuron; this collateral terminates on an NMDA receptor-bearing GABAergic neuron which synapses back onto the PC/RS pyramidal neuron. Increased firing of the PC/RS neuron automatically causes increased activity in this inhibitory loop and provides a means by which the pyramidal neuron can regulate its own firing. Inactivation or under-functioning of this inhibitory feedback loop would compromise this control, causing the pyramidal neuron to fire excessively and in a poorly modulated manner, thereby disturbing mental functioning subserved by the PC/RS neuron and its projections. While inactivation of this loop by NMDA antagonists would lead to reduced control over the firing of the PC/RS neuron, it is unclear whether this inactivation contributes to the injury of the PC/RS neuron.

Other comments

Ongoing work in our laboratory suggests that at least four different populations of NMDA receptors, one for each of the NMDA receptor-bearing GABAergic neurons in the two different excitatory limbs and one for each of the two accessory control arms of the circuit, need to be hypofunctional in order to induce the acute neurotoxic reaction.

In other words, it requires that both the excitatory input channels be disinhibited simultaneously (and thereby simultaneously hyperactivated) for acute physical injury of the pyramidal neuron to occur. Preventing hyperactivation of any one of the excitatory input receptors (non-NMDA-glutamatergic, m_3-muscarinic), either by blocking one of these two receptors or by restoring inhibitory tone over one of the excitatory limbs, prevents the neurotoxic reaction from occurring (Table 17.1).

Inspection of the circuit reveals that Glu plays two key but opposing roles in this neurotoxic process. By stimulating the NMDA receptors Glu prevents the circuit from wreaking destruction on itself. However, by stimulating the non-NMDA receptor on the vulnerable pyramidal neuron, Glu also is partially responsible for causing degeneration of the vulnerable pyramidal neurons. Studies aiming to confirm the role of Glu and NRHypo in neuropsychiatric disorders will have to be conducted by methods that separately evaluate the status of Glu at NMDA and non-NMDA receptors.

NRHypo hypothesis of schizophrenia

NRHypo causes psychosis

A variety of NMDA antagonists (e.g., ketamine, PCP, CPP, CPP-ene, CGS19755, CNS 1102), when administered to humans, cause a psychotic state (Domino and Luby, 1981; Javitt and Zukin, 1991; Kristensen *et al.*, 1992; Herrling, 1994; Krystal *et al.*, 1994; Grotta *et al.*, 1995; Malhotra *et al.*, 1996; Muir *et al.*, 1997; Newcomer *et al.*, 1999). In-depth studies of two of these agents, PCP and ketamine, indicate that these drugs, more faithfully than other psychotomimetics (e.g., amphetamines, LSD), can mimic a broad range of symptoms seen in schizophrenia, including impairments in cognitive function (Domino and Luby, 1981; Krystal *et al.*, 1994; Malhotra *et al.*, 1996; Newcomer *et al.*, 1999). Because of these differences among the different psychosis-producing agents, NMDA antagonists have been described as "schizophrenomimetic" and are being intensively studied in the hope of gaining insight into the pathophysiology of schizophrenia.

NRHypo-induced disinhibition: a common denominator of the neurotoxic and psychotic effects of NMDA antagonists

Several classes of GABAergic agents – benzodiazepines (Olney *et al.*, 1991), barbiturates (Olney *et al.*, 1991), and GABAergic anesthetics, including halothane, isoflurane, and propofol (Ishimaru *et al.*, 1995; Jevtovic-Todorovic *et al.*, 1997), block the neurotoxic action of NMDA antagonists, presumably by restoring GABAergic inhibition to the excitatory limbs in the relevant neural circuitry (Figure 17.3). Similarly anesthesiologists have long reported that the psychotomimetic effects of NMDA antagonists, which they term "emergence reactions," can be prevented by benzodiazepines (White *et al.*, 1982; Reich and Silvay, 1989) and barbiturates (Magbagbeola and Thomas, 1974). The ability of GABAergic agents to block both the neurotoxic and psychotomimetic effects of NMDA antagonists has led us to propose that the complex polysynaptic disinhibition mechanism that underlies the neurotoxic action of NMDA antagonists also underlies their psychotomimetic effects (Olney and Farber, 1995). The same core disinhibition mechanism is proposed to underlie both the mental and morphological changes associated with the NRHypo state. Mild disinhibition and consequent mild overstimulation would produce cognitive and behavioral disturbances, while persistent and severe disinhibition would result in morphological changes. Consistent with this proposal several classes of agents besides GABAergic agents which prevent the neurodegenerative changes (Table 17.1) also attenuate the psychosis associated with the NRHypo state (Malhotra *et al.*, 1997; Newcomer *et al.*, 1998; Anand *et al.*, 1997).

In addition, we have found that pre-pubertal rats are insensitive to the neurotoxic effects of NMDA antagonists, even following very high doses, but during adolescence they become increasingly sensitive to the neurotoxic action so that by late adolescence even low doses of an NMDA antagonist can cause neurotoxicity (Farber *et al.*, 1995). Similarly, anesthesiologists have found that the ability of ketamine to produce emergence reactions is dependent on the age of the patient. Infants and children prior to puberty rarely develop an emergence reaction when exposed to these agents, whereas up to 50 percent of post-pubertal teenagers and adults develop these symptoms (Reich and Silvay, 1989; Marshall and Longnecker, 1990). The similar age dependency profile of these two NRHypo-induced phenomena reinforces the proposal that a common pathophysiology underlies the neurotoxic and psychotomimetic effects.

NRHypo-induced disinhibition and schizophrenia

Given the similarity of the mental disturbances seen with NMDA antagonists and in schizophrenia we have further proposed that mental signs and symptoms seen in schizophrenia are produced by NRHypo (as others have also argued), and that these symptoms are a direct result of NRHypo-induced disinhibition of a polysynaptic circuit analogous to the one described in the rat (Olney and Farber, 1995; Farber *et al.*, 1999a). Consistent with this proposal is the observation that patients with schizophrenia appear to be more sensitive to NMDA antagonists than healthy humans, and following exposure to these drugs can even develop a prolonged recrudescence of the acute psychotic state lasting for days to months (Luby *et al.*, 1959; Ban *et al.*, 1961; Lahti *et al.*, 1995).

Schizophrenia typically has an age of onset in late adolescence or early adulthood. In order to account for growing evidence of an early developmental lesion in patients with schizophrenia, investigators have postulated that unknown maturational events must occur in the brain during adolescence that allows for the expression of symptoms. As discussed above, NRHypo-induced neurotoxicity and psychosis has a similar age of onset in adolescence, suggesting similar dependence on maturational events. Given this similar age of onset profile, we propose that a common mechanism, namely an NRHypo-induced disinhibited state, underlies the neurotoxic and psychotomimetic effects of NMDA antagonists as well as the signs and symptoms of schizophrenia, and that similar maturational events allow the onset of both. Consistent with this proposal, patients with schizophrenia develop antibodies to HSP-72 (Schwarz *et al.*, 1999), a protein that is expressed in response to the NRHypo state (Sharp *et al.*, 1991; Olney *et al.*, 1991).

Neurodegeneration and schizophrenia

The proposal that the same mechanism (NRHypo) underlies the neurotoxic and psychotomimetic action of NMDA antagonists, and that this mechanism is also operative in schizophrenia, does not imply that NRHypo-induced damage to the adult brain causes the psychosis or even that such damage will necessarily occur in the majority of people with schizophrenia. Rather, we believe that in many cases dysfunction in the NRHypo circuit may lead to mental changes without causing histological damage. Our reasoning is based on the observation that induction of histological damage requires the simultaneous disinhibition of converging excitatory projections plus the loss of other control elements. This signifies that a patient with schizophrenia would need to have pervasive abnormalities affecting multiple components of the circuit in order to sustain histological damage. In contrast, disruption in only one or two parts of the circuit would be sufficient to cause hyperactivation and excess firing of the pyramidal neuron, with consequent mental disturbances.

In addition, even if all parts of the circuit were in a NRHypo state it is not necessarily so that neurodegeneration would result. A mild NRHypo state, by slightly increasing stimulation of the vulnerable corticolimbic neuron, could produce psychosis without over-stimulating the neuron to such an extent that it degenerates and dies. Only in the cases where the NRHypo state and associated excitatory transmitter release were unremittingly severe on a chronic basis would neurodegenerative changes occur. Thus we believe that NRHypo neurodegeneration could explain the cognitive degeneration that is seen in a subset of patients with schizophrenia.

Regarding the relative absence of scarring in the brains of patients with schizophrenia and the dogma that ongoing brain damage in the adult brain should produce glial scarring, it has been observed that when neurons are deleted from the adult brain by the NRHypo mechanism, only a transient and very limited gliotic response occurs (Fix *et al.*, 1995). Thus, NRHypo is an excellent mechanism to explain how brain damage can occur during adulthood in schizophrenia with consequent cognitive deterioration but little or no residual scarring in the brain.

Abnormal neural development in schizophrenia

Consistent with evidence for genetic factors and early brain insults in schizophrenia (Lyon *et al.*, 1989), it is proposed that a pathological process leading to NRHypo in certain brain circuits might occur *in utero* and remain clinically silent until adolescence. As noted above, even though NMDA antagonists can instill a profound NRHypo state in either the human or rat brain during various stages of development, they do not elicit either a psychotic or neurotoxic response until after puberty. This signifies that the immature brain could harbor a lesion that renders the NMDA receptor system hypofunctional and this defect would not be expressed symptomatically until a critical stage in adolescence when unknown maturational changes make it possible for the NRHypo state to trigger a pathological reaction. We postulate that as the maturation process occurs in each particular brain region, mental functions subserved by that brain region would become dysfunctional. Since each region would mature at different times, one would see the unfolding of the full syndrome over time as each specific clinical sign appears in succession. For example, relatively subtle impairments in cognitive functions, such as attention and memory, might occur in early adolescence while the more obvious psychotic signs would not become present until later in adolescence. In addition some brain regions and their associated mental functions may be more sensitive to disruption by NRHypo so that different symptoms may present at different times even if brain maturation is relatively homogeneous. In those individuals with a particularly severe and persistent form of NRHypo, secondary structural brain changes accompanied by progressive clinical deterioration would also occur. Thus, two types of structural brain changes are postulated. One would originate *in utero* as a function of genetic, environmental factors, or both, altering the relevant normal circuit to produce NRHypo. The other would originate in adolescence as the brain matures and becomes vulnerable to the NRHypo state.

There are a number of potential mechanisms that could occur early and produce an NRHypo condition in the pre-adolescent brain. Candidates include loss of NMDA-receptor bearing GABAergic neurons, dopaminergic hyperactivity, hypofunctional NMDA receptors, and abnormalities in one or more of the other neurotransmitter systems in the NRHypo circuit. In addition, abnormal neuronal migration resulting in an abnormal arrangement or "miswiring" of the circuit could lead to an NRHypo-equivalent disinhibited state. Of these candidate mechanisms, the loss of NMDA-receptor bearing GABAergic neurons is of particular interest, given that deficiencies in the GABAergic system exist in postmortem brain tissue of patients with schizophrenia (Benes *et al.*, 1991; Woo *et al.*, 1998). Because the GABAergic neuron in the feedback loop normally regulates the firing pattern of cerebrocortical pyramidal neurons and aberrant firing of these neurons could cause considerable mental dysfunction, we propose that this population of GABAergic neurons might be differentially abnormal in patients with schizophrenia. Indeed, an excellent formula for psychosis would be loss of GABAergic inhibition simultaneously in this feedback loop and in any one of the two main excitatory inputs to the pyramidal neuron, since this would cause the pyramidal neuron to be persistently hyperactive at the same time that it has lost feedback control over its firing on other neurons.

Potential inconsistencies of the NRHypo schizophrenia hypothesis

An apparent inconsistency between the NRHypo animal model and schizophrenia is that female rats and humans are more sensitive than males to the neurotoxic and psychotomimetic effects, respectively, of NMDA antagonists, whereas in schizophrenia there are either no gender differences or a slightly earlier age of onset for males. However, the greater sensitivity of females to NMDA antagonists pertains only to PCP and its structural analogs and is most likely explained by these drugs being metabolized in the liver less efficiently by females than males. In contrast male rats are slightly more sensitive than females to the neurotoxic effects of competitive NMDA antagonists (Olney *et al.*, unpublished observations), presumably because these agents are not PCP analogs and are not subject to the same sex-linked differential in hepatic metabolism. Thus if the hepatic metabolic variable is corrected for, the male/female sensitivity ratio may be quite similar for the NRHypo animal model and the human disease.

A potential shortcoming of the NRHypo hypothesis as formulated herein is that it does not explain why D_2 dopamine receptor antagonists are often therapeutically beneficial in schizophrenia. However, as discussed elsewhere (Olney and Farber, 1995), D_2 dopamine receptors may interact with the NRHypo circuitry at points that remain to be identified, and such interaction could easily explain the observed therapeutic efficacy of D_2 dopamine antagonists in terms of their ability to correct or counteract the NRHypo state.

Summary of the NRHypo schizophrenia hypothesis

Given the schizophrenia-like quality of the mental disturbances caused in normal humans by NMDA antagonists, the exquisite sensitivity of schizophrenic patients to the psychotogenic actions of these drugs, the fact that the usual onset of susceptibility to schizophrenia and to the neurotoxic and psychotic effects of NRHypo has the same age-dependency profile, and substantial evidence that the basic mechanism underlying the neuropsychotoxic effects of NMDA antagonists is an NRHypo-induced disinhibitory state, we hypothesize that in the brains of patients with schizophrenia there is a comparable NRHypo-induced condition that causes a complex network disturbance featuring disinhibition as the core principle. We further postulate that in schizophrenia the NRHypo condition is instilled in the brain early in life but remains quiescent until maturational changes occur in adolescence that permit the disinhibition mechanism to disrupt mental functions and produce the symptoms of schizophrenia. Finally we propose that if the NRHypo state is severe enough it will cause ongoing neurodegenerative changes and progressive cognitive deterioration.

Fundamental features of schizophrenia accommodated by this hypothesis are: (1) the occurrence of structural brain changes during early development that have the potential for producing subsequent clinical manifestations of schizophrenia; (2) a quiescent period in infancy and adolescence before clinical manifestations are expressed; (3) onset in late adolescence of psychotic symptoms; (4) involvement of D_2 dopamine receptors in some cases but not others, which would explain why some but not all patients are responsive to typical antipsychotic therapy; (5) ongoing neurodegenerative changes and cognitive deterioration in some patients.

NRHypo hypothesis of AD

Over the past decade major strides have been made in discovering important genetic abnormalities in AD. Mutations on four different chromosomes, each of which can promote amyloidopathy, have now been identified as etiologic factors in familial AD and the role of apoE genotype as a risk factor in sporadic AD has been established. While recent research has elucidated the basic neurochemistry of β-amyloid, and it is clear that abnormal deposition of β-amyloid in the brain occurs early in AD, it is not at all clear how β-amyloid deposition contributes to the neurodegenerative events in AD. Based on evidence that a severe degree of NRHypo is present in the human AD brain, that NRHypo can induce an AD-like pattern of neurodegeneration in animal brain and that β-amyloid deposition may contribute to a worsening of the NRHypo state, we propose NRHypo as a missing link that can help explain major aspects of the pathophysiology of AD that have heretofore remained elusive.

Similarities between the neuronal degeneration seen in NRHypo and in AD

There are important similarities between the overall pattern of NRHypo neurodegeneration (Figure 17.2) and the pattern that has been described in the AD brain by various researchers. The PC/RS cortex, which is the brain region most vulnerable to NRHypo degeneration, is selectively affected early in the course of AD (Reiman *et al.*, 1996;

Minoshima *et al.*, 1997). The PC/RS cortex has also been shown to be markedly atrophic late in the disease (Braak *et al.*, 1992). In contrast, neurodegeneration in the anterior cingulate cortex is less severe in both the AD brain (Brun and Englund, 1981) and NRHypo animal model (Figure 17.2). While it is difficult to make precise anatomical comparisons between the rodent and human brain, the transentorhinal area, considered among the earliest and most severely affected regions in the human AD brain (Braak and Braak, 1995), is roughly homologous to the perirhinal cortex in rat brain, which is second only to the PC/RS cortex in its sensitivity to NRHypo neurodegeneration (Figure 17.2). Other brain regions preferentially affected in both the AD brain and NRHypo model include portions of the parietal, temporal, entorhinal, amygdaloid, subicular, hippocampal and insular cortices. A mild but transient microglial and astrocytic response accompanies the neurodegeneration seen with NRHypo. However, consistent with the known pathology of AD, a robust phagocytic response is conspicuously absent (Fix *et al.*, 1995).

The neurons primarily involved in neurofibrillary tangle (NFT) formation in the AD brain are distributed widely throughout cortical and limbic brain regions, but in each region these neurons tend to be pyramidal or multipolar neurons and in certain cortical regions they are distributed in a bilaminar pattern. This fits the description of the subpopulation of neurons primarily affected in the NRHypo neurodegenerative syndrome (Corso *et al.*, 1997). Interneurons in the cerebral cortex are also occasionally involved, but the most prominently affected neurons are medium-sized pyramidal or multipolar neurons in each region.

The tortuosity of dendritic processes in the NRHypo model is accompanied by a parallel pattern of tortuosity of the microtubular cytoskeleton within the distorted dendrite. This suggests that changes in the external configuration of the dendrite are due to cytoskeletal changes within the dendrite. The cytoskeleton of the injured neurons appears to be undergoing both degenerative and regenerative processes, but the repair effort is not very successful. As described above, the mechanism of injury involves simultaneous hyperactivation of the neuron through a muscarinic (m_3) cholinergic receptor and a glutamatergic (non-NMDA) receptor (Figure 17.3). Second messenger systems associated with non-NMDA receptors are not known at this time, but it is well known that the m_3 muscarinic receptor is coupled to a phosphoinositide/Ca^{2+}/protein kinase C second messenger system that mediates protein phosphorylation functions. Thus the erratic pattern of microtubule tortuosity and disarray, together with apparent efforts at microtubule regeneration, may reflect abnormal hyperactivity of this m_3-linked second messenger system and consequent disruption of its protein phosphorylation functions. Since other second messenger systems probably are also hyperactivated simultaneously in the same neuron, these systems may also contribute to the cytoskeletal disruption pattern.

We are just beginning to study the possible relationship between this cytoskeletal disruption process in the NRHypo model and NFT, its potential counterpart in the AD brain. Thus far, we have not detected ultrastructural evidence of paired helical filaments. However, even if such evidence cannot be found, this might signify species specificity of this particular abnormality without disqualifying the NRHypo degenerative process as a valid model of the mechanism giving rise to NFT in human AD brain. Thus, our findings in the rat are consistent with the conclusion that NRHypo by itself can produce many of the neuropathological features of AD.

NRHypo is present in aging and to a greater degree in AD

At least four different laboratories studying three different non-human species (mice, rats, monkeys) have reported that the NMDA receptor transmitter system becomes markedly hypofunctional with advancing age (Gonzales *et al.*, 1991; Wenk *et al.*, 1991; Magnusson and Cotman, 1993; Saransaari and Oja, 1995). In addition it was recently shown in humans that a more severe degree of NRHypo is present in the AD brain than in age-matched normal controls (Ulas and Cotman, 1997). Thus, in the aging human brain the stage may already be set for widespread corticolimbic neurodegeneration to occur. All that is required to explain why it occurs to a more severe degree in the AD brain than in the "normal" aging brain is to identify one or more adjunctive conditions peculiar to the AD brain that may serve as catalysts or promoters of the NRHypo state. Because we have seen no evidence of amyloidosis or amyloid plaque formation in the NRHypo animal model, we propose that genetic or other predisposing factors peculiar to the AD condition are primarily responsible for the amyloidopathy in AD brain and that when amyloidopathy occurs alongside NRHypo the pathological process known as AD develops. How then does amyloidosis interact with NRHypo to produce the histopathological features of AD?

AD neurodegeneration: a multi-stage process

A major tenet of our proposal is that in AD the brain after having first gone through a stage of NMDA receptor hyperfunction (NRHyper; i.e. typical Glu excitotoxicity) enters a stage where the NMDA receptor system becomes hypofunctional in AD. This hypothesis, consistent with the bulk of available data (e.g. Gómez-Isla *et al.*, 1996), assumes that the pattern of massive cortical neurodegeneration in AD tends to follow the pattern of NFT formation, and that the neurons that display NFT at the time of autopsy are injured neurons that would be destined to die slowly and leave behind neurofibrillary debris. However, this hypothesis also assumes that there is a separate pattern of neuronal degeneration that is less massive and that corresponds with the pattern of amyloid deposition. Finally, there is a third type of neurodegeneration which occurs as cholinergic neurons, bereft of trophic support from their cerebrocortical target neurons, involute and die. Our hypothesis conceives of the neurodegenerative events in AD occurring in multiple separate stages, by multiple separate mechanisms and according to multiple separate patterns. These have been difficult to tease apart because the stages have a significant degree of temporal overlap and the patterns have significant spatial overlap.

Stage I (amyloidogenic/NRHyper stage)

We propose that the first neurodegenerative stage entails the deposition of low concentrations of amyloid in the brain. β-amyloid alone is not toxic except at very high concentrations, whereas at substantially lower concentrations β-amyloid causes cultured neurons to become hypersensitive to Glu or NMDA excitotoxicity (Koh *et al.*, 1990; Gray and Patel, 1995; Patel, 1995). We propose, therefore, that predisposing factors (e.g. apoE4 genotype in sporadic AD, amyloidogenic mutations in familial AD) promote amyloidosis. Amyloidosis, in conjunction with other factors that may contribute to β-amyloid's augmentation of neuronal sensitivity to Glu (i.e. oxidative stress, defects in energy metabolism), increases the sensitivity of NMDA receptors so that even normal concentrations of Glu can trigger abnormal currents, which on a chronic low-grade excitotoxic basis can destroy NMDA receptor-bearing neurons. As these neurons degenerate, amyloid plaques may form and incorporate portions of the degenerating neurons and other neural and glial processes in the immediate environment. The pattern of these early neurodegenerative and reactive events will follow the pattern of distribution of the specific neurons vulnerable to this amyloid/NMDA receptor-mediated neuropathological process. We postulate that it may not be a very conspicuous pattern of neuronal loss because it may be restricted to a subpopulation of NMDA receptor-bearing neurons, including those in the NRHypo circuit that control the release of transmitters onto vulnerable pyramidal neurons (Figure 17.3). In Stage I, the neurodegenerative process may produce few if any symptoms because it is limited to a small population of neurons. However, loss of these neurons and their NMDA receptors would serve to increase the NRHypo burden in the aging brain.

Stage II (NFT/NRHypo stage)

The second stage commences when the NRHypo state, produced by the destruction of the NMDA receptor-bearing neurons, has become severe enough to substantially unleash the disinhibition syndrome in which many primary cerebrocortical and corticolimbic neurons are pathologically hyperstimulated through several signal transduction pathways at the same time. At this point, cognitive disturbances should become clearly evident. We propose that pyramidal neurons in many cortical and limbic brain regions will be affected. As the Stage II process progresses, these neurons will slowly degenerate and die. While these neurons are degenerating, we propose that at least some of them will develop NFT on the basis of excessive activation of second messenger pathways associated with muscarinic and/or non-NMDA glutamate receptors. These second messenger systems are coupled to kinases or other possible factors relevant to protein phosphorylation. Hyperactivation of these systems provides a rational explanation for NFT formation which is believed to result from hyperphosphorylation of microtubule-associated proteins. In Stage II, therefore, neurodegeneration occurs as a network disturbance. The pattern of degeneration is determined by the pattern of connections within the network, and by the failure of inhibition over certain excitatory pathways within the network, causing specific cortical and limbic neurons innervated by these excitatory pathways to degenerate. This provides a rational explanation for the pattern of cortical degeneration seen in AD.

Stage III (cholinergic deficit stage)

As the corticolimbic pyramidal neurons die in Stage II they are no longer able to provide the needed trophic support to their afferent neurons. Deprived of needed trophic factors these other neurons, which project to the pyramidal neurons that have undergone NRHypo neurodegeneration in Stage II, begin to involute and die. Since basal forebrain cholinergic neurons provide one of the key inputs to the neurons that degenerate as a consequence of NRHypo, these cholinergic neurons would be one of the main groups of neurons to die in Stage III. Evidence to date indicates that the degeneration of basal forebrain cholinergic neurons, and the resultant cholinergic deficit state, is a late event in the overall sequence of neurodegenerative events that occur in AD (Doucette *et al.*, 1986; Davis *et al.*, 1999). In addition, consistent with the NRHypo model, Miller *et al.* (1993), using magnetic resonance spectroscopy, observed that early in AD there is an apparent loss of non-cholinergic neurons but no loss of cholinergic neurons. Thus, while cholinergic loss commonly is assumed by many to occur early in the illness, the bulk of available evidence suggests that cholinergic loss occurs relatively late.

In summary, we propose that in contrast to schizophrenia, the brain of someone afflicted with AD develops normally and then functions normally throughout most of the adult life. As the brain ages the NMDA receptor system becomes progressively hypofunctional. In those individuals destined to develop AD, other abnormalities (e.g. amyloidopathy and oxidative stress) interact to increase the NRHypo burden and the brain enters into a severe and persistent NRHypo state which leads to widespread neurodegeneration and accompanying cognitive deterioration.

Comparison of the NRHypo models of schizophrenia and AD

Since the time of Emil Kraeplin and Alois Alzheimer neuroscientists have delineated the aberrations in mental functioning seen in schizophrenia (dementia praecox) from those seen in Alzheimer's Dementia. It thus seems counter-intuitive that a common mechanism (NRHypo) could underlie the pathophysiology of both illnesses. However, each illness has its own set of specific abnormalities, and the differences between these specific abnormalities could account for the disparate clinical picture (Table 17.2). The first difference is that the NRHypo state originates at different times. In schizophrenia

Table 17.2 Differences in the NRHypo models of AD and schizophrenia

	Schizophrenia	*AD*
Onset of NRHypo	Early in life	Late in life
Elements of Circuit Involved	a) excitatory limbs and accessory control elements minimally involved b) feedback loop maximally involved	a) excitatory limbs and accessory control elements maximally involved b) feedback loop minimally involved
Degree of NRHypo	Usually mild and non-progressive	Severe and progressive
Accessory Abnormalities	Hyperdopaminergia Abnormal neuronal migration Miswiring of the circuits Abnormalities in other receptors or neurotransmitters in the circuit	Amyloidosis Unfavorable ApoE genotype Oxidative stressors Disturbance in energy metabolism

the condition is instilled early in life (probably *in utero* in most cases), whereas in AD NRHypo occurs late in life. Given the maturational processes that occur in the brain during childhood and adolescence, it is likely that introduction of the NRHypo condition early in the life of an individual destined to develop schizophrenia would have a drastically different effect on the functioning of the CNS than it would if it were to occur *de novo* late in life.

Second, each illness has its own set of specific transmitter or neurochemical abnormalities that would coexist with the NRHypo state. For example, in AD excessive deposition of amyloid, disturbances in energy metabolism, oxidative stressors, and apoE genotype are all potential contributors to the pathological process. In schizophrenia, abnormalities such as hyperdopaminergic tone, early loss of GABAergic interneurons, abnormal neuronal migration, miswiring of circuits or abnormal receptors would contribute to the pathogenesis of the illness.

The third and last major difference is that the specific elements in each circuit which are abnormal or missing are different. In schizophrenia, the main abnormality is proposed to lie in the feedback control loop with an additional mild abnormality in one or more of the excitatory limbs or accessory control elements. These abnormalities alone would be sufficient to lead to abnormal and uncontrolled output from the vulnerable pyramidal neurons resulting in mental dysfunction characteristic of schizophrenia. However, since the disturbances in many cases are not pervasive or severe enough, the vulnerable pyramidal neurons typically will sustain only limited or no histological damage. Thus it is unlikely that most patients with schizophrenia would develop NFTs or neurodegeneration.

In contrast, in AD the main problem is postulated to lie with the excitatory inputs to the pyramidal neurons and the two accessory control elements. The abnormalities are severe, pervasive, and persistent enough that pyramidal neurons develop structural damage and ultimately die. The hyperactivation of these neurons in AD, as they are abnormally stimulated through multiple signal transduction pathways, is proposed to not typically translate into schizophrenia-like symptoms early in the disease because the feedback loop is intact, allowing pyramidal neurons to maintain control over their firing onto other neurons. However, as the disease progresses a point will be reached where the feedback loop is unable to appreciably prevent abnormal output from the pyramidal neuron, resulting in florid mental disturbances (e.g. psychosis). Consistent with this proposal psychosis is a frequent manifestation of AD (Drevets and Rubin, 1989; Rosen and Zubenko, 1991; Deutsch *et al.*, 1991; Farber *et al.*, 1999b), occurring in over 60 per cent of the subjects who come to autopsy (Farber *et al.*, 1999b).

Treatment implications

In attempting to apply knowledge gained from studying the NRHypo drug model to the clinical management of schizophrenia, it is important to remember that this model, although useful as a research tool, does not perfectly mimic the human disease process. We propose that the NRHypo state of schizophrenia is instilled in the brain by a pathological event early in life and remains quiescent until late adolescence when maturational changes in brain circuits allow the NRHypo mechanism to begin expressing its psychoneuropathological potential. If this were the case, the early pathological event will have caused permanent damage in brain circuits and numerous compensatory changes of an unknown kind will have occurred during development. Attempting to correct the dysfunction associated with these complicated permanent changes in the brain of a schizophrenia patient by pharmacological manipulations that effectively counter the acute effects of NMDA antagonist drugs in the normal adult rat brain is not likely to have a high success rate. For example, if NMDA receptor-bearing GABAergic neurons are deleted from the brain in early life this would instill a permanent NRHypo state, but attempting to correct this state by applying agents in adulthood that interact with the receptor systems for the missing GABAergic neurons would be futile since these receptor systems will undoubtedly be absent or pathologically altered as a consequence of the early lesion. Indeed, while benzodiazepines given adjunctively with antipsychotic medications have a defined role in the treatment of schizophrenia, monotherapy with benzodiazepines has minimal if any antipsychotic action. Thus, although the NRHypo model suggests pharmacological methods that could protect the brain against the adverse consequences of disinhibited activity in corticolimbic circuits, these methods were developed in a model based upon the acute induction of NRHypo in adulthood and at least some may not be directly applicable to treatment of a schizophrenia syndrome of developmental origin.

However, in AD where the pathological process is initiated in the setting of a normally developed adult CNS, it is more likely that pharmacological methods for preventing NRHypo neurotoxicity could be directly translated into effective treatments. If the NRHypo mechanism does underlie, at least in part, NFT formation and neurodegeneration in AD, then use of one of these agents might curtail the development of morphological changes in AD and potentially slow the progression of the illness. Early evidence indicates that some of these agents might be useful in AD (Fastbom *et al.*, 1998; Tekin *et al.*, 1998). Moreover, if NRHypo is operative in AD, then these agents would also be expected to ameliorate the psychosis seen in AD. Developing new treatments for psychosis in AD is important since traditional treatment with antagonists of the D_2 subtype of dopamine receptor is not overwhelmingly beneficial and carries significant risks (Schneider *et al.*, 1990). The current debate about the use of D_2 antagonists in dementia is of added interest given that antagonism of D_2 receptors does not appreciably ameliorate the psychosis or morphological changes associated with the NRHypo state (Farber *et al.*, 1993; Krystal *et al.*, 1999). If psychosis in AD is a more mild manifestation of a process that eventually can result in NFTs and neurodegeneration, then its presence in a patient could signify that the individual is prone to developing NFTs and neurodegeneration. Such a relationship might underlie the finding that the presence of psychosis is associated with a more rapid cognitive decline (Mayeux *et al.*, 1985; Stern *et al.*, 1987; Drevets and Rubin, 1989; Rosen and Zubenko, 1991; Levy *et al.*, 1996). Institution of appropriate treatment with a protective agent in such individuals might slow any further decline associated with NFT formation and neurodegeneration.

Summary

Here we have described a novel excitotoxic process in which hypofunctional NMDA receptors cease driving GABAergic neurons which cease inhibiting excitatory transmitters in the brain. These disinhibited excitatory transmitters then act in concert slowly to hyperstimulate neurons in corticolimbic brain regions. We have discussed how such an abnormality could exist in the brains of individuals with schizophrenia or AD and could account for the clinical stigmata of the two disorders. In addition, we have highlighted how other disorder-specific factors would account for the differences in the clinical presentation of AD and schizophrenia. In an animal model, pharmacological methods have been developed for preventing the overstimulation of these vulnerable corticolimbic pyramidal neurons, and at least some of these methods may be applicable for treating AD and possibly schizophrenia.

Acknowledgements

Supported in part by NIDA Scientist Development Award for Clinicians DA 00290 (NBF), NIMH Independent Scientist Award MH 01510 (JWN), MH 53363 (JWN) from NIMH, DA 05072 (JWO) from NIDA and AG 11355 (JWO) from NIA.

References

Adams, B. and Moghaddam, B. (1998) 'Corticolimbic dopamine neurotransmission is temporally dissociated from the cognitive and locomotor effects of phencyclidine', *Journal of Neuroscience* **18**: 5545–5554.

Allen, H.L. and Iversen, L.L. (1990) 'Phencyclidine, dizocilpine, and cerebrocortical neurons', *Science* **247**: 221.

Anand, A., Charney, D.S., Berman, R.M., Oren, D.A., Cappiello, A. and Krystal, J.H. (1997) 'Reduction in ketamine effects in humans by lamotrigine', *Society for Neuroscience Abstracts* **23**: 1755.

Ban, T.A., Lohrena, J.J. and Lehmann, H.E. (1961) 'Observations on the action of Sernyl – A new psychotropic drug', *Canadian Psychiatric Association Journal* **6**: 150–157.

Benes, F.M., McSparren, J., Bird, E.D., SanGiovanni, J.P. and Vincent, S.L. (1991) 'Deficits in small interneurons in prefrontal and cingulate cortices of schizophrenic and schizoaffective patients', *Archives of General Psychiatry* **48**: 996–1001.

Braak, H. and Braak, E. (1995) 'Staging of Alzheimer's disease-related neurofibrillary changes', *Neurobiology of Aging* **16**: 271–284.

Braak, H., Braak, E. and Bohl, J. (1992) 'Retrosplenial region involvement in Alzheimer's disease', *Neurodegeneration* **1**: 53–57.

Brun, A. and Englund, E. (1981) 'Regional pattern of degeneration in Alzheimer's disease: neuronal loss and histological grading', *Histopathology* **5**: 549–564.

Choi, D.W. (1988) 'Glutamate neurotoxicity and diseases of the nervous system', *Neuron* **1**: 623–634.

Choi, D.W. (1992) 'Excitotoxic cell death', *Journal of Neurobiology* **23**: 1261–1276.

Corso, T.D., Sesma, M.A., Tenkova, T.I., Der, T.C., Wozniak, D.F., Farber, N.B. and Olney, J.W. (1997) 'Multifocal brain damage induced by phencyclidine is augmented by pilocarpine', *Brain Research* **752**: 1–14.

Davis, K.L., Mohs, R.C., Marin, D., Purohit, D.P., Perl, D.P., Lantz, M. and Austin, G. (1999) 'Cholinergic markers in elderly patients with early signs of Alzheimer disease', *Journal of the American Medical Association* **281**: 1401–1406.

Deutsch, L.H., Bylsma, F.W., Rovner, B.W., Steele, C. and Folstein, M.F. (1991) 'Psychosis and physical aggression in probable Alzheimer's disease', *American Journal of Psychiatry* **148**: 1159–1163.

Domino, E.F. and Luby, E.D. (1981) 'Abnormal mental states induced by phencyclidine as a model of schizophrenia', in E.F. Domino (ed.) *PCP (Phencyclidine): Historical and Current Perspectives*, Ann Arbor: NPP Books, pp. 401–418.

Doucette, R., Fisman, M., Hachinski, V.C. and Mersky, H. (1986) 'Cell loss from the nucleus basalis of Meynert in Alzheimer's disease', *Canadian Journal of Neurological Sciences* **13**: 435–440.

Drevets, W.C. and Rubin, E.H. (1989) 'Psychotic symptoms and the longitudinal course of senile dementia of the Alzheimer type', *Biological Psychiatry* **25**: 39–48.

Ellison, G. (1994) 'Competitive and non-competitive NMDA antagonists induce similar limbic degeneration', *Neuroreport* **5**: 2688–2692.

Farber, N.B., Price, M.T., Labruyere, J., Nemnich, J., St Peter, H., Wozniak, D.F. and Olney, J.W. (1993) 'Antipsychotic drugs block phencyclidine receptor-mediated neurotoxicity', *Biological Psychiatry* **34**: 119–121.

Farber, N.B. and Olney, J.W. (1995) 'α_2 adrenergic agonists prevent MK-801 neurotoxicity', *Neuropsychopharmacology* **12**: 347–349.

Farber, N.B., Wozniak, D.F., Price, M.T., Labruyere, J., Huss, J., St Peter, H. and Olney, J.W. (1995) 'Age specific neurotoxicity in the rat associated with NMDA receptor blockade: potential relevance to schizophrenia?' *Biological Psychiatry* **38**: 788–796.

Farber, N.B., Kim, S.H. and Olney, J.W. (1997) 'Costimulation of muscarinic and non-NMDA glutamate receptors reproduces NMDA antagonist neurotoxicity', *Society for Neuroscience Abstracts* **23**: 2308.

Farber, N.B., Newcomer, J.W. and Olney, J.W. (1999a) 'Glycine agonists: what can they teach us about schizophrenia?' *Archives of General Psychiatry* **56**: 13–17.

Farber, N.B., Rubin, E.H., Newcomer, J.W., Kinscherf, D.A., Miller, J.P., Morris, J.C., Olney, J.W. and McKeel, D.W., Jr. (2000) 'Increased neocortical neurofibrillary tangle density in subjects with Alzheimer Disease and psychosis', *Archives of General Psychiatry* **57:** 1165–1173.

Fastbom, J., Forsell, Y. and Winblad, B. (1998) 'Benzodiazepines may have protective effects against Alzheimer's disease', *Alzheimer Disease and Associated Disorders* **12**: 14–17.

Fix, A.S., Wightman, K.A. and O'Callaghan, J.P. (1995) 'Reactive gliosis induced by MK-801 in the rat posterior cingulate/retrosplenial cortex: GFAP evaluation by sandwich ELISA and immunohistochemistry', *Neurotoxicology* **16**: 229–239.

Giovannini, M.G., Camilli, F., Mundula, A. and Pepeu, G. (1994) 'Glutamatergic regulation of acetylcholine output in different brain regions: a microdialysis study in the rat', *Neurochemistry International (Oxford)* **25**: 23–26.

Gómez-Isla, T., Price, J.L., McKeel, D.W., Jr., Morris, J.C., Growdon, J.H. and Hyman, B.T. (1996) 'Profound loss of layer II entorhinal cortex neurons occurs in very mild Alzheimer's disease', *The Journal of Neuroscience* **16**: 4491–4500.

Gonzales, R.A., Brown, L.M., Jones, T.W., Trent, R.D., Westbrook, S.L. and Leslie, S.W. (1991) 'N-methyl-D-aspartate mediated responses decrease with age in Fischer 344 rat brain', *Neurobiology of Aging* **12**: 219–225.

Gray, C.W. and Patel, A.J. (1995) 'Neurodegeneration mediated by glutamate and beta-amyloid peptide: a comparison and possible interaction', *Brain Research* **691**: 169–179.

Grotta, J., Clark, W., Coull, B., Pettigrew, L.C., Mackay, B., Goldstein, L.B., Murphy, D. and LaRue, L. (1995) 'Safety and tolerability of the glutamate antagonist CGS 19755 (Selfotel) in patients with acute ischemic stroke. Results of a phase IIa randomized trial', *Stroke* **26**: 602–605.

Hargreaves, R.J., Rigby, M., Smith, D., Hill, R.G. and Iversen, L.L. (1993) 'Competitive as well as uncompetitive N-methyl-D-aspartate receptor antagonists affect cortical neuronal morphology and cerebral glucose metabolism', *Neurochemical Research* **18**: 1263–1269.

Hasegawa, M., Kinoshita, H., Amano, M., Hasegawa, T., Kameyama, T. and Nabeshima, T. (1993) 'MK-801 increases endogenous acetylcholine release in the rat parietal cortex: a study using brain microdialysis', *Neuroscience Letters* **150**: 53–56.

Herrling, P. (1994) 'D-CPPene (SDZ EAA 494), a competitive NMDA antagonist. Results from animal models and first results in humans', *Neuropsychopharmacology* **10** (3S/Part 1): 591S.

Horvath, Z.C., Czopf, J. and Buzsaki, G. (1997) 'MK-801-induced neuronal damage in rats', *Brain Research* **753**: 181–195.

Ikonomidou, C., Qin, Y.Q., Labruyere, J. and Olney, J.W. (1996) 'Motor neuron degeneration induced by excitotoxin agonists has features in common with that seen in the SOD-1 transgenic mouse model of amyotrophic lateral sclerosis', *Journal of Neuropathology and Experimental Neurology* **55**: 211–224.

Ishimaru, M., Fukamauchi, F. and Olney, J.W. (1995) 'Halothane prevents MK-801 neurotoxicity in the rat cingulate cortex', *Neuroscience Letters* **193**: 1–4.

Javitt, D.C. and Zukin, S.R. (1991) 'Recent advances in the phencyclidine model of schizophrenia', *American Journal of Psychiatry* **148**: 1301–1308.

Jevtovic-Todorovic, V., Kirby, C.O. and Olney, J.W. (1997) 'Isoflurane and propofol block neurotoxicity caused by MK-801 in the rat posterior cingulate/retrosplenial cortex', *Journal of Cerebral Blood Flow and Metabolism* **17**: 168–174.

Jevtovic-Todorovic, V., Olney, J.W. and Farber, N.B. (1998a) 'Lamotrigine prevents NMDA antagonist neurotoxicity', *Society for Neuroscience Abstracts* **24**: 745.

Jevtovic-Todorovic, V., Todorvic, S.M., Mennerick, S., Powell, S., Dikranian, K., Benshoff, N., Zorumski, C.F. and Olney, J.W. (1998b) 'Nitrous oxide (laughing gas) is an NMDA antagonist, neuroprotectant and neurotoxin', *Nature Medicine* **4**: 460–463.

Kim, S.H., Olney, J.W. and Farber, N.B. (1998) 'Clonidine prevents NMDA antagonist neurotoxicity by activating α_2 adrenergic receptors in the diagonal band', *Society for Neuroscience Abstracts* **24**: 462.

Kim, S.H., Price, M.T., Olney, J.W. and Farber, N.B. (1999) 'Excessive cerebrocortical release of acetylcholine induced by NMDA antagonists is reduced by GABAergic and α_2-adrenergic agonists', *Molecular Psychiatry* **4**: 344–352.

Koh, J.Y., Yang, L.L. and Cotman, C.W. (1990) 'Beta-amyloid protein increases the vulnerability of cultured cortical neurons to excitotoxic damage', *Brain Research* **533**: 315–320.

Kristensen, J.D., Svensson, B. and Gordh, T., Jr. (1992) 'The NMDA receptor-antagonist CPP abolishes neurogenic "wind-up pain" after intrathecal administration in humans', *Pain* **51**: 249–253.

Krystal, J.H., Karper, L.P., Seibyl, J.P., Freeman, G.K., Delaney, R., Bremner, J.D., Heninger, G.R., Bowers, M.B., Jr. and Charney, D.S. (1994) 'Subanesthetic effects of the noncompetitive NMDA antagonist, ketamine, in humans. Psychotomimetic, perceptual, cognitive, and neuroendocrine responses', *Archives of General Psychiatry* **51**: 199–214.

Krystal, J.H., D'Souza, D.C., Karper, L.P., Bennett, A., Abi-Dargham, A., Abi-Saab, D., Cassello, K., Bowers, M.B., Jr., Vegso, S., Heninger, G.R. and Charney, D.S. (1999) 'Interactive effects of subanesthetic ketamine and haloperidol in healthy humans', *Psychopharmacology* **145**: 193–204.

Lahti, A.C., Holcomb, H.H., Medoff, D.R. and Tamminga, C.A. (1995) 'Ketamine activates psychosis and alters limbic blood flow in schizophrenia', *Neuroreport* **6**: 869–872.

Levy, M.L., Cummings, J.L., Fairbanks, L.A., Bravi, D., Calvani, M. and Carta, A. (1996) 'Longitudinal assessment of symptoms of depression, agitation, and psychosis in 181 patients with Alzheimer's disease', *American Journal of Psychiatry* **153**: 1438–1443.

Luby, E.D., Cohen, B.D., Rosenbaum, G., Gottlieb, J.S. and Kelley, R. (1959) 'Study of a new schizophrenomimetic drug – Sernyl', *Archives of Neurology and Psychiatry* **81**: 363–369.

Lyon, M., Barr, C.E., Cannon, T.D., Mednick, S.A. and Shore, D. (1989) 'Fetal neural development and schizophrenia', *Schizophrenia Bulletin* **15**: 149–161.

Magbagbeola, J.A. and Thomas, N.A. (1974) 'Effect of thiopentone on emergence reactions to ketamine anaesthesia', *Canadian Anaesthesia Society Journal* **21**: 321–324.

Magnusson, K.R. and Cotman, C.W. (1993) 'Age-related changes in excitatory amino acid receptors in two mouse strains', *Neurobiology of Aging* **14**: 197–206.

Malhotra, A.K., Pinals, D.A., Weingartner, H., Sirocco, K., Missar, C.D., Pickar, D. and Breier, A. (1996) 'NMDA receptor function and human cognition: the effects of ketamine in healthy volunteers', *Neuropsychopharmacology* **14**: 301–307.

Malhotra, A.K., Adler, C.M., Kennison, S.D., Elman, I., Pickar, D. and Breier, A. (1997) 'Clozapine blunts N-methyl-D-aspartate antagonist-induced psychosis: a study with ketamine', *Biological Psychiatry* **42**: 664–668.

Marshall, B.E. and Longnecker, D.E. (1990) 'General anesthetics, in L.S. Goodman, A. Gilman, T.W.

Rall, A.S. Nies, and P. Taylor (eds) *The Pharmacological Basis of Therapeutics*, New York: Pergamon Press, pp. 285–310.

Mayeux, R., Stern, Y. and Spanton, S. (1985) 'Heterogeneity in dementia of the Alzheimer type: evidence of subgroups', *Neurology* **35**: 453–461.

Miller, B.L., Moats, R.A., Shonk, T., Ernst, T., Woolley, S. and Ross, B.D. (1993) 'Alzheimer disease: depiction of increased cerebral myo-inositol with proton MR spectroscopy', *Radiology* **187**: 433–437.

Minoshima, S., Giordani, B., Berebt, S., Frey, K.A., Foster, N.L. and Kuhl, D.E. (1997) 'Metabolic reduction in the posterior cingulate cortex in very early Alzheimer's disease', *Annals of Neurology* **42**: 85–94.

Moghaddam, B., Adams, B., Verma, A. and Daly, D. (1997) 'Activation of glutamatergic neurotransmission by ketamine: a novel step in the pathway from NMDA receptor blockade to dopaminergic and cognitive disruptions associated with the prefrontal cortex', *Journal of Neuroscience* **17**: 2921–2927.

Muir, K.W., Grosset, D.G. and Lees, K.R. (1997) 'Effects of prolonged infusion of the NMDA antagonist aptiganel hydrochloride (CNS 1102) in normal volunteers', *Clinical Neuropharmacology* **20**: 311–321.

Newcomer, J.W., Farber, N.B., Selke, G., Melson, A.K., Jevtovic-Todorovic, V. and Olney, J.W. (1998) 'Guanabenz effects on NMDA antagonist-induced mental symptoms in humans', *Society for Neuroscience Abstracts* **24**: 525.

Newcomer, J.W., Farber, N.B., Jevtovic-Todorovic, V., Selke, G., Kelly Melson, A., Hershey, T., Craft, S. and Olney, J.W. (1999) 'Ketamine-induced NMDA receptor hypofunction as a model of memory impairment in schizophrenia', *Neuropsychopharmacology* **20**: 106–118.

Olney, J.W. (1990) 'Excitotoxic amino acids and neuropsychiatric disorders', *Annu. Rev. Pharmacol. Toxicol.* **30**: 47–71.

Olney, J.W. and Farber, N.B. (1995) 'Glutamate receptor dysfunction and schizophrenia', *Archives of General Psychiatry* **52**: 998–1007.

Olney, J.W., Labruyere, J. and Price, M.T. (1989) 'Pathological changes induced in cerebrocortical neurons by phencyclidine and related drugs', *Science* **244**: 1360–1362.

Olney, J.W., Labruyere, J., Wang, G., Wozniak, D.F., Price, M.T. and Sesma, M.A. (1991) 'NMDA antagonist neurotoxicity: mechanism and prevention', *Science* **254**: 1515–1518.

Olney, J.W., Wozniak, D.F. and Farber, N.B. (1997) 'Excitotoxic neurodegeneration in Alzheimer's disease: new hypothesis and new therapeutic strategies', *Archives of Neurology* **54**: 1234–1240.

Patel, A.J. (1995) 'Pretreatment with a sublethal concentration of β-amyloid$_{25-35}$ potentiates neurodegeneration mediated by glutamate in cultured cortical neurons', *Alzheimer's Research* **1**: 41–44.

Paxinos, G. and Watson, C. (1986) *The Rat Brain in Stereotaxic Coordinates* (2nd edn), New York: Academic Press.

Price, M.T., Farber, N.B., Labruyere, J., Foster, J. and Olney, J.W. (1994) 'Tracing the circuitry that mediates NMDA antagonist neurotoxicity', *Society for Neuroscience Abstracts* **20**: 1532.

Reich, D.L. and Silvay, G. (1989) 'Ketamine: an update on the first twenty-five years of clinical experience', *Canadian Journal of Anaesthesia* **36**: 186–197.

Reiman, E.M., Caselli, R.J., Yun, L.S., Chen, K., Bandy, D., Minoshima, S. and Osborne, D. (1996) 'Preclinical evidence of Alzheimer's disease in persons homozygous for the epsilon 4 allele for apolipoprotein E', *New England Journal of Medicine* **334**: 752–758.

Rosen, J. and Zubenko, G.S. (1991) 'Emergence of psychosis and depression in the longitudinal evaluation of Alzheimer's disease', *Biological Psychiatry* **29**: 224–232.

Rothstein, J.D., Van Kammen, M., Levey, A.I., Martin, L.J. and Kuncl, R.W. (1995) 'Selective loss of glial glutamate transporter GLT-1 in amyotropic lateral sclerosis', *Annals of Neurology* **38**: 73–84.

Saransaari, P. and Oja, S.S. (1995) 'Dizocilpine binding to cerebral cortical membranes from developing and ageing mice', *Mechanisms of Ageing &d Development* **85**: 171–181.

Schneider, L.S., Pollack, V.E. and Lyness, S.A. (1990) 'A meta-analysis of controlled trials of neuroleptic treatment in dementia', *Journal American Geriatric Society* **38**: 553–563.

Schwarz, M.J., Riedel, M., Gruber, R., Ackenheil, M. and Muller, N. (1999) 'Antibodies to heat shock proteins in schizophrenic patients: implications for the mechanism of the disease', *American Journal of Psychiatry* **156**: 1103–1104.

Sharp, F.R., Jasper, P., Hall, J., Noble, L. and Sagar, S.M. (1991) 'MK-801 and ketamine induce heat shock protein HSP72 in injured neurons in posterior cingulate and retrosplenial cortex', *Annals of Neurology* **30**: 801–809.

Sharp, F.R., Butman, M., Wang, S., Koistinaho, J., Graham, S.H., Sagar, S.M., Noble, L., Berger, P. and Longo, F.M. (1992) 'Haloperidol prevents induction of the hsp70 heat shock gene in neurons injured by phencyclidine (PCP), MK801, and ketamine', *Journal of Neuroscience Research* **33**: 605–616.

Sharp, F.R., Butman, M., Koistinaho, J., Aardalen, K., Nakki, R., Massa, S., Swanson, R.A. and Sagar, S.M. (1994) 'Phencyclidine induction of the hsp70 stress gene in injured pyramidal neurons is mediated via multiple receptors and voltage gated calcium channels', *Neuroscience* **62**: 1079–1092.

Sharp, J.W., Petersen, D.L. and Langford, M.T. (1995) 'DNQX inhibits phencyclidine (PCP) and ketamine induction of the hsp70 heat shock gene in the rat cingulate and retrosplenial cortex', *Brain Research* **687**: 114–124.

Stern, Y., Mayeux, R., Sano, M., Hauser, W.A. and Bush, T. (1987) 'Predictors of disease course in patients with probable Alzheimer's disease', *Neurology* **37**: 1649–1653.

Tekin, S., Aykut-Bingol, C., Tanridag, T. and Aktan, S. (1998) 'Antiglutamatergic therapy in Alzheimer's disease – effects of lamotrigine', *Journal of Neural Transmission* **105**: 295–303.

Ulas, J. and Cotman, C.W. (1997) 'Decreased expression of N-methyl-D-aspartate receptor 1 messenger RNA in select regions of Alzheimer brain', *Neuroscience* **79**: 973–982.

Wenk, G.L., Walker, L.C., Price, D.L. and Cork, L.C. (1991) 'Loss of NMDA, but not GABA-A, binding in the brains of aged rats and monkeys', *Neurobiology of Aging* **12**: 93–98.

White, P.F., Way, W.L. and Trevor, A.J. (1982) 'Ketamine – its pharmacology and therapeutic uses', *Anesthesiology* **56**: 119–136.

Woo, T.U., Whitehead, R.E., Melchitzky, D.S. and Lewis, D.A. (1998) 'A subclass of prefrontal gamma-aminobutyric acid axon terminals are selectively altered in schizophrenia', *Proceedings of the National Academy of Sciences of the United States of America* **95**: 5341–5346.

Wozniak, D.F., Dikranian, K., Ishimaru, M., Nardi, A., Corso, T.D., Tenkova, T.I., Olney, J.W. and Fix, A.S. (1998) 'Disseminated corticolimbic neuronal degeneration induced in rat brain by MK-801: potential relevance to Alzheimer's disease', *Neurobiology of Disease* **5**: 305–322.

Chapter 18

Modulators of the $GABA_A$ receptor

Novel therapeutic prospects

Christian Thomsen and Bjarke Ebert

Introduction

In a number of clinical conditions hypoactivity of the inhibitory GABA system has been hypothesised as the underlying mechanism of the pathology in question. These conditions include epilepsy, anxiety, stress, sleep disorders and pain. In contrast, schizophrenia has been suggested to be related to a hyperactivity of the GABAergic system (Theobald *et al.*, 1968; Tamminga *et al.*, 1978). However, although positive modulators of the $GABA_A$ receptor complex such as benzodiazepines are very effective in a number of circumstances, there is a general consensus that unselective benzodiazepines produce so many side effects that compounds substituting for presently used drugs are needed (Costa and Guidotto, 1996). In addition to allosteric modulators of the $GABA_A$ receptor complex (e.g., benzodiazepines, barbiturates and steroids) alternative strategies for enhancing GABAergic functioning includes direct GABA agonism and inhibitors of GABA's re-uptake or metabolism. Elevation of the synaptic concentration of GABA will activate both the ionotropic $GABA_A$ receptors and the metabotropic $GABA_B$ receptors. For a review of the therapeutic aspects of $GABA_B$ receptor intervention the reader is referred to recent reviews (Bowery and Enna, 2000; Malcangio and Bowery, 1995). Inactivation of the inhibitory neurotransmitter GABA in synapses of the central nervous system (CNS) is primarily mediated by re-uptake into presynaptic terminals of neurons and into glial cells via high affinity sodium-dependent GABA uptake carriers (Nelson, 1998).

Benzodiazepines act by modulating the $GABA_A$ receptor complex, thereby increasing the opening probability in the presence of a fixed concentration of GABA. The consequence is a leftward shift of the dose response curve to GABA and other $GABA_A$ agonists (Haefely *et al.*, 1975; Polc *et al.*, 1982; Choi *et al.*, 1977, 1981). In the absence of GABA, benzodiazepines do not produce any effect on the channel opening, thereby making benzodiazepines highly use-dependent, and function as a receptor amplifier. However, high degrees of $GABA_A$ receptor activation invariably produce an immediate desensitisation (Mierlack and Farb, 1988) leading to a reduction in maximum response and thus a reduced effect of GABA. This reduced sensitivity to GABA following application of benzodiazepines for longer time intervals may be one of the mechanisms underlying the abuse problem with benzodiazepines and the need of increasing doses during a prolonged treatment period with benzodiazepines (Haefely, 1986). In order to address this problem with abuse potential, a number of groups have tried to develop selective partial benzodiazepine agonists. However, one major obstacle for the development of selective compounds is the presence of a plethora of *in vivo* active receptors. $GABA_A$ receptors are formed as a pentameric assembly of different families of receptor subunits. The assembly, which in most receptors includes two α subunits, two β subunits and a γ: ϵ or δ subunit, determines the pharmacology of the functional receptor. The binding site for benzodiazepines is located at the interface between the α and γ subunit, whereas the binding site for GABA and other $GABA_A$ agonists is located at the interface between the α and β subunit (described in detail in Chapter 9 by Olsen and Macdonald). At present, it is believed that approximately twenty different receptor subunit combinations can account for up to 90–95 per cent of the $GABA_A$ receptors (McKernan and Whiting, 1996), leading to the conclusion that these receptor combinations are the most interesting from a drug development point of view.

It can, however, equally well be hypothesised that targeting of the large populations will result in the disturbance of larger populations of neurones, thereby producing more side effects. In order to address this question in more depth a detailed knowledge of the significance of every subunit combination at different types of neurons is needed. These kinds of studies are only in the initial phase. However, a wealth of data seems to suggest that even highly abundant and widespread receptor subunits have a very distinct expression pattern, raising the possibility that some degree of region selectivity may be obtained. This may also apply to other strategies for modulating the GABAergic system, such as inhibition of GABA uptake. While uptake of GABA may be inhibited by the GABA uptake inhibitor, nipecotic acid, its lipophilic derivatives such as (*R*)-*N*-(4,4-bis(3-methyl-2-thienyl)but-3-en-1-yl) nipecotic acid (tiagabine) are much more potent (Braestrup *et al.*, 1990). Such potent and selective GABA transport inhibitors, which are a prerequisite for understanding the roles of GABA transporters in physiology and pathophysiology, have only been identified for the first GABA transporter to be cloned (GAT-1) (Guastella *et al.*, 1990). However, with the molecular cloning of four different GABA transporters which show a unique distribution in the CNS (Borden *et al.*, 1992; Guastella *et al.*, 1990; Ikegaki *et al.*, 1994; Liu *et al.*, 1993; Yamauchi *et al.*, 1992) it should ideally be possible to target GABA transport processes more specifically for certain diseases involving GABAergic transmission. In the case of the human and rat homologues, the four subtypes have been termed GAT-1 (which equals mouse GAT1), GAT-2 (=mouse GAT3), GAT-3 (=mouse GAT4) and the betaine/GABA transporter (BGT-1) (=mouse GAT2) (see Chapter 14). For example, while GAT-1 is abundantly expressed in most regions of the brain, GAT-3 shows a more restricted pattern of distribution in the rodent CNS (Ikegaki *et al.*, 1994).

Receptor functional considerations for novel modulators of the GABAergic system

The clinical action of unselective benzodiazepines includes sedation, muscle relaxation, anxiolyis, anticonvulsion and memory impairment (e.g. Haefely *et al.*, 1981). Very recently it has been demonstrated that different $GABA_A$ receptors contribute to the different actions of the unselective benzodiazepines (Rudolph *et al.*, 1999; Löw *et al.*, 2000). The binding site for benzodiazepine at $GABA_A$ receptors is probably located at the interface of the α and β subunits. The $\alpha_{1-2-3 \text{ and } 5}$ containing receptors respond to diazepam, whereas α_6 containing receptors do not respond to diazepam or zolpidem (Wafford *et al.*, 1996; Hadingham *et al.*, 1996), but to compounds like abecarnil and bretazenil (Wafford *et al.*, 1996). One single amino acid (histidine in $\alpha_{1-2-3 \text{ and } 5}$ and arginine in $\alpha_{4 \text{ and } 6}$) is the basis for this difference in sensitivity towards diazepam. By introducing a single point mutation (histidine mutated to arginine) in $\alpha_{1-2-3 \text{ and } 5}$ it has been possible to investigate the pharmacological consequences of every type of benzodiazepine receptor (Rudolph *et al.*, 1999; McKernan *et al.*, 2000; Löw *et al.*, 2000). These data suggest that α_1 containing receptors are heavily involved in the sedative actions of diazepam, whereas α_2/α_3 containing receptors are important for different types of anxiolytics (Löw *et al.*, 2000). Whereas α_6 receptors may have a hypnotic effect, α_6 knock-out mice did not show any altered sensitivity towards benzodiazepines, barbiturates or ethanol (Homanics *et al.*, 1999). This very crude way of attributing different pharmacological actions to specific $GABA_A$ receptor subunit combinations has provided some insight into the different actions of benzodiazepines and thus established a paradigm for the development of novel compounds with more specific actions. However, it has become evident that in one region several different receptor combinations at different cell types can exist at one neurone, adding to the complexity (Fritschy *et al.*, 1998; Mellor *et al.*, 2000). Thus, receptor combinations appearing at the soma may be different from those at the dendrites (Mellor *et al.*, 2000; Hájos *et al.*, 2000; Miles *et al.*, 1996). Our understanding of this receptor heterogeneity at the single cell level is as yet sparse and the pharmacological consequences remain to be explored; however, the potential for targeting specific areas by GABAergic drugs may be the challenge for medicinal chemists in years to come. The GABA binding site is located at the interface between the α and β subunit (Smith and Olsen, 1994; Amin and Weiss, 1993; Sigel *et al.*, 1992; Ebert *et al.*, 1997), raising the possibility that several different subunit combination-dependent micro-binding domains/receptors may be formed. However, as amino acids contributing to the binding site appear to be conserved within the α and β subunit families (Amin and Weiss, 1993) the consequence is that the actual binding site is conserved independent of the receptor configuration. Therefore receptor affinity appears to be independent of receptor combination (Ebert *et al.*, 1997). However, the functional consequences of receptor binding are highly dependent on the receptor assembly (Ebert *et al.*, 1994, 1997). So, as is the case for benzodiazepines, only functional characterisation of ligands may provide a basis for understanding the *in vivo* pharmacological consequences of the drug. Thus, functional selectivity is obtainable for a number of compounds like the GABA agonists imidazole-4-acetic acid (IAA) and 4,5,6,7-tetrahydroisoxazolo[5,4-c]pyridin-3-ol (THIP) (see Figure 18.1).

As illustrated in Figure 18.1, IAA and THIP display a highly subunit dependent potency and maximal response resulting in a functional selectivity whereby the compounds at some combinations may act as agonists and, at the same time, as antagonists or low efficacy partial agonists at other combinations. Our understanding of the determinants at this functional selectivity is still very limited, so a strategy for the development of these compounds with a predefined receptor subtype selectivity profile is still based on trial and error. A very important aspect of this problem is the selection of subunit combinations relevant for the prediction of *in vivo* activity. An example of the consequences of variations of subunits is the effect of a substitution of the δ subunit for a γ_{2S} subunit in a receptor complex containing $\alpha_4\beta_3$.

The α_4 containing receptors exist predominantly in the thalamic area (Sur *et al.*, 1999a). Recent studies (including Mody, 2000) have indicated that some of these receptors may be located extrasynaptically, making them a potentially very interesting

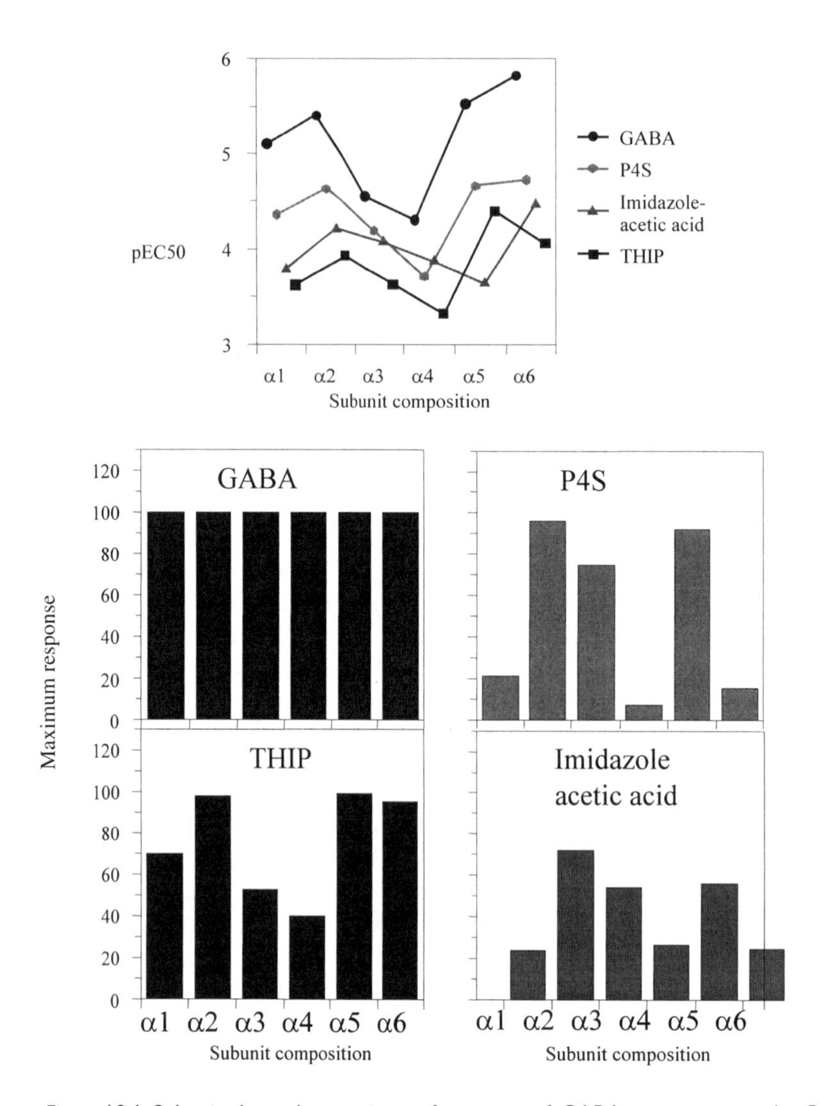

Figure 18.1 Subunit dependent activity of a series of GABAergic compounds. Oocytes were injected with cDNA encoding for $\alpha_x\beta_3\gamma_{2s}$ and concentration response curves were constructed for the compounds. Maximum response at every receptor combination at every oocyte was determined using 3 mM GABA. *Top:* pEC50 values. *Bottom panels:* Maximum response as function of α subunit.

interesting drug target. When $\alpha_4\beta_3\gamma_{2S}$ receptors are compared with $\alpha_4\beta_3\delta$ containing receptors in *Xenopus* oocytes, THIP is approximately ten times more potent at δ containing receptors than at γ_{2S} containing receptors. Furthermore, at δ containing receptors THIP appears to have a completely different pharmacological profile than at other receptor combinations (Wafford, 2000). So despite the fact that the GABA binding site is located at the interface between the α and β subunits, the pharmacological profile of the receptor is determined by the interaction of all subunits present (e.g., Ebert *et al.*, 1994). An additional example illustrating the complete unpredictability of the functional consequences of modifications of subunit compositions relates to the $GABA_A$ receptors in the cerebellar granule cells. As described in Chapter 8, cerebellar granule cells contain $\alpha_1\beta_{2/3}\gamma_2$, $\alpha_6\beta_{2/3}\gamma_2$ and $\alpha_1\alpha_6\beta_{2/3}\gamma_2$ containing receptors. A comparison of the activity of piperidine-4-sulphonic acid (P4S) at these three different combinations reveals that at $\alpha_1\beta_2\gamma_2$ and $\alpha_6\beta_{2/3}\gamma_2$ containing receptors P4S is a partial agonist with a maximal response of 10–15 per cent of that of GABA, whereas at $\alpha_1\alpha_6\beta_3\gamma_2$ containing receptors, the maximum response to P4S is 75 per cent of that of GABA and the potency is different from that at the other receptor combinations (Ebert *et al.*, unpublished). This result is not only surprising, it is also in contrast to data for P4S at $\alpha_1/\alpha_5/\alpha_1\alpha_5$ or $\alpha_1/\alpha_3/\alpha_1\alpha_3$ containing receptors, where co-expression of two different α subunits resulted in an α_1-like maximum response and a significantly lower potency than at α_1/α_5 or α_1/α_3 containing receptors (Ebert *et al.*, 1994). In general, coexpression of different α subunits results in novel pharmacology (see e.g. Verdoorn, 1994), underlining the complexity of the system.

New subtype selective modulators of the GABAergic system

The development of selective benzodiazepines is in deep contrast to what has been achieved with compounds interacting directly with the GABA recognition site at the $GABA_A$ receptors. In general these compounds were synthesised prior to the recognition of the existence of $GABA_A$ subunits and are *very* unselective. In a series of studies characterising the subunit-dependent receptor affinity for a number of compounds, including agonists, partial agonists and competitive antagonists, it has been shown that the traditional selectivity based on affinity drug design in the case of $GABA_A$ agonists/partial agonists is virtually impossible (Ebert *et al.*, 1997). Rational design of novel drugs based on the pharmacological profile in *in vitro* systems seems therefore still to be based on relatively vague grounds. However, in the case of benzodiazepines, a number of novel approaches may show that, despite the very complex situation with subunit combinations being localised in most CNS areas, it is possible to develop selective compounds with novel and very promising profiles in *in vivo* studies. Using an elegant combination of molecular biology, transgenic animals and structure – activity relationships determined in functional assays, the group at Merck, Sharpe and Dohme has been able to develop benzodiazepine-like compounds with agonist activity at α_2/α_3 containing receptors and antagonist activity at $\alpha_1/\alpha_5/\alpha_6$ containing receptors (McKernan *et al.*, 2000). In behavioural studies one of these compounds (L813.417) was acting as an anxiolytic devoid of the normal sedatory and amnesic effects usually ascribed to these compounds in higher doses (McKernan *et al.*, 2000; McKernan *et al.*, unpublished). It is very interesting that it is possible to develop benzodiazepine drugs with agonist activity at some receptor combinations and at the same time antagonist activity at other receptors. Since no endogenous ligand for the benzodiazepine receptor has been identified, and because the functional consequences of a benzodiazepine antagonist like L813.417 at $\alpha_1/\alpha_5/\alpha_6$ containing receptors seem to be minimal, benzodiazepine ligands with antagonist action at some receptor types, and at the same time agonist activity at other receptor types, appear to be a paradigm within which to obtain functional selectivity *in vivo*. This *fait accompli* may prove to be the solution to selectivity problems normally encountered during development of novel drugs; however, in order to obtain this information it is necessary to characterise the compounds in a vast array of functional assays, thereby changing the affinity screening approach which has been used so far. An important issue will therefore be the development of high throughput functional assays for the GABAergic receptors, stressing the importance of the development of novel electrophysiological techniques or alternative techniques (Simpson *et al.*, 2000; Smith *et al.*, 2000; Alder *et al.*, 1998). It may, however, be argued that selective compounds may not be the solution to the quest for novel clinically better drugs, since most diseases are multi-receptor imbalances and a selective effect is usually compensated for by the system.

With tiagabine and related potent and highly selective inhibitors of GAT-1 (Borden *et al.*, 1994b; Thomsen *et al.*, 1997) much information on the physiological roles and therapeutic prospects of GAT-1 has been obtained with the introduction of tiagabine for the indication of generalised convulsive epilepsy (Bialer *et al.*, 1999; Loiseau, 1999; Suzdak and Jansen, 1995). In addition to epilepsy, other therapeutic possibilities for this GAT-1 inhibitor are now emerging. A few compounds with moderate potency and selectivity for GAT-3 and BGT-1 have been identified (Borden *et al.*, 1994a; Falch *et al.*, 1999; Thomsen *et al.*, 1997; Dhar *et al.*, 1994) to provide preliminary suggestions to the preclinical potential of these subtypes (see also Chapter 14).

Epilepsy

The term 'epilepsy' refers to a family of neurological conditions characterised by seizures that result from the uncoordinated electrical activity of a large number of brain cells. It is now thought that epilepsy is not a disease in itself but rather a symptom of many different pathological conditions affecting the brain either directly or indirectly as a consequence of metabolic disturbances. Some of these are of genetic origin, others have unknown causes or are the result of either head injuries, vascular lesions, infections or alcohol/drug abuse (Dreifuss and Ogunyemi, 1992). Seizures are divided into two groups: partial seizures with a more or less defined locus of onset, and generalised seizures for which no locus is defined. These types of seizures are further sub-classified into various symptomatic categories (Dreifuss and Ogunyemi, 1992). Examples of generalised seizures are absence and tonic–clonic seizures. Absence seizures are characterised by abrupt cessation of ongoing behaviour, associated with a blank stare and followed by a sudden return to normal behaviour after 30–60 seconds. Clonic seizures are characterised by shock-like contractions of muscles with periods of muscle relaxation, whereas tonic seizures are characterised by sustained muscle contractions. A generalised tonic–clonic seizure involves muscle groups throughout the body and is associated with loss of consciousness.

In neuropharmacological terms, the major causative factor in epilepsy is believed to be an imbalance between excitatory glutamatergic neurotransmission and inhibitory GABAergic transmission (Meldrum, 1992; Bradford, 1995). The balance between excitatory and inhibitory events is organised on cellular, synaptic and circuitry levels. At the cellular level, depolarising inward currents (Na^+ and Ca^{2+}) are balanced by outward K^+-currents and inward Cl^--currents. At the synaptic level, excitatory transmission (e.g., mediated by glutamate) is balanced by inhibitory transmission mediated by, for example, GABA. The third factor is the circuitry between excitatory projecting neurons and inhibitory interneurons. In general terms, blockade of excitatory transmission (e.g., with an NMDA antagonist) or enhancement of inhibitory transmission (e.g., with a benzodiazepine) will lead to protection from seizures. However, in some cases facilitation of GABAergic transmission or blockade of glutamatergic transmission may lead to seizures because the primary effect of the drug is on inhibitory interneurons rather than on the projecting neuron (Gale, 1992). Mechanisms by which anticonvulsant drugs may act by interfering with GABAergic transmission include direct GABA agonism, allosteric modulation of GABA channel function (benzodiazepines, steroids and barbiturates), inhibition of GABA's uptake or metabolism, and finally activation of $GABA_B$ receptors (Bradford, 1995; Lin and Kadaba, 1997). At least theoretically, GAT inhibitors may be more advantageous than the other approaches because inhibition of GABA uptake will be most effective in areas of ongoing intense activity in GABAergic synapses.

Tiagabine (Gabitril) has been on the market for three years for adjunct treatment of partial seizures with or without secondary generalisation. Therefore much information has been obtained about the therapeutic potential of selective inhibition of GAT-1 for epileptic seizures. Tiagabine is effective as add-on treatment for simple partial, complex partial and secondarily generalised tonic-clonic seizures (Bialer *et al.*, 1999; Loiseau, 1999; Suzdak and Jansen, 1995). The efficacy against partial seizures was also predicted from pre-clinical studies in which tiagabine inhibited amygdala-kindled seizures (Dalby and Nielsen, 1997). Currently, monotherapy studies for partial seizures and add-on treatment for paediatric partial seizures are ongoing with tiagabine (Bialer *et al.*, 1999; Schacter, 1999). An alternative indication within epileptic seizure types is the potential use of tiagabine against myoclonic seizures suggested by its ability to reduce such convulsions in photosensitive baboons (Smith *et al.*, 1995). Owing to the different distribution of GAT subtypes in the CNS, future developments in the use of GAT inhibitors within the indication of epileptic seizures may be derived from compounds targeting other subtypes than GAT-1, or perhaps more likely from compounds targeting different GAT subtypes in addition to GAT-1. GAT-1 is abundantly expressed in most regions of the brain, including the cerebral and cerebellar cortices, hippocampus (CA3), piriform cortex, thalamus, substantia nigra, thalamus and superior colliculus (Guastella *et al.*, 1990; Ikegaki *et al.*, 1994). GAT-3 is predominantly found in the thalamus, hypothalamus, pons, medulla, central gray, some deep cerebellar nuclei, substantia nigra and in the inferior colliculus (Ikegaki *et al.*, 1994). While GAT-1 and GAT-3 is found exclusively in the brain, GAT-2 and BGT-1 are also found in peripheral tissues (e.g., kidney, liver, heart) (Yamauchi *et al.*, 1992; Lopez-Corcuera *et al.*, 1992; Borden *et al.*, 1995a, b) and is believed to be involved in non-synaptic functions such as osmoregulation (Yamauchi *et al.*, 1992; Borden *et al.*, 1995a). Accordingly, the most significant contribution to overall GABA uptake in the CNS is likely to be mediated by GAT-1. In regions that are sparse in GAT-3, such as the cerebral and cerebellar cortices and hippocampus, GABA uptake may exclusively be mediated via GAT-1. In some midbrain regions (e.g., thalamus and inferior colliculus) GABA uptake is likely to be mediated by both GAT-1 and GAT-3 (Dalby, 2000; Ikegaki *et al.*, 1994). Interestingly, these areas are very important for the initiation and spread of seizure activity (Gale, 1992). When inhibition of $[^3H]$GABA uptake is measured in membranes from the rat inferior colliculus selective GAT-1 inhibitors such as tiagabine or NNC711 inhibits ~70 per cent of the uptake at low nM concentrations, while much higher concentrations of NNC711 or tiagabine (i.e., mM) are required to inhibit the remaining fraction

(Dalby *et al.*, 1997). This NNC711-resistant fraction was inhibited by compounds (NNC 05–2045 and SNAP-5114) which were subsequently shown to be relatively selective for GAT-3 (Thomsen *et al.*, 1997). In microdialysis studies application of tiagabine has been shown to lead to increases in GABA levels in various brain regions, including the substantia nigra, globus pallidus, hippocampus and thalamus (Fink-Jensen *et al.*, 1992; Richards and Bowery, 1996; Dalby, 2000). Interestingly, SNAP-5114 increases GABA levels in the thalamus but not in the hippocampus (Dalby, 2000), a region which has very low levels of GAT-3 (Ikegaki *et al.*, 1994). The anticonvulsant profiles of NNC 05–2045 and NNC 05–2090 are distinct from tiagabine when compared in preclinical models of epileptic seizure activity (excluding the DBA/2 model of sound-induced seizures in which all are effective) (Table 18.1). Selective GAT-1 inhibitors such as tiagabine potently inhibit PTZ-induced seizures in mice and reduce convulsions in amygdala kindled rats but are ineffective against electrically induced seizures (MES-test) (Table 18.1). On the other hand, NNC 05–2045 and NNC 05–2090 are effective anticonvulsants in the MES test but not against clonic convulsions induced by PTZ, and do not affect amygdala kindled afterdischarges (Table 18.1). Accordingly, it may be speculated that a mixed GAT-1/GAT-3 inhibitor may be effective against several types of seizures, including primary generalised seizures with tonic convulsions for which tiagabine has proved less effective.

Anxiety disorders

Low GABA function is proposed to be an inherited biological marker of vulnerability for development of mood disorders (Petty, 1995). Thus, increasing GABAergic functioning may be beneficial for the treatment of anxiety. This approach has been followed by a number of groups developing benzodiazepines or benzodiazepine-like compounds. For a review on benzodiazepine-like compounds in relation to anxiety, see Teuber *et al.* (1999). Increasing the synaptic content of GABA is another mechanism to increase GABAergic functioning in the brain. Tiagabine has been shown to be effective in preclinical models of anxiety, including the Vogel water lick suppression test (Nielsen, 1988), the elevated plus maze (Dalvi and Rodgers, 1996; Schmitt and Hiemke, 1999) and the open field test (Schmitt and Hiemke, 1999). In line with these findings, improvements in mood and overall adjustment have been reported in patients receiving low doses of tiagabine as monotherapy (Dodrill *et al.*, 1998). Whether GAT-1 inhibitors are effective in patients suffering from anxiety disorders remains, however, to be demonstrated. In a case report a beneficial effect of tiagabine was observed in two patients suffering from refractory bipolar disorder (Schaffer and Schaffer, 1999). However, an open trial with eight patients suffering from acute mania with bipolar disorders did not confirm these initial observations (Grunze *et al.*, 1999).

Sleep

GABAergic compounds acting at the barbiturate site, the neurosteroid site or the benzodiazepine site have been used as hypnotics for several decades (Haefely, 1989). The action of these compounds is an increased efficacy of the GABAergic synapse, resulting in a leftward shift of the dose response curve to GABA. Most of the used hypnotics will interact with all of the different subtypes of receptors. However, as α_4 and δ containing

Table 18.1 Subtype selectivity of GABA uptake inhibitors and their anticonvulsant activity in rodents

	GAT-1 K_i (μM)	*GAT-2* K_i (μM)	*GAT-3* K_i (μM)	*BGT-1* K_i (μM)	*DBA/2 mice*		*MES*	*PTZ-seizures*		*Kindling*	
					Clonic (μ*mol/kg*)	*Tonic* (μ*mol/kg*)	*Tonic* (μ*mol/kg*)	*Clonic* (μ*mol/kg*)	*Tonic* (μ*mol/kg*)	*Focal* (μ*mol/kg*)	*Gener.* (μ*mol/kg*)
Tiagabine	0.11	>100	>100	>100	1	1	133	2	5	36	6
SNAP-5114	>30	20	6.6	22	110	110	>198	NT	NT	NT	NT
NNC 05–2045	27	14	6.1	1.6	17	6	30	242	31	>242	>242
NNC 05–2090	19	41	15	1.4	26	19	73	NT	NT	>242	222

Sources: The data are from Dalby and Nielsen (1997a, 1997b), Dalby *et al.* (1997), Thomsen *et al.* (1997), Dalby (2000).

receptors do not bind benzodiazepines the effect of benzodiazepines at these receptor types is absent (Wafford *et al.*, 1996). It is generally agreed that the sedative effect of benzodiazepines is mediated primarily via α_1 containing receptors (Rudolph *et al.*, 1999; McKernan *et al.*, 2000), whereas side effects related to amnesia may be α_5 mediated which is primarily hippocampal (Collinson, 2000). However, even short acting α_1 selective compounds like zalepon, zopiclone and zolpidem do produce memory impairment and hangover the following day (Landolt and Gillin, 2000), suggesting that even an α_1 selective compound may produce unwanted side effects by mode of action. The reason for this side effect profile may well be that as a consequence of the high degree of GABA receptor activation during the presence of benzodiazepine several other systems are influenced, resulting in a general inhibition of the CNS and, thus, an acute modification of the general function of the CNS. In line with this hypothesis are EEG measurements during sleep. A normal sleep pattern involves a very complex variation of different degrees of sleep, ranging from light sleep via deeper sleep stages to REM sleep. Present understanding of sleep quality and the relation to EEG patterns is still very limited; however, not only the REM sleep but also the transitions between the different sleep stages are important (Landolt and Gillin, 2000). The effect of benzodiazepines at the sleep micro architecture is not only at the onset of REM, which is delayed by benzodiazepines, but also the number of transitions are significantly reduced – probably as a consequence of a general depression of the excitatory systems (Lancel *et al.*, 1996). A series of studies by Lancel and co-workers have shown over the last years that this perturbation of the sleep micro architecture may not be the case with compounds acting directly at the GABA recognition site (Lancel *et al.*, 1996; Lancel and Faulhaber, 1996), Quite interestingly, THIP (Gaboxadol) appears to improve the quality of sleep, measured using behavioural parameters without affecting the onset of REM sleep, measured with EEG (Lancel and Faulhaber, 1996; Lancel and Langebartels, 2000). Following dosing with THIP, persons did not experience the drug hangovers or the problems with attention reported for benzodiazepines (Faulhaber *et al.*, 1997). Similar results have been obtained with muscimol and with the GABA-uptake inhibitor tiagabine (Lancel *et al.*, 1998), suggesting that the functional consequence of a direct acting agonist is different from those of the modulators. One explanation for this difference may be a combination of the functional receptor profile and the distribution of receptors essential for modulating the sleep architecture, whereby benzodiazepines are distinctly different from GABA agonists.

In sleep, the focus has been on the induction of sleep – the hypnotic effect. As shown above, the strategy invariably leads to problems with dependence and tolerance making the use of benzodiazepines especially a very questionable strategy. In light of the effects with THIP, muscimol and tiagabine it may be time to focus on the quality instead of the quantity of the sleep. The hypnotic effect of most compounds probably arises, as described above, from a massive activation of $\alpha 1$ containing receptors. However, the functional receptor profile of the GABA agonists is more unselective. Thus, GABA will activate all subtypes with different potency and yielding a regional selectivity, whereby receptor combinations with high affinity will determine the overall pharmacological profile. In case of the exogenous THIP the consequences are more profound: since THIP acts with different degrees of maximum response at different receptor combinations (Ebert *et al.*, 1997), the functional selectivity is more marked.

Premenstrual dysphoric disorder

Premenstrual dysphoric disorder (PMDD) is thought to be a consequence of the rapid drop in progesterone levels, and especially progesterone metabolites, which act as positive modulators of the GABAergic activity (Gallo and Smith, 1993). Other studies have shown a correlation between PMDD and a reduced level of GABA (Halbreich *et al.*, 1996). The resulting reduction of GABAergic activity in animal models can be compensated via modulation with neurosteroids or by blockade of the expression of α_4 subunits, suggesting that PMDD symptoms may in part be attributable to alterations in expression of $GABA_A$ receptor subunits as a result of progesterone withdrawal (Smith *et al.*, 1998). Furthermore, illustrating the complex interaction between different neurotransmitter systems, novel clinical studies have indicated that selective serotonin reuptake inhibitors like fluoxetine are able to modulate the PMDD (Dimmock *et al.*, 2000). Furthermore, tiagabine has been shown to improve mood in monotherapy studies (Dodrill *et al.*, 1998). However, in contrast to depression, these effects have a very fast onset and are probably mediated via the release of neurosteroids induced by serotonin. Neurosteroids have been developed for the treatment of PMDD and other indications; however, side effects have resulted in discontinuation of most of these compounds (e.g., dehydroepiandrosterone from Neurocrine Biosciences Inc., press release).

The main difference between benzodiazepines and GABA agonists is that benzodiazepines are inactive at α_4 and δ containing receptors, whereas $GABA_A$ agonists will act irrespective of the subunit composition (e.g., Ebert *et al.*, 1997). In relation to sleep and probably a number of other disorders, including anxiety and pre-menstrual dysphoric disorder, modulation of the thalamic areas may play a key role. In these areas a high abundance of $\alpha_4\beta_3\delta/\gamma_2$ containing receptors are found, making interaction with these receptors particularly interesting. Furthermore, novel data seem to suggest that these receptors are particularly susceptible to direct activation by $GABA_A$ agonists (Wafford, 2000). With the large density of α_4 containing receptors located exstrasynaptically (Sur *et al.*, 1999a; Sassoè-Pognetto *et al.*, 2000; Mody, 2000) only a relatively low level of activation at the individual extrasynaptic receptors will sum up to a significant inhibition of the neuron, raising the possibility that highly functional selective compounds can be developed for these receptors. In relation to this hypothesis is the effect of the neuroactive steroids with direct effect at the $GABA_A$ receptor. Although neurosteroids like alfaxalone and 3α-5α-dihydroxyprogesterone are interacting with all types of GABA receptors, data with $\alpha_4\beta_3\delta$ containing receptors indicate that the potency and efficacy at the receptors are higher than at other types of $GABA_A$ receptors (Wafford, 2000). This may allow the development of functional selective neurosteroids acting primarily at $GABA_A$ receptors in the thalamic area. However, a series of studies have shown that prolonged application of neurosteroids as hypnotics, as in the case with benzodiazepines, results in compensatory mechanisms which ultimately lead to dependence (Lancel *et al.*, 1997). It can therefore be questioned if it will be possible to develop neurosteroids devoid of dependence liability. At present, this question remains unaddressed. However, since most benzodiazepines and neurosteroids have been developed with the aim of inducing sleep as soon as possible, the compounds have been evaluated in models where the acute hypnotic effect is the parameter measured. One could speculate that if lower doses of neurosteroids were used, a lower level of activation would appear and thus a more functional selectivity would be obtained. In case of the neurosteroids this would imply an α_4 selective profile with the possibility of more subtle modulation and subsequently less risk of compensatory mechanisms. This still remains to be evaluated.

Analgesia

The analgesic action of benzodiazepines and barbiturates has been known for decades but the development of tolerance and the associated side-effects (see above) limit the usefulness of these drugs for the indication of pain. More recently, inhibitors of GABA uptake have been shown to be effective in various antinociceptive tests, including the mouse tail immersion test (Zorn and Enna, 1985), against electrical-induced pain in mice (Swedberg, 1994), the rat paw pressure threshold test (Ipponi *et al.*, 1999) and in the mouse hot plate test (Ipponi *et al.*, 1999). Interestingly, the antinociceptive effects of tiagabine are completely blocked by pre-treatment with a $GABA_B$ antagonist (Ipponi *et al.*, 1999), suggesting that the effects are mediated via activation of $GABA_B$ receptors through spill-over of GABA in the synaptic cleft (Scanziani, 2000).

Memory/learning

Positive modulation of the GABAergic system is associated with negative effects on memory and learning, which is clear from the vast evidence resulting from treatment with benzodiazepines over the decades (Landolt and Gillin, 2000). Thus, increasing the synaptic content of GABA would also be expected to induce impairment of memory and learning. However, this has not been observed to any significant extent in the clinic with the selective GAT-1 inhibitor, tiagabine (Kalviainen, 1998), which makes it an attractive alternative to conventional epilepsy treatment particularly in intellectually disabled patients.

Accordingly, at least in theory, negative modulation of the $GABA_A$ receptor complex with, for example, inverse benzodiazepines would have a beneficial effect on memory and cognition. An inverse agonism mediated via the benzodiazepine receptor corresponds to a rightward shift of the GABA dose response curve and can as such be seen as a functional competitive antagonist of GABA. Obviously such compounds need to target-specific subtypes of the $GABA_A$ receptor complex owing to the anxiogenic and proconvulsive nature of the previously identified inverse agonists (e.g., FG7142 and DMCM) (Stephens *et al.*, 1987). Such an example of benzodiazepine ligands with novel pharmacological activity are the relatively selective inverse α_5 agonists (Collinson, 2000; Atack, 2000). The α_5 receptors are predominantly expressed in the rodent and human hippocampus at interneurons controlling the activity of glutamatergic neurons (Sur *et al.*, 1999b; Howell *et al.*, 2000) suggesting a pivotal role in relation to memory and learning. This is further highlighted by the fact that α_5 knock-out mice do not display impaired memory and learning (Whiting, 2000). A single amino acid at the benzodiazepine binding site distinguishes the α_5 subunit from other α subunits, thereby allowing medicinal chemists to develop compounds which selectively interact with the α_5 subunit. Two different approaches illustrating the complex situation in the development of selective compounds have been followed by the group at Merck, Sharpe and Dohme. In one series, the development of a selective compound based on affinity was achieved, L655.708 (Atack, 2000), and in the other series the focus was on the functional selectivity, L792.782 (Collinson, 2000). The latter example showed that despite the fact that the compound acted with equal affinity at the different receptor populations the functional consequences were different. Thus L792.782 acted as an inverse agonist at α_5 containing receptors and at the same time as a low efficacy partial agonist at other receptors. In a series of electrophysiological experiments it was shown that compounds inhibiting α_5 containing $GABA_A$ receptors in the hippocampus were able to potentiate Long Term Potentiation, which is a correlate of memory at the synaptic level. Selective inverse α_5 benzodiazepine agonists like L792.782 and L655.708 have since been developed and shown in animal models to act as cognitive enhancers devoid of other side effects (Collinson, 2000; Atack, 2000). This finding is very interesting and opens up a whole new avenue of cognitive modulators – an area which due to the very controversial aspects of memory enhancers probably, too, will draw the attention of ethical committees. Similar effects may arise from the development of selective ligands at ϵ-containing receptors. Receptors containing the ϵ subunit are expressed primarily in the hippocampus and hypothalamus and do not respond to benzodiazepines (Whiting *et al.*, 1997; Thompson *et al.*, 1998). Inverse agonists at this receptor could therefore act as cognitive enhancers. However, due to the lack of selective compounds available, this still remains to be addressed.

Schizophrenia

Dysfunction in GABAergic neurotransmission may be implicated in the pathophysiology of schizophrenia owing to the important role of GABA as neurotransmitter in, for example, the prefrontal cortex (for a review, see Benes, 2000). The activity of excitatory glutamatergic neurons has been shown to be altered in patients suffering from schizophrenia and, more importantly, selective antagonists have been shown to induce psychological conditions reminiscent of schizophrenic symptoms (for reviews, see Benes, 2000; Duncan *et al.*, 1999) which lead to the so-called glutamate hypo-function of schizophrenia. Because GABA also has a profound role in controlling the activity of excitatory transmission in the prefrontal cortex, alterations in GABA transmission may also be implicated in schizophrenic disorders. Although the evidence is more indirect for the GABAergic system, overstimulation by the GAT inhibitor CI-966 produces psychotomimetic effects similar to those of schizophrenia (Sedman *et al.*, 1990). In brains from schizophrenic patients a decrease in the expression of GAT-1 mRNA has been observed (Ohnuma *et al.*, 1999) which, however, paradoxically was accompanied by a decrease in the levels of GABA in the brains. In GABAergic neurons projecting to excitatory pyramidal neurons a 40 per cent decrease in the density of GAT-1-immunoreactivity has been observed in schizophrenic subjects compared to normal controls and subjects with other psychiatric disorders (Woo *et al.*, 1998). Whereas these observations may suggest an involvement of GAT-1 in the pathophysiology of schizophrenia it should be noted that such symptoms have never been observed with tiagabine (Dodrill *et al.*, 1998; Loiseau, 1999). Thus, the alterations observed in the levels of GAT-1 expression may also be secondary to the disease process and/or perhaps related to antipsychotic treatment. Accordingly, clozapine and other antipsychotic drugs have been shown directly to affect GABAergic functioning in rats (Squires and Saederup, 1998), and this is likely to interfere with the measurements of GABA markers in schizophrenic patients. Therefore the concept of down-regulating

GABAergic activity for treatment of schizophrenia (e.g., by enhancing GABA uptake or decreasing its function with inverse benzodiazepine agonists) is highly speculative and should also be considered with caution.

Stroke

In several models of ischaemia, including the gerbil global ischaemia model, and in the middle cerebral artery or rat four vessel occlusion models, a marked increase in the extracellular levels of GABA has been observed in the rodent brain using micro-dialysis (Hagberg *et al.*, 1985; Phillis *et al.*, 1994). The elevated levels of GABA after an ischaemic episode may either be related to increased synaptic transmission, also in GABAergic neurons, and/or to reversal of the GABA transporter similar to what has been observed with the glutamate transporter (Rossi *et al.*, 2000). Selective inhibitors of GAT-1 such as tiagabine or CI-966 further potentiated the increase in GABA after an ischaemic episode (Inglefield *et al.*, 1995) and protected hippocampal CA1 from necrotic cell death in several models (Johansen and Diemer, 1991; Inglefield *et al.*, 1995; Phillis, 1995; Yang *et al.*, 2000). However, it cannot be ruled out that these effects are related to the hypothermic effects of tiagabine as previous suggested by Inglefield *et al.* (1995). Similar concerns should be kept in mind regarding the proposed anti-ischaemic effects of benzodiazepines (for a review, see Green *et al.*, 2000).

Future perspectives

Activation of GABAergic neurons is the focus in a number of approaches to treatment of CNS diseases. As has been discussed, epilepsy, anxiety, sleep and pain all share a hypoactive GABAergic neurotransmitter system as part of the key pathological mechanism. The use of selective and unselective compounds in animal models and in clinical studies has shown that GABAergic intervention in a number of cases does have ameliorating effects. However, in most cases these are accompanied by side effects or fading response in the long-term studies. Whether it is possible to develop benzodiazepine agonists targeting specific subunit combinations which in clinical studies are devoid of side effects and/or lack tachyphylaxis remains to be demonstrated. Based on these findings the naive approach would be to use selective $GABA_A$ receptor benzodiazepine agonists or inverse benzodiazepine agonists in disease states where an inhibition of GABAergic activity is desirable. Whereas it has been possible to develop selective inverse α_5 benzodiazepine agonists devoid of side effects in animal studies, this appears not to be the case with the competitive antagonists. The reason is that the competitive antagonists are unable to distinguish between the different receptor complexes, wherefore the selectivity is based on the agonist affinity combined with the actual concentration. Data for GABA indicate that only a relatively small variation in potency as a function of the subunit composition is seen, making functional selectivity hardly obtainable. Furthermore, during agonist activation the concentration of GABA in the synapse will rise from μM to mM within milliseconds, making the competitive $GABA_A$ agonist inactive. Competitive antagonists in the GABA system at the synaptic level will therefore in general be bound to the receptor when the GABA concentration is low and thus produce side effects, whereas it will dissociate from the receptor during hyperactivity and thus be inactive. At extrasynaptic receptors the situation may be different: since the time-dependent concentration profile at these sites is mainly a consequence of diffusion, the rise and decay in concentration is much slower. Although the inhibition of the GABA response under these two situations is highly dependent on the dissociation rate of the antagonist from the receptor, it is a challenge for the medicinal chemist to develop compounds with the desired kinetic profile. The task facing drug researchers in this area is to understand the complex interacting mechanisms underlying the action of currently used drugs and, based on this, to develop compounds with selectivity profiles so that interaction with several different receptor types are possible at the same time. Functional selectivity profiles in contrast to specific action will probably be in focus in the coming years.

Acknowledgement

John Atack and Keith A. Wafford, MSD Research Laboratories, are thanked for constructive comments and valuable discussions.

References

Alder, L., Smith, A.J., Priestley, T., Silk, J., Adkins, C., McKernan, R.M. and Atack, J.R. (1998) 'Use of a 96 well format ^{36}Cl flux assay to measure $GABA_A$ ion channel function in stably expressed recombinant receptors: constitutive channel opening in the absence of agonist', *Br. J. Pharmacol.* **123**: 199P.

Amin, J. and Weiss D.S. (1993) '$GABA_A$ receptor needs two homologous domains of the β-subunit for activation by GABA but not by pentobarbital', *Nature* **366**: 565–569.

Atack, J. (2000) 'Properties of L-655,708, selective for α-subunit containing $GABA_A$ receptors', Presentation at GABA2000 meeting 23 to 29 July. Organisers Peter Gage and Istvan Mody.

Benes, F.M. (2000) 'Emerging principles of altered neural circuitry in schizophrenia', *Brain. Res. Brain. Res. Rev.* **31**: 251–269.

Bialer, M., Johannessen, S.I., Kupferberg, H.J., Levy, R.H., Loiseau, P. and Perucca, E. (1999) 'Progress report on new antiepileptic drugs: a summary of the fourth Eilat Conference (EILAT IV)', *Epilepsy Res.* **34**: 1–41.

Borden, L.A., Dhar, T.G., Smith, K.E., Branchek, T.A., Gluchowski, C. and Weinshank, R.L. (1994a) 'Cloning of the human homologue of the GABA transporter GAT-3 and identification of a novel inhibitor with selectivity for this site', *Receptors and Channels* **2**: 207–213.

Borden, L.A., Murali Dhar, T.G., Smith, K.E., Weinshank, R.L., Branchek, T.A. and Gluchowski, C. (1994b) 'Tiagabine, SK&F 89976-A, CI-966 and NNC-711 are selective for the cloned GABA transporter GAT-1', *Eur. J. Pharmacol. Mol. Pharmacol.* **269**: 219–224.

Borden, L.A., Smith, K. E, Hartig, P.R., Branchek, T.A. and Weinshank, R.L. (1992) 'Molecular heterogeneity of the γ-aminobutyric acid (GABA) transport system. Cloning of two novel high affinity GABA transporters from rat brain', *J. Biol. Chem.* **267**: 21,098–21,104.

Borden, L.A., Smith, K.E., Gustafson, E.L., Branchek, T.A. and Weinshank, R.L. (1995a) 'Cloning and expression of a betaine/GABA transporter from human brain', *J. Neurochem.* **64**: 977–984.

Borden, L.A., Smith, K.E., Vaysse, P.J., Gustafson, E.L., Weinshank, R.L. and Branchek, T.A. (1995b) 'Re-evaluation of GABA transport in neuronal and glial cell cultures: correlation of pharmacology and mRNA localization', *Receptors and Channels* **3**: 129–146.

Bowery, N.G. and Enna, S.J. (2000) 'Gamma-aminobutyric acid (B) receptors: first of the functional metabotropic heterodimers', *J. Pharmacol. Exp. Ther.* **292**: 2–7.

Bradford, H.F. (1995) 'Glutamate, GABA and epilepsy', *Prog. Neurobiol.* **47**: 477–511.

Braestrup, C., Nielsen, E.B., Sonnewald, U., Knutsen, L.J., Andersen, K.E., Jansen, J.A., Frederiksen, K., Andersen, P.H., Mortensen, A. and Suzdak, P.D. (1990) '(R)-N-[4,4-bis (3-methyl-2-thienyl)but-3-en-1-yl]nipecotic acid binds with high affinity to the brain gamma-aminobutyric acid uptake carrier', *J. Neurochem.* **54**: 639–647.

Choi, D.W., Farb, D.H. and Fischbach, G.D. (1977) 'Clordiazepoxide selectively potentiates GABA conductance in spinal cord cell cultures', *Nature* **269**: 342–344.

Choi, D.W., Farb, D.H. and Fischbach, G.D. (1981) 'Clordiazepoxide selectively potentiates GABA conductance in spinal cord and sensory neurons in cell cultures', *J. Neurophysiol.* **45**: 621–631.

Collinson, N. (2000) 'Cognition enhancing effects of a $GABA_A$ α_5 selective inverse agonist', Presentation at GABA2000 meeting 23 to 29 July. Organisers Peter Gage and Istvan Mody.

Costa, E. and Guidotto, A. (1996) 'Benzodiazepines on trial: a research strategy for their rehabilitation', *Trends Pharmacol. Sci.* **17**: 192–200.

Dalby, N.O. (2000) 'GABA-level increasing and anticonvulsant effects of three different GABA uptake inhibitors', *Neuropharmacol.* **39**: 2399–2407.

Dalby, N.O. and Nielsen, E.B. (1997a) 'Comparison of the preclinical anticonvulsant profiles of tiagabine, lamotrigine, gabapentin and vigabatrin', *Epilepsy Res.* **28**: 63–72.

Dalby, N.O. and Nielsen, E.B. (1997b) 'Tiagabine exerts an anti-epileptogenic effect in amygdala kindling epileptogenesis in the rat', *Neurosci. Lett.* **229**: 135–137.

Dalby, N.O., Thomsen, C., Fink-Jensen, A., Lundbeck, J., Søkilde, B., Man, C.M., Sorensen, P.O. and Meldrum, B. (1997) 'Anticonvulsant properties of two GABA uptake inhibitors NNC 05-2045 and NNC 05-2090, not acting preferentially on GAT-1', *Epilepsy Res.* **28**: 51–61.

Dalvi, A. and Rodgers, R.J. (1996) 'GABAergic influences on plus-maze behaviour in mice', *Psychopharmacology* **128**: 380–397.

Dhar, T.G., Borden, L.A., Tyagarajan, S., Smith, K.E., Branchek, T.A., Weinshank, R.L. and Gluchowski, C. (1994) 'Design, synthesis and evaluation of substituted triarylnipecotic acid derivatives as GABA uptake inhibitors: identification of a ligand with moderate affinity and selectivity for the cloned human GABA transporter GAT-3', *J. Med. Chem.* **37**: 2334–2342.

Dimmock, P.W., Wyatt, K.W., Jones, P.W. and O'Brien, S (2000) 'Efficacy of selective serotonin-reuptake inhibitors in premenstrual syndrome: a systematic review', *Lancet* **356**: 1131–1136.

Dodrill, C.B., Arnett, J.L., Shu, V., Pixton, G.C., Lenz, G.T. and Sommerville, K.W. (1998) 'Effects of tiagabine monotherapy on abilities, adjustment, and mood', *Epilepsia* **39**: 33–42.

Dreifuss, F.E. and Ogunyemi, A.O. (1992) 'Classification of epileptic seizures and the epilepsies: an overview', *Epilepsy Res. Suppl.* **6**: 3–11.

Duncan, G.E., Sheitman, B.B. and Lieberman, J.A. (1999) 'An integrated view of pathophysiological models of schizophrenia', *Brain Res. Rev.* **29**: 250–264.

Ebert, B., Thompson, S.A., Saounatsou, K., McKernan, R., Krogsgaard-Larsen, P. and Wafford, K.A. (1997) 'Differences in agonist/antagonist binding affinity and receptor transduction using recombinant human γ-aminobutyric acid type A receptors', *Mol. Pharmacol.* **52**: 1150–1156.

Ebert, B., Wafford, K.A., Whiting, P.J., Krogsgaard-Larsen, P. and Kemp, J.A. (1994) 'Molecular pharmacology of γ-aminobutyric acid type A receptor agonists and partial agonists in oocytes injected with different alpha, beta and gamma receptor subunit combinations', *Mol. Pharmacol.* **46**: 957–963.

Fabienne, L., Wieser, H.-G., Yonekawa, Y., Aguzzi, A. and Fritschy, J.-M. (2000) 'Selective alterations in $GABA_A$ receptor subtypes in human temporal lobe epilepsy', *J. Neurosci.* **20**: 5401–5419.

Falch, E., Perregaard, J., Frølund, B., Søkilde, B., Buur, A., Hansen, L.M., Frydenvang, K., Brehm, L., Bolvig, T., Larsson, O.M., Sanchez, C., White, H.S., Schousboe, A. and Krogsgaard-Larsen, P. (1999) 'Selective inhibitors of glial GABA uptake: synthesis, absolute stereochemistry, and pharmacology of the enantiomers of 3-hydroxy-4-amino-4,5,6,7-tetrahydro-1,2-benzisoxazole (exo-THPO) and analogues', *J. Med. Chem.* **42**: 5402–5414.

Faulhaber, J., Steiger, A. and Lancel, M. (1997) 'The $GABA_A$ agonist THIP produces slow wave sleep and reduces spindling activity in NREM sleep in humans', *Psychopharmacol.* **130**: 285–291.

Fink-Jensen, A., Suzdak, P.D., Swedberg, M.D., Judge, M.E., Hansen, L. and Nielsen, P.G. (1992) 'The γ-aminobutyric acid (GABA) uptake inhibitor, tiagabine, increases extracellular brain levels of GABA in awake rats', *Eur. J. Pharmacol.* **220**: 197–201.

Fritschy, J.M., Johnson, D.K., Mohler, H. and Rudolph, U. (1998) 'Independent assembly and subcellular targeting of $GABA_A$-receptor subtypes demonstrated in mouse hippocampal and olfactory neurons in vivo', *Neurosci. Lett.* **249**: 99–102.

Gale, K. (1992) 'Subcortical structures and pathways involved in convulsive seizure generation', *J. Clin. Neurophysiol.* **9**: 264–277.

Gallo, M.A. and Smith, S.S. (1993) 'Progesterone withdrawal decreases latency to and increases duration of electrified prod burial: a possible rat model of PMS anxiety', *Pharmacol. Biochem. Behav.* **46**: 897–904.

Green, A.R., Hainsworth, A.H. and Jackson, D.M. (2000) 'GABA potentiation: a logical pharmacological approach for the treatment of acute ischaemic stroke', *Neuropharmacol.* **39**: 1483–1494.

Grunze, H., Erfurth, A., Marcuse, A., Amann, B., Normann, C. and Walden J. (1999) 'Tiagabine appears not to be efficacious in the treatment of acute mania', *J. Clin. Psychiatry* **60**: 759–762.

Guastella, J., Nelson, N., Nelson, H., Czyzyk, L., Keynan, S., Miedel, M.C., Davidson, N., Lester, H.A. and Kanner, B.I. (1990) 'Cloning and expression of a rat brain GABA transporter', *Science* **249**: 1303–1306.

Hadingham, K.L., Garrett, E.M., Wafford, K.A., Bain, C., Heavens, R.P., Sirinathsinghji, D.J. and Whiting, P.J. (1996) 'Cloning of cDNAs encoding the human γ-aminobutyric acid type A receptor 6 subunit and characterization of the pharmacology of α_6-containing receptors', *Mol. Pharmacol.* **49**: 253–259.

Haefely, W. (1986) 'Biological basis of drug-induced tolerance, rebound and dependence. Contribution of recent research on benzodiazepines', *Pharmacopsychiatry* **90**: 353–361.

Haefely, W.E. (1989) 'Pharmacology of the benzodiazepine receptor', *Eur. Arch. Psychiatry. Neurol. Sci.* **238**: 294–301.

Haefely, W., Kulcsar, A., Möhler, H., Pieri, L., Polc, P. and Schaffner, R. (1975) 'Possible involvement of GABA in the central actions of benzodiazepines', *Adv. Biochem. Psychopharmacol.* **14**: 131–151.

Haefely, W., Pieri, L. and Schaffner, R. (1981) 'General pharmacology and neuropharmacology of benzodiazepine derivatives', in F. Hoffmeister and G. Stiller (eds) *Handbook of Experimental Pharmacology 55II*, Berlin: Springer, pp. 13–262.

Hagberg, H., Lehmann, A., Sandberg, M., Nystrom, B., Jacobson, I. and Hamberger, A. (1985) 'Ischemia-induced shift of inhibitory and excitatory amino acids from intra- to extracellular compartments', *J. Cereb. Blood Flow. Metab.* **5**: 413–419.

Hajos, N., Nusser, Z., Rancz, E.A., Freund, T.F. and Mody, I. (2000) 'Cell type- and synapse-specific variability in synaptic $GABA_A$ receptor occupancy', *Eur. J. Neurosci.* **12**: 810–818.

Halbreich, U., Petty, F., Yonkers, K., Kramer, G.L., Rush, A.J., Bibi, K.W. (1996) 'Low plasma γ-aminobutyric acid levels during the late luteal phase of women with premenstrual dysphoric disorder', *Am. J. Psych.* **153**: 718–720.

Homanics, G.E., Harrison, N.L., Quinlan, J.J., Krasowski, M.D., Rick, C.E., DeBlas, A.L., Mehta, A.K., Kist, F., Mihalek, R.M., Aul, J.J. and Firestone, L.L. (1999) 'Normal electrophysiological and behavioral responses to ethanol in mice lacking the long splice variant of the γ_2 subunit of the γ-aminobutyrate type A receptor', *Neuropharmacol.* **38**: 253–265.

Howell, O., Atack, J.R., Dewar, D., McKernan, R.M. and Sur, C. (2000) 'Density and pharmacology of α_5 subunit-containing $GABA_A$ receptors are preserved in hippocampus of Alzheimer's disease patients', *J. Neurosci.* **98**: 669–675.

Ikegaki, N., Saito, N., Hashima, M. and Tanaka, C. (1994) 'Production of specific antibodies against GABA transporter subtypes (GAT1, GAT2, GAT3) and their application to immunocytochemistry', *Mol. Brain. Res.* **26**: 47–54.

Inglefield, J.R., Perry, J.M. and Schwartz, R.D. (1995) 'Post ischemic inhibition of GABA reuptake by tiagabine slows neuronal death in the gerbil hippocampus', *Hippocampus* **5**: 460–468.

Ipponi, A., Lamberti, C., Medica, A., Bartolini, A. and Malmberg-Aiello, P. (1999) 'Tiagabine antinociception in rodents depends on $GABA_B$ receptor activation: parallel antinociception testing and medial thalamus GABA microdialysis', *Eur. J. Pharmacol.* **368**: 205–211.

Johansen, F.F., and Diemer, N.H. (1991) 'Enhancement of GABA neurotransmission after cerebral ischemia in the rat reduces loss of hippocampal CA1 pyramidal cells', *Acta Neurol. Scand.* **84**: 1–6.

Kalviainen, R. (1998) 'Tiagabine: a new therapeutic option for people with intellectual disability and partial epilepsy', *J. Intellect. Disabil. Res.* **42**: 63–67.

Krogsgaard-Larsen, P., Falch, E., Larsson, O.M. and Schousboe, A. (1987) 'GABA uptake inhibitors: relevance to antiepileptic drug research', *Epilepsy Res.* **1**: 77–93.

Lancel, M. and Faulhaber, J. (1996) 'The $GABA_A$ agonist THIP (Gaboxadol) increases non-REM sleep and enhances delta activity in the rat', *Neurorep.* **7**: 2241–2245.

Lancel, M. and Langebartels, A. (2000) 'γ-aminobutyric aid_A ($GABA_A$) agonist 4,5,6, 7-tetrahydroisoxazolo[4,5-c]pyridin-3-ol persistently increases sleep maintenance and intensity during chronic administration to rats', *J. Pharmacol. Exp. Ther.* **293**: 1084–1090.

Lancel, M., Cröenlain, T.A.M. and Faulhaber J. (1996) 'Role of $GABA_A$ receptors in sleep regulation. Differential effects of muscimol and midazolam on sleep in rats', *Neuropsychopharmacol.* **15**: 63–74.

Lancel, M., Faulhaber, J., Deisz, R.A. (1998) 'Effect of the GABA uptake inhibitor tiagabine on sleep and EEG power spectra in the rat', *Br. J. Pharmacol.* **123**: 1471–1477.

Lancel, M., Faulhaber, J., Schiffelholz, T., Romeo, E., Di Michele, F., Holsboer, F. and Rupprecht R. (1997) 'Allopregnanolone affects sleep in a benzodiazepine-like fashion', *J. Pharmacol. Exp. Ther.* **282**: 1213–1218.

Landolt, H.P. and Gillin, J.C. (2000) '$GABA_A$ α_1 receptors: involvement in sleep regulation and potential of selective agonists in the treatment of insomnia', *CNS Drugs* **13**: 185–199.

Lin, Z. and Kadaba, P.K. (1997) 'Molecular targets for the rational design of antiepileptic drugs and related neuroprotective agents', *Med. Res. Rev.* **17**: 537–572.

Liu, Q.-R., Lopez-Corcuera, B., Mandiyan, S., Nelson, H. and Nelson, N. (1993) 'Molecular characterization of four pharmacologically distinct γ-aminobutyric acid transporters in mouse brain', *J. Biol. Chem.* **268**: 2106–2112.

Loiseau, P. (1999) 'Review of controlled trials of gabitril (tiagabine): a clinician's viewpoint', *Epilepsia* **40**: 14–19.

Lopez-Corcuera, B., Liu, Q.-R., Mandiyan, S., Nelson, H. and Nelson, N. (1992) 'Expression of a mouse brain cDNA encoding novel γ-aminobutyric acid transporter', *J. Biol. Chem.* **267**: 17,491–17,503.

Löw, K., Crestani, F., Keist, R., Benke, D., Brunig, I., Benson, J.A., Fritschy, J.M., Rulicke, T., Bluethmann, H., Möhler, H. and Rudolph, U. (2000) 'Molecular and neuronal substrate for the selective attenuation of anxiety', *Science* **290**: 131–134.

Malcangio, M. and Bower, N.G. (1995) 'Possible therapeutic application of $GABA_B$ receptor agonists and antagonists', *Clin. Neuropharmacol.* **18**: 285–305.

McKernan, R.M., Rosahl, T. W, Reynolds, D.S., Sur, C., Wafford, K.A., Atack, J.R., Farrar, S., Myers, J., Cook, G., Ferris, P., Garrett, L., Bristow, L., Marshall, G., Macaulay, A., Brown, N., Howell, O., Moore, K.W., Carling, R.W., Street, L., Castro, J. L, Ragan, C.I., Dawson, G.R. and Whiting, P.J. (2000) 'Sedative but not anxiolytic properties of benzodiazepines are mediated by the $GABA_A$ receptor α_1 subtype', *Nat. Neurosci.* **3**: 587–592.

McKernan, R.M. and Whiting, P.J. (1996) 'Which $GABA_A$-receptor subtypes really occur in the brain?', *Trends Neurosci.* **19**: 139–143.

Meldrum, B. (1992) 'Excitatory amino acids in epilepsy and potential novel therapeutics', *Epilepsy Res.* **12**: 189–196.

Mellor, J. R, Wisden, W. and Randall, A.D. (2000) 'Somato-synaptic variation of $GABA_A$ receptors in cultured murine cerebellar granule cells: investigation of the role of the α_6 subunit', *Neuropharmacol.* **39**: 1495–1513.

Mierlack, D. and Farb, D.H. (1988) 'Modulation of neurotransmitter receptor desensitisation: chlordiazepoxide stimulates fading of the GABA response', *J. Neurosci.* **8**: 814–820.

Miles, R., Toth, K., Gulyas, A.I., Hajos, N. and Freund, T.F. (1996) 'Differences between somatic and dendritic inhibition in the hippocampus', *Neuron* **16**: 815–823.

Mody, I. (2000) 'Synaptic actication of GABA-Rs', Presentation at GABA2000 meeting 23 to 29 July. Organisers Peter Gage and Istvan Mody.

Möhler, H., Fritschy, J.M., Lüscher, B., Rudolph, U., Benson, J. and Benke, D. (1996) 'The $GABA_A$ receptors. From subunits to diverse functions', *Ion Channels* **4**: 89–113.

Nelson, N. (1998) 'The family of Na+/Cl− neurotransmitter transporters', *J. Neurochem.* **71**: 1785–803.

Nielsen, E.B. (1988) 'Anxiolytic effect of NO-328, a GABA uptake inhibitor', *Psychopharmacol.* **96**: S42.

Ohnuma, T., Augood, S.J., Arai, H., McKenna, P.J. and Emson, P.C. (1999) 'Measurement of GABAergic parameters in the prefrontal cortex in schizophrenia: focus on GABA content, $GABA_A$ receptor α_1 subunit messenger RNA and human GABA transporter-1 (hGAT-1) messenger RNA expression', *Neuroscience* **93**: 441–448.

Olsen, R.W., Bureau, M., Houser, C.R., Delgado-Escueta, A.V., Richards, J.G. and Möhler, H. (1992) 'GABA/benzodiazepine receptors in human focal epilepsy', *Epilepsy Res.* **8**: 383–391.

Petty, F. (1995) 'GABA and mood disorders: a brief review and hypothesis', *J. Affect. Disord.* **34**: 275–281.

Phillis, J.W. (1995) 'CI-966, a GABA uptake inhibitor, antagonizes ischemia-induced neuronal degeneration in the gerbil', *Gen. Pharmacol.* **26**: 1061–1064.

Phillis, J.W., Smith-Barbour, M., Perkins, L.M. and O'Regan, M.H. (1994) 'Characterization of glutamate, aspartate, and GABA release from ischemic rat cerebral cortex', *Brain Res. Bull.* **34**: 457–466.

Polc, P., Bonetti, E.P., Schaffner, R. and Haefely, W. (1982) 'A three-state model of the benzodiazepine receptor explains the interactions between the benzodiazepine antagonist Ro 15–1788, benzodiazepine tranquilizers, beta-carbolines, and phenobarbitone', *Naunyn Schmied. Arch. Pharmacol.* **321**: 260–264.

Richards, D. A and Bowery, N.G. (1996) 'Comparative effects of the GABA uptake inhibitors, tiagabine and NNC-711, on extracellular GABA levels in the rat ventrolateral thalamus', *Neurochem. Res.* **21**: 135–140.

Rossi, D.J., Oshima, T. and Attwell, D. (2000) 'Glutamate release in severe brain ischemia is mainly by reversed uptake', *Nature* **403**: 316–321.

Rudolph, U., Crestani, F., Benke, D., Brünig, I., Benson, J.A., Fritschy, J.M., Martin, J.R., Bluethmann, H. and Möhler, H. (1999) 'Benzodiazepine actions mediated by specific γ-aminobutyric $acid_A$ receptor subtypes', *Nature* **401**: 796–800.

Scanziani, M. (2000) 'GABA spillover activates postsynaptic $GABA_B$ receptors to control rhythmic hippocampal activity', *Neuron* **25**: 673–681.

Schachter, S.C. (1999) 'A review of the antiepileptic drug tiagabine', *Clin. Neuropharmacol.* **22**: 312–317.

Schaffer, L.C. and Schaffer, C.B. (1999) 'Tiagabine and the treatment of refractory bipolar disorder', *Am. J. Psychiatry* **156**: 2014–2015.

Schmitt, U. and Hiemke, C. (1999) 'Effects of GABA-transporter (GAT) inhibitors on rat behaviour in open-field and elevated plus-maze', *Behav. Pharmacol.* **10**: 131–137.

Sedman, A.J., Gilmet, G.P., Sayed, A.J. and Posvar, E.L. (1990) 'Initial human safety and tolerance study of a GABA uptake inhibitor, CI-966: potential role of GABA as a mediator in the pathogenesis of schizophrenia and mania', *Drug Dev. Res.* **21**: 243–252.

Sigel, E., Baur, R., Kellenberger, S. and Malherbe, P. (1992) 'Point mutations affecting antagonist affinity and agonist dependent gating of $GABA_A$ receptor channels', *EMBO J.* **11**: 2017–2023.

Simpson, M.D., Slater, P., Deakin, J.F., Gottfries, C.G., Karlsson, I., Grenfeldt, B., and Crow, T.J. (1998) 'Absence of basal ganglia amino acid neuron deficits in schizophrenia in three collections of brains', *Schizophr. Res.* **31**: 167–175.

Simpson, P.B., Woollacott, A.J., Pillai, G.V., Maubach, K.A., Hadingham, K.L., Martin, K., Choudhury, H.I. and Seabrook, G.R. (2000) 'Pharmacology of recombinant human $GABA_A$ receptor subtypes measured using a novel pH-based high-throughput functional efficacy assay', *J. Neurosci. Methods* **99**: 91–100.

Smith, A.J., Alder, L., Silk, J., Adkins, C., Fletcher, A.E., Scales, T., Kerby, J., Marshall, G., Wafford, K.A., McKernan, R.M. and Atack, J.R. (2001) 'Effect of alpha subunit on allosteric modulation of ion channel function in stably expressed human recombinant $GABA_A$ receptors determined using ^{36}Cl-ion flux', *Mol. Pharmacol.* **59**: 1108–1118.

Smith, G.B. and Olsen, R.W. (1994) 'Identification of a [^{3}H]muscimol photoaffinity substrate in the bovine γ-aminobutyric $acid_A$ receptor α subunit', *J. Biol. Chem.* **269**: 20,380–20,387.

Smith, S.E., Parvez, N.S., Chapman, A.G. and Meldrum, B.S. (1995) 'The γ-aminobutyric acid uptake inhibitor, tiagabine, is anticonvulsant in two animal models of reflex epilepsy', *Eur. J. Pharmacol.* **273**: 259–265.

Smith, S.S, Gong, Q.H., Hsu, F.C., Markowitz, R.S., French-Mullen, J.M., Li, X. (1998) '$GABA_A$ receptor α_4 subunit suppression prevents withdrawal properties of an endogenous steroid', *Nature* **392**: 926–930.

Squires, R.F. and Saederup, E. (1998) 'Clozapine and several other antipsychotic/antidepressant drugs preferentially block the same "core", fraction of $GABA_A$ receptors', *Neurochem. Res.* **23**: 1283–1290.

Stephens, D.N., Schneider, H.H., Kehr, W., Jensen, L.H., Petersen, E. and Honore, T. (1987) 'Modulation of anxiety by beta-carbolines and other benzodiazepine receptor ligands: relationship of pharmacological to biochemical measures of efficacy', *Brain Res. Bull.* **19**: 309–318.

Sur, C., Fresu, L., Howell, O., McKernan, R.M. and Atack, J.R. (1999a) 'Autoradiographic localization of α_5 subunit-containing $GABA_A$ receptors in rat brain', *Brain Res.* **822**: 265–270.

Sur, C., Farrar, S.J., Kerby, J., Whiting, P.J., Atack, J.R. and McKernan, R.M. (1999b) 'Preferential coassembly of α_4 and δ subunits of the γ-aminobutyric $acid_A$ receptor in rat thalamus', *Mol. Pharmacol.* **56**: 110–115.

Suzdak, P. (1993) 'Lipophilic GABA uptake inhibitors: biochemistry, pharmacology and therapeutic potential', *Drugs of the Future* **18**: 1129–1136.

Suzdak, P.D. and Jansen, J.A. (1995) 'A review of the preclinical pharmacology of tiagabine: a potent and selective anticonvulsant GABA uptake inhibitor', *Epilepsia* **36**: 612–626.

Swedberg, M.D. (1994) 'The mouse grid-shock analgesia test: pharmacological characterization of latency to vocalization threshold as an index of antinociception', *J. Pharmacol. Exp. Ther.* **269**: 1021–1028.

Tamminga, C.A., Crayton, J.W. and Chase, T.N. (1978) 'Muscimol: GABA agonist in therapy in schizophrenia', *Am. J. Psychiatry* **135**: 746–747.

Teuber, L., Watjens, F. and Jensen, L.H. (1999) 'Ligands for the benzodiazepine binding site: a survey', *Current Pharm. Des.* **5**: 317–343.

Theobald, W., Busch, O., Kunz, H.A., Krupp, P., Stenger, E.G. and Heinemann, H. (1968) 'Pharmakologische und experimentalpsychologische untersuchungen mit 2 inhaltsstoffen des fliegenpilzes (amanita muscaria)', *Arzneim. Forsch.* **18**: 311–315.

Thompson, S.A., Bonnert, T.P., Whiting, P.J. and Wafford, K.A. (1998) 'Functional characteristics of recombinant human $GABA_A$ receptors containing the epsilon-subunit', *Toxicol. Lett.* **100**: 233–238.

Thomsen, C., Sørensen, P.O. and Egebjerg J. (1997) '1-(3-(9H-carbazol-9-yl)-1-propyl)-4-(2-methoxyphenyl)-4-piperidinol, a novel subtype selective inhibitor of the mouse type II GABA-transporter', *Br. J. Pharmacol.* **120**: 983–985.

Uhl, G.R. and Hartig, P.R. (1992) 'Transporter explosion: update on uptake', *Trends Pharmacol. Sci.* **13**: 421–425.

Verdoorn, T.A. (1994) 'Formation of heteromeric γ-aminobutyric acid type A receptors containing two different α subunits', *Mol. Pharmacol.* **45**: 475–480.

Wafford, K.A. (2000) 'Properties of a stable cell line expressing human $\alpha_4\beta_3\delta$ $GABA_A$ receptors', Presentation at GABA2000 meeting 23 to 29 July. Organisers Peter Gage and Istvan Mody.

Wafford, K.A., Thompson, S.A., Thomas, D., Sikela, J., Wilcox, A.S. and Whiting, P.J. (1996) 'Functional characterization of human γ-aminobutyric $acid_A$ receptors containing the α_4 subunit', *Mol. Pharmacol.* **50**: 670–678.

Whiting P. (2000) 'α_5 containing receptors: what do they do and what might they be good for?', Presentation at GABA2000 meeting 23 to 29 July. Organisers Peter Gage and Istvan Mody.

Whiting, P.J., Bonnert, T.P., McKernan, R.M., Farrar, S., Le Bourdellès, B., Heavens, R.P., Smith, D.W., Hewson, L., Rigby, M.R., Sirinathsinghji, D.J., Thompson, S.A. and Wafford, K.A. (1999) 'Molecular and functional diversity of the expanding GABA-A receptor gene family', *Ann. NY Acad. Sci.* **868**: 645–653.

Whiting, P.J., McAllister, G., Vassilatis, D., Bonnert, T.P., Heavens, R.P., Smith, D.W., Hewson, L., O'Donnell, R., Rigby, M.R., Sirinathsinghji, D.J., Marshall, G., Thompson, S.A., Wafford, K.A. and Vasilatis, D. (1997) 'Neuronally restricted RNA splicing regulates the expression of a novel $GABA_A$ receptor subunit conferring atypical functional properties', *J. Neurosci.* **17**: 5027–5037.

Woo, T.U., Whitehead, R.E., Melchitzky, D.S. and Lewis, D.A. (1998) 'A subclass of prefrontal γ-aminobutyric acid axon terminals are selectively altered in schizophrenia', *Proc. Natl. Acad. Sci. USA* **28**: 5341–5346.

Xu, C.W., Yi, Y., Qiu, L. and Shuaib, A. (2000) 'Neuroprotective activity of tiagabine in a focal embolic model of cerebral ischemia', *Brain Res.* **874**: 75–77.

Yamauchi, A., Uchida, S., Kwon, H.M., Preston, A.S., Robey, R.B., Garcia-Perez, A., Burg, M.B. and Handler, J.S. (1992) 'Cloning of a Na^+- and Cl^--dependent betaine transporter that is regulated by hypertonicity', *J. Biol. Chem.* **267**: 649–652.

Yang, Y., Li, Q., Wang, C.X., Jeerakathil, T. and Shuaib, A. (2000) 'Dose-dependent neuroprotection with tiagabine in a focal cerebral ischemia model in rat', *Neuroreport* **11**: 2307–2311.

Zorn, S.H. and Enna, S.J. (1985) 'GABA uptake inhibitors produce a greater antinociceptive response in the mouse tail-immersion assay than other types of GABAergic drugs', *Life Sci.* **37**: 1901–1912.

Index